WATER TREATMENT PLANT DESIGN

American Water Works Association
American Society of Civil Engineers

Third Edition

McGRAW-HILL
New York San Francisco Washington, D.C. Auckland Bogotá
Caracas Lisbon London Madrid Mexico City Milan
Montreal New Delhi San Juan Singapore
Sydney Tokyo Toronto

Library of Congress Cataloging-in-Publication Data

American Water Works Association, American Society of Civil Engineers.

   Water treatment plant design—3rd ed.
      p.   cm.
   Includes index.
   ISBN 0-07-001643-7
   1. Water treatment plants—Design and construction.
I. American Water Works Association. II. American Society of Civil Engineers.
TD434.W38   1997
628.1'62—dc21   97-26057
                                                                                                                 CIP

McGraw-Hill

A Division of The *McGraw·Hill* Companies

Copyright © 1998 by American Water Works Association and American Society of Civil Engineers. All rights reserved. Printed in the United States of America. Except as permitted under the United States Copyright Act of 1976, no part of this publication may be reproduced or distributed in any form or by any means, or stored in a data base or retrieval system, without the prior written permission of the publisher.

Copyright © 1990 by The McGraw-Hill Companies, Inc.

The first edition was published in 1969 by the American Water Works Association, Inc.

    5 6 7 8 9   BKM BKM   0 9 8 7 6 5 4 3

ISBN 0-07-001643-7

The sponsoring editor for this book was Larry Hager, the editing supervisor was Carol O'Connell of Graphic World Publishing Services, and the production supervisor was Sherri Souffrance. It was set in Times Roman by Graphic World, Inc.

  This book was printed on recycled, acid-free paper containing a minimum of 50% recycled de-inked fiber.

Information contained in this work has been obtained by The McGraw-Hill Companies, Inc. ("McGraw-Hill") from sources believed to be reliable. However, neither McGraw-Hill nor its authors guarantee the accuracy or completeness of any information published herein, and neither McGraw-Hill nor its authors shall be responsible for any errors, omissions, or damages arising out of use of this information. This work is published with the understanding that McGraw-Hill and its authors are supplying information but are not attempting to render engineering or other professional services. If such services are required, the assistance of an appropriate professional should be sought.

CONTENTS

Preface   viii
Acknowledgments   ix

Chapter 1. The Challenge of Water Treatment Plant Design   1

Today's Challenges / 2
Development of Water Treatment Design Projects / 4
The Purpose of This Text / 4
Bibliography / 5

Chapter 2. Master Planning and Treatment Process Selection   7

The Master Plan / 7
Water Quality Goals / 9
Treatment Options / 13

Chapter 3. Design and Construction   21

Preliminary Design / 21
Final Design / 24
Construction / 28
Plant Start-Up / 30
Bibliography / 30

Chapter 4. Intake Facilities   31

Intake Features / 31
Types of Intake Systems / 37
Intake Design / 41
Racks and Screens / 56
Bibliography / 59

Chapter 5. Aeration and Air Stripping   61

Uses of Aeration / 61
Types of Aeration Equipment / 61
Principles of Air Stripping and Aeration / 69
Design of Aeration Processes / 71
Pilot Testing / 83
Bibliography / 85

Chapter 6. Mixing, Coagulation, and Flocculation — 87

Definitions / 87
The Coagulation Process / 88
Design of Chemical Mixing / 90
Flocculation Process Design / 98
Process Monitoring and Control / 107
Bibliography / 109

Chapter 7. Clarification — 111

Conventional Clarification Design / 111
High-Rate Clarification / 132
Solids Contact/Slurry Recirculation Units / 137
Sludge Blanket Clarification / 140
Dissolved Air Flotation / 145
Contact Clarification / 149
Bibliography / 151

Chapter 8. High-Rate Granular Media Filtration — 153

Mechanism of Filtration / 153
Design Considerations / 154
Filter Design Criteria / 184
Filter Operation and Maintenance / 187
Bibliography / 190

Chapter 9. Slow Sand and Diatomaceous Earth Filtration — 193

Slow Sand Filtration / 193
Diatomaceous Earth Filtration / 207
Bibliography / 219

Chapter 10. Oxidation and Disinfection — 221

Regulatory Framework / 222
Chemical Oxidation Treatment / 228
Chlorination / 230
Chlorine Dioxide Systems / 249
Ozone Disinfection Systems / 254
Bibliography / 278

Chapter 11. Lime Softening — 281

Water Hardness and Softening Treatment / 281
Lime Softening Processes / 286
Special Lime Softening Design Considerations / 292
Future Trends in Softening / 295
Bibliography / 296

CONTENTS   v

Chapter 12. Ion Exchange Processes   299

    The Ion Exchange Process / *299*
    Ion Exchange Resins / *302*
    Cation Exchange Processes / *304*
    Anion Exchange Processes / *308*
    Demineralization / *318*
    Ion Exchange Equipment Design / *318*
    Boundaries between Suppliers, Consultants, and Users / *330*
    Future Trends in Ion Exchange Treatment / *330*
    Bibliography / *331*

Chapter 13. Membrane Processes   335

    Types of Membrane Processes / *335*
    Membrane System Components and Design Considerations / *341*
    Membrane Unit Design / *356*
    Other Membrane Process Design Considerations / *368*
    Bibliography / *375*

Chapter 14. Activated Carbon Processes   377

    Characteristics of Activated Carbon / *377*
    Design of Powdered Activated Carbon Facilities / *380*
    Design of Granular Activated Carbon Facilities / *385*
    Regeneration of Granular Activated Carbon / *408*
    Bibliography / *414*

Chapter 15. Chemicals and Chemical Handling   417

    Receiving and Storing Process Chemicals / *417*
    Chemical Feed and Metering Systems / *433*
    Designing for Safety and Hazardous Conditions / *456*
    Recent Trends in Chemical Handling and Use / *466*
    Bibliography / *467*

Chapter 16. Hydraulics   469

    Hydraulic Design / *469*
    Head Loss Types and Calculations / *473*
    Hydraulic Design Hints / *478*
    Ancillary Hydraulic Design / *481*
    Bibliography / *482*

Chapter 17. Process Residuals   485

    Types of Process Residuals / *485*
    Process Residual Disposal Methods / *492*
    Design Considerations and Criteria / *496*

Residuals Handling Considerations / *520*
Bibliography / *529*

Chapter 18. Architectural Design — 531

The Role of the Architect in Water Treatment Plant Design / *531*
Facilities Design / *535*
Construction Alternatives and Building Material Selection / *544*
Design Standard Promulgating Organizations and Organizations Offering
  Design Recommendations / *547*

Chapter 19. Structural Design — 553

The Design Process / *553*
Structural Design Considerations / *554*
Design of Basins, Vaults, Large Conduits, and Channels / *562*
Buildings and Superstructures / *563*
Bibliography / *565*

Chapter 20. Process Instrumentation and Controls — 567

Purpose of Instrumentation and Controls / *568*
Types of Instrumentation for Water Treatment / *570*
Control System Design Considerations / *584*
Design of Water Treatment Plant Control Systems / *592*
Bibliography / *603*

Chapter 21. Electrical Systems — 605

Electrical System Reliability / *605*
Elements of Electrical System Design / *611*
Electrical System Testing and Maintenance / *617*
Bibliography / *619*

Chapter 22. Design Reliability Features — 621

Reliability and Redundancy Concepts / *621*
Design Concepts for System Reliability / *623*
Reliability Design Principles and Practices / *626*
Operations and Maintenance for Plant Reliability / *631*
Bibliography / *632*

Chapter 23. Site Selection and Plant Arrangement — 635

Water Treatment Plant Site Selection / *635*
Arrangement of Water Treatment Plant Facilities / *655*
Bibliography / *670*

Chapter 24. Environmental Impact and Project Permitting — 673

Environmental Issues Associated with Plant Construction / 673
Environmental Issues Associated with Plant Operations / 677
Project Permitting / 681
Integration of Environmental Issues and Permitting into Plant Design, Construction, and Operation / 686
Bibliography / 688

Chapter 25. Operations and Maintenance — 689

Operations and Maintenance Considerations during Plant Design / 689
Preparing a Plant for Start-Up / 698
Bibliography / 703

Chapter 26. Construction Costs — 705

Level of Estimates / 705
Estimating Methodologies / 707
Special Cost Considerations / 710
Finalizing the Cost Estimate / 712
Bibliography / 714

Chapter 27. Operator Training and Plant Start-Up — 715

Training and Start-Up Considerations / 715
Design-Phase Training / 716
Construction-Phase Training, Start-Up, and Post-Start-Up / 719

Chapter 28. Pilot Plant Design and Construction — 729

Pilot Plant Studies / 730
Pilot Plant Design / 731
Pilot Plant Construction / 746
Bibliography / 750

Appendix A. Properties and Characteristics of Water Treatment Chemicals — 751

Appendix B. Abbreviations Commonly Used in the Water Industry — 777

Index   788

PREFACE

Water treatment plant design has, in recent years, developed into an increasingly complicated procedure. The design is generally executed by a design team made up of engineers and architects from several different specialties who must coordinate with owners, several levels of regulatory authorities, and public health officials to provide an efficient, reliable, and long-lasting facility. To do this, designers must not only be aware of successful past designs, but also stay abreast of new technology that reduces costs, improves reliability, and provides the best possible water quality. *Water Treatment Plant Design* was written to assist professionals in the field responsible for the design and construction of today's water treatment plants.

The first version of *Water Treatment Plant Design* was published in 1939 as a manual of engineering practice for the American Society of Civil Engineers (ASCE). In 1969, the manual assumed book form and was updated to include a discussion of developments in pretreatment and filtration processes. The 1969 edition was the result of a joint effort between committees of the ASCE, the American Water Works Association (AWWA), and the Conference of State Sanitary Engineers (CSSE).

The second edition was produced in 1990 through a joint effort of the AWWA and the ASCE. The material for each chapter was prepared by one or more authors and reviewed by a joint committee of AWWA and ASCE members.

This third edition is a joint AWWA and ASCE effort and is essentially a complete rewrite of the previous edition. The information presented in this book has been prepared as a guide and represents a consensus of opinion of recognized authorities in the field. A steering committee made up of members from both associations agreed on a greatly expanded list of chapters for the text to be of maximum value to designers. Chapters have been reviewed by the joint steering committee and by volunteer experts from engineering firms and water systems located throughout North America.

Water Treatment Plant Design has been written as a companion to AWWA's *Water Quality and Treatment*. The two books are intended to complement each other; *Water Treatment Plant Design* provides information on equipment, planning, design, and construction, and *Water Quality and Treatment* emphasizes the theory behind the unit processes.

The members of the steering committee for the third edition were:

AWWA	*Thomas J. Lane*, Malcolm Pirnie, Inc., Mahwah, N.J.
	Stephen R. Martin, Camp Dresser & McKee Inc., Cambridge, Mass.
	Leonard C. Rodman, Black & Veatch, Kansas City, Mo.
ASCE	*George P. Fulton*, Hazen & Sawyer, P.C. New York, N.Y.
	John E. Spitko, CH2M Hill, Inc., Philadelphia, Pa.
	David R. Zima, PSC Engineers & Consultants, Inc., Limerick, Pa.

ACKNOWLEDGMENTS

The following persons are authors and coauthors of chapters of the third edition:

Chapter 1	*Thomas J. Lane,* Malcolm Pirnie, Inc., Mahwah, N.J.
Chapter 2	*George P. Fulton,* Hazen & Sawyer, P.C., New York, N.Y.
	C. Michael Elliott, Stearns & Wheler, Cazenovia, N.Y.
Chapter 3	*R. Gary Fuller,* HDR Engineering, Inc., Denver, Colo.
Chapter 4	*Steven N. Foellmi,* Black & Veatch, Irvine, Calif.
Chapter 5	*John E. Dyksen,* Malcolm Pirnie, Inc., Mahwah, N.J.
Chapter 6	*George M. Wesner,* Consulting Engineer, San Clemente, Calif.
Chapter 7	*John R. Willis,* Camp Dresser & McKee Inc., Cambridge, Mass.
Chapter 8	*Stephen R. Martin,* Camp Dresser & McKee Inc. Cambridge, Mass.
Chapter 9	*Mark Choreser,* Stearns & Wheler, LLC, Cazenovia, N.Y.
	Michael Y. Broder, Hazen & Sawyer, New York, N.Y.
Chapter 10	*James C. Hesby,* Black & Veatch, Las Vegas, Nev.
Chapter 11	*Michael B. Horsley,* Black & Veatch, Kansas City, Mo.
	Kevin Tobin, Metropolitan Utility District, Omaha, Neb.
Chapter 12	*Michael C. Gottlieb,* Resin Tech, Inc., Cherry Hill, N.J.
	Peter Meyer, Resin Tech, Inc., Cherry Hill, N.J.
Chapter 13	*Robert A. Bergman,* CH2M Hill, Inc., Gainesville, Fla.
Chapter 14	*Richard D. Brady,* Corrao-Brady Group, San Diego, Calif.
Chapter 15	*Jerry L. Anderson,* CH2M Hill, Inc., Dayton, Ohio
	Craig E. Brackbill, Metropolitan Water District of Southern California, San Dimas, Calif.
	Richard E. Hubel, American Water Works Service Company, Voorhees, N.J.
	Vance G. Lee, HDR Engineering, Inc., Phoenix, Ariz.
Chapter 16	*Peter J. Barthuly,* Barthuly Hydraulics & Pumping Engineering, Inc., Burlington, Mass.
Chapter 17	*Jerry S. Russell,* John Carollo Engineers, Phoenix, Ariz.
	Brian E. Peck, John Carollo Engineers, Santa Ana, Calif.
Chapter 18	*Philip M. Zimmerman,* Malcolm Pirnie, Inc., White Plains, N.Y.
Chapter 19	*Ralph Eberts,* Black & Veatch, Los Angeles, Calif.
	T. Albert Chen, Black & Veatch, Los Angles, Calif.
Chapter 20	*John C. Nourse,* Westin Engineering Inc., San Jose, Calif.

Chapter 21   *Daniel E. Honeycutt,* Jordan, Jones & Goulding Inc., Atlanta, Ga.
Chapter 22   *John E. Spitko,* CH2M Hill, Inc., Philadelphia, Pa.
Chapter 23   *Darryl Corbin,* Montgomery Watson, Dallas, Tex.
Chapter 24   *Phillip C. Kennedy,* Camp Dresser & McKee Inc., Cambridge, Mass.
             Jane W. Wheeler, Camp Dresser & McKee Inc., Cambridge, Mass.
Chapter 25   *Richard A. Rohan,* Metcalf & Eddy, Inc., San Diego, Calif.
             Steve Baker, Montgomery Watson, San Diego, Calif.
             Eugene Nelms, JMM Operational Services, Inc., Edmonton, Alberta
Chapter 26   *Ron Cilensek,* CH2M Hill, Inc., Chesapeake, Va.
Chapter 27   *Thomas Arn,* Malcolm Pirnie, Inc., Chesapeake, Va.
             James Gasser, Malcolm Pirnie, Inc., White Plains, N.Y.
Chapter 28   *Robert A. Stoops,* Camp Dresser & McKee Inc., Providence, R.I.
Appendix A   *Jerry L. Anderson,* CH2M Hill, Inc., Dayton, Ohio
Staff Advisor—*George C. Craft,* AWWA, Denver, Colo.
Technical Editor—*Kathleen A. Faller,* AWWA, Denver, Colo.
Technical Editor—*Harry Von Huben,* Lake Bluff, Ill.

The following persons provided review of one or more chapters of the third edition: David J. Carlson, Mark Carlson, Ian A. Crossley, Patrick L. Daigle, C. Michael Elliott, Herbert D. Fiddick, Daniel J. Guillory, Leland L. Harms, Ron Henderson, Phillip Kennedy, Steve Lavinder, Tony Meyers, O. J. Morin, Walter J. O'Brien, Paul Osborne, Steven J. Randke, William F. Reeves, Dave Reves, Donald R. Stevens, Scott Trusler, Ian Watson, Stancil B. Weill, Jane Wheeler, David R. Wilkes, and Herbert G. Zeller.

CHAPTER 1
THE CHALLENGE OF WATER TREATMENT PLANT DESIGN

When water treatment engineering first evolved in the early part of the twentieth century, its main goal was to ensure that infectious organisms in drinking water supplies were removed or inactivated. Chlorination and filtration practices were applied with tremendous success to the point that major death-causing waterborne disease outbreaks in the United States were virtually eliminated by the 1930s.

As a result, for engineers trained in the 1960s, 1970s, and 1980s, both education and industry belief was that all concerns of microbiological contamination in surface waters could be eliminated by providing filtration (with suitable pretreatment) to produce water of sufficient clarity (turbidity less than 1.0 or 0.5 ntu), and then chlorinating. Groundwater was thought to be already filtered, requiring only chlorination to maintain a distribution system residual. Any additional treatment was generally considered necessary only to address non–health-related parameters, such as excessive hardness or water discoloration caused by iron and manganese.

The principal challenge to water treatment engineers in the 1960s and 1970s was engineering cost-effectiveness: how to accomplish these simple treatment goals at the lowest total cost to the water utility. Thus, in these decades many new techniques and processes were developed to clarify surface water economically. These developments included improvements to sedimentation basin designs; high-rate clarification processes such as tube settlers, plate settlers, and dissolved air flotation; high-rate filtration processes; and proprietary pre-engineered or package equipment integrating flocculation, settling, and filtration processes.

In the 1970s and 1980s a new drinking water concern arose: the potential long-term health risks posed by trace amounts of organic compounds present in drinking water. A wave of regulations ensued with new maximum contaminant levels (MCLs) established for total trihalomethanes (TTHMs), pesticides, and volatile organic chemicals (VOCs). This trend continues today. In response to this concern and resulting treatment needs, water treatment engineers have successfully devised new methods of water treatment to remove organic compounds. These methods, such as air stripping, activated carbon adsorption, and enhanced coagulation, have been the primary focus of water treatment engineering over the last 15 years.

Recently, the old concern about microbiological contamination is reemerging as the primary focus of water treatment engineers. The main driving forces behind this development have been:

- The promulgation of the Surface Water Treatment Rule and Total Coliform Rule by the U.S. Environmental Protection Agency (USEPA) and the monitoring and enforcement actions that have occurred since 1989, when they went into effect.
- Recent documented cases of contamination of drinking water supplies by waterborne diseases, mainly giardiasis and cryptosporidiosis, caused by cysts rather than bacteria.

New approaches and processes are evolving to address these concerns. These approaches include renewed emphasis on source water protection, optimizing plant performance, and recycle stream management, plus consideration of new technologies, especially membrane treatment and ozonation.

TODAY'S CHALLENGES

Engineers who design water treatment systems today face many challenges. The most important of these are described as follows.

Integrated Treatment Systems

Traditional treatment engineering has focused on the treatment plant as the sole vehicle for controlling drinking water quality. The engineer's role was to characterize the quality of the source water to enter the plant and devise treatment facilities to produce water meeting drinking water standards. The point of measurement for drinking water standards was the finished water exiting the plant.

Today's engineer must view the water treatment plant as only a major component in a multistep treatment process. This process includes consideration of the path that the water travels upstream of the plant in the watershed and the elements of the water transmission and distribution system downstream of the plant. Changing water quality must be managed in each of these steps, and new regulations require that drinking water standards be met at the customer's tap.

Regulatory Uncertainties

The definition of "safe" drinking water, which remained relatively fixed in the 1950s, 1960s, and 1970s, now seems to be constantly changing or under review as the water utility industry grapples to understand the potential health effects of trace amounts of an increasing variety of chemical compounds and infectious organisms. Today's treatment system engineer, in addition to addressing current drinking water standards, must anticipate potential future requirements. A water system designed today must be designed with sufficient flexibility to be modified to meet these potential requirements.

Regulatory uncertainties extend to other environmental concerns important to water treatment plant design, including waste management practices and chemical storage and feed operations.

New Technologies

The state of the art of water treatment plant design is continually changing as new technologies emerge, offering new unit processes for water treatment or making currently used processes more efficient or economical. In addition, advances in computer technology and building materials are rapidly changing and improving the support systems associated with water treatment plants.

Multidiscipline Teams

A water treatment plant engineering design team traditionally consisted simply of a small group of civil engineers. This single-discipline team performed the majority of design work for virtually all plant components. Support disciplines of architects and structural, electrical, and mechanical engineers were used to execute the basic decisions made by the design team.

Today, the complexity of project and regulatory requirements dictates that a far more multidisciplined approach be used. Typically, a small group of civil engineers remains as the "project" engineers, but this group uses the expertise and resources of many different specialists to execute the design. In addition to traditional design support disciplines, these may include:

- Process engineers
- Plant operations specialists
- Instrumentation and control engineers
- Health and safety specialists
- Environmental scientists
- Specialists in environmental permitting and public participation

Major design decisions today are no longer made unilaterally by the project team. Instead, a consensus is reached after participation by members of the design team and by individuals outside the team, including owners, operators, regulatory agencies, and the general public.

Project Delivery

The traditional procedure for construction of a new water treatment plant was for engineering design and specification to be prepared by an engineering firm or the owner's in-house staff. Bids were then taken and the contract awarded to the lowest responsible bidder. The design team then usually monitored the construction to see that the design intent was carried out, and after construction was completed, the facilities were operated by the owner.

Today, a number of changes and variations to this traditional approach are being considered or implemented. Two of the principal alternatives are:

- Design-build approaches, in which one entity is responsible for both design and construction
- Privatization approaches, in which the facility is owned by a private entity providing treatment service for the water utility

In addition, a global marketplace for water treatment engineering is evolving. Ideas and practices are being exchanged among countries all over the world. In North America, there is increasing consideration of European treatment practices, technologies, and firms.

DEVELOPMENT OF WATER TREATMENT DESIGN PROJECTS

A water treatment design project passes through many steps between the time when the need for a project is identified and the time that the completed project is placed into service. The period before construction commences can generally be divided into the following phases:

1. Master planning—treatment needs and feasible options for attaining those needs are established in a report. In subsequent phases, this report may be periodically updated to adjust to both system and regulatory changes.
2. Process train selection—viable treatment options are subjected to bench, pilot, and full-scale treatment investigations. This testing program provides background data sufficient in detail to enable decisions on selecting the more advantageous options for potential implementation. These tests provide design criteria for major plant process units.
3. Preliminary design—a "fine-tuning" procedure in which feasible alternatives for principal features of design, such as location, treatment process arrangement, type of equipment, and type and size of building enclosures, are evaluated. In this phase, preliminary designs are prepared in sufficient detail to permit development of meaningful project cost estimates. These estimates help in evaluating and selecting options to be incorporated into the final design and allow the owner to prepare the required project financial planning.
4. Final design—preparation of contract documents (drawings and specifications) that present the project design in sufficient detail to allow for gaining final regulatory approvals, obtaining competitive bids from construction contractors, and actual facility construction.

Many technical and nontechnical individuals must be involved, not only during the four phases of project development, but also between these phases to ensure that a project proceeds without undue delay. In addition to the engineer's design staff and the owner, these may include public health and regulatory officials, environmental scientists, and the public affected by both the proposed construction and the future water supply services to be provided.

The process train selection phase is only briefly covered in this book. Theory and procedures needed for this phase are the focus of *Water Quality and Treatment*. It is important that the interface between phase 1 and phase 2 and between phase 2 and phase 3 be carefully coordinated to allow uninterrupted continuity of design. In other words, viable options developed for consideration in phase 1, master planning, should provide a base for developing unit process test studies in phase 2. The process train selected in phase 2 provides the basis for phase 3, preliminary design, in which other factors influencing design are included in the evaluations before criteria for final design are developed and finalized.

Careful coordination of the various phases and entities involved provides the owner and the engineer with the opportunity to develop the most advantageous treatment solutions and designs, and helps avoid pitfalls in the schedule and decisions that might add to the cost of the project.

THE PURPOSE OF THIS TEXT

Water Treatment Plant Design is intended to serve as the primary reference for engineers who take on today's challenges of water treatment plant design. It covers the organization and execution of a water treatment plant project from planning and permitting through design, construction, and start-up.

The book is aimed at "project" engineers and managers: those professional engineers who lead the group of specialists who make up the design team. Generally, these individuals are graduates of civil or environmental engineering programs and are registered professional engineers.

For certain topics, especially the practical application of water treatment unit processes, this book aims to be an authoritative reference to design engineers. For other topics, only a general discussion of major concepts and issues is provided, and the reader is referred to more specialized references for detailed information.

Many books in circulation address the subject of water treatment engineering. As a joint publication of the American Society of Civil Engineers (ASCE) and the American Water Works Association (AWWA), Water Treatment Plant Design attempts to present an industry consensus on current design practices.

Organization of This Text

For the convenience of readers, chapters in this book have been organized as follows:

- Chapters 1, 2, and 3 examine the general preliminary and final design phases and engineering needs during project construction and initial operation.
- Chapters 4 through 14 address design practices for the major categories of unit processes applicable to water treatment plants.
- Chapters 15 through 21 cover the support systems associated with the design of water treatment facilities.
- Chapters 22 through 28 discuss general topics essential to developing a successful water treatment plant project.

Because of the rapid changes that the water treatment engineering industry is undergoing, many chapters include a discussion on future trends. These sections attempt to inform the reader of new developments not yet in general practice, but that may soon become so.

BIBLIOGRAPHY

A listing is provided at the end of each chapter of references specific to the chapter subject. In addition, a number of other texts are cited that the design engineer should be aware of for reference and additional details.

Relationship with *Water Quality and Treatment*

The AWWA publication *Water Quality and Treatment* is intended to be a companion reference to *Water Treatment Plant Design* and to serve as the primary reference for process engineers. *Water Quality and Treatment* provides details of water quality goals, with an emphasis on unit processes, standards, and the evaluation and selection of treatment process trains.

The two committees who have supervised the latest editions of these two books have attempted to eliminate overlap of subject matter and have tried to arrange the contents of the books to complement each other.

Other Design References

Other design references currently available include the following:

Kawamura, Susumu. *Integrated Design of Water Treatment Facilities.* New York: John Wiley and Sons, 1991.

Montgomery, James M. *Water Treatment Principles and Design.* New York: John Wiley and Sons, 1985.

Sanks, Robert L. *Water Treatment Plant Design for Practicing Engineers.* Ann Arbor, Mich.: Ann Arbor Science, 1978.

CHAPTER 2
MASTER PLANNING AND TREATMENT PROCESS SELECTION

Because the master plan is the first of many steps leading to a final water treatment plant design, this chapter introduces the principal issues that provide the basis for design through its many phases.

THE MASTER PLAN

A master plan is the orderly planning of a water system's future improvement program. The initial step in preparing any water system design is updating the system's master plan. Many states and state utility commissions require all water systems to have an active master plan that anticipates system additions and improvements for as many as 20 years into the future. The master plan for treatment should be periodically updated to reflect the improvements needed to compensate for changing system requirements imposed by facility wear, customer requirements, and changing water quality regulations.

Master Planning Issues

Master planning for water treatment facilities is often incorporated into the long-range capital improvement program for the water system. The master plan identifies present and future needs and direction for developing the water system's facilities. Some specific items that should be covered are:

- Identification of existing system components and service area
- Long-range projections of the area to be served by the water system
- Planning periods for the various water system facilities
- Present and future water demands
- Regulatory requirements for the ultimate approval and operation of the system
- Evaluation of alternative sources of supply

In addition, technical, environmental, institutional, financial, and operations and maintenance issues related to developing the recommended plan should be identified.

This text is concerned with the design of water treatment plant facilities and with those aspects of master planning related to the physical features of design. Discussion of master planning, therefore, is essentially limited to those treatment considerations outlined by the chapter subjects. In addition to process and facility design issues, the scope also includes:

- Site and facilities arrangement
- Environmental impact
- Construction costs
- Operations and maintenance (scope and costs)

A principal difficulty in master planning is that meaningful background data must be available to determine the future direction for water treatment without benefit of the more detailed information available from subsequent phases of design. That problem is somewhat simplified where design is to rehabilitate an existing plant because much of the data, especially operations and maintenance costs, is provided by past experience. In this case, a background of site and environmental experiences would also exist to guide future direction.

The major master planning difficulties are encountered in the design of a completely new facility. In this situation, background process and cost data are usually obtained either from nearby operating facilities, from the experience of other water supplies with a similar water source, or from references such as those published by the U.S. Environmental Protection Agency (USEPA) and the AWWA on typical treatment costs. Other factors such as the influence of site location and environmental impacts would need to be developed from local knowledge.

The most important considerations in preparing a master plan are:

- To provide general guidelines for future water treatment action.
- To develop all possible alternatives for further evaluation unless background data from existing experiences are so overwhelming that final decisions on treatment are obvious.
- To enable a liberal use of contingencies in developing cost estimates, with the magnitude of contingencies reflecting the confidence in the cost base. It is important that the owner not be led into quick acceptance of a treatment program that eventually turns out to be too costly. If the available information is questionable, it is best to delay program discussion until more meaningful background information is available.
- To include, as much as possible, potential features limiting site locations and environmental impacts in the early determination to resolve which alternatives should obviously be eliminated and what difficulties may be encountered with other options.

Principal Influences on the Master Plan

The principal influences and controls in generating water treatment process options in a master plan include the following:

1. Drinking water regulations, including those that are current, pending, and anticipated
2. Treatment options and limitations that produce quality consistent with regulations
3. Choices available where more than one treatment method or treatment train may be equal in cost and other features

Every treatment method development should be analyzed in the above order. Rules and regulations establish what must be accomplished in treated water quality, and often more than one method of treatment may accomplish the required result.

The potential options and the limitations of each method are developed in the next step. Limitations may include factors such as cost, operating control, and size of facilities. Where several viable options are available, different treatment trains are evaluated, taking advantage of the multiple choice to determine the most advantageous option. Comparisons here may include factors such as best fit on the site, appearance, ease of operations and maintenance, and vulnerability to upset.

When alternative source waters are available, the potential sources (surface water and groundwater) and specific intake or well locations should be evaluated to determine the water quality characteristics of each source. The vulnerability of source water to future contamination or water quality deterioration should receive particular consideration. The variability of the source water quality should be investigated because extreme water quality conditions often dictate treatment requirements. Once a new water supply is chosen, a protection program should be implemented (either watershed or groundwater protection) to maintain the integrity of the supply.

Planning Periods for the Master Plan

Major capital projects, such as large water treatment plants, generally require many years for planning, design, and construction. This, together with the expected long life of these facilities, results in exceptionally long design periods for water systems. Master planning studies often develop the water supply and treatment needs for 30 to 50 years into the future.

Based on complexity, expendability, and cost, the various components of the water filtration plant are sized to meet the needs of varying periods of time. More difficult and expensive facilities, such as intake tunnels and major structures, are often designed for the life of the facility, which can be as much as 50 years or more. Other facilities, such as process treatment units, are often initially planned with a first-phase design of 10 to 15 years, with a plan to allow for future increments of expansion to accommodate the full life of the project.

Equipment such as pumps and chemical feed systems have an expected life of 10 to 15 years. Therefore they are designed for shorter-term capacities with allowances for replacement to meet future needs of the facility.

WATER QUALITY GOALS

The Safe Drinking Water Act (SDWA) and its subsequent amendments provide the basic rules for water quality produced by a treatment system. But the design engineer cannot work solely from federal requirements because they are only minimum standards. Individual states have the option of making the standards more stringent or of expanding the basic regulations to include other quality standards, so it is important to work closely with state officials when considering process options and design details.

The USEPA is in a continuous process of modifying and expanding drinking water regulations under the SDWA and its amendments. This has become a complex process that has involved difficult scientific issues, as well as the political considerations that are inevitable to regulatory processes. A number of rule-making proceedings are involved in this process. Some of these have been completed and others are in varying stages of development. These rules tend to fall into one of four categories:

- Finalized rules that are in effect. The rules have become established regulations. Where relevant, state public water supply enforcement agencies must incorporate these rules into their own regulations and determine how they will be administered.
- Finalized rules that are not yet in effect. These rules have completed the promulgation process, and provisions are known with certainty. An effective date for these rules, however, has not been reached and they have yet to formally become a component of the established SDWA regulations. At this stage of rule development, state primacy agencies may be developing procedures for incorporating the rules into their respective state regulations and are assessing options for administering the rules once they become effective.
- Proposed rules. These rules have reached an intermediate stage that reveals specific USEPA intent. However, the provisions are at a proposed level that allows for comment from interested parties. The USEPA is required to formally respond to all comments and may make modifications before promulgation of the final rule, depending on the availability of additional information and the impact of comments. It is significant to note that most rules have been modified during the time between the proposed and final stage as a consequence of this process.
- Rules under development. These rules have not been proposed, and USEPA intent is not fully developed. In some cases, draft rules may be developed; these have no formal status and the USEPA has flexibility for changing the drafts without going through a formal response process to outside parties.

It is essential that the design engineer be aware, and stay abreast, of both the established rules and those in stages of development.

Finalized Regulations

Finalized rules that are presently in effect include the Surface Water Treatment Rule (SWTR); the Total Coliform Rule; the Phase I, II, and V Contaminant Rules; and the Lead and Copper Rule. Rules that have been finalized provide two levels of standards to be regulated—primary and secondary standards.

Primary Drinking Water Standards are health-related criteria that require mandatory enforcement by state primacy agencies. Existing primary standards, as established by rules currently in effect as of December 1995, are described in Table 2.1.

Secondary Drinking Water Standards include criteria that are intended for control of aesthetic factors. Unlike primary standards, parameters developed as secondary standards are established as guidelines that are strongly recommended, but not required. They are generally enforced at the discretion of the state primacy agency. Table 2.2 summarizes existing secondary regulations. Although parameters governed by secondary standards are not health related, they can have a significant effect on customer acceptance and can be a source of considerable customer complaint. Among these are water color and tastes and odors in the drinking water. In general, the secondary standards are observed in drinking water treatment more by customer needs than by regulatory control.

Significant Proposed Rules

The more significant of the ongoing procedures that will affect future standards and ultimately add to existing primary and secondary standards include proposals for the Information Collection Rule (ICR), Disinfectant-Disinfection By-Product Rule (DDBPR), and

TABLE 2.1 Presently Enforceable USEPA Primary Drinking Water Standards

Inorganics
A list of some 17 chemical elemental or compound materials. It is noted that traditional treatment would have little or no effect in the reduction of many of the materials in this list.

Synthetic organic contaminants
A list of over 30 materials, for the most part manufactured for use as herbicides and pesticides. Many of these would require more than the usual filtration treatment to achieve the required reduction.

Volatile organics
A group of over 20 materials, essentially solvents used in manufacturing and other activities for cleaning.

Contaminant	Maximum contaminant level in mg/L (except as noted)
Disinfection by-products	
Total trihalomethanes (four-quarter average)	0.10
Turbidity	
Not to exceed 5% of monthly samples	0.5 ntu
Microbiological	
Total coliform bacteria	More than 5% of samples positive
Disinfectant residual/heterotrophic plate count	More than 5% of samples with less than detectable residual and HPC 500/ml
Radionuclides	
Radium 226 and 228	5 pCi/L
Gross alpha radioactivity	15 pCi/L
Beta and photon radioactivity	4 rem/yr*

*The contribution of this category of radionuclides is to be computed. Tritium at 20,000 pCi/L and strontium 90 at 8 pCi/L, respectively, are computed to give no more than 4 rem/yr.

Enhanced Surface Water Treatment Rule (ESWTR). All of these have a widespread impact on water utilities.

In the case of the DDBPR, the USEPA recognizes a substantial dilemma because excessive control of disinfection by-products could compromise efforts for providing disinfection and other treatment to control waterborne disease. Waterborne disease is a potentially serious risk for drinking water, and there is concern that the capability to control this risk may be compromised by overly restrictive regulation of disinfectants and disinfectant by-products. Therefore there is a degree of uncertainty in the drinking water industry for dealing with this issue.

Existing regulation of DDBPs is presently confined to a total trihalomethane (TTHM) standard of 100 μg/L, averaged over four consecutive calendar quarters. In the proposed DDBPR, the USEPA aims to impose lower levels for TTHM, with the addition of haloacetic acid (HA) and bromate as new by-product categories to be regulated. Regulations are also proposed to limit the allowable level of certain disinfectant residuals, including chlorine dioxide, chlorite, chlorine, and chloramine. A number of other by-products have also been discussed for future regulation. It should be noted that several ozone by-products are also being considered for future regulation. These include several additional by-products at varying levels of concern: aldehydes, epoxies, peroxides, and iodate.

TABLE 2.2 Presently Enforceable USEPA Secondary Drinking Water Standards

Contaminant	Maximum contaminant level
Aluminum	0.05 to 0.2 mg/L*
Chloride	250 mg/L
Color	15 color units
Copper	1 mg/L
Corrosivity	Neither corrosive nor scale forming
Fluoride	2.0 mg/L
Foaming agents	0.5 mg/L
Iron	0.3 mg/L
Manganese	0.05 mg/L
Odor	Three threshold odor numbers
pH	6.5 to 8.5
Silver	0.1 mg/L
Sulfate	250 mg/L
Total dissolved solids	500 mg/L
Zinc	5 mg/L

*Selected level for aluminum depends on the discretion of the state primacy agency.

The principal issues addressed in the DDBPR proposed in the Federal Register of July 29, 1994, are summarized in Table 2.3. As indicated, the principal contaminants of concern were bromate, chlorites, HA, and TTHM. As proposed, the MCL of the latter two contaminants would be reduced in two stages. Best available technology (BAT) was proposed for the reduction or control of formation of all four contaminants. Through control of ozonation and chlorine dioxide disinfection, satisfactory levels of bromate and chlorites might be achieved. TTHM and HA levels, however, would be controlled by enhanced coagulation or enhanced softening to remove precursors that react with chlorine.

In one significant development related to goals for the SDWA, the USEPA had considered an interim emergency rule to respond to recent concerns for *Cryptosporidium*, but it has elected to use the voluntary Partnership Program for Safe Water as an alternative. This voluntary program is presently being implemented with an enrolled membership of over 300 utilities. The focus of this program is to ensure a response to concerns for *Cryptosporidium* through optimizing the filtration process for effective cyst removal by particle filtration mechanisms. A key treatment goal is the production of a filtered water turbidity of 0.1 ntu.

General guidelines have also been proposed for the ESWTR, the objective of which is essentially the removal and inactivation of *Cryptosporidium* cysts. Technology to control these cysts has not advanced sufficiently to propose a specific BAT with confidence. Although cyst particle removal through stringent filtration can substantially reduce *Cryptosporidium*, the disinfectant most capable of achieving inactivation in an efficient manner appears to be ozone.

The two rules proposed in the same Federal Register were actually at odds. In this same period of time, tests conducted for treatment of a large Northeast water supply indicated that it would take three times the coagulant dosage for optimum particulate removal in rapid sand filtration to reduce the organic precursor content sufficiently to produce water with satisfactory TTHM levels after chlorination. This increased dose would produce water with higher particulate content, especially in the 3 to 15 μm range considered significant for cyst reduction.

It is emphasized that the primary purpose of this discussion of new proposed rules is to illustrate the challenge presented to the design engineer in establishing treatment goals.

TABLE 2.3 Proposed USEPA Primary Drinking Water Standards for the Disinfectant/Disinfection By-Product Rule

Contaminant	Maximum contaminant level, mg/L		
Stage 1			
Bromate	0.010		
Chlorite	1.0		
Haloacetic acid (five) (HA)	0.060		
Total trihalomethanes (TTHM)	0.080		
Stage 2			
HA	0.030		
TTHM	0.040		
Required removal by enhanced coagulation (subpart H systems using conventional treatment)			
Source water TOC (mg/L)	Percent TOC removals for source water alkalinity (mg/L)		
	0 to 60	>60 to 120	>120
>2.0 to 4.0	40.0	30.0	20.0
>4.0 to 8.0	45.0	35.0	25.0
>8.0	50.0	40.0	30.0

The rules as proposed will, no doubt, be changed in some manner to permit treatment acceptable for by-product control and cyst removal and inactivation. The question that remains is how to prepare a design that must be completed before the rules are finalized.

A design must be prepared to produce water quality in accordance with the enforceable standards, but with accommodation to permit changes and additions in the future for new rules under consideration. The principal concerns in this regard would be to allow for such changes and additions in both space and hydraulic flow allowances to minimize future problems. In some instances, it may be possible to predict acceptable future additions such as a change to ozone disinfection to inactivate *Cryptosporidium* cysts before rapid sand filtration to reduce by-product precursors. In other situations, addition of granular activated carbon (GAC) adsorption or biologically activated carbon media for filters or in separate treatment steps might provide for by-product reduction while maintaining optimum particulate removal conditions.

In preparing any water treatment plant design, it may be considered almost a certainty that some treatment changes will be required in the future. It is important, therefore, that at least space and hydraulic flow accommodation be provided for future changes.

TREATMENT OPTIONS

As illustrated in Table 2.4, many treatment options are available to the designer to achieve the desired water quality results.

Process Alternatives

Detailed unit process tests and comparisons should be conducted under the guidelines of the companion volume to this text, *Water Quality and Treatment*. Bench and pilot scale treatment operations would also be conducted in the water quality and treatment phase.

TABLE 2.4 Most Common Drinking Water Treatment Processes

Water quality parameter	Process components

Turbidity—particulate reduction	Filtration
	• Rapid sand—conventional
	  Coagulation
	  Flocculation
	  Sedimentation
	  Filtration
	• Rapid sand—direct mode
	  Coagulation/flocculation
	  Filtration
	• Slow sand filtration
	• Diatomaceous earth filtration
	• Membrane filtration

Bacteria, viruses, cyst removal	Partial reduction—filtration (above) Inactivation—disinfection • Chlorine • Chloramine • Chlorine dioxide • Ozone
Color	Coagulation/rapid sand filtration Adsorption • Granular activated carbon (GAC) media • Powdered activated carbon (PAC) addition • Synthetic resins (ion exchange) Oxidation • Ozone • Chlorine • Potassium permanganate • Chlorine dioxide
Taste and odor control	Oxidation • Ozone • Chlorine • Potassium permanganate • Chlorine dioxide BAC adsorption
Volatile organic reduction	Air stripping GAC adsorption Combination of above
THM and HA control	Precursor reduction • Enhanced coagulation • GAC adsorption • Biologically activated carbon (BAC) media—preozonation By-product removal • GAC adsorption • Air stripping (partial)
Iron, manganese reduction/sequestering	Filtration of precipitators formed by preoxidation Coagulation/filtration of colloids Polyphosphate sequestering agents

TABLE 2.4 Most Common Drinking Water Treatment Processes *(Continued)*

Water quality parameter	Process components
Hardness reduction	Cold lime softening
	Ion exchange
Inorganic, organic chemical reduction	Ion exchange
	Biologically activated carbon (BAC) media adsorption
	Membrane filtration

There are essentially five methods of filtration that may be considered, depending on the limitations of the water to be treated:

- Conventional rapid sand filtration is used for high-turbidity situations.
- Direct filtration might be limited to waters of lower turbidity.
- Slow sand and diatomaceous earth (DE) filtration may be considered for particulate removal of almost any source water of 5 ntu (nephelometric turbidity units) or less.
- Depending on the nature of the particles contributing to turbidity, some waters of up to 10 ntu may be treated using slow sand or DE filtration.
- It is difficult to define the limitations of membrane filtration using only turbidity. Several membrane types are available, each of which can remove particles of different sizes, as well as certain dissolved constituents.

Treatment Train Alternatives

In developing a water treatment train, the multiple treatment capabilities of the different methods and materials should all be considered to both simplify and reduce the cost of facility construction and operation. Multiple capabilities of the different options in Table 2.4 are grouped in Table 2.5. This information indicates why rapid sand filtration with coagulation is the most common type of water treatment in use. This type of filter plant reduces content particulate matter, pathogens, disinfection by-product precursors, and color. If a filter is fitted with a GAC medium, the feature of taste and odor control is added, and color and precursor reduction is enhanced. With the addition of preozonation, the effective life of the GAC medium is increased because organics are removed more through biological action. In addition, preozonation, or application of another oxidant, conditions dissolved iron and manganese for removal in the filter.

Options and supplementary features that may be considered in developing baseline filtration trains are illustrated in Figures 2.1 through 2.3. On the schematics, basic treatment facilities are shown in solid boxes, and options are dashed. Alternatives both in fitting out facilities and in operations are listed below the particular facility.

Figures 2.1 through 2.3 are schematics of the four baseline treatment trains included in USEPA rules. There are two alternatives for rapid sand filters (Figure 2.1). In the first or conventional mode, part of the coagulated solids are removed in a settling or separation step preceding the filters. In direct filtration, all coagulated solids are removed by granular media filters. The other baseline trains are slow sand filters with a single sand medium and diatomaceous earth filters using powdered media coated on a porous septum.

TABLE 2.5 Common Multipurpose Treatment Measures

Treatment measure	Quality improvement
Filtration (all)	Particulate reduction
	Bacteria, virus, cyst reduction
Coagulation, rapid sand filtration (additional)	Precursors and by-product reduction
	Color removal
Oxidation	Pathogen inactivitation
	Color removal
	Taste and odor control
	Iron and manganese reduction*
GAC media	Rapid sand filter particulate removal
	Color removal
	Precursor and by-product reduction
	Precursor reduction with preozonation
	Taste and odor control

*After filtration.

Figure 2.1 illustrates the flexibility and broad treatment capabilities of the options. Facility modifications or options may be adapted for many treatment requirements. For instance, different coagulants may be considered that can be afforded high to low energy flocculation to produce a floc size and density best suited to the particular settling device or filter medium. Small, light, and floatable solids may best be removed by dissolved air flotation rather than by settling. Using GAC media preceded by ozonation provides the added capability of organics reduction by biologically activating the media. Ozonation may also reduce or convert many organics to enhance disinfection by-product reduction and to reduce chlorine demand. The multiple advantages of the options shown in these figures will be found in more detail later in this text.

Figure 2.2 shows how an intermediate GAC layer or "sandwich" in the slow sand filter medium, along with ozonation, can be used to enhance organics reduction. Chlorine disinfection, ozone disinfection, and posttreatment alternatives would be the same for all baseline options.

As indicated in Figure 2.3, both facility and chemical addition can broaden treatment capabilities of DE filtration. Adding potassium permanganate to the flow followed by sufficient detention time provide for manganese reduction.

In spite of its multiple benefits in addition to particle and turbidity reduction, rapid sand filters may not be the most advantageous filter for many applications, especially for those systems with low turbidity and color source water. Slow sand and DE filter plants may be less costly and less complicated to operate. Membrane filtration technology is an option that may become more attractive in many applications in the future.

Treatment Comparisons and Evaluations

The treatment train options discussed above illustrate the flexibility of porous media filtration. There are many other facets of water treatment, such as membrane filtration, softening, demineralization, and air stripping, which are discussed later in the text. All unit process issues and options should be a part of the design procedures covered in the companion text *Water Quality and Treatment*.

MASTER PLANNING AND TREATMENT PROCESS SELECTION

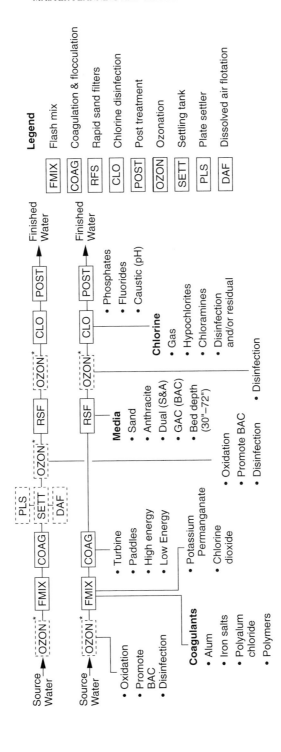

FIGURE 2.1 Baseline filtration options—rapid sand filters.

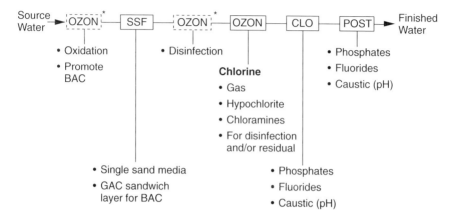

*Only single location used

Legend

SSF	Slow sand filters
CLO	Chlorine disinfection
POST	Post treatment
OZON	Ozonation

FIGURE 2.2 Baseline filtration options—slow sand filters.

The principal intent of these discussions is to emphasize that many treatment options and combinations of options available to the designer. It is essential that all viable options be investigated for each treatment application. It is also essential that issues other than treatment capability be investigated for each option and each treatment train. These other issues may include the following (not necessarily in order of importance):

- Construction cost
- Annual operation costs
- Site area required

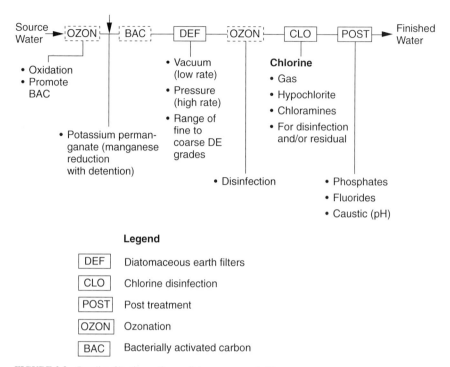

FIGURE 2.3 Baseline filtration options—diatomaceous earth filters.

- Complexity of operation (required capability of operating staff and laboratory monitoring)
- Operation risks (most common causes, if any, of treatment failure)
- Flexibility of plant arrangement for future changes

Consideration of all viable options would also be critical to provide a flexible facility arrangement in which additions and modifications may be made for future treatment requirements. Drinking water treatment design is not static; it is a dynamic, ever-changing process.

CHAPTER 3
DESIGN AND CONSTRUCTION

This chapter provides general guidelines for managing the design and construction phases of a water treatment project. The principal design and construction phases are preliminary design, final design, construction, and start-up.

PRELIMINARY DESIGN

Completion of the master planning and process selection phases establishes the design concepts for the water treatment project, forming the basis for preliminary design. The preliminary design phase must address a variety of design issues, concluding with concise recommendations for the principal components of the water treatment plant facilities.

Regulatory Agency Coordination

Regulatory agencies having approval authority over construction of a water treatment plant project may include local building departments, zoning commissions, and environmental commissions. In addition, there will be state and federal authorities having jurisdiction over water supply public health and environmental issues. Close coordination and communication with these agencies during the preliminary design phase is essential, because plant design will eventually have to be reviewed and approved by these agencies.

Most regulatory agencies have design standards or guidelines that should be reviewed to identify any potential conflicts the project poses. As part of this review, agency permits required for the project should be identified. Permits may be required for a number of plant design issues that could affect project implementation, such as site approval, storm water management, plant discharges to receiving streams, construction in waterways, environmental impacts, and water rights.

Although it may not be required, it is often beneficial to submit copies of preliminary engineering documents to regulatory agencies for review and comment to avoid major changes during final design.

Plant Siting Analysis

Selection of potential sites for a new water treatment plant must take into consideration a number of factors. Some of the principal ones are:

- Proximity of the plant site to the source water and to customers being served
- Consideration of finished water transmission requirements to interconnect the plant to the water distribution system
- Environmental and land use concerns
- Subsurface and geotechnical considerations
- Land availability, cost, and zoning
- Compatibility with surrounding developments
- Potential for flooding
- Availability of utilities
- Site topography and accessibility

Source water supply, treatment, and finished water transmission facilities must function as a complete system to provide a safe, reliable source of drinking water to the water distribution system.

Upgrading or Expansion of Existing Facilities

Many water treatment plant design projects involve modifying existing treatment facilities to provide increased capacity, to improve treatment performance, or to modernize and renovate equipment and structures. These projects require special considerations and investigations that are not required when designing a new plant on a new site.

Investigations made during preliminary design for these types of projects must answer questions such as:

- To what extent must the existing facilities be brought up to modern design standards and building codes as part of the project?
- How can the original design, as-built condition, current condition, and remaining useful life of existing facilities and structures be determined?
- How can the existing treatment facilities be kept in reliable service during the construction of new modifications?
- What safety or environmental hazards (such as lead paint, asbestos insulation, and underground petroleum spills) may be uncovered during the construction of the modifications?

Design decisions on these issues often require reasonable compromises that must be worked out among the design team, the owner, and regulatory agencies.

Pre-Engineered versus Custom-Built Equipment

When the treatment plant under design has a capacity of less than about 10 million gallons per day (mgd), it is usually possible to choose between custom-designed, pre-engineered, or "package" units. In custom-designed units, major process units usually involve reinforced concrete construction. The design engineer is responsible for all dimensions of the unit and the complete process and structural design. Equipment suppliers provide and are responsible for design and performance of specific equipment such as weirs, gates, sludge collectors, and other equipment located in the structural unit.

Pre-engineered units are generally made of steel or fiber-reinforced plastic. An equipment supplier takes responsibility for the full process, structural, and mechanical design of the unit. The design engineer generally is responsible only for the foundation design and items such as influent water quality, hydraulic and mechanical connections, and, if required, buildings or enclosures. Pre-engineered units are field assembled. Package units are partially or completely factory assembled and installed on the site.

Custom-designed process units tend to have higher construction costs but lower maintenance needs and longer useful lives. The best decision for a specific project can be made only after a careful analysis of the costs and benefits of each feasible alternative. In general, as the plant capacity becomes smaller, pre-engineered and package units become more cost-effective.

Financial Feasibility Evaluation

Financing for a water treatment plant project should be finalized during, or immediately after, the preliminary design phase. Preliminary estimated costs for plant facilities should be developed, including the implementation schedule, approximate cash flow, and operations and maintenance costs, that establish projected annual costs for the new facility. Potential sources of revenue for financing the project must then be investigated to establish a viable program to fund the planned facility. AWWA Manual M29, *Water Utility Capital Financing,* provides a comprehensive overview of traditional and innovative financing sources that should be evaluated by water utility management when determining long- and short-term strategies for meeting capital requirements.

Design Development

As part of the preliminary design phase, several key design components must be developed:

- Plant flow schematics indicating all treatment units and equipment for both primary treatment processes and waste handling systems
- Conceptual plant layouts showing the arrangement of treatment process units and other related facilities
- Plant hydraulic profile establishing operating water elevations through the plant at normal and peak flow rates
- Design parameters establishing treatment unit sizes and specific design criteria for minimum, average, and maximum conditions of the major treatment processes and waste handling systems
- Requirements for ancillary facilities including space needs for plant administration, laboratory, maintenance, chemical handling and storage, low and high service pumping, and water storage
- Plant control concepts including plant control logic and key control parameters

Ongoing Coordination and Cost Evaluations

During the preliminary design phase, design development should be coordinated with all support disciplines to obtain input on design. This coordination is necessary to make sure that preliminary engineering does not overlook any design issues that could affect design decisions. Although the process and civil disciplines are normally the primary focus of the preliminary engineering stage, the architectural, structural, mechanical, electrical, and instrumentation support disciplines should be included in these discussions.

Regular meetings should be held with the owner during preliminary design to discuss design issues and update the client on the process of design. Design disciplines pertaining to design issues being discussed should be involved in client meetings whenever possible.

Value engineering is often conducted before beginning final design to evaluate the cost-effectiveness and constructability of the preliminary design. At this stage of design, value engineering focuses only on major design elements, such as number and size of treatment units, plant layout, design parameters, and support facilities.

During preliminary design, a preliminary cost estimate for the water treatment plant project should be prepared to ensure that project cost is within the owner's budget established during feasibility studies. Cost estimates are important for (1) making decisions between viable plant alternatives, (2) developing the process train, and (3) planning the project's budgeting and financing. The final cost estimate should provide a breakdown of costs for each major component of the project. A construction contingency of 15% to 25% is usually included in the cost estimate at this stage of design to provide for unforeseen costs not included in a preliminary estimate.

FINAL DESIGN

The final design stage is initiated after completion of preliminary engineering, documentation, obtaining approvals of design decisions, and securing project financing.

Production of Design Documents

Final design involves preparing detailed drawings and specifications within the project framework established during preliminary design.

Drawings and Specifications. The final design drawings and specifications preparation must take into consideration a number of constraints and requirements not identified in the preliminary engineering stage. Treatment process design should be completed in more detail before undertaking the final design with the support disciplines. Critical design elements that should be completed early in final design are:

- Selecting process equipment to be installed, including equipment sizes, weights, utilities required to support the equipment, equipment control requirements, and provisions for variations in the requirements of equipment from different manufacturers
- Delineation of ancillary facilities required, including piping, valves, chemical systems, controls, pumps, and other equipment
- Finalization of major plant piping and structures, including pipe size and routing, treatment structure sizes and elevations, and support facilities size and functions

Project Control. It is important to maintain close communications with the client and regulatory agencies through all stages of design. Regular meetings should be held with clients to solicit their input and apprise them of project status. Members of the project operations staff should be involved in the meetings when possible to provide input on design decisions and to become familiar with the new plant facilities. During design, all disciplines should be kept informed of design decisions that could affect the design by any discipline. Regular design team meetings should be conducted for the designers to exchange information and coordinate conflicts between disciplines.

Schedules of project work tasks and project milestones should be prepared for all stages of design, with the schedule updated as changes are necessary. Reviews of the schedule should be conducted periodically to confirm that work tasks and milestones are on schedule. As part of schedule development, project budgets can be prepared based on the identified work tasks and schedule. Input from each design discipline should be obtained during budgeting to establish the needed work tasks, schedule, and budgets for each design stage. Previous project budgetary information can be a particularly valuable source of information for developing budgets for a new project. Manual M47, *Construction Contract Administration,* provides information on coordinating stages of design and maintaining communications between all parties involved in the construction project.

Design Standards. It is critical that facilities' details conform to both legally required and generally accepted design standards, including:

- Treatment design criteria published by state and federal agencies
- Design codes, such as building, electrical, fire, and plumbing codes
- Applicable laws and ordinances, such as Occupational Safety and Health Administration (OSHA) safety standards and Americans with Disabilities Act (ADA) requirements
- Design standards developed by local agencies

If the proposed design requires any deviations from legally required criteria, the changes must be discussed with the approving agencies and written variances or exemptions obtained. Any deviance from generally accepted practice should receive careful consideration and should be reviewed with the owner.

Coordination of Design Disciplines

Once final design is under way, it is essential that all design support disciplines be closely coordinated to avoid design conflicts and to make sure design information from all disciplines is received in a timely manner.

Civil/Mechanical Process Design. In developing the plant layout for construction of new facilities or modification of existing facilities, a number of design considerations should be addressed:

- Interfacing requirements with any existing facilities, including treatment units, piping, buildings, controls, power supply, and chemical feed systems
- Siting of new facilities to avoid construction conflicts and allow continual operation of existing facilities
- Providing ample site space for future additions or expansion to the facilities, including provisions for future tie-ins and equipment additions

Plant hydraulic design should be conducted early in the project, with the hydraulic profile developed to establish structure elevations at the plant site. Plant facilities should be arranged on the site to take advantage of available physical relief at the site, locating treatment structures to minimize site work where practical. Siting plant facilities should also take into consideration a number of other site factors, including:

- Topography
- Physical features or constraints
- Flood protection
- Drainage and storm water detention
- Piping between structures
- Vehicle access and parking
- Gas, power, and telephone utility routing
- Plant aesthetics and public acceptability
- Plant site buffer requirements
- Archaeological, historical, or cultural resources

The design should also address a number of factors that affect other design disciplines, including:

- Final dimensions of treatment structures
- Utilities requirements
- Equipment locations, dimensions, and operations requirements

Site Surveying. Topographic and property surveys should be made for the water treatment plant site to provide the baseline conditions for developing the plant location design and for obtaining many of the required permits. A permanent site benchmark should be established at the plant site located at a point that will not be disturbed by future construction, and a legal description of the property is needed.

A site plan is then prepared indicating property limits, site contours, physical features, buried utilities and pipelines, structures and buildings, roadways, easements and rights-of-way, and other features.

Geotechnical Investigation. The geologic conditions at the water treatment plant site should be established by a geotechnical investigation. Any existing subsurface information available should be collected first and reviewed to identify potential problems. The location of soil and rock strata should then be established by test borings and excavation of test pits. This information is vital to establish the bearing capacity of soils, to establish rock elevations, and to determine groundwater levels. Location of soil borings or test pits should be indicated on the site topographic plan, with the site survey coordinated to accurately locate the borings. Seismic conditions at the site may also have to be investigated to determine any special design considerations that need to be addressed. In addition, it may be necessary to conduct hazardous materials investigations to identify any potential site contamination or subsurface problems.

Architectural Design. Development of the architectural design of water treatment plant facilities must be closely coordinated with the process design because building requirements are normally based on the process requirements. Architectural concepts should be compatible with plant surroundings, with particular emphasis on minimizing building maintenance requirements. Architectural design is based on input from the other design disciplines, including:

- Treatment unit enclosure requirements
- Equipment space requirements
- Operator access requirements to plant facilities
- Public access requirements to plant facilities

Where practical, architectural concepts used for new plant additions or modifications should be compatible with existing plant buildings. Provisions for laboratory, office, maintenance shop, locker rooms, restrooms, storage, lunch rooms, meeting rooms, and control rooms are usually considered in the plant design.

Structural Design. Structural design is normally most affected by the civil, process, and architectural design decisions. The size and location of the treatment structures are particularly critical to structural design. If unusual or difficult geotechnical conditions are encountered, it may be necessary to change structure concepts to adapt to the subsoil constraints. As a result, early coordination with the geotechnical investigations is necessary to identify any unusual conditions.

Mechanical Design. Mechanical design establishes heating, ventilating, and air conditioning (HVAC) requirements for buildings. Chemical handling, storage, and laboratory areas normally require special attention because of their unique HVAC requirements. Mechanical and architectural designs must be coordinated closely to ensure that sufficient building space is allocated for the plumbing and HVAC equipment and ductwork.

Electrical Design. Electrical design should address a number of issues related to other design disciplines including:

- Power supply needs for all plant equipment
- Standby power requirements
- Energy conservation measures
- Interior and exterior lighting requirements

Adequate building space must be provided for electrical switchgear and motor control centers, with major electrical equipment often isolated to protect and adequately cool equipment.

Instrumentation Design. Instrumentation design is closely interrelated with process design to provide necessary control and monitoring of treatment processes. Automation of plant control should be considered, where practical, which requires particularly close coordination with process designers. A central control room is normally provided for most treatment plants, with plant operations monitored and controlled from that location.

Design Reviews and Final Documents

It is extremely important that the design team continually keep in mind various reviews that must be periodically made of final documents as they are developed and completed.

Owner Reviews. Throughout final design, the owner should be involved in key design decisions. Periodic meetings to discuss plans and specifications should be conducted to ensure that design meets the owner's needs. Plans and specifications should be submitted to the owner for review as project design proceeds, including intermediate and final documents. The owner's comments should be discussed with design disciplines and incorporated into the documents. Participation of the owner's operations and engineering staff is extremely important, because they will be the ultimate users of the facility.

Value Engineering. If value engineering is performed at this stage of the project, it focuses only on design details. It would not ordinarily consider major design concepts, because changing the design at this stage involves significant changes to plans and specifications. Issues usually addressed include materials of construction, equipment selection, and constructability.

Regulatory Agency Reviews. Final design plans and specifications must be submitted to various regulatory agencies for review and the review comments incorporated into documents where necessary. If possible, meetings should be conducted with regulatory agencies during final design to keep them informed of project status and to obtain their input on critical design issues.

Final Cost Estimates. A detailed cost estimate should be prepared as soon as practical during final design. The estimate should be completed for each design discipline indicating unit quantities and costs for all construction items. Cost estimates must be submitted for owner review. In addition, they must accompany applications for state or federal funding assistance.

Bid Documents. Final design plans and specifications can be formatted to allow construction using one of several contracting approaches. Normally, conventional bid documents are used that allow the owner to accept competitive bids and award the entire construction job, or job components, to the lowest qualified bidder. Under this approach all elements shown in the bid documents are the contractor's responsibility to construct, with the owner and engineer inspecting the project to ensure compliance with bid documents. In some instances, major equipment to be installed on the project is prepurchased by the owner and supplied to the contractor for installation. Prepurchased equipment is normally used to accelerate the construction schedule or to allow use of proprietary equipment.

Another bid approach gaining popularity is use of a design/build format. Under this approach, only preliminary engineering is completed for bidding, with final design prepared by the low, qualified bidder. The owner then monitors compliance with bid documents during construction, usually with assistance from the preliminary designer.

CONSTRUCTION

After completion of final design and incorporation of all pertinent review comments, the water treatment plant design can proceed into the construction phase. During construction of water treatment plant facilities, a number of design issues must be addressed to ensure the successful implementation of contract documents. These relate to bid administration, construction administration, design disciplines coordination, and owner and regulatory review of change orders.

Administration of Construction

Administration of a construction job includes bid administration, construction administration, and coordination of design disciplines.

Bid Administration. The time frame required for bidding varies depending on the size and complexity of the project but is normally about 30 to 60 days. Bid administration activities generally include printing the bid documents, conducting prebid meetings, answering bidder questions, preparing addendums to bid documents, bid evaluation and award recommendation, and contract preparation. Bid administration concludes with the award of the construction contract and the issuance of a notice to proceed to the selected contractor.

Construction Administration. The administration of project construction consists of a number of tasks that must be provided to oversee the work in progress. Some of the principal duties include:

- Field observation of the work as it progresses to verify the contractor's compliance with contract documents
- Clarification of contractor questions pertaining to the drawings and specifications
- Review of shop drawings submitted by the contractor
- Review of payment requests
- Monitoring compliance with the construction schedule
- Testing materials, such as concrete, for acceptability
- Coordination of field orders and change orders

Coordination of Design Disciplines. All design disciplines should be actively involved in construction administration. Members of the design team should make periodic visits to the construction site to monitor construction activities related to their discipline. As part of plant construction administration, it is necessary to coordinate modifications and changes to design with the various design disciplines.

Construction Approvals

No matter how carefully a construction job is planned and designed, there are almost always some changes that must be made during construction as a result of unexpected problems or a change of mind by the owner, designer, or regulatory agency.

Owner Approval of Change Orders. To facilitate timely processing of changes to contract documents, the owner should be apprised of potential or upcoming change orders as they become apparent. Owner approval is normally required for most change orders unless the designer has been designated as the client's contractual authority. Documentation should be provided for each change order to ensure that all parties have a clear understanding of the changes it will cause in the project cost and construction schedule.

Regulatory Approval of Change Orders. In many instances in which regulatory approval of the design has been obtained, change orders must be approved by the agencies to ensure that the project still complies with agency regulations. Regulatory approval of change orders may also be necessary when funding for construction is provided by a regulatory agency.

PLANT START-UP

The final phase of a water treatment plant construction project involves activities associated with start-up and initial operation of new facilities.

Operator Training

With any new or modified plant facility, it is important that plant staff receive training to operate the new equipment and systems. Operations and maintenance manuals and formal training should be provided by manufacturers for all new equipment and systems. Before start-up, equipment or systems manufacturers or suppliers should inspect equipment to verify proper installation, supervise any adjustments or installation checks, provide a written statement that equipment is installed properly and ready to operate, and instruct the owner's personnel on proper equipment operations and maintenance.

Equipment Start-Up

Starting up a new or modified treatment plant must be carefully coordinated to provide a manageable sequence of operation that demonstrates acceptable operation of each system and piece of equipment. Depending on the type of facility, individual or multiple systems of the plant may undergo individual start-up if they will not interfere with normal operations of existing facilities. For new facilities, it may be necessary to operate the entire treatment process at start-up due to the close interrelationship of all treatment systems.

BIBLIOGRAPHY

American Water Works Association. *Construction Contract Administration,* M47. Denver, Colo.: AWWA, 1996.

———. *Water Utility Capital Financing,* M29. Denver, Colo.: AWWA, 1997.

Corbitt, Robert A. *Standard Handbook of Environmental Engineering.* New York: McGraw-Hill, 1990.

Kawamura, Susumu. *Integrated Design of Water Treatment Facilities.* New York: John Wiley and Sons, 1991.

Water Environment Federation and American Society of Civil Engineers. *Design of Municipal Wastewater Treatment Plants.* Vol. 1. Alexandria, Va.: WEF and ASCE, 1991.

CHAPTER 4
INTAKE FACILITIES

Intakes are structures built in a body of water for the purpose of drawing water for human use. As discussed in this chapter, intake systems include the works required to divert and transport water from a supply source, such as a river, lake, or reservoir, to a shore well or pumping station. For small water supplies, the intake system may be relatively simple, consisting of little more than a submerged pipe protected by a rack or screen. In contrast, for major water supply systems, intake systems can be extensive, with diversion accomplished by intake tower structures or submerged inlet works. An intake system may also include transmission conduits, screens, pumping stations, and, in some instances, chemical storage and feeding facilities. For intakes located on rivers, jetties or low-head dams may also be required to ensure adequate submergence during low-flow periods.

INTAKE FEATURES

An intake system must possess a high degree of reliability and be able to supply the quantity of water demanded by a water utility under the most adverse conditions. Intakes are exposed to numerous natural and artificial perils, and it is important that the designer anticipate and make provision for them. Conservative structural and hydraulic design and careful consideration of intake location should be the rule.

The Purpose of Intakes

The purpose of an intake system is to reliably furnish at all times an adequate quantity of the best available water quality. Reliable intake systems are costly and may represent as much as 20% of the total water treatment plant investment. Pipeline construction associated with intakes may involve extensive underwater work and the use of specialized marine equipment. As a result of these and other factors, the cost of such work will be 2.5 to 4 times that of a similar land project.

New challenges must be considered in the design of intake facilities, including drinking water regulations that make it desirable to begin chemical treatment at the earliest point in the source water delivery process. In addition, problems associated with zebra mussels add new criteria to intake facility design.

Intake Components

Specific components that make up an intake facility are influenced by many factors, including characteristics of the water source, required present and future capacity, water quality variations, climatic conditions, existing and potential pollution sources, protection of aquatic life, water-level variations, navigation hazards, foundation conditions, sediment and bed loads, required reliability, and economic considerations. In general, an intake facility consists of some combination of the components shown in Figure 4.1.

Each intake system presents unique problems. Once the specific components are defined, the facility design must achieve the following:

- Provision against failure to supply water because of fluctuations in water level or channel instability
- Provision for water withdrawal at various depths where desirable and feasible
- Protection against hydraulic surges, ice, floods, floating debris, boats, and barges
- Location of the intake to provide water of the best available quality, to avoid pollution, and to provide structural stability
- Provision of racks and screens as required to prevent entry of objects that might damage pumps and treatment facilities
- Provision of adequate space for routine equipment cleaning and maintenance
- Provision of facilities for removing pumps and other equipment when major repairs are necessary
- Location and design to minimize damage to aquatic life
- Provision of space and facilities for receiving, storing, and feeding treatment chemicals if chemical treatment is to be practiced at the intake

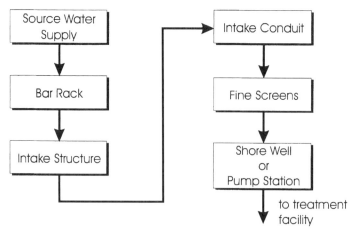

FIGURE 4.1 Component block diagram.

Types of Intakes

A variety of intake systems have been employed and can be generally divided into two categories: exposed intakes and submerged intakes (Table 4.1). Many variations of these principal intake types have been used successfully.

Intake towers are commonly used at lakes and reservoirs and are usually located in the deepest water that can be economically reached. Intake towers may also be incorporated into the dam that creates a reservoir and used as part of a river intake system. Water is conducted from the intake tower by a gravity pipeline or tunnel to the shore well or pumping station. Intake towers offer permanence, reliability, and flexibility in depth of draft, but their cost is substantial, and accessibility can at times be a problem.

The Metropolitan Water District of Southern California (MWD) owns the Eastside Reservoir, which will contain, at its maximum water level, about 800,000 acre-ft (986,200 m^3) of imported source water to serve as a water supply source for southern California. The tower, shown in Figure 4.2, is the structure by which water enters and exits the reservoir. The tower is 260 ft (79 m) tall, 100 ft (30 m) long in the direction of flow, and 80 ft (24 m) wide. It includes nine piping tiers, each equipped with two 84 in. (2.1 m) outlet pipes with butterfly control valves. Multiple tiers provide MWD with a way to selectively withdraw the best quality water as reservoir conditions change.

The tower is equipped with four movable wire cloth screens to prevent the entry of fish, and screens can be positioned over the intake ports selected for operation. Normal hydraulic withdrawal capacity of the intake tower is 1,100 mgd (4,200 ML per day) with maximum capacity during emergency drawdown of the reservoir of 4,500 mgd (17,000 ML per day). Chlorination is performed in the tower to control algae and mussel growth.

Siphon-well intakes are usually installed in rivers and consist of a shore structure that receives water from the river through a siphon pipe (Figure 4.3). The siphon-pipe inlet may be a submerged crib equipped with a trash rack or simply a screen section attached to the open pipe. Siphon-well intakes have a record of satisfactory service and are generally less costly than other types of shore intakes.

Elsinore Valley Municipal Water District draws source water from Canyon Lake for its 9 mgd (34 ML per day) water treatment plant. The intake for this facility, shown in Figure 4.4, consists of four 3 mgd (11 ML per day) capacity horizontal centrifugal pumps mounted on a floating platform anchored to the shore. The depth of water withdrawal is variable based on changeable suction piping length for each pump. A flexible hose is used for transmission of source water from the intake to the shore piping. An in-lake aeration

TABLE 4.1 Types of Intakes

Category	Design type	Remarks
Exposed	Tower (integral with dam)	Applicable to larger systems, more expensive
	Tower (lake interior)	Navigational impact
	Shore well	Design for floating debris or ice
	Floating or movable	Improved access for O&M
	Siphon well	Increased flexibility, provisions for expansion
Submerged	Plain end pipe or elbow	Applicable to smaller systems, less expensive
	Screened inlet crib	No navigational impact
		No impact from floating debris or ice
		Less flexibility
		Difficult O&M

34 CHAPTER FOUR

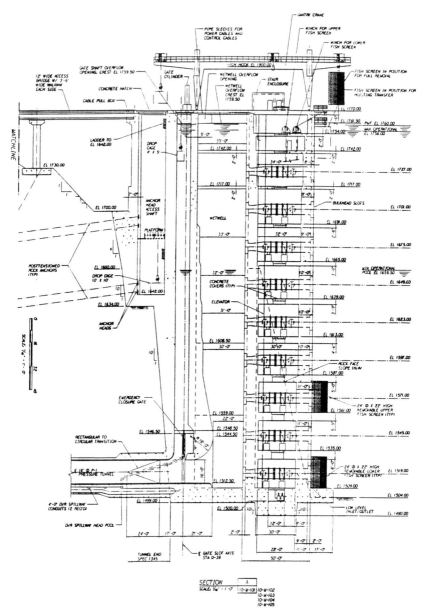

FIGURE 4.2 Eastside reservoir inlet/outlet tower.

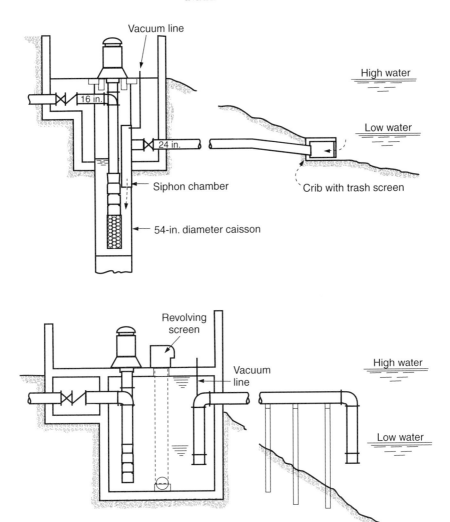

FIGURE 4.3 Siphon well intakes.

system is also used to minimize lake stratification and provide the best possible water quality to the plant for treatment. A design variation used where the water level varies is a movable carriage intake (Figure 4.5).

Submerged intakes may be constructed as "cribs" surrounding an upturned, bell-mouth inlet connected to an intake pipeline. The crib is often constructed as a timber polygon, weighted and protected by crushed rock. The intake conduit conveys water to the shore well, which may also be the source water pump station. The shore well is designed to dissipate surges and may contain either fixed or traveling screens. Submerged intake systems using the wood-crib arrangement have proven generally reliable on the Great Lakes when properly located. Other inlet configurations have also been successfully employed.

FIGURE 4.4 Floating intakes.

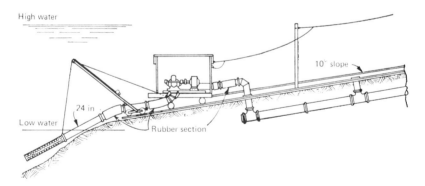

FIGURE 4.5 Intake pump on movable carriage.

Intake Capacity

Unlike water system elements such as pumping stations, basins, and filters, basic intake facilities usually cannot be readily expanded to provide additional capacity. For that reason, long-range water supply needs must be carefully considered when an intake is designed. In general, intake requirements to meet water needs 20 to 40 years in the future should be investigated. Although intake facility construction is generally costly, the incremental cost of providing increased capacity may be relatively small.

Intake Reliability and Redundancy

Reliability is essential, especially for waterworks that depend on a single intake system. Construction of duplicate intake structures is becoming increasingly common, including multiple inlet ports and screens, screen chambers, pump wetwells, and discharge conduits. Where impact from zebra mussels can be expected, redundancy takes on increased emphasis because of the need to provide periodic inspection and maintenance of intake components. For systems served by a single intake, failure of the intake works means failure of the supply, an emergency condition that, if not promptly corrected, can become a water supply disaster.

Intake Location

Selecting the appropriate location for an intake facility must include evaluating the major factors presented in Table 4.2. A thorough study should be made of water quality data to help determine the best location for an intake. Where insufficient data is available, a water sampling and testing program may be warranted.

TYPES OF INTAKE SYSTEMS

Intake systems are classified as either river intakes or lake/reservoir intakes because the circumstances, location, and types of structures used are generally quite different.

River Intakes

Water systems using rivers as a supply source can often combine an inlet structure and source water pumping in one facility. As an example, the bank intake system located on the Missouri River serving the Johnson County, Kansas, water system is shown on Figure 4.6.

TABLE 4.2 Intake Location Considerations

Criteria	Remarks
Water quality	Local surface drainage
	Wastewater discharge points
	Lake and stream currents
	Wind and wave impacts
	Water depth and variation
Water depth	Maximum available
	Adequate submergence over inlet ports
	Avoid ice problems
Silt, sand	Locate to minimize impact
Navigation	Outside shipping lanes, designed for accidental impact
Trash and debris	Provisions for unrestricted flowby
W.S. elevation	Maximum practical hydraulic gradient
Treatment facility	Close as practical to minimize conduit length
Cost	Least cost consistent with long-term performance and O&M requirements

FIGURE 4.6 Missouri River intake (shore structures).

This facility, part of a long-range water supply program, was designed for an ultimate capacity of 100 mgd (380 ML per day). Space was provided for six vertical, wet-pit pumps, each located in an individual cell and each protected by a removable bar rack and traveling screen. Source water enters through six rectangular ports located about 1 ft (0.3 m) above the bed of the river. Sluice gates enable each cell to be isolated for inspection, cleaning, or maintenance.

The Johnson County intake also includes presedimentation facilities at the water source. Heavy solids (sands and silts) are separated from the flow stream to reduce the load on pumps and pipelines, as well as to improve overall performance of the treatment plant.

Alternative intake systems should be considered because, under some circumstances, they may offer equally reliable service at lower cost. Alternatives that may prove useful in overcoming low-water and flood difficulties and bed-load problems include an exposed or submerged river inlet tower and shore pumping station, siphon intake, suspended intake, floating intake, and movable intake.

Figure 4.7 shows the Louisville (Kentucky) Water Company crib-type intake located on the Ohio River. The initial design capacity is 120 mgd (450 ML per day) but is expandable to 200 mgd (760 ML per day). The intake includes multiple screens mounted in a concrete crib structure constructed on the river bottom. Parallel source water conduits enable redundancy and increased hydraulic capacity. Provisions are included to backflush each source water conduit from the pump station. Capabilities for applying heavy doses of chlorine to control Asian clams and freshwater sponges to each conduit are also included.

The groundwater potential of a river valley is also worthy of consideration. Gravel-packed wells or horizontal groundwater collection systems may, on occasion, offer an economical alternative to difficult, costly intake construction and associated operating problems.

River Intake Locations. The preferred location for a shore intake system provides deep water, a stable channel, and water of consistently good quality. In general, the outside bank of an established river bend offers the best channel condition. The inside bank is likely to be troublesome because of shallow water and sandbar formation. The location should preferably be upstream of local sources of pollution. Considerable variation in water quality can result from the entrance of pollution from tributary streams above the proposed intake location, and water quality near one bank may be inferior to the quality at midstream or at the opposite shore.

Flood Considerations. It is essential to protect intake structures against flood damage. The intake structure must also be designed to prevent flotation and to resist the thrust of ice jams. Flood stages at the intake site should be considered carefully and a substantial margin of safety provided. Because of watershed and channel alterations, future flood stages may exceed those of the past, and so the designer should consider the possibility that the intake will be exposed to flood stages in excess of those of record.

Silt and Bed Load Considerations. Many streams carry heavy loads of suspended silt at times, and heavy material moves along the bed of many streams. The intake must be designed so that it will not be clogged by silt and bed-load deposits. Silt, sand, and gravel can also cause abrasion of pumps and other mechanical equipment, leading to severe problems at the treatment plant. To help prevent such deposits, jetties may be built to deflect the principal flow of the river toward and past the face of the intake.

Lake and Reservoir Intakes

Both tower intakes and submerged intakes are employed for water supplies drawing from lakes and reservoirs. A tower intake may be designed as an independent structure located some distance from shore in the deepest part of a reservoir. Access to these towers is

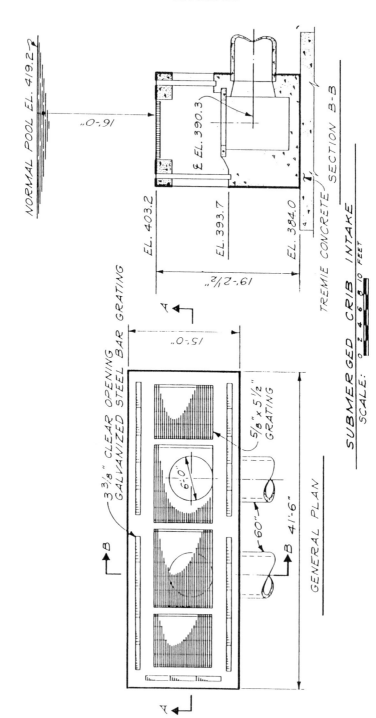

FIGURE 4.7 Ohio River intake (crib type).

provided by bridge, causeway, or boat. Towers must be designed to cope with flotation, wind, wave, and ice forces.

The Charlotte-Mecklenburg Utilities Department intake on Lake Norman is an in-lake tower facility (Figure 4.8). The intake has an initial capacity of 54 mgd (200 ML per day) and is expandable to an ultimate capacity of 108 mgd (400 ML per day), which should satisfy projected demand for the next 35 years. The intake uses passive stainless steel screens mounted on the exterior of the structure's inlet pipes to exclude entry of debris and fish. The screens are designed with air backwash for cleaning, and redundancy is provided with two 60 in. (1.5 m) source water conduits to shore facilities. This facility incorporates a special "gazebo" architectural concept to make the structure blend with the park setting.

Submerged intakes generally do not obstruct navigation and are usually less costly than exposed towers. If properly located and designed, submerged intakes experience a minimum of ice difficulties. They consist of a submerged inlet structure, an intake conduit, and a shore shaft or suction well. Shore intakes are occasionally built on lakes and reservoirs, but in general, their effectiveness is reduced by ice, sand, and floating debris.

The intake shown in Figure 4.9 has an initial capacity of 60 mgd (230 ML per day) but is expandable to the ultimate project capacity of 90 mgd (340 ML per day). This is a submerged type of intake consisting of three T-shaped, stainless steel passive screens located in Modesto Reservoir. Each screen is equipped for air backwashing for cleaning.

Lake and Reservoir Intake Location. The location of a lake or reservoir intake should be selected to obtain an adequate supply of water of the best possible quality, with consistent reliability, economical construction, and minimum effect on aquatic life. To avoid sediment, sand, and ice problems, a submerged intake's inlet works should be ideally located in deep water. Water of at least 50 ft (15 m) is desirable. To achieve this depth in lakes where shallow water extends for a long distance from shore, a long intake conduit may be required. In some locations, a shallower location may be acceptable if water of acceptable quality is available.

Water Quality Impacts on Intake Location. In the past, pollution has forced the abandonment of some lake intakes; therefore the pollution potential at proposed new intake sites should be carefully evaluated. When considering the pollution potential of a site, prevailing winds and currents are often significant. Review of seasonal water quality data provides further guidance in site selection. Special water quality surveys over a period of several years may be required to define the optimum location.

INTAKE DESIGN

Designing an intake system involves many different considerations of the capacity required, the best method of designing the intake structure, the best way to install the intake conduit, and how best to allow for problems such as ice formation and infestation by zebra mussels.

Design of Intake Capacity

Selecting design criteria and flow rates should reflect the longer planning period appropriate for major intake facilities. Hydraulic criteria to be evaluated are summarized in Table 4.3.

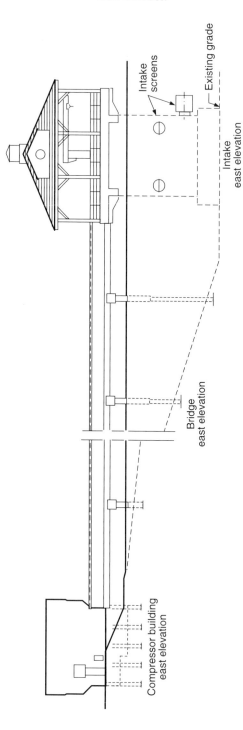

FIGURE 4.8 Lake Norman in-lake tower.

INTAKE FACILITIES 43

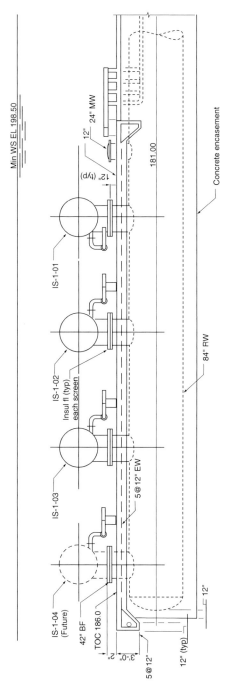

FIGURE 4.9 Submerged intake screens.

TABLE 4.3 Hydraulic Criteria

Flow criteria	Capacity	Remarks
Design flow	Q	Present design capacity
Minimum flow	0.10 to 0.20 Q	System specific
Ultimate flow	2.00 Q or higher	System specific
Hydraulic overload	1.25 to 1.50 Q	Evaluate for all design conditions

If the need for future increased capacity is likely, the potential for constructing an enlarged intake facility initially should be carefully evaluated. This may mean installing an oversized screen structure or parallel intake conduits (with one reserved for future capacity) to facilitate future expansion. Hydraulic analysis should also include calculation of a hydraulic overload condition to provide a safety factor against lost capacity caused by conduit fouling or for emergency conditions.

Submergence Design

Intake capacity is directly related to inlet submergence, and the designer should be aware of the substantial variations in water levels encountered in many lakes, reservoirs, and rivers. The intake system must have the capacity to meet maximum water demand during its projected service life, and intake capacity equal to this demand must be available during the period of minimum water level in the source of supply.

A conservative approach should be used to establish the lowest water level for intake system design. Lack of adequate submergence during periods of extreme low water results in greatly reduced capacity. Deeper than normal submergence should be considered for rivers subject to floating and slush ice, or where extremely low water may occur during the winter season as a consequence of ice jams.

On some rivers, it may be impossible to obtain adequate submergence under all expected operational conditions. For such situations, a low, self-scouring channel dam is required for a reliable intake system.

Cellular Design

Cellular or parallel component design of intake systems should be evaluated for major facilities. Cellular design divides the intake into two or more independent and parallel flow streams. This arrangement enhances reliability, provides flexibility, and simplifies maintenance activities. Individual cells can be taken out of service as required for inspection, cleaning, or repair.

Inlet Works and Ports

For intakes located on deep reservoirs and rivers, gated inlet ports may be provided at several depths for selective withdrawal of water at different depths. In rivers, submergence may govern, and ports must be placed at as low an elevation as practical. The lowest ports of an intake should be sufficiently above the bottom of the channel to avoid clogging by silt, sand, and gravel deposits. Port inlet velocities should be selected to minimize entrainment of frazil ice, debris, and fish. Factors that affect selection of intake port configuration and locations are summarized in Table 4.4.

TABLE 4.4 Intake Port Selection

Concern (to be avoided)	Required port location		
	Shallow	Intermediate	Deep
Organisms (requiring sunlight)		✓	✓
Warm water		✓	✓
Storm turbidity	✓	✓	✓
Plankton		✓	✓
Carbon dioxide	✓		
Iron and manganese	✓	✓	
Color	✓	✓	

TABLE 4.5 Location and Spacing of Ports

Criterion	Suggested location
Vertical spacing	10–15 ft (3–5 m) maximum
Depth above bottom	5–8 ft (1.5–2.4 m)
Depth below surface	Variable
Intermediate ports	Best water quality, avoidance of pollution
Ice avoidance	20–30 ft (6–9 m) below surface
Wave action	15–30 ft (5–9 m) below surface

Developing an intake port design should include an overall operational strategy identifying potential problems that may be encountered and the capability to minimize problems by selective withdrawal of water from the water source. It is likely that annual facility operation will require use of several intake ports to always draw the best water quality. A water quality monitoring system capable of defining water quality parameters throughout the depth of the water source should also be considered as an element in the intake facility design.

Experience has shown that the final number and spacing of inlet ports is affected by the specific conditions to be encountered, as shown in Table 4.5.

There is usually a wide variation of water quality with depth in stratified lakes. At Shasta Dam in California, turbid water behind the dam became stratified within several weeks after a major storm. The most turbid water settles near the bottom, and the water becomes progressively clearer toward the surface. These and other studies confirm the value of intake tower designs that include multilevel inlet ports. Occasional adjustment of the depth of draft often substantially improves chemical, physical, and biological water quality parameters. These improvements are reflected in enhanced treatment performance and reduced treatment costs.

Intake ports should be selected to achieve reliable delivery of water while minimizing the inclusion of unwanted material or contaminants. Table 4.6 provides a summary of design criteria.

The intake tower shown in Figure 4.10 is located in Monroe Reservoir, the water supply source for Bloomington, Indiana. Two intake cells, each equipped with three intake ports, provide variable depth withdrawal capability. Bar and traveling screens protect four vertical, wet-pit pumps. Design capacity of the intake system is 48 mgd (180 ML/d).

TABLE 4.6 Exposed Intake Design Criteria

Criterion	Remarks
Port velocity	0.20–0.33 ft/s (6–9 cm/s); 0.50 ft/s (15 cm/s) maximum
Ports	Multiple; three minimum
Water level variation	Design capacity at minimum level; operating deck above 500-year flood level

Various inlet designs for submerged intakes avoid sediment, sand, and ice problems. Inlets are best located in deep water with inlet velocities less than 0.5 ft/s (15 cm/s), preferably about 0.2 to 0.3 ft/s (6 to 9 cm/s). Inlet structures are usually constructed of wood or other nonferrous materials with low heat conductivity that are less susceptible to ice deposits. Wood cribs are usually polygonal, built of heavy timbers bolted together, weighted with concrete and crushed stone, and bedded on a crushed stone mat. The crib surrounds a bell-mouth pipe connected to the intake conduit. Riprap fill is placed on all sides of the crib. Wooden inlet cribs have provided many years of reliable service in the Great Lakes.

Figure 4.11 shows a section of the 315 mgd Milwaukee, Wisconsin, intake crib located in Lake Michigan, with a rated capacity of 315 mgd (1,200 ML per day). The crib is an octagonal, coated steel structure 11 ft (3.3 m) high and 52 ft (16 m) wide between parallel sides. It was floated into position and sunk by filling the air tanks with water. The horizontal baffle ensures relatively uniform flow through all parts of the intake screen. At the design flow rate, average velocity through the screen openings is 0.31 ft/s (9 cm/s). Mean water depth at the intake location is approximately 50 ft (15 m). The 108 in. (2.7 m) intake conduit extends 7,600 ft (2,300 m) to the pumping station located on the shore of the lake.

Other configurations for submerged intakes include hydraulically balanced inlet cones; screened, baffled steel cribs; and inlet drums. Hydraulically balanced inlet cones have been used at several Great Lakes intakes. The structure shown in Figure 4.12 consists of three groups of three equally spaced inlet cones connected to a cross at the inlet end of the intake conduit. This configuration provides essentially identical entrance velocities through all cones. The lower photograph shows a group of three inlet cones before placement in Lake Michigan.

Intake Conduit

The intake conduit, which connects the submerged inlet works with the shore shaft, may be a pipeline or a tunnel. Tunnels have a high degree of reliability, but they are usually too costly to construct for small water systems. A tunnel is often an economical choice for very large water systems. The selection of the design velocity for intake conduits is a balance between hydraulic headloss at high flow rates and the potential for sedimentation at low flow. Velocities in the conduit should be sufficient to minimize deposition. If very low flow rates may be experienced, provision of high-velocity backwash should be considered. Biological growths on the interior surface of the conduit may reduce its capacity, and this should be factored into the hydraulic design. Design criteria applicable to the design of intake conduits are presented in Table 4.7.

Subaqueous concrete pipe is generally used for intake conduits, with pipe laid in a trench in the lake or reservoir bottom. It is desirable to have approximately 3 to 4 ft (1 to 1.5 m) of cover over the top of the pipe, plus an additional protective top layer of crushed rock.

The allowable drawdown in the shore shaft is a critical factor in hydraulic design of the intake conduit. Drawdown results from friction loss in the conduit and equals the

INTAKE FACILITIES

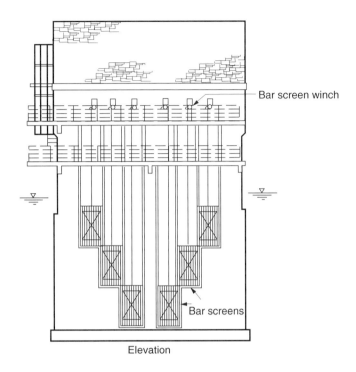

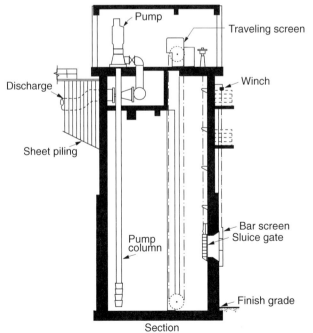

FIGURE 4.10 Multiple inlet gates.

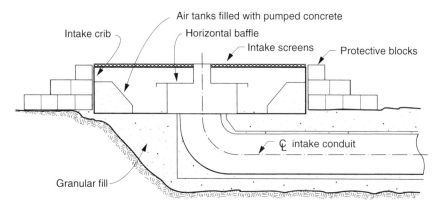

FIGURE 4.11 Lake intake crib.

difference between the lake or reservoir level and the level of water in the shore shaft. Drawdown should be limited to avoid excessive excavation for the shore shaft.

The conduit should be laid on a continuously rising or falling grade to avoid reduction in conduit capacity from air accumulating at high points. When an undulating conduit cannot be avoided because of the profile of the lake bottom, air release should be required at the high points.

Shore Well or Pump Station

The shore well serves as a screen chamber and a source water pump suction well. The depth of the shore shaft must be adequate to allow for drawdown when the intake operates at maximum capacity and the source water elevation is at its minimum. In addition, the well must provide an ample submergence allowance for the source water pumps. The well must also be capable of resisting and dissipating surges that occur when a power failure stops the source water pumps. Some shore wells are constructed with a shaft large enough to readily discharge the surge back to the supply source.

Many shore wells are equipped with fixed or traveling screens to remove objects large enough to cause pump damage. In general, traveling screens are preferable for all except small systems. Cellular construction of the shore well is advisable, and inlet control gates should be provided so that all or part of the shore well can be taken out of service for inspection and repair.

Flotation Considerations

Designing intake facilities requires careful evaluation of uplift forces to ensure that the structure is stable over the full range of water surface elevations that may be present. Two general methods for resisting the uplift forces are deadweight or foundation ties. The most common approach to stabilizing uplift forces is to design the structure with enough deadweight to resist all possible uplift force. Deadweight can be provided by tying the structure to a concrete mat or by filling cells of the structure with crushed rock or other heavy material. Foundation ties or rock anchors drilled into foundation rock are also occasionally used to tie down an intake structure. With either method, a safety factor against uplift of at least 1.2 should be provided.

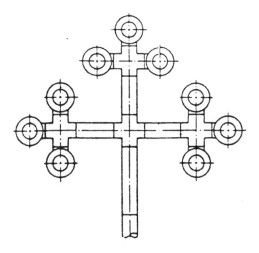

FIGURE 4.12 Hydraulically balanced inlet cones.

Geotechnical Considerations

Geotechnical investigations must be completed as the first step in designing an intake. Investigations usually include borings unless adequate data are already available from a previous installation in the same area. During the study phase, geotechnical data permit a more accurate comparison of the cost of viable alternatives to identify the most economical solution for the project. Geotechnical investigations should cover a sufficient area to account for adjustments in the intake location during final design. The cost of a

TABLE 4.7 Intake Conduit Design Criteria

Number of conduits	One minimum; two or more preferred
Velocity	1.5–2.0 ft/s (46–60 cm/s) at design flow; 3–4 ft/s (90–120 cm/s) maximum
Type of construction	Tunnel or pipeline
Slope or grade	Continuous to drain or to an air release valve

few additional borings during the initial study is much less than the cost of mobilizing a barge for additional borings if they are found necessary later in the design.

Provisions should also be included in construction specifications for diver inspections to verify that the subgrades are clean of loose and soft material before placing tremie concrete.

Mechanical Considerations

The mechanical design for an intake structure includes many of the routine considerations found in other, similar structures. Special mechanical design considerations for intakes include ventilation, insects, and flooding impacts.

Ventilation systems must be designed to accommodate heat generated by electric motors, motor control centers, and solar gain. Ventilation systems must also consider odorous gases that may be present, especially if screening material is temporarily stored in the building. With the trend toward more chemical treatment at the intake, ventilation must provide airflow required by applicable building codes for each stored chemical system. Ventilation of below-grade spaces that may occasionally be occupied for inspection or maintenance should also be provided.

Insects can be a serious problem at some intake sites, and insect screens on the air intake and exhaust louvers are one solution. However, large quantities of insects can block these screens, requiring increased maintenance by facility personnel. Another solution is to provide air filters with a large surface area on air handling units. Facility personnel may prefer the ease of changing a filter to the difficulty of cleaning louver screens.

Flooding conditions can affect the design of sanitary and storm drainage systems. If flooding is expected, backwater or knife gate valves should be provided in these systems.

Chemical Treatment Considerations

The intake facility provides the first opportunity in a water treatment system to impact, remove, or alter contaminants in the source water before subsequent treatment. A number of chemical treatment choices that may be provided at the intake should be considered in intake facilities design (Table 4.8).

The choice of chemical treatment to be applied at the intake must be coordinated with overall optimization of the water treatment process, including the control of disinfection by-product formation. Chlorine used to be the standard treatment to oxidize taste and odor compounds at the intake structure. Current practice is to avoid the early use of chlorine in the treatment process where the level of organics in the source water is highest because of the potential for forming chlorinated by-products such as trihalomethanes (THMs). Chlorine dioxide, potassium permanganate, and carbon are alternatives that avoid or minimize the formation of chlorinated disinfection by-products.

TABLE 4.8 Chemical Treatment Alternatives

Contaminant or criterion	Suggested treatment chemical
Debris and screenings	No treatment; discharge to water source or landfill
Taste and odor	Chlorine
	Chlorine dioxide
	Potassium permanganate
	Carbon
Coagulation	Alum or ferric chloride
	Cationic polymer
Zebra mussels	Chlorine and other oxidants

Ice Design Considerations

Intake systems located in regions with long, severe winters are subject to a variety of problems associated with ice in its various forms.

Surface Ice. In some locations surface ice and ice floes create a structural hazard to exposed intakes. On lakes, an accumulation of wind-driven ice floes near a shore intake can produce a deep, nearly solid layer of ice restricting or completely blocking intake ports. Under such conditions, reliable intake operation is virtually impossible, and water supplies obtained from lakes and reservoirs subject to severe ice problems are usually served by offshore intakes.

Ice jams can cause partial or complete blockage of river intakes. Jams below a river intake can also cause extremely high river stages, and an upstream jam can produce low water levels at the intake location, reducing its capacity.

Frazil and Anchor Ice. Ice starts to form when water temperature is reduced to 32° F (0° C) and water continues to lose heat to the atmosphere. For pure water, supercooling to temperatures well below the freezing point is necessary to start ice formation, but with natural water, the required supercooling is much less. Two types of ice formation are recognized: static ice and dynamic ice (Table 4.9).

Static ice forms in quiet water of lakes and river pools. Dynamic ice formation occurs in turbulent water such as areas of greatest flow in rivers and in lakes mixed by wind action. Frazil ice formed under dynamic conditions adversely affects hydraulic characteristics of intakes.

When natural water loses heat to the atmosphere and a condition of turbulence exists, uniform cooling of a large fraction of the water body occurs. If initial water temperature is slightly above the freezing point and cooling is rapid, a small amount of supercooling occurs, and small, disk-shaped frazil ice crystals form and are distributed throughout the turbulent mass. These small crystals are the initial stage of ice production. Other ice forms can develop from this initial ice production in sizable quantities.

Where there is little or no mixing, supercooled water and existing surface ice crystals are not carried to a significant depth, and the result is the formation of a layer of surface ice rather than a mass of frazil. Surface ice formation reduces heat loss from the water and usually prevents formation of frazil ice.

Two kinds of frazil ice have been identified, as shown in Table 4.10: active and inactive (passive). Freshly formed frazil crystals dispersed in supercooled water and growing in size are in an active state. When in this condition, they will readily adhere to underwater objects such as intake screens or rocks. Frazil ice production and adhesiveness are associated with

TABLE 4.9 Ice Formation Conditions

Type	Remarks
Static ice	Quiet waters
	Small lakes and river pools
Dynamic ice	Turbulent water
	Rivers in area of great flow
	Reservoirs with significant wind action
	Frazil ice formation occurs first
	Massive ice formations may follow

TABLE 4.10 Frazil Ice Characteristics

Type	Remarks
Active	Initial phase of formation
	Rapid growth in size
	Readily adheres to intake facilities
	Short-lived phase
	System clogging within a few hours
Inactive (passive)	Static or declining size
	Lost adherence and characteristics
	Less troublesome

the degree of supercooling, which is related to the rate of cooling of the water mass. Frazil ice particles remain in an active, adhesive state for only a short time after their formation. With the reduction of supercooling and the return of the water to 32° F (0° C), frazil crystals stop growing and change to an inactive, or passive, state. Passive frazil has lost its adhesive properties and is therefore less troublesome.

Frazil ice has been aptly called "the invisible strangler." When conditions favor its formation, the rate of buildup on underwater objects can be rapid; frazil accumulation can reduce an intake's capacity substantially or clog it completely in a few hours.

Some confusion exists concerning the relationship between frazil ice and anchor ice. It has been suggested that anchor ice occurs rarely and consists of sheetlike crystals that adhere to and grow on submerged objects. Accumulations of frazil ice may closely resemble anchor ice. Some investigators designate all ice attached to the bottom as anchor ice regardless of how it is formed. Anchor ice may form in place on the bottom and grow by the attachment of frazil crystals. On the other hand, according to Giffen (1973) ice crystallization and growth directly on the surface of a shallow intake structure in open water is commonly termed *anchor ice*. Anchor ice normally does not form at depths greater than 40 to 45 ft (12 to 14 m), although the depth associated with anchor ice formation depends on water turbidity.

Predicting Frazil Ice Formation. The climatological conditions that encourage frazil ice formation are:

- Clear night sky
- Air temperature 9.4° F (-7° C) or less
- Day water temperature 32.4° F (0.2° C) or less

- Cooling rate greater than 0.01° F (0.01° C) per hour
- Wind speed greater than 10 mph (16 km/h) at water surface

Frazil ice generally accumulates in the late evening or early morning hours and seldom lasts past noon. Conditions favorable to frazil ice formation vary considerably from site to site, making it difficult to use weather data alone as a forecasting tool.

Frazil ice formation can be illustrated graphically by plotting water temperature variation versus time, as shown on Figure 4.13. When the original constant cooling rate (A to B) of water undergoing supercooling deviates, frazil ice begins to form. The process illustrated in Figure 4.13 takes only a few minutes. The rate of ice production at point C is equal to the cooling rate divided by the product of the ice density.

Design Features. Research and experience on the Great Lakes and elsewhere indicate that location and design features of submerged intakes can reduce intake ice problems, but probably not completely eliminate them. Submerging lake intakes in deep water and sizing inlet ports for a velocity of 0.3 ft/s (9 cm/s) or less minimizes the amount of frazil ice transported downward to the structure. However, during winter storms, strong wind and wave action can carry ice crystals and supercooled water to considerable depths, making accumulation of ice on and around the intake likely. Some of the procedures to alleviate frazil ice problems at submerged intakes are summarized in Table 4.11.

Creating a quiet body of water at the intake location promotes surface ice formation. At the Billings, Montana, waterworks, frazil ice was a severe winter problem in the turbulent Yellowstone River. In this case, the solution was to enlarge an off-river intake channel into an earthen forebay with a detention time of approximately one hour. Surface ice formed on the nearly quiescent forebay. This insulating ice cover prevented the formation of additional frazil ice and provided the opportunity for river frazil carried into the forebay to combine with the surface ice and revert to a passive condition.

To prevent or at least minimize ice clogging, the structure can be built of low heat transfer materials with smooth surfaces not conducive to the accumulation of ice crystals. Metals such as steel are more susceptible to frazil ice formation because they have a high heat conductivity and act as a sink for the latent heat released when ice begins to form. This encourages ice buildup. In contrast, ice does not readily crystallize or grow rapidly on wood or plastic. Screens can be constructed using fiberglass-reinforced plastic with low thermal conductivity and a smooth surface. Any exposed metal surfaces can be coated with an inert material such as black epoxy paint to effect better thermal properties and to increase radiation heat gain.

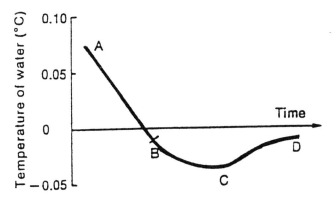

FIGURE 4.13 Frazil ice formation: water temperature sequence.

TABLE 4.11 Frazil Ice Mitigation

Criterion	Remarks
Location	Quiet waters away from turbulence Protected by ice cover Deep water
Inlet ports	Velocity Large bar rack opening (<24 in.) Low heat transfer materials Smooth surfaces Coating to improve thermal characteristics
Temperature	Heat to prevent ice adherence Raise local water temperature 0.1° C

Heated intake screens have been used successfully at power plant installations and may have application at waterworks intakes. Frazil ice does not adhere to objects whose temperature is slightly above the freezing point. For larger installations in cold climates, a prohibitive amount of energy would probably be required for heating intake screens.

Control Methods. An indication of icing problems at an intake is abnormal drawdown in the intake well. If the intake has a screen, excessive drawdown can rupture the screen.

Methods most commonly used to control frazil ice at an intake structure include injecting steam or compressed air at the intake opening and backflushing the intake with settled water. If backflushing is to be routinely practiced, provisions must be included when the intake is designed. It is sometimes possible to clear partially clogged intake ports by a method termed control drawdown, which involves throttling the intake-well pumps and maintaining reduced intake flow. Under some conditions, this flow may be sufficient to erode ice bridges at the ports and restore intake capacity.

If the plant has an alternate intake or sufficient storage capacity to operate without source water for a few hours, the easiest solution is to shut down the intake and wait for the ice to float off.

Fish Protection Consideration

Design of intake systems may affect aquatic organisms present in the river, lake, or reservoir. An effective approach for fish protection that also minimizes other contaminants is to use low entrance velocities.

A positive method to reduce fish entry into submerged intake facilities is the vertical velocity cap. Figure 4.14 illustrates a horizontal cap on the top of the intake structure, which forces all water to enter horizontally. Fish tend to swim against horizontal currents and avoid the intake.

Zebra Mussel Consideration

The zebra mussel *(Dreissena polymorpha)* is a small bivalve mollusk native to Europe that has alternating light and dark stripes on its shell. Zebra mussels average about an inch in length and may typically live for four to eight years. The mussel grows filament-like threads (byssus filaments) from the flat side of its shell that allow it to

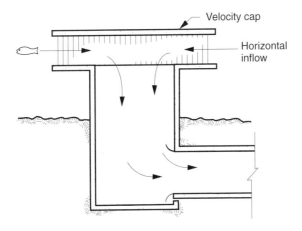

(Source: FMC Corporation)

FIGURE 4.14 Vertical velocity cap.

attach to hard surfaces such as the bottom of ships and the components of water facility intakes.

The zebra mussel is now well established in the Great Lakes region of the United States and Canada and proliferates under a relatively wide range of water conditions, as illustrated in Table 4.12. Projections indicate that they will eventually infest all freshwaters in two-thirds of the United States and the majority of southern Canada.

Zebra mussels have few natural predators, including some diving ducks and bivalve-eating fish such as the common carp and the freshwater drum. These predators appear to be of little help in controlling the zebra mussel population.

The magnitude of the zebra mussel threat can be illustrated by the rate at which the mollusk can reproduce. A single female can produce between 30,000 to 40,000 eggs per year. Although biologists estimate that only 5% of the laid eggs may actually survive, population increases can be huge in areas where the environment is conducive to growth. The rapid reproduction rate can result in a zebra mussel problem changing from nonexistent to a water supply emergency in a few months.

A range of design techniques have been investigated as a means to control or prevent zebra mussel infestation of water intake facilities, as shown in Table 4.13. The most commonly used control method involves combinations of mechanical removal and chemical treatment with chlorine and potassium permanganate applications to control recurrence. Chemical treatment typically consists of a chemical feed point at the intake facility entrance to distribute treatment chemical across the full flow stream of the inlet. Regulatory agencies restrict the discharge of treatment chemicals where they may enter the reservoir or river. Mechanical removal may be necessary to clean bar screens or pipelines infested by adult mussels. If this is done, care should be exercised in planning the removal process. Disposal of large volumes of shells and rotting mussel can be a problem, and the rotting tissue has the potential of causing serious taste and odor problems in drinking water.

A second bivalve, the quagga mussel, has been discovered in the Great Lakes region. This mussel is named after an extinct African relative to the zebra mussel and looks much like the zebra mussel to the untrained eye. The quagga tends to be slightly larger than the zebra and does not have a flat side. The quagga mussel appears to tolerate a higher salinity than does the zebra and appears to survive at greater water depths.

TABLE 4.12 Zebra Mussel Requirements for Reproduction

Water characteristics	Remarks
Water pH	Basic, 7.4 minimum
Calcium content of water	28 mg/L minimum
Water temperature	
Maximum	90° F (32.2° C)
Minimum	32° F (0° C)
Velocity of flow	5.0–6.5 ft/s (1.5–2.0 m/s)
Location	Prefer a dark location

TABLE 4.13 Zebra Mussel Control

Treatment technique	Remarks
Thermal	95° F for two hours 100% effective Repeat two or three times per year
Chemical	Oxidizing chemicals such as chlorine, bromine, potassium permanganate, ozone Continue for two or three weeks or apply continuously Nonoxidizing chemicals are effective and being developed
Coating of components	Silicone-based coatings prevent attachment 80% successful
Mechanical	Shovel or scrape High-pressure hose Sand blasting Pipeline pigging
Other methods	Ultrasound High pressure Electrocution Oxygen depletion

RACKS AND SCREENS

Racks and screens remove suspended particulates from water, including leaves, debris, and other sizable clogging material. Racks and screens are essential to provide protection to downstream conduits, pumps, and treatment works. Properly designed intake racks and screens can also minimize the effect on fish.

Racks and screens can be divided into two broad categories: coarse screens (racks) and fine screens. A summary of characteristics associated with each type is presented in Table 4.14.

Design Considerations

Head loss through racks and screens is an important consideration to be evaluated as part of the design process. Racks and screens should be designed to minimize head loss by providing sufficient flowthrough area to keep velocities low. The total area of clear openings in a screen should be 150% to 200% or more of the area or channel protected by the screen. The maximum head loss from clogging should be limited to 2.5 to 5.0 ft (0.8 to

TABLE 4.14 Rack and Screen Characteristics

Type	Remarks
Coarse screens (racks)	
Trash racks	Clear opening 3-4 in Inclined or vertical Manually or mechanically cleaned
Bar racks	Clear opening $\frac{3}{4}$ to 3 in. Inclined or vertical Manually or mechanically cleaned
Fine screens	
Traveling water screen	6 to 9 mm mesh cloth
Basket screen	$\frac{1}{8}$ in. to $\frac{3}{8}$ in. most common
Disk or drum screen	Vertical Water spray cleaning

1.5 m), and the screen should be designed to withstand the differential hydraulic load. Head loss for mechanically cleaned screens of all types can be held nearly constant with proper operator attention. Curves and tables for head loss through screening devices are available from equipment manufacturers.

Design of Racks

Coarse screens (also termed *trash racks* or *bar racks*) are commonly located at inlet ports to prevent entrance of large objects. Racks are generally constructed of $\frac{1}{2}$ to $\frac{3}{4}$ in. (13 to 19 mm) metal bars, spaced to provide 1 to 3 in. (25 to 75 mm) openings.

Coarse screens are typically installed vertically or at an incline of about 30 degrees from vertical. Both manual and automatic operation of the cleaning mechanism can be provided. Figure 4.15 illustrates one method for providing an economical grating

FIGURE 4.15 Coarse bar screen, mechanically cleaned.

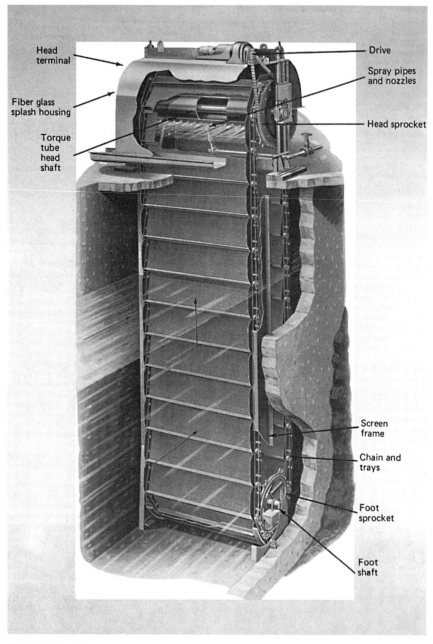

FIGURE 4.16 Traveling water screen.

system by mounting the rake on a traveling rail system so that a single rake can clean several coarse screens.

Design of Screens

Fine screens remove material too small to be deflected by coarse screens. Water velocity in net screen openings should be less than 2.0 ft/s (60 cm/s) at maximum design flow and minimum screen submergence. Hydraulically cleaned (backwashed) stationary screens have been used successfully at reservoir and river intakes. Numerous variations in design involving the use of backwashed, fixed screens have been used. It is generally recommended that velocity through stationary screen openings not exceed about 0.6 ft/s (18 cm/s).

Traveling water screens are commonly installed after coarse bar screens. Figure 4.16 illustrates a typical traveling water screen. Screens can be made to any desired opening, but $\frac{1}{8}$ to $\frac{3}{8}$ in. (6 to 9 mm) is the most common size. Operation includes flushing from behind the screen with water jets to remove accumulated material from the screen and washing it into a disposal trough. Screenings may be returned to the water source or disposed of in a landfill.

BIBLIOGRAPHY

Ashton, G. D. "River Ice." *American Scientist* 67(1):38, 1979.

Barnes, H. T. *Ice Engineering.* Montreal: Renouf, 1928.

Baylis, J. R., and H. H. Gerstein. "Fighting Frazil Ice at a Waterworks." *Engineering News-Record* 140(16):562, 1948.

Burdick, C. B. "Water Works Intakes." *Journal AWWA* 38(3):315, 1946.

Clasen, J., and H. Bernhardt. "Experiences of Quality Control of Raw Water Storage Reservoirs." *Aqua—Journal of Water Supply Research Technology* 38:256, 1989.

Coughlan, Jr., F. P. "Locating the New Intake." *Journal AWWA* 50(5):668, 1958.

Cunningham, J. W. "Waterworks Intakes and the Screening of Water." *Journal AWWA* 23(2):258, 1931.

Donnan, J. A., and C. A. MacKay. "The Economic Implications of the Zebra Mussel in the Great Lakes." *Water Pollution Control—Canada* 129:12,1991.

Driscoll, F. G. *Groundwater and Wells.* 2nd ed. St. Paul, Minn: Johnson Division, 1986.

Evans, G. P. and D. Johnson. *Detection of Pollution at Drinking Water Intakes, Environmental Protection: Standards, Compliance, and Costs.* Chichester, England: Ellis Horwood, Ltd.,1984, p. 239.

"Fine Screens." *Public Works* 126(4):c-12,1995.

Foulds, D. M., and T. E. Wigle. "Frazil—The Invisible Strangler." *Journal AWWA* 69(4):196, 1977.

Gifren, A. V. *The Occurrence and Prevention of Frazil Ice Blockage at Water Supply Intakes.* Toronto, Ontario: Research Branch Publication No. W 43, Ministry of the Environment, 1973.

Gillespie, D. D. "Novel Construction Concept for Raw Water Intake." *Civil Eng.* 41:43, 1971.

Goda, Takesha. "Water Quality Management in the Yodo River Basin." *Water Sci. Technol.* 23:65, 1991.

Hardenbergh, W. A. *Water Supply and Purification.* Scranton, Penn.: International Textbook, 1945, p. 325.

Heinzel, L. R. "Storm Effects on Turbidity in Trinity Project Waters." *Journal AWWA* 59(7):835,1967.

Howson, L. R. "Saginaw-Midland Supply Project Begins Operations." *Water Works Engineering* 102(5):415, 1949.

Howson, L. R., and Gerald Remus. "Best Features of Four Existing Plants Highlighted in New Detroit Water Facility." *Water and Wastes Engineering* 7(12):50, 1970.

Huber, F. "Coquitlam Lake Water Tunnel Upgrading—Design and Construction. A Case History." *Canadian Geotechnical Journal* 26(1):90, 1989.

Ihling, H. M. and D. P. Proudfit. "New Facilities Add 100 mgd to Milwaukee's Water Supply." *Civil Engineering* 33(9):31, 1963.

Klauber, Avery. "Investigators Hunt for Great Lakes Enemy. New York Sea Grant." *Nor'easter* 2:14, 1990.

Kubus, J. J. and D. A. Egloff. "Monitoring Water Quality Parameters from Municipal Water Intakes." *Water Pollution Control Journal* 54:1592, 1982.

Lamarre, Leslie. "Invasion of the Striped Mollusks." *EPRI Journal* 16:12, 1991.

Lamarre, Leslie. "Zebra Mussels: the Assault Continues." *EPRI Journal* 18:22, 1993.

Lee, G. F. and C. C. Harlin, Jr. "Effects of Intake Location on Water Quality. *Industrial Water Engineering* 2(3):36, 1965.

Lischer, V. C. and H. O. Hartung. "Intakes on Variable Streams." *Journal AWWA* 44(10):873, 1952.

Logan, T. H. *Prevention of Frazil Ice Clogging of Water Intakes by Application of Heat.* REC-ERC-74-15. Denver, Colo: U.S. Dept. of Interior, Bureau of Reclamation, September 1974. Distributed by National Technical Information Service, U.S. Dept. of Commerce, Springfield, Va.

Matisoff, G., et al. "Controlling Zebra Mussels at Water Treatment Plant Intakes." *AWRA Surface and Ground Water Quality Symposium,* Cleveland, Ohio: ARWA, p. 79.

McDonald, Adrian, and Pamela Naden. "Colour in Upland Sources: Variations in Small Intake Catchments." *Water Services* 91:121, 1987.

Merriman, Thaddeus, and T. H. Wiggin. "Collection of Water Intakes." In *American Civil Engineers' Handbook.* 5th ed. New York: John Wiley and Sons, 1930.

Monk, R. D., Terry Hall, and Mohammed Hussain. "Real World Design: Appropriate Technology for Developing Nations." *Journal AWWA* 76:68, 1984.

Peters, L. F. "Hydraulically Backwashed Well Screens Used as Intakes." *Water and Wastes Engineering* 6(3):52, 1969.

Richardson, W. H. "Intake Construction for Large Lakes and Rivers." *Journal AWWA* 61(8):365, 1969.

"Rise and Fall of a Pumping Station." *Water Works Engineering* 112(3):208, 1959.

San Giacomo, Richard, and M. E. Cavalcoli. "Opposing the Threat: Practical Application of Zebra Mussel Control." *Proceedings of the AWWA 1991 Annual Conference.* Denver, Colo: AWWA, p. 137.

Schuler, V. J. and L. E. Larson. "Improved Fish Protection at Intake Systems." *Journal of Environmental Engineering Division (ASCE)* 101(EE6):897, 1975.

Shaefer, V. J. "The Formation of Frazil and Anchor Ice in Cold Water." *Trans-American Geophysical Union* 31(6):885, 1950.

Tobin, Patrick, and George Wesner. "Siting of Water Intakes Downstream from Municipal Wastewater Facilities." AWWA Research Foundation et al. *Water Reuse Symposium* 1:192, 1979.

Tone, Ralph, and E. T. Conrad. "Centrifugal Pumps Ride the Rails of Unique Water Intake." *Public Works* 100(12):64, 1969.

Tredgett, R. G. and D. R. Fisher. "New Waterworks Intake for Hamilton, Ontario." *Water and Wastes Engineering* 7(2):32, 1970.

Whitlock, E. W. and R. D. Mitchell. "Hydraulically Backwashed Stationary Screens for Surface Water." *Journal AWWA* 50(10):1337, 1958.

Weiss, C. M., and R. T. Oglesby. "Limnology and Quality of Raw Water in Impoundments." *Public Works* 91(8):97, 1960.

Wolman, Abel. "Ice Engineering." *Journal AWWA* 21(1):133, 1929.

CHAPTER 5
AERATION AND AIR STRIPPING

Aeration processes have been used to improve water quality since the earliest days of water treatment. In this process, air and water are brought into intimate contact with each other to transfer volatile substances to or from the water. The removal of a gas from water is classified as desorption, or stripping. The transfer of a gas to water is called gas adsorption. The U.S. Environmental Protection Agency (USEPA) has identified air stripping as one of the best available methods for the removal of volatile organic chemicals (VOCs) from contaminated groundwater.

USES OF AERATION

Principal uses for aeration in water treatment include:

- To reduce the concentration of taste- and odor-causing substances and, to a limited extent, for oxidation of organic matter
- To remove substances that may in some way interfere with or add to the cost of subsequent water treatment. A prime example is removal of carbon dioxide from water before lime softening
- To add oxygen to water, primarily for oxidation of iron and manganese so that they may be removed by further treatment
- To remove radon gas
- To remove VOCs considered hazardous to public health

TYPES OF AERATION EQUIPMENT

Structures or equipment for aeration or air stripping may be classified into four general categories: waterfall aerators, diffusion or bubble aerators, mechanical aerators, and pressure aerators.

Waterfall Aeration

The waterfall type of aeration accomplishes gas transfer by causing water to break into drops or thin films, increasing the area of water exposed to air. The more common types are:

- Spray aerators
- Multiple-tray aerators
- Cascade aerators
- Cone aerators
- Packed columns

Spray Aerators. Spray aerators direct water upward, vertically or at an inclined angle in a manner that causes water to be broken into small drops. Installations commonly consist of fixed nozzles or a pipe grid located over an open-top tank.

Spray aerators are usually efficient with respect to gas transfer such as carbon dioxide removal or oxygen addition. However, they require a large installation area, are difficult to house, and pose operating problems during freezing weather.

Multiple-Tray Aerators. Multiple-tray aerators consist of a series of trays equipped with slatted, perforated, or wire-mesh bottoms. Water is distributed at the top, cascades from each tray, and is collected in a basin at the base. It is important to have even distribution of water from the trays to obtain optimum unit efficiency. Coarse media such as coke, stone, or ceramic balls ranging in size from 2 to 6 in. (5 to 15 cm) are used in many tray aerators to improve the efficiency of gas exchange and to take advantage of the catalytic effects of deposited manganese oxides.

A type of multiple-tray aerator, the crossflow tower, has been extensively used in water cooling applications. Water is allowed to fall over the tray area while air is either forced or induced to flow across the slats, perpendicular to the water path. Tray aerators are analogous to cooling towers, and the problems encountered in design and operation are similar. Tray aerators must be provided with adequate ventilation. If they are placed in a poorly ventilated building, performance will be impaired by contamination of inlet air with the compounds being removed.

Artificial ventilation is provided to some types of tray aerators by supplying air from a blower at the bottom of the enclosure. These aerators exhibit excellent oxygen absorption and carbon dioxide removal.

Cascade Aerators. With cascade aerators, increases in exposure time and area-volume ratio are obtained by allowing water to flow downward over a series of steps or baffles. The simplest cascade aerator is a concrete step structure that allows water to fall in thin layers from one level to another. The exposure time of air to water can be increased by increasing the number of steps, and the area-volume ratio can be improved by adding baffles to produce turbulence. In cold climates, these aerators must be housed, and adequate provisions must be made for ventilation. As with tray aerators, operating problems include corrosion and slime and algae buildup.

Cone Aerators. Cone aerators are similar to cascade aerators. They have several stacked pans arranged so that water fills the top pan and cascades down to each succeeding pan. A common type of commercial cone aerator is shown in Figure 5.1.

Packed Columns. Packed columns (also called packed towers or air strippers) are a relatively new development for drinking water treatment. They were developed primarily for

FIGURE 5.1 View of a typical cone aerator. *(Courtesy of Infilco Degremont, Inc.)*

removal of volatile compounds such as VOCs from contaminated water. The extremely large surface area provided by packing in a column, combined with forced air flowing counter to the flow of water, provides considerably more liquid-gas transfer compared with other aeration methods. A packed column consists principally of a cylindrical tower, packing material contained in the tower, and a centrifugal blower.

The quantity of air provided in relation to the quantity of water flowing through the column is known as the air-to-water ratio. This ratio is important in designing for removal of VOCs. A packed-column installation is shown in Figure 5.2.

Diffusion-Type Aeration

Diffusion- or bubble-type aerators accomplish gas transfer by discharging bubbles of air into water by means of air-injection devices. Compared with packed columns, diffused aeration provides less interfacial area for mass transfer but greater liquid contact time. On the other hand, packed columns provide a greater effective area but lower liquid volumes.

Diffuser Aerators.   The most common type of equipment for diffusion aeration consists of rectangular concrete tanks in which perforated pipes, porous diffuser tubes or plates, or other impingement devices are inserted. Compressed air is injected through the system to produce fine bubbles, which, on rising through the water, produce turbulence resulting in effective water-air mixing.

This type of aeration technique is often adapted to existing storage tanks and basins. If porous tubes or perforated pipes are used, they may be suspended at about one-half tank depth to reduce compression head. Porous plates are usually located on the bottom of the tank. Static tube aerators are also used in a variety of applications and provide adequate aeration when properly designed.

5.4 CHAPTER FIVE

FIGURE 5.2 View of a packed-column installation. *(Source: US Environmental Protection Agency.)*

Draft-Tube Aerators. A draft-tube aerator consists of a submersible pump that rests on the bottom of a basin and is equipped with an air intake pipe extending to above the water surface. The partial vacuum created by the pump pulls air through the tube and mixes it with water at the pump intake. Aerated water is then directed outward along the floor of the basin. This type of aeration is an inexpensive and relatively effective means of adding aeration to an existing basin.

In-Well Aeration. A variation of diffused aeration to remove VOCs from groundwater supplies is in-well aeration. This technique has been investigated by the North Penn Water Authority, a public water utility located in Lansdale, Pennsylvania. The Authority received a grant from the AWWA Research Foundation to conduct detailed tests on the feasibility of in-well aeration.

Aside from obvious advantages of this type of treatment system, several disadvantages exist. One disadvantage is the dissolution of large quantities of air into water, causing water to appear milky. For practical use of this type of aeration system, water needs some atmospheric contact time to allow the milky appearance to disappear before it is pumped into the distribution system. The study found this treatment method to have relatively low efficiency.

Mechanical Aeration

Mechanical aerators employ motor-driven impellers alone or in combination with air-injection devices.

Surface Aerators. Mechanical surface aerators are used extensively in wastewater applications for supplying oxygen to water. To a lesser extent, they are used to control taste and odor problems in water treatment and are commonly installed at a reservoir rather than at the treatment plant. They generally consist of an electric motor suspended on a float, with a driveshaft operating a propeller located a short distance below the water surface. The water is drawn up by the blade and thrown into the air in tiny droplets so that the water can pick up oxygen.

A variation in design is the surface aerator equipped with a draft tube extending below the propeller. With this design, water is drawn up from near the bottom of deeper basins.

Submerged Aerators. Submerged aerators operate in the reverse of surface aerators. The submerged blade draws water downward, and in the process, draws in air, which is diffused into the water. This type of aerator results in relatively calm water at the surface compared with surface aerators. Submerged aerators are best used for increasing dissolved oxygen levels.

Pressure Aerators. There are two basic types of pressure aerators, and the object of both is to aerate water that is under pressure. In one type, water is sprayed into the top of a closed tank while the tank is continuously supplied with compressed air. Aerated water leaves at the bottom of the tank. In the second type of pressure aerator, compressed air is injected directly into a pressurized pipeline and adds fine air bubbles to the flowing water.

With both systems, the higher the pressure used, the more oxygen dissolves into the water. Pressure aerators are primarily used for oxidizing iron and manganese for subsequent removal by settling, filtration, or both.

Emerging Technologies

Several types of proprietary equipment have recently been developed with improvements or advantages compared with standard aerator designs. Three examples are:

- Low-profile, diffused air system
- Maxi-strip system
- Rotor-strip unit

The principal advantage of these pieces of equipment is that they have much lower height requirements than standard equipment to achieve a given removal efficiency.

Low-Profile, Diffused Air System. The low-profile system is a multistage, diffused bubble air stripping device. The device differs from conventional systems in that it operates with a water depth of about 18 in. (46 cm) compared with 10 ft (3 m) or greater for conventional installations. Shallow water depth permits the use of regenerative blowers instead of compressors and reduces overall height of the device to less than four feet.

In this system, water flows by gravity through a series of completely mixed staged reactors. Individual modules are designed and fabricated to include 3, 4, 6, or 12 stages, depending on the level of treatment required. Each stage is separated by a baffle wall and includes separate diffuser heads. Air is blown into each stage, creating a turbulent mixture of diffused bubbles and water to provide efficient mass transfer.

Individual modules are designed to treat water at flow rates ranging from 23 to 150 gpm (1.45 to 9.46 L/s). Modules used in parallel can treat water at higher flow rates. This system also removes VOCs, gasoline components, and radon from contaminated groundwater and is particularly applicable to small water systems with low flow rates. A sketch of a typical device of this type is shown in Figure 5.3.

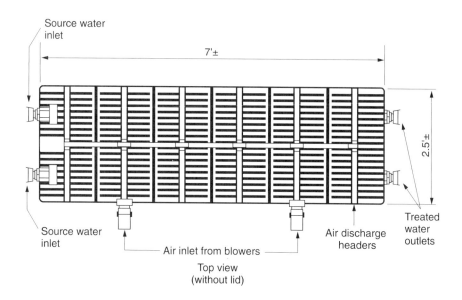

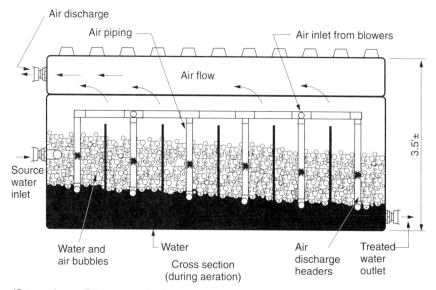

(Source: Lowry Engineering, Inc.)

FIGURE 5.3 Low profile, diffused air system. *(Source: Manufacturer's brochure.)*

Maxi-Strip System. The maxi-strip system uses an aeration device developed from a high-intensity turbulent mixer called the Turbo-Jet. The system is designed to increase the driving forces of air and water to improve mass transfer. Because larger surface areas and contact effectiveness are generated in a much smaller space than required by packed columns, less off-gas is produced for an equal degree of VOC removal.

Figure 5.4 illustrates a maxi-strip system consisting of a recirculation tank and several treatment units. One of these units is shown in Figure 5.5. The number of units required for a particular application is based on water flow rate and VOC concentration.

In operation, contaminated water is pumped from a well to the first series of maxi-strip units, from which treated water is discharged into a baffled recirculation tank. Water is then pumped to the next series of treatment units and subsequently back into a second cell of the recirculation tank. The process is continued through as many treatment units as required to achieve the desired degree of contaminant removal.

The standard maxi-strip unit is currently rated at 450 gpm (28.4 L/s). Higher concentrations of contaminants can be removed either by decreasing flow rates or by recycling some treated water back through the unit. Other unit sizes will be available in the future. Although experience is currently limited on VOC removal, these units appear to have good potential for assisting small systems with VOC-contaminated wells to meet current MCLs.

Rotor-Strip Unit. The rotor-strip is essentially a rapidly rotating drum filled with high-density packing material, with a surface area in excess of 8,000 ft^2/ft^3 (26,000 m^2/m^3). As shown in Figure 5.6, water is sprayed onto the packing inside the drum, and centrifugal force generated by the rotating drum throws the water outward through the media under

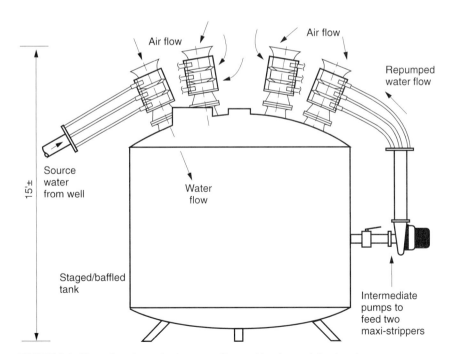

FIGURE 5.4 Illustration of a maxi-strip system. *(Source: Manufacturer's brochure.)*

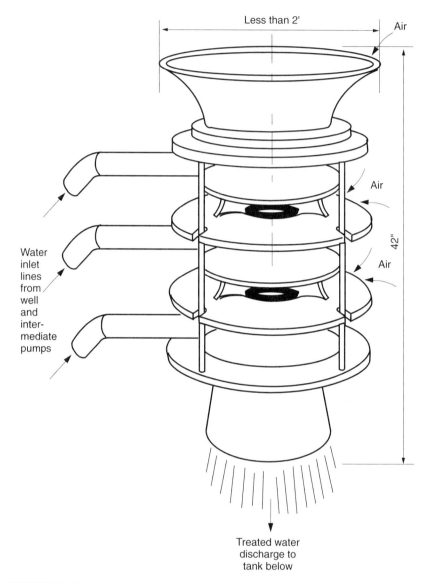

FIGURE 5.5 One unit of a maxi-strip system. *(Source: Manufacturer's brochure.)*

turbulent conditions. At the same time, air is forced through media countercurrent to the water flow.

The rotor-strip unit is currently available in sizes that operate at flow rates up to 700, 1,050, and 1,400 gpm (44, 66, and 88 L/s). Inlet water pressure is generally 15 to 30 psi (103 to 207 kPa) and discharge is atmospheric. One disadvantage of the design is that the device has more mechanical equipment that will require maintenance than other treatment units.

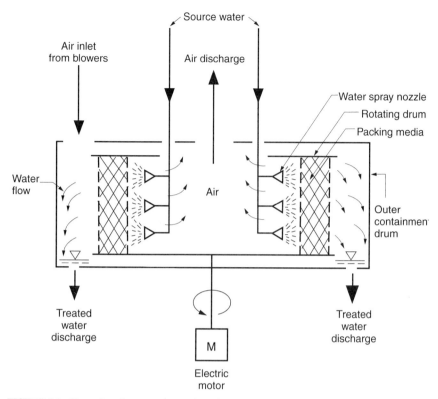

FIGURE 5.6 Illustration of a rotor strip aeration unit. *(Source: Manufacturer's brochure.)*

PRINCIPLES OF AIR STRIPPING AND AERATION

Transfer by aeration of a volatile material to or from water depends on a number of factors, including:

- Characteristics of the volatile material
- Water and surrounding air temperatures
- Gas transfer resistance
- Partial pressure of the gases in the aerator atmosphere
- Turbulence in gaseous and liquid phases
- Area-to-volume ratio
- Time of exposure

Equilibrium Conditions

The term *equilibrium* applied to gases dissolved in water signifies a steady-state concentration of dissolved substances. Aeration promotes the establishment of equilibrium between dissolved, volatile constituents in the water and the constituents in the air to which the water is exposed. For example, when water is exposed to air, oxygen and nitrogen dissolve in the water until a state of equilibrium is reached. The function of aeration is to speed up this natural process. True equilibrium may not be attained by aeration unless the air-water exposure period is relatively long. From a practical standpoint, however, it is generally not necessary to achieve absolute equilibrium.

Saturation Value

The concentration of a gas dissolved in a liquid at equilibrium is known as its saturation value. This value is an important characteristic of a dissolved gas. Saturation value is principally dependent on water temperature, partial pressure of the gas in the atmosphere in contact with the water, and presence of dissolved solids. The higher the partial pressure, the greater the dissolved gas concentration. This relationship is known as Henry's law. At a fixed partial pressure, the higher the temperature, the lower the solubility or saturation value of a gas. Gas solubility is also reduced by dissolved solids.

Saturation value has considerable practical and theoretical significance. It is the difference between the saturation value of a gas and its actual concentration in the water that provides the driving force for the interchange of gas between air and water. Water deficient in oxygen will absorb it when brought into contact with air, and the air-water equilibrium will be reached from the direction of oxygen deficiency. Prolonged aeration produces oxygen saturation.

On the other hand, if water contains more oxygen or, as is more commonly encountered, more carbon dioxide than the saturation amount, aeration brings about release of the gas. In this instance, equilibrium is approached from the direction of supersaturation. The final result of prolonged aeration, however, is the same—saturation.

Rate of Achievement

Equilibrium conditions are important in the aeration process, but of even greater significance to the design engineer is the rate of achievement of equilibrium. Equilibrium and rate of approach to it are not independent of one another. Under similar conditions, the further the air-water system is from equilibrium, the more rapid the interchange of gas in the direction of attaining equilibrium.

Significance of Films

Films at the air-water interface appear to have an important bearing on the rate of gas transfer. Both liquid and gas films can retard the rate of exchange of volatile material, but the liquid film is a more important factor in the transfer of gases of low solubility, such as oxygen and carbon dioxide.

Film resistance is influenced by many factors, but the most important are turbulence and temperature. High temperature and turbulence promote gas transfer by reducing film thickness. Increased temperature also increases the rate of molecular diffusion.

Rate of Transfer

The rate of transfer of a volatile substance from water to air is generally proportional to the difference between the existing concentration and the equilibrium concentration of the substance in solution. The relationship is expressed as follows:

$$M = K_L a (C_i^* - D_i)$$

where M = the mass of substance transferred per unit of time and volume, lb/hr/ft³ (kg/hr/m³)
K_L = the overall liquid mass transfer coefficient, ft/hr (m/hr)
a = effective area for mass transfer, ft²/ft³ (m²/m³)
C_i^* = liquid phase concentration in equilibrium with the gas phase concentration, lb/ft³ (kg/m³)
D_i = bulkhead liquid phase concentration, lb/ft³ (kg/m³)

The driving force for mass transfer is the difference between actual conditions in the air stripping unit and conditions associated with equilibrium between the gas and liquid phases. Equilibrium concentration of a solute in air is directly proportional to the concentration of the solute in water at a given temperature.

Henry's law states that the amount of gas that dissolves in a given quantity of liquid, at constant temperature and total pressure, is directly proportional to the partial pressure of the gas above the solution. Thus Henry's law constant can be considered a partition coefficient describing the relative tendency for a compound to separate, or partition, between the gas and the liquid of equilibrium; it indicates a contaminant's volatility and its affinity for the aeration process. Substances with high Henry's law constants are easily removed by aeration, and those with low constants are difficult to remove. Table 5.1 lists the Henry's law constants for several compounds. Vinyl chloride has an extremely large constant relative to the other VOCs.

The mass transfer coefficient, K_L, is a function of the compound being stripped from water, the geometry and physical characteristics of the air stripping system, and the temperature and flow rate (contact time) for the liquid. It also incorporates the diffusion resistance to mass transfer in both liquid and gas phases and is related to local gas and liquid phase mass transfer coefficients, k_g and K_L, respectively.

For most stripping applications in water treatment, the bulk of resistance to mass transfer resides in the liquid phase. As a result, air stripping process design should be based on maximizing the liquid mass transfer coefficient.

The effective area, a, represents the total surface area created in the air stripping unit by producing numerous fine water droplets or by forming minute gas bubbles. The effective area is a function of air stripping equipment. In terms of effective area, an optimum treatment system is one that includes a high surface area for mass transfer per unit volume.

The mass transfer coefficient, K_L, and the effective area for mass transfer, a, are usually evaluated as one constant, $K_L a$.

DESIGN OF AERATION PROCESSES

Air stripping equipment design has been developed extensively in the chemical industry for handling concentrated organic solutions. Procedures found in chemical engineering literature can be applied to water treatment for trace organics removal. Based on principles described in the last section, the design of an air stripping installation is primarily dependent on the following factors:

- Temperature of the water and the surrounding air
- Physical and chemical characteristics of the contaminant to be removed
- The ratio of air to water being provided in the process
- Contact time between the air and water
- The water surface area available for mass transfer

TABLE 5.1 Henry's Law Constants for Selected Compounds

Compound	Formula	Henry's Constant atm[a,d]
Vinyl chloride	CH_2CHCl	3.55×10^5
Oxygen	O_2	4.3×10^4
Toxaphene[b]	$C_{10}H_{10}C_{18}$[c]	3.5×10^3
Carbon dioxide	CO_2	1.51×10^3
Carbon tetrachloride[b]	CCl_4	1.29×10^3
Tetrachloroethylene[b]	C_2Cl_4	1.1×10^3
Trichloroethylene[b]	$CHClCCl_2$	5.5×10^2
Hydrogen sulfide	H_2S	5.15×10^2
Chloromethane[b]	CH_3Cl	4.8×10^2
1,1,1-Trichloroethane[b]	CCH_3Cl_3	4.0×10^2
Toluene[b]	$C_6H_5CH_3$	3.4×10^2 (at 25° C)
Benzene[b]	C_6H_6	2.4×10^2
Chloroform[b]	$CHCl_3$	1.7×10^2
Bromodichloromethane	$CHCl_2Br$	1.18×10^2[e]
1,2-Dichloroethane[b]	CH_2ClCH_2Cl	61
Dibromochloromethane	$CHClBr_2$	47[f]
Bromoform[b]	$CHBr_3$	35
Methyl tertiary butyl ether	$C_5H_{12}O$	22[g]
Ammonia	NH_3	0.76
Pentachlorophenal[b]	$C_6(OH)Cl_5$	0.12
Dieldrin[b]	$C_{12}H_{10}OCl_6$	0.0094

[a] Temperature 20° C except where noted otherwise.
[b] Computed from water solubility data and partial pressure of pure liquid at specified temperature.
[c] Synthetic; approximate chemical formula.
[d] Kavanaugh & Trussell, 1980, except where otherwise noted.
[e] Warner, Cohen, and Ireland, 1980.
[f] Symons et al, 1981.
[g] Zorgorske et al, 1996.

The first two factors are fixed by source water quality and location of the installation. The other factors can be varied with the type of aeration equipment used.

Design of Diffused Air Equipment

When diffused aeration is employed, air stripping is accomplished by injecting bubbles of air into the water. Ideally, diffused aeration is conducted counter to the flow of water. Untreated water should be entering at the top and treated water exiting at the bottom while fresh air enters at the bottom and exhausted air exits at the water surface. Gas transfer can be improved by increasing the basin depth, producing smaller bubbles, improving the contact basin geometry, and incorporating a turbine to produce smaller bubbles and increase bubble holdup.

Diffused-air aerators usually provide a longer aeration time than waterfall aerators, generally an advantage, but other factors influencing performance are the turbulence provided, air-volume ratio, and gas transfer resistance. Because of these factors, comparison between the two types of equipment cannot be made solely on the basis of aeration contact time.

Basin Design. Tanks used for the diffused-air process are usually made of concrete and are commonly 9 to 15 ft (3 to 5 m) deep and 10 to 30 ft (3 to 9 m) wide. The ratio of width to depth should not exceed 2 in order to achieve optimum mixing. Tank length is governed by the desired detention time, which usually varies from 10 to 30 min. Air diffusers are generally mounted along one side of the tank to impart a spiral flow to the water. A spiral flow pattern produces higher water surface velocities, which in turn promotes better gas transfer. In addition, with a spiral flow, a substantial number of bubbles do not escape immediately, but are carried across the basin where they are held in a more or less fixed position by the descending water.

Diffusers. Common types of diffusers are perforated pipes, porous plates or tubes, and various patented impingement or sparger devices. Compressed air is generally furnished by a rotary compressor sized to produce the correct volume and pressure. Diffusers produce small bubbles that rise through the water and cause turbulence and the opportunity for the exchange of volatile materials.

Diffusers are generally located near mid-depth in the tank, usually about the optimum efficiency point. Deeper location of the diffusers requires more pressure head, which increases compressor power costs. The amount of air required ranges from 0.01 to 0.15 ft^3/gal (0.0008 to 0.012 m^3/L) of water treated. Sufficient diffuser capacity must be provided to supply air at the required rate without excessive pressure loss. Some installations include lateral baffles to prevent short-circuiting.

Air pressure requirements depend on submerging diffusers and friction loss through piping. Power requirements vary from 0.5 to 2.0 kw/mgd (0.00013 to 0.00053 kw/m^2 per day), with the average about 1.0 kw/mgd (0.00026 kw/m^2 per day). When porous plates or tubes are used, air should be filtered to avoid clogging of the diffusers.

Diffuser-type aerators require less space than spray aerators and generally more than tray aerators. They have practically no head loss through diffusion units, and this is usually an important aspect in overall plant design. Aeration units have few cold weather operating problems, and there is no need to house them. In some instances, diffusion aeration basins are used to provide chemical mixing.

Design of Spray Aerators

Exposure time for each drop from a spray aerator depends on its initial velocity and trajectory. Drop size, and the resulting area-volume ratio, is a function of the dispersing action of the nozzle. The initial velocity V of a drop emerging from an orifice or nozzle appears in the formula

$$V = C_v \sqrt{2gh}$$

and the discharge by the equation

$$Q = C_d A \sqrt{2g}$$

where h = total head on the nozzles in ft
  g = acceleration from gravity in ft/sec^2
  A = area of the opening in ft^2
  C_v = coefficient of velocity
  C_d = coefficient of discharge ($C_d = C_v C_c$, where C_c is the coefficient of contraction)

Coefficients of velocity, contraction, and discharge vary with the shape and other characteristics of the orifice or nozzle.

The trajectory of the spray used in an aerator may be vertical or inclined. If the angle between the initial velocity vector and horizontal is zero, theoretical exposure time t of the water drops is given by the formula

$$t = 2C_v \sin \theta \sqrt{\frac{2h}{g}}$$

The sine of an angle of less than 90 degrees is less than 1.0, so a vertical jet gives the longest exposure time for a given value of h. But an inclined jet has the advantage of a longer path and less interference between falling drops. Wind also influences the path of the rising and falling drops, so an allowance must be made for its action.

Nozzle design is important in achieving optimum dispersion of water. Among special designs used are rifled nozzles, centrifugal (West Palm Beach) nozzles, Sacramento floating cones, impinging devices, and rotating reaction nozzles.

The size, number, and spacing of spray nozzles depend on the head of water being used, space available for aeration facilities, and interference between adjacent sprays. Theoretically, numerous small nozzles capable of producing atomized water would be the most efficient design. However, from a practical standpoint, very small nozzles should be avoided because of clogging and high maintenance requirements. Nozzles used in most spray aerators are 1.0 to 1.5 in. (2.5 to 3.8 cm) in diameter and have discharge ratings of 75 to 150 gpm (4.73 to 9.46 L/s) at about 10 psi (69 kPa). Nozzle spacing in most installations is between every 2 and 12 ft (0.6 and 3.7 m). The area allocated to spray aeration varies from 50 to 150 ft^2/mgd (106 to 318 m^2/m^3/s) capacity, although much larger areas have been used at some treatment facilities.

Because interior and exterior corrosion can be serious problems in aerator piping, corrosion-resistant materials should be used wherever possible.

Spray aerators providing a high area-to-volume ratio are spectacular to see. They are rarely housed, so ventilation presents no problem. Gas transfer between water drops and air proceeds rapidly, and spray-type aerators usually have a relatively high efficiency. In general, spray aerators remove more than 70% of dissolved carbon dioxide, and removals as high as 90% have been documented. Disadvantages of spray aerators are principally the relatively large space requirements, freezing problems in colder climates, short exposure time between water and air, and high head requirements.

Design of Multiple-Tray Aerators

Multiple-tray aerators are generally constructed with three to nine trays and a spacing of 12 to 30 in. (30 to 76 cm) between trays. Space required for an aeration unit ranges from about 25 to 75 ft^2/mgd (2 to 6 m^2/ML per day) capacity, with 50 ft^2/mgd (4 m^2/ML per day) being about average. Water application rates range from roughly 20 to 30 gpm/ft^2 (17 to 20 L/s/m^2). These aerators have excellent oxygen adsorption and carbon dioxide removal capacities.

Ventilation Requirements. Tray aerators are, in many respects, analogous to cooling towers, and the design is similar. Ventilation and water distribution must be carefully considered in connection with location and design.

Multiple-tray aerators are usually housed, particularly in colder climates. A good example of an enclosed but well-ventilated installation is the Allen substation aerator at

Memphis, Tennessee. Aluminum scroll panels are used to promote good cross-ventilation, and the roof is open except directly over the distributing trays. Carbon dioxide concentration in the source water exceeds 90 mg/L, and this aerator has consistently produced a 90% or greater reduction.

If a tray aerator must be enclosed and there is not sufficient natural ventilation, artificial ventilation must be provided. This is usually accomplished by supplying air with a blower at the bottom of the aerator so that it travels counter to water flow.

Important design considerations in designing tray aerators are the use of corrosion-resistant materials and methods of dealing with slime and algal growths. Aeration units are generally constructed using concrete, stainless steel, aluminum, and rot-resistant wood. Slime and algal growths may be controlled by treating the source water with chlorine or copper sulfate.

Carbon Dioxide Removal. Carbon dioxide removal by multiple-tray aerators can be approximated by the following empirical equation developed by Scott (1955):

$$C_n = C_c 10^{-kn}$$

where C_n = concentration of carbon dioxide in mg/L after passing through n trays
C_c = concentration determined originally in the distribution tray
n = the number of trays including the distribution tray
k = a coefficient dependent on ventilation, temperature, turbulence, and other characteristics of the installation; generally ranges from 0.12 to 0.16

Design of Packed Columns

The rate at which a volatile compound is removed by air stripping in packed tower aeration (PTA) depends on the following factors:

- Air-to-water ratio (A ratio)
- Height of packing in the column
- Available surface area for mass transfer
- Water loading rate
- Air and water temperatures
- Physical chemistry of the contaminants to be removed

The first four factors may be controlled in the design of an air stripping unit.

Air and Water Flow Requirements. Air flow required for a packed column depends on the Henry's law coefficient for the compounds to be removed from the water. Packing height is a function of the required VOC removal efficiency. In general, an increase in packing height results in higher VOC removal.

The air-to-water ratio used in a column is a function of water temperature and desired level of contaminant removal. This ratio determines the size of the blower, the primary component of operating costs for PTA systems. Air-to-water ratios typically range from 30:1 to 100:1. Water loading rate, the amount of water passing through the column, usually ranges from 25 to 30 gpm/ft^2 (17 to 20 L/sec/m^2). Column diameter is selected to accommodate the desired water loading on the column.

Packed tower aeration removal effectiveness usually increases with an increase in water temperature, but it has been found that heating the influent water to increase removal effectiveness is not generally cost-effective.

Column Design. The relationship between packing height and column performance is derived from the basic mass transfer relationship. In the following formula, packing height Z in feet, is related to the height of a transfer unit (HTU) in feet, and the number of transfer units (NTU):

$$Z = (HTU)(NTU)$$

The HTU is a function of the liquid-loading rate and $K_L a$. This relationship is expressed in the following equation:

$$HTU = \frac{L}{K_L a \, C_0}$$

where L = liquid flow, lb mole/hr/ft² (g mole/hr/m²)
C_0 = molar density of water in lb mole/ft³ (g mole/m³)

The number of transfer units is a function of column performance and the substance to be removed. This relationship is expressed in the following equation:

$$NTU = \frac{R_1}{R_1 - 1} \ln \frac{(X_i/X_o)(R_1 - 1) + 1}{R_1}$$

where X_i/X_o = the ratio of influent to effluent liquid phase concentration
R_1 = dimensionless stripping factor

$$= \frac{(H)(G)}{(p_t)(L)} \text{ where } G \text{ is the gas flow in lb mole/hr/ft}^2 \text{ (g mole/hr/m}^2\text{)}$$

The ability of a particular VOC to be stripped may be determined from the Henry's law constant. The higher the constant, the easier the VOC is removed by air stripping. The effect that the type of compound to be removed has on packing depth and air-to-water ratio is shown in Figure 5.7. An air-to-water ratio of about 20:1 is required to achieve 95% removal of trichloroethylene (TCE) with 15 ft (4.6 m) of packing medium that is 1 in. (2.5 cm) in diameter. For 95% removal of a less volatile compound such as 1,2-dichloroethane, an air-to-water ratio of about 120:1 is required for a column with the same size and depth of packing.

Desired removal efficiency also affects the design of a packed column. Figure 5.8 illustrates the relationship between air-to-water ratio and packing depth to achieve various efficiencies for TCE removal. About 6 ft (2 m) of 1 in. (2.5 cm) packing medium is required to achieve 80% removal of TCE with an air-to-water ratio of 20:1. To achieve 99% removal with the same packing and air-to-water ratio, about 20 ft (6.1 m) of packing would be required.

Water temperature must also be considered in designing a packed column. Most groundwater supplies have a water temperature of about 55° F (13° C). However, water temperature may be as low as 45° F (6° C) in northern regions and as high as 75° F (24° C) in some warm regions. The relationship between water temperature and removal efficiency is illustrated in Figure 5.9.

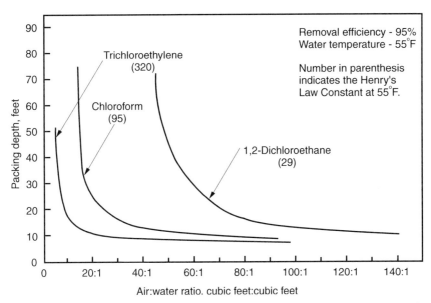

FIGURE 5.7 Effect of type of VOC on packed column design. *(Source: AWWARF/KIWA Report, 1983.)*

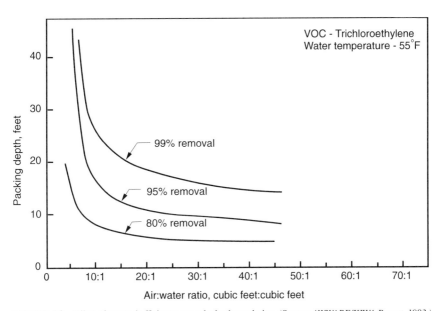

FIGURE 5.8 Effect of removal efficiency on packed column design. *(Source: AWWARF/KIWA Report, 1983.)*

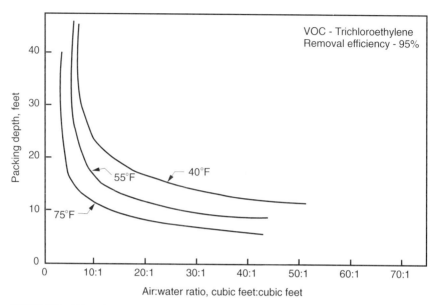

FIGURE 5.9 Effect of water temperature on packed column design. *(Source: AWWARF/KIWA Report, 1983.)*

Design Considerations. A diagram of a typical packed column installation is shown in Figure 5.10. In an installation to treat groundwater, the water is generally pumped directly from the wells to the top of the column. Treated water is collected at the bottom of the column in a clearwell, from which it is usually pumped directly to the distribution system. The principal facility elements of a packed column include the following:

- Column and internal parts
- Packing material
- Blower
- Clearwell
- Booster pumps

The column is basically a tank usually constructed from fiberglass-reinforced plastic, aluminum, stainless steel, or concrete.

A demister is usually installed at the top of the tank to prevent objectionable clouds of moisture from coming off the column. Near the top of the column, piping is installed to distribute influent water evenly over the top of the packing material.

As water flows downward through the packing, it tends to migrate to the column wall, and redistributors are installed at intervals to support the packing material and redirect the water back toward the center of the column. Four commonly used distributor styles are the orifice plate, trough, orifice headers, and spray nozzles (Figure 5.11).

Packing. Packing materials are designed to simultaneously provide a low pressure drop for air passing through the column and maximum air-water contact area. Packing pieces for "dumped packing" are randomly dumped into the column. They are available in various shapes of ceramic, stainless steel, and plastic materials. Plastic is most commonly used

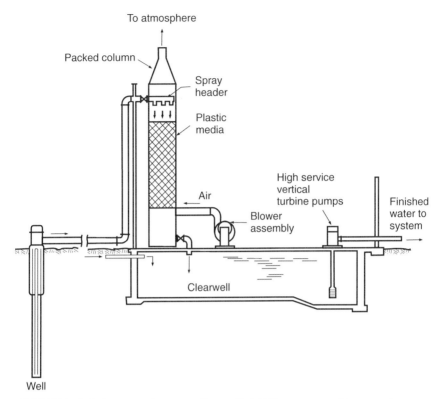

FIGURE 5.10 Packed tower aeration system. *(Source: Cook College operator training course.)*

in water treatment applications because it is durable, lightweight, and cost effective. Some common shapes are illustrated in Figure 5.12.

Fixed packing comes in prefabricated sheets mounted at intervals inside the column. Although the initial cost is higher for this type of packing, manufacturers claim higher transfer efficiency.

Air Blower. Airflow is provided at the base of the column by a centrifugal blower driven by an electric motor. Small towers are designed with blowers requiring a motor as small as 5 hp (3,700 W), but much larger blowers are required for larger units and when a high air-water ratio is required. Care must be taken in providing screens and locating the air inlet to prevent insects and airborne contaminants from being blown into the column.

Site Considerations. In general, water temperature in the column stays close to the temperature of the influent water. It has been demonstrated in several installations that there is no danger of freezing even under air temperatures well below freezing. In cold climates, blowers and pumping equipment are usually housed for protection from ice and snow. Housing this equipment also provides increased security, reduced noise, and reduced maintenance.

Site considerations include zoning restrictions, height restrictions, and noise restrictions. There are instances where residents have opposed installation of a column in a residential neighborhood because of potential noise, visual impact of the relatively tall tower, and the

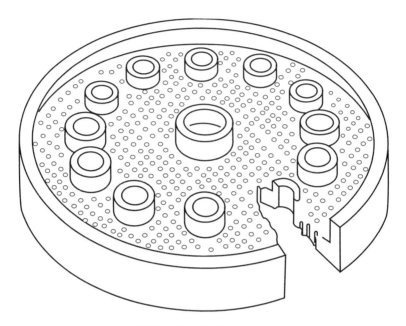

Orifice - type distributor

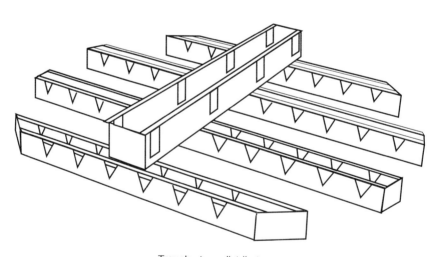

Trough - type distributor

FIGURE 5.11 Types of distributors. *(Source: USEPA workshop on emerging technologies for drinking water treatment, 1988.)*

FIGURE 5.12 Some common shapes of plastic packing. *(Source: US Environmental Protection Agency.)*

perceived inhalation danger of the off-gas. Where there is public opposition to installation of a column, other means of treatment may have to be considered or the source water may have to be piped to a more remote location for treatment.

Exhaust Emission Considerations. The emission of contaminated exhaust air from packed-column aeration systems creates potential air quality problems. The transfer of VOCs from water to air might be a concern depending on site or local regulations, type of VOC, proximity to human habitation, exposure of treatment plant workers, local air quality, local meteorological conditions, daily quantity of water to be processed, and the contamination level.

The emission rate must be evaluated in the context of applicable air quality regulations and other site-specific factors. Air emission regulations are expressed in terms of permissible emission rates (lb/day or lb/hr) or projected ground level concentrations (mg/m^3). If the treatment plant's emission rate is unacceptable, the column or plant process may be changed to bring the installation into compliance. Treatment options currently available to remove organics from off-gas include:

- Packed column design modification
- Thermal incineration of the gas
- Catalytic incineration of the gas
- Ozone destruction
- Vapor-phase carbon adsorption

The least costly method of achieving compliance with air regulations is usually to modify the packed column design to dilute the emissions. Possible modifications include increasing the tower height, airflow rate, and exhaust gas velocity. If these steps are insufficient to achieve compliance, a vapor-phase treatment component may be required.

Thermal incineration of packed-column off-gas has the disadvantage of high energy requirements. Catalytic incineration has lower temperature requirements but is currently not effective for removing low levels of chlorinated organics. Similarly, ozone destruction is being evaluated on a pilot scale at this time.

Granular activated carbon (GAC) adsorption is generally the most cost-effective method to remove low-level organics from packed column exhaust air. Vapor-phase adsorption is attractive because the vapor-phase mass transfer zone (MTZ) is much shorter than the liquid-phase MTZ, and the cross-sectional area requirement of the fixed bed is much smaller. Activated carbon usage is also less than that for liquid phase. A schematic of a vapor-phase GAC system is illustrated in Figure 5.13.

In operating a vapor-phase GAC system, the relative humidity of the off-gas should be lessened to prevent condensation of water vapor in the activated carbon pores by heating the air before it enters the GAC contactor. The competition of water vapor adsorption and gas-phase VOC adsorption onto GAC is minimized at an off-gas relative humidity of 40% to 50%.

Predicting contaminant breakthrough is a major concern with vapor-phase GAC systems because reliable methods of estimating the vapor-phase GAC bed life are not yet available. Possible approaches include monitoring GAC effluent air quality either continuously or intermittently, using a mass balance around the contactor, or combining these two approaches.

For example, a GAC bed from a pilot plant in Wausua, Wisconsin, treating off-gas containing TCE and PCE was regenerated with steam three times. The TCE capacity decreased from 80% to 60% of the initial capacity over the three cycles. This lessening in TCE capacity with successive adsorption/regeneration cycles was because of PCE buildup on the GAC. The PCE was not removed effectively under the existing regeneration conditions (100° C, 1 atm). Use of the equilibrium model indicated that regeneration with saturated steam 50° C above the boiling point of PCE (121° C) could improve PCE removal.

When considering installing a stripping column, as a first step, state and local officials should be contacted for information on air quality requirements.

Fouling of Packing. Packed column design must consider the possibility of scaling and fouling of the packing. Some installations experience few problems, but others have seri-

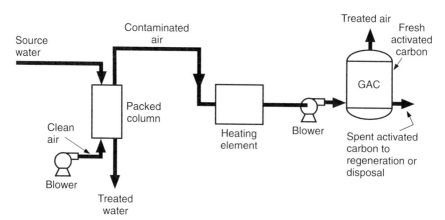

FIGURE 5.13 Schematic of vapor-phase GAC system. (*Source: AWWA WQD Committee Report, 1991.*)

ous problems. Three main causes of fouling are carbonate scaling, iron oxidation, and microbial action. Fouling gradually causes a decrease in airflow through the column and if not corrected can seriously reduce the column's performance.

Carbonate scaling of packing becomes a problem when influent water has a relatively high calcium carbonate hardness. Deposition occurs primarily because of pressure changes and a rise in the water's pH as CO_2 is released from the water. Carbonate scale deposits usually are a brittle, cementlike scale.

Ferrous iron in groundwater, another source of scaling, oxidizes easily in the presence of oxygen to form insoluble ferric compounds. The result is primarily iron hydroxide, which accumulates on the packing as a rust- or black-colored gel.

Microbial fouling is, in most cases, primarily from the presence of iron bacteria. Iron bacteria generally thrive in a dark environment, under aerobic conditions, and at temperatures between 40° F and 70° F (4° C and 21° C). These organisms derive energy from the oxidation of iron from the ferrous to ferric form. Some of the more common species of iron bacteria found in groundwater are *Gallionella, Crenothrix, Leptothrix,* and *Sphaerotilus.* Colonies of the bacteria can grow on the packing media, forming a slimy material that, if not controlled, can completely fill all void spaces in the packing.

One method of controlling fouling involves pretreating water with chlorine or permanganate and then filtering to remove oxidized solids. Another pretreatment method is to add chelating agents to inhibit formation of oxidation products. In any event, if the influent water has a high potential for fouling, the plant must provide facilities and a regular schedule of periodic cleaning of the buildup. Cleaning consists of circulating strong chlorine or acid solutions through the media.

PILOT TESTING

Pilot plant aeration studies are usually conducted to determine the effectiveness of an aeration system in removing the contaminants of concern. Testing is performed using a laboratory or bench-scale prototype unit. Municipalities, private consultants, and the USEPA use pilot aeration tests to determine the usefulness of aeration in removing various VOCs.

Pilot tests are sometimes performed on samples "spiked" with the contaminant to be studied. However, it is generally best to use water directly from the contaminated source, because subtle differences in both the physical and chemical composition of the water could have some effect on the ability of an aeration unit to remove a contaminant.

Diffused Air Pilot Studies

The main components of a diffused air pilot plant are the tank, diffuser, and blower. The system is usually operated in a countercurrent fashion. Influent water is piped through a rotameter to measure flow and dispersed through a liquid distributor at the top of the column. Water then flows down through the tank to be discharged at the bottom. Air is supplied by a compressor, piped through a rotameter to measure flow, and dispersed by a diffuser located at the bottom of the tank.

Packed Column Pilot Studies

A typical packed column pilot plant consists of the column with influent piping, valves, packing media, blower, and support structure, as shown in Figure 5.14. Influent is pumped

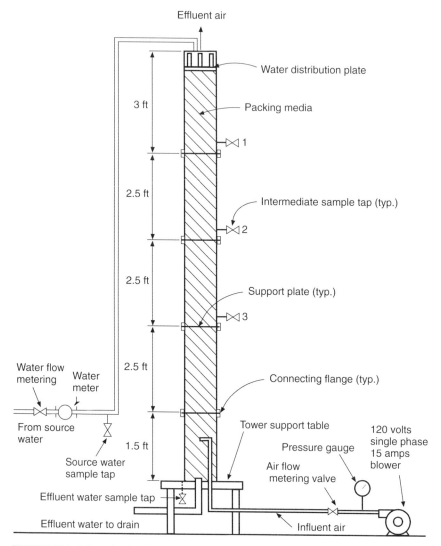

FIGURE 5.14 Schematic of a pilot air stripping column. *(Source USEPA technologies for upgrading existing or designing new drinking water treatment facilities, 1990.)*

through the metering valve and meter to the top of the column, where it is distributed by an orifice-type plate. Water trickles down through the packing and is discharged at the bottom of the column.

Support plates are placed at intervals in the column to provide intermediate support for media and to redistribute water and air flow. Air from a compressor or blower is controlled by a metering valve and is admitted at the bottom of the column. Taps for sampling water are located on the influent line, at three locations along the side of the column, and at the bottom of the column.

Pilot plant size should be small enough to allow it to be easily transported to test sites, yet large enough to allow "scale-up" procedures to be meaningful. The diameter of a pilot column is commonly 12 in. (30.4 cm).

It is critical for measurements of both water and air flows to be accurate in order to know the air-to-water ratio.

When using random packing in a column, the number of packing pieces per unit volume is ordinarily less in the vicinity of column walls. This packing distribution tends to produce a flow scheme in which water moves toward the walls and airflow is directed more toward the center of the column—a phenomenon referred to as channeling. Channeling is less pronounced when the diameter of the packing pieces (d_p) is smaller than one-eighth the column diameter (d_c). It is recommended that, if possible, the ratio $d_p/d_c = 1:15$.

Testing Program and Evaluation of Results

A testing program for operating a pilot plant and evaluating results is usually necessary before beginning to design full-scale treatment facilities. Decisions must be made on factors such as the number of operations to be made, loading rates, air-to-water ratio that must be provided, and, in the case of packed columns, the type of packing to be used. Length of a pilot test is usually based on a minimum time equal to the time required to reach equilibrium.

A critical part of operating a pilot plant is the sampling procedures. When samples are collected for VOC analysis, they must be collected without allowing any entrapped air to remain in the sample bottle. If even a very small bubble remains in the bottle, some of the volatile chemicals may leave the water, causing incorrect results when the sample is analyzed.

Other data that must be collected periodically during pilot plant operation are water and air temperatures, as well as chlorine, iron, and total organic carbon levels.

Once samples have been analyzed, test results can be evaluated. From these data, optimum full-scale operational and design conditions can be established. In a diffused air system, detention time, water depth, and air-to-water ratio are evaluated. In addition to these conditions, with a packed column system, KLa is determined and the best packing is chosen.

BIBLIOGRAPHY

Adams, J. Q., and R. M. Clark. "Evaluating the Costs of Packed Tower Aeration and GAC for Controlling Selected Organics." *Journal AWWA* 83(1):49, 1991.

Air Stripping for Volatile Organic Contaminant Removal. Denver, Colo: American Water Works Association, 1989.

American Water Works Association. *Water Quality and Treatment.* 4th ed. New York: McGraw-Hill, 1990.

Boyden, B. H., et al. "Using Inclined Cascade Aeration to Strip Chlorinated VOCs from Drinking Water." *Journal AWWA* 84(5):62, 1992.

Controlling Radionuclides and Other Contaminants in Drinking Water Supplies: a Workbook for Small Systems. Denver, Colo: American Water Works Association, 1991.

Dixon, K. L., et al. "Evaluating Aeration Technology for Radon Removal." *Journal AWWA* 83(4):141, 1991.

Dxombak, D. A., S. B. Roy, and H. J. Fang. "Air-Stripper Design and Costing Computer Program." *Journal AWWA* 85(10):63, 1993.

Jang, W., N. Nirmalakhandan, and R. E. Speece. *Cascade Air-Stripping System for Removal of Semi-Volatile Organic Contaminants: Feasibility Study.* Denver, Colo: American Water Works Association Research Foundation and American Water Works Association, 1989.

Kavanaugh, M. C., and R. R. Trussell. "Design of Aeration Towers to Strip Volatile Contaminants from Drinking Water." *Journal AWWA* 73(12): 684, 1980.

Langelier, W. F. "The Theory and Practice of Aeration." *Journal AWWA* 24(1):62, 1932.

Lamarche, P., and R. L. Droste. "Air-Stripping Mass Transfer Correlations for Volatile Organics." *Journal AWWA* 81(1):78, 1989.

Little, J. C., and R. E. Selleck. "Evaluating the Performance of Two Plastic Packings in a Crossflow Aeration Tower." *Journal AWWA* 83(6):88, 1991.

Scott, G. R. "Committee Report: Aeration of Water." *Journal AWWA* 44(4):873, 1955.

Staudinger, J., W. R. Knocke, and C. W. Randall. "Evaluation of the Onda Mass Transfer Correlation for the Design of Packed Column Air Stripping." *Journal AWWA* 82(1):73, 1990.

Symons, J. M., et al. "Treatment Techniques for Controlling Trihalomethanes in Drinking Water." U.S. EPA, Drinking Water Research Division, 1981.

VOCs and Unregulated Contaminants. Denver, Colo: American Water Works Association, 1990.

Warner, H. P., J. M. Cohen, and J. C. Ireland. "Determination of Henry's Law Constants of Selected Priority Pollutants." Wastewater Research Division, Municipal Environmental Research Laboratory, Cincinnati, Ohio, 1980.

Zogorski, J. S., et al. "Fuel Oxygenates and Water Quality: Current Understanding of Sources, Occurrences in Natural Waters, Environmental Fate, and Transport." Prepared for the Interagency Oxygenated Fuels Assessment, Office of Science and Technology Policy, Executive Office of the President.

CHAPTER 6
MIXING, COAGULATION, AND FLOCCULATION

Coagulation and flocculation may be broadly described as chemical and physical processes that mix coagulating chemicals and flocculation aids with water. The overall purpose is to form particles large enough to be removed by the subsequent settling or filtration processes. Particles in source water that can be removed by coagulation, flocculation, sedimentation, and filtration include colloids, suspended material, bacteria, and other organisms. The size of these particles may vary by several orders of magnitude. Some dissolved material can also be removed through the formation of particles in the coagulation and flocculation processes.

There are several excellent discussions on the theory of coagulation and flocculation in *Water Quality and Treatment* and other AWWA publications listed at the end of this chapter.

DEFINITIONS

Terms used in this chapter are defined as follows:

- *Coagulation:* the process in which chemicals are added to water, causing a reduction of the forces tending to keep particles apart. Particles in source water are in a stable condition. The purpose of coagulation is to destabilize particles and enable them to become attached to other particles so that they may be removed in subsequent processes. Particulates in source waters that contribute to color and turbidity are mainly clays, silts, viruses, bacteria, fulvic and humic acids, minerals (including asbestos, silicates, silica, and radioactive particles), and organic particulates. At pH levels above 4.0, particles or molecules are generally negatively charged. The coagulation process physically occurs in a rapid mixing process.

- *Mixing:* commonly referred to as flash mixing, rapid mixing, or initial mixing. The purpose of rapid mixing is to provide a uniform dispersion of coagulant chemical throughout the influent water.

- *Enhanced coagulation:* a phrase used by the U.S Environmental Protection Agency (USEPA) in the Disinfectants and Disinfection By-Products Rule. The Rule requires that the coagulation process of some water supplies be operated to remove a specified percentage of organic material from the source water, as measured by total organic carbon (TOC). Enhanced coagulation (removal of TOC) can be achieved in most cases by either increasing coagulant chemical dosage or adjusting the pH during the coagulation reaction.

- *Coagulant chemicals:* inorganic or organic chemicals that, when added to water at an optimum dosage, cause particle destabilization. Most coagulants are cationic when dissolved in water and include chemicals such as alum, ferric salts, lime, and cationic organic polymers.
- *Flocculation:* the agglomeration of small particles and colloids to form settleable or filterable particles (flocs). Flocculation begins immediately after destabilization in the zone of decaying mixing energy following rapid mixing, or as a result of the turbulence of transporting flow. In some instances, this incidental flocculation may be an adequate flocculation process. A separate flocculation process is most often included in the treatment train to enhance contact of destabilized particles and to build floc particles of optimum size, density, and strength.
- *Flocculation aids:* chemicals used to assist in forming larger, denser particles that can be more easily removed by sedimentation or filtration. Cationic, anionic, or nonionic polymers are most often used in dosages of less than 1.0 mg/L.
- *Direct filtration:* a treatment train that includes coagulation, flocculation, and filtration, but excludes a separate sedimentation process. With direct filtration, all suspended solids are removed by filtration. In the process sometimes called in-line filtration, flocculation occurs in the conduit between the rapid mixing stage and the filter, in the volume above the filter media, and within the filter media.
- *Solids contact clarifiers:* proprietary devices that combine rapid mixing, flocculation, and sedimentation in one unit. These units provide separate coagulation and flocculation zones and are designed to cause contact between newly formed floc and settled solids.

THE COAGULATION PROCESS

Coagulation reactions occur rapidly, probably taking less than one second. Principal mechanisms that contribute to the removal of particulates when coagulating chemicals such as alum or ferric chloride are mixed with water include chemical precipitation, reduction of electrostatic forces that tend to keep particles apart, physical collisions between particles, and particle bridging.

Several factors affect the type and amount of coagulating chemicals required, including the nature of suspended solids and the chemical characteristics of the influent water.

Coagulant Chemicals

The most commonly used coagulants are:

- Alum (aluminum sulfate), $Al_2(SO_4)_3 \cdot 14H_2O$ is the most common coagulant in the United States and is often used in conjunction with cationic polymers.
- Polyaluminum chloride, $Al(OH)_x(Cl)_y$, is efficient in some waters requiring less pH adjustment and producing less sludge.

- Ferric chloride, $FeCl_3$, may be more effective than alum in some applications.
- Ferric sulfate, $Fe_2(SO_4)_3$, is effective in some waters and more economical in some locations.
- Cationic polymers can be used alone as the primary coagulant or in conjunction with aluminum or iron coagulants.

Although alum is by far the most widely used coagulant chemical, ferric chloride or ferric sulfate form a better-settling floc in some waters and may be more consistently effective in removing natural organic matter.

Flocculation Aids

Floc formed in many waters with alum is light and fragile and somewhat difficult to settle. Polymers and other additives can often help form a floc that is more efficiently removed by settling and filtration. Typical additives used for flocculation aids are:

- High-molecular-weight anionic or nonionic polymers
- Activated silica
- Bentonite

These chemicals are normally added after applying coagulants, from 5 to 600 seconds after mixing. If the water to be treated with a flocculent aid is already in the flocculation stage, the chemical should be added so that it can be spread across the flocculation basin.

Chemical Selection

The selection of coagulant chemicals and flocculation aids for use in a particular plant is generally based on economic considerations along with reliability, safety, and chemical storage considerations. The best method of determining treatability, the most effective coagulants, and the required dosages is to conduct bench scale and, in some cases, pilot tests. Jar tests can be used to determine treatability and estimate chemical dosages. If possible, testing should cover all critical seasonal conditions. Pilot plant design and construction is discussed in Chapter 28.

When designing for coagulant application, as much flexibility as possible should be allowed to accommodate changing conditions. Several points of addition for coagulant chemicals, particularly polymers, should be provided in the rapid mixing and flocculation processes. The order of chemical addition is also important in almost all waters.

Sludge quantity and disposal are important considerations in selecting the coagulant to be used. Metal ion coagulants produce considerably larger volumes of sludge than polymers. The ability to predict the exact reaction and quantity of sludge that will be produced solely by the reaction formulas is limited. For this reason, predictions of treatability, chemical dosages, and sludge quantities must generally be determined by laboratory and pilot plant tests.

The coagulation process may, in some cases, be improved by preozonation. Ozone may significantly reduce coagulant requirements to the point where low residual solids (or filtration efficiency) make direct filtration feasible. Oxidation with air and chemical oxidants such as chlorine and potassium permanganate may also aid coagulation by oxidizing iron and manganese, which can aid floc formation.

Adjustment of pH

Control of pH and alkalinity is an essential aspect of coagulation. The optimum pH for coagulation varies but is generally within the following ranges:

- Alum: pH 5.5 to 7.5; typical pH 7.0
- Ferric salts: pH 5.0 to 8.5; typical pH 7.5

It may be necessary to adjust the pH of some source waters to achieve optimum coagulation. The pH is often lowered by adding carbon dioxide or an acid. Alum and ferric chloride consume alkalinity and can lower pH; however, reducing pH by adding more chemical than is required for coagulation should be avoided. In some source waters with low pH or low alkalinity, it may be necessary to add caustic soda or lime to raise pH and to offset the acidity of metal ion coagulants. A thorough discussion of the effects of pH on coagulation appears in *Water Quality and Treatment*.

If pH is lowered to improve coagulation, it may be necessary to raise the pH in the final effluent from the plant to provide a less corrosive finished water. The pH may be adjusted at one or more points in the treatment, including rapid mixing, prefiltration, and postfiltration.

For plants where only a small increase in pH is required, liquid caustic soda is most commonly used because of its ease of handling. When a large increase in pH is required, lime is normally the most economical choice. Lime, however, may add turbidity. If lime is used for postfiltration pH adjustment, it is generally best to use a lime saturator to ensure that no turbidity is added.

DESIGN OF CHEMICAL MIXING

Chemical mixing can be accomplished by several different types of equipment designed to mix the applied chemicals with the source water as quickly as possible.

Mixing Intensity

The intensity of agitation required for optimum rapid mixing and flocculation is measured by the G value. The G value concept developed by Camp and Stein in 1943 is widely used in designing rapid mixing and flocculation processes, and is defined by the equation

$$G = (P/\mu V)^{1/2}$$

where G = the root-mean-square velocity gradient, the rate of change of velocity, expressed in ft/s/ft
 P = power input, ft-lb/s
 μ = dynamic viscosity, lb-s/ft^2
 V = volume, ft^3

Equations are also available to calculate G for various types of mixing arrangements, and manufacturers of mixing and flocculation equipment provide information on G values for their equipment. Another parameter used in designing mixing systems is Gt, which is the dimensionless product of G and detention time in seconds.

TABLE 6.1 Water Viscosity and Water Temperature

Temperature, °C	Temperature, °F	μ, cP	μ, lb-s/ft²
0	32	1.792	3.75×10^{-5}
5	41	1.520	3.17×10^{-5}
10	50	1.310	2.74×10^{-5}
15	59	1.145	2.39×10^{-5}
20	68	1.009	2.10×10^{-5}
25	77	0.895	1.87×10^{-5}
30	86	0.800	1.67×10^{-5}

TABLE 6.2 Guidelines for Mixer Detention Times

Temperature, °C	Detention time factor
0	1.35
5	1.25
10	1.15
15	1.07
20	1.00
25	0.95
30	0.90

Temperature Effects on Mixing

Rapid mix and flocculation systems design is temperature dependent because water viscosity varies with temperature, as shown in Table 6.1. Guidelines for adjusting detention times in both rapid mix and flocculation basins are shown in Table 6.2.

Types of Rapid Mixing Systems

Coagulant chemicals can be mixed by several methods, including:

- Mechanical devices in a dedicated basin
- In-line blenders
- Hydraulic methods
- Air mixing

Mechanical Mixers. Propeller- or paddle-type mechanical mixers in a dedicated basin are the most commonly used rapid mix system in water treatment plants. A typical arrangement for this type of rapid mixer is illustrated in Figure 6.1. Another mixer arrangement without stators and with a turbine-type blade is illustrated in Figure 6.2.

Rapid mixers attempt to provide complete mixing by near-instantaneous blending throughout the entire basin. As a result, an incoming volume of water immediately loses

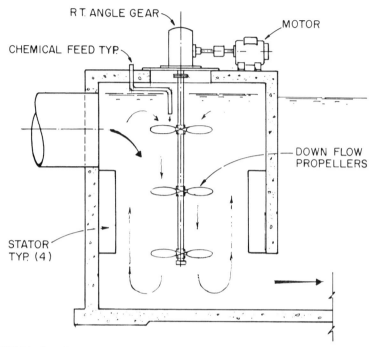

FIGURE 6.1 Propeller-type mechanical flash mixer.

its identity as it is mixed with water that entered the basin previously. A mixer operating in this manner is also called a backmix reactor because basin contents are always blended backward with incoming flow.

Mechanical mixers are generally propeller- or paddle-type devices. More than one set of propeller or paddle blades may be provided on a shaft. Stators (baffles near the blades of the mixer or on the wall of the basin) may be provided to maximize energy transfer to the fluid and to minimize residual velocities at the outlet. Mechanical mixers are often constructed with a vertical shaft driven by a speed reducer and electric motor. Propeller-type mixers can be arranged so that flow is directed in any direction. With propeller-type blades, coagulant chemical is generally directed to the eye of the propeller on the suction (upstream) side.

Mechanical mixers are not normally provided with variable-speed drives. If adjustments to energy input are necessary, they may be achieved by changing propellers or paddle blades or by mechanically adjusting the shaft speed.

Many treatment plant designs incorporate two or more rapid mix basins in series. The order in which coagulant chemicals are added is important in most waters, and more than one rapid mixer can provide the needed reaction time for each chemical. A plant under design in San Diego, California, uses three rapid mixers, as shown in the design criteria summarized in Table 6.3. Two other large water treatment plants in southern California use two vertical turbine mechanical mixers in series designed for G values of 440 to 670 s^{-1} with corresponding detention times of 30 to 10 s.

Typical design values for most mechanical rapid mix systems provide detention times of 10 to 60 s and G values of 600 to 1,000 s^{-1}.

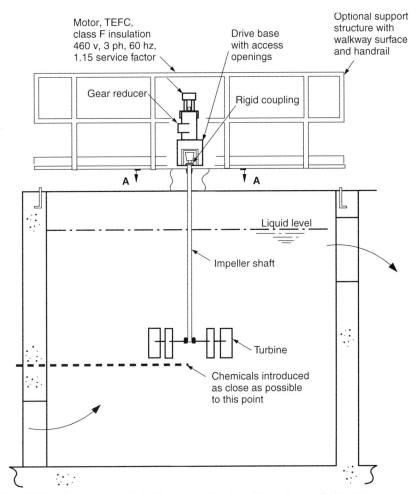

FIGURE 6.2 Turbine-type mechnical flash mixer. *(Courtesy of Eimco Process Equipment.)*

In-Line Blenders

Although the traditional complete-mix type of basin with a mechanical mixer has performed satisfactorily in many plants for years, recent experience has shown that in-line blenders often provide more efficient rapid mixing. Advantages of this type of mixer are that it can approach nearly instantaneous dispersion of chemicals. In-line blenders operate at short detention times (less than one second) and at high G values.

An important consideration is that a short detention time and high G value may be a disadvantage in waters requiring more reaction time and the use of more than one chemical for floc formation. One answer to this problem may be an in-line mixer used in conjunction with a mechanical rapid mix basin (for example, installing an in-line blender as a first stage, followed by rapid mix basins to provide more detention time). The design summarized in Table 6.3 uses this concept.

TABLE 6.3 Rapid Mixing Design Criteria from a 170 mgd Plant in San Diego, California

Number of stages: three

Type
 First stage: pump diffusion
 Second stage: mechanical
 Third stage: mechanical

Detention time
 First stage: 1 s
 Second stage: 30 s
 Third stage: 30 s

Number of basins: two
 Volume (each): 56,800 gal (214,900 L)
 Depth (each): 15 ft (4.6 m)
 Width (each): 22 ft (6.7 m)
 Length (each): 23 ft (7 m)

Mixing intensity, G
 First stage (in-line): 1,000 s^{-1}
 Second stage: (basin): 150 to 300 s^{-1}
 Third stage: (basin): 150 to 300 s^{-1}

In-Line Jet Mixers. Kawamura (1976) notes some problems with backmix-type rapid mixers and provides information on a design for several large water treatment plants. The system shown in Figure 6.3 was designed for an 82 mgd (310 ML per day) plant. The jet velocity at the nozzle is 24 ft/s (732 m/s) with a 950 gpm (60 L/s) capacity injection pump with a 10 hp motor. The G value is approximately 1,000 s^{-1}.

Advantages of this system are that either source water without added chemicals or partially destabilized source water can be used in the chemical injection system. A valve installed in the pump discharge line can control pumping rate and vary energy input for various plant flows and types of coagulating chemicals.

Chao and Stone (1979) presented a recommended design for a variation of an in-line jet mixing system illustrated in Figure 6.4. Instead of a single orifice directed either upstream or downstream with the flow, this design uses multiple jets that inject perpendicular to the flow in the pipe. With two rows of eight jets per row, the duration of mixing is about 0.5 s and the G value about 1,000 s^{-1}.

Mechanical In-Line Blenders. Mechanical in-line blenders provide rapid mixing of chemicals with water flowing in a pressure pipe. These devices consist of a propeller in the pipe and an electric drive system, as illustrated in Figure 6.5. These are normally proprietary items of equipment and can be specified to provide any required G value.

Static In-Line Blenders. Static in-line mixers, sometimes called motionless mixers, use energy of the flowing liquid to produce mixing. The design of this type of mixer attempts to create flow paths that result in consistent and predictable mixing performance. The units are available from more than one manufacturer and incorporate various arrangements of intersecting bars, corrugated sheets, and plates. A typical static mixer is shown in Figure 6.6.

Air Mixing. Air mixing can be a simple and reliable mixing method and has advantages where aeration of the source water is required anyway. In some instances, air mixing is incorporated into existing structures where it may not be convenient to install other mechanical equipment. It is especially applicable to deep conduits or vertical sections of piping.

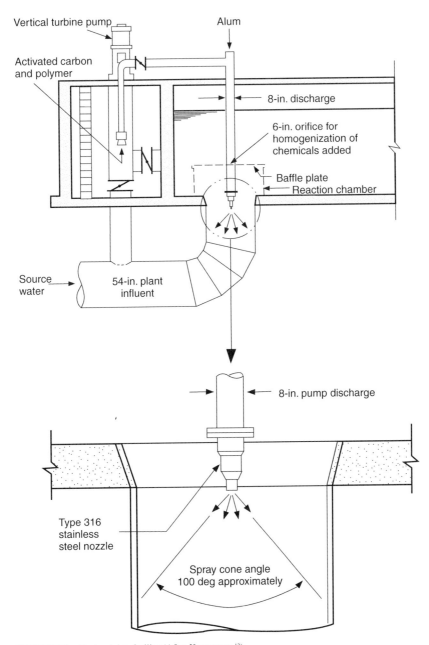

FIGURE 6.3 Flash mixing facility (After Kawamura [12]).

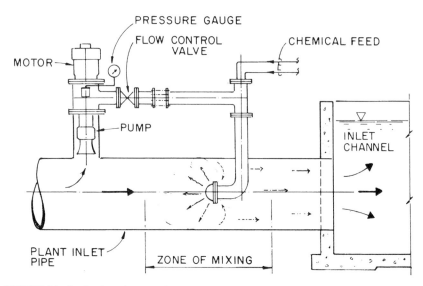

FIGURE 6.4 Section through pump mixer.

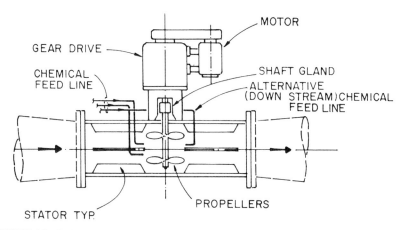

FIGURE 6.5 Typical in-line blender (mixer).

Energy applied in adding air may be computed as the volume of water displaced per unit time multiplied by the depth below the free surface, as shown in the following equation:

$$\text{hp} = Qh/528$$

where Q = free air discharge, ft^3/min
h = depth of air inlet nozzle below water surface, ft
hp = horsepower input

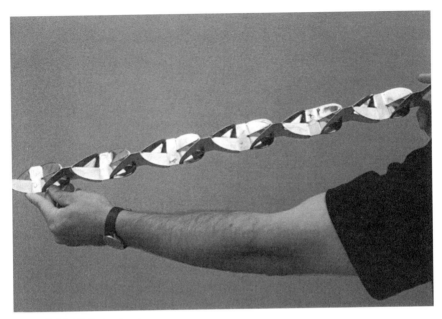

FIGURE 6.6 Typical static mixer.

Energy applied by air injection may be varied directly by adjusting air flow. Air mixing is not widely used, and before it is incorporated into a design, an inspection should be made to determine whether scum and floatable material would be a problem. Limited quantities of floating scum (or sludge) may be accepted onto filters, but certain coagulants and algae may increase scums.

Hydraulic Mixing. Hydraulic mixing can be achieved by using V-notch weirs, Parshall flumes, orifices, throttled valves, swirl chambers, and simple turbulence caused by velocity in a pipe, fitting, or conduit. Hydraulic mixing is a nonbackmix method that can sometimes be highly efficient. The principal problem is that energy input varies with the flow. However, if a plant has relatively constant flows, energy variations may not be a concern. Seasonal flow variations can sometimes be overcome by varying the number of plant mixing modules in operation to maintain more or less constant flows on those modules in operation.

Total head loss across a throttled valve used for mixing coagulant chemicals should not exceed 4 ft (3.2 m). If head loss exceeds this amount, coagulants should be added to the flow downstream of the valve in the zone of decaying energy because excessive confined energy may shear polymers.

The energy provided by a weir with an effective fall of 1 ft (30 cm) provides a G valve of 1,000 s^{-1} at 20° C. Such a weir mixer with a downstream baffle develops G values as a function of flow, as shown in Table 6.4. If the volume V where turbulence dissipates is assumed to be constant, G may vary significantly, but if turbulence volume is assumed to be proportional to the flow Q, there is a lesser variation in G.

Weir mixers require that coagulant chemicals be fed equally across the length of the weir at multiple points spaced at not more than the head distance of the weir. Because of maintenance problems with multiple orifice chemical feed manifolds and other practical

TABLE 6.4 Hydraulic Weir Mixing

Flow Q (percent of maximum)	G relative (V constant)	G relative (VQ)
1.0	1.0	1.0
0.9	0.92	0.97
0.8	0.83	0.93
0.7	0.74	0.88
0.6	0.65	0.84
0.5	0.47	0.73
0.35	0.42	0.71

considerations, weir mixers tend to be used on plants of less than approximately 40 mgd (151 ML per day) capacity.

FLOCCULATION PROCESS DESIGN

Building optimum size floc requires gentle mixing in the energy gradient range of 20 to 70 s^{-1} for a total period of approximately 10 to 30 min. Direct filtration requires a small, dense floc that can be formed at the higher end of the energy range. For settling in conventional basins and in units with settling tubes and settling plates, lower energy levels are applied to produce a large, dense floc that will resist breakup during contact with weirs and plates. Often, polymers are used to help form denser floc.

Floc begins to form within 2 s of coagulant addition and mixing. If high turbulence or shear is subsequently applied to the water, the formed flocs may be fragmented, and broken floc may not readily settle or re-form.

Optimum floc that is efficiently settled or filtered is usually formed under conditions of gradually reducing energy. In large plants, it may be difficult to distribute water to flocculation basins or filters without quiescent stages and high-energy stages. Conduits handling mixed water should minimize head losses, but may, on the other hand, include water jets or air mixing to maintain G at values of 100 to 150 s^{-1} before the water is transferred to the flocculation stage.

The gentle mixing process of flocculation is designed to maximize contact of destabilized particles and build settleable or filterable floc particles. It is desirable to maintain shear forces as constant as possible within the process. As a result, flocculator mechanisms tend to be slow and to cover the maximum possible cross-sectional area of floc basins.

It is desirable to compartmentalize the flocculation process by dividing the basin into two or more defined stages or compartments, as illustrated in Figure 6.7. Compartments prevent short-circuiting and permit defined zones of reduced energy input or tapered energy. To prevent short-circuiting, baffles are typically placed between each stage of flocculation. For mechanical (nonhydraulic) flocculation basins, baffles are designed to provide an orifice ratio of approximately 3% to 6% or a velocity of 0.9 ft/s (27 cm/s) under maximum flow conditions.

Incidental Flocculation

As coagulated water is transferred to flocculators in small plants, distances are short enough that incidental flocculation is negligible. But in large plants transfer may involve

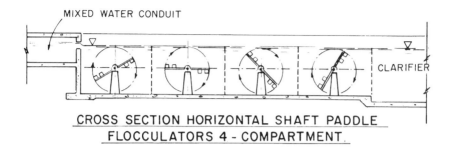

CROSS SECTION HORIZONTAL SHAFT PADDLE FLOCCULATORS 4 - COMPARTMENT.

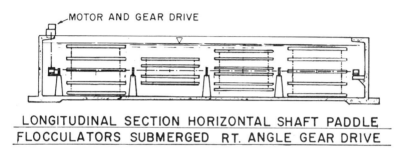

LONGITUDINAL SECTION HORIZONTAL SHAFT PADDLE FLOCCULATORS SUBMERGED RT. ANGLE GEAR DRIVE

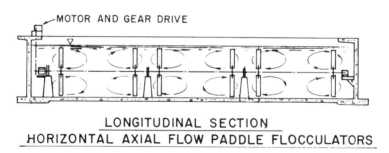

LONGITUDINAL SECTION HORIZONTAL AXIAL FLOW PADDLE FLOCCULATORS

FIGURE 6.7 Sections through horizontal shaft paddle flocculator.

distances of more than 100 ft (30 m) through low-velocity conduits, weirs, or other means of distributing water equally to each flocculation basin or compartment. This travel in large plants involves turbulence, and flocculation and incidental flocculation take place. If velocities or levels between the conduits and the flocculation basin are not limited, floc may be fragmented and plant efficiency impaired. Higher coagulant feed rates may be required to overcome fragile floc problems.

Flow splitting to distribute the flow to the flocculation basins is normally by weirs or orifices. Weirs at the postmixing stage of the process should be low-velocity, submerged weirs. Typical velocities in conduits from the mixer to flocculation basins are 1.5 to 3.0 ft/s (46 to 91 cm/s). Distribution channels between the mixer and flocculation basins are often tapered, either in width or depth, to maintain constant velocity.

TABLE 6.5 Typical Mixing and Flocculation Design Criteria for System Shown in Figure 6.8

Plant capacity	150 ft³/s
Design flow	97 mgd
Plant inlet pipe diameter	84 in.
Initial mixer	
Pump blender	
Pump rating, each	2 units
Energy input, G at 20° C	10 hp
Mixing zone	$1{,}000~\text{s}^{-1}$
Detention time	538 ft²
	3.6 s
Distribution channel	
Depth	10 ft
Width	6.5 to 1.5 ft
Maximum velocity	1.2 ft/s
Flocculation	
Number of basins	2
Compartments, each basin	4
Depth, average	16 ft
Compartment, width × length	15 × 80 ft
Volume, per basin	76,800 ft³
Detention time, total	18 min
Horizontal shaft paddles (each basin)	4
Maximum G, per compartment	$50~\text{s}^{-1}$
Maximum power, per compartment	2.0 hp
Chemical dosages	
Chlorine, refilter	5 mg/L
Alum, maximum, rapid mixer	20 mg/L
Cationic polymer, rapid mixer	2 mg/L
Nonionic polymer, second-stage flocculator	0.5 mg/L
Potassium permanganate, rapid mixer	2 mg/L

Typical design criteria for a large plant are summarized in Table 6.5 for the facilities illustrated in Figure 6.8.

Flocculation Time

Most modern plants provide approximately 20 min of flocculation time (at 20° C) under maximum plant flows. Some references recommend flocculation times of 30 min or longer. Older references (including the first edition of this book) do not define "nominal flows," and it appears that earlier texts based detention times on mean, or nominal, flows rather than maximum plant capacity.

In addition, temperature adjustments outlined in Table 6.2 should be considered. Design water temperature is the temperature most likely to be encountered under maximum flows.

For direct filtration plants, high-energy flocculation is typically in the range of 15 to 20 min detention. When clarification is required, lower energy input and detention times of 18 to 25 min are a guide. If compartmentalization is not provided, increased

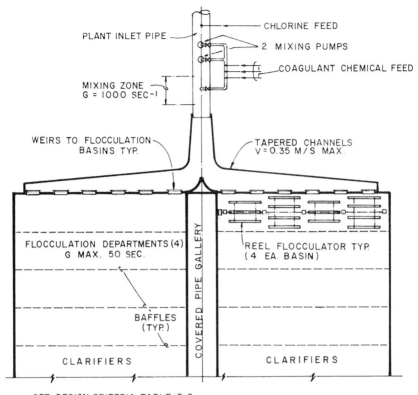

FIGURE 6.8 Partial plan for mixing and flocculation facilities [368 ML/day (97 mgd)].

detention times should be provided in addition to adjustments for water temperature. In all cases, pilot plant or full-scale tests and economic evaluation offer the most reliable indicators.

Energy Requirements

Energy input (G values) for flocculation in plants using metal ion and organic coagulant chemicals ranges from 20 to 75 s^{-1}. Typical G values and detention times for flocculation at 20° C are summarized in Table 6.6. Floc size and density should be matched to subsequent settling and filtration stages.

TABLE 6.6 Flocculation Design Criteria

Process	G, s^{-1}	Detention time, s	Gt
Distribution channels mixer to flocculator	100 to 150	Varies	—
High-energy flocculation for direct filtration	20 to 75	900 to 1,500	40,000 to 75,000
Conventional flocculation (presettling)	10 to 60	1,000 to 1,500	30,000 to 60,000

Types of Flocculators

Flocculation can be achieved by hydraulic methods or mechanical devices. Hydraulic methods are used most often in small plants. Mechanical flocculators cover a broad range of configurations.

Mechanical Flocculators. Mechanical flocculators are preferred by most design engineers in the United States because of their greater flexibility in varying G values and also because they have low head loss. Typical arrangements for horizontal shaft, reel-type flocculators, and vertical paddle units are shown in Figures 6.7, 6.8, and 6.9. Another type of horizontal shaft flocculator, illustrated in Figure 6.10, oscillates with a back-and-forth motion. One advantage of this type of unit is that it prevents water from rotating continually in the same direction around the shaft. Leakage from packing glands has occurred in some horizontal shaft units.

The type of mechanical flocculator influences the shape of flocculation compartments. Vertical flocculators are often associated with square compartments with maximum dimensions of approximately 20 ft (6 m) square and depths of 10 to 16 ft (3 to 5 m). Horizontal shaft, reel, or paddle flocculator compartments are often 20 to 100 ft (6 to 30 m) long and 10 to 16 ft (3 to 5 m) wide.

Between each zone or stage of mechanical flocculation, baffles should be designed to prevent short-circuiting. For these baffles, typical orifice areas should provide a velocity of approximately 1.0 to 1.5 ft/s (30 to 46 cm/s). Baffles are normally constructed of wood but may also be concrete, brick, or woven stainless steel strips.

Vertical flocculators are often higher-speed devices than horizontal shaft flocculators, and the proportion of volume of the compartment that receives energy from the vertical flocculators may be less. As a result, a wider range of energy is applied to the flow in vertical flocculators, and for a portion of the time, some of the flow may be subjected to a higher G. Vertical flocculators are more applicable to high-energy flocculation situations such as direct filtration.

Where uniform floc is required, low tip speed flocculators may be more suitable. Equipment manufacturers should be consulted to ensure that appropriate paddle designs are specified for large plants using vertical flocculators.

Vertical flocculators are often specified because they have no submerged bearings, are usually higher speed, and involve lower investments. High-speed flocculators, however, may not provide floc suitable for high-rate horizontal flow basins. Improved clarification may require increased coagulant doses or flocculent aids.

As a guide, for high-energy flocculators ($G = 50$ to 75 s^{-1}), maximum tip speed of mixer blades should not exceed 10 ft/s (30 m/s). For low-energy flocculators and paddle-type flocculators ($G = 20$ to 45 s^{-1}), blade tip speeds in the range of 1.0 to 2.5 ft/s (30 to 76 m/s) are appropriate. Some method of varying speed is normally provided.

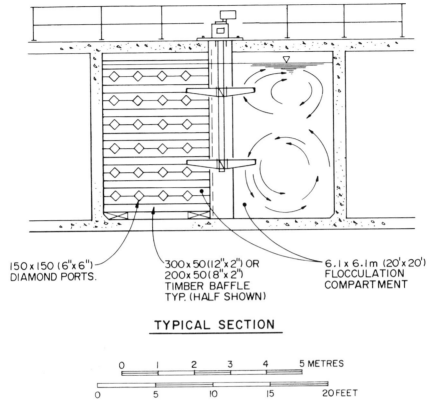

FIGURE 6.9 Vertical paddle flocculator compartment.

Variable-speed drives or provision to change pulleys or gears for different shaft speeds are valuable features that should be provided on mechanical flocculators. As a rule, only the upper 25% of the speed range requires adjustment, and such adjustment will provide a variation of 65% to 100% of maximum G. The G output of a flocculator does not normally have to be varied frequently, but it may require adjustment after installation or on a seasonal basis.

Hydraulic Flocculation. Hydraulic flocculation methods are simple and effective, especially if flows are relatively constant. The assumed flocculation volume is the total volume of each compartment, even though in some cases there may be reduced turbulence in portions of the compartments. The disadvantage of hydraulic flocculators is that G values are a function of flow that cannot be easily adjusted.

Energy may be applied to water by means of maze-type baffles or cross-flow baffles, as illustrated in Figure 6.11. For maze-type baffles, optimum plug flow conditions prevail, and excellent results can be obtained. At velocities in the range of 0.7 to 1.4 ft/s (21 to 43 cm/s), adequate flocculation may be achieved from turbulence caused by the 180 degree turn at each end of the baffle. For lower channel velocities, it may be necessary to provide an orifice at the end of each channel to induce higher energy input.

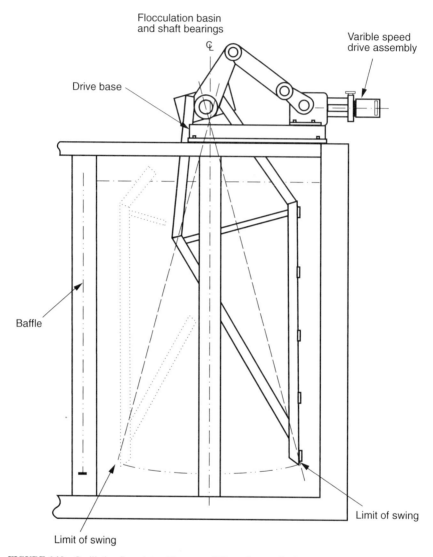

FIGURE 6.10 Oscillating flocculator. *(Courtesy of Eimco Process Equipment.)*

For cross-flow baffles, energy may be transmitted to the water in each compartment from the head loss across orifices in the entrance baffle. The G value in each compartment can be calculated from estimated head loss across baffles into each compartment with the following equation:

$$G = 62.5\, hv/t\mu$$

where hv = head loss entering compartment, ft
  t = detention time in compartment, s
  μ = viscosity, lb · s/ft^2

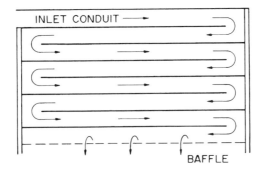

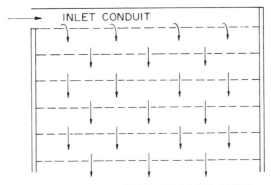

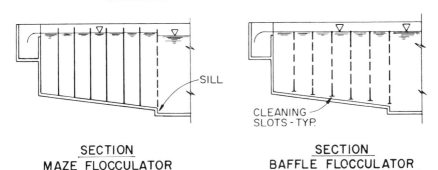

FIGURE 6.11 Plan and section of maze and baffle flocculators.

The head loss through orifices in baffles may be computed from the square-edged submerged orifice formula where the discharge coefficient may be assumed to be 0.8. Many texts give coefficients of 0.61 for square-edged orifices. In water treatment applications, head losses and velocities are lower than the velocities assumed in textbooks, less than 1.5 ft/s (46 cm/s), and coefficients are typically 0.8. If, in practice, head losses across baffles are greater than assumed, higher head losses and higher G values will result.

For cross-flow baffle flocculators, slots typically 4 to 6 in. (10 to 15 cm) high and 16 to 24 in. (41 to 61 cm) long should be provided in the bottom of baffles for cleaning purposes. Slots should be staggered to prevent short-circuiting. Normally the floors of hydraulic flocculation basins slope toward the discharge end to facilitate cleaning and to provide tapered energy.

Generally, flocculation basins are designed to be approximately the same depth as the adjacent clarifier.

End Baffles. Most flocculator systems require an end baffle between the flocculation zone and the clarifier, or some provision to prevent residual energy from the flocculation process from being transferred to the clarification stage. These baffles also minimize short-circuiting and reduce the effects of water temperature changes.

End baffles may be designed on the same basis as compartmentalization baffles described above. End baffles should not provide a barrier to removing sludge when the floc basin is cleaned, and limited openings in the bottom of the baffle are appropriate. Similarly, a small submerged section at the top of the baffle will allow scum to pass downstream.

Proprietary Designs. Several manufacturers provide proprietary designs that incorporate rapid mixing, flocculation, and settling in one unit. A typical configuration for this type of unit is illustrated in Figure 6.12. This equipment generally provides for upflow through the clarification stage, although in large-diameter units the flow may be more accurately described as radial.

The process may be designed to include intentional backmixing in the flocculation stage, and the clarification stage may be designed for "sludge blanket" conditions where flow from the flocculation zone moves upward through a layer, or blanket, of sludge in the settling zone. Designs may include a conical section in the mixing and flocculation stage and increased area in the upflow zone in the clarifier stage.

A single-unit solids contact type of flocculator-clarifier may have advantages for some applications. These units perform best in waters that can develop a dense, fast-settling floc, but the process may be difficult to control in some waters. Bench scale tests should be performed in evaluating this type of unit, and pilot tests are recommended for large plants.

Sludge Recirculation and Solids Contact. The intentional introduction of preformed floc or sludge into the mixing and flocculation stage is a feature of most proprietary water treatment equipment. High concentrations of suspended solids in the flocculation process (and in the sludge blanket of the clarifier) can provide improved efficiency in reducing particulates, colloids, organics, and certain ionized chemicals.

Recirculating sludge and reintroducing filter wash water into the mixing and flocculation stages of nonproprietary designs may improve efficiency and reduce chemical requirements. Sludge and wash water quality is an important consideration when determining the reuse of process residuals. Sludge recirculation may be difficult to optimize at the pilot plant stage of investigation. Providing flexibility for sludge recirculation and solids contact should be considered at the design stage of new plants.

Contact Flocculation. The ability of a coarse media bed to act as a flocculation system is well demonstrated. The time required to build optimum size and density of floc may be reduced where there is close contact with preformed flocs. The time factor is important for plants treating very cold water and for portable plants and pressure plants. Contact flocculation may be applicable for difficult high-suspended-solids water or water with low total dissolved solids that may not respond readily to metal ion coagulants.

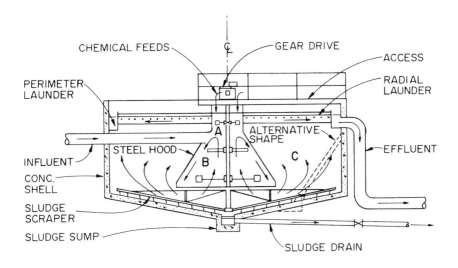

ZONE A – RAPID MIX
ZONE B – FLOCCULATION AND SOLIDS CONTACT
ZONE C – UPFLOW AND SLUDGE BLANKET ZONE
ALTERNATE SHAPE FOR SLUDGE BLANKET DESIGN

FIGURE 6.12 Typical proprietary design of solids contact reactor unit.

A typical application of the process is to pass coagulated water through a coarse media or gravel bed, either by gravity or under pressure. The gross detention time may be 3 to 5 min at 5° C. The G values should be tapered and in the 400 to 50 s^{-1} range. Flow may be either upflow or downflow, as illustrated in Figures 6.13 and 6.14. The system should provide for removing excess accumulated floc, usually by means of air scour, similar to filter air scour.

Pilot plant investigations should precede design, or the criteria for contact flocculation should be based on proven applications on similar water sources. Several manufacturers offer equipment or processes that use the principles of contact flocculation.

A new type of coarse media flocculator illustrated in Figure 6.15, using buoyant media instead of gravel, is being evaluated (Schulz et al., 1994). At this stage of development, the unit should not be considered unless tests on the source water to be treated can be conducted to develop design criteria.

PROCESS MONITORING AND CONTROL

Design dosages for chemicals should be based on experience with similar types of waters or, preferably, with jar tests and pilot tests. Design dosages of coagulants can be determined from jar tests. The effectiveness of coagulation and flocculation can also be monitored by several parameters, including turbidity, zeta potential, streaming current, particle counts, and bench and pilot tests (discussed in other chapters of this volume).

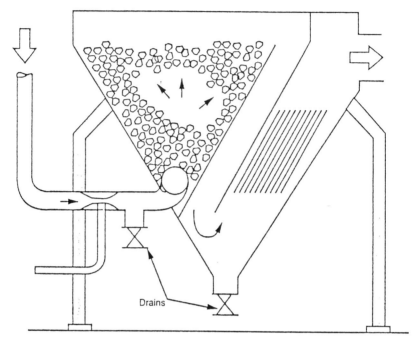

FIGURE 6.13 Upflow gravel-bed flocculator.

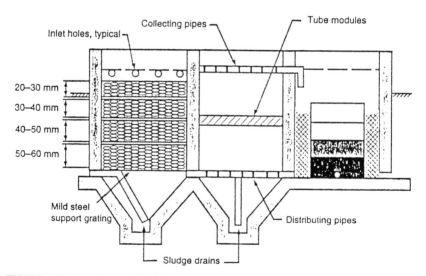

FIGURE 6.14 Downflow gravel-bed flocculator.

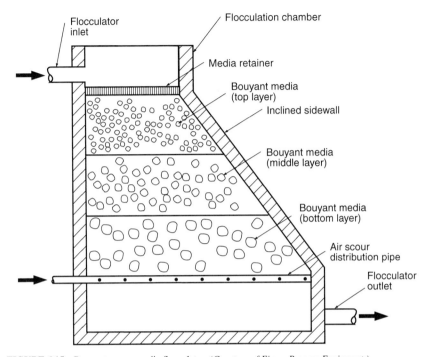

FIGURE 6.15 Bouyant course media flocculator. *(Courtesy of Eimco Process Equipment.)*

BIBLIOGRAPHY

American Water Works Association. *Water Quality and Treatment.* 4th ed. New York: McGraw-Hill, 1990.

Amirtharajah, A., M. M. Clark, and R. R. Trussel, editors. *Mixing in Coagulation and Flocculation.* AWWA Research Foundation Report. Denver, Colo.: American Water Works Association, 1991.

AWWA Coagulation Committee. "Committee Report: Coagulation as an Integrated Water Treatment Process." *Journal AWWA* 10(89):72, 1989.

Chao, J. L., and B.G. Stone "Initial Mixing by Jet Injection Blending." *Journal AWWA* 10(79):570, 1979.

Coagulation and Filtration: Back to the Basics. AWWA Annual Conference Seminar Proceedings. Denver, Colo.: American Water Works Association, 1981.

Crozes, Gil, et al. "Enhanced Coagulation: Its Effect on NOM Removal and Chemical Costs." *Journal AWWA* 1(95):78, 1995.

Culp, Wesner, and Culp R. L. *Handbook of Public Water Systems.* New York: Van Nostrand Reinhold, 1986.

James M. Montgomery, Consulting Engineers, Inc. *Water Treatment Principles and Design.* New York: John Wiley and Sons, 1985.

Kawamura, Susumu. "Considerations on Improving Flocculation." *Journal AWWA* 6(76):328,1976.

Kawamura, Susumu. *Integrated Design of Water Treatment Facilities.* New York: John Wiley and Sons, 1991.

Operational Control of Coagulation and Filtration Processes, M37. Denver, Colo.: American Water Works Association, 1992.

Schulz, C. R., et al. "Evaluating Buoyant Coarse Media Flocculation." *Journal AWWA* 8(94):51, 1994.

Selection and Design of Mixing Processes for Coagulation. AWWA Research Foundation Report. Denver, Colo.: American Water Works Association, 1994.

Technologies for Upgrading Existing or Designing New Drinking Water Treatment Facilities. EPA/625/4-89-023. Cincinnati, Ohio: U.S. Environmental Protection Agency, March 1990.

Upgrading Existing Water Treatment Plants. AWWA Annual Conference Seminar Proceedings. Denver, Colo.: American Water Works Association, 1974.

Vrale, Lasse, and R. M. Jorden. "Rapid Mixing in Water Treatment." *Journal AWWA* 1(71):52, 1971.

CHAPTER 7
CLARIFICATION

Clarification has more than one application in water treatment. Its usual purpose in a conventional treatment process is to reduce the solids load after coagulation and flocculation. A second application is removal of heavy settleable solids from source water from turbid sources to lessen the solids load on treatment plant processes, a process called plain sedimentation. Material presented in this chapter deals primarily with settling flocculated solids.

One way of designing the clarification process is to maximize solids removal by clarification, which generally requires lower clarifier loadings and larger, more costly units. Alternatively, the clarifier may be designed to remove only sufficient solids to provide reasonable filter run times and to ensure filtered water quality. This approach optimizes the entire plant and generally leads to smaller, less expensive facilities. Typical loading rates suggested in this chapter or by regulatory guidelines are generally conservatively selected to provide a high-clarity settled water.

Clarifiers fall into two basic categories: those used only to remove settleable solids, either by plain sedimentation or after flocculation, and those that combine flocculation and clarification processes into a single unit. The first category includes conventional sedimentation basins (Figure 7.1) and high-rate modifications such as tube or plate settlers and dissolved air flotation (DAF). The second category includes solids-contact units such as sludge blanket clarifiers and slurry recirculation clarifiers. Also included in this category is contact clarification in which flocculation and clarification take place in a coarse granular media bed.

CONVENTIONAL CLARIFICATION DESIGN

Most sedimentation basins used in water treatment are the horizontal-flow type in either rectangular, square, or circular design. Both long, rectangular basins and circular basins are commonly used; the choice is based on local conditions, economics, and personal preference. Camp (1946) states that long, rectangular basins exhibit more stable flow characteristics and therefore better sedimentation performance than very large square basins or circular tanks. Basins were originally designed to store sludge for several months and were periodically taken out of service for manual cleaning by flushing. Most basins designed today are cleaned continuously with mechanical equipment.

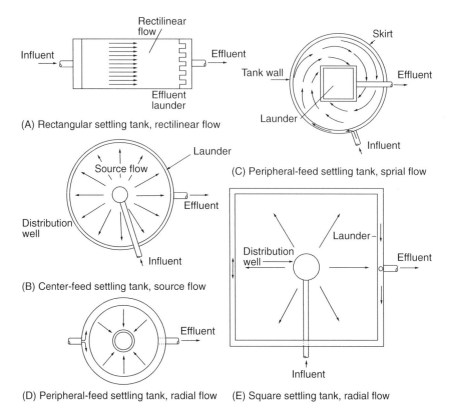

FIGURE 7.1 Typical conventional sedimentation tanks.

Sedimentation Theory

A complete discussion of sedimentation theory and its application can be found in *Water Quality and Treatment* (AWWA, 1990). The designer is urged to become familiar with this theory before selecting and designing the clarification process.

When flocculated particles enter a basin and begin to settle, particles' settling velocities change as particles agglomerate to form larger floc. Because the settling properties of flocculent suspensions cannot be formulated, a basin's performance cannot be accurately predicted. However, for new plants, settling rates can be estimated from batch settling data developed with laboratory jar tests. For expanding existing plants, settling rates can be derived from evaluating the performance of existing basins during various influent water quality conditions. These evaluations often allow for increasing rates for existing basins and establishing higher rates for new basins.

In an ideal continuous flow basin, sedimentation would take place as it does in the laboratory jar. However, in a real basin, wind, temperature density currents, and other factors cause short-circuiting, disruption of flow patterns, breakup of floc, and scouring of the settled sludge. The designer must learn as much as possible about the settling properties of the flocculated solids and then design basins to match these characteristics. When the designer does not have access to source water data, it is best to select design criteria known to have worked in similar applications, either from personal experience or from regulatory guidelines.

Design Approach

The primary approach in designing conventional sedimentation basins is to select a design overflow rate for maximum expected plant flow. This rate may be chosen based on all units being in service or on one unit being out of service, to allow for redundancy. When the overflow rate is selected, the designer should determine the number of units needed.

The type of sludge removal equipment should then be selected. This equipment may limit basin dimensions, which could affect the number of units. When the number of basins is selected, the designer should proceed to design inlet and outlet conditions and finalize dimensions to suit all design parameters and site conditions. The following are suggested guidelines for the various design parameters.

Overflow Rates. Hydraulic overflow rate is the primary design parameter for sizing sedimentation basins. This rate is defined as the rate of inflow (Q) divided by the tank surface area (A). Units are typically rated in gallons per day per square foot, gallons per minute per square foot, or cubic meters per hour per square meter. Acceptable overflow rates vary with the nature of the settling solids, water temperature, and hydraulic characteristics of the settling basin.

Typical design overflow rates for sedimentation of solids produced through alum coagulation/flocculation are shown in Table 7.1. For the ranges shown, higher rates are typical for warmer waters with heavier suspended solids. Rates higher than these may be applicable for warm waters greater than 20° C. Lower rates should be used for colder waters with lower turbidity or that are high in organic color or algae. After evaluating both cold and warm water loading rates, the design rate is based on whichever is more critical. Plant flow variations between cold and warm water periods often allow selection of higher rates for summer operation than typical suggested loading rates.

Overflow rates can also be selected based on pilot studies. Piloting of conventional settling basins is not especially reliable, but it is often done using tube settlers (see Chapter 28). Data from such studies, along with jar testing, are often useful in design. Pilot testing of other types of settling, especially proprietary processes, is useful and is recommended.

Detention Time. Detention time (i.e., flow rate divided by tank volume) is usually not an important design parameter. Many regulatory agencies (e.g., Great Lakes, 1992), however, still have a requirement for detention periods of four hours. It is likely that this detention requirement is a carryover from the days of manually cleaned basins designed to provide a sludge storage zone. These basins were often 15 to 16 ft (4.6 to 4.9 m) deep or greater and operated so that more than half the volume could be filled with sludge before being cleaned. Real detention time could vary from four hours when clean to less than two hours just before cleaning.

Modern designs with mechanical sludge removal equipment need not provide a sludge storage zone, and deep basins with long detention times are no longer required. Conventional basins with detention times of 1.5 to 2.0 hours provide excellent treatment.

TABLE 7.1 Typical Sedimentation Surface Loading Rates for Long, Rectangular Tanks and Circular Tanks Using Alum Coagulation

Application	gpd/ft^2
Turbidity removal	800 to 1200
Color and taste removal	600 to 1000
High algae content	500 to 800

Basin Depth and Velocities. In theory, basin depth should not be an important parameter either, because settling is based on overflow rates. However, in practice, basin depth is important because it affects flow-through velocity. Flow-through velocities must be low enough to minimize scouring of the settled floc blanket. Velocities of 2 to 4 ft/min (0.6 to 1.2 m/min) usually are acceptable for basin depths of 7 to 14 ft (2.1 to 4.3 m), the more shallow depths often used with multiple-tray basins. Single-pass basins are generally deeper to offset the effects of short-circuiting from density and wind currents.

Basin depth may also play a role in allowing greater opportunity for flocculent particle contact. Additional flocculation that takes place as particles settle allows for growth of heavier floc and the formation of a sludge blanket that may be less susceptible to resuspension. The formation of this blanket helps increase the solids content of the residuals withdrawn by removal equipment. The blanket can, however, also contribute to the creation of a density current along the bottom of the tank, causing floc carryover to the effluent.

Number of Tanks. One important choice to be made is the number of basins. The minimum, and by far the least costly plant, would have only a single settling basin. However, that would make for poor operation, because tanks must periodically be taken out of service for maintenance. Two tanks would partially offset this problem, but, unless plant flow can be reduced, the load on one tank could be excessive when the other is out of service. A minimum practical number of tanks would be three, allowing for a 50% increase on two tanks when one is out of service.

If design overflow rate is conservative, the three-tank approach is acceptable. In general, however, a minimum of four tanks is preferred. The number of tanks may also depend on the maximum size tank that can accommodate the selected sludge removal equipment or other factors such as site constraints.

Factors to consider in selecting the number of tanks are their relationships to the flocculation basins and the filters. Are units to be lined up as consecutive processes, or is each process to be a separate unit? When processes are to be consecutive, a decision on the minimum or maximum number of filters or floc basins may determine the number of settling basins. The designer should refer to the chapters on flocculation and filtration design, and approach design of these units as a common process.

Rectangular Basins

Long, narrow basins have been used for sedimentation for many years and will be in operation for years to come. In spite of new designs available, many new plants are still built with long, narrow basins.

Basin Dimensions. Rectangular basins are generally designed to be long and narrow, with width-to-length ratios of 3:1 to 5:1. This shape is least susceptible to short-circuiting—the hydraulic condition in a basin when actual flow time of water through the basin is less than the computed time. Short-circuiting is primarily caused by uneven flow distribution and density or wind currents that create zones of near-stagnant water in corners.

Basin widths are most often selected to match the requirements of the chosen mechanical sludge collection equipment. Chain-and-flight collectors, for example, are limited to about a 20 ft (6 m) width for a single pass, but it is possible to cover a wider basin in multiple passes. Traveling bridge collectors can be up to 100 ft (30 m) wide, limited only by the economics of bridge design and alignment.

Basin depths may be selected to provide a required detention time (though detention time is not a good design parameter) or may be selected to limit flow-through velocities

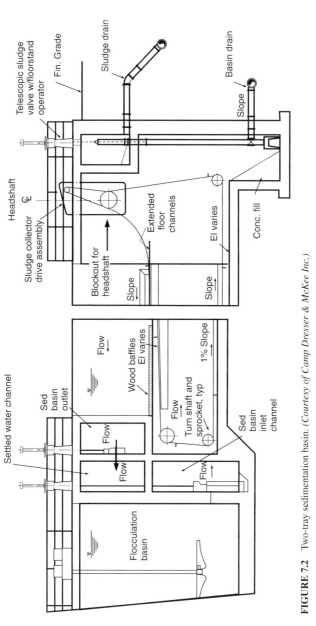

FIGURE 7.2 Two-tray sedimentation basin. (*Courtesy of Camp Dresser & McKee Inc.*)

and the potential for resuspension of settled floc. Basins with mechanical sludge removal are usually between 10 and 14 ft (3.0 and 4.3 m) deep.

Because settling is primarily based on area, multiple-tray basins have been developed as shown in Figure 7.2. A depth of about 7 ft (2 m) is typically provided between trays to allow access for cleaning and maintaining equipment.

Inlet Zone. A basin's effectiveness at any overflow rate can be greatly changed by short-circuiting. Short-circuiting reduces the actual area traversed by the flow, increasing the apparent overflow rate and reducing solids removal efficiency.

Many publications provide testimony to the importance of proper design of the basin inlet (Yee and Babb, 1985; Monk and Willis, 1987; Hudson, 1981; Kawamura, 1991). Good design of the inlet zone establishes uniform distribution into the basin and minimizes short-circuiting potential.

For long, narrow basins being fed directly from a flocculation basin, slots or a few individual inlets may suffice. To obtain uniform flow distribution through wider basins, perforated baffle walls should be provided; a typical arrangement is shown in Figure 7.3. For best results, flow from the flocculation basin should be in line with the basin axis.

Following hydraulic principles to ensure equal flow distribution, head loss through the perforations should be four to five times the velocity head of the approaching flow. The

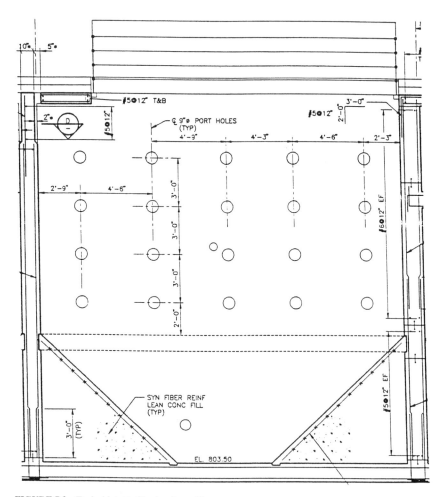

FIGURE 7.3 Typical inlet baffle showing orifices.

velocity gradient *(G)* should be equal to or less than that in the last flocculation compartment to minimize floc breakup (Hudson, 1981). The number of ports should be the maximum practical that will provide the required head loss. Port velocities typically must be about 0.7 to 1.0 ft/s (21 to 30 cm/s) for sufficient head loss. Ports should be arranged to cover as much of the basin's cross section as possible without creating high velocities in the sludge collection zone that might cause scouring action. Thus the lowest port should be about 2 ft (0.6 m) above the basin floor. Port spacing is typically 10 to 24 in. (25 to 61 cm) with a port diameter of 4 to 8 in. (10 to 20 cm).

Introducing flow across the entire inlet end of the basin reduces short-circuiting caused by density currents—created when water entering the top of the basin is colder and heavier than the water below. The influent settles quickly to the bottom, causing flow to move along the bottom of the tank and back across the top. When influent is introduced uniformly across the tank from top to bottom and side to side, water temperature remains the same, and density currents are less likely to form.

Outlet Design. Outlet design is also critical in reducing short-circuiting and scouring of settled solids. Outlet designs have undergone a number of transformations. Basins were originally designed with end weirs. This type of outlet causes an increase in horizontal and vertical velocity as flow is forced up the end wall to the weir, and the increased velocities cause considerable floc carryover by scouring settled floc and removing floc that has not had time to settle.

In an effort to reduce velocities and carryover, long-finger weirs extending well down the basin are used, with up to a third of the basin covered in some installations. By increasing the surface area over which flow is collected, vertical velocity is reduced. Regulatory agencies have promoted this type of design by requiring weir loading rates of 20,000 gal per day per foot of length (248,000 L/day/m) or less (Great Lakes, 1992).

Evaluation of the performance of long-finger weirs compared with end weirs showed that no benefit was achieved by the long weirs (Kawamuraand Lang, 1986). Bottom density currents still rise along the end wall causing floc carryover to the end of the weirs. Weirs are, however, somewhat effective in breaking up wind-induced surface currents. A single end weir could be used, but the problem of high velocities at the end wall still exists.

One approach to address this problem is to use a perforated end wall, similar to the inlet distribution wall, to maintain parallel hydraulic flow along the length of the basin (Monk and Willis, 1987). Velocity approaching the wall does not increase except near the ports because the entire basin cross section is used. Velocity remains low along the floor, reducing the potential for scouring. Flow uniformity approaching the end wall helps to ensure that flow covers the entire basin surface to achieve the design overflow rate. The approach to designing the perforated wall is the same as for the inlet. A small head loss must be taken to obtain uniform distribution. Velocities through the ports may be higher than for the inlet because the smaller floc carried over is less likely to be sheared.

The use of effluent baffle walls may not be effective in controlling density currents in the settled solids. Such density currents may be as high as 2.5 to 6 ft/min (0.8 to 1.8 m/min), much faster than the tank average flow-through velocity that could carry solids through a baffle wall (Kawamura, 1991). Control of density currents in the sludge blanket may best be accomplished by frequent solids removal at several locations (Taebi-Harandy and Schroeder, 1995). Traditional mechanical sludge removal equipment, however, does not typically provide for draw-off at several locations.

Manual Solids Removal. Although no longer used in many modern water treatment plants, basins can be designed for manual cleaning. In such cases basins must be designed to store sludge for a reasonable period of time. An extra depth of 4 to 5 ft (1.2 to 1.5 m) is usually provided, basin floors must slope to a drain, and adequate pressurized water must be available for flushing.

Basins can be designed with hoppers in the first half of the tank, where most of the sludge is likely to settle, and equipped with mud valves that can be frequently opened to waste the bulk of the sludge. This design reduces the frequency of removing the basin from service for complete cleaning. However, withdrawal of sludge from these hoppers should be frequent and controlled. Otherwise, if the solids become too thick, flow may "pipe" through to the drain, leaving the bulk of solids on the sides of the hoppers. Frequent removal of solids through these hoppers may also disrupt density currents and improve performance.

Mechanical Solids Removal.   Most modern sedimentation basins are designed to be mechanically cleaned using a variety of mechanisms, most of which are proprietary. These include systems that drag or plow sludge along the basin floor to hoppers and systems that rely on hydraulic or siphon action to withdraw solids. Because each system has different design requirements for basin dimensions and solids draw-off, it is important that the designer research the available equipment or systems around which to design the tanks. The following text discusses various types of equipment commonly used. It is not a complete discussion, and new devices are being developed all the time. A good reference for a list of currently available equipment and the names of manufacturers is the *Public Works Manual,* an annual publication of *Public Works Magazine.*

Traditional Equipment.   Traditional desludging equipment was mostly chain-and-flight drags made up of two strands of 6 in. (15 cm) pitch iron chain with wooden flights attached at 10 ft (3 m) intervals and operated at about 2 ft/min (0.6 m/min) to convey dense sludge to a hopper. Flights were usually made of redwood in lengths up to 20 ft (6 m). Cast iron or steel wearing shoes were attached to the wooden flights to prevent the wood from wearing, and these rode on steel T-rails cast into the concrete floor. A steel rail was also attached to the wall for the flight to ride on as it

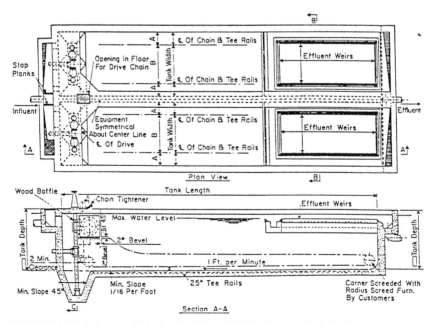

FIGURE 7.4   Typical rectangular basin with chain-and-flight collectors with sludge hoppers.

looped back. Figure 7.4 and Figure 7.5 show two typical types of rectangular basin equipped with chain-and-flight collectors.

Drags for continuous desludging have been popular with design engineers because they fit the long rectangular basin geometry and, although they desludge efficiently, they require considerable maintenance and basin downtime for chain, sprocket, and bearing repair.

New designs have been developed with ultrahigh-molecular-weight (UHMW) plastic or similar materials to replace the iron chain and sprockets. In addition, fiberglass flights have replaced the wooden boards, and plastic wearing strips are attached to the concrete floor and walls to replace iron rails. These new systems are corrosion free and require little maintenance. However, when first installed, plastic chain tends to stretch. This requires adjusting chain tension one or two times during the first year of operation. After that, the rate of creep reduces.

Circular Collector Equipment. Circular sludge collector units have been used in long, rectangular tanks to avoid using chain drag equipment. The circular units were generally installed at the influent end of the basin, and a transverse barrier wall was added to stop the density current and drifting of sludge toward the unscraped effluent end. In these cases, the circular mechanism "pushed" sludge to a circumferential hopper at the center pier from which it was automatically discharged as sludge underflow. The remainder of the basin was periodically cleaned manually.

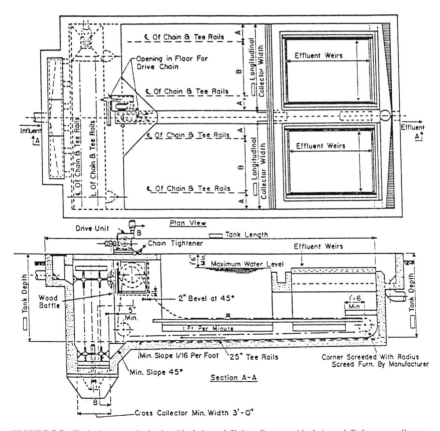

FIGURE 7.5 Typical rectangular basin with chain-and-flight collectors with chain-and-flight cross collectors.

Many of these units were not very effective, particularly where the circular desludging unit was placed only in the first quarter of a long basin with strong density currents. Sludge did not thicken where the unit was placed, which left the circular desludging mechanism with nothing to do, and the majority of the basin had to be manually cleaned. Recent designs have expanded basin coverage with multiple collector mechanisms to improve sludge removal and eliminate the need for manual cleaning.

In small rectangular basins, where corners not swept by the circular collector are relatively small, corner sweeps are generally not specified to minimize collector maintenance. When corner sweeps are not used, it is common practice to add steep corner fillets in basin construction to promote sludge movement to the area swept by the collector mechanism. However, in larger basins, where corner areas are large, or when sludge is of organic composition and must be continuously removed, corner sweeps are specified.

Corner sweeps can be subject to mechanical problems, and the larger the basin, the greater the likelihood of problems. Unfortunately, many corner sweeps fail or require an inordinate amount of maintenance because of poor mechanical details. The mechanical details of corner sweeps should be specified in great detail to obtain a high-quality, low-maintenance system.

In general, circular collector mechanisms are highly reliable and have low maintenance requirements because of their simplicity in providing positive sludge removal. Disadvantages include protecting their large submerged metal surfaces from corrosion and maintaining corner sweeps if they are required.

Carriage-Type Collectors. Oscillating-bridge collectors have become more popular during the past several decades. These top-of-wall running units span the width of one or more long, rectangular basins. One type cleans thick sludge by means of a single transverse vertical blade about 24 to 30 in. (0.61 to 0.76 m) deep that conveys dense sludge into a cross hopper. Another type uses a transverse suction header that discharges into a longitudinal trough along one side of the basin.

The transverse blade has an adjustable angle of attack and slides on flat bottom rails or sometimes on a heavy neoprene squeegee riding on the floor. The blade automatically adjusts to varying floor slopes and can be raised above the water level for maintenance. It pushes dense sludge at about 6 ft/min (1.8 m/min) toward the cross hopper on the cleaning run, and then it is hoisted about 3 ft (0.9 m) and travels at double speed back to the influent end to repeat.

Carriage units generally run on double-flanged iron wheels along heavy steel rails mounted on the long walls of the clarifier. Rubber tires running on top of concrete walls have been used occasionally in western Europe, but their use is discouraged in colder climates because of problems with snow and ice buildup. The units are traction driven with automatic compensation to prevent "crabbing" or, preferably, driven on cog rails located adjacent to steel rails on either side of the basin.

Power to drive and hoist motors is typically supplied through a flexible power cable reeled in and paid out by a cable reel. Motor-driven cable reels work best because simple spring motor reels often overstress the cable when the carriage is at the far end. Another problem is that they commonly do not have enough reserve force to reel in the cable at the near end, and they may run over and sever the loose, kinked cable. Several cleverly designed power reels have been developed, either using a backstay cable to power the reel as it retrieves the cable or synchronizing the cable reel to carriage travel and employing a spring motor slave reel core to compensate for minor variations in cable length.

Feed rails are sometimes used for power feed, but they are vulnerable to vandalism and sometimes burn out or carbonize because of the slow speed of the brushes. Another form of power supply is an overhead power cable festooned and sliding back and forth on a taut steel carrier cable running a few feet above the surface for the length of the basin. This type of power system is unsightly and generally suitable only for short basins.

CLARIFICATION   121

Cross Collectors/Cross Hoppers. A cross hopper is a trench, typically 3 or 4 ft wide by 2 or 4 ft deep (1 or 1.2 m wide by 0.6 or 1.2 m deep), running the width of one or more longitudinal sections of the sedimentation basin. Dense sludge falls into this cross trench and is scraped at about 2 ft/min (0.6 m/min) by chain-driven flights 8 in. (20 cm) deep, spaced 5 ft (1.5 m) on centers. These scraper flights deposit dense sludge into a deeper accumulating hopper at the end of the cross trench. The underflow is withdrawn hydraulically or by pumping from the hopper. Figure 7.6 shows one type of cross collector arrangement.

A helicoid screw is sometimes used in the cross trench in place of chain-driven flights. The screw turns slowly, paced to have a theoretical capacity of four times the volume of sludge to be actually moved, to minimize bearing wear. The bottom of the cross trench is filleted to accommodate the outside diameter of the screw. In screw cross collector applications, instead of propelling dense sludge all the way to one end, the flights of the helicoid screw may be opposed so that dense sludge is carried only half the trench length to the center point where the accumulating hopper is placed.

Some designers like to use the traditional steep-sided (60 degrees) hopper to remove sludge underflow, as illustrated in Figure 7.7. For basins greater than 10 ft (3 m) wide, more than one hopper must be used to keep the hopper depth within reason. Sometimes

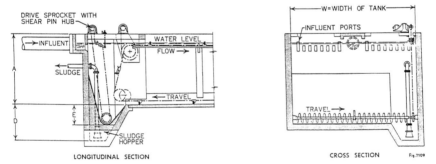

HELICAL SCREW TYPE CROSS COLLECTOR

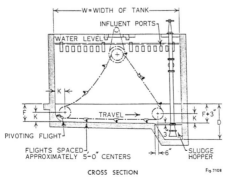

CHAIN AND FLIGHT TYPE CROSS COLLECTOR

FIGURE 7.6 Typical cross collector arrangements. *(Courtesy of FMC.)*

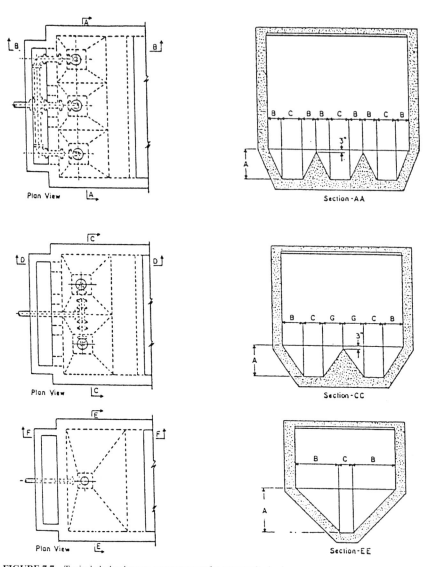

FIGURE 7.7 Typical sludge hopper arrangements for rectangular basins.

anchor sludge ("liver") forms even at a 60 degree slope, because the slope at corner intersections is far less than 60 degree and sludge accumulates by bridging. Multiple hoppers can be more expensive than cross hoppers and not as satisfactory in operation; if they are joined with a manifold to a common pipe, only one hopper will be served. Sludge must be removed independently from each hopper. This is often done by using telescoping sludge valves, air lift pumps, or individually valved outlets.

Indexing Grid System. The indexing grid system consists of a series of concave-faced, triangular blades rigidly connected to glide bars on the basin floor (Figure 7.8). The system operates beneath the sludge layer, sliding back and forth, indexing the sludge toward a sludge hopper at the end of the basin. The glide bars ride on top of ultrahigh-molecular-weight polyethylene wear strips anchored to the floor. The floor of the basin has no slope.

The stainless steel grid system is structurally reinforced by welded cross members to provide rigidity to the unit. Blade profiles are 2 in. (5 cm) tall, typically spaced 26 in. (66 cm) on center. The collector operates by gently pushing the grid system, including the sludge in front of each blade, in a forward direction at a speed of between 2 and 4 ft/min (0.6 and 1.2 m/min). The low profile height and slow operating speed help prevent resuspension of the sludge particles.

The system is driven by a low-pressure hydraulic unit using food-grade hydraulic fluid. The unit's speed can be adjusted to regulate sludge removal rates. The hydraulic unit is designed to oscillate the profiles. Stroke length of the actuator is longer than the spacing of the profiles to create an overlapping effect. When the hydraulic cylinder reaches the end of its stroke, the grid system reverses at a speed two to three times the forward speed. This allows the triangular profiles to wedge their way under the sludge.

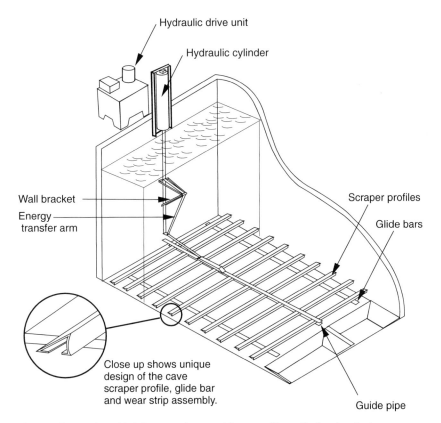

FIGURE 7.8 Indexing grid sludge removal system. *(Courtesy of Purac Engineering, Inc.)*

This oscillating movement of the collector with the concave profile is said by the manufacturer to provide thickening of the sludge.

The low clearance of 10 in. (25 cm) on the bottom of the basin allows the system to work well under tube and plate settler systems.

Track-Mounted Hydraulic Systems. As shown in Figure 7.9, one type of hydraulic removal system consists of a stainless steel header pipe with orifices sized and spaced for proper sludge removal. The collector pipe is attached to a pneumatically controlled drive assembly that travels on a stainless steel guide rail running the length of the tank. Collector pipes are generally a maximum width of 20 to 25 ft (6.1 to 7.6 m), and multiple units must be used to cover the width of wider tanks.

The collector pipe is attached to a sludge discharge pipe in the tank wall. The sludge discharge pipe contains a pneumatically actuated sludge valve located below water level. When the sludge valve is open, the water level in the basin creates a driving force to start flow into and through the collector system. The drive assembly is pneumatically powered to travel the length of the tank in both directions. As it moves along, sludge on the basin floor is picked up hydraulically at a travel speed of 1.5 ft/min (0.46 m/min). The number of times the collector traverses the tank must be determined from the volume of sludge produced and the flow capacity of the collector system. This is typically about 90 gpm (5.7 L/s), based on a differential head of 5 ft (1.5 m).

The system is fully automated through an electronic control system using a programmable logic controller to control how often the collector operates and the length of travel. For example, because most of the sludge typically accumulates in the first third of a rectangular basin, it is necessary to collect in that area on a frequent basis, with the collector traveling the full length of the basin less frequently. This method of operation avoids collecting a large volume of very low solids water.

Instrument-quality compressed air at 100 psi (689 kPa) is supplied from a compressor system to a local electric/pneumatic interface panel mounted at the basin. Air is provided to the drive assembly and pneumatic sludge valves by means of umbilical hoses from the control panel to the drive assembly and valves. At one end of the basin is an extractor assembly consisting of vertical guide rails and a removable winch assembly to lift the collector header out of the tank for maintenance.

Because these collectors do not require expensive sludge hoppers, they are a low-cost option for retrofitting manually cleaned basins. because they do not require access from the surface, they can also operate effectively beneath plate or tube settler systems.

Options to the pneumatically driven collectors include continuous stainless steel tapes or chains, powered by a motor mounted at the top of one end of the basin, that pull the collector pipe back and forth along the bottom-mounted rail.

Floating Bridge Hydraulic Systems. Floating suction header solids removal systems were developed to provide a less expensive means of retrofitting existing manually cleaned basins than chain-and-flight units. The units may also be a lower-cost approach for new construction.

The suction unit is mounted on massive floats built of closed-cell Styrofoam encased in fiberglass reinforced plastic (FRP). These floats are tied together as a rigid structure so that it is freestanding when the basin is drained. A header system supported by the floats draws dense sludge from the bottom of the basin, and by means of low-head siphon discharges it into a longitudinal trough attached to the side of the sedimentation basin. The cross header is broken into several subheaders, each carrying flow overhead to a control siphon freely discharging into the sludge trough. The effect of each lateral section can be observed and modulated. The siphon system is started with a portable low-differential suction source. Experience has shown that, for water treatment sludges, once started, the siphon can be maintained for months.

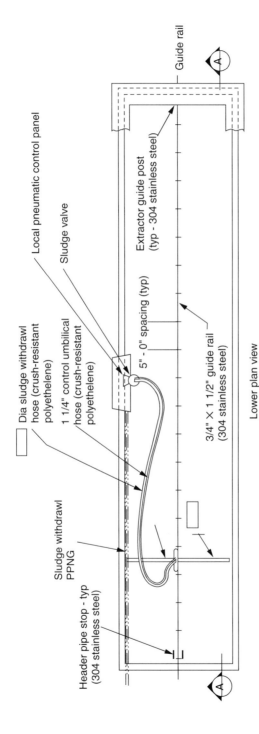

FIGURE 7.9 Track-mounted hydraulic sludge removal system. (*Courtesy of EMCO Process Equipment.*)

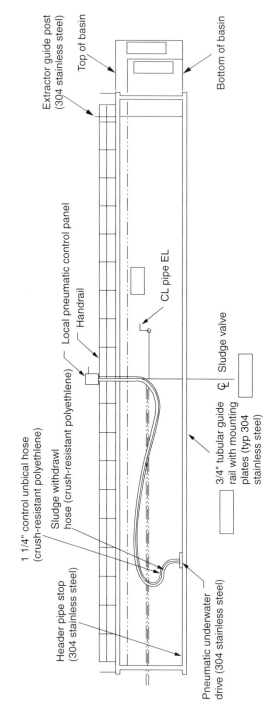

FIGURE 7.9 (Continued) Track-mounted hydraulic sludge removal system. (*Courtesy of ElMCO Process Equipment.*)

CLARIFICATION 127

The floating bridge system is towed back and forth, either by a single, center-mounted, stainless steel flexible tow cable or by two cables acting on either end of the rigid floating structure. In either case, cables are powered by a geared motor drive and idler sheave arrangement mounted at either end of the basin on top of the wall in an accessible location. Because the floating system eliminates friction and most wind problems, remarkably little power is required to tow the bridges, even in basins 200 ft (61 m) long or longer. Figure 7.10 shows a typical siphon desludging unit. Dense sludge sucked up by

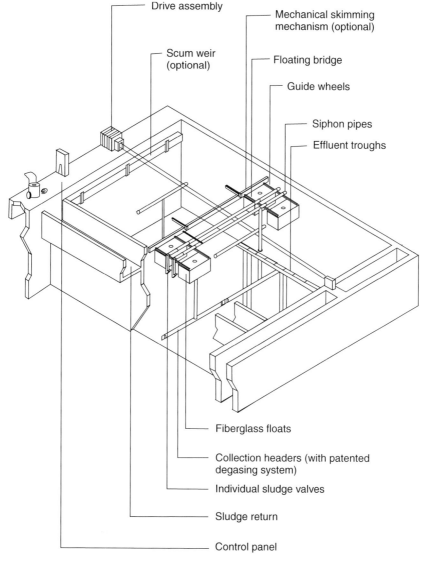

FIGURE 7.10 Floating bridge–type collector. *(Courtesy of Leopold.)*

the system is siphoned into the longitudinal trough. Valves control the amount of siphonage, with one valve for each siphon section forming the transverse header sludge pickup system.

Sludge suction pipes are either of light-gage, epoxy-coated aluminum, stainless steel, or PVC and are equipped with fluidizing vanes. In operation, the suction system travels up and down the length of the basin at about 6 ft/min (1.8 m/min) as often as is required to vacuum the dense sludge layer (compression zone). When there is not enough sludge to require continuous suction up and down the basin, siphon discharge is temporarily arrested by programming the longitudinal trough discharge gate to close. Water level in the trough then rises to equal the basin level. When the next programmed desludging cycle begins, the trough valve opens and the siphon continues from where it stopped, again discharging dense sludge into the trough.

Both carriage and floating bridge collection mechanisms are constructed with little submerged metal to minimize corrosion problems. However, these units can be used only in temperate climates where ice accumulation is not a problem. Some installations have also experienced drive synchronization problems and incomplete sludge removal.

Underflow Control. Underflow drawoff must be carefully controlled. If underflow is removed at too low a rate, dense sludge accumulates in the basin, creating a sludge blanket that is too deep. If underflow is removed too quickly for too long an interval, the drawoff "postholes," that is, less viscous liquid breaks through, and dense sludge accumulates in the basin, overloading the desludging equipment.

Drawing off underflow at regular intervals is best, either manually, with a programmed blowdown valve, or using a transfer pump. Such a program must be completely adjustable and programmed by the design engineer during early operation to fit plant operation. Manual drawoff control by guess and "gut feeling" often leads to operating problems.

New facilities should be designed with sludge viewing pits that permit the operator to observe the consistency of the sludge during blowdown. Direct observation provides a means of optimizing withdrawal rates and reduces excessive loading of sludge handling facilities.

Circular Basins

Circular sedimentation basins became more prevalent in water clarification when periodic manual cleaning of long, rectangular basins became unpopular. The top-drive circular mechanisms used for sludge cleaning have no bearings under water, resulting in longevity with little maintenance. In reasonable sizes—not exceeding 125 ft (38 m) in diameter—the circular center-feed clarifiers perform as well as long, rectangular basins provided there is a reasonably well-balanced radial flow from the center well with substantial water depth maintained at the center.

Some circular basins are designed for rim feed with clarified water collected in the center. However, most circular basins used today are the center-feed type. Included in this category are square tanks with center feed that are used for their feature of lower cost by means of common wall construction. A typical circular clarifier is shown in Figure 7.11.

Basin Dimensions. Circular basins, like rectangular basins, are designed based on surface overflow rates, and rates used are typically the same as for rectangular units. Circular basins may be of any diameter but are usually sized based on the commercially available standard sludge removal systems. Circular tanks have been built as large as 300 ft (91 m) in diameter but more typically are less than 100 ft (30 m) in diameter.

Alhough settling theory is based on overflow rate, side water depth is an important consideration. Adequate depth mitigates hydraulic instability caused by wind currents, ther-

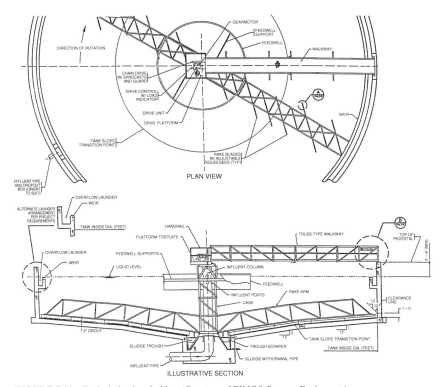

FIGURE 7.11 Typical circular clarifier. *(Courtesy of EIMCO Process Equipment.)*

mal currents, hydraulic scour, and random sludge blanket disturbances. Typical depths range from 10 to 15 ft (3 to 3.6 m). Because sludge is usually scraped to center hoppers, basin bottoms are sloped to the center. Large-diameter basins have two slopes, one steeper near the center to allow adequate depth to move the solids to central hoppers for removal.

Inlet Design. Flocculated water is usually introduced to the center of circular or square basins through a center riser into a circular feed well. Some clarifier designs allow the introduction of flocculated water into the side of the feed well. The intent of the feed well is to produce a smooth, radial flow outward toward the periphery of the basin.

The center feed amounts to a point source, because the feed well seldom represents more than 3% or 4% of basin area. For this reason, a great deal of flow mass is crowded into a small space and does not flow in an exactly radial pattern, leading to hydraulic imbalance and short-circuiting. This problem is accentuated by the tendency to design oversized basins that are expected to carry an 800 to 1,500 gpd/ft^2 (33,000 to 61,000 L/day/m^2) surface overflow rate; these basins are too shallow for an efficient center feed.

One questionable feed well design involves using a small-diameter circular skirt of about 1% of the basin area extending only 3 to 4 ft (0.9 to 1.2 m) below the surface. The feed into this well is either from four ports discharging horizontally from a pier riser or from a horizontal pipeline discharging horizontally into the well from just below the surface. In the four-port design, variation of flow rate is accommodated with a design exit flow of about 2 ft/s (61 cm/s). This, however, does not ensure equal egress from each port.

A more controllable feed design includes using a distribution well inside a large feed well that is about 3% to 4% of the basin area. This distribution well has multiple ports hooded with adjustable biased gates. The gates balance tangential feed discharges by imposing about a 4 in. (10 cm) head loss through the ports. This type of discharge causes the homogenized mass within the large feed well to rotate around the vertical axis at about 2 ft/s (61 cm/s). The well-distributed, fine-scale turbulence within and below the feed well encourages floc aggregation, and the overall slow rotation ensures that flow from the bottom of the skirt into the hindered sludge mass moving radially across the floor has equal displacement vectors. Some circular clarifiers include a flocculation zone in the center, as shown in Figure 7.12.

Density and displacement currents for circular basins are much the same as for long, rectangular basins. The vector system is influenced by well-flocculated influent mass sinking to the bottom adjacent to the feed well area, typically in the center one-third of the basin (about 10% of the total basin area). The vector system shows displacement radially along the bottom in the blanket zone and upwelling next to the peripheral wall. Clarified water generally flows across the surface toward the effluent.

Outlet Design. Clarified water collection must be uniform around the perimeter of the basin. This is accomplished by a circular trough around the perimeter with V-notch weirs or with submerged orifices. Some designs use a double-sided weir trough mounted inboard along at least 15% of the tank radius. This has the advantage of reducing wall flow disturbances and drawing overflow from a more widely distributed region to offset the effects of

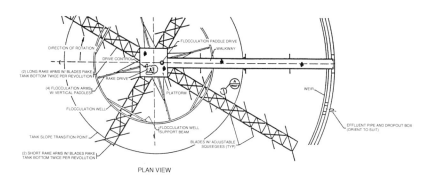

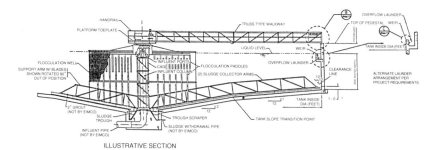

FIGURE 7.12 Typical flocculator clarifier. *(Courtesy of EIMCO Process Equipment.)*

bottom density currents running up the peripheral wall. Inboard weir troughs also partially break up wind-current stirring. Troughs should have small-diameter holes in the bottom to reduce buoyant uplift forces when they are empty.

Some designers prefer orifice troughs to overflow weir troughs because less floc breakup occurs. Others point out that the velocity gradient in a weir trough is no greater than in an orifice trough. Weir troughs are far easier to adjust for equal linear overflow, but if not properly adjusted they have greater variation in flow than improperly adjusted orifices. Submerged orifice troughs reduce passage of floating trash to the filters and permit variation in basin water depths during operation. This capability is useful when balancing differences in plant inflow and discharge rates, such as when multiple filter washing occurs.

Regulatory agencies sometimes stipulate that weir rates should not exceed around 20,000 gpd per linear foot (248,000 L/m) of weir. Flood (1961) found that weir overflow rates several times this value could be used if the weirs were well distributed over a substantial portion of the surface. Placing a double-sided weir trough 1 ft (0.3 m) away from the peripheral wall satisfies the regulatory requirements, but still draws overflow from a narrow band of surface immediately in the path of the upwelling peripheral flow.

Sludge Removal. Sludge is removed from circular basins using circular collection equipment powered by a center turntable drive and plows that move sludge into a center sludge hopper.

Turntable Drives. The tried and true, relatively trouble-free drive for both bridge-supported and pier-supported circular collectors is the sealed-turntable drive with the gear and pinion running in oil. Properly lubricated and with automatic condensate overflow, these drives operate for years without major repair. Typical turntable drives rotate on renewable bearing strips, and the gear is split so that the ball bearings and strips can be replaced without dismantling the remainder of the equipment.

These drives are protected by an indicator and overload circuit breaker device actuated by the thrust of the primary worm gear driving the pinion and turntable gear. The indicator senses the torque load exerted on the collection arms by the sludge and turns off power if the load exceeds a preset limit.

Sludge Hopper and Bottom Slopes. Because a circular sludge hopper surrounding a central pier holds the greatest volume, it is preferable to the older-style offset hopper design. A pair of heavy stirrups reach down from the arms of the circular scraper to move dense sludge around the hopper to the outlet to prevent buildup of anchor sludge and grit. The sludge drawoff pipe should never be less than 6 in. (15 cm) in diameter, and should be designed so that a rotor rodder, or "go-devil," can be placed into the line from outside the basin in case of clogging. In lime-softening plants, this line should be given a short purge of clear water after each blowdown cycle to flush out residual slurry.

The slope of the basin bottom is important, especially when there is heavy or sticky sludge. Plow blades keep the bottom free of adhesions, literally plowing extremely dense sludge and grit to the center hopper. Otherwise, the thixotropic sludge flows along the bottom to replace the blown-down underflow. As the dense sludge approaches the basin center, the plow blade spacing reduces, and the shorter radius results in reducing tangential blade velocities.

In large basins, a second set of arms is typically employed to cover the center half (25% of the basin area) because the blade movement at this point is extremely slow. Deep blades formed into spiral sections bridge the main and auxiliary arms to push the crowded sludge into the hopper.

In basins larger than about 80 ft (24 m) in diameter, it is advisable to use a double bottom slope. The double bottom slope is essential for basins more than 125 ft (38 m) in diameter. The two-slope design gives steeper slopes (greater hydraulic gradient) at the sludge hopper without excessive basin depth. The greater center depth dissipates scouring currents and is needed because of the concentration of influent energy in this relatively small region.

Square Basins

Basins larger than 30 ft (9 m) square are typically equipped with central sludge feed systems with corner sweeps, always in pairs, to clean sludge from corners. Corner sweeps eliminate the need for larger corner fillets, which are generally unacceptable except in very small basins. Corner sweeps should be avoided in basins larger than 100 ft (30 m) square because of structural problems and wear associated with large cantilevered corner sweep units.

Hydraulic problems usually occur with larger square basins. Radial density and displacement currents impinge on the peripheral walls at various angles, drift toward the corners, meet, and may cause a rising "corner floc" phenomenon that often contributes turbidity to effluent.

Where inboard weir troughs are used in square tanks, they are designed to cut across the corner to avoid rising corner floc. Where radial troughs are used (as is typical with upflow basins), they are always arranged to straddle the corner for the same reason. It is best not to run a single peripheral weir around the walls of a square basin because of the corner floc effect.

HIGH-RATE CLARIFICATION

High-rate clarification refers to all processes that can be loaded at higher rates than is typically used in designing conventional clarifiers. The principal types of units currently being used are:

- Tube settlers
- Plate settlers
- Solids contact units
- Sludge blanket clarification
- Dissolved air flotation
- Contact clarification

Tube Settlers

Tube settlers take advantage of the theory that surface overflow loading, which can also be defined as particle settling velocity, is the important design parameter. Theoretically, a shallow basin (i.e., short settling distance) should be effective. By using several shallow parallel tubes, surface area can be greatly increased and low flow-through velocity maintained in each tube to reduce scouring.

The first tube settlers were introduced in the 1960s by Microfloc. Typical tubes are 2 in. (5 cm) square, reducing the settling distance from several feet in a conventional basin to 2 in. (5 cm) or less inside the tube. The large effective settling surface results in a low overflow rate compared with the area of the tubes, which allows for a smaller basin, shorter detention time, and increased flow rates.

When tubes are placed at a 60-degree angle, they provide efficient settling and allow for settled solids removal from the tubes by gravity. As flocculated water rises through the tube, solids settle to the inclined surface, where they gradually gain mass and weight and eventually slide down the incline. In this way, a countercurrent flow pattern is developed. As the solids fall from the bottom of the tubes they settle to the floor of the basin, where they can be removed by conventional sludge collection equipment. Localized velocities caused by thermal currents are damped by the tubes. Likewise, surface wind currents have little effect because settling occurs within the tubes.

Overall depth of a tube clarifier is usually the same as a conventional basin. This is necessary to provide room: below the tubes for sludge collection equipment, for uniform flow approaching tube inlets, for the tubes themselves, and for uniform flow distribution through the tubes up to the collection launders. Figure 7.13 shows a typical tube settler installation.

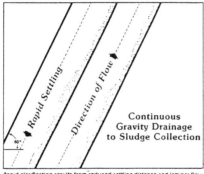

Rapid clarification results from reduced settling distance and laminar flow conditions.

FLOW DIAGRAM

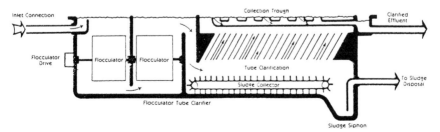

FIGURE 7.13 Typical tube settler. *(Courtesy of Wheelabrator Engineered Systems Inc—Microfloc.)*

Design Criteria. Tube settler clarifiers are designed on the basis of the total projected surface area of tubes. A loading rate of about 0.5 gpm/ft^2 (1.2 m/h) is typical, but the loading rate may range from 0.4 to 0.8 gpm/ft^2 (1.0 to 2.0 m/h). A 0.5 gpm/ft^2 (1.2 m/h) rate is equivalent to a loading rate of about 2 gpm/ft^2 (4.9 m/h) over the top area of tubes. Rates ranging from 1.0 to 3.0 gpm/ft^2 (2.4 to 7.3 m/h) over the top area may be used, depending on the settling characteristics of the flocculated solid (Neptune Microfloc, Inc., 1980).

Inlet Conditions. For tube settlers to operate with uniform loading, the hydraulics of the influent and effluent are very important. Influent turbulence adversely affects settling efficiency in two ways. First, high velocities do not allow even flow distribution into the tubes, important for ensuring equal loading on the tubes. Second, sludge falling from the tubes must be able to settle to the bottom of the basins. High velocities below the tubes break up and shear the falling floc, causing it to be resuspended and carried into the tubes. This overloads the tubes and affects operating efficiency.

To avoid inlet turbulence effects, tube settlers must be placed to create a stilling zone between the inlet and the settler modules. This stilling zone is usually at least 25% of the total basin area.

A minimum depth of 10 ft (3 m) is generally provided below the tubes to create low velocities approaching the tubes to allow sludge to settle without breaking. This depth also allows for access to sludge collection equipment.

Effluent Design. Flow leaving the tubes must be collected uniformly across the basin to equalize flow through the tubes. Flow is usually collected through submerged orifices in pipe laterals or launders, or in some cases by overflow weirs into the launders. A clear space of about 2 to 3 ft (0.6 to 1 m) above the tubes must be provided for transition distribution of flow to the collection laterals or launders. Launders are spaced at not greater than 5 ft (1.5 m) centers.

Solids Removal. Settled solids collect on the floor below the tubes and the plain settling area ahead of the tubes. These solids can be removed with the same types of equipment used in conventional basins. Although those types that travel along the basin floor are most applicable, tube modules can be designed and placed to allow for carriage-type collectors and circular-type units. Sludge equipment selection should be made before the basin is designed to accept the tubes.

Plate Settlers

Plate settlers were developed to improve the efficiency of conventional rectangular settling basins by taking advantage of the theory that settling depends on the settling area rather than detention time. Plate settlers date back to an English patent in 1886 (Purac) and were used primarily in the mining and mineral industries to separate heavy particles from slurries. In the late 1950s they were developed in Europe for treating drinking water.

Theory. Based on sedimentation theory, shallow basins provide the same settling as a deep basin. However, horizontal shallow basins are subject to the scouring action of the flow-through velocity, which lowers removal efficiency. Also, if horizontal shallow basins are stacked to reduce plant area, there is no easy method of sludge removal. For these reasons, plate settlers are designed to be vertically inclined, similar to tube settlers, to allow settled solids to slide down the inclined surface and drop into the basin below. Distance between plates is designed to provide an upflow velocity lower than the settling velocity of the particles, allowing particles to settle to the plate surface. The effective settling area is the horizontal projected area of the plate, calculated by multiplying the plate area by the cosine of the angle of the plate to the horizontal. Total settling area is the sum of the effective areas of each plate.

Design Criteria. The primary design criterion for plate settlers is the surface loading rate for each plate. Typical loading rates range from 0.3 to 0.7 gpm/ft^2 (0.7 to 1.7 m/h), depending on the settling characteristics of the solids, water temperature, and desired effluent quality. These loading rates allow for overall basin loadings from 2 to 6 gpm/ft^2 (5 to 15 m/h), several times that for conventional basins. This criterion allows for much smaller basins in new construction or for up-rating existing basins.

Basin Dimensions. Plate settlers are typically manufactured in modules. Dimensions of modules and plates vary by manufacturer and are proprietary. These proprietary dimensions require different basin geometries, and the system designer must work with manufacturers to establish appropriate dimensions for new construction or to decide how plates may be installed in existing basins. Some manufacturers provide standard-width plates of varying heights to fit custom designs.

The basic dimensions provided by each manufacturer are primarily established by how each manufacturer approaches the problem of flow control. Basin dimensions may also be controlled by how the sludge is to be removed. Sludge removal devices are typically the same as those used for conventional basins except those that travel on or above the surface of the basin. Two typical plate settler installations are shown in Figures 7.14 and 7.15.

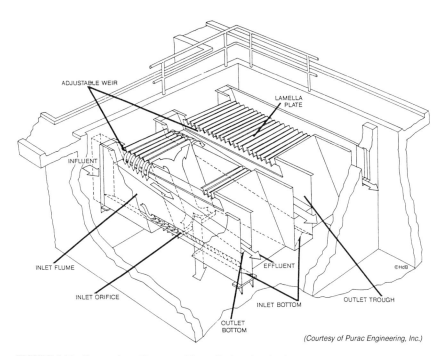

(Courtesy of Purac Engineering, Inc.)

FIGURE 7.14 Plate settlers. *(Courtesy of Purac Engineering, Inc.)*

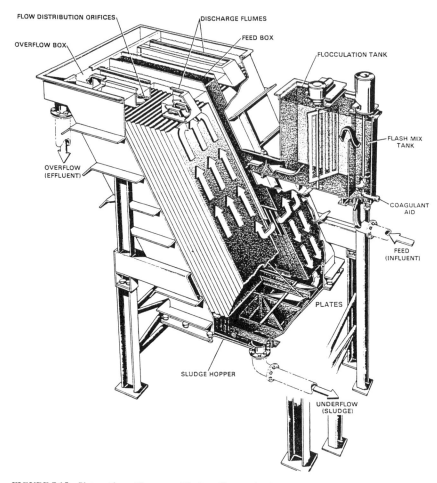

FIGURE 7.15 Plate settlers. *(Courtesy of Parkson Corporation.)*

Inlet Design. Inlet flow distribution is critical for efficient performance of plate settlers. In fact, the approach to distribution is a major difference between manufacturers. Some manufacturers feed the plates from side channels to prevent interference between the incoming flow and the falling settled solids. In one case the flow to each plate is controlled by head loss through an influent orifice. In another, flow is distributed from the side channel through slots and controlled by head loss taken through control orifices as flow leaves the plates.

A third approach provides a large space below the plates, similar to tube settler design, to minimize velocities and to design the plate outlet to induce sufficient head loss to ensure good flow distribution. The designer should evaluate each approach as it relates to the water to be treated and decide if one is more advantageous than another.

Outlet Design. Proprietary designs also provide for collection of clarified water in different ways. One manufacturer uses the upper half of the side inlet channel to collect clarified water laterally at the top of each plate section. Another manufacturer places a collection launder along the top of the plates with a control orifice located above each plate to induce enough head loss to ensure good flow distribution.

Other manufacturers and custom designs have uniformly spaced launders with weirs or submerged orifices that collect the flow along the entire surface. Launders should be spaced on the order of 6 ft (1.8 m). Submerged orifices should be designed to create sufficient head loss to ensure good flow distribution. An orifice velocity of 1.5 to 2.5 ft/s (46 to 76 cm/s) will generally be adequate.

Solids Removal. Settled solids that slide down the plates collect uniformly at the bottom of the basin, and they must be periodically removed. Typical removal equipment would be chain-and-flight collectors or bottom track units. Circular equipment with drive units above the plates has also been used.

SOLIDS CONTACT/SLURRY RECIRCULATION UNITS

Solids contact units, slurry recirculation, or sludge blanket types are designed to provide more efficient flocculation and greater opportunity for particle contact within the blanket, which also acts partially as a filter. The hydraulic design also provides for more uniform flow and is less subject to short-circuiting. For these reasons, solids contact units can handle three to four times the hydraulic loading of conventional basins. However, density currents created by differences in water temperature between the incoming flow or direct sunlight on the basin surface can be disruptive by causing the blanket to "boil" with resulting rising of the floc particles into the effluent. In the most severe situations the entire sludge blanket can be lost.

Solids contact units combine flocculation and sedimentation functions into a single basin. Some units operate with chemical feed directly to the inlet pipe, but a separate rapid mixer may provide better coagulation for turbidity or color removal applications. A large volume of previously settled solids is recirculated to the mixing zone to act as nuclei to form additional floc and to make more complete use of coagulation chemicals. Recirculation rates vary with the application, and may be up to 12 times influent flow for softening and up to 8 times influent flow for turbidity removal.

Slurry recirculation units were developed primarily for softening applications, where they have been extensively employed for many years. They have also been used in applications for turbidity or color removal.

Operation and Design Criteria

The basic theory of solids contact units is that contact of newly formed coagulation particles with previously formed floc enhances floc formation, creates more opportunity for particles to make contact, allows for larger floc development, and allows higher loading rates. Providing this recirculation within a single basin, compared with pumping settled solids to a flocculator in a conventional plant, reduces equipment requirements and lowers facilities costs. In addition, recirculating settling floc within the basin is less destructive to the floc than recirculating after it has settled and had time to thicken and further agglomerate. A typical slurry recirculation unit is shown in Figure 7.16.

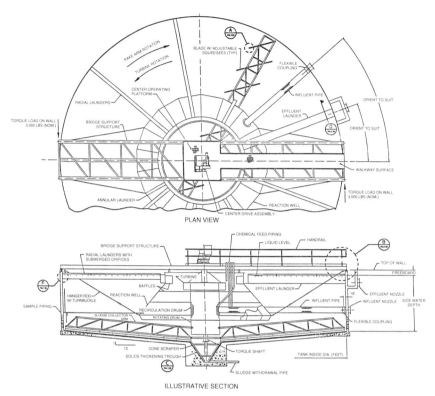

FIGURE 7.16 Typical solids contact slurry recirculation unit. *(Courtesy of EIMCO Process Equipment.)*

Surface Loading Rate. The surface loading rate is the primary basis for sizing the clarification zone. Loading rates may range from 0.5 to 1.5 gpm/ft² (1.2 to 3.7 m/h) for color and turbidity removal, with higher rates for turbid warm waters and lower rates for colored colder waters. Rates for softening are often higher. Surface settling rates used for solids contact units typically are higher than for conventional sedimentation in similar applications.

A number of factors affect settling efficiency, and loading rates should be adjusted accordingly. Specific gravity of particles in the settling zone often ranges from as low as 1.02 for alum floc to as high as 2.65 for sand particles. The designer should consider the nature of the flocculated material in selecting the loading rate.

Temperature is also an important consideration. Higher-temperature water allows for more efficient chemical coagulation and flocculation and has a lower viscosity, exerting less drag on settling particles. For these reasons, higher surface loading rates may be used for warm waters.

Flocculation/Contact Zone. A detention time of 30 minutes or more is provided in the reaction wells for units designed to remove turbidity to ensure complete chemical reactions and adequate time for flocculation. Shorter or longer times may be appropriate depending on water quality and temperature. Selection should be based on an understanding of flocculation principles. The designer should refer to Chapter 6 on flocculation for additional information.

Basin Dimensions. Slurry recirculation units are center-fed units with uniform collection of clarified water across the surface. To provide good flow distribution, units are generally circular with a diameter up to about 150 ft (46 m). Square tanks can take advantage of common wall construction, but they generally have problems associated with sludge removal from corners.

Side water depth, especially center depth, is an important design parameter to ensure adequate space for the mixing zone, uniform distribution of slurry to the basin clarification zone, and transition flow vertically to effluent collection troughs. Basin depth is usually established by proprietary equipment manufacturers' requirements.

Influent Design. Influent design is based on requirements for mixing chemically treated source water with previously formed flocculated and settled solids. To achieve this mix, flow is introduced into a recirculation drum or zone at the center of the unit.

Flow is then introduced to a flocculation-reaction zone in a center well separated from the clarification zone by a hood. The design of this area varies depending on the manufacturer. The hood is designed to control flow uniformly to the clarification zone. The designer should work with the manufacturer to select equipment most suitable to the application and to obtain necessary dimensions for basin structural design.

Effluent Design. Clarified water is collected in radial launders to maintain a vertical upflow within the settling area to help reduce short-circuiting. Launders are designed with low loading rates, typically about 20 gpm per linear foot (248 L/min/m), which sets the spacing. Spacing is usually about 15 to 20 ft (4.6 to 6.1 m).

Launders are designed with submerged orifices to overcome the problems of trying to level many weirs to the same elevation. Small changes in orifice elevation does not greatly affect flow distribution, but small changes in weir elevations have a major impact on flow.

Solids Removal. Most slurry recirculation units use a rake rotating around a center column to plow settled solids toward the center, where they are deposited in an annular hopper for discharge to waste. Some solids are picked up by the recirculating flow to be mixed with influent. The rake consists of a steel truss with squeegees and is supported from the rotating recirculation drum. Rake arms must be designed to withstand the torque caused by the weight of the solids being moved and by friction forces. The rake is driven by a center drive assembly similar to equipment for conventional circular clarifiers.

Equipment

Basin equipment consists of a conical reaction well, recirculation drum, influent pipe, and influent baffle. The equipment is generally mounted on a bridge extending across the basin and designed to act as a walkway for access to the center drive unit. Dual drives are provided, one to drive the sludge collection rakes, the other to drive the turbine mixer used to recirculate slurry and provide flocculation.

In large-diameter tanks, equipment is supported from a center column, and a bridge is provided only to access drive units. Some large-diameter tanks use peripheral drive units with rakes supported from the walls and a center column.

The Accelator Design

The Accelator (Figure 7.17) is a special design of solids contact basin developed by Infilco (Vincent, 1991) and is now a trademark of Infilco-Degremont Inc. This unit

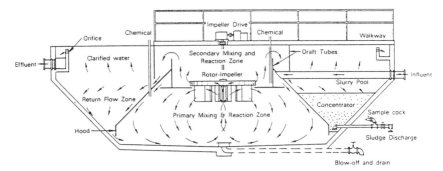

FIGURE 7.17 Accelator solids contact unit. *(Courtesy of Infilco Degremont Inc.)*

was initially used for softening applications but has widespread use for turbidity and color removal.

Influent, to which coagulating chemicals have been added, is introduced to a primary mixing zone where it is mixed with previously formed slurry. A large turbine mixer provides recirculation and mixing inside a central hooded area that separates the mixing zone from the clarification zone. Solids settling in the clarification zone drop below the edge of the hood and are picked up in the recirculating flow within the hood. The turbine impeller circulates slurry to a secondary zone located above the primary zone where continuous flocculation is taking place. Flocculated water overflows the top of the secondary zone baffle and flows downward and out along the slopes of the mixing zone hood.

In the clarification zone, solids settle while clarified water rises vertically to collection launders. Part of the clarification zone area contains hoppers with valved outlets. Solids settling over these hoppers can be retained and are periodically discharged, usually by a timed blow-off valve.

The volume of solids removed can be controlled by the number of hoppers in service. Solids not removed in the hoppers are drawn back under the hood to the mixing zone. The original Accelator did not have mechanical sludge removal equipment. This limited the diameter because of the depth required to maintain proper dimensions for the mixing zone and sludge collection hoppers. Depths greater than 20 ft (6 m) tend to be uneconomical compared with other types of slurry recirculation units.

Larger-diameter Accelerators operate with the same basic mixing zones but use mechanical scrapers to move sludge to a central hopper (Webb, 1993). As sludge moves under the hood of the mixing zone, some of it is resuspended, and the rest is periodically discharged to waste. The scraper may be structurally a part of the hood, supported by a separate scraper drive, or supported at the center by columns extending to the floor. Some Accelator designs use a peripheral drive and support the scraper mechanism from a traveling bridge that extends from the outer wall to a center column and rotates around the column.

The center mixing turbine operates at low tip speeds (variable-speed drives are typically used) to prevent floc particle shearing. The turbine circulates flow at pumping rates of up to 10 times the basin design flow rate.

SLUDGE BLANKET CLARIFICATION

Sludge blanket clarification is a variation of solids contact clarification in which coagulated water flows up through a blanket of previously formed solids. As the small coagulated particles enter the blanket, they contact particles within the blanket and flocculation takes place. Flocculated particles grow in size and become a part of the blanket. The sludge blanket grows in thickness and is suspended by the flow velocity passing through the blanket.

A blanket depth of several feet is required for efficient clarification. When the blanket depth has reached design depth, the top of the blanket is above the level of sludge removal hoppers. Because there is no upward flow in the hopper area, the portion of the blanket next to the hopper settles into the hopper. Cohesion in the sludge blanket helps pull solids into the hopper as the blanket settles, creating a flow of solids along the top surface of the blanket. Sludge is periodically removed from the hopper by gravity.

Design Criteria and Application

Sludge blanket clarification may be used for applications where flocculent suspensions are formed, such as to remove turbidity, color, organic matter, tastes and odors, and iron and manganese. The process may be applied to highly turbid waters if the turbidity is colloidal in nature. However, heavy suspended solids should be removed by presedimentation because they may not be supported by the upward flow in the blanket and may settle out to create a maintenance problem. Waters with high levels of algae may be difficult to treat because algae may float to the surface, carrying flocculated solids with them. Otherwise, sludge blankets can be an efficient removal process operating at much higher loading rates than conventional clarifiers.

For a sludge blanket to perform efficiently, it must be designed to provide uniform upward flow, ideally with equal velocities across the entire cross section. This ideal is approached by introducing flow across the bottom of the basin from uniformly spaced distribution laterals, each with uniformly spaced orifices. Clarified water is then collected in equally spaced launders of either the overflow weir or submerged orifice type.

One of the first applications of sludge blanket clarification was the pyramid type, shown in Figure 7.18, developed by Candy in the 1930s. In this unit, coagulated water is introduced to the bottom of the pyramid and flows upward through the sludge blanket with a reducing velocity resulting from the expanding area of the pyramid. Clarified water is collected in uniformly spaced launders, and sludge overflows to a hopper.

Because the hydraulic design reduces short-circuiting, and because the sludge blanket is heavier than the settling floc particles in a conventional basin, the surface loading rate can be increased. Recent application of plates and tubes with the sludge blanket has allowed even higher loading rates.

Pulsed Blanket Clarifier

The pulsed blanket clarifier (Hartman and Jacarrino, 1987), as shown in Figure 7.19, consists of the following subsystems:

- Vacuum chamber
- Inlet distribution system
- Effluent collection system
- Sludge extraction
- Drain, washdown, and sample system
- Controls

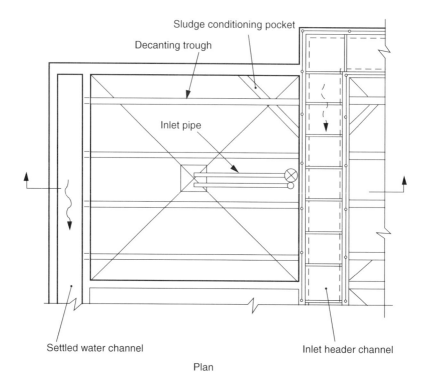
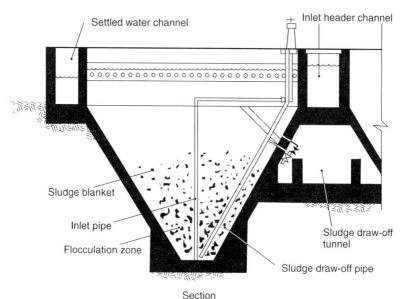

FIGURE 7.18 Pyramid-type sludge blanket clarifier.

CLARIFICATION

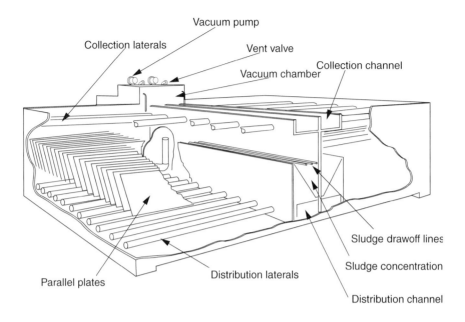

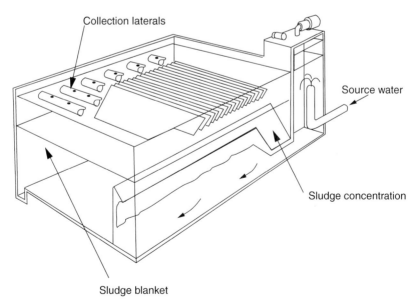

FIGURE 7.19 Cross-section of the Superpulsator sludge blanket clarifier. *(Courtesy of Infilco Degremont Inc.)*

The treatment cycle begins when previously mixed influent water and coagulant chemicals enter the vacuum chamber. With the vacuum pump running and the vacuum release valve closed, pressure is reduced inside the chamber, causing the water level to rise to a predetermined level. This level is controlled by a timer that actuates the vacuum release valve. When the vacuum is released, the water level drops rapidly, causing a surge of water to flow into the distribution system. As the water level in the vacuum chamber drops to the hydraulic grade line of the influent water, the head is dissipated and the surge slows. A timer then closes the vent valve, and the vacuum pump, which runs continuously, begins evacuating the air to repeat the cycle. This rising and falling of the water level in the vacuum chamber creates the rhythmic, controlled pulsing for which the pulsed blanket clarifier is known. Pulsations create and maintain the sludge blanket in homogeneous suspension critical for efficient solids removal. Pulsations also aid in flocculation created by the turbulence from the high-velocity flow leaving distribution pipe orifices.

As water enters the sludge blanket and passes upward through it under the forces of succeeding pulsations, the sludge blanket performs the double tasks of flocculation and filtration: it serves as a medium that agglomerates the newly formed floc, and it helps suspended matter and colloidal particles adhere to the floc.

Water exits the sludge blanket into the clarification zone. Because the flow rate is lower in the clarification zone than in the blanket, particles that escape the blanket settle back toward the blanket. In the Superpulsator clarification is enhanced by inclined plates that increase the settling area and efficiency of particle removal. At the top of the clarification zone, clarified water is collected in submerged orifice laterals or launders.

As solids are removed in the blanket and clarification zone, blanket volume increases and some solid must be removed to maintain stable conditions. A zone of sludge concentrators collects and concentrates excess solids. As these excess solids increase the volume of the blanket, they spill over into the concentrators. When the concentrator is full, thickened sludge is withdrawn through sludge removal headers usually on a timed basis by automatically controlled sludge valves.

Design Criteria. Pulsed blanket clarifiers are sized based on surface loading rate in the clarification zone. The area over the sludge concentrators is not included in this calculation. Typical design loading rates recommended by the manufacturer are as follows, but lower rates may be required by some regulatory agencies and for some water quality conditions (Webb, 1993):

Type of unit	Loading rate
Pulsed blanket	1.0 to 1.25 gpm/ft^2 (2.4 to 3.05 m/h)
Pulsed blanket with plate settlers	2.0 to 4.0 gpm/ft^2 (4.9 to 9.8 m/h)
Pulsed blanket with plate and tube settlers	3.0 to 5.0 gpm/ft^2 (2.3 to 12.2 m/h)

Basin Dimensions. Basin dimensions are set to provide uniform flow distribution. A typical Superpulsator consists of two 16 ft (4.9 m) wide blanket sections with a central 10 ft (3 m) wide source water influent chamber, sludge concentrator, and effluent section. Basin water depth is 16 ft (4.9 m). Basin length varies, limited by hydraulic considerations.

Inlet Design

Flow to the basin is controlled through the inlet distribution system. Flow from the vacuum chamber typically enters a conduit serving two halves of the basin. From this conduit, laterals extend to either side to provide uniform flow across the basin. Laterals are spaced at about 3.5 ft (1.1 m) on center. Orifices in these laterals further serve to equally distribute flow across the basin. Orifices are designed to provide a certain entrance velocity based on the vacuum chamber hydraulic head to ensure even flow distribution and create an energy level to enhance flocculation in the mixing zone as flow enters the basin. Baffles above the inlet laterals further ensure uniform flow to the blanket.

Outlet Design. Clarified water is collected at the clarifier surface in uniformly spaced laterals with submerged orifices. The lateral spacing is about 6.5 ft (2 m). Orifices are designed to induce enough head loss to aid in maintaining uniform upflow velocities in the blanket. Laterals discharge to an effluent channel located above the sludge concentrators. Submerged orifices are used because flow through them is not as affected by the pulsing action of the water surface as it would be over a weir.

Sludge Removal. As the sludge blanket builds in volume, it rises above the level of an overflow weir to sludge concentrators, which are hoppers located between the two halves of the basin. Sludge flows into concentrators where it is allowed to partially thicken. It is periodically drawn off through timer-controlled valves, usually by gravity, to a sump from which it may be pumped or flow by gravity to sludge handling facilities. No mechanical equipment is used, which is one of the attractive features of this type of clarifier.

DISSOLVED AIR FLOTATION

In flotation, the effects of gravity settling are offset by the buoyant forces of small air bubbles. These air bubbles are introduced to the flocculated water, where they attach to floc particles and then float to the surface. Flotation is typically sized at loading rates up to 10 times that for conventional treatment. Higher rates may be possible on high-quality warm water.

Dissolved air flotation is an effective alternative to sedimentation or other clarification processes. Modern DAF technology was first patented in 1924 by Peterson and Sveen for fiber separation in the pulp and paper industry (Kollajtis, 1991). The process was first used for drinking water treatment in Sweden in 1960 and has been widely used in Scandinavia and the United Kingdom for more than 30 years.

Previous uses of the process in the United States have been to thicken waste-activated sludge in biological wastewater treatment, for fiber separation in the pulp and paper industry, and for mineral separation in the mining industry. Only recently has this process gained interest for drinking water treatment in North America. It is especially applicable when treating for algae, color, and low-turbidity water. The first use in the United States was at New Castle, New York, in a 7.5 mgd (28 ML per day) plant that began operation in 1993. A typical DAF unit is shown in Figure 7.20.

Theory and Operation

Effective gravity settling of particles requires that they be destabilized, coagulated, and flocculated by using metal salts, polymers, or both. The same is true for DAF. In gravity settling the flocculation process must be designed to create large, heavy floc that settles to the bottom of the basin. In DAF, flocculation is designed to create a large number of smaller floc particles that can be floated to the surface.

For efficient flotation, flocculated particles must be in contact with a large number of air bubbles. Three mechanisms are at work in this air/floc attachment process:

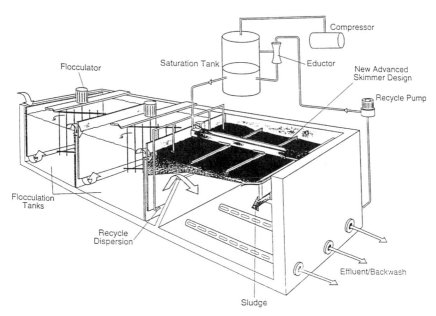

FIGURE 7.20 Typical dissolved air flotation tank. *(Courtesy of Purac Engineering, Inc.)*

- Adhesion of air bubbles on the floc surface
- Entrapment of bubbles under the floc
- Absorption of bubbles into the floc structure

The size of air bubbles is important. If bubbles are too large, the resulting rapid rise rate will exceed the laminar flow requirements, causing poor performance. If bubbles are too small, a low rise rate will result and tank size may need to be increased.

In a typical DAF tank, flocculated water is introduced uniformly across the end of the tank, near the bottom, into the recycle dispersion zone. Recycle is continuously introduced through a distribution system of proprietary nozzles, valves, or orifices. When the recycle flow pressure is suddenly decreased from its operating pressure of 60 to 90 psi (414 to 620 kPa) to atmospheric pressure, saturated air within the recycle stream is released in the form of microbubbles with a size range of 10 to 100 μm, and averaging around 40 to 50 μm. These microbubbles attach to flocculated material by the mechanisms described previously, causing flocculated material to float to the surface.

At the surface, the bubble-floc forms a stable and continuously thickening layer of float, or sludge. If left at the surface, the float can thicken to as much as 3% to 6% dry solids. This can be an advantage if solids are to be mechanically dewatered, because solids may be suitable for dewatering without further thickening, or the thickening process can be reduced. Sludge thickness depends on the time it is allowed to remain on the surface and the type of removal system employed.

Recycle System

About 5% to 10% of clarified water is recycled into the air saturation system by recycle pumps. Recycle water is pumped through an eductor that introduces air in the line from the air cushion of the saturation tanks. The air-water mixture is then discharged into the bottom of the saturation tank. The eductor increases the efficiency of air-to-water transfer by increasing contact time between the two media.

Air is supplied from an air compressor controlled by the saturation tank water level. Operating efficiency of this type system is in the 80% to 85% range and may be improved by using packed bed saturation tanks to increase the air-water interface. Packed bed saturators are more expensive, but their greater efficiency may make them more economical in large plants. Saturated water flows from the saturation tank through a system of headers to the proprietary recycle dispersion system, where the pressure is dropped to atmospheric and microbubbles are created.

Design Criteria

The size of DAF tanks is based on the surface loading rate. Standard practice has been to design in the range of 4 to 5 gpm/ft^2 (10 to 12 m/h), although higher rates may be possible. The use of pilot plant studies is recommended for rates higher than standard (Grubb, Arnold, and Harvey, 1994).

Design should allow for a recycle rate of 5% to 10% of plant flow at a dissolved air pressure of 60 to 90 psi (414 to 620 kPa). Recycle rate and air pressure should be adjustable to allow for process optimization.

The recycle dispersion system must be selected to evenly distribute recycle flow across the width of the unit into the flocculated water flow.

Basin Dimensions

Basin dimensions may be flexible; the length-to-width ratio is not important. However, a maximum length of about 40 ft (12 m) is recommended, because in this distance, all of the air bubbles would have typically risen to the surface. Tank depths are usually around 10 ft (3 m). If tanks are too shallow, forward velocities could carry air bubbles into the effluent. Tanks could be deeper, but there is no added benefit for increasing depth.

Influent Design

As noted, above flocculated water must be introduced uniformly along the width of the tank near the bottom. This can be done by means of a continuous slot or through uniformly spaced pipes. Recycle flow must also be uniformly distributed. When pipes are used for flocculated water, recycle flow is sometimes introduced into pipes just before the outlet.

Effluent Design

Clarified water is collected from the tank bottom by being forced under an end baffle wall or by pipe laterals on the basin floor. In either case, flow must then be discharged over a control weir to maintain a stable water surface in the clarifier. The weir may be adjustable to fine-tune the water surface level. This is especially important when scrapers are used to remove the float.

Floated Solids Removal

Floating solids (called float) are removed from the surface by mechanical skimming or by hydraulic flooding. Mechanical skimming equipment must be maintained, but it allows the float to be removed at a dry solids content of 3% to 6%. At this concentration it may be possible to eliminate or at least reduce thickening equipment required for mechanical dewatering.

Flooding eliminates mechanical equipment in the flotation tank but results in a dilute solids stream of around 0.5% or less. This float removal method could create a need for larger thickening equipment.

Mechanical Skimming. Three common types of mechanical skimming equipment are the chain-and-flight skimmer, oscillating skimmer, and rotating skimmer.

Chain-and-Flight Skimmer. A typical mechanical skimming device is the chain-and-flight type, similar to those used to scrape settled solids in gravity sedimentation basins. In this case, however, flights are equipped with rubber squeegees or nylon brushes that drag the float along the water surface and up onto a dewatering beach. As solids are dragged up the beach, water is squeezed out and drains away, further thickening the float. The float then drops off the beach into a hopper from which it flows by gravity or is pumped to waste or thickening.

Solids content of the float is controlled by how often it is removed and by how much excess water is dragged along with the solids. Adjusting the flight depth into the water is critical to the amount of excess water and also to treatment efficiency. If the moving flights are set too deep, they can create flow patterns in the basin that will cause poor particle removal and floc carry-through to the effluent. If the flight is set too shallow, it can disrupt the adhesive forces in the float blanket. This disruption causes solids to settle back into the tank and floc carry-through.

The skimmer is usually operated on a timer, with the frequency of operation based on solids production and the desired level of thickening. All mechanical components are above water, making maintenance easier than for gravity sedimentation basins.

Oscillating Skimmer. This skimming device consists of a carriage-and-blade assembly that is pushed and pulled across the surface by a gearmotor. Blade spacing is typically 5 to 7 ft (1.5 to 2.1 m), and the mechanism covers two-thirds of the flotation area. The lead carriage, closest to the sludge beach, is equipped with a profile-duplicating arm that allows the skimmer blade to exactly duplicate the shape of the beach. The beach can be either flat or curved.

All moving components are located above water, which allows easy inspection and maintenance. The gear motor oscillates a drive rod back and forth in 15 in. (38 cm) increments while engaging the skimmer carriages. When the carriage reaches the effluent end and the profile blade deposits the float into the hopper, the blades retract above the water surface. The carriage is then mechanically reversed to return to the influent end in an indexing manner.

The skimmer unit is constructed of stainless steel and plastic. High-hardness plastics provide low operating friction, allowing the mechanism to operate efficiently. The profile duplicating blade is typically a nylon brush that aids in sludge dewatering. The brush allows the sludge to thicken to a slightly higher solids percentage while the blade pauses inherently during operation on the sludge beach. Maintenance is minimal, consisting of checking the plastic wear blocks and regular servicing of the gear motor.

Rotating Skimmer. Another type of mechanical skimming device consists of a rotating shaft with curved or straight blades attached. The shaft is mounted at the effluent end of the basin above a sludge beach. As the blades rotate, they pull a portion of the float blanket onto the beach and into a hopper. The general flow of water toward the effluent causes the blanket to flow up to the beach, where the rotating blades continually remove a portion.

Hydraulic Removal. Float can be hydraulically removed from the surface of a DAF unit by flooding. This is accomplished by partially closing the basin effluent gate, which causes increased head loss and a small rise in the water surface level. This increase in level brings the surface up above a weir that discharges to a sludge drain. Part of the flow through the basin is now diverted over this weir, dragging the float blanket along with it.

Adhesion of the sludge particles helps keep the blanket intact as it moves across the surface to the weir. Water is trickled down the basin walls to break adhesion to the wall so that the blanket can move freely. Float removal is relatively frequent to prevent the blanket from becoming too thick because a heavy blanket does not flow well. A solids content of about 0.5% is best.

When most of the blanket is removed, usually in about 10 min or less, the effluent gate reopens and operations return to normal. This mode of float removal eliminates all mechanical equipment in the basins. The disadvantage is that the removed solids are diluted and may require more extensive thickening for subsequent mechanical dewatering.

CONTACT CLARIFICATION

Contact clarification resembles filtration more than clarification. Coagulated water flocculates within the contact medium, may stick to the medium grains, and builds in size, eventually clogging the medium, at which point it is backwashed. Because this process is more filtration than gravity settling, it is capable of operating at rates of up to 20 times that for conventional settling on low-solids source waters requiring low coagulant dosages. High-solids waters and high-color waters that require high coagulant dosages quickly clog the medium, resulting in excessive backwashing. The process is generally marketed as a two-stage process, along with filters in package plants, but may be used separately to upgrade existing plants

Roughing filters have been used for as long as filtration has been around. The contact clarification process, however, gained popularity with the development in the mid-1980s of the adsorption clarifier by Microfloc. Similar processes are now also marketed by Infilco Degremont, Inc., and Roberts Filter Co. The adsorption clarifier is shown in Figure 7.21.

Application

The fact that clarification results from filtration limits the solids levels that can be removed by contact clarification without causing excessive cleaning. This process is most effective for source water having low turbidity, color, iron, and manganese. On such waters, it can be an economical alternative to more conventional types of clarification. It is especially applicable to upgrading overloaded direct filtration systems.

Design Criteria

Contact clarifiers are rated on the basis of surface loading rate and typically operate in the range of 8 to 10 gpm/ft2 (19.5 to 24.4 m/h). To maintain effective flocculation, units must also be selected to operate in the range of 50% to 100% of design capacity.

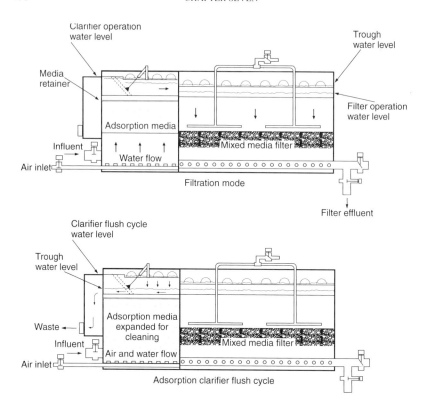

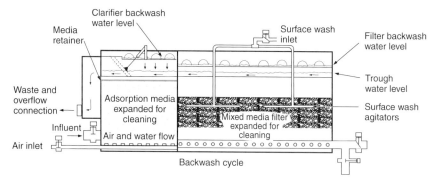

FIGURE 7.21 Adsorption clarifier. *(Courtesy of Wheelabrator Engineered Systems Inc.—Microfloc.)*

Operation

Source water is chemically treated by the addition of metal salts or polymers in a rapid mixer, usually of the static mixer type. Chemically treated water is then introduced to the clarifier in an upflow or downflow mode, depending on the proprietary process selected. Clarified water is then filtered in a typical granular medium filter.

Inlet Conditions

Proprietary units are designed as upflow units. The adsorption clarifier uses a buoyant medium held in place by a screen. Chemically treated source water is introduced under the medium through a pipe lateral. Other types of units use a medium heavier than water that must be supported by a media-retaining support system typical of gravity filters. In all cases, source water flows up through the medium as it would for any upflow filter.

Outlet Conditions

Clarified water is collected above the medium in an effluent trough similar to backwash troughs in a filter and then flows to the filter through an outlet valve.

Sludge Removal

Because solids are removed by filtration, they must be removed by backwashing or flushing. The flushing cycle is initiated after a preset head loss has been reached or a preset time has passed.

The cycle is started by closing the effluent and influent valves. The medium is first air scoured. Air is necessary in the adsorption clarifier to allow buoyant media to expand during the flushing cycle. Air reduces the apparent density of the water, causing media to lose buoyancy and opening pore spaces so that accumulated material can be flushed out. Air is also necessary with other types of media to assist water in scouring accumulated solids.

Source water is used to flush the bed, taking advantage of hydraulic head already available. Waste washwater is collected in the effluent trough but is directed to a waste drain instead of the filter. When the cycle is completed, the air is turned off, the waste drain valve is closed, the effluent valve is opened, and clarified flow once more is directed to the filter. No mechanical equipment is required for sludge removal, an advantage of these units.

BIBLIOGRAPHY

Camp, T. R. "Sedimentation and the Design of Settling Tanks." *Transactions of ASCE* 3(2285):895, 1946.

Camp, T. R. "Studies of Sedimentation Basin Design." *Sewage and Industrial Wastes* 25(1):1, 1953.

Dissolved Air Flotation: Field Investigations. American Water Works Association Research Foundation Report. Denver, Colo.: American Water Works Association, 1994.

American Water Works Association, *Water Quality and Treatment,* fourth edition. New York: McGraw-Hill, 1990.

Dissolved Air Floatation: *Laboratory and Pilot Plant Investigations.* American Water Works Association Research Foundation Report. Denver, Colo.: American Water Works Association, 1992.

Flood, F. L. *Sedimentation Tanks.* Seminar Papers on Wastewater Treatment and Disposal. Boston: Boston Society of Civil Engineers, 1961.

Great Lakes Upper Mississippi River Board of State Public Health and Environmental Managers. *Recommended Standards for Water Works.* 1992.

Grubb, T. R., S. R. Arnold, and P. J. Harvey. "Recent Applications of DAF Pilot Studies, Retrofits and Conversions." Presented at AWWA Annual Conference, June 1994.

Hartman, R. H., and R. F. Jacarrino. "Superpulsator Clarifier for Potable Water Treatment." Presented at AWWA Annual Conference, 1987.

Hazen, A. "On Sedimentation." *Transactions of ASCE* LIII:63, 1904.

Hudson, H. E. *Water Clarification Processes Practical Design and Evaluation.* New York: Van Nostrand Reinhold, 1981.

Ives, K. J. "The Inside Story of Water-Treatment Processes." *Journal of Environmental Engineering* 121(12):846, 1995.

Kawamura, S., and J. Lang. "Re-evaluation of Launders in Rectangular Sedimentation Basins." *Journal WPCF* 58(12):1124, 1986.

Kawamura, S. *Integrated Design of Water Treatment Facilities.* New York: John Wiley and Sons, 1991.

Kollajtis, J. A. "Dissolved Air Flotation Applied in Drinking Water Clarification." Presented at AWWA Annual Conference, 1991.

Monk, R. D. G., J. F. Willis. "Designing Water Treatment Facilities." *Journal AWWA* 79(2):45, 1987.

Neptune Microfloc, Inc. *60* Tube Application Guidelines for Water Treatment.* Technical Memorandum No. 1, Revision 2. Corvallis, Ore.: Neptune Microfloc, Inc.,1985.

Neptune Microfloc, Inc. *FTC and TC.* Technical Memorandum No. 14, Revision 0. Corvallis, Ore.: Neptune Microfloc, Inc., 1980.

Operational Control of Coagulation and Filtration Processes, M37. Denver, Colo.: American Water Works Association, 1992.

Taebi-Harandy, A., and E. D. Schroeder. "Analysis of Structural Features on Performance of Secondary Clarifiers." *Journal of Environmental Engineering* 121(12):911, 1995.

Vincent, J. "Effective Clarification and Softening with the Accelator Unit versus Conventional Treatment Methods." Presentation to the New York State Sixth Annual Water Treatment Technical Conference, Saratoga Springs, N.Y., April 9 and 10, 1991.

Webb, J. E. "Super Flocculation and High-Rate Clarification for Water Treatment." Presented at North Carolina Operators Association, Sept. 29, 1993.

Yee, L. Y., and A. F. Babb. "Inlet Design for Rectangular Settling Tanks by Physical Modeling." *Journal WPCF* 57(12):1168, 1985.

CHAPTER 8
HIGH-RATE GRANULAR MEDIA FILTRATION

Filtration, as it applies to water treatment, is the passage of water through a porous medium to remove suspended solids. According to Baker (1948), the earliest written records of water treatment, dating from about 4,000 BC, mention filtration of water through charcoal or sand and gravel. Although a number of modifications have been made in the manner of application, filtration remains one of the fundamental technologies associated with water treatment.

Filtration is needed for most surface waters in order to provide a second barrier against the transmission of waterborne diseases. Although disinfection is today the primary defense, filtration can assist significantly by reducing the load on the disinfection process, increasing disinfection efficiency. The Surface Water Treatment Rule (SWTR) recognizes three categories of granular filtration techniques:

- Rapid sand
- Slow sand
- Diatomaceous earth

This chapter covers the design of the first category of filters. However, in this instance the term *rapid sand* includes not only sand, but also other types of filter media such as crushed anthracite coal and granular activated carbon. Chapter 9 covers the other two categories of granular filtration techniques.

MECHANISM OF FILTRATION

Removing suspended solids by high-rate granular media filtration is a complex process involving a number of phenomena. Attempts to develop theories that quantitatively predict solids removal performance with sufficient precision and versatility to be of use in practical filter design have met with relatively little success. Consequently, filter media selection is often an empirical process. In current high-rate granular media filtration techniques, solids removal occurs primarily as a two-step process (Cleasby, 1972). During the initial transport step, particles are moved to the surfaces of media grains or previously captured floc.

Transport is believed to be caused largely by hydrodynamic forces, with contact occurring as stream lines converge in pore restrictions. The second step is particles' attachment to either grain or floc surfaces. Electrokinetic and molecular forces are probably responsible for the adherence of particles on surfaces within the bed (O'Melia and Crapps, 1964; Craft, 1966; O'Melia and Stumm, 1967). Physical straining through the surface layer of solids and biological growth (schmutzdecke) is the principal filtration mechanism of a slow sand filter, but it is generally a minor means of solids removal in high-rate granular media filters.

DESIGN CONSIDERATIONS

A number of interrelated components are involved in the overall design of a high-rate granular media filtration system:

- Pretreatment system
- Filter media
- Filtration rates
- Depth of the filter box
- Mode of operational control
- Filter washing system
- Filter arrangements
- Underdrain system
- Filter performance monitoring
- Auxiliaries

These components are discussed in detail in the following sections.

Pretreatment

Effective operation of a high-rate granular media filtration system requires pretreating the source water. The nature, as well as the quantity, of suspended material in the pretreated water is critical to filter performance.

Unflocculated water can be difficult to filter regardless of the type of medium in use (Cleasby, 1972; Hsiung, Conley, and Hansen, 1976). However, the work of Robeck, Dostal, and Woodward (1964) with dual-media filters showed that if the applied water is properly coagulated, filtration at rates of 4 or 6 gpm/ft^2 (10 or 15 m/h) produce essentially the same filtered water quality as filtration at a rate of 2 gpm/ft^2 (5 m/h). Subsequent investigations have shown similar results for mixed-media filters (Laughlin and Duvall, 1968; Westerhoff, 1971; Conley, 1972).

Chemicals used in conjunction with high-rate granular media filtration are limited primarily to metal salts or cationic polymers as primary coagulants. Primary coagulants are ideally fed into rapid mixing basins preceding flocculation. Whether clarification is also required depends on the quantity of suspended solids and algae in the source water. Primary coagulants are intended to produce agglomerations of natural and chemical solids. Nonionic or anionic polymers are often added with the coagulant as a coagulant aid to assist in strengthening and growth of these agglomerations during flocculation. These same polymers can also be added as a filter aid to the filter influent water or to the washwater to increase the strength of adhesion between media grains and floc in coarse-to-fine filters.

Pretreatment may also include aeration or introducing an oxidant if an objective of water treatment is to remove iron or manganese.

A filter aid polymer can improve floc capture, provide better filtered water quality, and increase filter runs with higher head loss before turbidity breakthrough. Filter aid polymers are not generally used with fine-to-coarse filters because they promote rapid surface clogging. Filter aids are often fed in dilute liquid form to allow dispersion without mechanical agitation just before filtration.

Filter aid polymer doses to gravity filters are usually low (0.02 to 0.05 mg/L). Doses required for pressure filters may be higher because of the higher operating head losses normally employed. Because water viscosity increases with decreasing temperature, breakthrough as a result of floc shearing is more likely at lower water temperatures. Consequently, increased polymer doses and a longer contact time before filtration may be required in cold weather.

Assuming that adequate coagulation is feasible, the designer must decide whether clarification is desirable. In the past, settling has been provided before high-rate granular media filtration when turbidities exceeded roughly 10 ntu (Culp and Culp, 1974). The increased storage capacities of dual- and mixed-media filters have made filtration of water with higher turbidities practicable. The primary advantage of providing direct filtration is eliminating capital and operating costs associated with clarification. The higher solids load on the filter will, however, shorten run times and increase the portion of product water required for filter washing. Although the point at which advantages outweigh disadvantages varies with local conditions, a number of investigators have suggested conditions that would justify consideration of direct filtration (Conley, 1965; Cleasby, 1972; Hutchison, 1976; Culp, 1977). It is imperative that pilot studies be conducted to determine the feasibility of direct filtration for each application.

Filter Media

Although the selection of filter media type and characteristics is the heart of any filtration system, selection is usually based on arbitrary decisions, tradition, or a standard approach. Pilot plant studies using alternative filter media and filtration rates can determine the most effective and efficient media for a particular water.

In drinking water applications in North America, the most commonly used filter media are natural silica sand, garnet sand or ilmenite, crushed anthracite coal, and granular activated carbon (GAC). Selecting appropriate filter media involves a number of design decisions concerning source water quality, pretreatment, and desired filtered water quality. Filter media cleaning requirements and underdrain system options depend on the filter configuration and filter media selected.

Media variables the designer can control include bed composition, bed depth, grain size distribution, and, to a lesser extent, specific gravity. In addition to media design characteristics, media quality can be controlled to some extent through specifications covering, where applicable, hardness or abrasion resistance, grain shapes, acid solubility, impurities, moisture, adsorptive capacity, manner of shipment, and other such factors. Suggested criteria and a discussion of the applicability of these parameters can be found in the AWWA Standards for Filtering Material (B100) and Granular Activated Carbon (B604).

In the United States, granular media have been traditionally described in terms of effective size (E.S.) and uniformity coefficient (U.C.). The E.S. is that dimension exceeded by all but the finest 10% (by weight) of the representative sample. It is also referred to as the "10% finer" size. The U.C. is the ratio of the "60% finer" size to the E.S. Common practice in Europe is to express media sizes as the upper and lower limits of a range. These limits may be expressed either as linear dimensions or as passing and retaining sieve sizes (i.e., 1.0 to 2.0 mm or $-10 + 18$ mesh).

Filter beds may be classified as graded fine-to-coarse, ungraded, graded coarse-to-fine, or uniformly graded, depending on the distribution of grain sizes within the bed during filtration. Transition from the ungraded media of a slow sand filter to the fine-to-coarse high-rate granular media filter resulted from dissatisfaction with the low loading rates and laborious cleaning procedure characteristic of slow sand filters. Filters with uniformly graded or coarse-to-fine beds are now operated at higher loading rates and for longer run times than are feasible with conventional rapid sand filters.

Ungraded Media. The slow sand filter is a primary example of an ungraded bed. Because slow sand filters are not backwashed, no hydraulic grading of the media occurs. Distribution of the various grain sizes in the bed is essentially random. Typical slow sand filter beds contain 2 to 4 ft (0.6 to 1.2 m) of sand with an E.S. of 0.2 to 0.35 mm and a U.C. not exceeding 3.0.

Fine-to-Coarse Media. Fluidization and expansion of rapid sand filter beds during backwashing results in accumulating fine sand grains at the top of the bed and coarse grains at the bottom. Consequently, filtration occurs predominantly in the top few inches, and head loss increases relatively rapidly during operation. This sand medium typically has an E.S. of 0.35 to 0.60 mm and a U.C. of 1.3 to 1.8. Grains passing a no. 50 sieve (0.3 mm) or captured on a no. 16 sieve (1.18 mm) are normally limited by specifications to very small portions of the medium. Bed depths are typically 24 to 36 in. (0.6 to 0.9 m).

Single-medium anthracite beds have been used in the same basic configuration as rapid sand beds. Because anthracite is more angular than sand, the porosity of an anthracite bed is higher than that of a sand bed containing media with the same E.S. The porosity of a sand bed is generally 40% to 45% whereas a typical anthracite bed has a porosity of 50% to 55%. Consequently, anthracite does not perform in exactly the same manner as sand of equivalent size. Because of the lower specific gravity, anthracite beds are also easier to fluidize and expand than sand beds.

Coarse-to-Fine Media. In a coarse-to-fine bed, both small and large grains contribute to the filtering process. The presence of fine media in a filter is desirable because of the relatively large surface area per unit volume that fine media provide for particle adhesion. Fine media are instrumental in achieving the best-quality filtered water. Coarse media, when placed before fine media in the filtering sequence, decrease the rate of head loss buildup and increase available storage capacity in the bed.

Tests by Oeben, Haines, and Ives (1968) and Craft (1971) demonstrated that sand media placed in a downflow reverse-graded (coarse-to-fine) alignment exhibit filtering performance superior to that of the same media in the conventional fine-to-coarse alignment. Better use of the entire bed, manifested as lower head loss and longer run time, was achieved by the reverse-graded beds without decline in filtered water quality. Reverse grading of the beds used in these studies was accomplished, however, by physically transferring backwashed media to another filter vessel, a method not applicable to full-scale operations. Attempts in the United States to approximate coarse-to-fine filtration have been directed almost entirely toward the use of dual-media and mixed-media (i.e., triple-media) beds.

Dual-media beds normally contain silica sand and crushed anthracite coal. Triple-media beds contain an additional layer of garnet or ilmenite sand. Beds with three or more media types that intermix after backwashing have been patented as mixed-media filters and are proprietary technology (Rice and Conley, 1967). Specific gravities of materials used in filtration are roughly as follows:

- Silica sand, 2.55 to 2.65
- Anthracite coal, 1.5 to 1.75

- Garnet, 4.0 to 4.3
- Ilmenite, 4.5

A typical dual-media bed contains 6 to 12 in. (0.15 to 0.3 m) of silica sand (E.S. 0.4 to 0.55 mm) overlain by 18 to 30 in. (0.46 to 0.76 m) of anthracite (E.S. 0.8 to 1.1 mm). A typical mixed-media filter bed contains 3 to 4 in. (5 to 10 cm) of garnet (E.S. 0.15 to 0.35 mm), 6 to 9 in. (0.15 to 0.3 m) of silica sand (E.S. 0.35 to 0.5 mm), and 18 to 24 in. (0.5 to 0.6 m) of anthracite (E.S. 0.8 to 1.1 mm).

The degree to which media layers are intermixed in the bed depends on the sizes and shapes of the media used, the nature of the backwashing procedure, and the specific gravities of the different media. Disagreement exists over whether distinct layers or intermixed layers are most desirable. If layers mix completely, the purpose of using more than one medium would be defeated. If no mixing occurs, individual fine-to-coarse layers would result, and the possibility of rapid clogging at interfaces is raised.

Proponents contend that in a properly designed mixed-media filter, a gradual decline in pore sizes from top to bottom of the bed is established after backwashing. The original argument can be traced to Conley and Pitman (1960), Conley (1961), and Camp (1961, 1964) in the early 1960s. Brosman and Malina (1972) concluded that a slightly mixed bed was superior to a distinctly layered bed in terms of head loss development, filter run time, and filtered water turbidity. Cleasby and Sejkora (1975), however, disagree that superior performance can be attributed to interfacial intermixing in and of itself; rather, it is a result of differences in the media sizes required to construct mixed and separated beds. They found that to provide a relatively sharp interface in a dual-media bed, fairly coarse sand was required. The resulting bed would not provide the same filtered water quality as a bed using finer sand that mixed more readily with the coal.

The anthracite coal and silica sand used in dual-media filters inevitably results in some intermixing of layers. In a triple-media bed, intermixing of silica sand and garnet sand normally occurs more readily than mixing of silica sand and coal. Cleasby and Woods (1975) suggest that, as a rule of thumb, the ratio of the average particle size of coarse silica grains to the size of coarse garnet grains should not exceed 1.5 to ensure that some garnet remains at the bottom of the bed. They also suggest that a ratio of coarse coal grain size to a fine silica sand grain size of about 3 results in a reasonable degree of mixing in dual- or mixed-media beds. Brosman and Malina (1972) found that anthracite-sand filter media with a size ratio at the interface of less than 3:1 exhibits little mixing and that the zone of mixing increases linearly as the size ratio increases above 3:1.

In a number of U.S. installations, taste and odor removal and filtration have been combined in a single unit using GAC (Hager, 1969; Hansen, 1972; Blanck and Sulick, 1975; McCreary and Snoeyink, 1977). GAC is sometimes added to existing rapid sand units from which some sand has been removed. GAC depths of 12 to 48 in. (0.3 to 1.2 m) over silica sand layers of 6 to 18 in. (0.14 to 0.5 m) have been reported. Typically, GAC with an E.S. of 0.5 to 0.65 mm has been used. This technique is usually applicable only where taste and odor and not turbidity are of primary concern. If turbidity levels are high, GAC pores become rapidly plugged, and carbon life is quickly reduced. If both turbidity and taste and odor are significant problems, GAC beds should be preceded by conventional granular media filtration. If carbon adsorption is desired to remove organics, the depth of GAC that can be provided in a converted gravity filter is likely to be too shallow to provide adequate contact time.

In Europe and the former Soviet Union, upflow and biflow filters are commonly employed to achieve coarse-to-fine filtration (Hager, 1969). Hamann and McKinney (1968) reported that upflow filters commonly used in the former Soviet Union are relatively deep (6.5 to 8.5 ft of sand) and that the E.S. ranged from 0.5 to 2.0 mm. The primary difficulty associated with upflow filters is breakthrough resulting from bed lifting as head loss increases. Grids are commonly installed above the sand to discourage lifting.

In the biflow filter, water is introduced simultaneously at the top and the bottom of the bed. Russian biflow filters typically contain 5 to 5.5 ft (1.5 to 1.7 m) of sand. Filtered water is withdrawn at an intermediate point 1.5 to 2 ft (0.5 to 0.6 m) below the top surface. During filtration, the head on the upper bed aids in preventing expansion of the lower bed. The loadings on the upper and lower portions of the filter are not equivalent. The lower bed is a coarse-to-fine bed, although the finest grains in the upflow filter are as coarse as the coarsest grains in the downflow bed. Variations observed in filtered water quality between the upper and lower portions of the filter are attributable to differences in grain sizes and media depth.

Uniform Media. Uniformly graded deep-bed filters used in Europe use relatively coarse media, ranging from 0.5 mm to as much as 6.0 mm. The U.C. is typically 1.2 to 1.3, but values as high as 1.5 may be found. Greater media depth is substituted for the lack of fine media in the bed. Such a substitution requires more vigilant operation and increased chemical usage to avoid breakthrough. Depths of 4 to 6 ft (1.2 to 1.8 m) are common, and in some cases media depths reach 8 ft (2.4 m). Filters of this type are not expanded during backwash, and stratification of grain sizes does not occur. These filters are generally designed to use air or air/water backwash.

There are many possible combinations of filter media size, d, and depth, L. Montgomery (1985) presents a methodology for determining the optimum relationship between these two variables. The relationship between L and E.S. d_e (10% finer) of many high-rate filters is shown in Figure 8.1. In this figure, the average E.S. for a dual- or mixed-media filter was computed as a weighted average. Data in the figure indicate that as the media used become coarser, the required depth is increased, and as the media become finer, the depth required is reduced.

Filtration Rates

Slow sand filters, designed for filtration rates of 3 to 6 million gallons per acre per day at a rate of 0.05 to 0.10 gpm/ft^2 (0.1 to 0.2 m/h), were initially replaced by rapid sand filters that operated at rates of 1 to 2 gpm/ft^2 (2.4 to 5.0 m/h). The 2 gpm/ft^2 (4.9 m/h) rate became widely accepted as an upper limit in U.S. water supply practice for many years. It has subsequently been demonstrated that dual-media and mixed-media, as well as single-medium (sand or anthracite), filters can be successfully operated at much higher rates.

A number of investigators found dual- and mixed-media filters to operate successfully at rates from 3 to 8 gpm/ft^2 (8 to 20 m/h) in a variety of locations (Conley, 1961, 1965; Robeck, Dostal, and Woodward, 1964; Dostal and Robeck, 1966; Laughlin and Duvall, 1968; Tuepker and Buescher, 1968; Rimer, 1968; Westerhoff, 1971; Kirchman and Jones, 1972). The quantity of evidence of the practicality of high-rate filtration was such that in 1972 the AWWA Committee on Filtration Problems concluded that it had been amply demonstrated that filters could be designed and operated to produce water of acceptable quality at flows substantially higher than the rate of 2 gpm/ft^2 (5 m/h) once considered the maximum. Over the last 15 years, a number of pilot-scale and full-scale deep-bed uniformly graded anthracite filters have been operated reliably at rates of 10 to 15 gpm/ft^2 (24 to 37 m/h).

Average filtration rates of roughly 2 to 7 gpm/ft^2 (5 to 17 m/h) are reported for the upflow, biflow, and deep-bed filters discussed previously (Hamann and McKinney, 1968; Jung and Savage, 1974).

Many regulatory agencies will not approve rates in excess of 2 gpm/ft^2 (5 m/h) without successful pilot-scale testing. The designer should make every effort to obtain approvals for operation at higher rates.

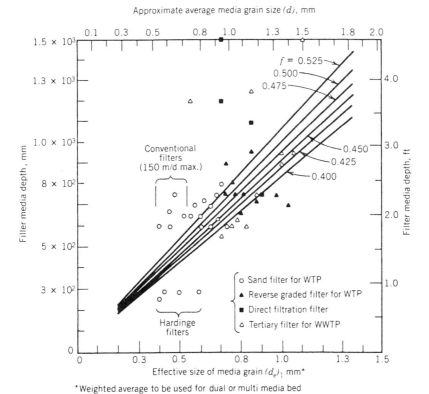

FIGURE 8.1 Relationship between depth and size of media. *(Source: Montgomery 1985.)*

Filter Operational Control

Filtration process control is critical to successful operation. Decisions regarding control methods must be made early during the design because they affect the physical layout of the filtering facilities. Filter control may be predicated either on head loss through the filter bed or on the rate of filtration. In either case, smooth transition during changes in filtration rate is highly desirable. The adverse effects of sudden flow surges on filtered water quality have been well documented (Cleasby, Williamson, and Baumann, 1963; Tuepker and Buescher, 1968).

Control methods are generally tested most severely by removal of filters for washing and their return to service. Of the choices open to the designer, the mode of control is perhaps the most controversial. Two basic modes of gravity filter control are commonly found: constant-rate and declining-rate. With the constant-rate mode, there are three ways to operate a filter: (1) a rate-of-flow controller in the filtered water piping; (2) influent flow splitting with the water level over the filter maintained at a constant level; and (3) influent flow splitting with the water level varying during the filter run.

Constant-Rate with Rate-of-Flow Controller. With this type of control, water levels in all filters and the filter influent channel are maintained at a constant level. Plant flow is proportioned equally among the operating filters by means of a flow-measuring device (e.g., venturi meter) and modulating valves incorporated in the effluent piping of each filter. Each filter controller receives a signal from the venturi meter and modulates the valves to ensure that each filter is filtering an equal portion of the influent flow.

As the water level in the influent channel rises or falls because of filter media clogging, filters being taken out of service for washing or maintenance, or variations in the plant flow, a level element in the filter influent channel signals this movement to the controller, which, in turn, modulates the flow through the other filters. At the start of each filter run, the valve is almost closed to dissipate the surplus head caused by a high water level over the filters. As the filter media clogs, the water level rises and is sensed by the level element, and the controller compensates accordingly by opening the valves to recover an equal amount of head. When one of the filters is taken out of service for washing, the remaining filters in service must pick up additional flow. Operating variation in the water level is usually 6 in. (15 cm). Figure 8.2 shows a typical arrangement of this method of filter control.

Constant-Rate with Constant Water Level and Influent Flow Splitting. This method incorporates individual inlet weirs in the influent channel entrance to each filter. Channel and weir lengths should be generously sized to ensure equal flow splitting. A level element in each filter sends a signal to a filter controller to maintain a constant level of water, accomplished by modulating a valve located in the effluent piping of each filter. At the start of the filter run when the head loss through the media is minimal, the valve is almost closed to dissipate surplus head. As head loss in the filter media increases, the water level rises, increasing the driving head. The level element signals this rise to the controller, which, in turn, commands the valve to open further, reducing head loss across the valve and maintaining a constant water level and flow. Figure 8.3 shows a typical arrangement of this method of filter control.

Constant-Rate with Varying Water Levels and Influent Flow Splitting. This method is similar to influent flow splitting with constant level except that there are no level elements, controllers, or modulating valves. The water level at the start of the filter run is just above the top of the bed. Filters of this type normally discharge over an effluent weir, eliminating the possibility of bed dewatering. The relatively high discharge elevation requires an unusually deep filter box to provide filtering head. From 5 to 6 ft (1.5 to 1.8 m) of

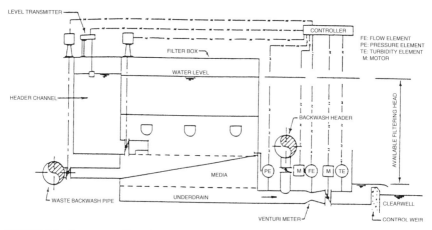

FIGURE 8.2 Constant-rate filter with rate-of-flow controller. *(Source: Monk 1987.)*

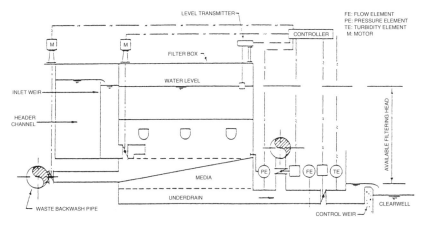

FIGURE 8.3 Constant-rate filter with influent splitting and constant water level. *(Source: Monk 1987.)*

additional depth is typical. The filtration rate is determined by plant influent flow. The level in each filter rises as necessary to accept an equal portion of influent and indicates head loss. When the level rises to a fixed upper limit, filter washing is initiated. Figure 8.4 shows a typical arrangement of this method of filter control.

Declining-Rate Control. Declining-rate filters are equipped with effluent weirs rather than rate controllers. Flow is distributed on the basis of the relative conditions of the beds. Assuming that influent piping losses are roughly the same for all filters, a uniform operating water level in all filters is achieved. The filtration rate then becomes the highest in the cleanest bed and lowest in the dirtiest bed. In each bed, the filtration rate decreases as solids accumulate. An orifice plate or other simple flow-limiting device is used on each filter effluent line to limit maximum flow rate. To determine which bed is in greatest need of washing, some type of effluent rate indication must be provided. Advantages claimed for declining-rate filters include higher water production for a given run length and improved filtered water quality. A principal disadvantage is that, after a filter has been backwashed, it immediately operates at a high rate, which can cause high turbidity to pass through for a period of time. Figure 8.5 shows a typical arrangement of this method of filter control.

Filter Media Washing

As the amount of solids retained in the filter media of a rapid sand filter increases, bed porosity decreases. At the same time, head loss through the bed and shear on captured floc increase. Before the head loss builds to an unacceptable level or turbidity breakthrough occurs, washing is required to clean the bed.

Failure to clean filter media adequately can lead to a multitude of problems. Initially, mudballs form and accumulate in the bed, causing clogging. Then the clogged areas contract as head loss increases. This shrinkage opens cracks in the filter media surface and sometimes at the filter walls. Cracks can cause short-circuiting of the bed during filtration, with subsequent decline in filtered water quality. Clogged areas also contribute to channeling of washwater, which can lead to bed upset. The mechanisms by which washing problems lead to filter failures are discussed in greater detail in Cleasby (1972).

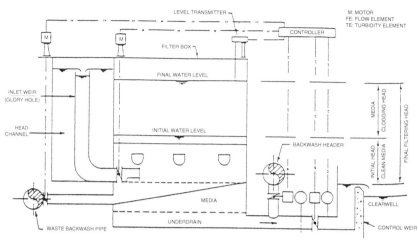

FIGURE 8.4 Constant-rate filter with influent splitting and varying water level. *(Source: Monk 1987.)*

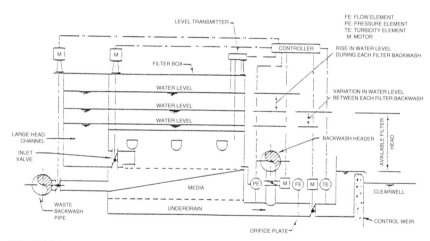

FIGURE 8.5 Declining-rate filter. *(Source: Monk 1987.)*

The selection of a washing technique is closely tied to filter media and underdrain selection. In current practice, washing normally includes upflow water flushing. The rate and duration of water flushing are variable, however, and may be supplemented with air scour or surface water wash. Operational sequencing of combined washing systems and the source of the washwater introduce additional variations.

Washwater Source. Washwater source options include the following:

- flow bled from high-service discharge and used directly for washing or to fill an above ground washwater tank that is subsequently used for gravity washing
- gravity flow from a separate elevated finished-water storage tank
- direct pumping from a sump or belowground clearwell

Bleeding flow from a high-service discharge main results in energy loss because of the pressure reduction required before washing. For direct washing, a pressure-reducing valve or orifice is placed in the washwater supply line. For bleeding flow to fill a washwater tank, an altitude valve or other level control device is used to control the water level in the tank. In either case, the washwater supply line is often sized to restrict the maximum amount of water that can be delivered. Both options avoid provision of separate washwater pumps. Direct washing also avoids construction of a washwater tank but presents greater difficulty in controlling washwater flow. Because of the large pressure drop often involved in supplying washwater by high-service bleeding, the potential for cavitation in or following head-dissipating devices in the supply line is significant.

If elevated finished-water storage is not available to provide head for filter washing, washwater may be pumped to a separate washwater storage tank or directly to the filters. Use of a washwater tank permits pumping at a lower rate. Tank storage volume must be sufficient to permit filter washing at the maximum wash rate while the pump operates at the minimum run times.

A number of proprietary filters are available that obtain washwater by means other than those previously listed. One design uses vertical steel tanks divided into upper and lower compartments. Sufficient filtering head is provided so that following downflow filtration in the lower compartment, filtered water flows through a pipe into the upper tank. When terminal head loss in the filter bed is reached, washwater flows from the upper tank back through the filter.

Some filter control systems permit gravity flow washing of a filter using effluent from the filters remaining in service. Such filters are called self-backwashing filters. They do not use pumps or piping for backwashing, but instead, all the filters discharge into a common channel. A filtered water weir controls the water level in the channel so that the water level is always higher than the filter washwater troughs or side weir. This difference in level must be sufficient to provide the head needed to deliver adequate water for backwashing. Also, there must be a sufficient number of filters in operation to meet the demand for backwash water. Figure 8.6 shows the configuration of a typical self-backwashing filter. To provide the required backwash driving head, the filter box must be substantially deeper than those required for more conventional types of filter backwash systems. But, because no

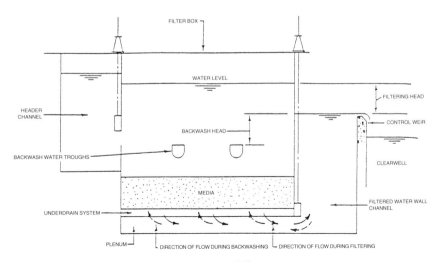

FIGURE 8.6 Self-backwashing filter. *(Source: Monk 1987.)*

equipment is involved, the capital cost is the lowest. One disadvantage of this design is that there is no way to control backwash rate.

Washing Methods. The AWWA Subcommittee on Backwashing of Granular Filters has acknowledged four basic backwash methods:

- Upflow water wash without auxiliary scour
- Upflow water wash with air scour
- Upflow water wash with surface wash
- Continuous backwash

The application normally dictates the method selected. Filter bed expansion during upflow water washing results in media stratification. Air washing results in filter media mixing. If stratification is desired, air scour must be avoided or must precede fluidization and expansion with water.

Upflow Water Wash without Auxiliary Scour. Upflow water wash alone may be sufficient in some filters receiving low solids loadings. In the absence of auxiliary scour, washing in an expanded bed occurs as a result of the drag forces on the suspended grains. Grain collisions do not contribute significantly to washing (Camp, Graber, and Conklin, 1971; Cleasby, Stangl, and Rice, 1975; Cleasby et al., 1977).

Maximum shear on the grains theoretically occurs (for typical filter sand) at a bed expansion of 80% to 100% (Cleasby et al., 1977). The increase in shear with increasing bed porosity is relatively slight beyond the point at which expansion begins. Optimal expansion may be less than 20% (Johnson and Cleasby, 1966). Normally, when water wash is applied exclusively, an expansion of 20% to 50% is used. Water wash at a sufficient rate to substantially expand (10% or greater) a granular bed is generally referred to as high-rate water wash. Water wash incapable of fully fluidizing a bed (i.e., less than 10% expansion) is generally referred to as a low-rate wash.

Experience in the United States with high-rate water wash used alone is extensive. It is generally successful for applications that filter iron precipitates from groundwater or remove color from otherwise high-quality surface water. The relatively weak cleaning action of water wash without auxiliary scour of some type, however, generally renders it unsuitable for filters removing large quantities of suspended solids or for applications where polymers are used.

High-rate water wash tends to stratify granular media. In multimedia beds, this action is essential and beneficial, but it is not required for uniformly graded single-medium beds. In single-medium beds, high-rate water wash results in movement of the fine grains to the top of the bed, which has a negative effect on head loss and filter run length.

Upflow Water Wash with Air Scour. There are numerous approaches to using auxiliary air scour in backwashing filters. Air scour has been used alone and with low-rate water backwash in an unexpanded bed or slightly expanded bed. Each procedure takes place before either low- or high-rate water wash.

Air scour provides effective cleaning action, especially if used simultaneously with water wash. Cleaning is attributable to high interstitial velocities and abrasion between grains. On the other hand, air wash has substantial potential for media loss and gravel disruption if not properly controlled. Use of air scour can significantly reduce the quantity of water required for backwashing filters.

If more than one filtering medium is used and stratification of the bed is desired, high-rate water wash must follow air scour. In a single-medium bed, if a low-rate wash can adequately remove scoured solids, high-rate wash can be avoided.

If air scour occurs simultaneously with water wash, air flow must usually be stopped before washwater overflow into the washwater collection troughs to prevent media loss. For this reason, the permissible duration of air washing is short unless the concurrent water wash rate is low or the filter box is very deep.

Experience indicates that air scour essentially eliminates mudball formation. Difficulties have arisen, however, from failure to remove scoured solids from filter surfaces. Contributing factors probably include low water-washing rates, long horizontal-travel distances to backwash troughs, and a necessary lag between termination of air scour and initiation of higher-rate water wash.

Upflow Water Wash with Surface Wash. Surface wash systems have been widely used for many years. Fixed systems distribute auxiliary washwater from equally spaced nozzles in a pipe grid. Rotary systems have pipe arms that swivel on central bearings. Nozzles are placed on opposite sides of the pipes on either side of the bearing and the force of the water jets provides the thrust required to rotate the pipe arms.

Rotary systems are preferred because they generally provide better cleaning action, lower water requirements, and less obstruction for filter access. Possible problems with rotating surface wash units include failure to rotate, failure to clean in corners, abrasion of concrete walls near the point of closest passage of the arm, and locally high velocities caused when passing under washwater collection troughs. Either type of system may fail to provide auxiliary scour where it is most needed. This can be especially true in multimedia beds if substantial removals are occurring at media interfaces.

Surface wash systems are typically suspended about 2 in. (5 cm) above the surface of the unexpanded filter bed. Systems have also been placed in the unexpanded bed of dual- and mixed-media filters where there is an accumulation of filtered materials deep within the filter bed. Dual-arm rotary systems that have one arm above and one arm below the unexpanded surface are also available. Nozzle plugging with media has been a problem with the submerged units.

Rotary surface wash systems may have either straight or curved pipe arms and generally have nozzle diameters of $\frac{1}{8}$ to $\frac{1}{4}$ in. (3 to 6 mm). Single-arm units typically operate at 50 to 100 psi (345 to 690 kPa) and discharge from 25 to 160 gpm (2 to 10 L/s) depending on length. Standard units are available up to approximately 14 ft (4 m) in diameter. Some models induct air into the washwater jets. A typical surface agitator and arrangements are shown in Figure 8.7.

Advantages of auxiliary surface wash include proven effectiveness in alleviating dirty filter problems, improved cleaning (when compared with water wash alone) without a great change in system complexity, and possibly lessened danger of gravel upset if the quantity of washwater introduced through the underdrain is reduced.

Because surface wash systems constitute a possible connection between filtered and unfiltered water, backflow prevention devices must be provided in supply lines.

Continuous Backwash. An alternative to the automatic control of standard filters is the use of continuous backwashing filter beds, which eliminate the need to remove the beds from service for washing. Beds are divided into a series of narrow, contiguous cells, each containing its own underdrain system that allows it to be washed independently from remaining cells. Washing is accomplished by means of a traveling hood suspended above the bed. As the hood travels across the bed, each cell is isolated, and a small backwash pump draws clean water from the filter effluent and reverses the flow through that particular cell. Water is removed by a second washwater pump located in the traveling hood and discharged to waste.

The wash cycle time is controlled by preset adjustable timers to permit optimization of the automatic operation feature. Media depth varies with each application but is typically 30 to 36 in. (0.8 to 0.9 m).

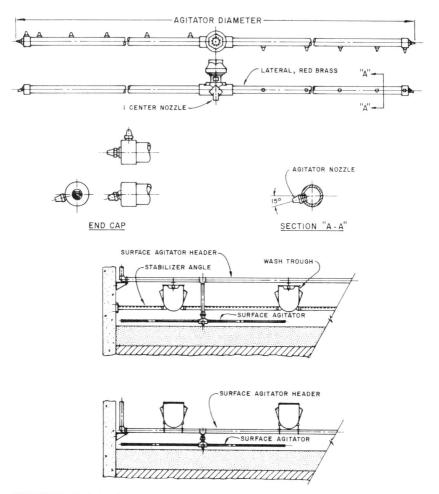

FIGURE 8.7 Typical surface agitator and arrangements. *(Source: F. B. Leopold Co.)*

In addition to automatic washing features, these filters have the capability of producing relatively constant washwater flow. In a properly sized system, this constant flow can eliminate the need to provide washwater equalization facilities and permits direct recycle to the plant headworks.

Wash Rates. In the United States, wash rates are expressed as volumetric flow per unit surface area (gpm/ft^2). In Europe, wash rates are expressed as the equivalent water rise velocity (ft/s, ft/min, in./min, mm/s, or m/h). Wash rates are generally variable and depend on washwater temperature, filter media characteristics, and washing method. Water viscosity decreases with increasing temperature. Consequently, as washwater temperature rises, drag forces on media grains are reduced and higher wash rates are required to achieve bed expansion. Each degree Celsius increase in water temperature requires roughly a 2% increase in wash rate to prevent a reduction in bed expansion. Filter wash systems should be designed for the warmest washwater temperature that will be encountered.

Filter media characteristics also affect washing rate. Rate requirements increase with increasing grain size and density. Also, angular grains are more easily expanded than round grains. In filters using more than one type of filter medium, sizes of each type of medium must be selected carefully to ensure proper positioning after water wash. Recommended size ratios for dual- and mixed-media beds were discussed previously. Figure 8.8 displays the effect of media size on the water wash rate required to achieve 10% bed expansion for three common filter media. Figure 8.9 shows the effect of water temperature on the viscosity of water and on the wash rate for silica sand and anthracite coal.

Characteristic washing rates and durations vary for each washing method discussed previously. The suitability of a washing method is related to influent water quality, filter media characteristics and bed configuration, and underdrain design. Consequently, not all washing methods are applicable in all cases, and different methods may or may not yield similar results in a particular case.

Upflow Water Wash without Auxiliary Scour. When water wash is used alone, a high-rate wash is employed. Generally a wash rate of 15 to 23 gpm/ft^2 (37 to 56 m/h) is applied. After the water level in the filter has been lowered to the top of the washwater collection trough, the wash starts. It usually lasts from 3 to 15 min. A low-rate water wash is used at the end of the wash cycle in multimedia filter beds to restratify the filter media.

Upflow Water Wash with Air Scour. Three variations of air and water wash were discussed previously. The first, air scour alone followed by low-rate water wash, is commonly applied in Great Britain to single-medium sand filters with 0.6 to 1.2 mm E.S. media. After the water level in the filter is lowered to below the washwater overflow, air is injected at 1 to 2 ft^3/min/ft^2 (0.3 to 0.6 m^3/min/m^2) for 3 to 5 min. Water wash of 5 to 7.5 gpm/ft^2 (12 to 18 m/h) follows. Bed expansion and stratification are not achieved, although relatively cool water temperatures may result in fluidizing upper sand layers. Problems with gravel disruption have not been experienced if air and water are applied separately (Cleasby et al., 1977).

Air scour alone followed by high-rate water wash can be applied to dual-media or multimedia filters, because bed stratification occurs during water wash. This method has been

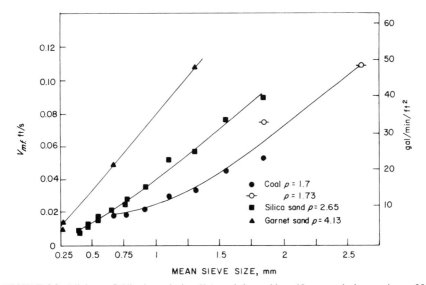

FIGURE 8.8 Minimum fluidization velocity (V_{mf}) needed to achieve 10 percent bed expansion at 25 degrees C. *(Source: Cleasby and Baumann 1974.)*

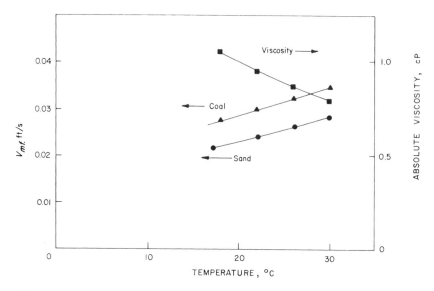

FIGURE 8.9 Effect of water temperature on V_{mf} of sand and coal, and on absolute viscosity of water. *(Source: Cleasby and Baumann 1974.)*

used in the United States with air scour at 2 to 5 ft^3/min/ft^2 (0.6 to 1.5 m^3/min/m^2) followed by high-rate water wash at 15 to 23 gpm/ft^2 (37 to 56 m/h).

Concurrent air scour and water wash is generally limited to the deep, coarse-grained filters common in Europe. For 1 to 2 mm E.S. media, air-scour rates of 2 to 4 ft^3/min/ft^2 (0.6 to 1.2 m^3/min/m^2) are used with a water flow of 6.3 gpm/ft^2 (15.4 m/h). For 2 to 6 mm E.S. media, 6 to 8 ft^3/min/ft^2 (1.8 to 2.4 m^3/min/m^2) and 6.3 to 7.5 gpm/ft^2 (15.4 to 18.3 m/h) are used. Concurrent air and water wash typically lasts 5 to 10 min and is followed by water wash alone for another 5 to 10 min. The rate of final water wash is generally one to two times that used with air scour. In some installations, concurrent air and water wash is used to improve transport of solids to the washwater collection troughs rather than to increase scour in the bed (Harris, 1970).

Upflow Water Wash with Surface Wash. Combined surface and water wash usually involves three phases. After the water surface level is lowered in the bed, surface wash is activated and operated alone for 1 to 3 min. Low-rate water wash is then applied simultaneously for an additional period of roughly 5 to 10 min. Termination of surface wash precedes a final phase (1 to 5 min) during which a higher washwater rate is used to expand the bed 20% to 50%. This usually requires a washwater rate of 15 to 23 gpm/ft^2 (37 to 56 m/h). Washwater flow during surface agitation is usually limited to that required to expand the bed only slightly. If anthracite makes up the top filtering layer, bed expansion above the surface-wash system may be desirable to reduce the likelihood of media loss. Rotary surface-wash systems typically add 0.5 to 2.0 gpm/ft^2 (1.2 to 5 m/h) to the washwater flow. Fixed-nozzle systems typically deliver 2 to 4 gpm/ft^2 (5 to 10 m/h).

Filter Arrangements

Filters can be configured in a number of ways in the overall plant layout. It is important to develop a layout that is the least costly and is operationally optimized (Begin and Monk,

1975). Amirtharajah (1982) shows how a minimum-cost filter design is obtained by differentiating the cost function in terms of length-to-width ratio as a variable.

Configuration of Filters. Filters are normally placed next to each other along one or both sides of a pipe gallery. This approach provides the most compact arrangement and also simplifies filter operation and maintenance. If possible, areas for future expansion should be provided at one end of the row (or rows) of filters, and piping in the gallery should be installed with blind flanges at the ends to make future filter additions easier.

In larger plants, placing filters in rows on opposite sides of a pipe gallery is common practice. In smaller plants, a single row of filters results in simpler construction. Typical filter configurations are shown in Figure 8.10.

Location of the filtered water clearwell under the pipe gallery, and common walls between filtered and unfiltered water, should be avoided to prevent the possibility that leakage through the walls may contaminate filtered water. A false floor should be provided in the washwater gullet to prevent a common wall from existing between unfiltered water in the gullet and filtered water in the underdrain system. Drainage should be provided for spaces beneath false floors of this type.

Many designers favor construction of the conduits connecting pretreatment basins and filters in such a manner that floc destruction because of turbulence and high velocities is minimized. Drops, bends, and long runs should be minimized or avoided. The need to avoid turbulence, however, is not universally accepted (Cleasby, 1972). Low velocities may be dictated by head loss between basins and filters.

In warm climates, filters may be placed outdoors with precautions for controlling algae formation in the filter box. One method is to provide for shock chlorination of filter influent or washwater. In colder climates, filters are normally housed to prevent ice formation. Filter control consoles are usually provided adjacent to each filter or each set of filters to allow for local control of the backwash operation. Consoles should be provided with climate-controlled housing.

Number of Filters. From a cost point of view, one filter is the most ideal. Practically, however, four filters are the minimum number that should be used to allow for filter washing and the occasional need for a filter to be out of service for maintenance, without resulting in unreasonable rate increases in the other filters. If self-backwashing filters are used, allowing for the lowest rate of operation and for one filter out for maintenance, the remaining filters must be able to produce sufficient water to wash one filter effectively. At a filtration rate of 4 gpm/ft^2 (10 m/h) and a 16 gpm/ft^2 (39 m/h) wash rate, a minimum of six filters is required. If a lesser filtration rate or a higher backwash rate is used, even more filters are needed.

Size of Filters. The size of individual gravity filters is determined by plant capacity, filtration rate, and the number of filters desired. Hydraulic considerations and the effect of removing a filter from service limit maximum filter size. Additional considerations include the maximum area to which washwater or air scour can be evenly distributed, the maximum span length of washwater collection troughs, and available sizes of surface wash equipment, if used. Single gravity filters of up to 4,500 ft^2 (420 m^2) have been reported (Clark, Viesman, and Hammer, 1977), but units less than half this size are more typical, even in large plants. Large filters may be divided into two sections by using a central gullet, permitting half the filter to be washed at a time, although influent and effluent piping are usually shared.

Pressure filters are usually limited by shipping constraints. The largest standard units typically available are 12 ft (3.7 m) diameter tanks. This limits vertical filters to about 113 ft^2 (10 m^2) of filter medium. Horizontal filters are normally not longer than 40 ft (12 m). Larger units of both types can be specially fabricated on site.

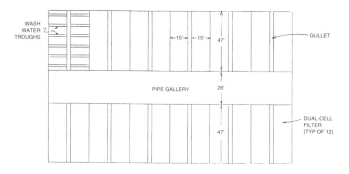

Los Angeles (CA) Aqueduct Water Filtration Plant (300 mgd)

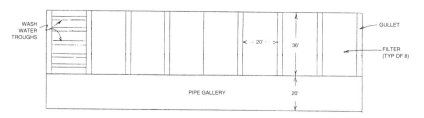

Hemlock Water Filtration Plant; Rochester, NY (48 mgd)

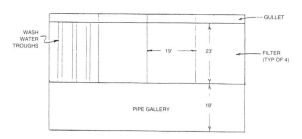

Attleboro (MA) Water Treatment Plant (12 mgd)

FIGURE 8.10 Typical filter configurations. *(Source: Camp Dresser & McKee.)*

Capital cost of filters can generally be minimized by designing for the minimum number of filters consistent with size limitations. Plant expansions are usually accomplished by adding filters of the same size as existing units, because installing larger filters may require extensive changes in the filter wash system.

Depth of Filter Box. A number of factors should be considered in designing a filter box, including:

- Available head
- Depth of water over the filter media surface
- Fixed head losses
- Head losses through the filter media
- Shape of head-loss curves
- Rate of filtration
- Elevation of the filtered water effluent control weir (if used)

Available head may be restricted by site conditions and plant layout, by the designed maximum length of filter run, by optimizing head loss, or by terminating turbidity breakthrough. Ideally, allowable head loss is determined from pilot plant studies. Fixed head losses through piping, venturi meters, and throttling valves can be calculated from manufacturers' literature and by hydraulic analyses. Head losses through filter media are determined by pilot plant studies or are calculated from Darcy's formula.

Once these factors are known, water depth and weir elevation can be established to achieve a cost-effective design. Objectives in designing a filter box should be to minimize the cost of construction and to avoid the possibility that the filter will develop negative pressures. Monk (1984) presents the methodology required to determine optimum filter box depth.

Underdrain Systems

An underdrain system has two purposes: to collect water that passes through the filter media and to distribute washwater (and air, if used) uniformly across the filter bed. Support gravel is required when openings in the underdrain system are larger than the filter medium directly above it. Although the support gravel or other support method does not contribute to particulate matter removal, it aids in distributing washwater. For this reason, it should be considered part of the underdrain system. Uneven distribution of washwater can displace support gravel, eventually requiring removal of the filter media to be regraded or replaced.

Four basic types of underdrain systems are common: pipe laterals, blocks, false bottom, and porous bottom.

Traditionally, the greatest difficulty in underdrain design has been providing a barrier to the finest medium that does not clog during filtration or filter washing. Early attempts to use fine screens or strainers were largely unsuccessful, leading to the use of gravel layers below filter sand. The position of gravel layers may, however, be disrupted during filter washing. Jet action, which is discussed in greater detail elsewhere (Cleasby, 1972), causes sand and gravel mixtures to be more easily disrupted than gravel alone. If auxiliary air scour is used, even greater gravel disturbance may occur. Fine gravel, usually placed at the sand-gravel interface, is most easily dislocated. A possible solution to this problem is the use of gravel in a coarse-to-fine-to-coarse, or "hourglass," gradation, which has been shown to be highly stable at high washwater rates (Cleasby, 1972). Fine media penetrate the upper coarse gravel layer without apparent ill effect.

Mixed-media beds with very fine garnet at the bottom of the bed are generally constructed with a layer of coarse garnet on top of the silica support gravel. Coarse garnet prevents leakage of the fine garnet and also helps stabilize the underlying silica gravel.

European-type deep-bed filters use relatively coarse and uniformly graded media. As a result, bed stratification is not required and air scour presents less of a hazard to proper bed operation. Also, the use of strainers is more likely to be feasible because of the larger permissible openings. Consequently, false-bottom underdrains with nozzles designed for both air and water distribution and without support gravel are commonly used in deep-bed filters.

Pipe Laterals. Pipe lateral underdrains were once popular because of their relatively low cost and adaptability for use in pressure filters. Problems with relatively high head loss and poor washwater distribution resulted in a general decline in their use. They are still encountered, however, when older filters are upgraded.

Pipe underdrain systems generally consist of a centrally located manifold pipe to which smaller, equally spaced laterals are attached. Lateral pipes usually have one or two rows of $\frac{1}{4}$ to $\frac{3}{4}$ in. (6 to 19 mm) diameter perforations on their bottom sides. The lateral pipes may be fitted with nozzles as illustrated in Figure 8.11. Guidelines for lateral design include the following ratios:

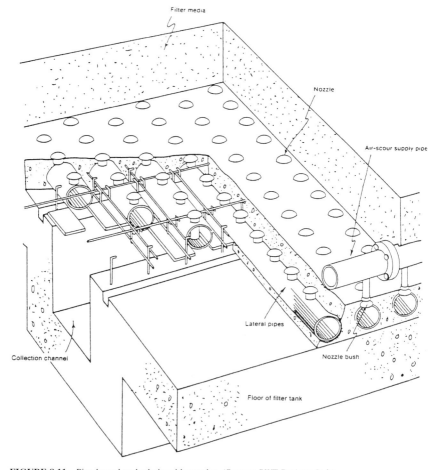

FIGURE 8.11 Pipe lateral underdrain with nozzles. *(Source: PWT Projects Ltd.)*

- Total area of orifices (surface area of bed)—0.0015 to 0.005:1
- Cross-sectional area of lateral (total area of orifices served)—2 to 4:1
- Cross-sectional area of manifold (total area of laterals served)—1.5 to 3:1

Orifices are normally spaced at 3 to 12 in. (8 to 30 cm) and laterals at roughly the same spacings as the orifices. Approximately 18 in. (45 cm) of support gravel is required to cover a lateral network. Three to five graded layers are usually involved, with sizes varying from $1\frac{1}{2}$ to $\frac{1}{8}$ in. (38 to 3 mm). The bottom layer should extend 4 in. (10 cm) above the highest washwater outlet.

Blocks. A commonly used block underdrain consists of vitrified clay blocks with $\frac{1}{4}$ in (6 mm) diameter dispersion orifices located across the top of each block. Support gravel is required with this type of underdrain system. The size and arrangement of these blocks and typical support gravel layers are shown in Figure 8.12. In mixed-media applications, the third gravel layer is replaced by garnet of similar size. This type of underdrain system is suitable only for water washing. However, auxiliary air scour may be provided by adding an air piping grid at the sand-gravel interface, as shown in Figure 8.13.

Another type of block underdrain is designed for concurrent air/water wash. Blocks are constructed of polyethylene and consist of a primary feeder lateral (lower) and a secondary compensating lateral (upper), shown in Figure 8.14. Small control orifices open from the feeder lateral directly into the compensating lateral. Washwater and air are admitted to and flow through the feeder lateral and rise to discharge from the control orifices into the compensating lateral. The triangular shape of the primary lateral distributes incoming washwater and air uniformly along its length. Support gravel is typically used with this type of underdrain, graded in an hourglass configuration. As a replacement for support gravel, an integral media support (IMS) cap made of plastic beads sintered together may be installed on top of the plastic block underdrain, as shown in Figure 8.15.

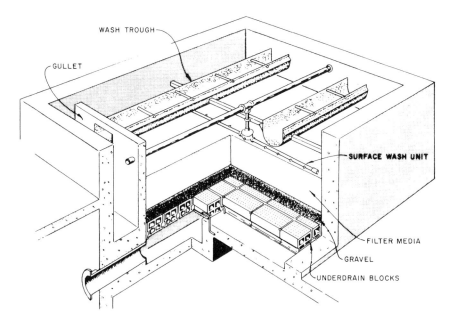

FIGURE 8.12 Typical gravity filter. *(Source: F. B. Leopold Co.)*

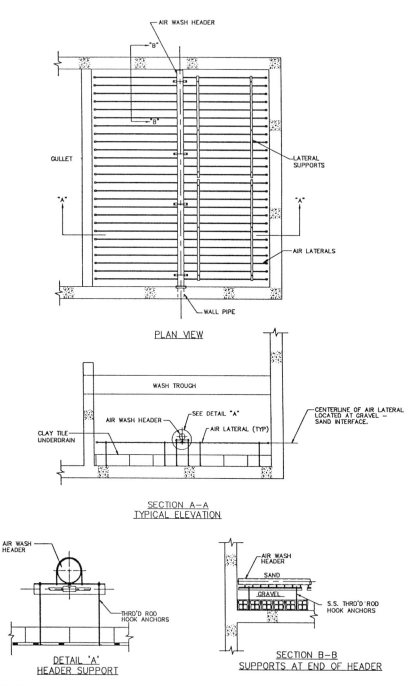

FIGURE 8.13 Vitrified clay tile underdrain with auxiliary air scouring system. *(Source: Roberts Water Technologies, Inc.)*

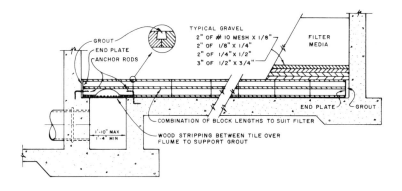

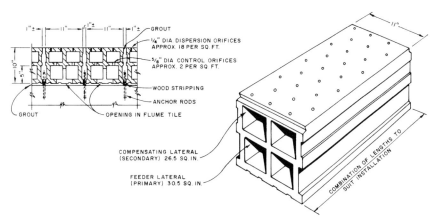

FIGURE 8.14 Plastic block underdrain designed for use with air/water wash. *(Source: F. B. Leopold Co.)*

False Bottoms. One of the most widely used false-bottom underdrains is constructed of precast or cast-in-place reinforced concrete supported on concrete sills. This underdrain system contains uniformly spaced inverted pyramidal depressions. Unglazed porcelain spheres are placed in the depressions to distribute flow. Each depression is filled and leveled with 1 to $1\frac{1}{2}$ in. (25 to 38 mm) gravel before placement of overlying gravel support layers. A typical arrangement including the gravel layers is shown in Figure 8.16. The last silica gravel layer should be replaced by coarse garnet in a mixed-media filter bed.

Other false-bottom underdrains have impervious bottoms penetrated by nozzles. Nozzle-type underdrains are used primarily in filters employing air/water wash systems. Fine openings in the nozzles eliminate the need for support gravel, which, accordingly, reduces filter box depth. Other varieties of false-bottom underdrains are constructed of concrete, polyethylene, or tile blocks; monolithic concrete; or steel plates. Nozzles are equipped with plunge pipes for air wash and are usually constructed of stainless steel, plastic, or brass. Plunge pipes usually adjust to allow for leveling after installation. Nozzle orifices are sometimes smaller on the filter side to prevent clogging

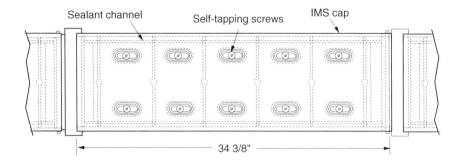

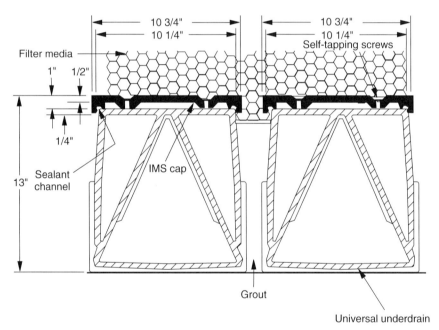

FIGURE 8.15 Integral media support (IMS) cap designed for use without support gravel. *(Source: F. B. Leopold Co.)*

during filtration, even though such a configuration can contribute to clogging during filter washing. Some failures resulting from plugging and breakage have been experienced with this type of underdrain. A typical false-bottom underdrain using monolithic concrete is shown in Figure 8.17.

Porous-Bottom Underdrains. Porous-bottom underdrains constructed of porous aluminum oxide plates have been used in both block and false-bottom configurations. They

HIGH-RATE GRANULAR MEDIA FILTRATION

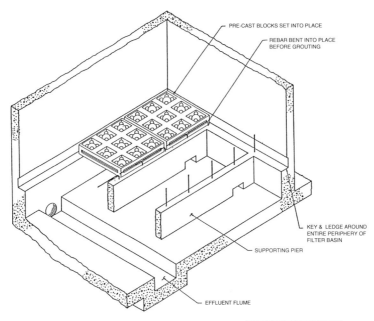

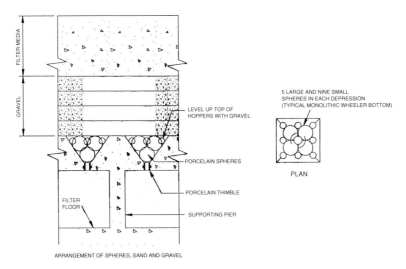

FIGURE 8.16 Concrete wheeler bottom with gravel support. *(Source: Roberts Water Technologies, Inc.)*

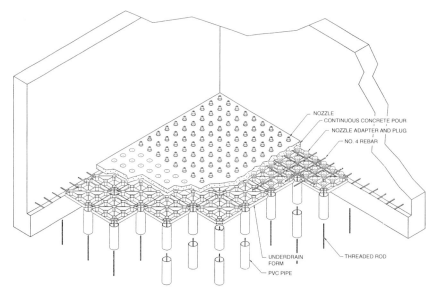

FIGURE 8.17 Underdrain layout. *(Source: Infilco Degremont, Inc.)*

are constructed of plates mounted on concrete or steel piers or on clay tile saddles to form blocks. Very small pore sizes make porous-bottom underdrains susceptible to plugging and therefore unsuitable for use in softening or iron-and-manganese removal plants or other plants where plugging by chemical deposition may occur. They may also clog with rust or debris during backwashing.

Additional problems that may occur include breakage because of the brittle nature of the porous material and failure of caulked joints between plates. Porous bottoms have been used successfully in a few locations, but they are less widely accepted than the block or false-bottom underdrain systems discussed previously.

Filter Performance Monitoring

Filter design should include instrumentation to monitor filtered water turbidity and particle count, filtration rate, head loss, backwash rate, and filter run length. If auxiliary air scour backwash is used, airflow monitoring should also be included. Pilot filters may help determine primary coagulant, coagulant aid, and filter aid dosage.

Turbidity. Turbidity is the most common measure of filter performance. Turbidity is an indication of the amount of suspended material in a water sample. The nephelometric technique, the current standard in the United States, compares the intensity of light scattered by a water sample with a standard reference suspension under the same conditions. Turbidimeters can be either discrete sample or on-line type. On-line turbidimeters include a light source to illuminate the water sample, one or more photoelectric detectors, and an analog indicator of turbidity based on the intensity of light scattered 90 degrees to the path of the incident light. Turbidity is reported as nephelometric turbidity units (ntu).

Each filter's effluent turbidity should be monitored and recorded continuously using an on-line turbidimeter to detect variances from normal operation immediately. Observations of the effects of fluctuations in source water quality, rate changes, equipment malfunctions, chemical feed variations, filter washing, and other such occurrences contribute to the operator's understanding of the plant's performance and increase the ability to deal with such situations.

Turbidity measurement is sometimes used to automatically initiate a filter wash cycle or to actuate an alarm whenever the filter effluent reaches a preset maximum turbidity level. Most water treatment plants establish a filtered water turbidity goal (e.g., 0.1 ntu) well below current regulatory standards.

Turbidity of the waste washwater and filter-to-waste water can also be monitored to assess the performance of filter washing and ripening, respectively. A different type of turbidimeter, capable of measuring higher turbidity levels, monitors waste washwater.

Sample piping should be designed so that it does not collect air bubbles that can distort readings (Letterman, 1994). Air bubble traps, available from most turbidimeter suppliers, can be used for this purpose. On-line turbidimeters should be located where they are readily accessible because they require periodic cleaning and calibration to ensure accurate readings.

Particle Counting. Particle counting is rapidly gaining acceptance to monitor filter performance (Lewis, Hargesheimer, and Ventsch, 1992). Particle counters are instruments that can quantify and size particles in water by light-scattering techniques and can be either discrete sample or on-line type. Particles ranging in size from approximately 1 to 500 μm can be quantified by particle counters. Particle counters can provide a direct measurement of the number of particles in a particular size range.

Particle counters are particularly useful in determining the log removal of particles in the *Giardia* and *Cryptosporidium* size ranges. *Giardia* cysts generally fall in the 5 to 15 μm range, and *Cryptosporidium* oocysts fall in the 4 to 7 μm range; however, the correlation between particle counts and *Giardia* and *Cryptosporidium* concentration has not been fully established. Filtration theory indicates that 1 to 3 μm particles are the most difficult to remove, but Moan et al. (1993) found earliest breakthrough of 3 to 7 μm particles in test filters. The 3 to 7 μm size range includes cysts and oocysts currently of primary concern.

Current standards require a 2 to 2.5 log removal of *Giardia* cysts by a filtration process, depending on the type of filtration. Future regulations may also include a *Cryptosporidium* oocyst log removal requirement. For very clean source waters, particle counters may not demonstrate proper log removals because of the low required filtered water particle counts. Even in these situations, fluctuations in filtered water particle counts can still give an excellent indication of filter performance.

Engineers and plant operators should be aware that, at the present time, no standard method of analysis has been established, so particle counts can vary from instrument to instrument. Sensors associated with particle counters may have different operating principles such as particle size detection limit, particle concentration range, sample flow rate, or pressure requirements. Bends and flow-altering devices should be avoided upstream of particle counters because they can vary performance. Chemical and physical cleaning of particle counters is required and should be considered in their installation. The operator should be aware that stable particle counts are difficult to obtain when samples are switched periodically between filter effluent lines.

Filtration Rate. Flow-measuring devices are recommended for monitoring flow through individual filters. Filtration rates can be monitored and controlled by comparing the metered flow rate through a filter to its surface area. Meters can be combined with modulating valves to automatically control filtration rates. The most common type of measuring device is a venturi tube because it can be easily checked in the field with manometers. It is often impossible to provide the ideal length of pipe preceding the flow-measuring device, particularly when it is mounted in the filter's effluent piping. However, this is usually not a concern if the total plant flow is being split equally between all operating filters. Local and remote indicating and recording devices are usually included. Orifice plates can also be placed in filter effluent lines to limit maximum filtration rates. If flow splitting by weirs or other such device is employed, the filtration rate may be determined by dividing the total plant flow by the number of operating filters.

Head Loss. Head loss in a filter bed is a valuable indicator of filter bed condition and may be used to automatically activate filter washing. Head loss through the filter media is normally monitored by differential pressure-cell devices that measure the water pressure above and below the filter media.

Aside from head loss developed by particle retention within the filter media, operating head loss depends on the filtration rate, the clean bed head loss through the filter media, and head losses through the filter underdrain system and the effluent rate controller. Terminal head loss is the difference between the static head "available" between the water elevations in the filters and the filtered water effluent control weir, less the operating head losses through the clean media, underdrain system, effluent piping connections and bends, and the effluent rate controller. Modern plants typically have a terminal head loss of 8 to 10 ft (2.4 to 3 m), while many older plants have significantly less. Clean bed head losses range from 1 to 2 ft (0.3 to 0.6 m) depending on media specifications and filtration rate.

Filters should be washed when terminal head loss is reached, otherwise turbidity breakthrough may occur. Also, a vacuum can result if head loss at any level in the filter bed exceeds static head. This situation is referred to as negative head and can cause air binding of the filter media. When pressure in the filter bed drops below atmospheric levels, dissolved gases are released from the water being filtered. Gas bubbles trapped in the bed further increase head loss and aggravate the problem. They may also result in media displacement during filter washing.

This problem is particularly acute when filtering with insufficient water depth over the media or when surface waters are saturated with atmospheric gases because of rising temperatures in the spring. Remedies for air binding in gravity filters include increasing washing frequency, maintaining adequate static head above the media surface, and keeping the clearwell water level above the top of the filter media to keep it submerged. Pressure filters normally discharge well above atmospheric pressure and are not subject to air binding.

The head loss sensor connection to the filter box should be located approximately 4 in. (10 cm) above the top of the washwater collection trough to prevent washwater from entering the sensor. A sediment trap with drain installed on the sensor line will capture any sediment that may enter the line. The end of the sensor should be turned up, keeping a full column of water in the line at all times to minimize air entrainment. A fine mesh stainless steel screen installed on the end of the sensor will prevent clogging with filter media. However, this screen requires periodic cleaning to prevent a buildup of material that may cause a false head loss reading.

Another valuable monitoring method is to measure head loss at points within the filter bed by installing several pressure taps at various depths of the filter bed. These pressure taps can be connected to transparent tubes, creating a piezometer board. The pressure taps can be monitored and recorded continuously for better observation and control (Monk and Gagnon, 1985). Although this unit is a good performance monitoring tool, it is not necessary to install one on each filter.

Filter Wash Rate. Because filter wash flow requirements may vary with the seasons because of differences in water temperature and pretreated water quality, operator knowledge of the filter washwater rate in use at a particular time is essential. Flow tubes are usually employed as the monitoring device and are usually matched with a downstream flow-control device to control the washing rate. Recording meters are generally not necessary, but a totalizing device is desirable to determine the overall volume of water used in washing. An alarm can be provided that is actuated if the wash rate exceeds a predetermined maximum.

Length of Filter Run. Filter performance is often judged by the length of the filter run, but too long a filter run may not be good for filter operation. Long filter runs make washing a filter much more difficult because of particulate matter compaction in the filter media. In addition, a long filter run indicates that the filter is not working at its most cost-effective capacity.

Rapid sand filters are generally operated with run lengths between 12 and 72 hours, typically with 24-hour runs. Pressure filters may have somewhat longer filter runs than gravity filters if they can be operated at higher head losses without turbidity breakthrough.

Pilot Filters. Pilot filters are bench-scale models of full-scale plant filters that can be used to determine optimum coagulant dosage. Coagulated water is diverted to pilot filters from the full-scale pretreatment units. Monitoring the pilot filter effluent turbidity provides an indication of the adequacy of coagulant feed. Pilot filters greatly reduce the lag time in coagulant feed system adjustment and improve plant performance. Parallel pilot filters are usually provided to ensure continuous control. Because of the nature of connecting piping to pilot filters, higher filter aid polymer dosages are usually required than in the full-scale plant. The effect of increasing the polymer dose is to shorten filter run times. Consequently, pilot filters are generally not used to predict run lengths or polymer dosage. In most cases, however, these variations do not affect determination of optimum coagulant dosage.

Filter Operation and Control

Several types of filtration operation and control are used. Many plants have filter control consoles located immediately adjacent to the filters they serve to observe filters for malfunction during backwashing. Remote operation from the plant's central control console is also practiced in some plants. This allows a single operator to wash filters and still observe other plant processes.

In the past, all major valves were controlled by individual manual controls, and all filter operations were operator directed. However, advances in sensing and control equipment have made the use of remote automatic or semiautomatic control commonplace. In semiautomatic operation, filter washing is initiated by the operator but consists of a predetermined sequence that requires no additional attention.

Fully automatic filters are washed without operator input on the basis of loss of head in the filter bed, filtered water turbidity, or a fixed maximum run time. Automatic systems permit operation of all filters from a central location, reducing personnel requirements. However, remote automatic operation may not permit the operator to observe the wash cycle directly. Automatic systems should allow easy modification of the washwater and auxiliary scour rates and sequence by the operator. Process instrumentation for monitoring and alarming to indicate when problems arise is essential for all modes of operation.

Auxiliaries

Auxiliary equipment that the engineer must consider in filter design include piping and conduits supplying and removing water from the filter, valves used to control filter operations, and washwater troughs.

Conduits. Filters require hydraulic connections for influent water, filtered water, washwater supply, auxiliary scour, washwater drain, and (if used) a connection to allow filtered water to be wasted. Influent water is usually delivered to a gravity filter through the washwater gullet. Influent to a pressure filter is generally distributed by a tapped pipe serving as a manifold or by a baffle plate. Influent conduits should be designed to deliver water to the filters with as little disturbance as possible. Free fall or turbulence, which can disturb the filter media, is undesirable. Delivering influent beneath the water surface in the filter or baffling the incoming stream prevents media disturbance. Necessary measures depend on the control strategy used.

Typical piping serving a gravity filter is shown in Figure 8.18. Influent conduits should be sized to limit velocities to 2 ft/s (0.6 m/s). This may result in the use of an influent flume rather than a pipe in large plants. Hydraulic considerations generally result in velocities of 3 to 6 ft/s (0.9 to 1.8 m/s) in washwater and filtered water piping. At higher velocities, head losses often become excessive, and undesirable effects such as water hammer are more

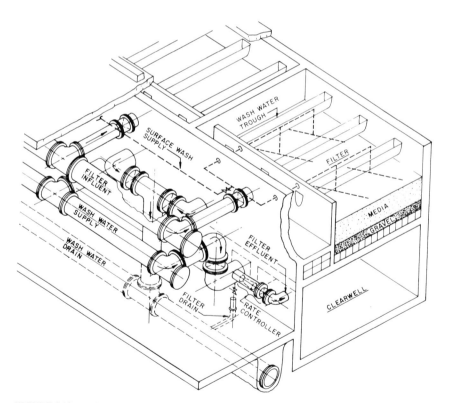

FIGURE 8.18 Typical gravity filter piping. *(Source: Camp Dresser & McKee.)*

likely to occur. Cement-lined ductile iron or steel pipe or stainless steel pipe is commonly used for filter piping. Flanged, grooved-end, or mechanical joints and connections should be used for ease of maintenance.

Design considerations include avoiding high points and including air release valves in washwater supply piping, ensuring adequate backpressure within the throat of the filtered water venturi meter, and providing accessible piping to permit proper maintenance.

Valves. A typical filter is equipped with five valves: influent, filtered water, washwater supply, washwater drain, and surface wash or air wash supply. A filter-to-waste valve may also be included; however, waste lines constitute a potential cross-connection and must be equipped with air-gap protection against backflow from the drain to the filter.

Rubber-seated butterfly valves are most common in filter pipe galleries because of their relatively short laying length. In larger plants, the influent and washwater drain may be gates rather than valves. These valves and gates can be operated either manually or automatically from a local filter operating console or from a remote central control console. The filter rate control valve and its actuator must provide for stable and accurate flow control. It typically should operate between 0 and 45 degrees open. It should also automatically close during power failure to prevent dewatering the filter media. Valve placement should permit easy access for maintenance, with valves and gates equipped with easily distinguishable position indicators.

Valve operating systems may be hydraulic, pneumatic, or electric. Hydraulic systems were developed first but are generally no longer installed in new plants because of problems with leakage and with plugging of orifices in the lines by deposition from the fluid. Pneumatic or electrical systems are used in most new construction. Pneumatic systems are generally less expensive, but they require oil- and moisture-free air. Electrical systems offer greater reliability, but initial cost is usually higher. In the event that maintenance is required, electrical controls may require greater technical skill to service than pneumatic controls. Flow control valves must be carefully designed to provide required accuracy and to avoid cavitation.

Washwater Troughs. In the United States, washwater troughs are suspended at even spacings above gravity filter beds to provide uniform removal of washwater during backwashing. These same troughs also normally distribute influent flow uniformly across the filter media's top surface. This limits horizontal travel required and equalizes static head on the underdrain system. In contrast, European designs often feature narrow beds with overflow walls on one or both sides, but not suspended over the media. In these designs, tilting side weirs, horizontal water jets, and a procedure allowing influent water to enter the filter on the side opposite the overflow wall are sometimes used to aid the movement of scoured solids to waste. This method is termed *cross wash.* Spacing troughs in U.S. practice is usually at 5 to 7 ft (1.5 to 2 m) centers to limit horizontal travel distances to 2.5 to 3.5 ft (0.8 to 1 m).

If troughs are placed too close to the surface of the unexpanded bed, media may be lost during backwash. The design elevation of the weir edge of the trough may be determined by adding the depth required for maximum bed expansion (usually 50%) and the overall depth of the trough, plus a small margin of safety of 6 to 12 in. (0.15 to 0.3 m). If air scour is practiced, additional care must be taken. In a conventional trough, simultaneous use of washwater and air must be stopped when the level of water rises to the bottom of the trough, or media will be lost. In a limited number of plants, special baffled plates on either side of the trough minimize the loss of media during concurrent water/air wash.

French (1981) has given the following criterion for the location of the top of the troughs based on trough spacing:

$$H = 0.34S$$

where H = height of the top edge of the trough above the fluidized bed
  S = center-to-center spacing of the troughs

Troughs are usually made of fiberglass reinforced plastic (FRP), stainless steel, or reinforced concrete. Troughs made of FRP and stainless steel usually have semicircular bottoms, and concrete troughs have V-shaped bottoms. Trough bottoms should not be flat because froth and sludge tend to accumulate and fall back onto the surface of filter media. Typical trough cross sections, including two with special baffled plates for concurrent air/water wash, are shown in Figure 8.19.

The required cross-sectional area of the trough for a given washwater flow can be estimated from Figure 8.20. A more rigorous analysis can be obtained by referring to derivations provided by Fair, Geyer, and Okun (1968) and Brater and King (1976).

After troughs are installed in a filter, weir edges must be leveled to uniformly match a still water surface at the desired overflow elevation. It is critical that troughs be properly supported both vertically and horizontally so that their weirs remain absolutely level during backwashing. Center supports are typically used whenever trough length exceeds 14 ft (4.3 m).

FILTER DESIGN CRITERIA

The first step for the design engineer when designing filtration facilities should be to review all current federal, state, and local laws and regulations that may be applicable. The second step should be to review all applicable standards prepared by various organizations and associations.

Standards Set by Regulatory Agencies

The first design criterion to be considered for a filtration process is its ability to meet applicable water quality standards. The U.S. Environmental Protection Agency (USEPA) has established the Surface Water Treatment Rule (SWTR) as the controlling standard for filtration. The SWTR requires a public water supplier using a surface water or groundwater source deemed "under the influence of surface water" to achieve 3 log (99.9%) removal or inactivation of *Giardia* and 4 log (99.99%) removal or inactivation of viruses.

Credit for log removal is given to filtration processes based on their type, with the remaining required log removal to be achieved by disinfection. Conventional filtration is usually given a 2.5 log credit for *Giardia* removal, and direct, slow sand, and diatomaceous earth filtration are usually given a 2 log credit.

Maximum turbidity levels are established for filtration performance standards. For conventional and direct filtration, filtered water turbidity must be less than 0.5 ntu in 95% of samples collected in a month. For slow sand and diatomaceous earth filtration, less than 1.0 ntu must be achieved in 95% of the samples. A level of 5 ntu must not be exceeded at any time regardless of the type of filtration employed. These water quality limits must be considered in designing filtration facilities, as well as design of the entire treatment train.

In addition to drinking water quality standards, most states have adopted minimum standards for designing filtration facilities, either with specific criteria or reference to other design standards. The most widely known and adopted standard is *Recommended Standards*

HIGH-RATE GRANULAR MEDIA FILTRATION   **185**

a. Fiberglass reinforced plastic trough.   b. Reinforced concrete trough.

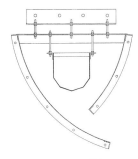

c. Engineered Stainless Steel Design (ESSD^{TM}) wash trough with Type II MULTIWASH^R separator baffles. (*Source: General Filter Co.*)

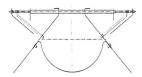

d. ScourGuard^R baffled wash trough. (*Source: EIMCO Process Equipment Co.*)

FIGURE 8.19 Typical washwater trough cross sections.

for Water Works, commonly known as the "Ten States Standards" (1992). This standard contains specific criteria on filter size, arrangement, allowable filtration rates, structural details, hydraulics, materials of construction, filter washing, and control systems. In some instances, states require pilot testing of treatment processes to establish their validity and as design criteria.

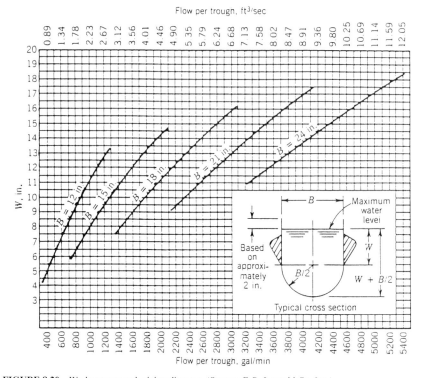

FIGURE 8.20 Washwater trough-sizing diagram. *(Source: F. B. Leopold Co. Inc.)*

Standards that limit the discharge of waste washwater to surface water sources have also been established by the USEPA. Waste discharges to a waterway usually require the implementation of a federal National Pollution Discharge Elimination System (NPDES) permit. If a sanitary sewer is available, local sewer use ordinances (based on wastewater treatment plant discharge limits set by an NPDES permit) usually require pretreatment of the waste washwater. State or local regulations may also require recycling of waste washwater within the water treatment plant. State regulatory officials should be consulted during the planning phase to determine discharge or treatment requirements for waste washwater.

Standards Recommended by Others

The American Water Works Association (AWWA) has published various manuals and handbooks that provide useful and up-to-date information for planning and designing filtration facilities. In particular, the AWWA Research Foundation has published many research reports and design manuals related to filtration, generally specific to a particular type of filtration or new advances in filtration technology.

The AWWA standards for filtering materials are a particularly useful tool because they establish minimum quality standards for the purchase and installation of filter media of various types.

NSF International has established a certification program for all direct and indirect drinking water additives. Standard 60 relates to drinking water treatment chemicals, such as coagulants and coagulant aids. Standard 61 relates to drinking water system components to ensure that these additives are compatible with drinking water systems, including filter media and coatings and linings for valves and piping systems.

Standards such as building, fire, and plumbing codes and other local ordinances must be considered in the design of filters and their housing. However, most codes and ordinances do not specifically mention water treatment plants, so local building code and fire department officials should be consulted during the planning phase to obtain clarification of their specific requirements. Occupational Safety and Health Administration (OSHA) standards for worker safety must also be considered.

FILTER OPERATION AND MAINTENANCE

When designing filtration equipment, the engineer must keep in mind how the facilities will be operated and maintained, including common filter behavior and filter operating problems.

Common Filter Behavior

In between filter backwashes, filters "ripen" improving their ability to remove flocculated particulate matter from the water, until turbidity breakthrough, terminal head loss, or the next scheduled filter wash occurs. Filters remove floc and larger particles by straining at the top surface of the filter media and smaller particles by the mechanisms of transport (i.e., interception, sedimentation, and diffusion) and attachment to the deeper filter media. In general, increasing the filtration rate tends to decrease the time to reach terminal head loss or breakthrough. Increasing the E.S. of the filter media tends to decrease the time to reach breakthrough but increases the time to reach terminal head loss. Montgomery (1985) includes a thorough discussion of filtration theory and mathematical models for particle removal and hydraulics.

Filter Ripening. Clean filter media require a period of conditioning before particle attachment mechanisms take effect. The mechanisms of particle transport apply during and after filter ripening, but the mechanisms of particle attachment require attraction to the filter media of opposite charges or coagulation with charge-neutralized particles previously attached to the filter media. Consequently, inadequate particle and turbidity removal may occur for a period of time after a filter is put into service following backwash. This time period typically lasts from 5 to 30 min. Amirtharajah (1980) thoroughly explores and describes the filter ripening process. Moran et al. (1993) found that for smaller particles, ripening continues well into the filter run, whereas for larger particles, ripening occurs early in the filter run, but removal efficiency decreases afterward.

Methods successfully used to minimize the problem of filter ripening include a short period of wasting the first portion of filtered water (i.e., filter-to-waste), a slow increasing of the filtered water flow rate after a backwash, and adding a filter aid to the washwater supply.

Filter Efficiency The computation of filter efficiency relates the effective filtration rate to the operating filtration rate in the following equation:

$$\frac{Re}{Ro} = \frac{\text{UFRV} - \text{UBWV}}{\text{UFRV}}$$

where Re = effective filtration rate
Ro = operating filtration rate
UFRV = unit filter run volume
UBWV = unit backwash volume

As an example, a filter operating at 40 gpm/ft^2 (98 m/h) for 48 hours and requiring 300 gal/ft^2 (12,200 L/m^2) for an adequate backwash would have an effective filtration rate of 3.9 gpm/ft^2 (9.5 m/h) and an efficiency of 97%.

Net Filter Production. A filter operating at 4.0 gpm/ft^2 (10 m/h) for 48 hours and requiring 300 gal/ft^2 (105 L/m^2) for an adequate backwash would have a net filter production of 4 gpm/ft^2 × 60 min × 48 hours − 300 gal/ft^2 = 11,220 gal/ft^2 (3,946 L/m^2). Properly designed and operated filters should exhibit net filter production volumes ranging from 7,500 to 12,500 gal/ft^2 (305,500 to 509,200 L/m^2). A net filter production volume of 5,000 gal/ft^2 (203,700 L/m^2) or less could indicate inadequate pretreatment, filter clogging algae in the influent water, excessive fines or mudballs in the filter media, mineral precipitates in the underdrains, air binding, or hydraulic restrictions causing inadequate head between filters and clearwell.

Common Filter Problems

Common filter problems include inadequate pretreatment or filter washing, gravel bed upset, air binding, restart after shutdown, and filter media replacement.

Inadequate Pretreatment. The pretreatment process (i.e., coagulation, flocculation, and clarification) in a conventional plant generally should produce pretreated waters with turbidities no greater than about 4 ntu. Pretreated waters with turbidities much greater than 4 ntu are indicative of floc carryover that tends to cause short filter run lengths. Conversely, pretreated waters with turbidities of 1 ntu or less in a conventional plant may result in inefficient filter operation or inadequate particulate removal.

In a direct filtration plant, pretreated water turbidity typically exceeds source water turbidity. The flocculation process should be operated to minimize floc size and allow penetration of the floc deep within the filter media, effectively using its entire depth.

Jar tests or on-line pilot filters, zeta potential, or streaming current instrumentation can greatly assist the operator in optimizing the coagulation process and ensuring adequate pretreatment for effective filtration.

Inadequate Filter Washing. Inadequate filter washing can result in poor filtered water quality and mudball formation. Cracks can occur in filter media when compressible solids remaining from previous filter runs pull filter media together and away from the filter box wall. Pretreated water can then travel through the cracks and bypass much of the filter media. Mudballs result from residuals remaining from previous filter runs sticking to filter media and forming agglomerations that grow too large to reach the washwater collection troughs during washing. As they grow heavier, mudballs can sink to create impassable regions within the filter media, typically at the anthracite-sand or sand-gravel interface. The impassable regions result in higher effective filtration rates, poorer filtered water quality, and shorter filter runs.

Air scour and surface wash systems can prevent the formation of cracks and mudballs, but previously formed mudballs may have to be removed manually or by soaking filter media with acidified water. Some plants include waste washwater turbidimeters as an operational tool to monitor the waste washwater quality and to minimize the volume of washwater usage.

Support Gravel Upset. Filter media support gravel upset may occur from operational errors such as washing a dry or drained filter. The initial rate for filling the filter should not exceed 5 gpm/ft^2 (12.2 m/h). Otherwise, rising water can compress air within the filter media pores to short-circuit the washwater, channel through the filter media, and disturb the support gravel.

Opening the washwater rate control valve too fast can cause gravel upset. Also, trapped air in the washwater header piping or in the underdrain system may be released in an uncontrolled manner creating a visible boiling action at the top of the fluidized media. If the support gravel is disturbed, it requires manual regradation after the overlying filter media are removed.

Improper specifications for the gradation and thickness of gravel layers can also create support gravel upset. Refer to *AWWA Standard B100—Filtering Material* and consult with filter media and filter underdrain suppliers for guidance on support gravel for various filter media and underdrain combinations. Design errors relating to the washwater supply or rate controller, improper programming of an automatic filter wash sequence, and improper installation of the gravel layers can also lead to support gravel upset.

Air Binding. Filter influent waters, particularly from surface water sources, typically contain significant concentrations of dissolved gases. Depending on water temperature, the dissolved gas concentration may reach saturation point. Surface water supplies typically reach saturation point during algae blooms, during seasonal changes when temperatures increase, or where there is significant cascading and aeration of source water.

When head loss exceeds the available head at some elevation within the filter media, pressure falls below atmospheric, and air escapes from solution. Air binding occurs when the accumulation of air bubbles blocks the water's path. An excessive effective filtration rate and significantly increased head loss result. Air binding most often occurs a few inches into a single-medium filter or just below the anthracite-sand or GAC-sand interface in a dual-media filter. Air escapes from the media upon closure of the filtered water rate control valve, and the resulting agitation can cause loss of media. By design, air binding does not occur in pressure, declining-rate, or self-backwashing filters.

Recommended Standards for Water Works (1992) recommends a minimum water depth of 3 ft (1 m) above filter media in gravity plants. Modern plants that operate at relatively high filtration rates typically require a water depth of 5 feet (1.5 m) or more to prevent air binding within the media. The designer should consider the rate of head loss development along with the desired filter run length to develop design criteria for the filter media, filter box depth, and freeboard.

Restart After Shutdown. Water treatment plants sometimes operate with individual filters removed from service during low-demand periods. Additionally, some plants shut down at night or cycle the filters on and off based on distribution storage tank levels. Some plants place individual filters or entire plants on standby for periods of time and even drain the filters when they have sufficient redundancy or more than one plant. Preferably, all filters should be operated continuously at a reduced filtration rate. Continuous operation prevents formation of undesirable biological activity caused by water standing in the filter box and avoids loss of ripened filter media in a drained filter. In any case, inactive filters require special measures for restart.

Filters that are shut down with standing water require a brief washing or period of filter-to-waste to ensure adequate filter performance before filtered water flows to the clearwell. The restart wash or filter-to-waste period should be programmed into the centralized control system for the plant.

Drained filters require manual restart with close supervision, including gradual refilling with washwater to prevent gravel upset or media separation, a complete backwash, and a longer than normal period of filter-to-waste to ensure adequate performance before opening the filtered water control valve.

Filter Media Replacement. Rapid sand filters with dual and mixed media can lose 5% to 7% of the media per year from air binding, excessive washing, air scour, or surface wash. The loss can be particularly serious when there is cold water, mismatched dual or mixed media, low washwater collection troughs, and leakage through the support gravel or underdrains. Filter performance and the effectiveness of surface wash decline when media loss exceeds about 20% of the original depth (Kawamura, 1991).

Lost media should be replaced with appropriately specified material following the guidelines presented earlier to ensure that the entire bed of media approximates original specifications. Replacement media for dual-and mixed-media filters requires special attention. If a layer of fines has accumulated on the surface of the filter, it may be wise to scrape off a thin layer of top media before adding new media. After replacing the new media on top of the filter, the operator should initiate two complete filter wash cycles to fluidize the media, wash off fines, and ensure restratification.

BIBLIOGRAPHY

Amirtharajah, A. "Initial Degradation of Effluent Quality during Filtration." *Journal AWWA* 72(9):518, 1980.

Amirtharajah, A. "Design of Granular-Media Filter Units." In *Water Treatment Plant Design.* Ann Arbor, Mich.: Ann Arbor Science, 1982.

AWWA/ANSI. *Standard for Filtering Material—B100.* Denver Colo.: American Water Works Association, 1996.

AWWA/ANSI. *Standard for Granular Activated Carbon B604.* Denver Colo.: American Water Works Association, 1994.

Baker, M. N. *The Quest for Pure Water.* Lancaster, Pa.: Lancaster Press, 1948.

Begin, E. E., and R. D. G. Monk. "Improvements to New Bedford's Water Supply." *Journal of the New England Water Works Association* 89(2):145, 1975.

Blanck, C. A., and D. J. Sulick. "Activated Carbon Fights Bad Taste." *Water and Wastes Engineering* 12(9):71, 1975.

Brater, E., and H. King. *Handbook of Hydraulics.* 6th ed. New York: McGraw-Hill, 1976.

Brosman, D. R., and J. F. Malina, Jr. *Intermixing of Dual Media Filters and Effects on Performance.* Austin, Tex.: University of Texas Center for Research in Water Resources, 1972.

Camp, T. R. "Discussion: Experience with Anthracite-Sand Filters." *Journal AWWA* 53(12):1478, 1961.

Camp, T. R. "Theory of Water Filtration." *Journal Sanitary Engineering Division, American Society of Civil Engineers* 90(8):1, 1964.

Camp, T. R., S. D. Graber, and G. F. Conklin. "Backwashing of Granular Water Filters." *Journal Sanitary Engineering Division, American Society of Civil Engineers* 97(12):903, 1971.

Clark, J., W. Viesman, Jr., and M. Hammer. *Water Supply and Pollution Control.* 3rd ed. New York: EP-DUN-Donnelly, 1977.

Cleasby, J. L. "Filtration." In *Physicochemical Processes for Water Quality Control.* Edited by W. J. Weber, Jr. New York: Wiley-Interscience, 1972.

Cleasby, J. L., J. Arboleda, D. E. Burns, P. W. Prendiville, and E. S. Savage. "Backwashing of Granular Filters."*Journal AWWA* 69(2):115, 1977.

Cleasby, J. L., and E. Baumann. *Wastewater Filtration: Design Considerations.* EPA 625/4-74-007. Washington D.C.: U.S. Environmental Protection Agency Office of Technology Transfer, 1974.

Cleasby, J. L., and G. D. Sejkora. "Effect of Media Intermixing on Dual Media Filtration. *Journal Environmental Engineering Division, American Society of Civil Engineers,* 101(8):503, 1975.

Cleasby, J. L., E. W. Stangl, and G. A. Rice. "Developments in Backwashing of Granular Filters." *Journal Environmental Engineering Division, American Society of Civil Engineers* 101(10):713, 1975.

Cleasby, J. L., M. W. Williamson, and E. R. Baumann. "Effect of Filtration Rate Changes on Quality." *Journal AWWA* 55(7):869, 1963.

Cleasby, J. L., and C. F. Woods. "Intermixing of Dual Media and Multimedia Granular Filters." *Journal AWWA* 67(4):197, 1975.

"Committee Report. State of the Art of Water Filtration." *Journal AWWA* 64(10):662, 1972.

Conley, W. R. "Experience with Anthracite-Sand Filters." *Journal AWWA* 53(12):1473, 1961.

Conley, W. R. "Integration of the Clarification Process." *Journal AWWA* 57(10):1333, 1965.

Conley, W. R. "High Rate Filtration." *Journal AWWA,* 64(3):205, 1972.

Conley, W. R., and R. W. Pitman. "Test Program for Filter Evaluation at Hanford." *Journal AWWA* 52(2):205, 1960.

Craft, T. F. "Review of Rapid Sand Filtration Theory." *Journal AWWA* 58(4):428, 1966.

Craft, T. F. "Comparison of Sand and Anthracite for Rapid Filtration." *Journal AWWA* 63(1):10, 1971.

Culp, R. L. "Direct Filtration." *Journal AWWA* 69(7):375, 1977.

Culp, G. L., and R. L. Culp. *New Concepts in Water Purification.* New York: Van Nostrand Reinhold, 1974.

Dostal, K. A., and G. G. Robeck. "Studies of Modifications in Treatment of Lake Erie Water." *Journal AWWA* 58(11):1489, 1966.

Fair, G., J. Geyer, and D. Okun. Water and Wastewater Engineering. Vol. 2. New York: John Wiley and Sons, 1968.

French, J. A. "Flow Approaching Filter Washwater Troughs." *Journal Environmental Engineering Division, American Society of Civil Engineers* 107(2):359, 1981.

Great Lakes–Upper Mississippi River Board of State Sanitary Engineers. *Recommended Standards for Water Works.* Albany, N.Y.: Health Education Service, 1992.

Hager, D. G. "Adsorption and Filtration with Granular Activated Carbon." *Water and Wastes Engineering* 6(8):39, 1969.

Hamann, C. L., and R. E. McKinney. "Upflow Filtration Process." *Journal AWWA* 60(9):1023, 1968.

Haney, B. J., and S. E. Steimle. "Upflow Filter for Potable Water Production." *Journal Environmental Engineering Division, American Society of Civil Engineers* 101(8):489, 1975.

Hansen, R. E. "Granular Carbon Filters for Taste and Odor Control." *Journal AWWA* 64(3):176, 1972.

Harris, W. L. "High Rate Filter Efficiency." *Journal AWWA* 62(8):515, 1970.

Hsiung, A. K., W. R. Conley, and S. P. Hansen. "The Effect of Media Selection on Filtration Performance." Paper presented at Spring Meeting of Hawaii Section, AWWA, April 1976.

Hutchison, W. R. "High Rate Direct Filtration." *Journal AWWA* 68(6):292, 1976.

Johnson, R. L., and J. L. Cleasby. "Effect of Backwash on Filter Effluent Quality." *Journal Sanitary Engineering Division, American Society of Civil Engineers* 92(2):215, 1966.

Jung, H., and E. S. Savage. "Deep Bed Filtration." *Journal AWWA* 66(2):73, 1974.

Kawamura, S. "Design and Operation of High-Rate Filtration. Part 1." *Journal AWWA* 67(10):535, 1975.

Kawamura, S. *Integrated Design of Water Treatment Facilities.* New York: John Wiley and Sons, 1991.

Kirchman, W. B., and W. H. Jones. "High Rate Filtration." *Journal AWWA* 64(3):157, 1972.

Laughlin, J. E., and T. E. Duvall. "Simultaneous Plant-Scale Tests of Mixed Media and Rapid Sand Filters." *Journal AWWA* 60(9):1015, 1968.

Letterman, R. D. "What Turbidity Measurement Can Tell Us." *Opflow AWWA,* 20(8), 1994.

Lewis, C. M., E. E. Hargesheimer, and C. M. Ventsch. "Selecting Particle Counters for Process Monitoring." *Journal AWWA* 84(12):46, 1992.

McCreary, J. J., and V. L. Snoeyink. "Granular Activated Carbon in Water Treatment." *Journal AWWA* 69(8):437, 1977.

Monk, R. D. G. "Improved Methods of Designing Filter Boxes." *Journal AWWA* 76(8):54, 1984.

Monk, R. D. G. "Design Options for Water Filtration." *Journal AWWA* 79(9):93, 1987.

Monk, R. D. G., and A. P. Gagnon. "A New Method of Filter Monitoring." *Public Works* 116(10):68, 1985.

J. M. Montgomery Consulting Engineers. *Water Treatment Principles and Design.* New York: John Wiley and Sons, 1985.

Moran, D. C., M. C. Moran, R. S. Cushing, and D. F. Lawler. "Particle Behavior in Deep-Bed Filtration. Part l: Ripening and Breakthrough." *Journal AWWA* 85(12):69, 1993.

Oeben, R. W., H. P. Haines, and K. J. Ives. "Comparison of Normal and Reverse Graded Filtration." *Journal AWWA* 60(4):429, 1968.

O'Melia, C. R., and D. K. Crapps. "Some Chemical Aspects of Rapid Sand Filtration." *Journal AWWA* 56(10):1326, 1964.

O'Melia, C. R., and W. Stumm. "Theory of Water Filtration." *Journal AWWA* 59(11):1393, 1967.

Rice, A. H., and W. R. Conley. United States Patent 3,343,680. 1967.

Rimer, A. E. "Filtration through a Trimedia Filter." *Journal Sanitary Engineering Division, American Society of Civil Engineers* 94(6):521, 1968.

Robeck, G. G., K. A. Dostal, and R. L. Woodward. "Studies of Modifications in Water Filtration." *Journal AWWA* 56(2):198, 1964.

Tuepker, J. L., and C. A. Buescher, Jr. "Operation and Maintenance of Rapid Sand and Mixed Media Filters in a Lime Softening Plant." *Journal AWWA* 60(12):1377, 1968.

Westerhoff, G. P. "Experience with Higher Filtration Rates." *Journal AWWA* 63(6):376, 1971.

CHAPTER 9
SLOW SAND AND DIATOMACEOUS EARTH FILTRATION

Although rapid sand filters predominate, the two types of filters covered in this chapter may be effective in many applications where source water quality permits. In general, the combined costs of constructing and operating slow sand and diatomaceous earth (DE) filters may be considerably less than the cost of rapid sand filtration plants for the same capacity.

The principal mechanisms for separation of solids in all porous media filters are attachment and straining or entrapment. Because of the relationship of the somewhat large pores in rapid sand filter media compared with particulates, the mechanism for separation is attachment. In the case of slow sand and DE filters, however, the pore-particulate size relationship results in more substantial separation by entrapment.

During the initial operation period of slow sand filters, the separation of organic matter and other solids generates a layer of biological matter on the surface of the filter media. Once established, this layer is the predominant filtering mechanism. Solids are removed from water by a DE filter as the source water flows through a precoat layer of powderlike DE.

For all practical purposes, most solids are separated at the surface of the media in both actions. Because of the small pore size of the media, particulate separation is ideal for removing the cysts of *Giardia* and *Cryptosporidium*. In most situations, neither type of filter requires previous conditioning of the raw water.

In general, application of slow sand and DE filtration should be limited to source waters with turbidity levels less than 5 ntu. Where particulates are dominant (rather than organic matter) both types of filters may be used with water of up to 10 ntu turbidity. If either type of filter is used on water with higher turbidity, filtrate quality will generally be acceptable, but the more rapid buildup of solids on the filter results in rapid loss of head and shortened length of filter runs.

SLOW SAND FILTRATION

Slow sand filtration was the first type of porous media filtration used in water treatment. The first recorded installations occurred in Scotland and England in the early 1800s. By the mid-1800s, legislation was passed in London, England, requiring filtration of water to be consumed. The first recorded installation of slow sand filtration facilities in the United States was in Poughkeepsie, New York, in 1872. Subsequent development of rapid-rate filtration then slowed the pace of construction of slow sand plants in the United States in the early 1900s.

Around 1980, interest in using slow sand filtration was rekindled as the U.S. Environmental Protection Agency (USEPA) conducted research to develop treatment options that are simple to operate for use by small communities and that produce high-quality effluent. Research reconfirmed that, at recommended filter rates and with appropriate media and source water quality, slow sand filtration can produce a low turbidity effluent and can effectively remove microbiological contaminants.

When the USEPA passed the Surface Water Treatment Rule (SWTR) in 1989, further pressure was placed on communities that were not filtering surface water supplies to add filtration. Slow sand filters were rated along with rapid sand and DE filters as baseline treatment in the regulations. As a result, slow sand filtration has once again become a treatment method routinely considered in evaluating filtration options in many U.S. communities. Abroad, it is used to provide safe drinking water to many poor or rural communities. A major portion of the city of London water supply is treated by slow sand filtration.

Renewed interest in the slow sand process has generated new research into improving treatment performance. The focus has been on expanding use of slow sand filtration in treating poorer-quality source water, especially with higher turbidity and organic content.

Slow Sand Treatment Mechanism

As its name implies, slow sand filtration is accomplished by passing water at a relatively low rate through a sand media. The filtration rate is on the order of 1/100th of the rate used in a typical rapid sand filter.

Because of the relatively low filter rate, head loss across the bed occurs gradually over a much longer period of time. Average filter run length is normally between 45 and 60 days. In some newer installations, filter run lengths in excess of six months and even greater than one year have been reported.

Slow sand filtration accomplishes its treatment primarily through biological activity, with the bulk of this activity taking place on the surface of the sand bed. A layer develops on the sand surface that is called "schmutzdecke," an accumulation of organic and inorganic debris and particulate matter in which biological activity is stimulated. It has been found that some biological activity also extends deeper into the bed, where particulate removal is accomplished by bioadsorption and attachment to the sand grains.

Source Water Quality Considerations

In considering whether slow sand filtration is an appropriate treatment method, source water quality must be carefully evaluated. If source water quality data are not available, pilot testing of the source water is essential to determine the applicability of the slow sand treatment option. Table 9.1 lists source water quality parameters with recommended limits (Collins and Spencer, 1991).

Turbidity. Both the level and type of turbidity in source water must be considered. In general, most existing slow sand plants successfully treat source water turbidity of less than 10 ntu (Slezak and Simms, 1984), which is recommended for an upper limit in designing new facilities. Also of some importance is the stability of the water. Slow sand facilities operate more efficiently if source turbidity is relatively constant and generally ≤5.0 ntu.

TABLE 9.1 Source Water Quality Limitations of Slow Sand Filters

Parameter	Recommended limit
Turbidity	5 to 10 ntu*
Algae	200,000/L†
True color	15 to 25 Platinum color units
Trihalomethanes	50 μg/L
Nonpurgeable dissolved organic carbon	2.5 mg/L
UV absorbance‡	0.080 cm^{-1}
Dissolved oxygen	>6 mg/L
Phosphorous (PO$_4$)	30 μg/L
Ammonia	3 mg/L
Iron	<1 mg/L
Manganese	<1 mg/L

*Presence of clay particles must be determined.
†Dependent on identification of algal species and assumes covered filter.
‡Absorbance at wavelength of 254 nanometers.

Of equal importance is the nature of particulates. Source waters that normally contain clay particulates or that pick up clay after storm events will cause problems for slow sand filters. This difficulty for slow sand filters occurs because clay penetrates deep into the bed or may even carry through the filter, causing an immediate problem of elevated filtered water turbidity and a long-term problem of filter clogging and reduced length of filter runs.

Algae. In a few instances, it has been found that the presence of certain types of algae actually enhances the filtration process by providing greater surface area for biological activity. In general, however, the presence of algae in the source water reduces filter run lengths. Table 9.2 presents a list of commonly found algal species divided into categories related to their effect on filter performance (Collins and Spencer, 1991). Filter-clogging species are detrimental to filter performance, while filamentous species may actually enhance filter performance by providing greater surface area. Floating species would not result in direct clogging of the filter, but may shorten run lengths based on poorer-quality raw water.

Algae may be present in source water delivered to the filter and may also occur in an uncovered filter bed open to sunlight. In general, it is prudent to reduce algal content in source water to as low a level as possible to limit its effect on filter performance. Observation of algal growths, as well as identification, will aid with assessing the need for pretreatment, such as copper sulfate, and in determining when filter run lengths may be shortened. Some researchers have suggested the measurement of chlorophyll at concentrations of 5 mg/m^3 as a limit in source water (Cleasby et al., 1984).

Color. Color in treated water is currently categorized by USEPA as a secondary contaminant in drinking water supplies, with the focus being aesthetic concerns. As identified by Christman and Oglesby (1971) the yellow to brown color of many source waters can be the result of microbial breakdown of lignins from woody plants. True color removals of 25% or less were reported by Cleasby et al. (1984). Other research has indicated a removal range between 15% and 20% for total organic carbon (Fox et al., 1984; Collins et al., 1989).

TABLE 9.2 Algal Species Classification for Slow Sand Filtration

Filter clogging	Filamentous	Floating
Tabellaria	*Hydrodictyon*	*Protococcus*
Asterionella	*Oscillaria*	*Scenedesmus*
Stephanodiscus	*Cladophora*	*Synura*
Synedra	*Aphanizomenon*	*Anabaena*
	Melosira	*Euglena*

When evaluating the applicability of slow sand filtration for a specific source water, a review of historical trihalomethane (THM) data can reveal whether the expected low removal efficiency of aquatic organic substances by the process is a concern. Where historical color and THM data are unavailable, a sampling program can be initiated to aid in evaluating whether slow sand filtration is an appropriate treatment method.

Iron and Manganese. Slow sand filters remove iron and manganese through precipitation on the sand surface in a scalinglike action, but an upper limit of 1 mg/L of iron is suggested to avoid forming an iron precipitate that could clog filters. A similar limit for manganese would also appear to be acceptable. Collins et al. (1989) showed that iron precipitate on a slow sand filter enhanced the removal of organic precursors.

Dissolved Oxygen. The presence of dissolved oxygen in source water is critical for stimulating a healthy schmutzdecke for proper slow sand filter operation. Some slow sand plants use aeration of the water as a pretreatment. Reduction of dissolved oxygen levels commonly occurs following algal blooms so that the importance of dissolved oxygen in the source water is another reason to control algal growth in the source. Potential problems resulting from dissolved oxygen deficiencies include tastes and odors, redissolving of precipitated metals, aesthetics, and increased chlorine demand (Ellis, 1985).

Nutrients. The proper operation of the schmutzdecke is somewhat dependent on the presence of sufficient concentrations of carbon, nitrogen, phosphorous, and sulfur. Carbon and sulfur (in sulfate form) are prevalent in most source waters. However, protected reservoir systems may have limited concentrations of nitrogen and phosphorous present.

It has been reported that, for every 1 mg of carbon removed by the schmutzdecke, 0.04 mg of nitrogen and 6 μg of phosphorous are required (Skeat, 1961). Slow sand filters have also shown the ability to remove up to 3 mg/L of ammonia from source water under the right conditions. Ammonia can be used as a source of nitrogen for the filter.

Effluent Water Quality

Slow sand filtration has been shown to be effective in achieving removal of *Giardia* and viruses. Effluent turbidities in the range of 0.1 to 0.2 ntu are typical for high-quality source waters. Removal of organic substances is generally in the range of 15% to 25%. Recent research has focused on improving removal because of disinfection byproduct formation considerations.

Typical treatment performance of conventional slow sand filtration plants is listed in Table 9.3. Limited data are available on removal capabilities with respect to *Cryptosporidium,* but more research is being initiated.

TABLE 9.3 Typical Treatment Performance of Conventional Slow Sand Filters

Water quality parameter	Treatment performance or reduction capacity
Turbidity	<1.0 ntu
Coliforms	1 to 3 log units
Enteric viruses	2 to 4 log units
Giardia cysts	2 to 4+ log units
Total organic carbon	<15% to 25%
Biodegradable dissolved organic carbon	<50%
Trihalomethane precursors	<25%

Design of Slow Sand Filters

The slow sand filter is relatively simple in arrangement, having only three basic elements in addition to a control system. Typical of any filter design, the complete train includes clearwell storage, disinfection, and posttreatment. Figure 9.1 presents a cross-sectional view of a typical filter bed.

Filter Box Design. The filter box contains all the filtering components of the system. These include source water storage (above the sand bed), filter sand, underdrain system, and, in some cases, facilities for collecting wastewater generated during the cleaning process. The box floor and sides are generally constructed of concrete. Roof designs for covered filters vary and may include wood truss, steel, precast concrete, or cast-in-place concrete.

If the filter unit is to be covered, the height of the box must be adequate to provide for the depth of sand and support media, underdrain system, source water storage above the media, and headroom for cleaning and resanding operations. The filter box area is determined by the unit rate of flow and required supply flow.

Slow sand filters may also be uncovered. There are currently many operating facilities that are uncovered in the U.S. Pacific Northwest, Europe, and South America. Besides lower initial cost, an advantage of uncovered filters is the far greater ease of using mechanical equipment for cleaning and maintenance. If filters are to be covered, major considerations include providing headroom for equipment during cleaning and repair, lighting, and ventilation.

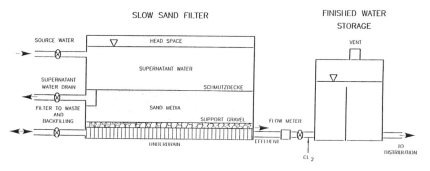

FIGURE 9.1 Typical covered slow sand filter installation.

A 2 to 3 ft (0.6 to 0.9 m) freeboard depth should be added over the normal water surface to provide for fluctuations in water depth within the filters without reaching ceiling height. This also provides room to install permanent lighting to improve the efficiency of cleaning and resanding. Generous headroom also allows the use of larger mechanical equipment within the filters, which can significantly reduce the time required for cleaning or resanding operations, particularly in large filters.

Filter Sand. Guidelines for filter sand characteristics and proper media depth vary between those of the International Research Center (IRC) and the *Recommended Standards for Water Works* (commonly known as the 10-State Standards). The IRC manual recommends sand with an effective size of 0.15 to 0.30 mm with a uniformity coefficient between 3 and 5. The 10-State Standards recommend an effective size of 0.30 to 0.45 mm and a uniformity coefficient of ≤ 2.5.

A finer effective size may improve particulate removals but generally results in shorter filter run lengths. Media that are too large allow deeper bed penetration and may even result in filter breakthrough or clogging. A deeper penetration of particles in the filter bed also means that more sand must be removed during a scraping cycle. It has been suggested that a better approach is to increase the depth of the sand rather than to reduce media size if a more conservative design is desired.

Pilot testing of the process using different media sizes provides data on removals and filter run lengths and can serve as a basis for media selection.

Sand depth should generally be between 18 and 35 in. (460 and 890 mm), but some plant operators have reported satisfactory treatment with sand depth as low as 12 in. (300 mm). Most slow sand plants in the United States are designed with a minimum sand depth of 30 in. (760 mm).

If filters are cleaned by manual scraping, about $\frac{1}{2}$ in. (1 cm) of sand is removed during the scraping. Final sand depth should be determined based on cleaning method, anticipated filter run lengths, number of scrapings desired before resanding, sand availability and expense, and impact of downtimes on plant capacity. The minimum depth before resanding should be 18 in. (460 mm).

Underdrain and Support System. One common type of underdrain consists of a manifold and perforated laterals installed below the sand bed. Most new designs use a plastic piping system for filter underdrains. Piping material must be certified for contact with potable water. Typical lateral sizes range from 4 to 8 in. (100 to 200 mm) with the underdrain system header in the range of 8 to 16 in. (200 to 400 mm). Figure 9.2 is a view of an installation in progress of a perforated PVC underdrain system.

The underdrain system must be designed to cause minimal head loss within the system. Head losses through the individual perforations of the laterals must be a fraction of head loss through the lateral itself to provide a balanced flow across the system. The design engineer should refer to hydraulic textbooks for guidance with respect to piping manifold system designs.

Other underdrain systems use prefabricated plastic or clay filter blocks or a false floor of concrete blocks or brick with gravel media above. Because of the large area of a slow sand filter, the prefabricated type is normally expensive to install and is used infrequently. The hydraulics of a false floor system must be similar to that of the piped system.

Gravel support media usually consist of multiple layers of graded gravel. The gravel layers are coarsest on the bottom and become finer with each layer. Gravel supports the sand, and the fine layer prevents sand from migrating down to clog underdrain openings. The 10-State Standards recommend gravel support layers similar to those required for a rapid sand installation with a media depth between 18 and 24 in. (460 and 610 mm) and gravel sizes in a range from $\frac{3}{32}$ to $2\frac{1}{2}$ in. (2 to 64 mm) in a five-layer system.

FIGURE 9.2 Underdrain installation.

Source Water Storage. Source water storage is the depth and volume of water overlying the sand surface within the basin. This volume varies in different designs (between 3 and 24 hours of plant capacity). As storage capacity increases, there are some benefits with respect to equalizing source water quality, sedimentation of larger particles, and even biological action within the water column itself.

The primary purpose of the water level above the sand, however, is to provide the driving head across the filter bed. A typical terminal head loss for a slow sand filter is in the range of 4 to 5 ft (1.2 to 1.5 m). Therefore typical depths of water above the sand should range between 6 and 7 ft (1.8 and 2.1 m) to provide for the additional driving force required for head loss through the clean sand bed and through the piping systems. If a filter is to be covered, the height of the filter box above the sand is governed primarily by space requirements for cleaning and resanding, so provision of a 6 to 7 ft (1.8 to 2.1 m) depth of water can easily be accomplished.

Access for Cleaning and Maintenance. Early sand filters were constructed either with no covers (roofs) or with earthen embankments over cast-in-place concrete roofing systems. Filters constructed with a roof typically used access hatches or ports spaced at intervals above the bed surface. Cleaning the beds was cumbersome and normally accomplished either by manually hoisting the scraped schmutzdecke up through the access way or by hydraulic transfer of a slurry of removed material through a piping system.

New covered filter installations typically use some sort of structure with an access ramp into the basins or "ship's doors" to allow direct access into the filter box at the elevation of the sand surface. Installation of access ramps is generally more costly (when installed with a covered filter) but may result in reduced maintenance when compared with a ship's door. An access ramp and filter are depicted in Figure 9.3.

A typical ramp is constructed of concrete (or earth for uncovered installations). Ramp widths vary, but consideration should be given to the width of equipment expected to be used in the box. The slope should be 1 ft (0.3 m) vertical to 5 ft (1.5 m) horizontal or shallower. A steeper slope could result in equipment slipping on the ramp surface when sand or other debris is tracked on it.

FIGURE 9.3 Access ramp and filter.

A cross section of a typical plant using a filter entrance structure is shown in Figure 9.4. Entrance doors at the top of the ramp must be corrosion resistant because of the prevalent high humidity from the water within the filters.

Ship's doors are constructed of steel or aluminum. Steel is stronger but generally requires more maintenance, such as painting, and corrodes faster than aluminum. Design of aluminum doors must consider anticipated loads carefully. Hinge lubrication is critical with either material to ensure smooth operation.

Lighting. Older covered slow sand installations relied on portable or minimal lighting when working within the filters or used natural light through skylights or from open hatches on the roof. If funds are available, a permanent lighting system increases visibility and eliminates hazards associated with portable lighting.

Because of high-humidity conditions, the lighting design must incorporate corrosion-resistant materials such as aluminum and stainless steel. Lighting units must include integral gaskets and methods of sealing wire penetrations to minimize moisture within the units. Conduit systems must use corrosion-resistant materials.

Lighting units must be mounted in a manner that minimizes damage from high water levels or from operating cleaning equipment. Lighting levels on the order of 10 foot-candles (100 lumens/m^2) provide sufficient light to allow efficient and safe operations.

Ventilation. Little consideration was given to ventilation in older covered slow sand filters. Usually ventilation consisted of open hatches with portable ventilators installed during maintenance operations. Although portable units increase ventilation, the actual rates are generally below those required by code when engine-driven equipment is being operated inside the filter unit.

For newer installations, ventilation rates for motorized equipment must be designed to protect operators and should be similar to those provided in parking garages. Rates must be set according to expected emissions of motorized equipment operated within the basin. The number of air changes required is generally governed by state or federal regulations.

Filter Roofing System. Selecting a filter roofing system must consider capital expense, long-term reliability, headroom, availability of material, and site conditions. Cast-in-place systems are generally the most expensive but are the most durable. Precast concrete slabs are less expensive than cast-in-place but have a slightly shorter life span. Systems con-

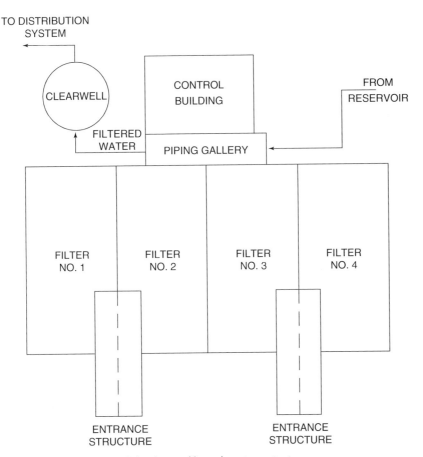

FIGURE 9.4 Typical slow sand plant layout with exterior entrance structures.

structed of wood or steel experience significant deterioration or corrosion problems and require replacement within a relatively short time. Obviously, the replacement expense versus initial capital expense must be evaluated in design.

Methods of covering the roof system include shingles, metal roofing, single-ply membrane, built-up roofing, and earthen fill. Again, each system has specific capital costs, and not all are usable with each roof support system. An earth cover system can help blend a facility into its surrounding but requires a strong support structure to carry the load.

Filter-to-Waste System. The design of a slow sand filter should provide the ability to discharge filtered water to waste. The filter-to-waste system allows the operator to check effluent quality after a filter cleaning and before bringing the filter back on line. It also provides a method to "cleanse" filter media after resanding or other major reconstruction work.

The filter-to-waste system must be designed to avoid a potential cross-connection. The ideal system includes an air gap on the filter effluent to the drain pipe. However, many regulators allow a hard piped connection to a drain system, with valving to direct water to finished water storage or to waste. In the latter case, it is normally required to install the filter-to-waste line at an elevation below the filtered water line and to provide an air gap at the ultimate discharge point to minimize potential contamination.

Filter Draining. At some point in the filter cycle, head losses increase to such a level that filter production is unacceptably low, and the filter must be taken off-line for cleaning. Because the rate is so low through the filter and the hydraulics between adjoining filters may not allow significant drop in filter level to occur, some form of separate drain should be provided.

This drain should generally be installed to discharge filter supernatant either back to the source or to a lagoon. The drain can also be connected to the filter-to-waste system through an air gap, or it may be installed as a separate system. Cross-connection to the filtered water system must be considered when designing this drain. After the supernatant is discharged, full drainage of the filter box can be accomplished through the filter and into the filter-to-waste system. Figure 9.5 shows a typical piping gallery for a slow sand facility.

Flow Control. Flow through a slow sand filter can be controlled on either the influent or effluent side. One method of influent control provides a constant level over the filter, and declining-rate filtration results. This means that, as the head loss increases across the bed, the filter rate is reduced. In the second method, the filtered water level is adjusted to increase as filter head loss increases to provide the same flow, or constant-rate filtration.

Effluent is controlled by either a control valve or by fluctuating finished water storage levels in response to head loss changes across the filter bed. The intent is to maintain a constant filter rate by adjustments in available head as head loss changes.

With either system, finished water storage must ensure that water levels are maintained at a minimum of 1 ft (0.3 m) above the top of the filter sand to avoid problems associated with negative pressures within the sand bed. Control systems should seek to maintain constant flow through the plant and minimize the surge of filter starts and stops that could affect effluent quality.

FIGURE 9.5 Piping gallery.

Storage of finished water after slow sand filtration serves two functions. First, it can provide a method of maintaining submergence of the filter media under all conditions. This limits problems that could develop from air binding within the sand bed. In addition, storage provides contact time for disinfection after filtration. The volume and form of storage vary according to system requirements.

Filter Rates. Typical filter rates are usually in the range of 0.04 to 0.10 gpm/ft^2 (0.09 to 0.24 m/h). Filter rates can be established through pilot testing. In some cases, it has been demonstrated that higher rates are possible while maintaining acceptable effluent quality and filter run lengths. However, based on concerns about microbiological contaminants, it is prudent to design filter rates conservatively to allow flexibility for future increased treatment requirements and fluctuating demands for water quantity and quality.

One design approach is to allow a moderate to low rate, even when one filter is off-line and the treatment system is under maximum water demand conditions. This allows some flexibility to meet future increased demands without capital expenditures and provides a buffer under poor source water quality conditions. This approach also increases initial capital expenditures because of larger filter area requirements.

Some cost savings may be gained by allowing on-line filters to operate at high rates when one unit is off-line, but research has shown that even though the schmutzdecke has been established, an increase in filter rate can have an adverse impact on treated water quality. The higher rate also results in reduced length of filter runs.

A compromise approach is to provide moderate flow rates when all units are in service. Higher rates result with one unit off-line, but only for short periods. This provides flexibility when dealing with water quality and demand issues and reduces capital expenditures.

Automation. Slow sand filtration plants can require much simpler automation than other filtration processes. But to some degree, the level of automation is driven by regulations and the client. For automation, the design must consider:

- State and federal regulations
- Initial cost and operating costs
- Level of complexity desired by client and operator
- Work shifts to be used
- Source water quality fluctuations
- Record-keeping needs and requirements
- Critical plant parameters and equipment

The recommended minimum location points for recording and monitoring include:

- Source water turbidity
- Source water flow
- Filter level
- Filtered water flow (individual)
- Filter head loss
- Filtered water turbidity (individual or combined)
- Chlorine residual (before and after storage)
- Finished water flow
- Finished water storage level

Other monitoring points may be specific to each site's requirements and should be discussed with the client.

Pilot Testing

The most advantageous way of determining the applicability of a treatment process to a specific source water and the expected performance that may be expected is through pilot studies. In fact, reliable prediction of treatment results is difficult without piloting. The results of studies provide background data to establish design parameters and estimate filter run lengths. Data can also be used as an introduction of the process for orienting operations staff.

Planning is critical to the successful implementation of the piloting program. The protocol must establish:

- Pilot plant design and construction
- Conditions for runs
- Duration of tests
- Run parameters
- Data to be collected
- Laboratory analyses required

Some suggested water quality parameters to be monitored on the source and filtered water include:

- Turbidity
- Color
- Temperature
- Algae
- Bacteria
- Dissolved oxygen
- Iron and manganese
- Particle counts
- Filter head loss
- Flow rate
- Filter run length

It is important to collect all samples in accordance with established procedures to minimize potential for contamination. Whenever possible, filter sand used for pilot studies should be the same sand that will be used in the full-scale plant. It has also been suggested that sand be installed in a "dirty" condition (similar to that delivered in construction) to determine the washing-out period during plant start-up. When tests are completed, pilot filters can be moved into the full-scale plant for future research.

Pilot plant construction varies. Previous designs used many types of materials, including fiberglass or concrete pipe and PVC pipe. The intent is to provide a watertight container with room for installing sand and support media, underdrain piping, sample ports, flow measurement equipment, and source water storage. Typical pilot plant setups are 8 to 12 ft (2.4 to 3.7 m) high. The diameter of the pilot should be as large as possible to approximate future flow rates and facilitate operation and cleaning.

New Developments

Improvements and innovations to expand slow sand filtration treatment capabilities are discussed in the following.

Roughing Filters. Roughing filters serve as a pretreatment method to reduce sediment loading to the sand filters. Many varieties are currently in use; in general they consist of gravel-type media with different gradations. The filters commonly are designed in stages and for either vertical upflow or downflow or horizontal flow.

Recent designs used gravel filter material that decreases in size with flow direction. The gravel size range is between 0.2 and 2 in. (5 and 50 mm) and flow velocities are in the range of 0.02 and 0.08 ft/min (0.3 to 1.5 m/h). Pardon (1991) showed that greater than 90% removal can be obtained for particles 10 μm and greater, and 72% removal of particles between 2 and 5 μm, through use of vertical roughing filters. Roughing filters are cleaned by flushing at high rates.

Other pretreatment methods for extremely turbid waters use single-stage gravel filters located next to or within the source (mostly rivers or canals). These filters operate similarly to roughing filters. Cleaning is normally accomplished manually because most of the removal is at the top surface of the filter.

Preozonation. Concern with disinfection by-products in finished water has increased the need to improve organic precursor removals through the treatment process. Research has been conducted on ozone as a preoxidant ahead of slow sand filtration as a means of improving organics removal.

In general, ozone use is greater in Europe than in the United States, and many plants in Europe have ozone preoxidation and GAC adsorption before slow sand filtration. Malley et al. (1991) reviewed past research and performed pilot studies to evaluate the effect on treatment performance of preozonation before slow sand filtration. Ozonation converts nonbiodegradable organic matter to biodegradable forms to enhance biological activation of the filter media.

In this ozone treatment scheme reductions occurred in ultraviolet (UV) absorbance and trihalomethane formation potential (THMFP). It was also observed that ozone enhanced conditions in the filter water columns for removal of other objectionable matter. However, the breakdown of organic matter also reduced filter run times. The studies also showed that ozonation by-products were removed through the biological slow sand process.

Using ozone ahead of slow sand filtration may allow many communities to meet new disinfection by-product regulations. However, treatment improvements may result in increased operating and maintenance costs relating to shorter filter run times and operating costs of the ozone system.

Granular Activated Carbon. Adding granular activated carbon (GAC) to slow sand filter media was initially tested in England at the Thames Water Utilities. Thames Water currently operates seven slow sand treatment facilities with a combined capacity of about 700 mgd (2,500 ML per day). Because of pesticide levels in the source waters and strict regulations for pesticide removals, the utility determined that adding a GAC treatment step could allow water quality goals to be met.

To avoid the relatively high cost of constructing GAC adsorbers, the utility explored installing GAC within the filter bed. The "sandwich" bed used a 3 to 8 in. (75 to 200 mm) layer of GAC installed 4 to 6 in. (100 to 150 mm) below the sand surface. The performance of the system was compared with conventional slow sand filtration with respect to head loss, color removal, TOC removal, and THMFP. The results are presented in Table 9.4. Chlorine demand was also reduced with the use of GAC, and pesticide levels were reduced to below standards.

TABLE 9.4 Comparisons of the Operation of Slow Sand and GAC Sandwich Filters

Parameter	Conventional slow sand	GAC sandwich filter
Cleaning frequency	30 days	No change
Color removal	20%	50%
Total organic carbon reduction	20%	35% to 40%
Total trihalomethane formation potential reduction (24-hour contact time)	130 μg/L	60 g/L

Based on test results, Thames Water is planning to use a three-year cycle for regeneration of the GAC and to install the GAC sandwich throughout their treatment facilities.

Filter Fabric. In research conducted in the United States by Collins et al. (1989) and in England by Graham et al. (1991) a synthetic, nonwoven fabric installed on the sand surface was shown to increase filter run lengths while maintaining effective treatment. The benefit is that removal of suspended particles occurs on the fabric with the intent of simplifying the cleaning process.

When specifying a fabric, parameters to be considered include porosity, specific surface area, and fabric thickness. Graham et al. (1991) showed that a fabric thickness between $\frac{3}{4}$ and $1\frac{1}{4}$ in. (20 and 30 mm) was necessary to allow adequate fabric removal and cleaning and that fabrics with a specific surface area between 4,000 and 4,500 ft^2/ft^3 (13,000 and 15,000 m^2/m^3) provided optimal results. They also demonstrated that use of fabric meeting this criteria increased filter run times by 400% over a conventional filter.

Issues to resolve before using filter fabric involve the mechanism for removal and cleaning of the fabric, especially for large installations. Further pilot studies may also be warranted to determine the applicability of this modification with source waters of varying quality.

Filter Harrowing. The process of harrrowing filters was initially developed as a filter cleaning method at the West Hartford, Connecticut, slow sand filtration facility. The plant currently has 22 slow sand beds in operation with a capacity of 50 mgd (189 ML per day). Beds vary in size between $\frac{1}{2}$ and $\frac{3}{4}$ acre (0.2 to 0.3 hectares). To reduce the time and expense for cleaning, the utility developed the process of "harrowing" filters in the 1950s. In the process, a tractor with a mounted spring-tooth harrow operates within the bed while water about 6 in. (15 cm) deep is flowing across the sand surface. The harrow breaks up the top of the filter surface and water carries away the debris. The process removes the accumulated source water particulates while maintaining an active biological material in the top several inches of the sand.

The process creates wastewater as the bed is being cleaned. The volume generated depends on source water quality, filter area, and cross-flow velocity and depth and is generally in the range of 40 to 60 gal/ft^2 (163 to 244 L/m^2) of filter area.

The experience at West Hartford has shown that using the harrowing process reduces cleaning time to one quarter of the time required for conventional cleaning. The harrowing process requires the following facilities that may not be provided for a conventionally cleaned slow sand filter:

- Access ramp for harrowing equipment to enter the filter
- Harrowed water influent distribution system

- Harrowed wastewater collection system
- Holding lagoon for the harrowed wastewater

The filter box must be designed with enough headroom to permit tractor operation, and support columns should be minimized to improve maneuverability. Figure 9.6 shows a harrow tractor operating in a filter basin. Influent and waste headers can consist of either perforated piping or channels with adjustable weirs.

Wastewater lagoons should be designed to carry one to two harrowings with the ability to allow time for settling and to discharge the decanted water in stages. Because wastewater has generally not received previous chemical pretreatment, discharge requirements are usually less rigorous.

The process is now currently used at several operating plants in the United States to improve plant performance and reduce cleaning times. Designers of larger-capacity plants should consider using this cleaning process.

DIATOMACEOUS EARTH FILTRATION

DE filtration has been used effectively for drinking water treatment since 1942 when it was adopted as a standard method for the U.S. Army. The DE filter was selected because of its portability and effectiveness in removing *Entamoeba histolytica* cysts. These cysts are pervasive in some parts of the world and are difficult to control with disinfectants alone. DE filters are commonly called precoat filters because of the precoat of the filter leaves that initiates every operating cycle. Although DE is the most common precoat material used, other precoat material such as ground perlite performs as well in many applications.

For many years, the type of equipment available limited the use of DE filters for municipal drinking water treatment. The use of stainless steel and plastics in the fabrication of equipment has significantly changed the performance capability of the filters by improving their ease of operations and maintenance.

FIGURE 9.6 Harrow tractor.

DE is mined from the fossilized remains of microscopic plants called diatoms deposited in what were the beds of ancient oceans. A powdered medium is manufactured from the diatomite deposits that is almost pure silica. One of the more common diatomite media used for drinking water treatment has a mean particle size of 22.3 μm with 80% of the particles ranging in size from 5 to 64 μm. This medium, when deposited on the filter septum, has an average pore size of about 7.0 μm. Specifications for DE material is covered in AWWA standard B101.

DE Filter Operation

As illustrated in Figure 9.7, DE filter operations occur in three steps:

1. A precoat of about $\frac{1}{8}$ in. (3 mm) is deposited on the filter.
2. After the precoat has been deposited, filtering begins, and at the same time a small amount of DE material (called body feed) is added to the source water to maintain the porosity of the media.
3. Particulates in the source water are trapped in the precoat layer until maximum head loss is reached, at which time the filter run is terminated and media material is cleaned from the septa.

Porosity Control. The principal requirement for maintaining effective DE filter runs is to maintain the porosity of the filter cake. Source water solids generally vary in size and are a mixture of relatively inert matter and solids that are predominantly organic. If source water is filtered through the precoat alone, the buildup of solids and compression of the accumulated cake quickly reduce filter cake porosity, and head loss increases at an exponential rate. This may be avoided by adding body feed to the source water in sufficient amounts to produce a constant flow versus head loss relationship (Figure 9.8).

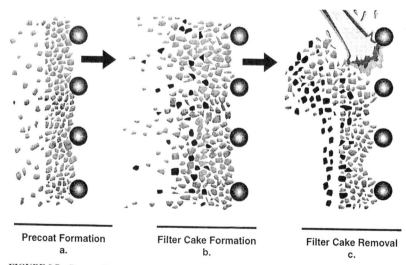

| Precoat Formation | Filter Cake Formation | Filter Cake Removal |
| a. | b. | c. |

FIGURE 9.7 Precoat filtration cycles.

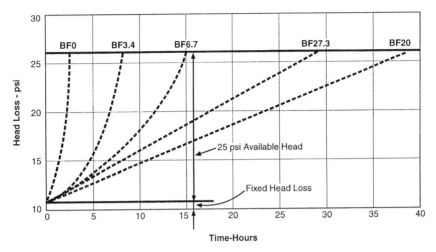

FIGURE 9.8 Effect of varying body feed addition.

Although the rate of flow does not affect effluent quality or turbidity breakthrough, the flow rate for precoat filters should generally be limited to about 2 gpm/ft^2 (7.8 m/h). The shape of the precoat filter head loss curve that reflects both feed and flow conditions is, therefore, an important feature to control effective filter run performance.

Supplementary Treatment. Supplementary measures may be added to the basic DE filter process to enhance the filtration process and also to expand the process to remove some nonparticulate constituents.

Natural color in source water supplies can be caused by either organic or mineral matter. Color can result from the decay of plant matter or from the solubilization of iron in the soil, and in many instances the mineral and organic matter may be bound together. Therefore color can be present in either particulate form or in solution. Particulate color consists mostly of negatively charged colloids, and, even though the precoat media has low pore size, charged colloids pass through unless the charge is neutralized. The use of a strong oxidant such as ozone has been demonstrated to be effective in conditioning color for removal.

When color is particulate rather than dissolved, DE filters reduce source water color of about 25 color units (CU) and less to below 5.0 CU. With source color between 25 and 60 CU, filter effluent is generally no higher than 10 CU. Supplemental treatment such as pre-ozonation or alum-coated media may be required to improve removal of particulate color and to reduce dissolved color.

Dissolved iron may be precipitated by aeration or by adding a strong oxidant so that the iron may be removed as a particulate in DE filtration. The use of magnesite (magnesium oxide) has been found to facilitate removal of some forms of iron. Magnesite mixed along with body feed is held for about 10 min to form a negatively charged suspension of magnesium oxide (MgO) that gradually undergoes hydration and solution.

Manganese may be removed in DE filtration with potassium permanganate (KMnO$_4$) added to the body feed, followed by flow detention. Detention time is important and should be determined in bench and pilot tests. The rate of KMnO$_4$ addition and body feed rate depend on the amount of manganese in the source water and other water quality characteristics.

Where iron and manganese are both present in source water, supplementary conditioning must usually be accomplished in separate steps, with iron treatment preceding the $KMnO_4$ addition. When there is a large amount of iron to be removed, it may be necessary to have two filters in series, with iron conditioning preceding the first filter and the addition of $KMnO_4$ and detention between the first and second filters.

Filter Design Considerations

The principal elements of a flat-leaf DE filters are shown in Figure 9.9. Several options are available in designing each of these elements, as well as in the integrated assembly design.

Types of DE Filters. Two basic groups of DE filters are available. If source water is to be forced through the filter under pressure, the containment vessel must be closed, as illustrated in Figure 9.9. Filters operated under a vacuum, on the other hand, may be open vessels.

Although there is theoretically no limitation to what pressure may be applied to a pressure-type filter, practical considerations of pumping costs have limited head loss to a maximum of 35 psi (241 kPa). Most systems used for drinking water filtration are designed for a maximum head loss of 25 psi (172 kPa).

Filter Construction. Pressure filters are always constructed as cylindrical pressure vessels mounted either vertically or horizontally. Most units fabricated today are made of stainless steel for the shell and most internal parts. The type of stainless steel used depends on the corrosivity of the water being treated.

Vacuum filters are built as rectangular tanks. Because of the low differential heads they are subject to, vacuum filter containments and internal parts, except certain structural

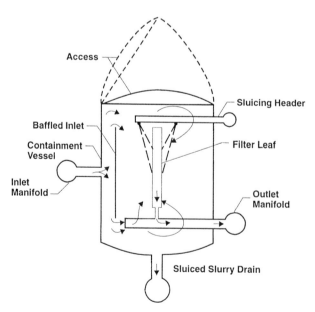

FIGURE 9.9 Principal elements of a flat leaf filter.

supports, are most often fabricated of plastics for their chemical resistance and reduced maintenance. For larger units, the containment vessel may be constructed of concrete.

Filter Elements. Inlet water flow is introduced to a DE filter through the containment wall, fitted with an internal baffling device to prevent disturbance of the filter cake. Filter cake may be cleaned from the filter by scraping, vibration, hydraulic bumping (surging), or manually hosing down the septa from the top of an open vessel.

Many arrangements of filter elements are available, constructed in both a tubular and flat form, with the flat (or "leaf") design being by far the most common. Some typical filter element designs are illustrated in Figure 9.10. Filter elements may be mounted either horizontally or vertically, and may be either fixed in position or able to rotate.

Vertical mounting of the leaves is used almost exclusively for water treatment applications. Most pressure and vacuum filters constructed today have fixed leaves mounted by means of spigot-type "push-on" outlets installed in sockets on a manifold and sealed by O-rings or flat gaskets. The outlet manifold is usually located below the leaves to provide them with support while allowing gravity to assist in seating the push-on connections.

Variations for both leaf connections and manifold location are selected depending on operating conditions and requirements for inspection and maintenance. Fixed-leaf pressure filters may also be divided into retracting shell and retracting bundle types for internal access. Both options may be used for any size filter, but the retracting bundle type is generally preferred for larger units. As illustrated in Figure 9.11, the retracting shell or tank design has the shell mounted on wheels on rails. An electric motor or hydraulic piston opens and closes the unit.

In the retracting bundle design, the shell head is suspended from an overhead monorail. The bundle, attached by the manifold and frame to the head, is retracted by means of the monorail, which is usually motor driven. Internal rails attached to the shell support the end of the leaf bundles when the shell is opened.

Thorough cleaning of the septa at the end of a filter run is important in maintaining peak efficiency. Most units used for water treatment sluice the cake with water sprays, which creates a slurry that can be easily handled and treated and does not require opening the filter vessel. Fixed-leaf filters are usually cleaned with high-pressure spray jets mounted on oscillating spray heads, with single or multiple jets directed between the filter leaves.

Rotating filter leaves usually have a stationary spray header, and coverage is obtained as the leaves rotate past the sprays. Open filters may be cleaned manually using high-pressure

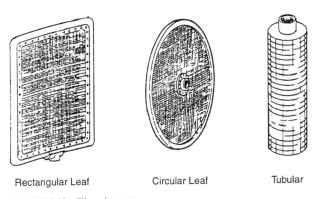

Rectangular Leaf   Circular Leaf   Tubular

FIGURE 9.10 Filter elements.

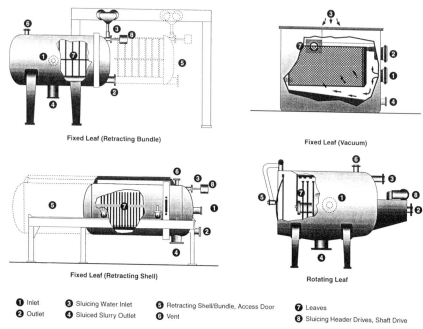

FIGURE 9.11 Common vertical leaf filters.

sprays and may require covers over the units to contain the spray. Additional devices that may assist in the complete removal of the cake slurry from the filter containment include spray jets in the invert of the vessel or an air scour to suspend the material before the vessel is drained.

Filter Leaf Design

A section of a typical flat leaf showing the principal construction elements appears in Figure 9.12. The flat filter leaf with a broad surface and limited thickness should be designed with the following goals:

- Leaf and outer frame must be stiff enough to resist warping under the force exerted at maximum differential pressure.
- The unit must have a backing screen to prevent the cloth septum from flexing under gradually increasing pressure.
- The path provided for filtrate flow must not restrict flow and create minimal head loss through the leaf.

It is essential that the filter cake remain undisturbed. An adequate filter leaf design prevents the possibility of cake movement from warping of the frame or flexing of the septum.

Central Drainage Chamber. There are three basic types of filter leaf drainage chambers: heavy wire mesh with wire spacing up to 1 in. (2.5 cm); expanded metal sheets that

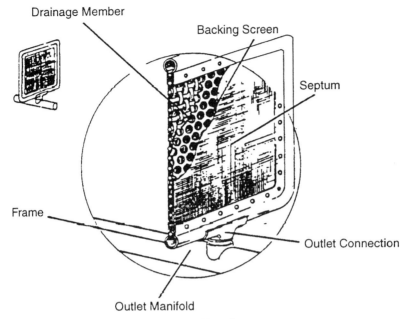

FIGURE 9.12 Typical construction of flat leaf filter element.

provide a deeper chamber with increased rigidity; and the Trislot, a proprietary design having thin metal bars with welded transverse round or wedge-shaped wires.

Backing Screen. The backing screen is an intermediate screen used when the irregular surface of the central drainage chamber may permit flexing of the cloth septa under conditions of varying pressure.

Filter Septum. Filter septa material are cloth weaves made with either stainless steel wires or plastic monofilaments. The principal purpose of the septa is to retain the precoat, which must bridge the openings in the weave. Because openings in the weave are larger than the major portion of particulates in precoat material, the precoat is retained by bridging. The cloth septum must be uniformly woven to produce an even precoat that reduces the extent of the recirculation required to deposit the material. The weave must also be designed so that it sluices cleanly, drops the cake freely, and resists plugging and damage.

One of the more common wire cloths for water filtration is a standard 24 × 110 Dutch weave. Another type of weave is the Multibraid, composed of bundles of wire in both directions. This weave is less vulnerable to the entrapment of particles and blinding than the standard weave. Woven wire cloth may also be "calendered," which involves passing the cloth through compression rollers to flatten the rounded wire at the surface of the weave. Calendering improves precoat retention characteristics and generally strengthens the cloth against rough treatment.

Plastic cloth is used predominantly for vacuum-type DE filters and is available in a variety of weaves using either polyester or polypropylene monofilament. Plastic cloth may be supplied as a bag to envelop a filter leaf or as a cloth caulked into a leaf frame.

Binding Frame Closures. The binding frame surrounds the filter leaf to prevent leakage around the septum. The outside frame is also the principal structural element to provide rigidity and prevent warping. Depending on the shape, the outside binder may also collect flow from the central chamber and supplement flow routing to the outlet nozzle.

Vacuum Filter Leaves. Vacuum filter leaves used for drinking water treatment are often made of plastic. As shown in Figure 9.13, the central drainage chamber and outlet spigot are molded in a single piece, usually of high-impact styrene. Ridges or other raised patterns provide the required flow path. The raised pattern is spaced so that intermediate screens are not required. The septum, in the form of an envelope with zipperlike closures, is sealed at the bottom outlet by a gasket that also provides tight closure for the manifold connection. Figure 9.14 shows two ways that filter leaves may be mounted in a vacuum filter with supplementary support to improve rigidity.

Outlet Connections. Fixed-leaf filter leaves usually have spigot-type outlet connections made of castings machined to fit into the sockets of the outlet manifold. Figure 9.15 shows two types of outlet construction. The central drain hub shown in the figure is used for rotating leaves and is of two-piece construction, clamped to the center of the circular leaf by bolts. The leaf outlet connection must be of sufficient size to allow full flow of the filtrate collected by the leaf at a minimum head loss. As leaf size and loading increase, the distribution of flow within the leaf and the transition to the outlet connection increase in importance.

Comparison of DE Filter Systems

If DE filtration has been selected for use in a water treatment system, a selection must be made between vacuum and pressure filter options. The length of filter run expected from either type of filter should be determined in pilot testing preceding design evaluation. The two types of filter systems may be evaluated in regard to costs, including the following items:

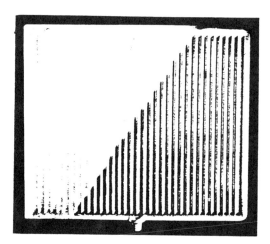
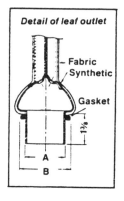

FIGURE 9.13 Vacuum filter leaf.

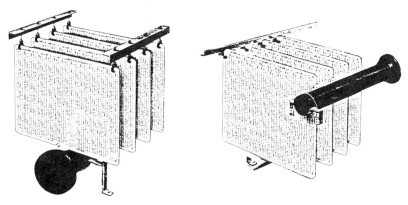

FIGURE 9.14 Vacuum leaf mountings.

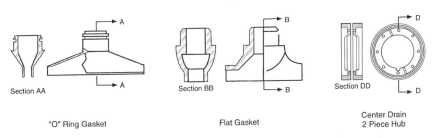

FIGURE 9.15 Outlet connections.

- Capital cost of installation of the complete filter system and ancillary equipment
- Capital costs of structures to house the equipment
- Average length of filter runs
- The cost of DE material required for operation
- Pumping power costs

Pressure Filter Systems. The principal elements included in a pressure filter system are shown in Figure 9.16. In the filtering mode, source water is pumped into the filter shell, and a baffle plate distributes flow and prevents it from eroding the cake on the septa. A flow-measuring device installed in the filter outlet line activates a valve that maintains a constant flow rate.

In the precoat cycle, the source water inlet and filtrate outlet valves remain closed. If the system includes a precoat tank as shown in the schematic, the system is first filled with previously filtered water to a depth sufficient to reach the top of the shell. The pump then recirculates the water as precoat slurry is added. In an alternative arrangement, a recirculation tank is not used. With level controls in the filter shell, the unit is first partially filled with water, then filled completely along with the addition of precoat slurry.

Precoat accumulates on the septa as the slurry is recirculated. Because the flow tends to seek the path of least resistance, recirculation tends to concentrate on those leaf areas not previously covered, until a reasonably uniform precoat thickness is eventually formed.

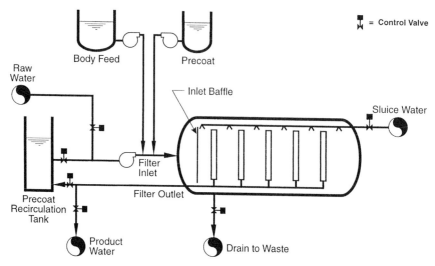

FIGURE 9.16 Pressure system schematic.

After full cover is achieved, the remaining suspended precoat material is intercepted by the filter and the turbidity of recirculated flow quickly clears up. At this time, in a simultaneous operation, the recirculation valve closes and the filter inlet and outlet valves open to initiate the filter cycle.

When the maximum available head differential is reached, the filter run is terminated and the filter is cleaned. In the cleaning cycle the inlet valve is closed, the waste valve is opened, and after the unit is fully drained, sluice water is turned on to flush away the filter cake.

The principal advantages of the pressure filter system are as follows:

- The ability to operate at application rates of up to 3 gpm/ft^2 (7.8 m/h) requires fewer filter units to process a given amount of water.
- The enclosed filter units generally make a neater installation with minimum housekeeping maintenance.
- A wide range of sizes of pressure units are available, from small up to about 3,000 ft^2 (280 m^2).

The principal disadvantages of the pressure-type precoated filter system are the relatively high initial cost of the filter units and somewhat higher power costs for operation because of the higher operating pressure.

Vacuum Filter Systems. As illustrated in Figure 9.17, a vacuum precoat filter system operates essentially the same as a pressure filter except that the filter is open to the atmosphere and the pump pulls the flow of water through the filter cake. Vacuum-type DE filters operate at flow rates of from 0.5 to 1.0 gpm/ft^2 (1.2 to 2.4 m/h) of leaf area, with 0.75 gpm/ft^2 (1.8 m/h) about average. Total head available for operating the filter system is limited by the suction capability of the pump, generally about 18 in. (460 mm) of mercury.

Because the vacuum filter is an open unit and operations can be easily observed, precoat and cleaning cycles are simplified. Sluicing may be done by fixed or rotating sprays, or the leaves can be manually cleaned with a high-pressure hose.

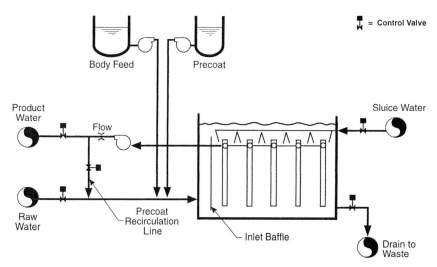

FIGURE 9.17 Vacuum filter system schematic.

One limitation of vacuum filters sometimes occurs during operation with cold water. In winter some water is saturated with air, and if the velocity through the cake generates a low enough hydraulic pressure, the air is released from solution to form bubbles that "pop" particles from the precoat, producing turbidity. However, many vacuum filters operated with cold water do not experience this problem.

Principal advantages of the vacuum filter system are as follows:

- For small plants, the initial cost is about one-third the cost of a stainless steel pressure filter of the same capacity.
- It is easy to observe the precoat and cake accumulation because the unit is open.
- Precoat, filter, and sluicing cycles may be operated manually because the unit is open.
- Operation and maintenance of the units generally requires fewer skills than does pressure filter operation and maintenance.

The principal disadvantages are as follows:

- The relatively low filtration rate requires more or larger units than for pressure units.
- Filter units must be protected from the weather because wind, rain, or ice disturb the filter cake, and sunlight promotes algae growth.

Ancillary Facilities

DE is a fine powder that has abrasive properties when handled both in dry form and in slurries. Special precautions must be taken to protect plant staff against dust, and equipment and pipelines must be protected against erosion. Because of its abrasive properties, DE breaks down in size under turbulent operating conditions.

DE Delivery. DE can be delivered in 50 lb (22.7 kg) plastic-lined paper bags, in 900 lb (408 kg) woven plastic bags, or in bulk truck or rail car loads. The 50 lb (22.7 kg) bags are usually stacked about 12 to a pallet and are handled with a forklift truck. Bags are then individually lifted by plant staff and discharged through a bag breaker equipped with a dust collector that empties into a dry feeder hopper or into a slurry tank.

The 900 lb (408 kg) bags are delivered, two to a pallet, in an "over-and-under" stack. Special forklift trucks and unloading frames are used for handling and unloading the bags. Bags are transferred from the pallet with a sling to a monorail and then moved into place over a discharge hopper. A spout, tucked into a flap in the bottom of the bag, is unrolled and attached to a discharge spout, and the bag is then unloaded with the assistance of a vibrator mounted on the holding frame. Discharge may be to a pneumatic conveyor for dry bulk storage or to a wetting chamber for pumping and storage as slurry. Because bags are completely enclosed, dust collectors are not required for unloading facilities.

Bulk delivery by truck or rail car may be made directly to storage silos with a pneumatic system similar to equipment used for handling bulk carbon or lime.

Slurry Storage. For small operations, DE slurry is usually stored in fiberglass-reinforced plastic tanks and, for large volumes, in reinforced concrete basins. All storage units must be equipped with continuously operating mixers to keep DE in suspension. Depending on slurry concentration, it may be difficult to resuspend DE once it has been allowed to settle. In order to limit attrition of DE in the turbulence of mixing, slurries should be used within three days.

Slurry Conveyance. Steel pipelines convey slurry either by gravity or by pumping. The flow must be continuous, and at the end of a conveying cycle, the pipeline must be completely flushed. The line should not be flushed to the point of delivery unless flushing water volume is measured and considered in determining the concentration of DE in the slurry at the point of use. Another method used is to deliver flushing water to a holding tank when DE concentration is determined before supplementary DE is added for use in mixing a new batch of slurry.

For DE concentrations of 4% or less, case-hardened centrifugal pumps may be used for slurry transfer. Abrasion-resistant chemical feed metering pumps are also acceptable for these concentrations. Thicker slurries must be conveyed with either peristaltic-type pumps or rubber-lined centrifugal pumps. To minimize breakdown of DE material, pipeline velocities should be limited to no more than 8 ft/s (2.4 m/s), but must be maintained at over 3 ft/s (0.9 m/s) to prevent settling.

Waste Disposal. Because it is predominantly inert, waste DE slurry may be easily treated by thickening with polymers and dewatering on belt filters. DE slurry may also be settled and concentrated in lagoons.

At large installations, DE may be recovered from wastes for reuse as body feed, but it should never be used for precoat. Virgin DE should be used for precoat to provide a final barrier for any pathogens that may pass through the upper levels of the cake. DE recovery saves costs of DE purchases and reduces the costs of waste disposal.

In the recovery process, separation of the lighter source water solids from the more dense DE is accomplished by pumping wastes through a series of vertical hydrocyclones. The heavier DE material gravitates to the bottom discharge to exit into a collector tank, while most of the source water solids and the finer fraction of the DE remain in the center core of the unit. The finer material is then removed hydraulically from the top of each unit. Because of the erosive nature of DE slurry, it is necessary to use hydrocyclones constructed of ceramic material and pumps designed for abrasion resistance.

A typical recovery system with connections to the filtration process is shown in Figure 9.18. Recovery rates generally range between 85% and 90%. Because of the possible

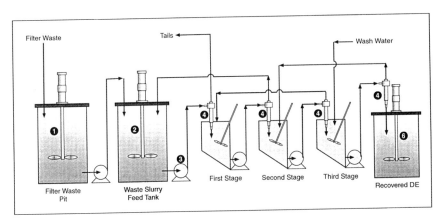

FIGURE 9.18 Schematic of a diatomaceous earth recovery system.

carryover of some bacteria, viruses, and cysts that have been separated in precoat filtration, it is advisable to disinfect the recovered material before use as body feed. One simple method of disinfection is the application of a heavy dose of potassium permanganate to the material before it is reused. Where *Cryptosporidium* cysts may exist, ozonation may be the only suitable method of disinfection.

System Controls.   Most pressure filter systems are provided with automatic operating systems similar to those for rapid sand filters. Vacuum filter systems may also be automated or operated manually. Where automation is used, the precoat cycle, body feed, and washing cycles may be initiated and performed automatically. The washing cycle may be automatically initiated when maximum head loss is reached, or the staff operator may be alerted by an alarm so that the cycle may be manually initiated.

A flow-measuring device on the outlet of pressure filters controls system flow with a throttling valve on the product water line. Most installations use a supply pump that operates at a fixed head, and the flow is controlled by the throttling valve. In some installations, it may be advantageous to use a variable-speed supply pump instead of the control valve to maintain required flow. For vacuum filters, a level control may be added to maintain a relatively fixed water level in the filters while the flow device controls outlet flow.

BIBLIOGRAPHY

Slow Sand Filtration

Allen, R. E. "A Method of Cleaning Slow Sand Filters Developed at the West Hartford, Connecticut, Water Treatment Plant." Slow Sand Filtration Workshop, University of New Hampshire, October 1991.

Biologically Enhanced Slow Sand Filtration for Removal of Natural Organic Matter. American Water Works Association Research Foundation. Denver, Colo.: American Water Works Association, 1993.

Calgon Carbon Corporation. *GAC Sandwich Filter.* Pittsburgh: Water Services, February 1994.

Christman, R. F., and R. T. Oglesby. "Microbiological Degradation and the Formation of Humus." In *Lignin: Occurrence, Formation, Structure, and Reactions.* Edited by K. V. Sarkanen and C. H. Ludxig. New York: Wiley-Interscience, 1971.

Cleasby, J. L., D. J. Hilmoe, and C. J. Dimitracopoulos. "Slow Sand and Direct In-Line Filtration of a Surface Water." *Journal AWWA* 76(12):44, 1984.

Collins, M. R., T. T. Eighmy, J. M. Fenstermacher, and S. K. Spanos. *Modifications to the Slow Sand Process for Improved Removals of Trihalomethane Precursors*. American Water Works Association Research Foundation Report. Denver, Colo.: American Water Works Association, 1989.

Ellis, R. V. "Slow Sand Filtration." *CRC Critical Reviews in Environmental Control* 15(4):315, 1985.

Fox, Kim, M. R. Collins, and N. J. D. Graham. *Slow Sand Filtration: An International Compilation of Scientific and Operational Developments*. Denver, Colo.: American Water Works Association, 1994.

Graham, N. J. D., T. S. A. Mbwette, and L. Di Bernardo. "Fabric Protected Slow Sand Filtration: a Review." Slow Sand Filtration Workshop, University of New Hampshire, October 1991.

Huisman, L., and W. E. Wood. *Slow Sand Filtration*. Geneva, Switzerland: World Health Organization, 1974.

Leland, D. E. "Slow Sand Filtration—Piloting Requirements." Slow Sand Filtration Workshop, University of New Hampshire, October 1991.

Malley, J. P., M. R. Collins, and T. T. Eighmy. "The Effects of Preozonation on Slow Sand Filtration." Slow Sand Filtration Workshop, University of New Hampshire, October 1991.

Manual of Design for Slow Sand Filtration. American Water Works Association Research Foundation. Denver, Colo.: American Water Works Association, 1991.

Mbwette, T. S. A. *The Performance of Fabric Protected Slow Sand Filters*. Doctoral Dissertation, University of London, United Kingdom.

McNair, D. R., R. C. Sims, D. L. Sorenson, and M. Hulbert. "Schmutzdecke Characterization of Clinopilolite-Amended Slow Sand Filtration." *Journal AWWA* 79(12):74, 1987.

Pardon, Maurice. "Removal Efficiency of Particulate Matter through Vertical Flow Roughing Filters." Slow Sand Filtration Workshop, University of New Hampshire, October 1991.

Recommended Standards for Water Works. Albany, N.Y.: Health Education Services, 1992.

Skeat, W. O. *Manual of British Water Engineering Practice*. 3rd ed. "The Biology of Water Supply," Chapter 20. Cambridge, England: Institution of Water Engineers, W. Heffer and Sons Ltd., 1961.

Slezak, L. A., and R. C. Sims. "The Application and Effectiveness of Slow Sand Filtration in the United States." *Journal AWWA* 76(12):38, 1984.

Spencer, C. H., and M. R. Collins. "Water Quality Limitations to the Use of Slow Sand Filtration." Slow Sand Filtration Workshop, University of New Hampshire, October 1991.

Wegelin, M. "Roughing Filters for Surface Water Pretreatment." Slow Sand Filtration Workshop, University of New Hampshire, October 1991.

Whipple, G. C. *The Microscopy of Drinking Water*. 3rd ed. "Purification of Water Containing Algae," Chapter 17. New York: John Wiley and Sons, 1914.

Diatomaceous Earth Filtration

Precoat Filtration, M30. Denver, Colo.: American Water Works Association, 1995.

Schuler, P. F., and M. M. Ghosh. "Diatomaceous Earth Filtration of Cysts and Their Particulates Using Chemical Additives." *Journal AWWA* 82(12):67, 1990.

CHAPTER 10
OXIDATION AND DISINFECTION

In 1854 Dr. John Snow established that water could be a mode of communication for dreaded diseases such as cholera. At first, slow sand filtration and the use of uncontaminated water supplies were the only means employed to prevent the spread of waterborne diseases. In the 1870s Louis Pasteur and Robert Koch developed the germ theory of disease, and innovations began to occur.

In 1881 Koch demonstrated in the laboratory that chlorine could kill bacteria, and by 1890 the first electrolytic chlorine generation plant was built in Germany. In 1905 continuous chlorination was used for the first time in Lincoln, England, to arrest a typhoid epidemic. The first regular use of disinfection in the United States was at the Bubbley Creek Filtration Plant in Chicago in 1908, about the same time that Dr. Harriette Chick first advanced her famous theory of disinfection.

Some of the physical constants for the three most common disinfecting agents, chlorine, ozone, and chlorine dioxide, are listed in Table 10.1. It should be noted that all three normally exist as gases, although chlorine dioxide liquefies at a temperature near 10° C. Chlorine is available as a compressed liquid, but ozone and chlorine dioxide must be manufactured on site (ozone because it decomposes, chlorine dioxide because it is dangerous to store in a concentrated compressed form).

TABLE 10.1 Physical Constants for Common Disinfecting Agents

Name	Symbol	Molecular weight	Solubility in water at 1 atm and 25° C, g/L	Boiling point, °C	Melting point, °C	Heat of vaporization, cal/g
Chlorine	Cl_2	70.91	7.29	−34.5	−101	68.7
Ozone	O_3	48.00	0.006*	−112	−192	54.0
Chlorine dioxide	ClO_2	67.45	8.0†	10.9	−59	96.6

*190 O_3 by weight.
†Assumes equilibrium with 10% ClO_2 gas phase.

The first use of ozone for disinfection was at Nice, France, in 1910. Since that time, disinfection has become an accepted water supply practice throughout the world. Chlorination has been the dominant method employed, but ozonation has been widely used also, particularly in France, Germany, and Canada. There has also been increasing use of chlorine dioxide as a disinfectant in the United States and Europe.

Oxidants are used in water treatment to accomplish a wide variety of treatment objectives besides disinfection, including mitigation of objectionable tastes and odors, removal of color, removal of iron and manganese, and oxidation of organic chemicals. Oxidation of contaminants in water by means of aeration is covered in detail in Chapter 5. Information on the chemicals used in water treatment is provided in Appendix 1. Information on chemical handling and chemical feed equipment is covered in Chapter 15.

For additional information on the theory and chemical reactions involved in oxidation and disinfection, refer to the companion text *Water Quality and Treatment*.

REGULATORY FRAMEWORK

A number of regulations have been promulgated by the U.S. Environmental Protection Agency (USEPA) under the Safe Drinking Water Act (SDWA) that affect how or when a water treatment system performs oxidation or disinfection. In addition to the requirements under the Primary and Secondary Regulations, several specific rules have a great impact on oxidation and disinfection process and are described in the following sections.

Secondary Regulations

The National Secondary Drinking Water standards apply to drinking water contaminants that may adversely affect the aesthetic qualities of water, such as odor and appearance. These qualities have no known adverse health effects, and thus secondary regulations are not mandatory. However, the qualities listed in the Secondary Standards do seriously affect acceptance of water by the public, and for this reason, compliance with the limits established by USEPA is strongly recommended. In addition, some states have enacted more stringent regulations that require compliance with some of the maximum contaminant levels (MCLs) in Secondary Standards.

The contaminants most commonly treated for removal by oxidation under the Secondary Standards are taste and odor, iron, and manganese. These oxidation processes are discussed in more detail later in this chapter.

The Surface Water Treatment Rule

The Surface Water Treatment Rule (SWTR) was promulgated by USEPA on June 29, 1989. This regulation applies to every public water system in the United States that uses surface water as a source. It also applies to groundwater systems that the state determines might become contaminated by surface water; these systems are labeled "groundwater under the direct influence of surface water," or GWUI.

The purpose of the regulation is to protect the public from waterborne diseases that are most commonly transmitted by contamination of surface water. Because it is difficult to monitor for particular microorganisms, such as *Giardia lamblia* and viruses, the SWTR emphasizes treatment techniques as the condition for compliance instead of having MCLs for microorganisms. Because of the wide variety of water qualities, local conditions, and methods of treatment, the rule does not prescribe a particular method of treatment but instead offers several alternatives. Any of these methods may be used by a water system to meet the overall goal, which is removal or inactivation of essentially all disease-causing organisms.

To ensure that water quality goals are met, the SWTR contains many operation and monitoring requirements. Studies indicate that *Giardia* cysts, viruses, and *Cryptosporidium* are among the most resistant waterborne pathogens; therefore water systems that attain adequate removal or inactivation of these organisms will, to the best of current knowledge, provide adequate protection from other waterborne disease organisms.

Disinfectant Contact Requirements. Most water systems using a surface water source must use sedimentation and filtration to ensure adequate removal of microorganisms. Under the SWTR requirements, only water systems with extremely low-turbidity source water may be allowed to operate without filtration, and then it is under very stringent operating and monitoring conditions. All surface water and GWUI systems, whether they provide filtration or not, must practice disinfection under highly specific conditions.

The effectiveness of a disinfectant in killing or inactivating microorganisms depends on:

- The type of disinfectant used
- The disinfectant residual concentration (abbreviated C)
- The time the water is in contact with the disinfectant (abbreviated T)
- Water temperature
- The pH of the water, which has an effect on inactivation if chlorine is used

The residual concentration C of a disinfectant in milligrams per liter (mg/L) multiplied by the contact time T in minutes is called the CT value. The CT values required by the SWTR to guarantee the necessary reduction in microorganisms by various disinfectants may be obtained from tables in publications referenced at the end of this chapter. Each water system's treatment must be sufficient to ensure that the total process of removal plus disinfectant CT achieves at least 99.9% (3-log) inactivation or removal of *Giardia* cysts and 99.99% (4-log) inactivation or removal of viruses. Source waters that are particularly vulnerable to microbial contamination may require greater log reductions, at the discretion of the primacy agency.

Credit for physical removal of pathogenic organisms is given to properly operated filtration processes, as indicated in Table 10.2. The remaining log inactivation is required to be achieved by the disinfection process. Application of the CT concept is discussed further under the design consideration section for each oxidant.

TABLE 10.2 Surface Water Treatment Rule Disinfection Requirements

Process	Log removals	
	Giardia cysts	Viruses
Minimum log removal inactivation	3	4
Conventional treatment credit	2.5	2
Remaining for disinfection	0.5	2
Direct filtration credit	2	1
Remaining for disinfection	1	3

Approval of Lower CT *Values.* The *CT* values presented in the tables provided by USEPA are generally considered to be conservative. Each primacy agency may allow lower *CT* values for individual systems based on on-site studies showing that adequate inactivation is achieved under all flow and raw water conditions. Protocols and requirements are extensive but may be justified for systems that have unusual circumstances warranting the studies.

Single Point of Disinfection. Systems with only one point of disinfectant application may calculate the *CT* that is being achieved by the entire system by measuring the disinfectant residual at a point before the treated water reaches the first customer. The multiplication of this residual concentration *(C)* and the contact time *(T)* through all basins and piping from the application point to the measurement point will provide a conservative *CT* value. If the value meets the required criteria, this is the simplest method of calculation.

Greater credit for the inactivation being achieved by a system using a single point of disinfection may be obtained by "profiling" the system to determine the inactivation credit for each of several contact "sections" in the system. In this method, the disinfectant residual is measured at several points in the treatment train and used along with the detention time in the respective section to determine the proportion of the total required inactivation that is being achieved by that section.

The calculated inactivation being achieved by each section divided by the total required inactivation is called the inactivation ratio for the section. If the inactivation ratios for all sections are then totaled and the result is equal to or greater than 1, the required total inactivation is being achieved.

Multiple Disinfectants or Application Points. Systems that apply disinfectant at more than one point will have to profile the system by computing the inactivation ratio for each section between application points. Some systems may also find it advantageous to divide the treatment train into additional sections between the disinfection application points to achieve the greatest *CT* credit.

The inactivation ratio for each section is calculated using the disinfectant residual at the end of the section, the contact time for the section, and the required *CT* for the disinfectant being used. The sum of the inactivation ratios for all sections will then determine the total inactivation being achieved by the entire treatment system.

Variations in Peak Hourly Flow. The determination of inactivation credit in each disinfection section of a system is to be determined under the conditions of peak hourly flow. However, some systems with large reservoirs may find that peak hourly flow does not occur at the same time in all sections.

To simplify determination of peak hourly flow, USEPA suggests that *CT* values for all sections be calculated during the hour of peak flow through the last section. This is best determined by a flowmeter immediately downstream of the last section.

Determination of Contact Time. The contact time *T* used in calculating *CT* values is the time it takes water to move from the disinfectant application point to the point at which the residual is measured.

Contact Time in Pipes. The time during which water is in contact with a disinfectant while flowing through pipes is straightforward. It assumes that water moves in a relatively uniform manner between two points and can be calculated on the basis of uniform plug flow as

$$T = \frac{\text{Internal volume of the pipe}}{\text{Peak hourly rate through the pipe}}$$

Contact Time through Reservoirs. Under most conditions, water does not move through reservoirs, tanks, and basins in a uniform manner. Therefore the time *T* used to compute *CT* in reservoirs depends on the design of the reservoir, such as the shape, inlet and outlet design and locations, and the baffling. In general, reservoirs with a large length-to-width ratio and with good inlet and outlet baffling minimize short-circuiting and provide the most uniform flow.

The contact time used to calculate the *CT* is the detention time at which 90% of the water passing through the reservoir is retained within the reservoir—in other words, the time it takes for 10% of the water to pass through the reservoir. This detention time, or contact time, is designated as T_{10}. The value of T_{10} for a reservoir at various flow rates may be determined experimentally by tracer studies or theoretically by approximation.

The most accurate method of determining contact time through reservoirs is by experiments using tracer chemicals such as chloride or fluoride. The studies are performed by feeding controlled amounts of the tracer chemical at the reservoir inlet and making repeated analyses of samples collected at the outlet. Unfortunately, the detention time under various flow rates is not a linear function, so it is recommended that tracer studies be performed using at least four flow rates that span the normal flow range. This information can then be used to construct a curve of detention time versus flow rate that can be used to determine T_{10} at any flow with fair accuracy.

Under certain conditions, the state primacy agency may allow the contact time for a reservoir to be determined by an approximation. The method involves multiplying the theoretical contact time (plug flow) of a reservoir by a rule-of-thumb factor that takes into consideration the reservoir design. Examples of reservoirs with poor, average, and superior baffling conditions are shown in Figures 10.1, 10.2, and 10.3. The shaded areas on the figures indicate areas with little or no flow (dead space) in both a horizontal and vertical perspective, which causes much of the flow to short-circuit directly from the inlet to the outlet. Table 10.3 summarizes the baffling conditions and the proportion of T_{10} to the theoretical contact time for each classification.

The detention time in reservoirs may also vary with the water level, the depth of sludge, and whether the water level is rising or falling, so these factors must be included in the *CT* calculations.

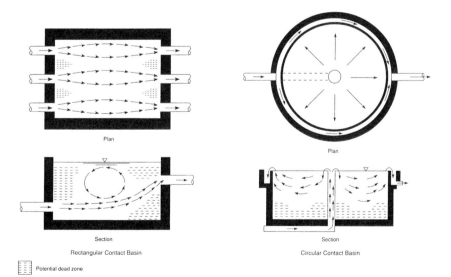

FIGURE 10.1 Examples of poor baffling conditions in basins. *(Source: Guidance Manual.)*

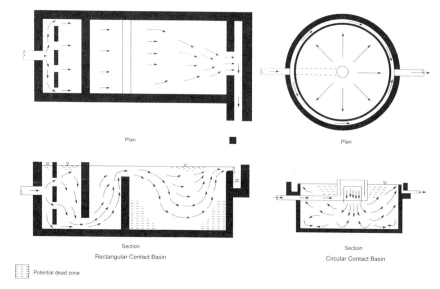

FIGURE 10.2 Examples of average baffling conditions in basins. *(Source: Guidance Manual.)*

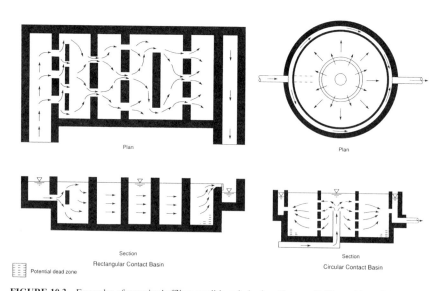

FIGURE 10.3 Examples of superior baffling conditions in basins. *(Source: Guidance Manual.)*

Contact Time through Filters. Filters usually have good flow distribution, so they can be considered as having conditions similar to superior baffling. The T_{10} for filters can therefore be calculated by subtracting the volume of filter media, gravel, and underdrains from the reservoir volume, dividing by the flow rate, and multiplying by the superior factor of 0.7.

TABLE 10.3 Baffling Classifications

Baffling condition	T_{10}/T	Baffling description
Unbaffled	0.1	No baffling, agitated basin, very low length-to-width ratio, high inlet and outlet velocities
Poor	0.3	Single or multiple unbaffled inlets and outlets, no baffles
Average	0.5	Baffled inlet or outlet with some intrabasin baffles
Superior	0.7	Perforated inlet baffle, serpentine or perforated intrabasin baffles, outlet weir or perforated launders
Perfect (plug flow)	1.0	Very high length-to-width ratio, perforated inlet, outlet, and intrabasin baffles

Disinfectant Monitoring Requirements. The SWTR requires that the disinfectant residual of water entering the distribution system must be continuously monitored by water systems serving a population of more than 3,300. The residual cannot be less than 0.2 mg/L for more than 4 hours during periods when water is being served to the public. Any time the residual falls below this level the water system must notify the state. Systems serving a population of less than 3,300 may take grab samples on an ongoing basis in place of continuous monitoring. Samples must be collected at least as frequently as given in Table 10.4.

The disinfectant residual must be measured at the same points in the distribution system and at the same time as total coliforms are sampled. Disinfectant residuals must not be undetectable in more than 5% of the samples each month for any two consecutive months that water is served to the public.

Water systems must submit special reports to the state detailing the monitoring required by the SWTR.

Proposed SWTR Changes. Proposals have been made to add new requirements to the SWTR in what is being called the Enhanced SWTR. One of the major concerns that must be addressed is what to do in response to several major outbreaks in surface water systems caused by *Cryptosporidium*. The steps necessary to provide protection against the organism at reasonable cost to water systems is still not fully determined.

It is expected that data collected under the new Information Collection Rule (ICR) will help experts define more fully what additional treatment requirements should be enacted. The ICR requires larger water system to monitor for certain microbial agents, viruses, and disinfection by-products and to conduct studies to assist in developing new surface water treatment and disinfection rules.

Regulation of Disinfection By-Products

Although chlorine has been primarily responsible for almost eradicating waterborne disease in most of the world, it has been found that free chlorine will react with precursors—organic substances that occur naturally in most raw water—to produce trihalomethanes (THMs). Trihalomethanes are a family of organic chemicals, with the principal one, chloroform, known to be a carcinogen to some animals, and other THMs are suspected carcinogens. In 1979 the USEPA established a regulation that was initially applicable only to water systems serving a population of over 10,000, with an MCL of 0.1 mg/L total THMs.

Since the THM regulation was enacted, research has shown that there are other by-products from chlorination and other disinfectants that may have harmful effects on humans, as well as possible harmful effects of the disinfectants themselves. The issues

TABLE 10.4 Required Disinfectant Residual Sampling Frequency for Small Systems

Population served	Samples per day*
<500	1
501 to 1,000	2
1,001 to 2,500	3
2,501 to 3,300	4

*Samples must be taken at intervals prescribed by the state.

involved in establishing additional regulation of disinfectant and disinfection by-products (D-DBP) are complex, so the regulations will be implemented in stages over a period of years.

Designers of water treatment systems must stay abreast of the development of the new D-DBP regulations because they will undoubtably have a major impact on the choice of disinfectants, the location of disinfectant application, and the design of other parts of the treatment process.

Groundwater Disinfection Rule

Federal regulations require all surface water and GWUI systems to practice disinfection and maintain a chlorine residual in the water entering the distribution system, but there is no federal requirement for groundwater systems to practice disinfection. Many states have gone beyond the federal requirements and currently require all or certain classes of public water systems to practice chlorination.

The USEPA is currently developing a groundwater disinfection rule that focuses on source protection to prevent waterborne disease. Part of the development of the rule will involve research efforts to determine whether groundwater is naturally purified or requires chemical disinfection.

The final rule's treatment requirements will probably apply to all systems using groundwater. Under the new requirements, systems having a protected groundwater supply and a history of good compliance with microbiological standards will probably be allowed a variance to avoid chlorination at the discretion of the state.

CHEMICAL OXIDATION TREATMENT

Oxidation-reduction (redox) reactions form the basis for many water treatment processes addressing a wide range of water quality objectives. These may include removal of iron, manganese, sulfur, color, tastes, odor, and synthetic organics (herbicides and pesticides). A redox reaction consists of two half-reactions: the oxidation reaction, in which a substance loses, or donates, electrons; and the reduction reaction, in which a substance gains, or accepts, electrons. An oxidation reaction and a reduction reaction must always be coupled because free electrons cannot exist in solution and electrons must be conserved.

Oxidizing agents, or oxidants, used in water treatment include chlorine, chlorine dioxide, permanganate, oxygen, and ozone. The appropriate oxidant for achieving a specific water quality objective depends on a number of factors, including raw water quality, specific contaminants, and local chemical and power costs. For critical applications, the designer should insist on bench or pilot-scale evaluations of treatment alternatives to select the best approach and determine appropriate design criteria.

Iron and Manganese Removal

Dissolved iron and manganese, when present at levels exceeding 0.1 mg/L and 0.05 mg/L, respectively, may stain clothes and plumbing fixtures, may foul water softeners, and may impart offensive tastes, odors, or color to the water. Removal is facilitated through oxidation of the soluble divalent ions to the insoluble trivalent forms, usually written as Fe(III) and Mn(III). These insoluble forms precipitate in subsequent sedimentation or filtration steps. The oxidation reactions for iron and manganese with various oxidants are presented in Tables 10.5 and 10.6 along with the stoichiometric oxidant requirement, alkalinity used, and sludge produced.

Small well water systems with excessive levels of iron and manganese often apply an oxidant, provide a period of detention for the reaction to take place, and then remove the precipitated iron and manganese with a pressure filter. Figure 10.4 illustrates such a system.

Taste and Odor Control

Most objectionable tastes and odors that occur in raw water, particularly those of organic nature, can be mitigated by judicious application of a preoxidant. Surface waters in particular are prone to taste and odor problems from the presence of algae, other odor-causing organisms, and decaying vegetation.

The most well-known and common odor-causing compounds associated with algae are methylisoborneol (MIB) and geosmin. Both are produced by actinomycetes and various blue-green algae and are particularly resistant to oxidation. Because of the competing reactions with an oxidant and the potential for creating new taste and odor problems, bench or pilot-scale studies should be conducted to establish design criteria and select the appropriate oxidant for specific taste and odor problems. Particularly tough applications may require both oxidation and an adsorption step to lower tastes and odor to acceptable levels.

Potassium permanganate is often used for taste and odor control. Figure 10.5 presents a diagram of a potassium permanganate feed system.

TABLE 10.5 Oxidation of Manganese

	Reaction	Oxidant, mg/mg Mn^{2+}	Alkalinity used, mg/mg Mn^{2+}	Sludge,* lb/lb (kg/kg) Mn^{2+}
A.	Oxygen $2MnSO_4 + 2Ca(HCO_3)_2 + O_2 = 2MnO_2 + 2CaSO_4 + 2H_2O + 4CO_2$	0.29	1.80	1.58
B.	Chlorine $Mn(HCO_3)_2 + Ca(HCO_3)_2 + Cl_2 = MnO_2 + CaCl_2 + 2H_2O + 4CO_2$	1.29	3.64	1.58
C.	Chlorine dioxide $Mn(HCO_3)_2 + 2NaHCO_3 + 2ClO_2 = MnO_2 + 2NaClO_2 + 2H_2O + 2CO_2$	2.46	3.60	1.58
D.	Potassium permanganate $3Mn(HCO_3)_2 + 2KMnO_4 = 5MnO_2 + 2KHCO_3 + 2H_2O + 4CO_2$	1.92	1.21	2.64

*Sludge weight based on MnO_2 as the precipitate. It is highly probable that portions of the sludge will consist of MnOOH and $MnCO_3$.

TABLE 10.6 Oxidation of Iron

Reaction	Oxidant, mg/mg Fe^{2+}	Alkalinity used, mg/mg Fe^{2+}	Sludge,* lb/lb (kg/kg) Fe^{2+}
A. Oxygen $4Fe(HCO_3)_2 + O_2 + 2H_2O = 4Fe(OH)_3 + 8CO_2$	0.14	1.80	1.9
B. Chlorine $2Fe(HCO_3)_2 + Ca(HCO_3)_2 + Cl_2 = 2Fe(OH)_3 + CaCl_2 + 6CO_2$	0.64	2.70	1.9
C. Chlorine dioxide $Fe(HCO_3)_2 + 2NaHCO_3 + ClO_2 = Fe(OH)_3 + NaClO_2 + 3CO_2$	1.21	2.70	1.9
D. Potassium permanganate $3Fe(HCO_3)_2 + KMnO_4 + 2H_2O = 3Fe(OH)_3 + MnO_2 + KHCO_3 + 5CO_2$	0.94	1.50	2.43

*Sludge weight based on $Fe(OH)_3$ as the precipitate. It is highly probable that portions of the sludge will consist of $FeCO_3$.

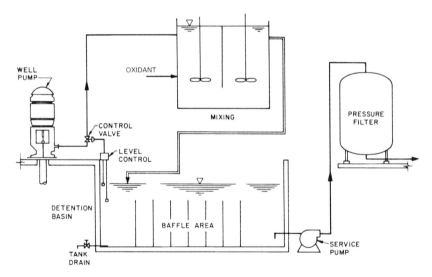

FIGURE 10.4 Chlorination, detention, and filtration (iron and manganese removal).

CHLORINATION

Although a good deal of work on modeling disinfection has recently been done, the principal disinfection theory in use today is still the Chick-Watson theory, which was developed by Dr. Harriette Chick and refined by H. E. Watson in 1908. In simplified form, this theory states that the rate of destruction of pathogens by a disinfectant is proportional to the number of pathogens and the concentration of the disinfectant.

OXIDATION AND DISINFECTION

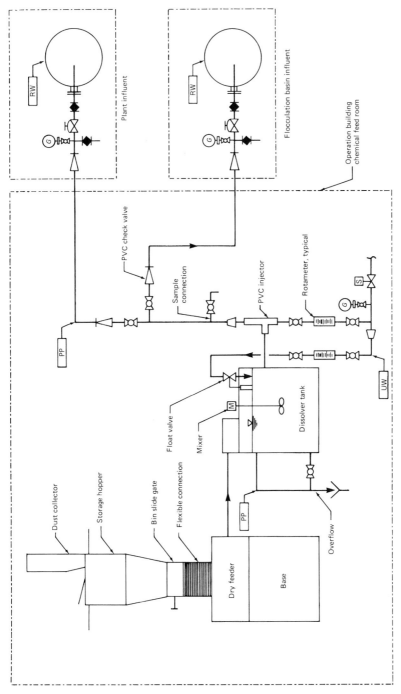

FIGURE 10.5 Potassium permanganate system. (*J. M. Montgomery Consulting Engineers.*)

At the present time, the maintenance of free chlorine residuals is the principal means by which water utilities ensure that the water they provide is properly disinfected. *Free residual chlorine* is a term used to refer to a chlorine residual that is not combined with ammonia or organic nitrogen.

Chlorine is most often available commercially in pressurized vessels containing both liquid and gas fractions, as sodium hypochlorite (household bleach), and as calcium hypochlorite. Gaseous chlorine is most often employed by larger water systems because of its significantly lower cost; however, transportation and storage of gaseous chlorine does impose a certain risk of serious accidents, and some utilities have switched to sodium hypochlorite in order to circumvent safety problems in densely populated areas.

This approach was further encouraged by the 1988 amendments to the Uniform Fire Code (UFC88), which classified gaseous chlorine as an acutely hazardous material (AHM). Many jurisdictions have adopted these amendments, which require installation of equipment for containment and scrubbing of gaseous chlorine leaks. On-site generation of chlorine gas or sodium hypochlorite is possible, and this approach is receiving increased interest. For greater detail, refer to White's *Handbook on Chlorination* (White, 1986), an extensive compendium that is widely used in the industry.

Chlorine Chemistry

When chlorine gas is dissolved in water, it quickly reacts to form hypochlorous acid (HOCl), which then disassociates to form a mixture of the acid and hypochlorite ion (OCl^-). Although both forms act as disinfectants, the acid is much more effective, so the effectiveness of free chlorine as a disinfectant depends primarily on the amount of hypochlorous acid available to react with the organisms. This fraction is most important because HOCl is nearly 1,000 times more effective than the ionized form, OCl^-.

The relative proportion of these two forms depends on the pH of the solution. The fraction of chlorine present as HOCl at various pH levels and at various temperatures is shown in Figure 10.6. Generally speaking, the HOCl species dominates at pH levels less than 7, and the OCl^- dominates at pH levels greater than 8. Between 7 and 8, the speciation of aqueous chlorine is highly pH dependent. It should be noted that, although pH is the dominant factor, changes in temperature also result in a modest change in the proportion.

Chloramination

If free chlorine is added to completely pure water, the free residual would be the same as the amount of chlorine added. However, if there are any contaminants in the water, the chlorine will react with them. The contaminants could be microorganisms, organic or inorganic compounds that can be oxidized (such as iron and manganese), or ammonia. The amount of chlorine used in the reaction with these contaminants is called the chlorine demand.

When chlorine reacts with ammonia, chloramine compounds are formed. The ammonia may be naturally occurring in the water or may be added in the treatment process to purposely form chloramines. Chloramines are useful disinfectants in some situations and are referred to as combined residual chlorine.

Understanding of the reactions between chlorine and ammonia is important to an adequate understanding of chlorination chemistry. These reactions are complex, and the temperature, pH mixing regimen, and Cl_2/NH_3 weight ratio all influence both the rate and the products of the reaction. When a small amount of chlorine is added ($Cl_2/NH_3 <4$),

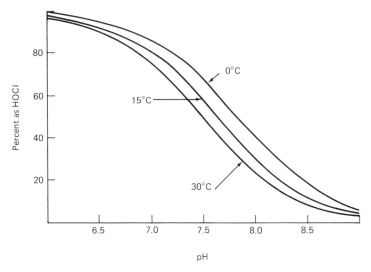

FIGURE 10.6 Influence of pH and temperature on fraction of aqueous chlorine as HOCl.

monochloramine is the dominant species formed. As additional chlorine is added, dichloramine and trichloramine are formed, along with other products such as NO_3 and N_2 gas. The following are some of the typical reactions that may occur:

Monochloramine: $NH_4^+ + HOCl = NH_2Cl + H_2O + H^+$

Dichloramine: $NH_2Cl + HOCl = NHCl_2 + H_2O$

Trichloramine: $NHCl_2 + HOCl = NCl_3 + H_2O$

Nitrogen: $2NH_4^+ + 3HOCl = N_2 + 5H^+ + 3Cl + 3H_2O$

Nitrate: $NH_4^+ + 4HOCl = NO_3^- + H_2O + 6H^+ + 4Cl^-$

These reactions give insight into the chlorine dose required to achieve free residual chlorination. Table 10.7 summarizes the theoretical chlorine dose required for these various chlorination reactions. It should be noted that a great deal of chlorine is required to form nitrate, and only slightly less to form trichloramine. The chlorine-to-ammonia weight ratio required for forming nitrogen, however, is even less than that required for converting all the ammonia to dichloramine. Saunier (1976) has shown that nitrogen is the principal species formed during the free residual reaction, but practical experience suggests that doses of 8 mg Cl_2/mg NH_3 are often required, and a dose of 9 to 10 mg Cl_2/mg NH_3 is recommended for design purposes. In practice, however, the NCl_3 concentration increases rapidly as the chlorine doses exceed the optimum, so excessive chlorine doses should be avoided. This is demonstrated by the free residual chlorine curves from Palin (1950) shown in Figure 10.7.

Combined residual chlorination was first used at Ottawa, Ontario, Canada. At the time, it was argued that chloramines had a germicidal action greater than that of chlorine alone, that their use could ensure that the water produced was free of taste and odors, that a much

TABLE 10.7 Chlorine Dose Required for Ammonia-Chlorine Reaction

Reaction	mg Cl_2/mg NH_3
Monochloramine (NH_2Cl)	4.2
Dichloramine ($NHCl_2$)	8.4
Trichloramine (NCl_3)	12.5
Nitrogen (N_2)	6.3
Nitrate (NO_3)	16.7
Recommended design dose for free residual reaction	9

more long-lasting residual was produced, and that the overall cost of disinfection could be reduced.

Many subsequent studies have shown that:

- The germicidal action of combined chlorine may be substantially less than that of free chlorine
- Combined chlorine is sometimes better from the taste and odor standpoint when the taste and odor of concern are the result of chlorine by-products
- The combined chlorine residual is indeed longer lasting than a free chlorine residual

Specific lethality data suggest that at pH of 7 or below, free chlorine is 200 times more effective in killing bacteria and viruses, 50 times more effective in killing spores, and 2.5 times more effective in killing cysts. For this reason, many of the water plants that currently use the chloramine process add chlorine first and maintain a free chlorine residual for some time before ammonia is added.

Chlorine Gas Systems

Gaseous chlorine is commercially available in 150 lb (68 kg) cylinders, 1-ton (1,016 kg) containers, in tank trucks, or in railroad tank cars. In all these vessels, liquid chlorine occupies approximately 85% of the volume when the product is delivered. This is to provide room for the expansion of liquid chlorine if the container should be heated.

Chlorine cylinders and containers should never be directly heated. As a safety precaution, their outlet valves are equipped with a small fusible plug that melts at approximately 158° F and releases some liquid chlorine to cool the cylinder before a more serious accident can occur.

Small to medium size water systems generally withdraw gas from the top of the container. The maximum withdrawal rate with this method is about 40 lb (18 kg) per day for a 150 lb (68 kg) cylinder and 400 lb (180 kg) per day for a 1-ton (1,016 kg) container. Higher feed rates can be obtained by connecting two or more cylinders or containers to feed simultaneously. The temperature of the chlorine feed room should be maintained at about 65° F (18° C).

Provision must also be made for a weighing device. This can be best accomplished by either a lever scale system or load cells. Load cells are more commonly used.

If the containers being used are 1 ton or larger and the withdrawal rates exceed those available with the direct evaporation method described earlier, chlorine evaporators may be used. Evaporators are available in capacities of 4,000, 6,000, and 8,000 lb (1,800, 2,700, and

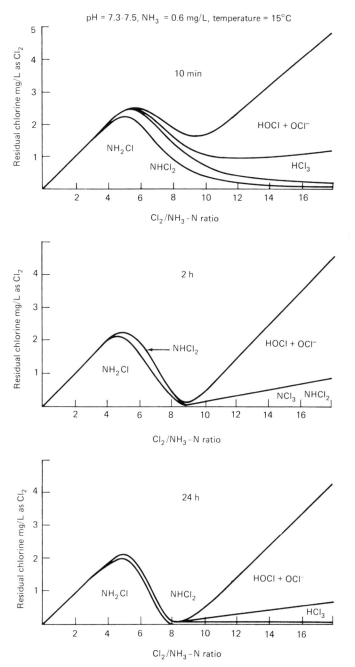

FIGURE 10.7 Free residual chlorine curves. *(Source: A. Palin, "A Study of the Chloro Derivatives of Ammonia," Water Engineering, vol. 54, December 1950, p. 151.)*

3,600 kg) per day. When an evaporator is used, liquid chlorine is withdrawn from the bottom of the container and transported to the evaporator, where it is converted to a gas. The most common type of evaporator uses an electric resistance heater in a hot-water bath surrounding a vessel in which the liquid chlorine is converted to gas.

The heat of evaporation of chlorine is very low, approximately 69 cal/g, compared with 540 cal/g for water. However, evaporators should be designed with extra capacity to ensure that the existing gas is superheated and will not recondense on the downstream side. When an evaporator is being used beyond its capacity, misting occurs. A chlorine gas filter should be installed on the exit gas line from the evaporators to remove impurities in the chlorine that would be detrimental to the chlorinator. Evaporators should be equipped with an automatic shutoff valve to prevent liquid chlorine from passing to the chlorinators.

When possible, all portions of the chlorine feed system that contain liquid chlorine should be designed and operated with all the liquid in the system as a continuous medium. To shut down the evaporator, it is necessary only to close the effluent valve on the evaporator. No other valves between the evaporator effluent valve and the liquid chlorine container should be shut. Long liquid chlorine lines should be avoided, but if they are unavoidable, chlorine expansion chambers should be provided. It should be emphasized that liquid chlorine has a high temperature-expansion coefficient. Unless expansion is permitted, the temperature increase in trapped liquid will result in pressure high enough to rupture the pipes.

Gas Feeders. A conventional chlorinator consists of an inlet-pressure–reducing valve, a rotameter, a metering control orifice, and a vacuum-differential regulating valve. A simple schematic is shown in Figure 10.8. The driving force for the system comes from the vacuum, which is created by the chlorine injector. The chlorine gas comes to the chlorinator and is converted to a constant pressure (usually a mild vacuum) by the influent-pressure–reducing valve. The chlorine then passes through the rotameter, where the flow rate is measured under conditions of constant pressure (and consequently constant density), and then through a metering or control orifice.

A vacuum differential regulator is mounted across the control orifice so that a constant pressure differential (vacuum differential) is maintained to stabilize the flow for a particular setting on the control orifice. The flow through the control orifice can be adjusted by changing the opening on the orifice. The control orifice has a typical range of 20 to 1, and the vacuum differential regulator has a range of about 10 to 1. Thus the overall range of these devices combined is about 200 to 1. On the other hand, a typical rotameter has a range of about 20 to 1. Thus the chlorinator should be selected based on design capacities, and the rotameter installed at any particular time should be appropriate for current demands. A large chlorine feed system is illustrated in Figure 10.9.

Chlorine Gas Piping. Chlorine gas flows in a vacuum from the chlorinator to the injector, and it is critical that the vacuum created by the injector be transmitted to the chlorinator without significant dissipation. As a consequence, the diameter of the chlorine vacuum lines should always be designed rather than arbitrarily selected. According to White (1986), lines should be sized to limit the total pressure drop over the pipe length to between 1.5 and 1.75 in. (3.8 and 4.4 cm) Hg under maximum injector vacuum levels. The following formula can be used to estimate pressure drop:

$$P = \frac{1.89\,L f W^2}{109\,p d^5}$$

where P = total pressure drop, in. Hg
L = length of line, ft

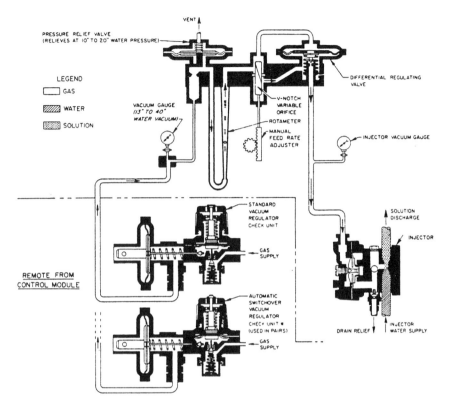

FIGURE 10.8 Flow diagram for conventional chlorinator. *(Courtesy of Wallace and Tiernan Division of Pennwalt Corp.)*

f = friction factor
W = chlorine flow, lb/day
p = chlorine density, lb/ft^3
d = inside pipe diameter, in.

The chlorine density can be estimated from Figure 10.10 if the pressure and temperature are known. For PVC gas lines, the friction factor can be estimated from the following formula once the Reynold's number is determined:

$$\log f = 1.75 - 0.722 \log N_{re}$$

The Reynold's number can be estimated by the following formula:

$$N_{re} = \frac{0.263W}{ud}$$

where u = viscosity of gas, cP
d = inside pipe diameter, in.

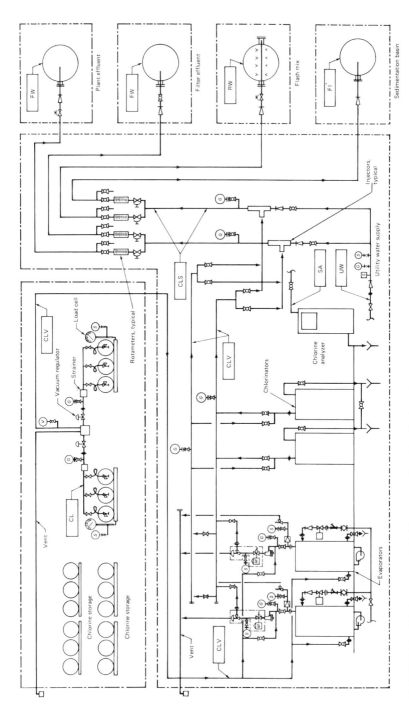

FIGURE 10.9 Chlorine system. (*J. M. Montgomery Consulting Engineers.*)

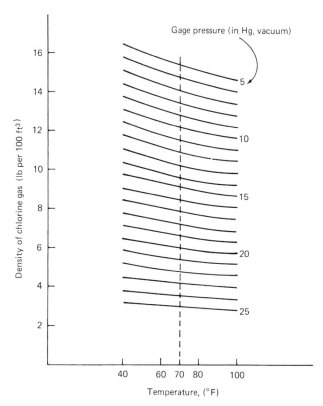

FIGURE 10.10 Density of chlorine gas under vacuum. *(Adopted from A. S. Ross and O. Mass, "The Density of Gaseous Chlorine," Canadian Journal of Research, vol. 18B, 1940, pp. 55–65.)*

In a 40° to 60° F (4° to 15° C) temperature range, the viscosity can be estimated by the following formula:

$$u = 0.0115 + 0.00003T(°F)$$

The design procedure is to select pipe sizes and calculate the pressure drop. If the drop is too large, a larger pipe diameter should be used.

Initial Mixing of Chlorine. It is important to distribute the chlorine dose uniformly across the flow cross section of the pipe or basin. Traditionally, good designs have accomplished this with open-channel or pipeline diffusers. It can be shown, however, that truly complete mixing across the flow cross section only takes place some distance downstream from diffusers of this type.

When free chlorine residuals are being used, this may not be a critical problem. But when water contains significant amounts of ammonia (10% of the chlorine dose or more), initial mixing takes on new importance. Under these conditions there is evidence that chlorine is more effective during the first few seconds after its addition. With ammonia present, failure to mix the chlorine solution rapidly with all the water can result in a "break-point" reaction with part of the water and little or no reaction with the remainder. The break-point reaction is not reversible. Thus part of the water receives excess chlorine, and the remainder receives little or none.

Consequently, good initial mixing becomes a much more important process step, and initial mixing devices should be designed to spread the chlorine throughout the flow cross section in the shortest possible time with the least possible backmixing. This can be accomplished by in-line mixers, by gas aspirator devices, by installing the chlorine diffuser in a hydraulic jump, or by installing a pump-diffuser system. Flash-mixing chambers such as those used for coagulants in a conventional water treatment plant are not satisfactory because they result in backmixing.

Materials for Handling Chlorine.   Dry chlorine gas will not attack carbon steel at normal temperatures; as a consequence, liquid chlorine is packaged in steel containers. On the other hand, like liquid oxygen, liquid chlorine will sustain combustion of steel once any portion of the steel-chlorine contact surface has been heated to the kindling point of 438° F (225° C). Because of this potential danger, heat should never be applied to a chlorine container, and all chlorine piping should be clearly labeled.

If a steel pipe containing liquid chlorine or even chlorine gas at reduced pressure is accidentally cut with a welder's torch, the pipe will ignite and continue to burn as long as there is a chlorine supply available. Small amounts of moisture will also cause chlorine to attack steel. Because a trace amount of moisture is unavoidable, some of the corrosion product ($FeCl_3$) is always found in chlorine containers and in chlorine lines.

The preferred materials for handling chlorine in the three basic parts of the system are listed in Table 10.8. The supply system includes the storage tank and all piping and fittings up to the pressure-reducing valve (PRV) on the chlorinator inlet, including the evaporator, if it is present. The vacuum system refers to all the piping, fittings, and other elements in which the chlorine gas is in a vacuum, including the area between the chlorinator inlet PRV and the injector. The chlorine water system includes all piping above and below water downstream of the injector.

All parts, from the chlorine supply to the feeders, should be carbon steel. Piping should be schedule-80 seamless-weld carbon steel. Reducing fittings should be used rather than bushings, and ammonia-type unions with lead gaskets should be used rather than ground-joint unions. All parts should meet Chlorine Institute standards. Mainline valves should be ball type or rising-stem type, made of cast iron. The ball type is preferred because it is easier to operate and because the lever indicates the position of the valve at a glance.

Piping systems can be assembled by welding or by threading, although welding is preferred. If threaded piping is used, Teflon tape should be specified as the thread lubricant. Only diaphragm-type pressure gauges should be installed in chlorine systems, and these should have silver diaphragms and Hastalloy-C housings.

TABLE 10.8 Materials Selection for Handling Chlorine

Portion of the chlorination system	Location	Form of chlorine	Acceptable materials
Supply system	Piping from storage through to the chlorinator inlet	Liquid chlorine Gaseous chlorine under pressure	Schedule-80 stainless steel Carbon steel Cast iron
Vacuum gas system	Chlorinator outlet to injector inlet	Gaseous chlorine under vacuum	Schedule-80 PVC Reinforced fiberglass
Chlorine water lines	Injector to diffuser	Chlorine solution	Schedule-80 PVC

Chloramine Systems

Between 1930 and 1940, a large number of plants installed facilities for using combined chlorine residuals. An AWWA survey of 36 states in 1940 showed that 2,541 supplies treated their water supply with chlorine and 407 used combined chlorine residuals. Thus, at its peak, combined residual chlorine was used in about one out of every seven water supplies. After 1940, the discovery of break-point chlorination, the difficulty of obtaining ammonia during World War II, and pressure from public health officials to use a more effective free chlorine residual caused combined residual chlorination to decrease in popularity.

Interest in combined residual chlorination again increased because chloramines do not react with the natural aquatic humus present in water to form trihalomethanes (THM). In many applications chloramines may be the most cost-effective solution to limit THM formation.

The requirements of the Surface Water Treatment Rule have again reduced use of chloramines for disinfection because it is such a weak disinfectant, particularly in inactivating viruses. To meet SWTR requirements, most systems must expose the raw water to free chlorine for a period of time before the addition of ammonia. If a system wishes to rely on chloramine without exposing the water to any free chlorine, it will have to show to the satisfaction of the state that the system is achieving the required inactivation. Chloramine performance is affected by water temperature, so tables of CT values at various temperatures must be consulted in considering the inactivation that can be obtained.

Chlorine/Ammonia Feed Ratio. In designing facilities for combined residual chlorination, the residual ratio of ammonia to chlorine should be considered. White (1986) recommends a ratio of 3 parts of chlorine to 1 part of ammonia. A survey of 24 utilities using combined residual chlorination revealed Cl_2/NH_3 ratios currently employed ranging from 1 to 4, the median being 3. As the Cl_2/NH_3 ratio gets higher, the cost of the ammonia required to maintain a given level of chlorine residual increases. For alkaline pH levels, the maximum long-term combined residual will be at a ratio of less than 5; for more acid pH levels, a higher ratio applies. Above this dose, substantial amounts of combined chlorine will be lost to nitrogen and nitrate ions. When the ratio is above about 4, chlorinous odors may be observed. Very low water ammonia ratios risk excessive corrosion of copper and brass elements of the distribution system.

Anhydrous Ammonia Systems. Ammonia is available on the commercial market in three useful forms: anhydrous ammonia, aqueous ammonia, and ammonia sulfate. Aqueous and anhydrous ammonia are the forms most commonly used.

Anhydrous ammonia is available as a compressed liquid in containers nearly identical to those used for chlorine. Most of the equipment used in connection with chlorine can also be used for feeding ammonia with minor modification. Chlorinators can be used as ammoniators through the use of a simple modification kit. Ammonia has a heat of vaporization of 328 cal/g and may be fed from the top of the cylinder in the same manner as chlorine, although not at the same rate. The capacity of a 1-ton ammonia cylinder by the evaporation method is about 84 lb (38 kg) per day. Like chlorine, ammonia liquids should be transported in black iron pipe with welded steel fittings.

Although anhydrous ammonia can be fed by an injector system, it is highly soluble in water, and simple direct-feed ammoniator designs are common. Although an ammonia injector–type system eliminates the need for transporting toxic ammonia under pressure to distant locations in the plant, ammonia is a base, unlike chlorine, and it will soften the water at the point of injection, producing a precipitate of calcium carbonate. This can cause severe scale problems in the injection system. Users of anhydrous ammonia report that carbonate precipitates also build up at diffuser ports when anhydrous ammonia is fed directly.

These problems are particularly troublesome in hard water, so ammonia injector systems may require that soft water or a special self-cleaning diffuser with rubber sleeve be used.

Aqueous Ammonia Systems. Aqueous ammonia is delivered in a solution that is 33% ammonia by weight. Aqueous ammonia is usually fed through PVC lines, although iron pipe is also acceptable.

In concentrated form, ammonia reacts chemically with copper, so under no circumstances should any brass, bronze, or other copper alloy be used in any ammonia feed system. The vapors above the solution in an aqueous ammonia tank are extremely potent. Therefore provision should be made for disposing of the displaced vapors in the tank in a safe manner, such as transfer back to the delivery vehicle, when the storage tank is being refilled. Pressure-release valves on these tanks should pass through a water-type scrubber before going to the atmosphere.

Aqueous ammonia is often fed through stainless steel diffusers with holes $\frac{1}{8}$ in. (3 mm) or larger, but designed with a significant back pressure. The diffusers should be carefully laid on a horizontal grade, because at low doses and low plant flows, poor distribution will result if variations of hydrostatic pressure occur.

Ammonia dosage generally is not critical, and a manual or flow-pace control system may be satisfactory. Some designs feed ammonia by gravity, drawing liquid ammonia from the bottom of the storage vessel through a flowmeter and regulating valve to the point of injection. The vapor pressure from the liquid ammonia is the prime mover in this instance. Metering pumps with Hypalon or Teflon surfaces may also be used to improve dosage control. The ideal control system would permit a feed rate proportional to the product of the flow and the chlorine residual.

When ordering aqueous ammonia, debris-free chemical should be specified, because aqueous ammonia is sometimes delivered with considerable debris present. Alternatively, facilities may be installed to strain undesirable debris from the product before it is used. Clogging with precipitated calcium carbonate is often reported to be a problem with diffusers using aqueous ammonia. Softening of the carrier water may be necessary to avoid excessive maintenance.

Ammonium Sulfate Systems. Solid ammonium sulfate is usually fed into a simple mixing tank using a gravity or volumetric feeder. Once mixed, the solution can be transported using the same methods described earlier for aqueous ammonia. If the local water is hard, scaling problems may occur, and softening of this carrier water should be considered.

Sodium Hypochlorite Systems

Sodium hypochlorite is used to feed chlorine by many small and medium-size water systems because of the greater ease of handling and much greater safety. In recent years, some larger water systems that are located in metropolitan areas have also changed from chlorine gas to the use of sodium hypochlorite because of an increased emphasis on safety where large amounts of gas are stored in or transported through urban areas.

If cost were the only criterion, liquid chlorine would always be chosen rather than sodium hypochlorite. Based on chemical costs only, hypochlorite would cost two to four times the equivalent quantity of gaseous chlorine. The recent UFC changes for gaseous chlorine as an AHM typically result in total net present worth for hypochlorite about twice that of gaseous chlorine.

Purchasing Hypochlorite. Sodium hypochlorite (liquid bleach) is formed by combining chlorine and sodium hydroxide. In some instances, it is made at the site, with both of these products generated electrolytically. In other instances, it is manufactured from chlorine and sodium hydroxide that have been separately shipped to the manufacturing site. The reaction that proceeds in this instance is as follows:

$$2NaOH + Cl_2 \rightarrow NaOCl + NaCl + H_2O$$

A slight excess of sodium hydroxide is often added to increase the stability of the chlorine in the product. When the hypochlorite is added to water, it hydrolyzes to form hypochlorous acid (HOCl), the same active ingredient that occurs when chlorine gas is used. The hypochlorite reaction slightly increases the hydroxyl ions (pH increase) by the formation of sodium hydroxide, whereas the reaction of chlorine gas with water increases the hydrogen-ion concentration (pH decrease), forming hydrochloric acid. In most waters, these differences are not significant, but when high chlorine doses are used in poorly buffered waters, these effects should be considered. They can be evaluated by calculation or by simple laboratory tests.

In the commercial trade, the concentration of sodium hypochlorite solutions is usually expressed as a percentage. The "trade percent" is actually a measure of weight per unit volume, with 1% corresponding to a weight of 10 g of available chlorine per liter. Common household bleach, at a trade concentration of 5.25%, has approximately 5.25 g/100 mL, or 52.5 g of available chlorine per liter. Hypochlorite available for municipal use usually has a trade concentration of 12.5% to 17%. These are approximate concentrations, and they should always be confirmed for a particular shipment by laboratory procedures.

Because increasing the concentration of any salt will lower the freezing point of a solution, the freezing points of various solutions of sodium hypochlorite are a function of their concentrations, with the more dilute concentrations approaching the freezing point of pure water.

Hypochlorite solutions are subject to degradation during storage, with chlorate and chlorite as the primary degradation products. These chemicals are of health concern in drinking water and are being considered by USEPA for future regulation.

The rate of degradation is accelerated by high hypochlorite concentrations, high temperature, the presence of light, low pH, and the presence of heavy metal cations such as iron, copper, nickel, and cobalt.

When purchasing bulk sodium hypochlorite, purchasing specifications should be used, delineating the acceptable ranges for available chlorine (12.5% to 17%) and pH (11 to 11.2), as well as maximum contaminant limits for iron (2 mg/L) and copper (1 mg/L). Specifications should also require that shipments be free of sediment and other deleterious particulate material. Shipments should be analyzed for the concentration of chlorine, the pH, and the concentration of metal contaminants.

Hypochlorite Storage. The storage of sodium hypochlorite must be carefully managed to limit degradation and the formation of chlorate. In that the rate of degradation is a function of temperature, and chlorine concentration, storage time should be limited to less than 28 days and temperatures to 70° F (21° C) or less. The chemical may be delivered warm—that is, at temperatures up to 85° F (30° C)—which must be considered in the system design. Alternative approaches to hypochlorite management include:

- Installation of immersion coolers in the storage tanks connected to a chilled water source to cool the chemical
- Installation of the storage tanks in a climate-controlled room to maintain the chemical at the desired temperature
- Provision of additional storage capacity to allow for dilution of the chemical on delivery (50% dilution will reduce the rate of decomposition by a factor of 4).

Materials used for bulk storage of sodium hypochlorite may be fiberglass, cross-linked rotational molded polyethylene, or lined carbon steel. Hypochlorite tanks should be vented, and provision should be made for sampling the contents. Connection to the delivery vehicle should be Hastalloy-C, titanium nipples securely braced to the tank, or schedule-80 PVC.

Schedule-80 PVC is the most commonly used material for hypochlorite piping. Valves may be plug valves made of steel, lined with PVC or polypropylene, PVC ball valves, or PVC diaphragm valves. Diffusers are generally made of PVC or Kynar. Because of the high salinity of hypochlorite solutions, the diffusers should be designed for high velocities to ensure mixing throughout the channel cross section. The rate of flow is most commonly controlled by a diaphragm valve. Steel valves lined with PVC, Kynar, hard rubber, or Saran are acceptable.

Hypochlorite Feed. Hypochlorite is highly volatile, so the feed systems must be carefully designed to prevent gas binding. Hypochlorite may be fed by gravity, may be pumped, or may be fed by eductor systems. Gravity systems are preferred because of the simplicity of their design and their reliability. In these installations, the difference in static head between the hypochlorite storage tanks and the point of application is used to provide gravity flow, and the flow is modulated using a flowmeter and a diaphragm valve. If a sufficient hydraulic gradient does not exist between a storage tank and the point of application, an intermediate constant-head tank can be used. In any instance, a constant-head tank will provide improved flow control.

Most small to medium-size water systems feed hypochlorite with positive displacement diaphragm metering pumps. The metering pumps should be located as close to the storage tanks as possible with the suction lines oversized. The pumps should be located well below the level of the liquid in the storage tank. For larger systems, centrifugal pumps have been used, with downstream modulating valves to control the flow rate, but they have not been found as satisfactory as gravity feed or the use of diaphragm pumps.

Eductor systems may be used for larger feed rates and use the same principle as chlorine gas injector systems. An eductor is used to draw a vacuum, which pulls the hypochlorite from the storage tank to the point of application, and the flow is controlled by a modulating valve.

On-Site Generation of Chlorine

Although on-site generation of disinfectants has been possible for many years, it has not generally been found to be practical. However, new concerns over accidental release of chlorine gas and the relatively high cost of scrubbers that may have to be installed by many plants to prevent release of gas into adjoining neighborhoods has raised new interest in on-site generation. Several systems are now available that use new technology to generate disinfectants and use only salt and electrical power. They can even be designed to operate unattended at a remote location and are intended to meet all USEPA standards for disinfection and maintaining a chlorine residual in the finished water.

Considerations that must be made in evaluating alternative on-site disinfectant generation systems include the operating cost of the process, the concentration of the brine produced, the availability of salt of sufficient quality, necessary conjunctive treatment processes, and the reliability of the process. The principal operating costs are the cost of salt, the cost of any on-site salt purification necessary, the cost of power, maintenance, and replacement of parts.

The concentration of brine is important because more dilute brine concentrations mean that more on-site storage is required. There is considerable variation in the maximum brine concentration that can be produced by the different processes. A commercial hypochlorite generation facility may include facilities for refining salt quality, evaporating and concentrating the product, and filtration. Hence the simpler municipal installations ordinarily produce much more dilute brine with a higher NaCl content.

Chlorination Control

Proper design of the control system for chlorination facilities is as important as any other aspect of their design. Methods currently used are manual setting based on flow and periodic or continuous residual measurements; continuous feedforward control based on continuous flow measurements and feedback control based on continuous residual measurements; and compound, closed-loop control based on continuous measurements of both flow and chlorine residual. Each of these methods is illustrated in Figure 10.11. Details of control-signal manipulation are not shown.

Control Methods. The manual control method is limited by the diligence of the operator. If either the flow or the chlorine residual changes, the operator must make adjustments; consequently, this method gives a wide range of performance. The feedforward (flow pacing) method is a significant improvement because the flow has been eliminated as a variable provided the operator is diligent in maintaining the control system. For water supplies where chlorine demand is very stable, feedforward control is preferred.

Feedback control is theoretically superior to feedforward control because the chlorine residual is being directly used to control chlorine addition. Unfortunately, this is not always the case, because chlorine residual analyses drift off course unless they are regularly maintained. Feedforward control is better than feedback control alone.

Whenever chlorine demand is variable, compound-loop control is preferred. This system is sometimes accomplished by controlling the differential vacuum regulator on the chlorinator with the flow signal and setting the chlorine gas-metering orifice with the signal from the residual analyzer. Alternatively, it can be provided by electronically adding both control signals and using the results to control the differential vacuum. The first approach allows the chlorinator to operate over a dynamic range of 200 to 1, whereas the differential vacuum will only allow a dynamic range (maximum feed/minimum feed) of 20 to 1. However, in most installations, a dynamic range of 20 to 1 is satisfactory.

With either feedback or compound-loop control, lag time is one of the principal design parameters. Lag time is the time between the moment when the chlorine is added to the effluent and the time when the residual-analyzer signal comes to the chlorinator. Lag time includes the transit time from the point where the chlorine is initially mixed until it reaches the sampling point, the transmission time between the sample point and the chlorine residual analyzer (in the sample line), and the analysis time.

The analysis time is usually a minor factor. If the lag time is too much longer than the response time of the analyzer, the level of the chlorine dose will sawtooth. White (1986) suggests that the lag time be maintained at an average level of 2 min, with a maximum of 5 min. Low-flow conditions should be considered. The following are some of the most common design errors: poor chlorine sample conditions, analyzer located too far from sampling point, and effluent chlorine dose paced to influent flow.

It should be understood from the beginning that the purpose of the chlorine residual analyzer discussed here is to control the chlorine dose. If continuous monitoring of the chlorine residual after the chlorine contact period is desired, another chlorine residual analyzer is required.

Sample Point Location. The principal consideration in locating the sampling point for the control analyzer is that there must be good mixing. If the sample is taken before adequate mixing has occurred, the result will be erratic readings unsuitable for control. For the

246   CHAPTER TEN

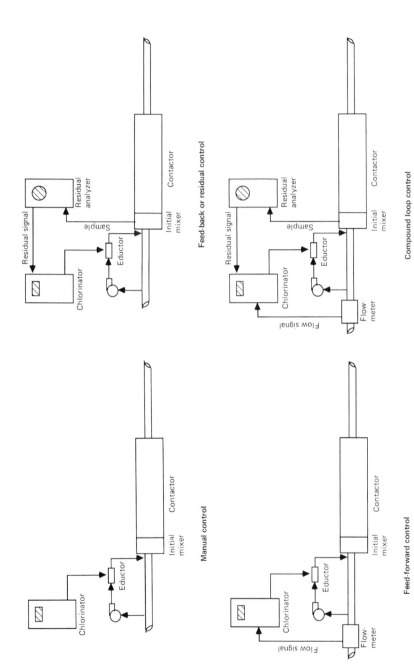

FIGURE 10.11 Alternative chlorination control methods.

majority of initial mixing designs, sampling should be provided immediately downstream of the initial mixing device. Ordinarily, chlorine residuals are stable enough for control measurements after just a few seconds of contact.

If no initial mixing device is present, the sampling point should be far enough downstream to ensure that good mixing has occurred. For turbulent flow, 10 pipe diameters is usually sufficient; however, low-flow conditions should be considered, and if adequate mixing cannot occur in a reasonable period of time, an initial mixing device will be necessary for control purposes.

Chlorine residual analyzers should always be located as near as possible to the sampling point, even if special housings are required. Sample lines should be designed for velocities of about 10 ft/s (3 m/s), and the transit time between the sampling point and the residual analyzers should be minimized. Finally, the chlorine dose should always be paced to the flow most representative of the point of addition. A common error in design is an arrangement in which the effluent chlorine dose is paced using influent flow measurements. Too many events occur between a plant's influent and its effluent, and such a design often results in an erratic chlorine dose and an unmanageable operating system.

Chlorine Residual Analysers

Two methods for continuous chlorine residual analysis are currently available: the automatic amperometric titrator and the ion-selective probe. In an automatic amperometric titrator, the cell has an indicating electrode made of copper concentrically mounted around a platinum reference electrode. Water flows into the space between the two, and a potential is imposed between the electrodes, resulting in a current flow that is proportional to the amount of chlorine in the sample. Ordinarily, a pH 4 buffer is used, and the free chlorine is measured. The use of a buffer with excess potassium iodide will cause the unit to titrate the total chlorine residual, and an excess of combined chlorine will interfere with attempts to measure the residual. One type of amperometric titrator is shown in Figure 10.12.

Chlorination System Design for SWTR Compliance

The following simplified steps can be taken to determine probable CT required and the probable credit:

1. Determine the total removal/inactivation required:
   - Determine whether the usual 3-log *Giardia* and 4-log virus removal/inactivation applies.
   - Determine whether higher removal/inactivation is required because of a vulnerable source.
2. Determine the credits that will be allowed for physical removal:
   - For conventional treatment—2.5-log *Giardia* and 2-log virus.
   - For direct filtration—2-log *Giardia* and 1-log virus.
3. Determine credits required for disinfection:
   - For conventional treatment—minimum 0.5-log *Giardia* and 2-log virus.
   - For direct filtration—minimum 1-log *Giardia* and 3-log virus.
4. Determine *CT* required for design conditions using tables provided in publications detailing the SWTR requirements. Note that the worst case would be a combination

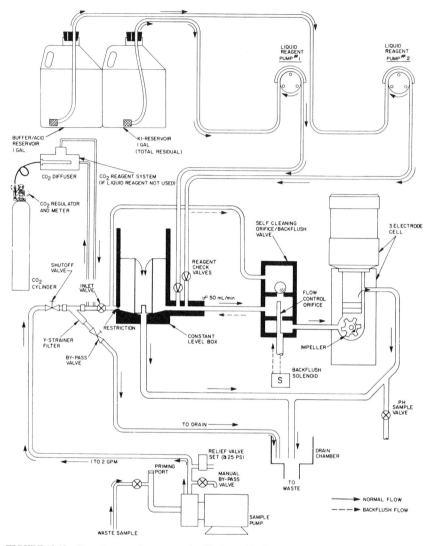

FIGURE 10.12 Flow diagram of amperometric chlorine residual analyzer. *(Courtesy of Wallace and Tiernan Division of Pennwalt Corp.)*

of high pH, low temperature, and high flow; however, these may not occur concurrently, and several combinations of representative conditions should be considered.

5. Using Figures 10.1, 10.2, and 10.3 determine the hydraulic efficiency of the basins that the water will pass through from the point of disinfectant application until it enters the distribution system (or at the first customer, if appropriate).
6. Using Table 10.2, compute the total detention time through the plant.

7. Multiply the detention time by the chlorine residual at the first customer, and compare it with the required *CT*.

If primary disinfection can be applied as the raw water enters the plant, there is usually not much problem in getting adequate detention time through the process basins in a full conventional plant. However, many plants now practice minimal prechlorination to control disinfectant by-product (DBP) formation, so the *CT* credit must be achieved in the clearwell or a separate contact basin with optimized hydraulic efficiency. The designer must be particularly careful when relying on a clearwell, or any treated water storage basin, for *CT* credit. Most existing clearwells are poorly baffled or have none at all. In addition, their basic function as useful storage volume, which may be withdrawn, may conflict with the need to maintain more than a minimum volume in order to satisfy contact time requirements.

CHLORINE DIOXIDE SYSTEMS

In certain circumstances, chlorine dioxide is an excellent choice among disinfectants. In particular, it is effective in destroying phenols, yet it does not form trihalomethanes in significant amounts. Chlorine dioxide's disinfectant properties are not adversely affected by a higher pH, as those of a free chlorine residual are. Consequently, chlorine dioxide is a much quicker disinfectant at higher pH levels. In western Europe, use of chlorine dioxide is increasing, particularly in Holland, Germany, France, and Switzerland, in regions where source water is of lower quality. In these locations, chlorine dioxide is used for disinfection, often as an adjunct to ozonation.

Chlorine Dioxide Disinfection

Chlorine dioxide does not dissociate or disproportionate as chlorine does at normal drinking water pH levels. Like chlorine, chlorine dioxide exerts a demand when it is first added to a water supply, which must be overcome if a persistent residual is to be maintained. Like chlorine, chlorine dioxide is photosensitive (light sensitive), and because it is a gas at temperatures above 11° C, its residuals are easily removed by aeration.

Although chlorine dioxide has recently been used at more treatment plants in the United States because it does not form trihalomethanes or haloacetic acids, there is still concern that there are other organic by-products of chlorine dioxide that are not yet well understood, and it may have other undesirable reaction products. Information presently available indicates that the reaction products include aldehydes, carboxylic acids, and ketones.

It is generally considered that chloro-organic by-products are not produced by reaction between chlorine dioxide and organic compounds, but may be present in practical applications as a result of free chlorine present in the chlorine dioxide solution. The principal inorganic by-products of chlorine dioxide reactions within water treatment are chlorite ion (ClO_2^-), chloride ion (Cl^-), and chlorate ion (ClO_3^-), in the order listed. Both chlorate and chlorite ions, particularly the chlorite ion, have been implicated in the formation of methemoglobin. Consequently, most European countries limit the level of chlorine dioxide that can be used, and USEPA has considered doing so in the United States as well.

The current USEPA recommendation is that the sum of chlorine dioxide, chlorite, and chlorate in the distribution system be maintained at less than 1.0 mg/L. The proposed Disinfectant-Disinfection By-Products Rule (DDBPR) (July 1994) provides stage 1 limits of 0.80 mg/L chlorine dioxide and 1.0 mg/L chlorite ion. No maximum contaminant level (MCL) has yet been proposed for the chlorate ion.

The basic design diagram of an automatic residual control chlorine dioxide system is shown in Figure 10.13.

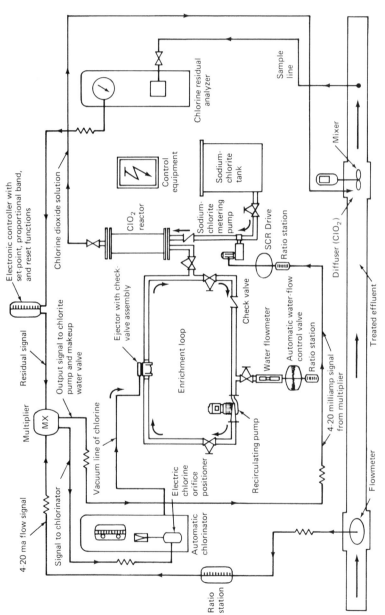

FIGURE 10.13 CIFEC automatic residual control chlorine dioxide system. *(Courtesy CIFEC, Paris, France.)*

Chlorine Dioxide Generation

Chlorine dioxide cannot be stored once it is generated because it is not safe. Numerous stimulants may cause the pure gas to explode, including an increase in temperature, exposure to light, changes in pressure, and exposure to organic contaminants. As a result, chlorine dioxide is usually generated on-site.

All chlorine dioxide for drinking water treatment is generated from sodium chlorite. Most generation techniques use the oxidative process, in which chlorine, either as a gas or in solution, is mixed with a sodium chlorite solution, $NaClO_2$. In addition to the desired formation of chlorine dioxide, chlorate ion may be formed in the generation system as an undesired by-product in a competing reaction when chlorine is in excess.

The goal in generating chlorine dioxide from chlorine and sodium chlorite is to maximize the chlorine dioxide yield, defined as the molar ratio of chlorine dioxide produced to the theoretical maximum. The term *conversion* is also used when referring to chlorine dioxide generation reactions; this is the molar ratio of the amount of chlorine dioxide formed to the amount of sodium chlorite fed to the system. For other reactions that produce chlorine dioxide, such as the hydrochloric acid–sodium chlorite reaction, yield and conversion will have different values.

Studies of the mechanism and kinetics of the chlorine–sodium chlorite reaction have shown that conditions favoring the formation of chlorine dioxide are those in which the reactants are present in high concentrations and the chlorine is present as either hypochlorous acid or molecular chlorine (Cl_2). Three methods for the generation of chlorine dioxide from chlorine and sodium chlorite are commercially available: the aqueous chlorine–sodium chlorite system, the gas chlorine–aqueous sodium chlorite system, and the solid sodium chlorite–gas chlorine system.

Aqueous Chlorine–Sodium Chlorite System. The earliest systems produced chlorine dioxide by simply pumping a sodium chlorite solution into a chlorine solution, followed by a short reaction time. Acceptable yields were achieved by feeding 200% to 300% more chlorine than the stoichiometric requirement. The chlorine dioxide solution from a generator of this type contains high levels of chlorine in addition to the chlorine dioxide.

With the discovery that potentially toxic chlorinated organics are generated by the reaction of chlorine and naturally occurring humic substances in water supplies, plus the growing interest in chlorine dioxide as a replacement for some chlorination practices, generation methods were sought that would produce a chlorine-free chlorine dioxide. One of the most common methods for chlorine dioxide generation currently in use that strives to meet this requirement is the pH-adjusted method.

The pH-adjusted system uses hydrochloric acid fed into the chlorine solution before reaction with the sodium chlorite. The acid feed serves to shift the chlorine solution equilibrium and the chlorine hydrolysis equilibrium, favoring hypochlorous acid and molecular chlorine. The acid feed must be carefully controlled so that the pH of the chlorine dioxide solution can be maintained between 2 and 3. Yield is reduced at both higher and lower pH values. Yields of more than 90% have been reported from the pH-adjusted system, with approximately 7% excess (unreacted) chlorine remaining in the solution.

Another modification that produces high yields of chlorine dioxide, with minimal amounts of chlorine remaining in the chlorine dioxide solution, requires that the chlorine solution used for generation have a chlorine concentration greater than 4 g/L. The exact relationship of excess chlorine required for 95% yield and initial chlorine solution concentration is shown in Figure 10.14.

Because this concentration of chlorine in solution is near the upper operating limit of commercial chlorine ejectors, and because these ejectors operate at constant water

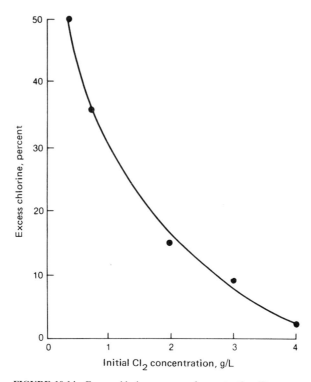

FIGURE 10.14 Excess chlorine necessary for greater than 95 percent conversion of chlorite to chlorine dioxide. *(Source: M. Aieta and J. Berg, "A Review of Chlorine Dioxode in Drinking Water Treatment," Journal AWWA, vol. 78, no. 6, June 1986, p. 62.)*

flow rates, the yield of this method of generation depends on the production rate, with lower production rates resulting in lower yields. This type of generator is normally operated on an intermittent basis to maintain high yield when less than maximum production capacity is required. Chlorine dioxide solutions in the 6 to 10 g/L concentration range are prepared and immediately diluted to about 1 g/L for storage and subsequent use as needed.

A schematic of the aqueous chlorine–sodium chlorite system is shown in Figure 10.15.

Gas Chlorine–Sodium Chlorite System. The most recent development in chlorine dioxide generator technology is a patented system that reacts chlorine gas with a concentrated sodium chlorite solution under vacuum. The chlorine dioxide produced is removed from the reaction chamber by a gas ejector, which is similar to the common chlorine gas vacuum feed system.

This generation technique produces chlorine dioxide solutions with yields in excess of 95%. The chlorine dioxide solution concentration is 200 to 1,000 mg/L and contains less than 5% excess chlorine, which is defined as the amount of unreacted chlorine remaining in the chlorine dioxide generator effluent. The system is operated on a continuous basis and achieves a high yield over the entire production range (Figure 10.16).

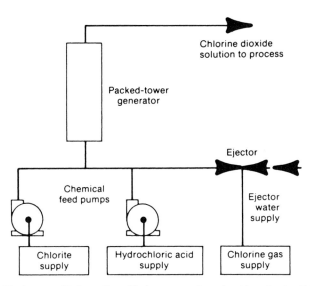

FIGURE 10.15 Aqueous chlorine-sodium chlorite system schematic with optional acid feed. *(Source: M. Aieta and J. Berg, "A Review of Chlorine Dioxide in Drinking Water Treatment," Journal AWWA, vol. 78, no. 6, June 1986, p. 62.).*

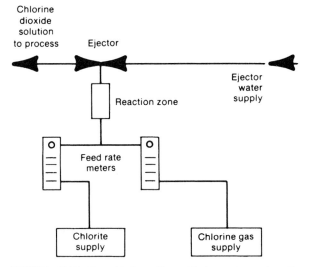

FIGURE 10.16 Gas chlorine-sodium chlorite system schematic. *(Source: M. Aieta and J. Berg, "A Review of Chlorine Dioxide in Drinking Water Treatment," Journal AWWA, vol. 78, no. 6, June 1986, p. 62.)*

Sodium Chlorite

Sodium chlorite is available as a solid, white-colored powder that is about 80% $NaClO_2$ by weight, and in the form of liquid solutions that are 25% to 32% $NaClO_2$ by weight. If granular sodium chlorite is used, it should be stored in a separate building equipped with sloped floors, drains, and facilities for hosing down spills. The building should be constructed of materials with maximum fire resistance, and it should be designed with the material's explosive potential in mind. Furthermore, $NaClO_2$ must be kept from contacting organic materials (leather boots, cloth, mops, etc.) because rapid oxidation reactions can ensue and cause fires.

OZONE DISINFECTION SYSTEMS

Although the first use of ozone in water treatment coincided with the first use of chlorine for disinfection, at the present time, ozonation is a common practice in only a small number of countries, notably France, Germany, and Switzerland. In many other countries, including the United States, England, and Japan, interest in ozonation in water treatment has increased substantially in the 1990s because of suspected carcinogenic properties of the trihalomethanes that are formed when organic compounds, naturally present in water, react with chlorine. By mid-1995 there were known to be at least 106 ozone systems in drinking water treatment service in the United States. At least 42 more were in design or under construction and scheduled to be in operation by the year 2000.

Ozone Chemistry

Ozone is a highly reactive gas formed by electrical discharges in the presence of oxygen. Its most distinguishing characteristic is a very pungent odor. In fact, the word "ozone" is derived from a Greek word that means "to smell." The use of this gas in water treatment requires an understanding of its physical and chemical behavior. The physical chemistry of ozone is important because a number of complex factors affect its solubility, reactivity, autodecomposition, and stability.

Ozone is an allotrope of oxygen. Substantial amounts of energy are required to split the stable oxygen-oxygen covalent bond to form ozone, and the ozone molecule readily reverts to elemental oxygen during the oxidation-reduction reaction. Ozone is more soluble in water than oxygen. The effect of temperature on ozone solubility is shown in Figure 10.17.

As illustrated in Figure 10.18, once ozone enters solution, it follows two basic modes of reaction: direct oxidation, which is rather slow and extremely selective, and autodecomposition to the hydroxyl radical. Autodecomposition to the hydroxyl radical is catalyzed by the presence of hydroxyl radicals, organic radicals, hydrogen peroxide, ultraviolet light, or high concentrations of hydroxide ion. The hydroxyl radical is extremely fast and nonselective in its oxidation of organic compounds, but at the same time, it is scavenged by carbonate and bicarbonate ions to form carbonate and bicarbonate radicals.

These carbonate and bicarbonate radicals are of no consequence in organic reactions. Furthermore, the hydroxyl radicals and organic radicals produced by autodecomposition become chain carriers and reenter the autodecomposition reaction to accelerate it. Thus low-pH conditions favor the slow, direct oxidation reactions involving O_3, and high-pH conditions or high concentrations of organic matter favor the autodecomposition route. High concentrations of bicarbonate or carbonate buffer, especially carbonate buffer, reduce the rate of autodecomposition by scavenging hydroxyl radicals. This means that ozone residuals last longer at low pH and in highly buffered waters.

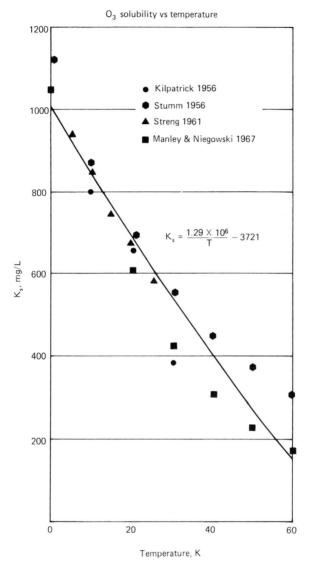

FIGURE 10.17 Effect of temperature on solubility of ozone.

The rate of autodecomposition is highly dependent on many factors, including pH, ultraviolet (UV) light, temperature, ozone concentration, and the presence of radical scavengers. Decomposition can be expressed assuming pseudo–first order kinetics and in terms of ozone residuals as given in the following equation:

$$C_t/C_0 = e^{-kt}$$

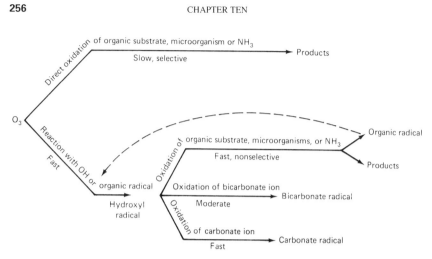

FIGURE 10.18 Reaction pathways of ozone in water. *(Source: J. Hoigne and H. Bader, "Role of Hydroxyl Radical Reactions in Ozonation Processes in Aqueous Solutions," Water Resources Bulletin, vol 10, 1976, p. 377.)*

where C_t = ozone concentration at time t, mg/L
C_o = ozone concentration at time 0, mg/L
k = decay rate constant, L/min
t = time, min

For design purposes, the decay rate constant must be determined for the water to be treated under expected operating conditions, including temperature, pH, and water quality.

Ozone Systems

Ozone systems are typically comprised of four basic subsystems: ozone generation, feed gas preparation, contacting, and off-gas disposal.

Ozone Generation. The basic configuration of an ozone generator (ozonator) is shown in Figure 10.19. An electromotive force (voltage) is impressed across two electrodes with a dielectric and discharge gap in between. Air is passed through the corona discharge between the two electrodes, and some of the oxygen in the air is converted to the ozone allotrope. Design principles suggest that the voltage necessary to produce ozone is a function of the product of the gap pressure and the gap width.

$$V = K_1 p g$$

where V = necessary voltage
p = gap pressure
g = gap width
K_1 = constant

In a similar sense, the yield of the ozonator is directly proportional to the frequency, the dielectric constant, and the square of the voltage applied and inversely proportional to the thickness of the dielectric.

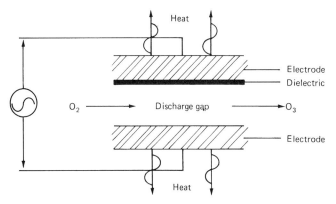

FIGURE 10.19 Basic ozonator configuration. *(Source: H. Rosen, "Ozone Generation and Its Relationship to the Economical Application of Ozone in Wastewater Treatment," in F. L. Evans III (ed.). Ozone in Water and Wastewater Treatment, Ann Arbor Science Publishers, Ann Arbor, Mich., 1972.)*

$$\text{Yield} = \frac{K_2\,(feV^2)}{d}$$

where f = frequency
V = voltage
e = dielectric constant
d = thickness of dielectric
K_2 = constant

This equation suggests some problems that are inherent in the design of ozonators. One is that the dielectric should be made as thin as possible. Very thin dielectrics are, however, more susceptible to failure. In the same sense, the yield is related to the square of the voltage, indicating that high voltages are desirable. On the other hand, dielectric failure also occurs when high voltages are used.

Equipment designs are currently available as low frequency (i.e., line frequency—60 Hz in the United States), medium frequency (i.e., 400 to 1,000 Hz), and high frequency (up to 2,000 Hz). The medium-frequency design has become the most common, with each manufacturer optimizing operating conditions to balance the interacting factors. Recent design improvements involving dielectric materials, gap width, and better methods of removing heat from the ozone cell have drastically improved the performance and capabilities of ozone generation.

Three basic types of ozone-generating systems are now in use: the Otto plate, the conventional horizontal tube, and the Lowther plate. Each of these designs is briefly sketched in Figure 10.20. The Otto plate was designed in 1905. Although inefficient, this design is still being used in parts of western Europe. Its principal disadvantages are that it is inefficient and that only low pressures can be used within the unit. High pressures are desirable so that the ozone can be bubbled through deep ozone contact chambers.

The tube-type generator is composed of a number of stainless steel tubes fitted into a large vessel and surrounded by cooling water. A concentric glass tube with a conducting coating on the inside is placed inside each stainless steel tube. A potential is applied between the inside coating of the glass tube and the outside steel tube, and air or oxygen is then passed through the gap in between. Variations of this design are by far the most common ozone generators in use today.

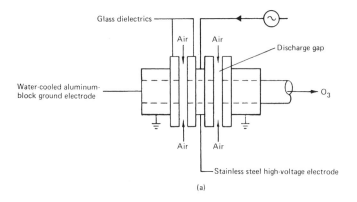

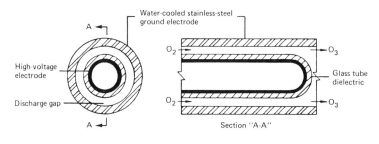

FIGURE 10.20 Alternative ozonator configurations. (a) Otto plate-type generator unit; (b) tube-type generator unit; (c) Lowther plate generator unit. *(Source: H. Rosen, "Ozone Generation and Its Relationship to the Economical Application of Ozone in Wastewater Treatment," in F. L. Evans III (ed.), Ozone in Water and Wastewater Treatment, Ann Arbor Science Publishers, Ann Arbor, Mich., 1972.)*

Discounting improvements in the tube-type generator, the Lowther plate unit is the most recent development. Whereas the other two units are usually water cooled, this unit is air cooled. It is made up of a gas-tight arrangement of an aluminum heat dissipator, a steel electrode coated with a ceramic dielectric, a silicone-rubber spacer, and a second ceramic-coated steel electrode with inlet and outlet. The silicone-rubber spacer sets the width of the discharge gap. Several of these units can be pressed together and manifolded to increase generator production. This design has been most effective in smaller capacities used for aquariums, cooling water, spas, and similar applications.

Feed Gas. Of all the component subsystems in an ozonation facility, feed gas preparation is the most complex, presents the most options to the designer, and consequently presents the most opportunities for error. While striving to achieve the functional objectives of feed gas preparation, the designer is often faced with conflicting demands in simplicity of equipment and operation and low power costs. The feed gas subsystem directly accounts for only 15% to 40% of the total power consumption for the entire facility.

The function of a feed gas preparation system is to supply gas to the ozone generators of appropriate quality and in quantities sufficient for the process. Design of a feed gas preparation system should address the following parameters of the gas:

- Moisture content
- Particulates and other contaminants
- Oxygen concentration
- Temperature
- Pressure
- Mass flow rate

Moisture Content. The most important gas quality parameter is moisture content. Moisture content of gases is usually expressed as dew point, which is the temperature at which moisture begins to condense out of the gas. To understand gas preparation systems, the designer should remember that the moisture-holding capacity of a gas decreases with decreasing temperature and with increasing pressures.

In ozone applications, the feed gas dew point must be extremely low. Excess moisture will adversely affect ozone production and may react with nitrous oxides within the ozone generator to form nitric acid, causing subsequent damage to the generator. Current practice is to design for $-60°$ C as the minimum acceptable dew point. Most systems operate at below $-80°$ C dew point.

Particulates and Other Contaminants. Particulates can potentially cause problems at a number of locations within an ozone system. Compressors with close running clearances must be protected from particles that would score the impeller, lobe, or rotor surfaces. Desiccant dryers must be protected from dust and hydrocarbons that would block the desiccant pores, thereby reducing the desiccant moisture-holding capacity. Because the ozone generator acts as an electrostatic precipitator, particulates and hydrocarbons will attach to the dielectric surface. In some cases, if particulates accumulate in the unit, hot spots will develop and eventually lead to dielectric failure. Fine bubble ozone diffusers must be protected from particulates that would eventually clog the diffuser.

Particulates may be inorganic or organic and may be present in the ambient air or generated from within the ozone system itself. Typical inorganic materials include sand, dust, lime dust, coal dust, construction debris, and moisture droplets. Common organic materials include pollen, cottonwood seeds, and the like. From within an ozone system, the most common particulate is desiccant dust.

In addition, hydrocarbons may be present in the atmosphere in large metropolitan areas and in the oil- and gas-producing areas of the country. Although not commonly used, lubricated-type compressors will also introduce unwanted hydrocarbons into the system.

Filters are the best method of controlling particulates and should be placed ahead of the compressor, ahead of the desiccant dryers, and after the desiccant dryers. Although filter rating is something of a black art, the goal should be a final feed gas entering the ozone generator free of all particulates larger than 0.3 μm.

Hydrocarbons in the feed gas can be controlled with coalescing and carbon-absorber filters. For installations with lubricated screw-type compressors, the coalescing filter will typically remove oil droplets larger than 0.05 μm. However, a carbon absorber is necessary for capturing and removing hydrocarbon vapors.

Oxygen Concentration. Because ozone generation is the conversion of molecular oxygen to ozone, increasing the oxygen concentration of the feed gas will increase the production of ozone. Increased oxygen concentration can be achieved through supplementing, or replacing, the air stream with a high-purity oxygen source using one of the following:

- Purchased liquid oxygen (LOX) delivered to the site and stored in refrigerated thermos tanks at 95% to 99%+ oxygen
- On-site generation of gaseous oxygen using cryogenic air separation technology at 90% to 99%+ oxygen
- On-site generation of gaseous oxygen using pressure or vacuum swing (PSA/VSA) technology at 90% to 95% oxygen

Temperature. At elevated temperatures, the rate of ozone decomposition increases. Consequently, feed gas temperatures should be relatively cool to avoid rapid decomposition of ozone as it is produced within the generator. Current practice limits entering gas temperature to below 90° F (33° C).

Pressure. The system pressure at the point of delivery from the feed gas preparation system to the generator is dictated by two factors. First is the pressure required to overcome all downstream pressure losses through the ozone contact basin, including those from control valves, the generator, line losses, diffusers, and the static head of the water above the diffuser.

The second is the design pressure at which the manufacturer has optimized generator performance. If the manufacturer's optimum design pressure exceeds the downstream losses, a pressure-maintaining valve will be placed downstream of the generator. Low-frequency generators typically operate at low pressures, 8 to 12 lb/ft^2 (55 to 83 kPa); medium-frequency generators operate at 18 to 25 lb/ft^2 (124 to 172 kPa).

Mass Flow Rate. Feed gas preparation systems must be designed to provide sufficient mass flow to achieve the design ozone production at the desired ozone concentration. The mass flow required can be calculated as follows:

$$\text{Required feed gas mass flow (lb/h)} = \frac{\text{Design ozone production (lb/d)}}{24 \times \text{Ozone concentration}}$$

The mass flow rate can be converted to a volumetric flow rate for compressor selection. However, although compressors typically used in ozone systems are constant-volume machines, the mass flow rate will vary with inlet air temperatures because of changes in inlet gas density. Because the lowest mass flow rate output from a constant-volume compressor occurs when the inlet temperature is highest, summer will usually be the critical condition for compressor selection.

Air Feed Systems Feed gas preparation systems can be classified according to the general operating pressure within the feed gas subsystem as follows:

- Low pressure—<15 psig (103 kPa)
- Medium pressure—15 to 65 psig (103 to 448 kPa)
- High pressure—>65 psig (>448 kPa)

Low-pressure systems are designed for very low pressure losses and usually operate as constant-volume systems (i.e., constant gas flow rate). This allows the use of rotary lobe–type blowers, which are reliable, low-maintenance units. Because of the high moisture-holding capacity of the low-pressure air, drying always includes both refrigerative and desiccant dryers, as presented on Figure 10.21. Desiccant dryers are limited to thermally regenerated types.

Medium-pressure systems are configured much like the low-pressure systems, as illustrated on Figure 10.22 except the rotary lobe blower is not an acceptable compressor type, and a pressure-reduction step is incorporated after the dryers. Many systems are being designed today for a broad range of system pressures to allow for additional pressure losses from dirty filters, aged desiccant, and fouled diffusers. Although "normal" system operating pressure is expected to be 22 to 27 psig (152 to 186 kPa), the systems are being designed to maintain operations at pressures up to 35 psig (241 kPa). Like low-pressure systems, both refrigerative and desiccant drying types will normally be provided in a medium-pressure system. Again, only thermally regenerated desiccant dryers can be used.

High-pressure systems can usually be differentiated from the other types by the use of multistage, positive-displacement compressors and pressure swing–type desiccant dryers, as presented on Figure 10.23. Pressure swing desiccant dryers will operate at pressures down to 65 psig (448 kPa), but with higher pressures, purge requirements for regeneration are reduced. Typical operating pressures are 85 to 100 psig (586 to 690 kPa). High-pressure systems have been used successfully in many small systems using less than 750 lb per day (340 kg per day) ozone production capacity. Only a few larger installations use high-pressure drying.

Gas Compression. Gas compression in feed gas preparation systems serves the purpose of providing air at mass flow rates required by the process and at pressures necessary to overcome all downstream losses. The five major types of compressors used in ozone systems are:

- Liquid ring
- Centrifugal
- Rotary screw
- Rotary lobe
- Reciprocating

Liquid Ring Compressors. Liquid ring compressors have been widely used in ozone systems. The available flow range and optimum operating discharge pressure fits well with a broad range of system sizes, and they are naturally oil free. They are positive displacement constant-volume machines; that is, they produce almost a constant flow regardless of downstream pressure. The only effective means of varying flow is to provide a bypass line from the discharge to the suction side.

Rotary Screw Compressors. Rotary screw compressors have been widely used in air feed gas preparation systems for ozone. They are a positive displacement machine, consisting of two interlocking screw-shaped rotors, available in either single-stage configuration for discharge pressures up to 50 psig (345 kPa) or two-stage for discharge pressures over 100 psig (690 kPa). They should preferably be nonlubricated as part of a total hydrocarbon control effort.

High-quality rotary screw compressors have a proven track record internationally and are used extensively in a wide variety of industrial applications, including such diverse installations as food processing and microchip manufacturing.

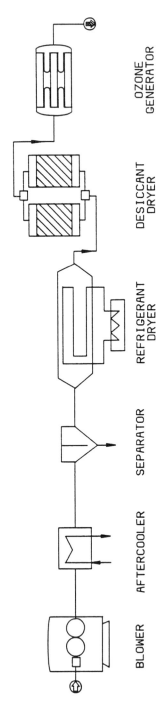

FIGURE 10.21 Low-pressure air preparation ozonation system.

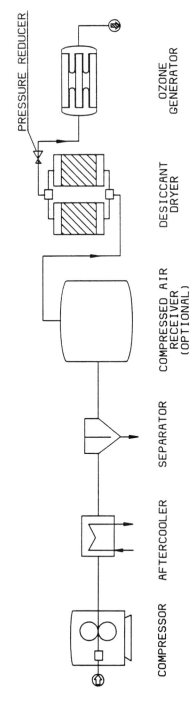

FIGURE 10.22 Medium-pressure air preparation ozonation system.

OXIDATION AND DISINFECTION   263

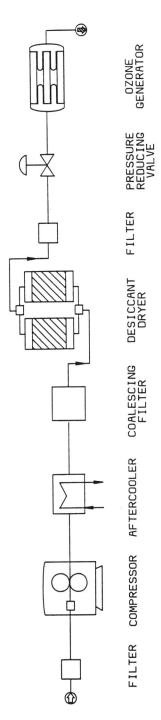

FIGURE 10.23 High-pressure air preparation ozonation system.

Rotary screw compressors can be readily used in constant- or variable-volume systems. In variable-volume systems, the compressor discharges into a receiver, which is maintained within a preset pressure range. When air in the receiver reaches maximum pressure, the compressor "unloads" by closing an inlet valve and opening a blowoff valve on the discharge.

The nonlubricated rotary screw compressors operate under dry conditions with relatively fine tolerances between the rotors. Although the air end casing is water cooled, the rotors must be replaced approximately every 40,000 hours at a cost of about 20% of the original purchase price. Recommended annual maintenance includes replacing the soft parts of the unloading valve.

Because they are not available in capacities less than 350 ft^3/min (9.9 m^3/min), nonlubricated rotary screw compressors are generally not suitable for ozone systems smaller than about 1,500 lb per day (680 kg per day).

Rotary Lobe Compressors. Rotary lobe blowers, with discharge pressures to about 15 psig (103 kPa), have been successfully used in low-pressure installations in which turbines are used for ozone dissolution. At least two manufacturers have also developed nonlubricated high-pressure rotary lobe compressors suitable for up to 35 psig (241 kPa) discharge pressure in a single-stage configuration and up to 110 psig (758 kPa) in a two-stage configuration. These have been used only on a limited basis in ozone systems, and one manufacturer has stopped marketing the compressor because of difficulties in maintaining quality control of the lobe castings.

The rotary lobe blower is ideally suited to constant-volume low-pressure systems and will see continued use in those applications.

Compressor Selection. The selection of a compressor for an ozone generation installation depends primarily on four considerations: required capacity, required discharge pressure, system type (whether variable or constant volume), and efficiency of power utilization. Required capacity is driven by the treatment process requirements and must take into account total required ozone production and design ozone concentration.

The selection of a design discharge pressure must be based on total system considerations, including desiccant dryer types. A conservative approach to estimating system pressure losses will allow the system to continue operating even when experiencing unexpected high losses from dirty filters or clogged diffusers. Because some compressors are more readily adapted to variable-volume control schemes than others, the proposed system operation should be carefully discussed with the compressor manufacturer. This will ensure appropriate use of each compressor type, resulting in optimum power utilization.

Aftercoolers. Heat generated in the gas stream during compression results in elevated compressor discharge temperatures. For low- to medium-pressure systems, the discharge temperature will be between 200° and 350° F (93° and 176° C) for most compressors. The liquid compressant of a liquid ring compressor will absorb most of that heat, and the discharge temperature will only be 10° to 15° F (5° to 8° C) above the compressant temperature. For all other compressors, temperature must be reduced using an aftercooler before the airstream enters any of the downstream dryers.

An aftercooler is typically an air-to-water shell and tube heat exchanger. Standard design provides for a 10° F (5° C) approach; that is, the exit air temperature is 10° F (5° C) above the entering cooling water temperature. High-efficiency units designed for 2° F (1° C) approach are available; they use thin-walled, internally finned tubes. Designs should provide for not greater than 100° F (38° C) exit air temperature.

Refrigerative Drying. Refrigerative drying utilizes physical cooling of the air to condense moisture out of the airstream, thereby lowering the dew point. Refrigerative drying used for ozone systems may be one of two types: direct expansion or chilled water–based systems. To avoid freezing of condensed moisture, refrigerative drying is limited to producing dew points no lower than the freezing temperature of water, 32° F (0° C). In actual practice, refrigerant-based dryers are rated for 35° to 38° F (3° to 4° C), and chilled water–based systems are typically rated for 50° F (10° C) dew point.

Although refrigerative drying is not able to achieve final feed gas dew point, it is a cost-effective process for removing over 80% of the incoming moisture and reducing the subsequent load on the desiccant dryers. For a typical system, the total annualized capital and operating costs can be estimated at $1.50 per pound of moisture removed for a direct expansion type of refrigerative dryer, and nearly $10 per pound of moisture removed for a desiccant dryer.

On the negative side, refrigerated dryers are relatively complicated and require specialized expertise for some maintenance tasks. With multiple component parts, including the condenser, hermetic compressor, and several control valves, refrigerated dryers require frequent maintenance attention for reliable long-term operation.

Direct Expansion Driers. Self-contained, direct expansion driers are available from several manufacturers that are provided exclusively for air drying. These dryers employ a conventional refrigeration cycle with a compressor, condenser, and heat exchanger for gas cooling. Condensers may be air or water cooled.

Because most refrigerated dryers are sold by manufacturers for use in high-pressure plant air systems, the standard units are rated for performance at 100 psig (690 kPa) and 100° F (38° C) inlet conditions and 3.5 to 5 psig (24 to 34 kPa) pressure drop through the dryer. For lower-pressure ozone gas preparation systems, dryer capacity must be derated to as little as one-third of the standard rating to maintain the design dew point and minimize pressure drop through the dryer.

Chilled Water Driers. Chilled water–based refrigerative drying has been used successfully where chilled water is being provided for generator cooling. In these systems, a separately mounted chilled water-to-air heat exchanger is used for gas cooling, with the chilled water being provided from a closed loop chilled water system. Design dried-air dew point from these systems is usually no better than 50° F (10° C), and the desiccant dryer must be sized accordingly.

Also available are self-contained, chilled water dryers provided exclusively for air drying. Compared with direct expansion dryers, chilled water dryers employ an extra heat exchanger. Although they normally use a water-glycol mixture, design outlet dew points are also limited to 50° F (10° C).

Desiccant Drying. Desiccant drying is the final step in the air preparation system to reduce moisture in the air and achieve the final design dew point. These dryers work by absorbing moisture vapor onto a solid surface—actually submicroscopic pores in the desiccant material. During regeneration, this absorbed moisture is driven off by high temperatures or system pressure reduction and carried out of the dryer bed by a purge gas stream.

Most desiccant dryers in use with ozone systems are of the dual-tower design. While one tower is in drying service, the other tower is being regenerated. All components are typically skid-mounted with a four-way or other switching valve to alternate flow between the chambers. Typical cycle times are 16 hours for thermally regenerated dryers (8 hours drying, 5 hours heating, and 3 hours cooling), and 4 minutes for heatless dryers (2 minutes each drying and purging).

One of the major advantages of solid absorbent is its ability to be regenerated. That is, the absorptive process is reversible; the absorbed moisture can be removed and the dryer placed back in service for another drying cycle. Although drying techniques are similar for all designs, dryers are usually identified by the mode of regeneration: pressure swing (heatless), internal heat, and external heat.

Proper regeneration of a desiccant dryer depends on several interrelated factors, including:

- Regeneration temperature
- Purge gas moisture content
- Regeneration direction
- Purge flow rate and duration
- Bed cooling

The total heat applied to the desiccant must be sufficient to bring the desiccant material and the absorber vessel to the generation temperature and to provide the heat of desorption. Because the desorption process is also a function of partial vapor pressure in the purge gas, higher temperatures are necessary for dryers using atmospheric air for regeneration when attempting to achieve comparable dew point performance.

Most desiccant manufacturers recommend that regeneration purge flow direction be countercurrent to the direction of drying. This ensures that the desiccant near the outlet of the bed is thoroughly regenerated. Because the dew point of the gas being dried is in equilibrium with the absorbent at the bed outlet, this results in the best dew point performance. Also, because most of the water is absorbed at the inlet of the bed, countercurrent regeneration minimizes the amount of water carried through the bed, thus minimizing hydrothermal aging. Finally, bed cooldown is especially important in ozone gas preparation because hot feed gas to the generators will adversely affect ozone production.

Pressure Swing Regeneration. For air preparation systems operating at pressures above 70 psig (483 kPa), the desiccant can be regenerated without heat. A portion (20% to 30%) of the dry process air is expanded to atmospheric pressure and passed through the bed countercurrent to the direction of drying. Because the dry low-pressure air has a lower vapor pressure than the adsorbate on the desiccant, moisture will desorb and be carried away by the purge gas.

The typical cycle time for heatless dryers is very short—on the order of two to three minutes for drying. Moisture loading is also kept very low—typically 2% to 3% of the total dynamic capacity. Consequently, the chamber size for a heatless dryer is not significantly different than for a heat regenerated dryer. Heatless dryers can reliably produce 100° F (38° C) pressure dew points, and this performance will not suffer from thermal aging of the desiccant material.

Because of their simplicity in operation and reliable performance, pressure swing dryers have been widely used in several industrial applications, including microchip manufacturing. Pressure swing dryers have been used extensively in moderate-size ozone systems.

Internal Heat Regeneration. Internal heat regenerated dryers rely on conduction and radiation to transfer heat to the desiccant and drive off the moisture. A small purge flow (8% to 12% of the dryer capacity) is passed over the bed countercurrent to the direction of drying to carry off this moisture.

A typical internal heat regenerated dryer has several vertical heater elements spaced throughout the desiccant bed. Each heater element should be enclosed in an individual sheath to allow space for thermal expansion and contraction of the heater element. Because of the difficulty in achieving even heating of the desiccant bed through this method, internal heat regenerated dryers have always been limited in size to approximately 1,000 ft^3/min (28 m^3/min) and smaller. More recently, many ozone system designers are avoiding these dryers altogether, citing uneven performance and premature aging of the desiccant.

External Heat Regeneration. External heat regenerated dryers rely on convection to heat the desiccant material and drive off the moisture. For these dryers, purge air is heated to the regeneration temperature and passed through the bed countercurrent to the direction of drying. Several variations on this basic theme have been developed, including atmospheric purge with dry gas sweep, closed loop, and atmospheric purge with closed loop cooling.

In the first scenario, a blower draws in ambient air, passes it over an externally mounted heater, passes it through the bed, and vents it to the atmosphere. To accomplish removal of additional residual moisture and improve dryer dew point performance, the desiccant bed is then purged with dry process air, also vented to the atmosphere.

In the closed loop system, a captive volume of air is circulated by a positive displacement compressor over an externally mounted heater, through the bed, through an aftercooler and separator to drop out the removed moisture, and then back to the compressor. The great advantage of this system is that no process air is used for regeneration purge flow. This is most important for constant-volume ozone systems where changing process gas flow rates will adversely affect system performance.

Atmospheric purge with closed loop cooling is a combination of the previous two systems. The atmospheric purge air is heated, passed through the desiccant bed as in the earlier alternative for the heating cycle, and a captive volume of gas is recycled in a coiled loop system for desiccant cooling.

Gas Filtration. Gas filtration is an important step at several locations in the process to protect the components of the gas preparation system, to provide the best feed gas quality for optimum long-term performance of the ozone generators, and to prevent plugging of the diffusion system, especially fine bubble diffusers.

Raw Gas Filtration. Raw gas filtration is used primarily to protect the compressor. Most inlet filters provided with the compressors are of the cleanable/replaceable, pleated cartridge type, similar to a high-quality automobile intake filter. Each type of compressor requires a different degree of protection. Many compressors, such as rotary screw and rotary lobe compressors, are typically provided with inlet filter/silencers as part of the prepackaged assembly. Because these compressors operate with very close running tolerances between lobes, filtration requirements are relatively stringent and the inlet filters are typically rated for less than 10 μm.

Coalescing Filter. The feed gas downstream of any refrigerative dryer that does not include a reheat cycle will be saturated and will be carrying liquid moisture. To protect the desiccant dryer from this moisture and from particulates in the airstream, the dryer must be preceded by a coalescing filter.

Coalescing filters provide an effective means of reducing the liquid and solid aerosol burden of the base stream. Unfortunately, there is no widely accepted, meaningful rating method. Table 10.9 presents the rating as stated by three different manufacturers for their coalescing filter. Most coalescing filters now employ inside-to-outside flow direction for optimum coalescing. In this arrangement, the coalesced moisture is collected on a final drainage layer on the outside surface of the filter and drips to the bottom of the housing for removal through an automatic drain trap.

Final Feed Gas. Final feed gas filtration is provided as a particulate after filter, skid mounted with the desiccant dryer. Particulate challenge to the after filter comes primarily from dusting of the desiccant material with an average particle size of 3 μm, and 99%+ of all particles are larger than 1 μm. To accommodate this loading, particulate after filters should be replaceable, pleated cartridge, with at least a 1.0 μm absolute rating. Life span is limited only by particulate loading, but six to nine months is a reasonable expectation.

High Purity Oxygen. High purity oxygen (LOX), purchased and delivered to the plant site, has gained in popularity because of its simple operation. This system has only two major components, the storage tank and the vaporizers.

Liquid Storage Tanks. Liquid oxygen storage tanks are double-walled thermos-type tanks designed to minimize evaporation losses, and they can be either vertical or horizontal. Although custom-made tanks can be fabricated, it is less expensive to purchase standard-sized tanks. If LOX is supplied from an outside commercial source, the storage tanks are normally rated at 200 to 300 psig (1,379 to 2,068 kPa). In special cases, pressurized storage tanks can be built to operate at 600 psig (4,137 kPa). Although the classification of the tanks by operating pressure varies from manufacturer to manufacturer, the following general guidelines can be used:

- Operating pressure < 50 psig (<345 kPa): low pressure
- Operating pressure = 200 to 300 psig (1,379 to 2,068 kPa): medium or high pressure
- Operating pressure = 600 psig (4,137 kPa): high pressure

TABLE 10.9 Coalescing Filter Performance Ratings

	Manufacturer		
	A	B	C
Aerosols			
Efficiency, %	95	—	—
Size, μm	0.009	—	—
Efficiency, %	100	99.985	99.999
Size, μm	0.75	0.3	—
Solids			
Efficiency, %	100	99.985	100
Size, μm	0.3	0.3	0.025

LOX storage tanks absorb ambient heat, which causes some of the stored liquid product to evaporate. The evaporated oxygen gas must be vented or routed to an ozone generator to prevent overpressurizing the tanks. Depending on tank size, about 0.2% to 0.5% per day of stored liquid oxygen will vaporize.

On-site storage capacity varies depending on the relative availability of LOX. Most installations should be provided with at least twice the expected delivery size, which is typically 5,000 lbs (2,268 kg) so that it will not be necessary to deplete the supply on hand before taking a full delivery.

Vaporizers. The stored LOX must be vaporized to be used. Vaporization is accomplished by using water, electricity, steam, liquid petroleum gas (LPG), natural gas, or an ambient-type vaporizer. The water-type vaporizer uses a water-to-LOX heat exchanger. Filtered and disinfected water should be used to prevent plugging of the vaporizer. To reduce water use, cooling water returned from the ozone generator can be used to supply the water-type vaporizer. To meet instantaneous oxygen demands with stored LOX, an uninterrupted water supply is required.

The electric vaporizer uses an immersion-type electrical heater submerged in a concrete basin filled with water. The heater raises the water temperature, and the heat is transferred to a submerged stainless steel or copper coil where LOX is vaporized to oxygen gas (GOX). This method requires about 80 kWh to maintain the safer temperature and about 120 kWh per ton of oxygen vaporized.

The steam-type vaporizer uses a steam-to-oxygen heat exchanger and requires approximately 350 lb (159 kg) of steam per ton of oxygen produced.

The ambient-type vaporizer is a radiator-type vaporizer using ambient heat to vaporize LOX. This type of vaporizer consumes the least amount of energy but requires more space. At 100,000 ft^3/min (2,830 m^3/min) vaporization capacity, the space required for the ambient-type vaporizer is about 36 by 14 ft (11 by 4 m). However, a fan can be provided to increase ambient airflow and reduce the size of the vaporizer. In addition, a heater is sometimes provided to prevent the vaporizer from freezing.

The LPG/natural gas vaporizer is a direct-fire vaporizer using LPG (such as propane) or natural gas to power a water heater and a water recirculation system. The heated water vaporizes LOX to GOX. This type of vaporizer can provide an uninterrupted oxygen supply from the storage tanks during plant power outage. If LPG is used as the primary fuel, a storage vessel is required. For safety reasons, this type of vaporizer should be located remote from the LOX storage tanks. The different types of vaporizers are compared in Table 10.10.

On-Site Generation of Oxygen. Oxygen can be generated on-site as it is needed using either the cryogenic separation or the adsorptive separation process.

Cryogenic Separation Process. Cryogenic air separation is the oldest of the air separation technologies. Since it was introduced in 1902, the technology has been improved many times and is the most widely used in both industrial and municipal applications to separate oxygen from air. Since its inception, it has been used in over 2,000 installations worldwide. Municipal applications in the United States alone include over 70 installations, mostly in the treatment of wastewater. It is a sophisticated process, offered only as a preengineered package. However, alternative components (e.g. air compression, cooling, prepurification, and system operating packages) are still available.

The cryogenic air seperation process uses the principles of gas liquefaction followed by fractional distillation to separate air into oxygen and nitrogen. Components of air will liquefy at a temperature below $-300°$ F ($-184°$ C) and at a pressure of 14.7 psia (101 kPa). Liquefaction is achieved by expanding the compressed air rapidly from a high-pressure stage to a low-pressure stage. This sudden expansion causes the air temperature to drop drastically and allows the air to liquefy.

TABLE 10.10 Comparison of Vaporizers

Type of vaporizer	Advantages	Disadvantages
Water-type vaporizers	Low space requirement Simple to operate Reliable	Require continuous supply of plant water Require pumps
Electric vaporizers	Low space requirement Simple to operate	High energy consumption Heating element prone to corrosion Heater must be continuously energized High cost
Steam-type vaporizers	Low space requirement Low cost	Require a continuous supply of steam
Ambient-type vaporizers	Low cost Low energy consumption	Large space requirement
Liquid petroleum gas (LPG) natural gas–type vaporizers	Does not depend on plant power supply Low space requirement	Require LPG or natural gas Noisy operation

Nitrogen and oxygen gases can be seperated from liquefied air by using their different boiling points at various pressures. The boiling points for oxygen and nitrogen are:

	Boiling points at different pressures		
	14.7 psia (101 kPa)	22 psia (152 kPa)	90 psia (620 kPa)
Nitrogen	$-320°$ F ($-196°$ C)	$-314°$ F ($-192°$ C)	$-286°$ F ($-177°$ C)
Oxygen	$-297°$ F ($-183°$ C)	$-290°$ F ($-179°$ C)	$-259°$ F ($-162°$ C)

As indicated in the table, nitrogen always has a lower boiling point than oxygen under the same pressure. If the temperature is maintained at the boiling point of nitrogen, the nitrogen will vaporize, while oxygen remains liquid. In addition, the boiling point of nitrogen at 90 psia (620 kPa) is higher than that of oxygen at 22 psia (152 kPa), so nitrogen gas at the high-pressure side of the condenser will condense and release heat to vaporize liquid oxygen at the low-pressure side. By using two separate distillation columns, one operated at the high pressure and the other at the low pressure, the nitrogen and oxygen can be separated.

In addition to the production of gaseous oxygen (GOX) and liquid oxygen (LOX), by-products such as gaseous nitrogen (GAN) and liquid nitrogen (LIN) can also be produced. Cryogenic air seperation is the only method that can produce GOX, LOX, LIN, and GAN at the same time.

A simplified schematic diagram of the cryogenic air separation system is presented in Figure 10.24. The major components of the system provide air compression, air purification, gas expansion, distillation, and product storage.

A cryogenic air separation system consumes about 260 to 340 kWh (936 to 1,224 MJ) of energy per ton (1,016 kg) of oxygen produced, depending on the plant capacity, with about 90% of it used by the main compressor. Because of its high power requirements, high-efficiency compressors and drives are required to reduce the total operating cost.

Turndown of a cryogenic air separation system is limited primarily by the main air compressor and the distillation column. The maximum turndown for a single main air compressor is about 70% to 75% of the rated capacity. Below that limit, the compressor will become unstable and must be vented. The turndown of a cryogenic system is also limited by the distillation column to about 50%. Below 50% of system

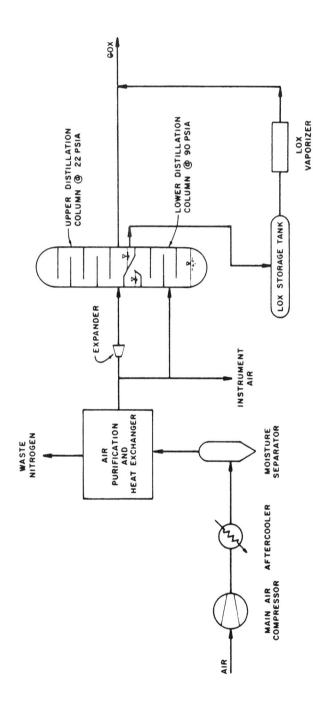

FIGURE 10.24 Typical schematic of cryogenic system.

capacity, GOX must be vented or LOX must be produced in order to maintain system stability.

The turndown range can be increased by using multiple trains. For a two-train system with each train rated at 50% of capacity, the turndown can extend to 30% of total rated capacity. However, because of high capital costs and operating complexity, the selection of the number of trains must be based on process requirements rather than turndown range.

The response rate to demand changes of a cryogenic system is relatively slow, normally about 2% to 3% per minute. Therefore, from 100% to 70%, it will take about 10 to 15 minutes to reduce the production rate. If one train is shut down but maintained in a cold condition (sometimes referred to as "bottled up") for less than 12 hours, the "cold start-up" time to reach full production is about 1 to 2 hours. If the system is maintained in cold condition for more than 12 hours, the cold start-up time will increase to 2 to 3 hours. If the cryogenic system is restarted from ambient temperature, LOX is required, and the "warm start-up" time is about 2 to 3 days.

Providing redundancy of equipment to ensure uninterrupted operation is a common practice for municipal agencies. Redundancy normally includes multiple trains of equal-capacity equipment, with one or two more trains on standby. However, manufacturers indicate that, with a sound operation and maintenance program, a cryogenic plant can be on line over 98% of the time without a second train. They also indicate that a standby cold box is not needed, because there are no rotating parts in this unit. However, this would require adequate LOX backup storage in the event of a cold box problem. Redundancy of rotating equipment and product storage tanks is desirable.

Noise levels from an air separation plant can be high unless special consideration is given during the design to reduce the noise. Noise levels from the main air compressors in general exceed 90 dbA (decibles A-weighted) at a distance of 3 ft (0.9 m). Therefore it is common to house this equipment in a sound-insulated enclosure. A separate silencer in addition to the inlet filter/silencer assembly may be needed to reduce the noise at the inlet of the compressors. Silencers should be provided at the outlet of the compressor surge vent line and the product and waste vent lines. Switching valves would generate a noise level over 100 dbA at a distance of 3 ft (0.9 m) and should be provided with an enclosure and silencer to reduce the noise.

Depending on the equipment's size, sheet metal enclosures with an acoustic treatment may be required for the booster compressors and gas product compressors. Most liquid product pumps are small and should not cause concern about the noise level. The cooling water pumps are usually centrifugal and should have noise levels below 85 dbA.

Depending on the size, the main air compressor may require a power supply of 480, 2,400, or 4,160 volts. The remaining equipment normally requires either a 120 or 480 volt supply. The main air compressor is started unloaded and may take 5 to 15 minutes to reach its operating status, depending on compressor size. If the standby unit is started before the duty unit is shut down to maintain uninterrupted air supply, the power requirement should include the standby unit.

The cryogenic plant requires large quantities of cooling water equal to 8 to 10 gpm (30 to 38 L/min) per ton (907 kg) of oxygen production.

Adsorptive Separation. Separation of air into high purity oxygen and nitrogen gas streams can be accomplished by preferential adsorption of nitrogen onto a solid adsorbent as air is passed through a column (or bed). Nitrogen is retained in the adsorbent bed while oxygen (being less preferentially adsorbed) passes through the column as the product gas at an oxygen purity between 90% and 95%.

When the column becomes saturated with nitrogen, oxygen production from that column is discontinued and the adsorbent is regenerated. Regeneration occurs by elevating the temperature or dropping the pressure in the adsorption column, which reduces the capacity of the adsorbent. The nitrogen is disturbed (released) from the adsorbent, and the highly concentrated nitrogen stream is purged from the system. At this point, the column is returned to the adsorption mode and production of enriched oxygen is resumed.

Adsorptive separation processes are classified by the method of regeneration. The adsorbent can be regenerated by raising the temperature or by decreasing the pressure. Raising the temperature of the bed in the regeneration process is known as thermal swing adsorption (TSA). When the pressure is lowered during regeneration, the process is called either pressure swing adsorption (PSA) or vacuum swing adsorption (VSA). The use of PSA or VSA technology for air separation is more popular than the TSA process because of lower capital and operating costs.

The difference between PSA and VSA systems is the operating pressure. A PSA system utilizes a compression step of between 30 and 60 psig (209 and 419 kPa) before the adsorbent beds, and the VSA system operates at between 0.5 and 15 psig (3 and 105 kPa) for adsorption. Regeneration takes place at near-atmospheric pressure in a PSA system; the VSA columns are regenerated under vacuum. Thus a pressure change occurs in both PSA and VSA systems to desorb nitrogen. One characteristic common to both PSA and VSA is the inability to produce liquid oxygen. This is in contrast to cryogenic air separation, which is based on the liquefaction process.

For typical municipal water treatment operations, separation capacity will be used during most of the year. Both VSA and PSA oxygen generation have the capability to operate at low utilization rates. Turndown can be nearly 100%, but with a substantial power penalty.

Most of the energy losses in an adsorptive system occur during the regeneration of the adsorbent bed. It is at this time that the energy expended to raise the pressure in a PSA system is lost (from depressurization of the column) or a pressure drop is induced in a VSA system (with a vacuum pump). Because the adsorption process utilizes multiple beds whose operation is set by the timing sequence for adsorption, desorption, and purging, provisions are made to alter the timing sequence during turndown operation. By extending the duration of each step, the regeneration step occurs less often and less energy is lost. This improves the energy efficiency of the process at turndown. These projections indicate that the VSA system would provide the most energy-efficient operation at capacity production. The lower energy requirements of the single-train VSA versus the single-train PSA continue until approximately 50% of production capacity, at which time the energy requirements are about equal. Accordingly, the magnitude of the energy savings offered by the VSA system decreases as the production rate is lowered.

The response time to changes in production rate is almost instantaneous for both types of adsorptive separation processes. If less oxygen is required, the flow rate through the system is reduced by adjusting a valve on the outlet of the column. The compressor on a PSA system would then go into unload mode to maintain the same pressure in the columns, which lowers the flow rate. For a system utilizing a preset volume of gas through the column to initiate regeneration, the microprocessor controlling the timing cycle for the beds would automatically adjust for the lower production rate. This allows quick response to changes in oxygen production.

The main sources of noise in a PSA separation process are the main air compressors. Noise emissions from these compressors can range from 80 to 90 dbA. Noise attenuation should be provided for these units. Typical installations are in an enclosed area, with inlet, surge, and purge silencers.

The sources of noise in a VSA system include the main air blower, the product compressor, the switching valves, and the vacuum pump used for regeneration. The vacuum pump is the major source of noise for this system, with noise levels above 90 dbA. Exhaust silencers are usually necessary.

Cooling water requirements range from 2 to 5 gpm (8 to 19 L/min) per ton (907 kg) of system capacity for a VSA process. Several manufacturers have suggested that high-quality water (total dissolved solids less than 200 ppm) be used as the seal water for the vacuum pumps. The manufacturer of the most popular vacuum pump used in VSA systems has indicated that city tap water is generally used for seal water, with no other special requirements. A PSA process requires between 4 and 5 gpm (15 to 19 L/min) per ton (907 kg) of system capacity, with the major portion being used by the compressor's intercooler and aftercooler.

Space requirements for VSA and PSA systems are roughly the same as for a cryogenic facility.

Ozone Contactors

Ozone contact basins provide for transfer of ozone gas into the liquid, promote ozone contact throughout the liquid, and serve to retain the ozonated liquid for a period of time as required to accomplish the desired reactions. The specific process objective and corresponding reactions should dictate contact basin design. Reactions that are rapid relative to the ozone mass transfer rate from gas to liquid phase are best served by contactors that promote the maximum transfer of ozone in the shortest period of time. For these applications, such as oxidation of iron, manganese, or simple organics, contact time is often less important, and contactors that rely on single points of application may be suitable. For reactions that are slow relative to the ozone mass transfer rate, such as disinfection or oxidation of complex organics (including the very persistent herbicides and pesticides), contact time is critical and favors contactors with extended detention time and multiple application points, such as the conventional multi-stage fine bubble diffuser design.

Factors Affecting Transfer Efficiency. The mass transfer of ozone into water has been described by the two-film theory of gas transfer. However, the calculations are complex and designers have usually avoided them in favor of conservative estimates for transfer efficiency based on past experience with similar designs. With the continued development and use of ozone technology, it becomes important for the designer to understand the basic factors that affect transfer efficiency, including contactor characteristics, feed gas characteristics, and source water characteristics.

Contactors have been developed in many configurations, and mass transfer will vary with any characteristic that affects the driving force between the gas and liquid. For the conventional fine bubble diffuser design, the essential factor is depth of water over the diffusers, with efficiency increasing with increasing depth. Additional factors of less importance include hydraulic detention time, liquid flow direction relative to gas flow direction, and number of stages.

Feed gas characteristics that influence mass transfer include ozone dose, feed gas concentration, and bubble size. Mass transfer efficiency will decrease with increasing ozone dose or bubble size but increase with increased ozone concentration. Recent developments in ozone generation technology with resulting ozone concentrations well over 10% have improved transfer efficiency. However, the corresponding decrease in gas flow rates may significantly change the hydrodynamics within the contact basin, making uniform contacting (distribution of the ozone throughout the liquid) much less certain.

Certain characteristics of the water itself can also influence the ozone transfer rate, including temperature, pH, and water quality. The solubility of ozone increases with increasing temperatures, and thus the transfer rate will increase at higher temperatures. As pH increases the transfer rate will increase. The presence of ozone-reactive materials, including organics, iron, or manganese, will increase mass transfer efficiency.

Common Types of Contactors. Types of contactors that are commonly used include:

- Conventional fine bubble
- Turbine
- Packed column
- Injectors
- Deep U-tube

The selection of contactor type depends on many factors, including site considerations and economics, but most often the contactor should be selected based on the specific treatment objective. Appropriate choices are indicated Table 10.11.

Fine Bubble Contactors. The fine bubble diffuser contactor consists of a series of over/under baffled cells. Ozone is applied to some or all cells through a grid of ceramic diffusers at the bottom of the basin. The fine bubble diffuser contact basin is the most widely used contactor. Its prevalence is justified by many factors, including:

- No moving parts
- Adaptable to both rapid and slow reactions
- Adaptable to high transfer efficiency
- Adaptable to high hydraulic efficiency (T_{10}/HRT where HRT is the theoretical hydraulic retention time)

In an effort to improve transfer efficiency, the depth of bubble contactors has increased over the years to the current practice of 20 to 22 ft (6 to 7 m) over the diffusers. Further increases in depth yield only marginal improvement in transfer efficiency, especially at low ozone doses.

Distribution of gas throughout the liquid must be carefully considered, especially in the design of disinfection applications where successful operation cannot be directly measured. Several guidelines have been developed to assist the designer in ensuring adequate contact between the ozone gas and the liquid:

- Maintain a gas-to-liquid ratio between 0.05 and 0.20. This guide comes from the chemical engineering field for effective transfer and contacting in a two-phase (liquid and gas) system. However, this may be difficult to maintain under a wide range of flow and dose conditions.

TABLE 10.11 Contactor Selection

Treatment objective	Suitable contactor choices
Primary disinfection	Multistage fine bubble diffuser type
Iron and manganese oxidation	Injectors, multistage fine bubble
Color removal	Multistage fine bubble, injectors, deep U-tube, turbine
Taste and odor control	Multistage fine bubble, deep U-tube
Algae removal	Turbine
Oxidation	Deep U-tube, injectors, multistage fine bubble

- Maintain a minimum gas floor loading rate of 0.12 ft³/min/ft² (0.037 m³/min/m²) This guide comes from studies on activated sludge systems and represents the minimum gas flow rate for proper mixing of a liquid mass.

- Maintain uniform floor coverage with diffusers spaced at no more than 3 ft (0.9 m). This guideline was developed in recognition of opportunistic flow of the liquid through areas low in or devoid of gas bubbles.

If followed, these guidelines will yield a design that effectively promotes uniform contacting and distribution of ozone throughout the liquid. In disinfection applications this means the designer can be assured that all of the liquid has been equally exposed to the disinfectant. These guidelines work well with conventional air feed systems operating at typical water treatment doses.

However, recent developments in ozone generation, resulting in ozone concentrations in excess of 10% when using high purity oxygen as the feed gas, have made it almost impossible to meet these guidelines when operating at typical disinfection doses of 1 to 3 mg/L. Solutions to these applications have included supplemental mixing with air or water injectors. One manufacturer recommends using more, small (4 in. [10 cm] diameter) diffusers with smaller bubbles (1 to 2 mm) to improve the floor coverage. Ultimately, the higher ozone concentration available from the latest generation of generators is forcing the industry to consider modifications, and alternatives, to the conventional multistage fine bubble diffuser contactor.

The search to optimize the hydraulic efficiency (T_{10}/T) in fine bubble contactors has been investigated by using tracer dye studies of existing basins, on computer modeling based on finite element analysis, and using computational fluid dynamic techniques. While complex in development, this work provides practical results for the designer.

Using D, L, and W to represent the depth, length in the direction of flow, and the width perpendicular to the direction of the flow, it has been shown that hydraulic efficiency is very closely related to D/L. While previous sources have recommended basin configurations of 1.5 to 1.0 to 1.0 *(D, W, L)*, the T_{10}/T can be improved by 50% or better if the D/L is increased to 4.0 or higher. Hydraulic efficiency appears to be independent of width. Consequently, to optimize T_{10} for contactors with a depth of 20 ft (6.1 m), the cell length should be less than 5 ft (1.5 m).

Appendix O of the SWTR Guidance Manual provides extensive discussion in evaluating CT for existing basins. For most applications, the designer may use the T_{10} method for estimating basin performance. However, the determination of CT for ozone contact basins is complicated by the time required to establish a residual concentration and the relatively rapid decay of ozone residual in water. Consequently, a cell-by-cell evaluation is necessary and the CT for an ozone contact basin is the sum of the CT values for all cells.

The designer may conservatively estimate the T_{10}/T for the entire contactor. D/L may be estimated by the following:

$$D/L = \frac{D \times \text{number of cells}}{\text{total contactor length}}$$ (excluding baffle thickness, and counting chimneys as cells)

The T_{10} for each cell may be estimated by allocating the total T_{10} based on a linear extrapolation by the following

$$T_{10} \text{ cell} = (V \text{ cell})(T_{10} \text{ total})(V \text{ total})$$

Each cell may be evaluated as follows:

First cell. In the first cell, the initial ozone demand is being satisfied, and the residual is just being established. This cell may be considered for flat inactivation credits in accordance with Table 10.12. Or C may be taken as the average for the contactor. For the designer, taking T_{10} as one-half the allocated T_{10}, and C as one-half the target outlet, C will give a conservative estimate of CT equivalent to assuming the residual is initially established mid-depth of the contactor.

Subsequent cells. For subsequent cells, T_{10} is taken as the allocated T_{10}. The value of C may be taken as the average for the cell, (that is, C value in + C value out, divided by 2), or as the C at the outlet.

For the designer, basin operation will usually be established to achieve a target residual in the outlet from the first cell of 0.1 mg/L or more, although long-established European practice has been to achieve a residual of 0.4 mg/L. In subsequent cells, additional ozone is applied to maintain the residual for the desired contact time, the last cell(s) provide for ozone decay, but may also be used for CT credit.

Diffuser contactors should be designed with 2 to 3 feet of headroom to provide for unimpeded gas flow to the off-gas exit. In addition, many operations will exhibit foaming within the basin, which should not be allowed to enter the off-gas piping.

Turbine Contactors. Turbine contactors have been used widely for water treatment around the world, including several installations in the United States. The aspirating turbine draws ozone gas into the contactor and mixes it with the water. Major advantages of this design include effective mixing, reduced opportunity for clogging from particulates or oxidation products, and high transfer efficiency without deep structures. Major disadvantages, which limit expanded use, include the additional energy input and limited turndown capability from fixed gas flow rates. Because the turbine contactor functions as a completely stirred reactor, in disinfection applications it should be followed by detention chambers or placed in multiple stages.

Packed Column Contactors. Packed column contactors have seen limited use in water treatment, but interest may increase with increasing ozone concentrations, which are less compatible with conventional designs. In this design, the reactor is filled with ceramic rings; the liquid flows down through the reactor while ozone gas is applied at the bottom. This design provides effective contacting for high liquid flow/low gas flow conditions as experienced with low doses of high ozone concentration. Hydraulic efficiency approaches plug flow characteristics. The primary disadvantages of packed columns are the high cost of the packing material and potential for scaling on the packing.

Direct Injection Contactors. Direct injection of ozone into the liquid stream has received renewed interest in the United States for several reasons, including the increasing number of smaller applications. In addition, direct injection offers potential advantages when operating with high ozone concentrations. Direct injection works on the venturi principal and involves pressurized water flow past a small orifice, which creates a partial vacuum, drawing in the gas. In most applications, this is conducted in a sidestream, which is subsequently blended with the remaining liquid flow in a static mixer. In small applications, this may all occur in a pipeline; for larger applications, injection would normally be followed by a reaction chamber. The primary

TABLE 10.12 First Cell Log Flat Inactivation Credits

Outlet ozone residual	Credit
0.1 mg/L	1 log virus
0.3 mg/L	0.5 log *Giardia*, 1 log virus

disadvantages of this system are the energy input required to move the liquid in the sidestream and poor turndown characteristics.

Off-Gas Disposal

One of the principal design problems in ozone contact systems is the disposal of off-gases from ozone contactors. Assuming that ozone contactors use from 90% to 100% of the ozone that is applied, the air exiting from the contactor may have ozone concentrations as high as 0.5% by volume. This compares with a threshold odor level of 0.05 ppm for ozone and an 8 h OSHA standard of 0.1 ppm.

To date, regulations have not been established on the levels of ozone that may be discharged to the atmosphere, but there is no question that large volumes of air containing 0.5% ozone cannot be casually discharged. Five principal methods of off-gas disposal may be considered:

- Reinjection
- Heating to cause autodecomposition
- Chemical reduction with a reducing agent
- Catalytic reduction with a metal oxide
- Dilution

Reinjection generally involves the construction of two ozone contact basins. The fresh ozone is introduced into the downstream contact basin, and the off-gases are then repumped and reinjected into the upstream contact basin. Given the efficiencies of ozone consumption in each contact stage and the loss of ozone during the repumping process, the ozone residual in the air exiting from the reinjection stage can be as low as 0.001% or 10 ppm. Thus reinjection alone does not completely solve the problem. Rather, reinjection must be used in tandem with some of the other techniques described.

Chemical reduction is another method for removing ozone residuals from off-gases. The chemical reduction could be accomplished by passing the off-gases from the ozone contact chamber in countercurrent flow with an ozone-specific reducing agent in a scrubber much like those used for removing fumes from industrial off-gases. The key to this method is the selection of an inexpensive reducing agent that is not also oxidized by the oxygen present in the air. No uniformly satisfactory reducing agent has been developed to date.

Ozone rapidly dissipates when it is heated. Consequently, in some designs the ozone contactor off-gases are heated to a temperature at which decomposition of the ozone is nearly instantaneous. Temperatures as high as 250° C have sometimes been indicated. The obvious disadvantage of this method is the amount of heat required. In some European designs, the hot air exiting from the ozone decomposer is recycled to a preheater to warm the air that is about to enter the decomposer. This reduces energy requirements but increases capital costs.

Most recent designs have used thermal/catalytic destruct units for off-gas treatment. These consist or a vessel containing the catalyst preceded by a heater. Catalytic reduction involves passing the ozone off-gases across a surface that catalyzes the decomposition of ozone to elemental oxygen. Most catalytic compounds shown to be effective for ozone reduction are proprietary and are based on iron or manganese oxides.

The catalytic reaction is very rapid, and empty-bed contact times are on the order of 1 min. The preheater primarily functions to eliminate moisture from the incoming gas, and are typically designed for a 20° F ($-8°$ C) temperature rise to reduce the relative humidity fron 100% to about 50%. The thermal/catalytic destruct systems are capable of reliably reducing off-gas ozone concentrations to well below 0.1 mg/L.

Construction Materials for Ozone Equipment

In designing systems for ozonation, the highly aggressive character of ozone should be kept in mind. All rubber, most plastics, neoprene, ethylene-propylene-diene rubbers (EPDM), and aluminum are unacceptable materials for use with ozone. The only acceptable materials are 316 stainless steel, 305 stainless steel, glass, Hypalon, Teflon, and concrete. There is some dispute about the usefulness of type 1 PVC. Manufacturers have often recommended type 1 PVC, but its quality does not seem to be uniform from place to place, and incidents of PVC failure occur regularly.

BIBLIOGRAPHY

Aieta, E. M., and J. D. Berg. "A Review of Chlorine Dioxide in Drinking Water." *Journal AWWA* 77(6):64, 1986.

Automated Measurement of Aqueous Ozone Concentration. Denver, Colo.: American Water Works Association Research Foundation, 1993.

Case Studies of Modified Disinfection Practices for Trihalomethane Control. Denver Colo.: American Water Works Association Research Foundation, 1990.

Chick, H. "An Investigation of the Laws of Disinfection." *Journal of Hygiene* 8:92, 1908.

Chlorination By-Products: Production and Control. Denver, Colo.: American Water Works Association Research Foundation, 1986.

Chlorine Dioxide: Drinking Water Issues. Denver, Colo.: American Water Works Association Research Foundation, 1993.

Controlling Disinfection By-Products. Denver, Colo.: American Water Works Association, 1993.

Development and Validation of Rational Design Methods of Disinfection. Denver, Colo.: American Water Works Association Research Foundation, 1995.

Disinfectant Residual Measurement Methods. Denver, Colo.: American Water Works Association Research Foundation, 1992.

Disinfection Dilemma: Microbiological Control versus By-Products. Denver, Colo.: American Water Works Association, 1993.

Duguang, Z., C. Ramirex Cortina, and M. Roustan. "Modeling of the Flow Patterns in Ozonation Chambers Contactor by Residence Time Distribution Studies and by Computational Fluid Dynamics Approach." In *Proc 12th World Congress of the International Ozone Association.* Vol 2. Tours: Instaprint S.A., 1995, pp. 215–226.

Experimental Methodologies for the Determination of Disinfection Effectiveness. Denver, Colo.: American Water Works Association Research Foundation, 1993.

Gordon, G., and A. Rosenblatt. "Gaseous Chlorine-Free Chlorine Dioxide for Drinking Water." In *Chemical Oxidation: Technology of the Nineties. Fifth International Symposium.* International Chemical Oxidation Association, Vanderbilt University, Nashville, Tenn., Feb. 15–17, 1995.

Guidance Manual of Compliance with the Filtration and Disinfection Requirements for Public Water Systems Using Surface Water Sources. U.S. Environmental Protection Agency. Denver, Colo.: American Water Works Association, 1990.

Henry, D. J., and E. M. Freeman. "Finite Element Analysis and t10 Optimization of Ozone Contactors." In *Proc 12th World Congress of the International Ozone Association.* Vol 2. Tours: Instaprint S.A., 1995, pp. 201–214.

Hill, A. G., and H. T. Spencer. "Mass Transfer in a Gas Sparged Ozone Reactor." In *Proceedings First International Symposium Ozone for Water and Wastewater Treatment.* Edited by R. G. Rice and M. E. Browning. Norwalk, Conn.: IOA, 1975, pp. 367–381.

Hoigne, J. "Mechanisms, Rates, and Selectivity of Oxidation of Organic Compounds Initiated by Ozonation of Water." In *Handbook of Ozone Technology and Applications.* Vol. 1. *International Ozone Technology and Applications.* Vol. 1. Edited by R. G. Rice and A. Netzer. International Ozone Association, 1982, pp. 341–379.

Identification and Occurrence of Ozonation By-Products in Drinking Water. Denver, Colo.: American Water Works Association Research Foundation, 1993.

Impact of Ozone on the Removal of Particles, TOC, and THM Precursors. Denver, Colo.: American Water Works Association Research Foundation, 1989.

In-Line Ozone and Hydrogen Peroxide Treatment for Removal of Organic Chemicals. Denver, Colo.: American Water Works Association Research Foundation, 1992.

Kilpatrick, M., et al. "The Decomposition of Ozone in Aqueous Solutions." *Water Research* 10:377, 1976.

Manley, T., and S. Niegowski. "Ozone." In *Encyclopedia of Chemical Technology.* 2nd ed. Vol. 14. New York: John Wiley & Sons, 1967, p. 410.

Masschelein, Willie J. *Chlorine Dioxide : Chemistry and Environmental Impact of Oxychlorine Compounds.* Ann Arbor, Mich.: Ann Arbor Science, 1979.

Morris, C. "The Acid Ionization Constant of HOCL from 5 to 35° C." *Journal of Physical Chemistry* 70:3798, 1966.

Morris, C. "Aspects of the Quantitative Assessment of Germicidal Efficiency." In *Disinfection in Water and Wastewater.* Ann Arbor, Mich.: Ann Arbor Science, 1975.

Ozone and Biological Treatment for DBP Control and Biological Stability. Denver, Colo.: American Water Works Association Research Foundation, 1994.

Ozone and Ozone-Peroxide Disinfection of Giardia *and Viruses.* Denver, Colo.: American Water Works Association Research Foundation, 1992.

Ozone Disinfection of Giardia *and* Cryptosporidium. Denver, Colo.: American Water Works Association Research Foundation, 1994.

Palin, A. "A Study of the Chloro Derivatives of Ammonia." *Water and Water Engineering* 54:151, 1950.

Pilot-Scale Evaluation of Ozone and Peroxone. Denver, Colo.: American Water Works Association Research Foundation, 1991.

Race, J. *Chlorination of Water.* New York: Wiley, 1918.

Rakness, K. L., R. C. Renner, B. A. Hegg, and A. G. Hill. "Practical Design Model for Calculating Bubble Diffuser Contactor Ozone Transfer Efficiency." *Ozone Science and Engineering* 10:173, 1988.

Roustan, M., and J. Mallevialle. "Theoretical Aspects of Ozone Transfer into Water." In *Ozonation Manual for Water and Wastewater Treatment.* Edited by W. J. Maschelein. New York: John Wiley and Sons, 1982, pp. 47–52.

Saunier, B. *Kinetics of Breakpoint Chlorination and of Disinfection.* Ph.D. Thesis in Civil Engineering, University of California–Berkeley, 1976.

Sources, Occurrence, and Control of Chlorine Dioxide By-Product Residuals in Drinking Water. Denver, Colo.: American Water Works Association Research Foundation, 1994.

Streng, A. "Solubility of Ozone in Aqueous Solution." *Journal of Chemical and Engineering Data* 6:431, 1961.

Taste-and-Odor Problems Associated with Chlorine Dioxide. Denver, Colo.: American Water Works Association Research Foundation, 1991.

Trussell, R. R. and M. D. Umphres. "The Formation of Trihalomethanes." *Journal AWWA* 70:604, 1978.

Von Huben, H. *Surface Water Treatment: the New Rules.* Denver, Colo.: American Water Works Association, 1991.

Water Chlorination Principles and Practices, Manual M20. Denver, Colo.: American Water Works Association, 1973.

Watson, H. "A Note of the Variation of the Rate of Disinfection with Change in the Concentration of the Disinfectant." *Journal of Hygiene* 8:536, 1908.

White, G. *Handbook of Chlorination.* 2nd ed. Princeton, N.J.: Van Nostrand, 1986.

CHAPTER 11
LIME SOFTENING

Cold lime softening uses chemical precipitation with lime and other chemicals to reduce a water's hardness and, in some cases, to enhance clarification before filtration. Hot-process softening is predominantly used in industrial applications and is not discussed here. Ion exchange softening is covered in Chapter 12, and membrane processes that may be used to soften water are covered in Chapter 13.

Designing plants with lime softening processes is somewhat different than the design of other types of water treatment plants. Factors responsible for these differences include types of chemicals used, the relatively large quantity of some of these chemicals, the special chemical handling considerations, and the nature of the corresponding chemical reactions. These factors influence process and equipment design and selection, plant layout, and other design considerations.

WATER HARDNESS AND SOFTENING TREATMENT

"Hardness" in water is the sum of the concentrations of multivalent ions, principally calcium and magnesium. Other ions that produce hardness include iron, manganese, strontium, barium, zinc, and aluminum, but these ions are generally not present in significant quantities. Hardness is generally expressed in terms of equivalent milligrams per liter (mg/L) of calcium carbonate. The sum can also be expressed in milliequivalents per liter. Another expression, used commonly in the past, is grains per gallons, where 17.1 milligrams per liter (as calcium carbonate) is equal to 1 grain per gallon.

Total hardness is usually defined as simply the sum of magnesium and calcium hardness in mg/L as $CaCO_3$. Total hardness can also be differentiated into carbonate and noncarbonate hardness. Carbonate hardness is the portion of total hardness present in the form of bicarbonate salts [$Ca(HCO_3)_2$ and $Mg(HCO_3)_2$] and carbonate compounds ($CaCO_3$ and $MgCO_3$).

Noncarbonate hardness is the portion of calcium and magnesium present as noncarbonate salts, such as calcium sulfate ($CaSO_4$), calcium chloride ($CaCl_2$), magnesium sulfate ($MgSO_4$), and magnesium chloride ($MgCl$). The sum of carbonate and noncarbonate hardness equals total hardness.

Acceptable Level of Hardness

Sawyer (1994) classified the degree of hardness as follows:

Hardness	mg/L as $CaCO_3$
Soft	0 to 75
Moderate	75 to 150
Hard	150 to 300
Very hard	Above 300

The degree of hardness acceptable for finished water varies with the consumer or industry served. In 1968, the American Water Works Association (AWWA) established a water quality goal for total hardness of 80 to 100 mg/L expressed as calcium carbonate, but current AWWA water quality goals do not include hardness.

Magnesium hardness of not more than 40 mg/L is also often proposed as a goal to minimize scaling at elevated temperatures in water heaters. The magnesium concentration present before precipitation of magnesium salt is a function of pH and temperature of the finished water. In recent years, many utilities using a softening process have allowed total hardness in finished water to approach 120 to 150 mg/L to reduce chemical costs and residuals production, and, in some cases, to produce a less corrosive water by increasing alkalinity.

Desired hardness is often stated as a policy by a utility that softens water. If a policy exists, the designer must provide a treatment plant design meeting that policy. If no policy exists but softening is desired, it is up to the designer to arrive at a reasonable hardness for design purposes. At present, total hardness of 120 mg/L or less and magnesium hardness of 40 mg/L or less appear to be acceptable design criteria for softening facilities for most applications.

In practice, total hardness produced and the pH of the treated water may be adjusted during operation by varying chemical feed to meet other treatment goals, including modification of hardness goals, stability and corrosivity requirements, and turbidity removal under varying source water quality conditions. Treated water pH of softening facilities may vary between 8.5 and 10.0, depending on all these factors.

Benefits of Softening

Potential benefits of softening water at a central treatment plant include the following:

- Reducing dissolved minerals and scale-forming tendencies
- Reducing consumption of household cleaning agents
- Removing radium 226 and 228
- Removing heavy metals
- Supplementing disinfection and reducing algal growths in basins
- Removing certain organic compounds and reducing total organic carbon (TOC)
- Removing silica and fluoride
- Removing iron and manganese
- Reducing turbidity of surface waters in conjunction with the hardness precipitation process
- Increasing the Langelier Saturation Index, useful for corrosion control under some conditions

The degree of removal of constituents usually depends on the treatment process used. Benefits to the consumer depend on source water quality and user requirements.

Softening Plants in the United States

Currently more than 1,000 domestic-use water plants in the United States use lime softening processes. Although water plants using lime softening are found in most sections of the United States, the majority of large plants are located in the midwestern states and in Florida.

Types of Softening Processes

The softening process generally consists of pretreatment (where required); single-stage, two-stage, or split-treatment softening; and filtration, supplemented by disinfection. Softening can be combined with iron and manganese removal, coagulation for turbidity removal, or both. As noted in Table 11.1, many large softening plants use both coagulation and softening in surface water treatment.

Lime softening processes used by public water supplies can be categorized into two general types:

1. The single-stage lime process is used when source water has high calcium hardness, and a portion of the hardness is removed to achieve the desired hardness. Magnesium hardness of the source water is low (generally less than 40 mg/L as $CaCO_3$) and is not removed.
2. The excess lime process is used when source water has high calcium and magnesium hardness. This process removes a portion of both calcium and magnesium hardness to achieve the desired hardness.

Where noncarbonate hardness must be removed to achieve desired hardness, the process requires adding alkalinity with soda ash and is generally called a lime–soda ash process.

Chemistry of Lime Softening

Water softening involves a number of complex and dynamic chemical interactions. The chemical reactions involved and calculating chemical feed requirements are discussed in detail in *Water Quality and Treatment*. The discussion that follows in this text simplifies the chemistry involved, highlighting only the predominant reactions.

TABLE 11.1 Summary of Domestic-Use Softening Plants

Softening process	Number of Plants	Average daily production, mgd	Total daily production, mgd
Softening	353	0.85	299
Coagulation, softening	265	8.50	2,246
Softening, iron removal	366	0.95	348
Coagulation, softening, iron removal	45	3.60	162

Lime, the primary chemical used for water softening, neutralizes carbon dioxide to convert bicarbonate to carbonate alkalinity and to precipitate calcium carbonate and magnesium hydroxide. Quicklime, CaO, is first slaked to produce calcium hydroxide:

$$CaO + H_2O = Ca(OH)_2 \tag{11.1}$$

Chemical Reactions. Reactions between calcium hydroxide and carbon dioxide and bicarbonate alkalinity are shown in Equations (11.2) and (11.3). The reactions first convert alkalinity present to carbonate alkalinity, which precipitates as insoluble calcium carbonate:

$$CO_2 + Ca(OH)_2 = CaCO_3 + H_2O \tag{11.2}$$

$$Ca(HCO_3)_2 + Ca(OH)_2 = 2CaCO_3 + 2H_2O \tag{11.3}$$

The optimum pH to produce minimum soluble calcium is about 10.3, depending on water temperature, total dissolved solids, and other factors affecting the solubility of inorganic compounds. In precipitating the calcium ion, two moles of calcium carbonate are formed for every mole of calcium ion removed from the water, as shown in Equation (11.3).

Magnesium hardness, present as magnesium bicarbonate, is removed in a stepwise fashion, as shown in Equations (11.4) and (11.5):

$$Mg(HCO_3)_2 + Ca(OH)_2 = CaCO_3 + MgCO_3 + 2H_2O \tag{11.4}$$

$$MgCO_3 + Ca(OH)_2 = CaCO_3 + Mg(OH)_2 \tag{11.5}$$

Magnesium hydroxide does not precipitate quantitatively, as suggested by Equation (11.5), because the solubility of magnesium hydroxide depends on pH. Generally a pH of 11.0 to 11.3 is necessary to reduce the magnesium ion concentration to low values. Unless the initial carbonate alkalinity exceeds the calcium concentration, considerable soluble calcium exists at this elevated pH. Excess hydroxide alkalinity must be converted to carbonate alkalinity to produce a water of minimum calcium hardness. This process, generally termed *stabilization* or *recarbonation,* requires carbon dioxide:

$$Ca^{2+} + 2OH^- + CO_2 = CaCO_3 + H_2O \tag{11.6}$$

Once calcium carbonate is formed, its properties are such that resolubilization takes place only at a very low rate. To remove noncarbonate hardness—calcium or magnesium hardness present in excess of the alkalinity—requires soda ash. Equations (11.7) and (11.8) illustrate noncarbonate hardness removal:

$$MgSO_4 + Ca(OH)_2 = Mg(OH)_2 + CaSO_4 \tag{11.7}$$

$$CaSO_4 + Na_2CO_3 = CaCO_3 + Na_2SO_4 \tag{11.8}$$

No softening occurs in Equation (11.7), as magnesium hardness is only exchanged for calcium hardness. Soda ash is used in Equation (11.8) to remove the calcium noncarbonate hardness either originally present or formed as a result of the reactions in Equation (11.7).

Chemical Requirements. These equations allow reasonably good approximations of the amounts of lime and soda ash required to soften a water. The lime required to remove carbonate hardness and magnesium can be calculated as shown in Equation (11.9):

$$\begin{aligned}CaO \text{ (lb/mil gal)} = \ & 10.6 CO_2 \text{ (mg/L)} \\ & + 4.7 \text{ [alkalinity (mg/L)} \\ & + \text{ magnesium hardness (mg/L)} + x]\end{aligned} \tag{11.9}$$

where CaO is 100% pure, CO_2 is expressed as CO_2, alkalinity is expressed as $CaCO_3$, and x is the required excess hydroxide alkalinity in mg/L as $CaCO_3$. The magnesium hardness shown is the amount to be removed by softening, and not the amount present. Desired excess alkalinity can be determined from the magnesium hydroxide solubility relationship; it is typically in the range of 30 to 70 mg/L and is often estimated at 50 mg/L, expressed as $CaCO_3$.

Equation (11.10) shows the calculation for the quantity of soda ash required to remove noncarbonate hardness:

$$Na_2CO_3 \text{ (lb/mil gal)} = 8.8 \text{ [noncarbonate hardness (mg/L)} - x] \qquad (11.10)$$

where Na_2CO_3 is 100% pure, noncarbonate hardness is expressed as $CaCO_3$, and x is the noncarbonate hardness left in the water.

Because CaO is usually 88% to 95% pure, results from Equation (11.9) must be divided by actual chemical purity. Soda ash is essentially pure, so no adjustment to the calculation is required. If $Ca(OH)_2$ is used instead of CaO, the required amount of CaO should be calculated by dividing by 56/74, the ratio of the molecular weights.

Use of Caustic Soda. Caustic soda, NaOH, can be used in place of lime or soda ash. Fewer residuals are produced, and caustic soda is easier to handle, store, and feed. Caustic soda is generally purchased as a 50% aqueous solution. Softening reactions with caustic soda are shown in Equations (11.11) through (11.15):

$$CO_2 + 2NaOH = Na_2CO_3 + H_2O \qquad (11.11)$$

$$Ca(HCO_3)_2 + 2NaOH = CaCO_3 + Na_2CO_3 + 2H_2O \qquad (11.12)$$

$$Mg(HCO_3)_2 + 2NaOH = Mg(OH)_2 + 2Na_2CO_3 + 2H_2O \qquad (11.13)$$

$$MgSO_4 + 2NaOH = Mg(OH)_2 + Na_2SO_4 \qquad (11.14)$$

$$CaSO_4 + Na_2CO_3 = CaCO_3 + Na_2SO_4 \qquad (11.15)$$

The sodium carbonate formed in Equations (11.11) through (11.13) is available to precipitate calcium noncarbonate hardness, as shown in Equation (11.15). A combination of lime and caustic soda can be used, the ratio depending on the calcium noncarbonate removal required. This combination provides some savings in chemical cost compared with the use of caustic soda alone because caustic soda is more expensive than lime. Using caustic soda may be a good option for low-alkalinity water, because alkalinity reduction with caustic soda is half that of lime softening. A disadvantage of using caustic soda is the increase in finished water sodium concentration.

Split-Treatment Softening. Split-treatment softening is used at several plants where source water turbidity is low and magnesium high, or where pretreatment or coagulation of bypassed flow is practiced. The lime dose required to treat the entire flow is added to the first basin, elevating the pH to between 11.0 and 11.3, removing calcium and almost all magnesium hardness. The settled water is then blended with a bypassed portion of source water to produce a pH of about 10.3, which precipitates only calcium carbonate hardness in the bypassed flow.

The principal advantage of split-treatment softening is to reduce or eliminate carbon dioxide stabilization of high-pH water, perhaps a significant cost savings in plant operation. An additional advantage is reducing residuals production because this process does not require excess lime or produce corresponding residuals. When designing split-flow systems for high iron or manganese content waters, the engineer should verify that the selected design and bypass flow will yield permissible iron and manganese concentrations in the finished water.

LIME SOFTENING PROCESSES

The three common lime softening processes are the single-stage lime process, the excess lime softening process, and the split-treatment process.

Pretreatment

The principal types of pretreatment used before lime softening are aeration and presedimentation.

Aeration. Aeration may be used to remove carbon dioxide from the source water before softening. This is usually only applicable to groundwaters where carbon dioxide concentrations are relatively high. Lime removal of carbon dioxide in source water adds to operations costs because of chemical expenses and increased calcium carbonate residuals, in accordance with Equation (11.2).

Induced-draft or open tray aeration is often used and may reduce the carbon dioxide level to 10 mg/L or less. Aeration also oxidizes iron and manganese that may be present. For some groundwaters containing substantial iron and manganese, clogging of aeration trays is a problem. The aerator should be designed to minimize clogging and provide ready access for periodic cleaning. The lime dosage required to react with carbon dioxide may be estimated using part of Equation (11.9), as follows:

$$\text{CaO (lb/mil gal)} = 10.6 \times CO_2 \text{ (mg/L)} \qquad (11.16)$$

where CO_2 is expressed as CO_2 (mg/L).

Reduced lime consumption and residuals production associated with aeration must be weighed against the capital cost, as well as the operating and maintenance costs of aeration equipment. Aeration is primarily used where carbon dioxide levels are high enough to justify the cost. Residuals produced by the reaction of lime and carbon dioxide are in accordance with Equation (11.2) and may be estimated as follows:

$$\text{Dry weight } CaCO_3 \text{ residuals (lb/mil gal)} = 19.0 \times CO_2 \text{ (mg/L)} \qquad (11.17)$$

where CO_2 is expressed as CO_2 (mg/L).

Aeration is often used to oxidize iron and manganese in iron and manganese removal plants, but the elevated pH of the softening process together with chemical oxidation, if needed, can remove iron and manganese without the need for aeration.

Presedimentation. Presedimentation is used primarily by those plants treating high-turbidity surface waters, such as sources on the Missouri and Mississippi rivers. Some of these plants use metal salt or polymer coagulants to enhance suspended-solids removal.

Presedimentation provides a more uniform water quality at the treatment plant, removes a major portion of the suspended solids with little chemical cost, and offers potential cost savings in residuals treatment and disposal. Because of the cost of dewatering and land disposal of these solids and the minor impact on highly turbid rivers, state and federal agencies have often allowed return of presedimentation residuals directly to the river.

Presedimentation also provides an opportunity for pretreatment and removal of taste and odors and other organic compounds with powdered activated carbon and oxidizing agents before the elevated pH associated with the softening process. In some cases, this provides more efficient and effective treatment.

Single-Stage Lime Process

For the single-stage lime process, only carbonate hardness in the form of calcium is to be removed, and only lime is added to the softening stage. The lime dosage required to react with calcium carbonate hardness may be estimated using part of Equation (10.9) as follows:

$$\text{CaO (lb/mil gal)} = 10.6 CO_2 \text{ (mg/L)} + 4.7 \text{ [alkalinity (mg/L)]} \quad (11.18)$$

where alkalinity is equal to the portion of calcium carbonate hardness to be removed.

For the single-stage lime–soda ash process, both carbonate and noncarbonate hardness are removed, and soda ash is added to the softening stage. The dosage of lime required to react with calcium hardness may be estimated using Equation (11.18), where alkalinity is the total alkalinity of the source water. The dosage of soda ash required may be estimated using Equation (11.10).

The lime or lime–soda ash process removes only calcium carbonate hardness down to a minimum of about 35 mg/L.

Recarbonation. Recarbonation with carbon dioxide is required to convert carbonate alkalinity to bicarbonate alkalinity and stabilize the water before filtration. An often used guideline is 40 mg/L of bicarbonate alkalinity in the finished water, although this could vary depending on stability requirements for the particular water. The reaction of carbonate alkalinity with carbon dioxide to produce bicarbonate alkalinity is as follows:

$$CO_2 + CaCO_3 + H_2O = Ca(HCO_3)_2 \quad (11.19)$$

The dosage of carbon dioxide required to react with the carbonate alkalinity to produce bicarbonate alkalinity may be estimated as follows:

$$CO_2 \text{ (lb/mil gal)} = 3.7 \times \text{carbonate alkalinity (mg/L)} \quad (11.20)$$

where carbonate alkalinity is that amount to be converted to bicarbonate alkalinity.

A flow diagram for the single-stage lime and the lime–soda ash processes is shown in Figure 11.1.

Conventional Softening Basins. Conventional softening basins are similar to conventional basins used for coagulation and clarification. They consist of rapid mixing, flocculation, and sedimentation.

Lime, soda ash, coagulant, and a polymer coagulant aid (if required) are added at the rapid mix basin. Mixing is usually accomplished by mechanical mixers, with a detention time of approximately 30 seconds.

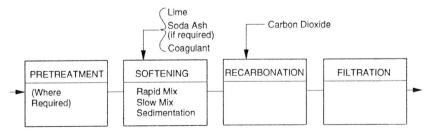

FIGURE 11.1 Single-stage lime process.

Flocculation typically consists of horizontal paddle wheel or vertical turbine flocculators with total detention times of 30 to 45 minutes and up to three-stage tapered flocculation. Flocculators are usually equipped with variable-speed drives. The flocculation basin should be designed to facilitate periodic cleaning because of residuals buildup.

Sedimentation loading rates are in the range of 0.4 to 1.0 gpm/ft^2 (0.98 to 2.4 m/h) with detention times of 2 to 4 hours. The higher loading rates are generally used for waters not requiring coagulation for turbidity removal. Continuous residuals collection equipment should be provided, and there must also be a means of draining the basin for periodic cleaning.

Recycling previously formed calcium carbonate residuals from the sedimentation basin to the rapid mix is beneficial in the softening process. Recycling accelerates the precipitation reactions, and the process more closely approaches true solubility when the mix is seeded with these previously formed crystals. Recycling residuals also allows precipitation to occur on the recycled residuals that serve as nuclei, reducing precipitation on the mechanical equipment.

In addition, recycling calcium carbonate residuals promotes growth of larger calcium carbonate crystals that settle and dewater more rapidly. A study has indicated that 50% to 100% residuals recycle, based on solids produced, controls particle size optimally for this type of process (Burris et al., 1976). This same study found that particle growth approached an equilibrium value after about four cycles.

Figure 11.2 illustrates the effect of residuals recycle on particle growth. Curve A is a sample of residuals taken before recycling was practiced. Curve B represents a sample of sludge taken after several cycles at 25% recycling based on solids produced. Curve C represents 300% sludge recycling. Curve D was a sample of sludge taken from a solids contact clarifier with an extremely high recycle rate operating at a high solids concentration. Residuals recycle for conventional basins may be accomplished by a separate basin sump and solids-handling pumps that recycle sedimentation residuals to the rapid mix.

Conventional softening basins are mostly found at older facilities. They provide a high degree of process stability, but the relative size and number of the basins, when compared

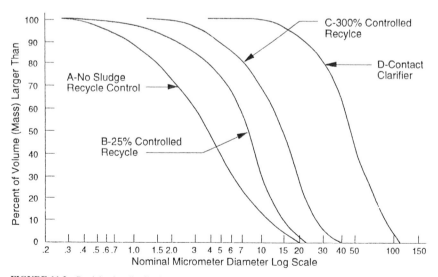

FIGURE 11.2 Particle-size distribution comparison.

with high-rate solids contact processes, substantially increases the initial cost. In addition, the high-rate solids contact processes are generally more effective in driving the softening reaction to completion and reducing chemical costs.

Solids Contact Softening Basins. Solids contact softening basins combine mixing, residuals recirculation, and sedimentation functions in one basin. Rapid mixing may be provided ahead of the solids contact unit but generally is not needed. Lime and other chemicals are applied directly to the mixing zone of the solids contact unit. This type of unit provides a high degree of continuous sludge recirculation and contact, and it produces more stable water and larger calcium carbonate crystals. The depth and density of the residuals in the basin can be controlled by wasting residuals and by the degree of recirculation. The mixing and recirculation zone is separated from the sedimentation zone by a conical baffle wall within the basin. An example of a solids contact unit is shown in Figure 11.3.

Effective solids contact units pull settled residuals from near the floor at the center of the basin with a large-diameter turbine or impeller and recirculate the residuals at a high rate with the water entering the basin. The maximum recirculation rate is typically 10 to 1 based on incoming flow. The mixer is provided with a variable-speed drive for adjusting the recirculation rate to meet treatment conditions.

Where multiple solids contact units are installed, small recirculation pumps, operated to recirculate solids from one unit to another, are often used during start-up of a unit that has been out of service for cleaning. This is especially helpful during periods of low water temperature.

Side water depth of solids contact basins usually varies from 14 to 18 ft (4.3 to 5.5 m) and depends on basin size and the equipment manufacturer's requirements. Contact time in the mixing zone is typically 20 to 30 minutes, measured by the volume of water within and directly under the baffle wall.

Surface loading rate in the sedimentation zone is generally measured 2 ft (0.6 m) below the water surface, based on the surface area between the baffle wall and the basin wall. Surface loading rates are usually in the range of 1.0 to 1.75 gpm/ft^2 (2.4 to 4.3 m/h). Where coagulation for turbidity removal is required, surface loading rates are generally in the lower end of this range or less. Where possible, design requirements should be based on successful plants using the same or similar source water.

For the single-stage lime and lime–soda ash process, residuals are composed primarily of heavy calcium carbonate crystals without the more gelatinous and lighter magnesium hydroxide component. The heavy calcium carbonate crystals create some difficulty in recirculating residuals. In any case, the manufacturer should be advised of the nature of the process so that the basin design will function properly. Equipment specifications should establish a minimum percent solids to be maintained in the mixing zone. A minimum percent solids of 1% by weight is recommended by the *Recommended Standards for Water Works* (commonly called the 10-State Standards).

Solids contact units offer a smaller footprint because of the combined, higher-rate process. They also provide a high degree of residuals recirculation and contact to complete the softening reaction and are typically used instead of conventional basins for softening. However, solids contact units can require more operator attention to monitor and control the solids inventory in the basin, and these units are more prone to upset from varying inlet water temperatures, flows, and turbidity conditions.

Recarbonation Basins. For the single-stage lime and lime–soda ash process, a recarbonation basin with a contact time of at least 20 minutes is typically provided to stabilize water before filtration. The contact basin should be designed for plug flow, and mixing should be provided ahead of the basin to introduce carbon dioxide.

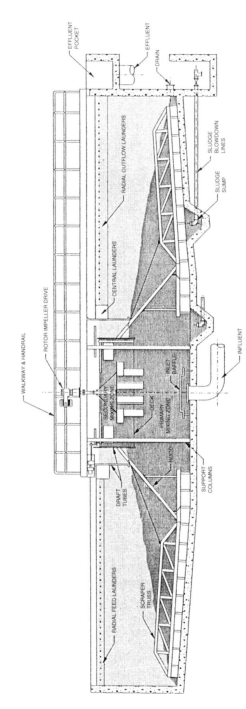

FIGURE 11.3 Solids contact clarifier.

Excess Lime Softening Process

In lime softening, the term *excess lime* refers to the removal of magnesium, in addition to calcium, by providing excess lime in the primary softening process. The excess lime process can generally remove calcium hardness to a level of about 35 mg/L and magnesium hardness to about 10 mg/L.

Single-Stage Excess Lime Process. The single-stage excess lime softening process is similar to Figure 11.1. The softening basin is a conventional or solids contact configuration as discussed under the single-stage lime process. Excess lime leaving the softening basin is reacted with additional carbon dioxide in the recarbonation basin to lower the pH of the water and to stabilize the water before filtration.

Although single-stage excess lime softening plants have been operated successfully, several problems are commonly encountered. It is more difficult to achieve process stability with excess lime softening in the typical recarbonation basin, resulting in deposition of calcium carbonate on the filter media. In addition, residuals buildup in the recarbonation basin creates greater cleaning requirements. Particulate loading to the filters is high during periods of process instability in the softening basin.

Two-Stage Excess Lime Process. A two-stage excess lime softening process is shown in Figure 11.4. The two-stage process can use either conventional or solids contact basin units for the first and second stages. The two-stage process has several advantages over the single-stage excess lime process. By limiting recarbonation at the second stage to enough carbon dioxide to react with the excess lime, additional softening by removal of calcium carbonate may be performed in the second stage.

The second stage provides additional process stability before filtration and is equipped with residuals collection equipment to eliminate residuals buildup. A coagulant added at the second stage further clarifies the water before filtration. The second stage also provides an opportunity for further treatment by adsorption or oxidation after primary softening and before filtration. After the second stage, carbon dioxide is added to convert carbonate alkalinity to bicarbonate alkalinity for stability, similar to single-stage treatment.

Loading rates for two-stage processes are similar to those for the single-stage lime process. For conventional units, residuals recirculation should be provided as discussed under single-stage lime processes.

Recarbonation. Recarbonation is required for the excess lime process to convert excess hydroxide alkalinity to carbonate alkalinity and also to convert carbonate alkalinity to bicarbonate alkalinity for stabilizing the treated water.

The reaction of carbon dioxide with hydroxide alkalinity is shown in Equation (11.6). The dosage of carbon dioxide can be estimated as follows:

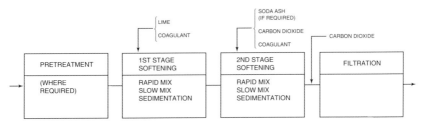

FIGURE 11.4 Excess lime softening process.

$$CO_2 \text{ (lb/mil gal)} = 3.7 \times [\text{hydroxide alkalinity (mg/L)}] \quad (11.21)$$

where hydroxide alkalinity is the excess hydroxide alkalinity required for magnesium removal.

In addition, carbon dioxide is required to convert carbonate alkalinity to bicarbonate alkalinity for stabilizing treated water in accordance with Equations (11.20) and (11.21).

Split-Treatment Excess Lime Process

Split-treatment softening is a variation of the two-stage excess lime softening process that can be advantageous under some conditions. A schematic of the process is shown in Figure 11.5.

In split-treatment softening, a portion of the flow to the first-stage softening process (generally 10% to 30%) is bypassed to the second stage. Carbon dioxide and bicarbonate alkalinity in the bypassed flow react with excess lime leaving the primary softening process both to soften the bypassed flow with excess lime and to stabilize water entering the secondary basin.

The net result of split treatment softening is a savings in lime and carbon dioxide feed requirements and a reduction in residuals production. In addition, the first-stage softening unit may be smaller because it receives reduced flow. Chemical reactions involved and calculating chemical feed requirements and flow split are covered in detail in *Water Quality and Treatment*.

Because a portion of the flow is not treated in the first stage of softening, it is important to determine whether adequate treatment of bypassed flow is provided in the second stage. Split-treatment softening is most often used for groundwaters but has also been used for some surface waters with pretreatment or coagulation of the bypassed flow.

SPECIAL LIME SOFTENING DESIGN CONSIDERATIONS

Many special design and operations and maintenance considerations are associated with lime softening processes, including problems with encrustation and removing, handling, and disposal of the large quantities of residuals resulting from treatment processes.

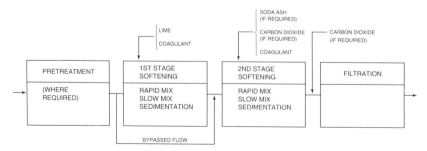

FIGURE 11.5 Split-treatment excess lime process.

Encrustation Problems

A significant problem in operations and maintenance of lime softening processes is encrustation of calcium carbonate on the process and hydraulic components. Encrustation can occur anywhere but is generally greater and harder near application points of lime and soda ash and where water velocity is increased, such as on mixing equipment. Encrustation can also occur in flumes and pipelines after lime or carbon dioxide addition or in split-treatment flow in a secondary process.

The burden of dealing with encrustation on equipment is usually placed on the basin equipment manufacturer. An allowance should also be made in designing pipelines and flumes, with additional hydraulic capacity and provisions for ease of cleaning. Open flumes are generally preferred over pipelines for this reason. Calcium carbonate buildup may be removed with high-pressure jet equipment or by mechanical means. Basins and pipelines are usually cleaned once per year during seasonal low-demand periods. If the plant has no low-demand periods, additional treatment capacity in the form of more treatment units will have to be provided.

Recirculating residuals to the point of addition of lime or, for second-stage processes, to the point of addition of carbon dioxide or split-treatment flow provides nuclei for calcium carbonate deposition and may reduce encrustation problems.

Additional information on sludge thickening, dewatering, and handling are provided in Chapter 17.

Residuals Removal and Handling

Residuals from softening reactions are dense when compared with coagulation residuals, with solids concentrations varying from 3% to 15% or greater by weight. Single-stage processes removing only calcium carbonate produce the highest-density residuals. Residuals from excess lime processes are generally less dense because of the influence of magnesium hydroxide, which is lighter and more gelatinous. For excess lime softening, an average residuals blowdown solids concentration of 5% by weight is often used as a rough guideline. For split-treatment softening with coagulation of the bypass flow, design consideration must be given as to whether to combine residuals from softening and coagulation units.

Residual scrapers for softening basins require higher-torque units than conventional treatment basins because of the high density of residuals. The design of the load on scraper mechanisms depends on the type of softening and the stage of the process. Equipment manufacturers and similar treatment facilities should be consulted for guidance.

Residuals blowdown from softening basins is usually controlled by a repeat cycle timer that allows periodic, timed blowdowns at full pipe flow. Backflushing piping with clear water after each blowdown must be provided to reduce the possibility of lines clogging from settled residuals. Provisions for sampling the residuals blowdown should be included. Some facilities have residuals blowdown viewing pits where the residuals can be observed and sampled.

Residuals pumps for recirculation or disposal are typically solids-handling centrifugal units, although positive displacement pumping units may be required for extremely dense residuals. Recirculation pumps should be provided with variable-speed drives.

Residuals Production

The quantity of residuals produced by the softening process can be estimated by performing a mass balance on the residuals-producing calcium and magnesium, which represent the hardness removed in the process. One or both of the substances are present as hardness in the water entering the plant. In addition, calcium is added to the plant flow by the addition of lime.

Some hardness also leaves the plant in the finished water. The rest of the hardness leaving the plant is in the residuals produced. The residuals are in the form of calcium carbonate for the calcium hardness removed and in the form of magnesium hydroxide for the magnesium hardness removed.

The calcium balance is as follows:

$$\text{Dry weight } CaCO_3 \text{ residuals (lb/mil gal)} = 20.9 \\ \times \text{ [(Ca in)} + \text{(Ca added by lime)} \\ - \text{(Ca out) in mg/L]} \quad (11.22)$$

where Ca is calcium in mg/L as calcium (calcium carbonate equivalent divided by 2.5).

For quicklime (CaO), Ca added by lime is $0.71 \times CaO$ in mg/L $\times$ percent purity/100. For hydrated lime [$Ca(OH)_2$], Ca added by lime is $0.54 \times Ca(OH)_2$ in mg/L $\times$ percent purity/100.

The magnesium balance is:

$$\text{Dry weight } Mg(OH)_2 \text{ residuals (lb/mil gal)} = 20.0 \\ \times \text{ [(Mg in)} - \text{(Mg out) in mg/L]} \quad (11.23)$$

where Mg is magnesium in mg/L as magnesium (calcium carbonate equivalent divided by 4.1). Impurities in lime are also a source of residuals produced in the softening process. The amount can be estimated as follows:

$$\text{Dry weight lime impurities (lb/mil gal)} = \text{lime dose in lb/mg} \\ \times (100 - \text{percent purity})/100 \quad (11.24)$$

Total dry weight solids of softening residuals for the excess lime treatment process is usually about 2.5 times the hardness removed, expressed in mg/L or lb/mg. The single-stage process produces total dry weight solids for softening residuals approaching 2.0 times the hardness removed.

Residuals are also produced by source water turbidity removed in the process, by the precipitation of iron and manganese that may be removed in the process, and by coagulants and powdered activated carbon that may be used in the process. Guidelines for determining the quantities of residuals produced by these items may be found in Chapter 17.

Filtration

The design of filters for softening plants is similar to conventional clarification plants. Many modern softening plants use dual-media filters consisting of 24 to 30 in. (0.6 to 0.75 m) or more of anthracite and approximately 12 in. (0.3 m) of sand, although other types of media have also been used successfully. Design loading rates typically range from 3 to 5 gpm/ft^2 (7.3 to 12.2 m/h). Available head should be 8 ft (2.4 m) or more at design rate, measured from the water surface in the filter to the filter effluent line as it leaves the filter.

Particulate loading on softening plant filters consists, to a large extent, of fine calcium carbonate that makes softening plant filters behave somewhat differently than coagulation plant filters. Softening plant filters are generally characterized by long filter runs, on the order of 36 to 48 hours or more, and are often washed based on time rather than head loss or turbidity breakthrough. However, effluent turbidity must be carefully monitored and controlled because it is not unusual for the fine calcium carbonate to pass through the filter. Filter aid polymer should be provided to assist in controlling turbidity breakthrough.

Water is typically under a calcium carbonate plating condition before filtration to maintain stability and to reduce corrosion in the distribution system. Therefore a polyphosphate is usually fed before the filters at a dosage of 0.25 to 1.0 mg/L to control encrustation of the filter media. Annual inspection of the media for cementation and inspection of the underdrain every 4 to 5 years through careful excavation of the media may also be performed to monitor encrustation.

Surface wash or air-water backwash is essential to assist filter backwashing. The supplemental agitation breaks up cementation of filter media and helps remove suspended solids in the backwash water. The design of backwash systems is also similar to conventional plants.

Pellet Reactors

Pellet reactors have been used for softening in the Netherlands for many years and have been installed at a number of locations in North America. A pellet reactor consists of an inverted conical tank where calcium carbonate crystallizes on a suspended bed of fine sand.

Advantages of the pellet reactor are its small size and low installation cost. Residuals consist of small pellets that dewater readily, minimizing residuals volume. However, pellet reactors should not be considered for systems high in magnesium content because of magnesium hydroxide fouling of the reactor.

Pellet reactor systems should be designed cautiously. Where the application is not proven, pilot testing is advisable. Design should be carefully coordinated with the equipment manufacturer.

Chemical Feed and Layout Considerations

The design of lime softening plants is influenced, to a large degree, by the need to handle and feed large quantities of lime. Lime solution readily encrusts solution pumps, pipelines, and troughs and presents major maintenance considerations. Pumping lime solutions should be avoided if possible; gravity flow in open troughs or hoses readily accessible for cleaning are preferable.

The lime feed system should be located as close to the point of feed as possible. On the other hand, feed systems for these chemicals, particularly lime, require frequent attention and should be located as close as practical to the operator's station. Because of these factors, it is generally preferable to bring the water to be treated as close as practicable to central chemical feed facilities readily accessible to the operator.

Provisions should be made to feed alum or ferric coagulants and coagulant polymer at the mixing facilities ahead of softening basins. It is often necessary to use coagulants to enhance clarification within softening units. Provisions should also be made to feed polyphosphate and filter aid coagulant ahead of the filters.

FUTURE TRENDS IN SOFTENING

With the technological advances in membrane process manufacturing, it is anticipated that lime softening may gradually be replaced by membrane processes for some applications. This option appears to be particularly applicable for smaller installations where the potential for reduced operator attention and remote monitoring of membrane processes has a significant cost impact. In addition, membrane processes produce residuals containing only the constituents removed from the source water, which may make membranes more attractive to regulatory agencies where residuals disposal to source waters is a consideration. Membrane processes may also remove regulated contaminants in conjunction with softening.

Lime softening uses large amounts of chemicals (primarily lime) and produces large amounts of residuals in comparison to coagulation treatment only. Because of the cost of lime, other chemicals, and residuals disposal problems, some utilities have considered reducing the degree of hardness removal. This idea has raised the question as to what degree of hardness removal for a particular utility is beneficial and cost-effective to the consumers. More attention to this area is expected in the future.

BIBLIOGRAPHY

Aieta, E. M., E. J. Singley, A. R. Trussell, K. W. Thorbjarnarson, and M. J. McGuire. "Radionuclides in Drinking Water: an Overview." *Journal AWWA* 79(4):144, 1987.

American Water Works Association Trace Inorganic Substances Research Committee. "A Review of Solid-Solution Interactions and Implications for the Control of Trace Inorganic Materials in Water Treatment." *Journal AWWA* 80(10):56, 1988.

Berg, G., R. B. Dean, and D. R. Dahling. "Removal of Poliovirus 1 from Secondary Effluents by Lime Flocculation and Rapid Sand Filtration." *Journal AWWA* 60(2):193, 1968.

Burris, M. A., K. W. Cousens, and D. M. Mair. "Softening and Coagulation Sludge—Disposal Studies for a Surface Water Supply." *Journal AWWA* 68(5):247, 1976.

Chaudhuri, Malay, and R. S. Engelbrecht. "Removal of Viruses from Water by Chemical Coagulation and Flocculation." *Journal AWWA* 62(9):563, 1970.

Corrosion Control by Deposition of $CaCO_3$ Films. A Handbook of Practical Application and Instruction. Denver, Colo.: American Water Works Association, 1978.

Internal Corrosion of Water Distribution Systems. 2nd ed. Cooperative research report. Denver, Colo.: American Water Works Association Research Foundation and DVGW–Technologiezentrum Wasser, 1996.

Introduction to Water Treatment. Vol. 2. *Principles and Practices of Water Supply Operations.* Denver, Colo.: American Water Works Association, 1996.

Liao, N. Y., and S. J. Randtke, "Removing Fulvic Acid by Lime Softening." *Journal AWWA* 77(8)78, 1985.

Liao, N. Y., and S. J. Randtke. "Predicting the Removal of Soluble Organic Contaminants by Lime Softening." *Water Research* 20(1):27, 1986.

Lime Handling, Application and Storage. Denver, Colo.: American Water Works Association, 1995.

Manual of Treatment Techniques for Meeting the Interim Primary Drinking Water Regulations. U.S. Environmental Protection Agency, 600/8-77-005, Washington, DC, 1978.

"Quality Goals for Potable Water: A Statement of Policy Adopted by the AWWA Board of Directors on January 28, 1968, Based on the Final Report of Task Group 2650P—Water Quality Goals, E.L. Bean, Chair," *Journal AWWA* 60(12):1317, 1968.

Randtke, S. J., et al. "Removing Soluble Organic Contaminants by Lime-Soda Softening." *Journal AWWA* 74(4):192, 1982.

Recommended Standards for Water Works. Albany, NY: Great Lakes–Upper Mississippi River Board of State Public Health and Environmental Managers, 1992.

The Removal and Disinfection Efficiency of Lime Softening Processes for Giardia *and Viruses.* Denver, Colo.: American Water Works Association, 1994.

Riehl, M. L., H. H. Weiser, and R. T. Rheins. "Effect of Lime Treated Water upon Survival of Bacteria." *Journal AWWA* 44(5):466, 1952.

Sawyer, C. N., P. L. McCarty, and G. F. Parkin. *Chemistry for Environmental Engineers.* New York: McGraw-Hill, 1994.

Sproul, O. J. *Critical Review of Virus Removal by Coagulation Processes and pH Modifications.* EPA-600/2-80-004, Washington, DC: EPA, 1980.

Waste Sludge and Filter Washwater Disposal from Water Softening Plants. Cincinnati, Ohio: Ohio Department of Health, 1978.

Water Quality and Treatment, A Handbook of Community Water Supplies. 4th ed. Denver, Colo.: American Water Works Association, 1990.

Wentworth, D. F., R. T. Thorup, and O.J. Sproul. "Poliovirus Inactivation in Water-Softening Precipitative Processes." *Journal AWWA* 60(8):939, 1968.

CHAPTER 12
ION EXCHANGE PROCESSES

Ion exchange processes are widely used in water treatment to remove objectionable ionic contaminants. The nature of ion exchange, as used by industry, has changed dramatically over the last 80 years. The first ion exchange process used inorganic zeolites mined from natural deposits as a cation exchanger for water softening. Modern synthetic polymer-based exchange media are used as cation and anion exchangers. They have largely displaced the natural zeolites, primarily because of higher capacities. However, some inorganic zeolites are still used in a few specialized applications.

Ion exchange processes primarily treat low mineral content waters to reduce undesirable ionic contaminants. By far the largest single application of ion exchange is for water softening, the process of removing calcium and magnesium ions from water. Ion exchange also reduces or removes potentially harmful ionic contaminants from drinking water supplies. Chemical processes use ion exchange for product purification and recovery, for specialty separations such as chromatographic separations by size and valance charge, and as catalysts.

A full discussion of ion exchange process theory can be found in *Water Quality and Treatment*.

THE ION EXCHANGE PROCESS

As its name implies, ion exchange is the exchange of ions from one phase to another. In water treatment the exchange of ions occurs between the solid phase of the ion exchanger and influent water. Ions such as calcium, magnesium, barium, copper, lead, zinc, strontium, radium, ammonium, fluorides, nitrates, humates, arsenates, chromates, uranium, anionic metallic complexes, phosphates, hydrogen sulfide, bicarbonates, sulfates, and many others are routinely removed from water by ion exchange resins.

It must be kept in mind that the ion exchange process works only with ions. Substances that do not ionize in water are not removed by ion exchange. Each type of ion exchange resin exhibits an order of preference for various ions. This can be stated quantitatively through selectivity coefficients. Equilibrium concentrations of ions in the resin phase and in the water phase can be calculated from the selectivity coefficient.

Each ion pair has a unique selectivity value for each ion exchange resin. Operating capacity and leakage data for ion exchange resins in the common ions in water are usually provided by the resin supplier.

In general, ion exchange resin is used in, and regenerated to, an ionic form that exchanges ions that are acceptable in treated water. As untreated water passes through the resin, the undesirable ion exchange for the unobjectionable ion occurs on the resin. For example, a cation exchange resin is regenerated with sodium chloride and operated in the sodium cycle. The resin exchanges sodium ions for all positively charged ions (cations), including hardness-causing calcium and magnesium ions, and the water is softened. Likewise, an anion resin can also be regenerated with sodium chloride and operated in the chloride cycle. It will exchange chlorides for the anions (e.g., bicarbonates, sulfates, nitrates) and the effluent water will have a chloride concentration equivalent to the concentration of anions in the source water.

In municipal and domestic applications, ion exchange resins are normally used in single beds to remove specific substances, such as hardness, nitrates, naturally occurring organics (color), and alkalinity. In most cases resins are regenerated with an exchange salt, so the total dissolved solids and pH of the treated water are not significantly altered. Acid and caustic regenerations are mostly used in deionization processes for industrial applications.

Although salt exchanges are the most common, other ion exchanges involve acids and bases. In the process of demineralization, for example, cation resin is used in the hydrogen form and exchanges hydrogen ions. This converts all incoming salts to their equivalent acids. After cation exchange, water passes through a hydroxide-form anion exchanger where anions are exchanged for hydroxides, neutralizing acids and converting them to form water molecules. In this process the net result is that an equivalent amount of water molecules is added to the water in exchange for the salts removed.

Resin Capacity

Ion exchange resins have a finite capacity. When this capacity is used up, the resins are exhausted and leakage of the unwanted ions begins to rise. Exhausted resin can be regenerated using a salt, acid, or base solution containing the ion whose "form" the resin will be operated in. This is passed through the resin bed in sufficient quantity and at a sufficiently high concentration to displace and replace the ions from the resin with the ions from the regenerant solution. The high concentration of the regenerant minimizes waste volumes and, in some cases, changes equilibrium relationships to make the regeneration more efficient.

Regeneration Reactions

Ion exchange is an equilibrium reaction, and regeneration is never 100% complete. There is always some concentration of the undesirable ionic constituent left in the product water as a result of the equilibrium relationships. The most complete exchange reactions are those in which the resulting products disappear, such as by H and OH neutralization, which drives the reaction to completion by forming water molecules.

When equilibrium is favorable, ion exchange occurs in a narrow band within the resin bed, and a large portion of the resin bed can be used before significant leakage occurs due to kinetic slippage. When equilibrium is not favorable, the exchange zone is wider and diffuse and can be as long as or longer than the entire resin bed. The leakage of the undesirable ion appears almost immediately and gradually increases throughout the service run. When the exchange zone is larger than the entire bed, the desired quality cannot be achieved, even in the initial portion of the service cycle.

The most common regenerant in drinking water applications is sodium chloride used in the softening process, dealkalinization, and barium, radium, uranium, selenium, arsenic, and nitrate removal. Potassium chloride, although more expensive, can be used with similar results in situations where low sodium levels are desired.

In demineralization, resins are regenerated with acids and bases. Cation exchange resins are usually regenerated with sulfuric or hydrochloric acid. Anion resins are regenerated with caustic soda, usually sodium hydroxide but sometimes potassium hydroxide. Weakly acidic cation resins can also be regenerated with waste or partially spent acids such as acid left over from regenerating strong acid resins. Weakly basic resins can be regenerated with weak bases such as sodium carbonate, ammonia, or spent caustic such as leftover from regenerating strong base resins.

In most cases weak acid and weak base resins are regenerated with the same strong acids and bases used to regenerate strong cation and strong anion resins, especially when they are used together. It is not unusual to use waste regenerants from strongly ionized resins to regenerate weakly ionized resins. In this manner, regeneration efficiencies can approach 100%, whereas 10% to 50% is normal for strongly ionized resin systems.

In general, for the ion exchange process to be effective, the volume of treated water must be greater than the volume of waste generated by the backwash regeneration and rinse cycles. As the total ionic strength of the solution increases, the exhaustion (service) cycle throughput is reduced proportionately. When the ionic concentration is greater than 500 mg/L, ion exchange may become impractical for some applications. At 1,000 mg/L, ion exchange is still an effective technology for softening and selective ion removal. At 10,000 mg/L, ion exchange is not practical for most applications except trace ion removal, usually involving special resins.

Source Water Quality

Ion exchange resins are generally limited to processing waters that are relatively free of oxidants, physical contaminants, or oily substances that might coat the resin beads. The organic polymers used to make ion exchange resins have upper temperature limits of about 300° F (149° C). Functional groups in strongly basic anion resins are thermally less stable than the polymer and have lower temperature limits; functional groups in cation exchanges are more stable than the polymer.

Point-of-Use versus Centralized Ion Exchange Systems

Where a specific contaminant must be removed, point-of-use (home-type) systems may have advantages over a centralized treatment system for small communities because of the lower capital costs involved. This particularly applies to hardness. Homeowners who dislike hard water can purchase a softener, while another homeowner on a salt-restricted diet might prefer not to have a softener or to limit its use to nonpotable outlets to avoid the increase in sodium caused by the hardness exchange.

Most homeowners do not soften water used for lawn sprinkling, which increases efficiency. Individual home systems in some cases may not have the same restrictions on waste discharge that centralized treatment facilities have. However, point-of-use units have the significant drawback that there is seldom any periodic monitoring of effluent quality, and the homeowner may not be aware that the softener has malfunctioned or that the treated water does not contain a lowered level of contamination.

Centralized treatment systems have advantages in economy of scale, allow for more convenient monitoring of effluent quality, and provide an easier pathway to collect and treat regenerant waste.

ION EXCHANGE RESINS

When salts dissolve in water, they separate into charged ions. Cations carry positive charges and anions carry negative charges. There are two classes of ion exchange resins: those that exchange cations and those that exchange anions. Each of these classes is further divided into those that are strongly ionized and weakly ionized, according to the nature of the functional groups. Almost all ion exchange materials can be classified in this manner. The characteristics of strongly ionized and weakly ionized resins are summarized in Table 12.1.

Most modern ion exchange resins consist of an organic polymer, chemically bonded to an acidic or basic functional group. Ion exchange polymers typically are polymerized polystyrene cross-linked with divinylbenzene. Other polymers used are acrylic or methacrylic acids, phenolformaldehyde, epoxypolyamine, and pyridine-based polymers. Acrylics and epoxys are the most widely used in this group, but they make up only a small percentage of ion exchangers produced commercially.

Resin Qualities

Ion exchange resins are homogeneous solids, and the exchange of ions from the liquid to the solid phase occurs at the surface of the beads. Ions then diffuse through the solvated polymer from one ion exchange site to another toward the center of the bead.

To create a balance between pressure loss and kinetics, the size of resins chosen for most water treatment applications is a range between 0.3 and 1.2 mm in diameter. This size range has been proven to provide the best combination of flow equalization, pressure drop, and kinetics. Resins smaller than 0.3 mm are used in specialty applications and have enhanced kinetic properties that make them superior to larger resins for certain types of slow chemical reactions. However, their small size makes them more expensive to produce, they require special containment equipment, and they create a higher pressure drop.

Gel and Macroporous Resins. The physical structure of commonly available ion exchange resins is classified as either gel or macroporous. In gel types, pores are in the copolymer structure and are due to molecular spacing.

Macroporous resins have actual physical pores in the resin beads and have high surface area with a higher surface-to-volume ratio than gel-type resins. Macroporous resins are currently made in two different ways. The original method was to agglomerate microspheres

TABLE 12.1 Characteristics of Ion Exchange Resins

Strongly ionized	Weakly ionized
Fully ionized	Ionization a function of pH
Fixed number of exchange sites	Number of exchange sites varies from low to high
Fast rate of neutralization ($T_{1/2} < 5$ min)	Slow rate of neutralization ($T_{1/2} > 10$ min)
Regeneration mass action controlled, excess regenerants required	Regeneration pH controlled, essentially 100% efficient
Operating capacity dominated by equilibrium factors	Operating capacity dominated by kinetic factors

into large spherical beads. A second process is based on extraction technology in which an inert substance is added to the reacting monomer mixture. After polymerization the inert substance is removed from the resulting copolymer beads, leaving behind discrete pores.

Historically, macroporous resins were originally made to provide physical stability to highly cross-linked cation exchange resins. By their nature, macroporous resins are more resistant to stresses brought on by changes in operating conditions. Highly cross-linked macroporous resins are better able to withstand severe and rapid changes in operating environments such as sudden and high flow rates, sudden temperature and pressure changes, high temperatures, and oxidative environments.

Increasing cross-linkage to gain thermal and oxidative resistance also decreases capacity and increases brittleness, leading to reduced physical stability. These effects are reduced when macroporosity is introduced. The volume of discrete pores provides stress relief but also reduces the amount of copolymer within a bead, reducing the volumetric capacity of the resin. This reduced volumetric capacity makes macroporous resins less desirable in strongly ionized resins in many bulk water treatment applications. However, macroporosity has been proven superior for high-capacity, weakly ionized resins because of the ability to accommodate the significant size change these resins undergo between the exhausted and regenerated ionic forms. Advantages of gel and macroporous resins are compared in Table 12.2.

Inorganic Zeolites. Inorganic zeolites are no longer commonly used in softening or in bulk deionization. They have useful selectivities that make them ideal to remove specific contaminants and are occasionally used for this purpose. Because they are not stable at all pHs, pH control is often required. The major categories of commercially available inorganic zeolites are green sand, activated alumina, and aluminosilicates.

Green sand and aluminosilicates are manufactured from mined deposits. Some aluminosilicates are also synthesized. Green sand is primarily used for iron and manganese removal. Activated alumina is a highly processed form of aluminum oxide that has high selectivity for fluoride, arsenic, and lead.

Adsorbents. In certain applications, nonfunctional polymers, especially those with macroporous structures, are used as adsorbents. They generally have limited or no ion exchange functionality. However, they have certain characteristics such as static charge and copolymer structure or pore size that make them useful for some types of chemical separations. They are not widely used in water treatment.

Resin Degradation

Ion exchange resins degrade both physically and chemically. In practical terms, the degradation of a resin can be defined as any change in the physical or chemical properties that

TABLE 12.2 Comparison of Advantages of Gel-Type and Macroporous Resins

Gel-type resins	Macroporous resins
Higher capacity	Better physical stability
Better kinetics	Better organic fouling resistance
Better operating efficiency	Better oxidation resistance
Lower cost	

adversely affects the operating performance or the ability of the resin to attain the desired results.

Most resin replacement occurs as a result of degradation, but a change in resin properties is not automatic grounds for replacement. Degradation must be evaluated in light of the way the resin is used. A physically degraded ion exchange resin that continues to perform well chemically without high pressure loss or flow channeling may be acceptable for certain low-flow applications.

Physical Degradation. Ion exchange resins degrade physically by breaking. New resins typically have over 90% whole, perfect, uncracked beads. Over time, resin beads develop cracks, which then fracture, reducing the average particle size. The void volume (space between beads) decreases as irregular-shaped fragments fill spaces between beads, increasing pressure loss. Filling void space can also lead to channeling (irregular flow distribution), and small particles may clog flow distributors.

Degradation from Oxidation. Resins are often exposed to oxidants such as dissolved oxygen and chlorine. Oxidation of resin particles occurs first on the outside of the beads and works its way inward. This results in an uneven distribution of cross-linking and swelling that creates an osmotic stress, a major cause of bead breakage in cation resins used in domestic water softeners. Both macroporous and gel types are subject to oxidative degradation.

Chemical Fouling. Ion exchange resin can become impaired by an ion exchange reaction that is irreversible, or by surface clogging that occurs when resin becomes fouled with oil, rust, or biological slime. Fouling can also be caused when a pretreatment clarification upset causes water-soluble polymers to pass onto the resin and coat the beads. When any of this happens, the resin is referred to as being fouled.

Cation resin fouling is most often due to chemical precipitants such as iron or barium. These can either coat the resin or form clumps of precipitated salts that cannot be removed by the normal regeneration process.

Anion resins are most commonly fouled by naturally occurring organic substances that, although exchangeable as ions, have such slow diffusion rates that they are not removed by the normal regeneration process unless special resins or special regeneration procedures are used. When fouling continues over a long period of time the resin becomes impaired and performance drops off. This condition is commonly referred to as organic fouling.

Proper regeneration techniques and chemical procedures reduce the tendency to foul. Some types of resins are more resistant to fouling than others, but there is usually a trade-off in other properties. The selection of the type of anion resin for a given application depends on factors such as water temperature ranges, operating efficiency capacity requirements, regeneration temperatures, and required quality levels.

CATION EXCHANGE PROCESSES

Softening is the most widely practiced of all ion exchange processes. Calcium and magnesium ions are referred to as hardness because they react with soap to form curds. This makes it hard to wash in waters containing higher levels of these chemicals. Ion exchange softening actually involves all cations, including copper, iron, lead, and zinc, which are all exchanged along with the calcium and magnesium.

Salt Selection

Any environmentally and economically acceptable salt can be used to regenerate the exhausted resins. A good regenerant should be inexpensive and contain an exchangeable ion that can effectively displace unwanted ions from the resin. It should be soluble so that it may be delivered at reasonably high concentrations in order to keep waste volumes low, and its co-ion should remain soluble when paired with displaced ions. Because most chloride salts are soluble and chlorides are nontoxic, it is not surprising that sodium chloride is the most common salt used to regenerate softeners. Potassium chloride, although more expensive, can be used with similar results.

When a strongly acidic cation exchange resin is regenerated with sodium chloride, it places the resin in the sodium form. The resin can then exchange its sodium ions for calcium, magnesium, and other cations in the source water. The result is a discharge water containing essentially all sodium cations.

Operating Capacity

Operating capacity of the resin is defined by the number of exchangeable ions it removes from the source water during each cycle. In the case of a water softener, calcium, magnesium, and other divalent ions constitute the exchangeable ions. There are several ways to express this. The most common expression of capacity in North America is kilograins (as $CaCO_3$) per cubic foot of resin. The higher the concentration of exchangeable ions, the lower the volume-based throughput capacity for a resin to achieve the same ion-based operating capacity.

The resin operating capacity also varies according to the regenerant level. The regenerant level is normally stated in the United States in terms of pounds of regenerant per cubic foot of resin. Higher dosages give higher operating capacities and lower hardness leakage (higher quality), but at reduced salt efficiency. The selection of regeneration levels is generally based on using the minimal amount of salt to operate the softener at the required degree of hardness leakage.

Operating efficiency depends on several factors, including choice of resin, regeneration level, method of regeneration, flow rate during the service cycle, and the quality level used to determine start and end of the service cycle.

A variety of process schemes can maximize regenerant effectiveness. These usually require more complex and costly equipment, essentially trading reduced chemical operating costs and a lower waste discharge for higher equipment costs.

Ammonia Removal

Ammonia is a colorless gas that dissolves readily in water. It is commonly present in most water supplies in trace quantities due to degradation of nitrogenous organic matter. It may also be present due to the discharge of industrial wastes. Ammonia reacts with water to form ammonium hydroxide, which dissociates into ammonium (NH_4^+) cations according to the following reaction:

$$NH_3 + H_2O \rightleftarrows NH_4^+ + OH^-$$

When present in concentrations of a few milligrams per liter (mg/L), the degree of ionization of ammonia is a function of the solution pH. At a pH below 8, ammonia is present primarily as the cationic ammonium ion. At a pH above 9.0, ammonia is essentially unionized and is not efficiently removed by ion exchange.

Strong-acid cation exchange resins, such as the kind typically used for softeners, exchange sodium for ammonium ions. As the cation resin exchanges for divalent hardness ions, hardness displaces the ammonia, which in turn displaces more sodium. As the service cycle continues, hardness forms a band in the upper portion of the resin and ammonia forms its own band just beneath the hardness.

If ammonia removal from a water source is required, ammonia must be included as an exchangeable ion in the system design. The service cycle is ended when ammonia levels rise in the effluent. If the service cycle is allowed to continue past ammonia breakthrough, hardness will continue to load on the resin and displace ammonia. Ammonia levels in the effluent could then reach concentrations equal to the total hardness plus ammonia concentrations in the source water.

Ammonia can also be removed by inorganic zeolites such as chabazite, mordenite, and clinoptilolite. Some zeolites are more highly selective for ammonia than hardness and are able to remove it throughout the acceptable pH range for drinking water (6 to 9) without danger of dumping. Not having to remove hardness gives them very high throughput capacities.

Barium Removal

Barium is an alkaline earth metal occasionally found in groundwater, primarily in Arizona, Texas, Michigan, Vermont, and Florida. Barium is present in trace amounts in some surface waters with the highest levels occurring in the lower Mississippi basin. Barium is highly toxic when ingested as a soluble salt—a dose of as little as 550 mg is considered lethal.

Barium Removal by Softeners. Barium is readily exchanged onto strong-acid cation exchange resins in softeners. Barium has a higher selectivity for cation resins than either calcium or magnesium and is removed throughout the entire softening cycle. It does not begin to leak until after hardness breakthrough. The increase in barium caused by hardness breakthrough, although small compared with the hardness leakage, may be sufficient to warrant termination of the service cycle. Therefore barium removal by ion exchange is achieved by ordinary softening.

When sulfate is present, barium solubility is limited. If sulfates are present in the source water, it is probable that barium is present only as a suspended solid and cannot be removed by ion exchange processes at levels above a few mg/L.

Because the barium form of resin is more difficult to regenerate than the hardness form, barium will tend to accumulate on the resin, especially at lower salt dose levels. During the initial exhaustion cycle, barium continues to load on the resin after hardness breakthrough by displacing previously exchanged calcium and magnesium ions. When the resin bed is regenerated, barium is less efficiently displaced from the resin than hardness ions. The ratio of barium to hardness left on the resin after regeneration will be substantially higher than in the influent water, and barium will be mostly at the bottom of the resin bed. The regenerant dose level should be high enough to prevent barium buildup in the resin bed.

If a significant amount of barium is exchanged onto the resin during the service cycle, there is the potential for barium fouling from precipitation of barium sulfate. Appreciable amounts of sulfate, either in the dilution water or in the salt itself, cause precipitation of the barium in the resin, which fouls the resin.

Barium-fouled resins can be cleaned, but the cleaning process is slow and involves corrosive chemicals such as hydrochloric acid. One method is to soak the resin bed in 10%

hydrochloric acid, converting sulfates to bisulfates that are more soluble. The process usually requires several hours and vigorous agitation before giving measurable improvement. It is often less expensive to discard and replace barium-fouled resins when performance drops below acceptable levels.

Combined Softening and Decationization. Waters containing hardness and appreciable levels of alkalinity can be partially demineralized and fully softened by having two columns of strong-acid cation resins operated in parallel. One column is in the hydrogen form and the other is in the sodium form. The column effluents are blended at a ratio determined by the composition of the source water. The overall effect is partial demineralization, alkalinity reduction, and softening. Through proper control of the blend ratio, pH can be maintained at an acceptable level.

Barium Removal by Hydrogen-Form Weak-Acid Dealkalizers. Barium can also be removed by using a weak-acid cation resin in either the hydrogen or sodium form. The use of the resin in the hydrogen form for softening is limited to removing hardness associated with alkalinity, usually limited to waters that have high alkalinity-to-hardness ratios. The effluent has a reduced and variable pH.

Initially the resin removes all cations plus those associated with alkalinity in exchange for hydrogen and produces a very low pH. But in a short time the resin ceases exchanging for monovalent ions such as sodium, then more gradually with hardness, and finally with barium. The pH rises gradually and the reaction of the hydrogen-form resin is limited to these divalent cations associated with the alkalinity content of the water. Alkalinity in the water is converted to carbon dioxide gas by the hydrogen ions exchanged and has to be removed. The most common methods of CO_2 removal are air stripping and deaerating heaters.

Hydrogen-form weak-acid resins combine readily with hydrogen ions and are efficiently converted back to the hydrogen form. Only about 20% excess regenerant above the theoretical dose is needed. The use of hydrochloric acid as a regenerant has the advantage that it can be used at high concentrations, producing lower waste volume. Hydrochloric acid regenerant concentrations of 6% to 10% are typical.

Concentrated hydrochloric acid is hazardous and corrosive and gives off toxic fumes, so proper safety precautions must be taken when using this chemical. Sulfuric acid cannot be used because it forms insoluble calcium and barium sulfate that would foul the resin bed.

A weak-acid resin can be used in the sodium form to remove barium and hardness. This involves a two-stage regeneration, first of acid, then an alkaline rinse to neutralize the bed with sodium carbonate or sodium hydroxide. Sodium bicarbonate can be used but is less effective.

Operating a weak-acid resin, whether in the hydrogen or sodium cycle, is more complex and usually more expensive than using salt-regenerated strong-acid cation resins. Other disadvantages are the need for acid-resistant materials, wastewater neutralization, and the need to deal with or strip CO_2 from the product water.

Radium Removal

Radium 226 and radium 228 are natural groundwater contaminants that usually occur at trace levels. Maximum contaminant levels for radium have been set by the U.S. Environmental Protection Agency (USEPA). Some public water systems with elevated radium levels in their groundwater and no other water source available have found it necessary to install radium removal systems.

A strong-acid cation exchange resin operated in the sodium cycle is an effective method of radium removal. Like barium, radium has a higher selectivity for cation exchange resins than hardness and is removed during the normal water softening cycle. In a manner similar to barium, on the first cycle of exhaustion, radium continues to load on the resin bed until well after hardness breakthrough by displacing all other ions previously loaded, including calcium and magnesium. This effect is valid only for the first cycle and can be misleading.

Point-of-use cartridges using ordinary softener resins on a one-time basis for radium removal provide relatively good radium removal for about 5 to 15 times (depending on the particular resin) as long as they produce softened water.

In regenerable systems, however, radium is much harder to exchange off the resin. More radium remains in the resin after regeneration as radium is pushed toward the exit (bottom) of the resin bed during the regeneration cycle. Hardness leakage reaches the radium-rich end of the resin bed as the softener becomes exhausted. The hardness that is less preferred by the resin displaces a small but significant amount of radium from the resin, causing radium levels to increase to unacceptable levels. In systems that are regenerated, it is necessary to limit the service cycle to the softener capacity for hardness. Because the amount of radium is insignificant compared with hardness, softener design calculations are made in the traditional manner.

ANION EXCHANGE PROCESSES

Sulfate, nitrate, selenium, arsenic, and uranium are all present as anions and can be removed by anion exchange. The most common process is to use salt-regenerated strongly basic anion exchange resins, operating in the chloride cycle. Equipment design is essentially the same for any of these substances. However, throughput capacities and equipment sizing are different because of the differences in relative affinities. Substances with high affinities can continue to load to higher concentrations on the resins by displacing other substances previously exchanged but with lower relative affinities.

The gel-type strongly basic resins, whose functional groups are based on either trimethylamine (called type 1) or dimethylethanolamine (called type 2), have the same order of affinities for common ions found in drinking water except for the hydroxide ion. Type 2 resin is somewhat less basic than type 1 so it combines more readily with the hydroxide ion. In demineralization, anion resins are regenerated with NaOH. In these applications, the type 2 resins give better regeneration efficiencies. When operated in the salt cycle, regeneration efficiency is about the same for both types.

The general order of affinities for both types of resins for the common ions found in water is:

phosphate → selanate → carbonate → arsenate → selonite → arsenite → sulfate → nitrate → chloride → bicarbonate

When strong-base resins are regenerated with NaCl they are commonly referred to as anion softeners because of their similarity to traditional cation-based water softeners. When resin pH is at or above 7, anion resins tend to give off fishy odors from the slow decomposition of their functional groups, which are amine based. The alkanol amine group of the type 2 resin gives off less offensive odors than the type 1 resins and is sometimes preferred in drinking water applications for this reason.

When a strong-base resin is operated in the chloride cycle, it exchanges chlorides for all incoming anions. When all exchangeable ions, in this case chlorides, have been used up, the resin bed contains only the ions it removed from the source water. These ions will have distributed themselves in order of their relative affinities.

The bicarbonate ion is the least strongly held and is positioned nearest the exit end of the vessel. If the vessel is operated as a dealkalizer, the service cycle would be ended at this point and the vessel regenerated with a salt solution. However, if the vessel is used for nitrate removal and there is no desire to remove bicarbonates, the service cycle can continue. Nitrates and other ions continue to be exchanged onto the resin bed and the bicarbonates are driven off. The service cycle ends when nitrates on the resin are pushed down to the discharge end of the vessel. Regardless of the application, the vessel, resins, and regenerations are identical, but operating capacity depends on the composition of untreated water and the specific treatment goal.

Dealkalization

Dealkalization is the removal of bicarbonates, carbonates, and hydroxides. In drinking waters, hydroxides can be controlled simply by injecting acid. This leaves bicarbonates and carbonates that are not amenable to acid injection because of the increased corrosivity at lower pH. Strongly basic anion exchange resins operated in the chloride cycle can remove bicarbonates and carbonates in exchange for chlorides. This also eliminates "temporary hardness," minimizing scale buildup in water heaters and boilers. Figure 12.1 shows a softener followed by a strong base dealkalizer.

Carbonates are usually a small percent of overall alkalinity. The bicarbonate ion that predominates has the lowest affinity of the common anions for the exchange resins. It is the first ion to appear in the effluent, leaving the exhausted bed filled with sulfates at the top, just above nitrates, which are above chlorides and the bicarbonates at the bottom. The chloride ion is extremely effective in regenerating the resin.

Because the bicarbonate ion is not as strongly held by the resin, operating capacity depends on the ratio of ion ratios in the inlet water during the service cycle as much as the total concentration of ions. Operating capacity for most gel type 2 anion exchange resins can be estimated at 12,500 grains per cubic foot of resin. Throughput capacities are calculated based on all anions in the inlet water including chlorides as being exchanged.

The sum of all the anions as their calcium carbonate equivalents must first be calculated and then divided by 17.1 to convert the number to grains per gallon. Bicarbonates are reduced to nearly 0 mg/L throughout most of the service cycle, and treated water pH is about 5. When alkalinity leakage reaches 10%, the service cycle is terminated.

pH Effects of Dealkalization. The pH of ordinary tap water is determined by the equilibrium between carbon dioxide, bicarbonate, and carbonate between the pH values of 5.0 and 10.6. In the absence of hydroxides, removal of bicarbonate and carbonates automatically drops the pH of tap water to 5.0.

When salt is used alone as a regenerant, only carbon dioxide remains unremoved by resin. This also means that pH becomes 5.0. This scenario can be changed by adding a small amount of caustic to the regenerant salt, usually 0.25 lb (0.11 kg) of NaOH to each 5.0 lb (2.3 kg) of salt, placing only a small amount of hydroxide on the resin. Its main effect is to convert the nonregenerated bicarbonate left on the resin to carbonates. Carbonates on resin combine with carbon dioxide to form bicarbonates held by resin. By this means, CO_2 is removed, pH is shifted back to 7.0, and the effluent is free of bicarbonates and carbon dioxide.

Dealkalization with Weak-Acid Resins Regenerated with Acid. Weakly acidic resins are primarily operated in the hydrogen form and are, by their nature, unable to react with cations except in the presence of a neutralizing agent such as alkalinity. They exchange hydrogen for cations, but only hardness and other divalent ions associated with alkalinity are exchanged. In this way, weakly acidic resins are able to work effectively as dealkalizers on certain waters. Because they are weakly acidic, these resins have extremely high affinities for hydrogen ions and are fully converted to the hydrogen form with little more than the stoichiometric amount of acid during regeneration.

The degree of exchange during the service cycle, or operating capacity, is limited by the amount of alkalinity, the ratio of divalent cations to alkalinity actually being exchanged, and total ionic concentration. The key factor to determining potential operating capacity to 10% alkalinity leakage is the hardness-to-alkalinity ratio. The base operating capacity can vary from about 10 to 60 kg/ft^2 (5.6 kg/m^2) of alkalinity as the hardness-to-alkalinity ratio varies from 0 to 1.2 or higher. The regeneration dose is calculated as 120% of operating capacity. Flow rate and temperature during the service cycle also affect operating capacity to a marked degree, possibly reducing operating capacity by 35% as the temperature drops from 70° to 35° F (21° to 2° C) or as the flow rate doubles from 2 to 4 gpm/ft^3 (27 to 53 L/min/m^3).

In the process, hardness and bicarbonates are converted to carbonic acid, and in effect the effluent is also partially demineralized. The degree of softening depends on the hardness-to-alkalinity ratio. Regeneration equipment and all process and tanks, piping, and valves must be acid resistant. Sulfuric acid is commonly used as the regenerant because it is less expensive, but calcium sulfate precipitation is a real concern.

When sulfuric acid is used, regenerant concentration is usually held between 0.5% and 0.75%. Hydrochloric acid can be used at higher concentrations, with 6% to 10% the common range, which gives much lower waste volumes. Hydrochloric acid is not as widely used because it is more expensive, is more corrosive, gives off hazardous fumes, and has higher equipment costs because of the increased corrosivity.

Effects of pH during Salt-Cycle Dealkalization. Weakly acidic resins have a small amount of strong acid capacity and convert all salts to acids in the beginning of the service cycle. Initial pH can be as low as 2 and rises gradually, but remains under 5 for over three-fourths of the service cycle. Carbon dioxide generated by the exchange reaction in some applications must be removed by aeration or vacuum degasification, or it will be converted back to biocarbonate when caustic is added to raise pH.

Dealkalization by Dual Strong-Acid Columns Operated in the Sodium and Hydrogen Cycles. In this scheme, two vessels, each loaded with strong-acid resins, operate in parallel, with one regenerated with salt and the other with an acid. The two effluents are combined in a ratio based on the composition of the untreated inlet water so that acid generated in the hydrogen cycle vessel is just sufficient to neutralize alkalinity in the effluent from the salt-regenerated vessel. This process produces partially demineralized, fully softened water with about zero alkalinity.

The sodium cycle vessel is sized and designed as a traditional salt-regenerated cation water softener. The hydrogen cycle vessel is typically designed to operate at the highest acid efficiency. Throughput capacity and leakage and blending ratios depend on the water analysis. When practical, the two vessels are sized to exhaust simultaneously.

Arsenic Removal

Arsenic occurs widely in the earth's crust, usually in the form of insoluble complexes with iron and sulfides. Another source of arsenic in drinking water supplies is its presence as a contaminant in groundwater, caused by its extensive use as a pesticide in the past. It also exists in soluble form, primarily as arsenites (AsO_3) and arsenates (AsO_4). Ingestion of as little as 100 mg of arsenic per day can cause severe poisoning in humans.

Arsenic is not detectable by normal human senses, so toxic levels can exist without warning. It is important that treated effluent from an arsenic removal system be properly

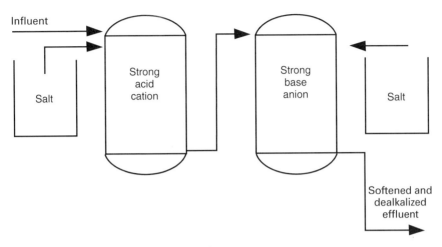

FIGURE 12.1 Strong base dealkalizing.

monitored to ensure that arsenic is removed and to prevent unsafe levels of arsenic in the treated water being delivered to the distribution system.

Removal Using Strongly Basic Ion Exchange Resins. Soluble forms of arsenic can be removed by strongly basic ion exchange resins or activated alumina. The process works best when arsenic is present in the arsenate form. Proper pretreatment and testing should be used to ensure that all arsenic is oxidized to the arsenate form.

In oxygen-deprived waters, such as in well water, arsenite species is usually 20% to 50% of total arsenic. Arsenite is a weak acid and at pH levels below 9 it is un-ionized and not readily removed by ion exchange. However, arsenate is a relatively strong acid and fully ionized at pH levels above 6. An effective way of converting arsenite to arsenate is with oxidation using chlorine or chloramine. This oxidation reaction occurs at a rapid rate with over 95% of the arsenite converted to arsenate with 1 mg/L chlorine in less than 5 s contact time. The reaction is insensitive to pH in the range of 6.5 to 9.5, but the reaction rate decreases substantially outside this range.

The presence of other oxidizable substances must be accounted for because they consume chlorine and may act as reverse catalysts for arsenite oxidation. For example, the presence of 5 mg/L of TOC will slow the reaction rate so that almost 50% of the arsenite remains unoxidized after 30 s, and more than 30 min may be required to reach 80% oxidation. The presence of chloramine has also been shown to reduce the rate of oxidation substantially. For this reason, the presence of both TOC and chloramines should be thoroughly evaluated at the pilot plant stage to ensure that this critical step is carried out effectively and that all arsenites are converted to arsenates.

The presence of iron or high levels of sulfate interferes with and impairs the removal efficiency of arsenic in any form because the sulfate ion is removed along with the arsenic, increasing the exchange load. Throughput capacities and resin volume calculations should include sulfates as exchangeable ions.

The operating capacity of strongly basic anion exchange resins is markedly higher for arsenate than for arsenite. Sulfates and chlorides are preferred over arsenites regardless of the type of medium. This means that they are removed before the arsenites and must be included as exchangeable ions that increase the loading. Their presence reduces the throughput capacity for arsenite proportionately. The arsenate ion, however, has a higher affinity for the strongly basic anion resins than sulfates or chlorides. This means that the presence of sulfate is less detrimental to the throughput capacity, and arsenate "dumping" cannot occur if the resin column is inadvertently overrun. This adds a measure of safety.

Anion exchange resins have the advantage over activated alumina of being regenerable in a single step with common salt in the same manner as a water softener, whereas activated alumina must be regenerated in two steps. As a result, strongly basic resins are more commonly used for arsenic removal, especially in small systems such as for individual homes.

Removal Using Activated Alumina. Activated alumina has been used successfully to remove both forms of arsenic, but it exhibits as much as an 80 times increase in throughput capacity for arsenate as for arsenite at the same concentrations on otherwise identical waters. It is necessary to control pH within a fairly narrow range and to carefully monitor water chemistry for success. Preoxidation of arsenic to the arsenate form and pH control to the 5.5 to 6.0 range are highly recommended.

Regeneration of activated alumina is a two-step process. Following the backwash cycle, it is first regenerated with sodium hydroxide and rinsed, next it is washed with a dilute acid such as sulfuric, and then the bed is rinsed down to the proper pH before being returned to service.

Sulfate Removal

High concentrations of sulfate in drinking water supplies cause diarrhea in humans and livestock. Sulfates can be removed by strong-base chloride-cycle anion exchange with removal rates typically between 95% and 100%. A salt dosage of 5 lb/ft^3 (8 g/m^3) is sufficient, and operating capacities of 20 to 25 kilograins of sulfate, expressed as calcium carbonate per cubic foot (0.03 m^3) of resin, are typical for gel-type strong-base resins. Sulfate displaces the other common ions, including chlorides and nitrates, during the exhaustion cycle. As a result, operating throughput is calculated based only on sulfate concentration in the source water.

Nitrate Removal

Nitrates are toxic, especially to infants and to ruminant animals such as cows. They occur naturally as a result of the decomposition of nitrogen-containing waste matter, and also as a result of farm runoff from fertilizer. Nitrates in drinking water are usually limited to 10 mg/L as nitrogen. At elevated levels nitrates can cause methemoglobinemia, also known as blue baby syndrome.

Nitrate Removal by Type 1 and Type 2 Resins. Nitrate is usually removed using type 1 or type 2 strong-base gel anion resins selective for nitrates over the common ions in tap water, with the exception of sulfates. Sulfate selectivity is reversed at the higher ionic concentrations used for regeneration. Salt at concentrations above 1% is an effective regenerant. The nitrate ion is held more strongly by the resin than chlorides during the regeneration and service cycles. During the service cycle, sulfates displace nitrates at high concentrations if the vessel is overrun (Figure 12.2).

Sulfate is more fully removed from the resin than nitrates. The typical strongly basic resins are about four times as selective for nitrates than for chlorides. During the service cycle, all anions are converted to chlorides at the entrance to the resin bed. At the exit end of the bed, chlorides displace some of the remaining nitrates and a small but significant

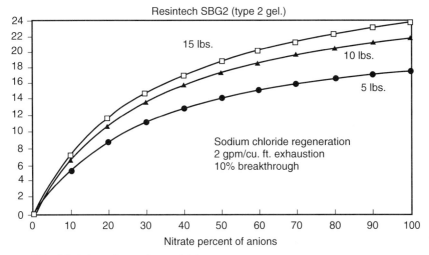

FIGURE 12.2 Nitrate removal capacity.

amount of nitrate leakage occurs as the service cycle begins. How much nitrate leakage occurs depends on the TDS concentration and the degree of regeneration. To ensure nitrate levels below 10 mg/L as nitrogen, about 10 lb/ft^3 (16 g/m^3) of NaCl or higher is often considered a minimum dose for cocurrent regenerated columns.

When nitrates in the effluent reach 20% to 30% or when nitrate levels reach 10 mg/L as nitrogen, the service cycle is terminated. If the resin bed is allowed to continue to process water, the sulfates will continue to load on the resin by displacing nitrates, increasing nitrates in the effluent to a level that could become as high as the sum of sulfate plus nitrates in the source water. In many waters, sulfate levels are high enough that nitrate levels could become toxic because of dumping.

Nitrate Removal by Selective Resins. Nitrate removal by so-called nitrate-selective anion exchangers is becoming widely practiced around the world, especially in areas where nitrates have been a problem for some time. These resins are similar in composition and structure to type 1 resins, but their functional groups are based on larger amine compounds than the trimethylamine used in ordinary type 1 resins.

The term *nitrate selective* is actually a misnomer. These resins are effective because they are less selective for multivalent ions such as sulfates. Because of this, sulfates are not able to cause massive dumping of nitrates when resins are overexhausted, as in the case of ordinary resins. For this reason, these resins offer more safety in high-sulfate waters. At levels of about 25% or fewer sulfates, standard resins have a higher operating capacity than selective types. Because of the larger size of the amine groups, selective resins have lower total capacities, making overall operating capacity smaller than for standard-type resins. When SO$_4$ levels are over 25%, selective resins give higher throughputs.

Sodium chloride is effective in regenerating both standard and selective types of resins. Average nitrate leakages are about the same at similar regenerant levels. The biggest difference is that sulfates are dumped by nitrates from the selective resins instead of the other way around in standard resins.

Fluoride Removal

Because bone char and activated alumina are both effective for fluoride removal, they are the preferred exchange media. They are highly selective for fluorides over other ions in water. Strongly basic anion exchange resins are not used because the fluoride ion is the least strongly held and is not well removed. Activated alumina is a semicrystalline inorganic adsorbent. Bone char is made from charred and steam-activated animal bones. Bone char is not as readily available or as widely used as activated alumina, but both successfully reduce fluoride levels.

Defluoridation with activated alumina is similar to other ion exchange processes. Alumina is generally used in a packed bed configured for downflow exhaustion and regeneration. Because of limited kinetics it is best to use bed heights of at least 3 ft (0.9 m). The pH of the water being treated should be adjusted to 5.5 to 6.0.

Unlike organic-based ion exchangers, the breakthrough curve is very gradual. Fluoride leaks from the column throughout the service cycle. The end point of the service cycle is the maximum fluoride concentration that is acceptable for the water's end user. Because of the ever-changing breakthrough pattern, it is difficult to blend or use bypass techniques to improve performance. A more practical approach is to store the entire exhaustion cycle and run the system until the average over the entire cycle meets or is just below the maximum target level.

Selenium Removal

The maximum contaminant level (MCL) for selenium is only 0.01 mg/L. When selenium is found in drinking water, it is usually as a divalent anion. The divalent selenate ion [SeO_4^{-2}, Se(+6)] is more preferred than the selenite [SeO_3^{-2}, Se (+4)] by strongly basic anion exchange resins. Operating capacity of a strongly basic anion exchange resin is greatest when all selenium is present in the +6 valence state. The selenate ion is more preferred than any ions commonly found in drinking water, including sulfate.

Operating Parameters. When a resin loaded with selenite is exhausted, sulfate continues to be loaded on the resin and displaces previously loaded selenite ions. As this happens, the concentration of selenite in the effluent can approach the sum of the sulfate plus selenite ions in the source water. The sulfate level in the source water is usually many times higher than the allowable limit of selenium. The danger exists of selenium dumping and appearing in the effluent at higher than original influent concentration if the resin bed is overrun. Dumping does not occur if the selenium is converted to the selenate ion before the ion exchange vessel because selenate is preferred over sulfate.

Oxidation of selenites to selenates occurs readily in the pH range of 6.5 to 7.5. A retention time of 5 min ensures that over 70% of the Se(+4) is converted to Se(+6) when the free chlorine level is maintained at 5 mg/L or more. Free chlorine is a much more effective agent than either potassium permanganate or hydrogen peroxide.

Over 99% of selenite (+4) can be converted to selenate (+6) in 15 minutes with a 5 mg/L free chlorine residual, but a 2 mg/L chlorine residual takes four times as long. Both pH and chlorine residual level must be controlled to maintain stable, effective operation. Each installation should be evaluated on its own to determine the necessary parameters for proper chlorination.

Even though the selenate ion has a higher affinity for the strongly basic resin than sulfates, their relative affinities are close. They are close enough that when sulfate breaks through, it causes an increase in the selenate level in the effluent. Selenates begin to rise gradually once sulfate breaks through, usually within 10% of the throughput at which sulfate breakthrough occurs. If the resin bed is continued in service, the selenium concentration will rise above maximum allowable levels. This could happen without notice unless the effluent is carefully monitored. For this reason it is considered standard practice to end the service cycle and regenerate the resin at or before sulfate breakthrough.

Operating Cycle. During the service cycle, all anions are loaded on the resin and exchanged for an equal amount of chlorides. During the early portion of the service cycle the effluent contains only chloride anions. But, as the cycle continues, selenates continue to load on the inlet portion of the resin and displace the previously loaded sulfates down the column. Sulfates do likewise to the nitrates and chlorides, which in turn displace bicarbonates that have the lowest relative affinity.

Bicarbonates appear first in the effluent, followed by chlorides and nitrates, and then sulfates. At these low concentrations, the divalent ions are much more strongly held. The presence of high levels of chlorides or bicarbonates in the inlet water has a negligible impact on the capacity of the resin for selenium. Selenium leakage remains low throughout the service cycle until sulfate breaks through. Sulfates displace a portion of the selenium at the bottom of the bed that is left over from the previous regeneration. Selenium levels rise soon after sulfate levels increase.

Regeneration. Because selenium is present only in trace quantities, its concentration alone has little effect on the resin's throughput capacity. It is prudent that any ion exchange system designed for selenium removal be designed to run as a sulfate removal system to a sulfate leakage end point. The system size of the water regeneration equipment for removing sulfates and selenium is the same as used for dealkalization, except for the monitoring equipment.

Relative affinities of divalent ions such as sulfates and selenates against monovalent chlorides drop at the higher ionic concentration during regeneration. To carry out a proper regeneration, resin should be regenerated with sodium chloride at concentrations of at least 5% and a flow rate that allows at least 30 min contact time. These divalent ions are easily displaced by chlorides.

Because of dilution effects from water remaining in the vessel, it is necessary to use enough of a salt dose at sufficient concentration to ensure that the concentration at the bed is high enough to properly reduce the affinities of the divalent ions. This can be achieved with a salt dose of 5 to 10 lb/ft^3 (8 to 16 g/m^3), injected at a concentration of at least 5% and a maximum of 15%. The selenium and sulfate are pushed from the resin bed, and any residual selenium is found at the bottom of the bed after the regeneration cycle. It will resist leakage until another divalent ion, such as sulfate, begins to leak through at the end of the next service cycle.

Uranium Removal

Above pH 6, uranium exists in drinking water primarily as a carbonate complex that is an anion and has a tremendous affinity for strongly basic anion exchange resins—the material generally used for removal. The process has been highly effective at pH levels of 6 to 8.2.

Higher pHs could result in uranium precipitation, which makes the problem one of physical removal. Lower pH levels change the nature of uranium to a nonionic or cationic species that prevents exchange reactions from occurring. Tests have shown that over 95% removal may be achieved at pH levels as low as 5.6. At a pH of 4.3, removal rate drops to 50% and run lengths (throughput capacities) are reduced by over 90%. It has also been shown that sudden changes in influent water pH to values below 5.6 can result in dumping of previously removed uranium. For these reasons, it is important to control inlet water pH above 6 at all times. In situations where pH cannot be maintained above 5.6, other treatment methods should be considered.

Throughput Capacity. The uranium carbonate complex has a relative affinity for strongly basic anion exchange resins over 100 times greater than common ions, including divalent ions such as carbonates and sulfates. These are the only ions able to offer competition. At pH levels of 6.0 to 9.0, the carbonate ion is negligible because it exists primarily as the bicarbonate species, which is monovalent. Only the sulfate ion exists to offer competition. Throughput for uranium removal can conservatively be estimated as being 100 times the calculated volume of throughput based on resin exhaustion by the sulfate content of the inlet water. To maintain this capacity over many cycles, it is necessary to use enough salt at sufficiently high concentrations to regenerate the resin back to the chloride form.

Regeneration. To regenerate the uranyl carbonate ion from the resin, the concentration of regenerant at the resin bed should be sufficiently high to reverse or reduce its relative affinity (compared with chloride) to workable levels and to use enough regenerant and contact time. Sodium chloride is the most common regenerant. Table 12.3 shows the difference in effectiveness of a 22 lb/ft^3 (35.2 g/m^3) sodium chloride dose rate and a type 1 gel anion resin.

At concentrations above 20%, a regenerant level of 15 lb/ft^3 (16 g/m^3) is enough to ensure greater than 90% uranium removal through successive service cycles. Leakage remains low through the service cycle even without complete regeneration because of the high selectivity at the low ionic concentrations during the service cycle. Leakage is normally well below 1% for sodium chloride regeneration levels of 15 lb/ft^3 (16 g/m^3), applied at concentrations of 10% or higher.

Other Regenerants. The chloride ion at neutral pH is the most commercially effective ion for regeneration of uranium from anion resins. Neutral salts are usually preferred because of environmental concerns and materials of construction considerations. Regeneration with pure hydrochloric acid, although not recommended because of the added expense for corrosion-resistant equipment, shows an even better efficiency than sodium chloride because of its low pH. This fact is mentioned here to point out that lower pH is acceptable during regeneration.

TABLE 12.3 Difference in Uranium Removal with Different Sodium Chloride Concentrations[*]

Percent sodium chloride concentration	Percent uranium removed from the resin
4	47
5.5	54
11	75
16	86
20	91

[*]Regenerant level is approximately 22 lb/ft^3 of type 1 gel resin.

Removal of Color, TOC, and Trihalomethane Precursors

Color, TOC, and trihalomethane (THM) precursors are all caused by naturally occurring organic matter that is the result of the natural decay of vegetable matter, primarily leaves, grasses, roots, and wood. The resulting soluble or colloidal dispersions of semisoluble matter are a mixture of fulvic and humic acids, tannins, and lignins. These are primarily aromatic hydrocarbon polymers with carboxylic groups that behave as anions.

Carboxylic acids are moderately to weakly ionized. Although they are exchanged as anions, they are kinetically slower and bulkier than inorganic ions. Therefore the ion exchange processing equipment and regenerant delivery systems must be designed for lower flow rates and longer regenerant contact times than systems designed for exchanging inorganic ions.

The molecular weight distribution of these substances varies widely from about 200 to 80,000 or more, depending on the source, age, and environmental conditions. In general, over 80% of naturally occurring substances that cause TOC have molecular weights under 10,000, larger than inorganic ions but still small enough to enter the gel phase of ion exchange resins. TOC values are typically expressed as mg/L as carbon. This can be converted to approximate ion exchange concentrations as calcium carbonate equivalents by multiplying the TOC values, mg/L as C, by 1.5 to get approximate mg/L as calcium carbonate.

The lower cross-linked, more highly porous "standard" gel resins designed for inorganic exchange are somewhat better then the standard 8% divinylbenzene (DVB) type, but still lack enough gel-phase porosity to be used strictly for exchanging these substances. Some relatively new special grades of resins with high gel-phase porosity are far more successful in this kind of service.

Macroporous resins were the first types widely used for organics removal because it was initially believed that surface adsorption played a significant role in the removal process. It has since been shown that the best performing macroporous resins are those that have the highest level of gel-phase porosity. It has also been shown that gel-phase porosity is the most significant predictor of performance. Because over 95% of naturally occurring TOC matter is removed only by ion exchange and not adsorption, macroporous structures cannot offer advantages unless combined with proper gel-phase properties.

The loading capacity and degree of breakthrough are affected by competing ion ratios, especially sulfates to TOC. Once sulfates begin to leak, about 50% of the influent TOC appears in the effluent. If complete TOC removal is required, the service cycle should be terminated when sulfates leakage increases. If the service cycle is run past the sulfate break, organics with the highest affinities will continue to load by displacing those with lower affinities and also sulfates. They will accumulate on the resin, making it more difficult to regenerate the resin, which could lead to premature fouling.

Naturally occurring TOC substances have affinities for resins that are similar to sulfates. Theoretically, they can be regenerated from a strong-base resin in the same type of process as used for sulfates, but the regeneration process is limited by kinetics. Salt regeneration can be made more effective by using 10% warm brine and adding 1% to 2% sodium hydroxide to it. A dose level of 10 lb/ft^3 (16 g/m^3) applied over a 2 h contact time at 70° F (21° C) is the minimum recommended. Shorter contact times produce proportionately lower organic removal, and longer contact times add only about 3% per hour to the 2 h organic removal amount. TOC operating capacity is calculated based on sulfates and TOC as the exchanging ions.

DEMINERALIZATION

Demineralization, also called deionization, is the result of exchanging all positively charged ions for the hydrogen ion and all negatively charged ions for hydroxide ions. Hydrogen and hydroxides react to form water molecules.

In general, demineralization requires cation and anion exchange. Several varieties of each kind can be selected depending on the specific water analysis, quality requirements, and operating conditions. Cation and anion resins can be in separate vessels (separate beds or two beds) or in a single vessel (mixed beds). In mixed beds, reaction products disappear by forming water molecules that drive the exchange reactions and produce a greater level of purity.

Mixed beds of cation and anion resins give the effect of multiple stages of two-bed systems. Most mixed-resin systems consist of strongly acidic and strongly basic cation and anion resins. Mixed beds are most widely used for polishing water previously deionized by a membrane process or by a two-bed demineralizer. The main advantage of mixed-bed demineralization is that of improved water quality.

ION EXCHANGE EQUIPMENT DESIGN

This section covers the basics of how ion exchange systems are sized, requirements for ancillary systems, and construction materials selection. These factors vary depending on the type of equipment and equipment suppliers' preferences.

Flow Rates

Most ion exchange systems are limited by flow rate for physical reasons, such as pressure loss through the resin bed, rather than by any particular chemical requirements. In general, the flow rate through an ion exchange unit should not exceed about 20 gpm/ft^2 (49 m/h). At higher rates the pressure exerted on the resin bed, together with thermal and osmotic stresses, is sufficiently large to begin breaking some of the resin beads. In general, the maximum pressure loss that should be allowed across anion and cation beds is 20 psi (138 kPa). It is common practice to allow for a 5 to 10 psi (34 to 69 kPa) increase in pressure drop across the resin bed over time. To keep below these pressure losses, the design value for the flow rate across a resin bed should be about 15 gpm/ft^2 (37 m/h).

Minimum flow rate is that which the liquid distributors can accommodate. The maximum practical turndown is about 5:1. A liquid distributor designed for a flow rate of 5 gpm (0.32 L/s) will not give good distribution at flow rates below 1 gpm (0.06 L/s). Some distributors can do somewhat better than this, and other distributors, such as those used in some of the less expensive commercial tanks, do not do this well. The other potential problem at very low flow rates is that of relatively high levels of leachables (from the resins and tank lining) entering the product water, especially when new resins are placed in service.

Regeneration Frequency and Media Depth

Resin volume requirements are generally selected based on the user preference for cycle times between regenerations, within limits defined by flow rate and bed depth requirements. Regeneration frequencies are usually kept at less than two or three times per 24 hours. The service cycle times can normally vary from a minimum of approximately eight hours to a maximum of several days, depending on process, volume of resin used, and mix of ions in the feed water. Systems that cycle more frequently are generally less reliable.

On the other hand, systems that regenerate too infrequently can encounter difficulties due to ions migrating deeply into the resin beads, foulants that harden into the resin over time, and bacterial growth that occurs in stagnant resin beds. For the purpose of initial design of a system, a reasonable starting point is to use a flow rate of 10 gpm/ft^2 (24 m/h) and a resin bed depth of 5 ft (1.5 m).

Resin bed depths of less than 24 in. (60 cm) are generally not recommended, primarily to avoid inefficient operation. Even though the theoretical height of an ion exchange zone is usually less than 6 in. (15 cm), the "average" height of an exchange zone is distorted by imperfections in the distribution and collection systems within a tank, and a considerable fraction of the total resin bed may be lost to these imperfections. Bed depths greater than 6 ft (1.8 m) are generally avoided because of concerns of exceeding the pressure drop limitations of the resin.

There are several different methods for regenerating resins. The method chosen largely determines how efficiently the ion exchange resins will operate and how complex the regeneration will be.

Degasification

One factor in selecting a demineralizer design is whether to use a degasifier. Carbon dioxide can be removed by a forced draft degasifier or by a vacuum deaerator between the cation and anion exchangers. This removal reduces the ionic load on the anion vessel. Degasifiers complicate the design for several reasons:

- They may make the anion exchanger much smaller than the cation exchanger. This creates potential problems in flow rates and regeneration times.
- In most cases, two or more trains of exchangers share a common degasifier, but the simultaneous regeneration becomes more complicated when the degasifier has to stay in service while a particular train is regenerating.
- Rinse recycle becomes far more complicated in cases with a degasifier. In fact, it is generally not practical to rinse recycle the cation exchanger, although in many cases, even with a degasifier the anion regeneration can finish with rinse recycle.
- Although the reason for using a degasifier is to reduce operating costs, the cost analysis should consider that a certain amount of caustic may be required to neutralize waste acid from the cation regeneration.

Co-Flow Exchangers

Co-flow ion exchange is the oldest of all designs, the best known, and the simplest. Although co-flow designs are inherently less efficient and do not provide water quality that is as good as other designs, they are a forgiving design that can be used in dirty water applications with high turbidity.

Because the regenerant flow is in the same direction as the service flow, ions at the top of the bed have to be pushed downward all the way through the resin bed before they can be purged. This makes the co-flow method somewhat inefficient. It also leaves a small portion of exchanged ions remaining in the resin at the bottom of the bed where they can cause leakage in subsequent service cycles. This phenomenon is slight in softeners, but it is particularly noticeable in nitrate removal salt-exchange units where the nitrate leakage at the beginning of the cycle is significantly higher than the leakage in midcycle.

The exchanger consists of a tank that contains a bed of ion exchange resin. At the top is an upper distributor that applies water over the surface of the resin bed, and at the bottom is an underdrain collector. Source water enters at the upper distribution and flows down through the resin bed (Figure 12.3).

During regeneration the regenerant chemical also flows downward through the resin bed. In smaller units the regenerant chemical is introduced through the same upper distributor. In larger units there is generally a separate regenerant distributor located just above the resin bed. The advantage of a regenerant distributor is that it saves water during regeneration.

Co-flow exchangers have freeboard or empty space between the upper distributor and the resin bed that allows for resin bed expansion during the backwash portion of the regeneration cycle. During backwash, the inlet distributor becomes the outlet collector.

Regenerating a co-flow ion exchanger consists of four steps:

1. *Backwash.* During backwash a flow of water is introduced through the underdrain and flows up through the resin bed at a rate sufficient to expand the resin bed about 50%. The purpose is to relieve hydraulic compaction, move the finer resin material such as resin fragments to the top of the bed, and remove any suspended solids from the bed that have accumulated during the service cycle.

2. *Chemical injection.* A dilute solution of regenerant chemical flows down through the resin bed, stripping the ions that were collected during the service cycle off the resin and restoring the resin into what is called the regenerated form.

3. *Displacement rinse or slow rinse.* The displacement rinse is conducted at or close to the same flow rate as the injection step. Its purpose is twofold. First, it slowly pushes chemical through the resin bed, allowing necessary contact time to complete the regeneration process. Second, it is an efficient way of removing the bulk of the regenerant solution from the exchange tank.

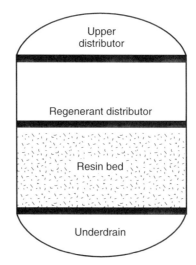

FIGURE 12.3 Co-flow exchanger.

ION EXCHANGE PROCESSES   **321**

4. *Fast rinse.* This is the final rinse step, usually conducted at or close to the service flow rate. Its purpose is to flush the last traces of regenerant chemical out of the exchanger and prepare the exchanger for return to service.

Split-Flow Exchangers

Split-flow (sometimes called co-counter) exchangers look similar to co-flow exchangers except that they have a regenerant collector, usually located about one-third of the way down in the resin bed (Figure 12.4). The regenerant collector is always screened and must be braced similar to a mixed-bed interface collector because of the hydraulic forces on the collector during service and regeneration cycles. Most of the regenerant comes in through the bottom and flows upward through the bottom portion of resin, but some of the regenerant is introduced through the top and flows downward through the top portion. Both portions of regenerant chemical exit through the regenerant collector.

The regeneration of a split-flow exchanger generally consists of the following steps:

1. *Subsurface backwash.* During this step, water is brought in through the regenerant collector and flows upward through the top portion of the resin bed to flush most of the dirt that might be trapped in the top portion of the resin bed. Subsurface backwash also relieves compaction of the upper resin bed and reduces stresses against the regenerant collector. Backwash time is generally limited to just a few minutes because the main purpose is fluffing the bed and relieving hydraulic compaction.
2. *Regenerant injection.* In this step the diluted regenerant is split into two portions. One is fed into the top of the bed and flows downward, and the other is fed into the underdrain and flows upward. The upward flow should not exceed 1 gpm/ft^2 (2.4 m/h) because any flow in excess of this may tend to fluidize the bottom portion of the resin bed.
3. *Displacement rinse or slow rinse.* This step is usually carried out by adding dilution water following chemical injection, but it is sometimes performed at a higher rate to

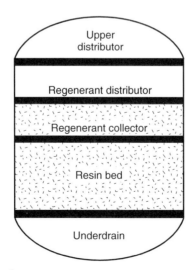

FIGURE 12.4 Split-flow exchanger.

minimize regeneration time. The volume of displacement water depends on whether or not there is a recycle rinse. If the bed is rinsed to waste, only about two bed volumes of displacement water are employed. If the final rinse is performed in a recycle manner, it is prudent to plan to use at least three bed volumes and possibly as many as six bed volumes of displacement water.

4. *Final rinse.* Final rinse can be conducted in a straight-through-to-waste manner or by recycling through the accompanying exchanger (cation to anion or anion to cation). Rinse recycle is a water-saving feature commonly employed in countercurrent regenerated two-bed demineralizers. The volume of rinse required depends on the water quality required at the beginning of the service cycle. Cation resins require less water than anion resins. Cation rinse volumes should be about 6 to 10 bed volumes, and anion rinse volumes about 6 to 15 bed volumes. The rinse water flow rate most commonly used is the same as the service cycle flow rate. However, it should not be less than about 2 gpm/ft^2 (4.9 m/h) and not higher than the service flow rate.

Countercurrent Regeneration Exchangers

Various methods of countercurrent regeneration have been devised to overcome the limitations of co-flow regeneration (Figure 12.5). Countercurrent regeneration is primarily used in demineralization, and, except for occasional use for nitrate removal, it is almost never used in salt cycle exchange. In all countercurrent regeneration methods, the dilute regenerant solution is introduced in a flow direction opposite the service flow, so that the least used portion of the resin (polishing zone) bed contacts the freshest regenerant first. This results in low leakages even at the lowest regenerant dose levels, with much higher regeneration efficiency. Countercurrent systems save chemical operating costs and reduce waste discharge. These advantages are offset by higher capital costs and a more complex operating system. Maintenance costs are also usually higher with countercurrent regeneration.

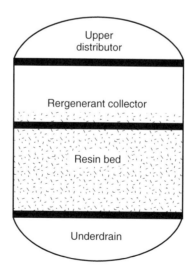

FIGURE 12.5 Countercurrent regeneration exchanger.

The most common form of countercurrent regenerated ion exchange system uses downflow service and up-flow regeneration. Water block countercurrent regeneration is similar to split-flow regeneration, except that instead of splitting regenerant flow between the top and bottom, the regenerant all goes through the bottom.

Because the up-flow regenerant tends to expand and fluidize the resin bed, there must be some means of keeping the bed packed during regeneration. The simplest method is to bring water in from the top of the bed. Air is used instead of water as the blocking flow in some systems.

Water-block and air-block exchangers both require a regenerant collector located just below the surface of the resin bed. It is necessary to bury the regenerant collector in the resin bed to a depth of approximately 6 in. (15 cm) to prevent fluidization of the resin during the upward regeneration. This layer of resin is considered inactive and is not included in the ion exchange capacity or throughput calculations.

The required flow rate of the blocking water must be sufficient to prevent excessive movement of the resin and must be compatible with the chemical solution employed during regeneration. This means that blocking water used for anion regeneration must be free of hardness.

Another method of preventing bed fluidization is called the air-block system. Water above the resin bed is drained to the regenerant collector, and pressurized air is brought in through the top of the tank. An air dome prevents water buildup above the regenerant collector and prevents the resin bed from expanding and fluidizing. The air-block method of countercurrent regeneration is more complicated than the water-block method. The regeneration procedure is complicated by the necessity to drain down and then refill the bed before and after chemical injection and displacement.

Another complication of air-block units is that a portion of the regenerant tends to flow up above the regenerant collector, and some fluidization of the resin inevitably occurs. The air dome cannot keep the water level or chemical level from rising above the bed. To the extent that the regenerant solution rises above the collector, the regenerant does not displace out of the bed. In systems where high-quality water is required, it is necessary to rinse to waste before the rinse recycle. The regeneration of a countercurrent regeneration exchanger consists of the following steps:

1. *Subsurface backwash.* Subsurface backwash is performed at the normal backwash rate for a reduced period of time by bringing water in through the regenerant collector and out of the top of the tank.
2. *Drain to bed level.* Air is usually brought in through the vent and is used to pressure drain the water to the level of the regenerant collector.
3. *Chemical injection.* The regeneration chemical is added at the same flow rate, concentration, and time as in other designs. The limitation of upward regenerant flow rate is about the same as for other upwardly regenerated units.
4. *Chemical displacement.* Displacement is normally employed at the dilution water rate, although it is sometimes advantageous to displace at a higher rate. The requirement is at least three bed volumes.
5. *Refill.* The vessel is generally refilled from the top, through the rinse water valve.
6. *Optional rinse-to-waste.* Use of the optional rinse-to-waste step depends on water quality requirements. If used, it should amount to approximately three to five bed volumes.
7. *Rinse recycle.* Rinse recycle is 6 to 10 bed volumes for the cation and 10 to 15 bed volumes for the anion, normally performed at the service rate. The volume depends on the end point desired.

Packed-Bed Design

A packed-bed demineralizer consists of a tank containing an upper distributor and an underdrain. The resin bed fills the entire tank except for the amount of freeboard required to allow for the resin swelling and contracting between the regenerated and exhausted form (Figure 12.6). There is no separate regenerant collector. Although the direction of the service flow rate may be upward or downward, most of the packed-bed designs built to date have up-flow service and down-flow regeneration.

An important factor in the choice of demineralizer is the size required. The packed-bed design offers the smallest possible size. The co-flow design offers the widest potential variety of size and does not require that the size be fitted exactly to the resin volume. One possible problem with all countercurrently regenerated units is that the resin volume is fixed by the vessel size and must be close to an exact volume. Resins expand and contract during service and exhaustion and are shipped by weight rather than volume. It is common practice to label and purchase resins by volume, so it is necessary to fit resin volume to the tank. If an unknown amount of resin is blindly put into a tank that is going to employ countercurrent regeneration, the chance of failure is high.

Regeneration

When the capacity of a resin is exhausted and the leakage of undesirable ions rises to an unacceptable point, the resin must be regenerated. The regeneration process reverses the exchange reaction under controlled conditions, leaving the resin in the desired ionic form.

Salt Regeneration. Sudden changes in ionic concentration cause ion exchange resin to rapidly change size, shocking the resin and perhaps breaking beads. Yet higher ionic concentrations are more effective regenerants and are more efficient. So it is necessary to control the rate of change and total concentrations during regeneration. For salt-regenerated exchangers, the best range of salt concentration is between 8% and 14%, with 10% being

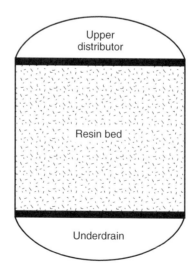

FIGURE 12.6 Packed-bed exchanger.

the most common. See Table 12.3, page 316, for the uranium removal differences with various concentrations of sodium regeneration.

Acid and Caustic Cycles. For acid-base exchanges, the concentrations used are generally somewhat lower than for salt regenerations. The typical concentration for strong acids is in the range of 2% to 10%, but concentrations of 0.5% to 1.0% are used with sulfuric acid regeneration of weak acid resins in high-calcium waters. For caustics, the typical concentration range is 4% to 6%.

Where calcium hardness is present, sulfuric acid must be used at controlled low concentrations to prevent calcium sulfate precipitation. Multiple regeneration steps with varying concentrations are sometimes used with sulfuric acid to reduce waste volumes and maintain regenerant efficiency. Although contact times as short as 10 to 15 min have been employed successfully, in most cases a contact time of 30 to 40 min is used for cation resins and weakly basic anion resins. Strongly basic resins may require 60 min injection time and heated regenerant where silica removal is critical.

In many units, after the introduction of regenerant chemical, a slow rinse or displacement rinse pushes the regenerant chemical through the resin and out at approximately the same flow rate used during chemical introduction. This slow rinse helps ensure adequate chemical contact time and helps decrease the rinse requirement. The final rinse step purges the last traces of chemical from the ion exchanger and prepares the system for the next service cycle.

Rinse volume requirements vary considerably depending on the type of resin used, adequacy of the internal design, age of the resin, and presence of foulants. Co-flow systems generally require from 7 to 10 bed volumes of final or fast rinse. Rinse requirements can vary from a minimum of about 2 bed volumes to over 20 bed volumes, depending on the type of system and end point requirements.

Rinse recycle is sometimes used to save wastewater volume in demineralizers. When effluent purity reaches about the same level as the source water, the rinse is recycled to the front end of the system rather than discharged to waste. Because effluent quality is changing rapidly, there is very little load on the resins. Rinse recycle can generally be used when the regenerant concentration has dropped to less than about 0.1%.

Rinse recycle cannot be used with salt-regenerated units because there is no mechanism to remove salt left in the rinse water. Although final rinse waters can be saved for backwash or regenerant dilution, this is usually not practical except on large salt-regenerated systems.

External Regeneration. External regeneration is performed in a vessel other than the service vessel. This type of regeneration is widely practiced by the service exchange business and also in condensate polishing demineralizers in the electric power industry. Because resin becomes mixed as it is transferred from the service vessel, externally regenerated units are always regenerated in a down-flow fashion. Resins used in mixed beds must be separated before regeneration so that the acid regenerant can contact the cation portion of the resin and the sodium hydroxide regenerant can contact the anion portion of the resin. The wrong chemical contacting the wrong resin can cause fractured resin beads from osmotic shock and precipitation of various salts.

Vessel Design

The ion exchange vessel must contain ion exchange resin in a way that allows the liquid to flow through while keeping the media bed packed. The design should allow the various chemical solutions used during regeneration to flow through the media properly. Most ion exchange systems use pressurized tanks, although a few gravity-flow ion exchange systems are in use. Gravity-flow systems are generally applicable only to very large units and systems that have sufficient surface area to allow necessary flow rates without large pressure drops.

Tank Design. There are two basic choices of tank materials for pressure vessels: fiberglass and steel. Fiberglass tanks are generally less expensive, but because of the way the fiberglass tank is made, there are fewer options for accessories and connections in the system design. Fiberglass tanks are lightweight and have an advantage over steel tanks in that they are inherently noncorrosive. However, they are not available in large sizes and are generally limited to tank diameters of 8 ft (2.4 m) and less. They are not generally used for pressures above 150 psi (1,034 kPa) or temperatures above 100° F (38° C).

Steel tanks have far more design options, and they can be made in almost any configuration. However, steel tanks are heavier than fiberglass tanks, and, because of concerns about corrosion, they are almost always lined and therefore more expensive. For high-temperature applications, stainless steel is sometimes preferred over carbon steel. The high cost of building a stainless steel vessel, though, makes it prohibitively expensive for all but a few specialty applications.

To ensure proper flow of liquid through the resin bed, water must enter the tank in such a fashion as to provide minimum turbulence and laminar flow into the resin bed. Any protuberance creates a disturbance and an area of maldistribution. The resin bed should present a flat surface so the water can flow straight down through it and into the underdrain collector. The path length should be relatively equal for all portions of the bed.

Because most pressure vessels have dished heads in the larger tanks, the dish creates a disturbance that distorts flow and reduces resin efficiency. Large tanks generally benefit from the addition of either flat, false bottoms, or subfills such as concrete or sometimes gravel that fill the bottom head and provide a flat surface for the resin bed.

Tank Equipment. At a minimum, the tank must be equipped with an inlet distributor for the water to flow into the tank and an underdrain collector to allow the water an exit. It is also generally advisable to have a separate connection to add and remove ion exchange resin.

For tanks larger than about 3 ft (0.9 m) diameter, it is necessary to have an access way, so that a worker can climb inside the tank to install or repair internal distributors and collectors. Larger openings are preferable to small openings. Access located on the side of the tank can create a distribution problem by the distortion of the tank.

View ports are a feature used in more sophisticated ion exchange systems. They allow the operator to visually inspect resin to make sure resin has not been lost from excessive backwash or problems with the underdrain collector. View ports also provide the opportunity to monitor resin during regeneration and backwash cycles to ensure that the resin bed remains packed and does not fluidize when it is not supposed to.

Underdrain. If the underdrain is not located close to the bottom of the tank, rinse requirements increase significantly because of the difficulty of removing regenerant chemical that settles below the collector. The underdrain must also have sufficient open area to collect water without resin fines clogging openings. For example, slotted pipe–type collectors sometimes used in less expensive, smaller systems have a small amount of open area highly prone to plugging.

One common system consists of a lateral pipe with drilled orifices pointed toward the bottom of the vessel. A layer of coarse plastic mesh protector wraps over the lateral, and a layer of plastic or stainless screen covers the mesh protector. Another typical system is similar to the first except that drilled orifices are covered with a continuous wedge wire sleeve. These two lateral systems are among the most widely used and are considered by some to be superior to strainer designs.

Strainers are reasonably good collectors, but they have too much open area to be inherently good distributors. The lack of pressure drop resulting from the open area inhibits equal distribution at the low flows used during regeneration. A second problem is that, in order to obtain reasonable open area in the strainer, it is generally over 2 in. (5 cm) high. The extra height creates an area of poor distribution toward the bottom of the strainer. Various manufacturers have tried to overcome this problem, with some success, by adding check valves within the strainers, putting shrouds over the strainers, or using a large number of small strainers.

Distributors. By far the most important factor in the efficiency and quality produced by an ion exchange system is the design of the internal distributors and collectors. The inlet distribution system must allow the water and chemical to enter the vessel and spread out equally throughout the entire tank diameter with equal velocity. Velocity must be sufficiently low to prevent disturbing the resin bed below. Distributor design is much more crucial in tanks with limited freeboard. Underdrain collectors must be located as close to the bottom of the vessel as possible, with the least amount of resin underneath the collector, and should have point collection spacing of 6 to 12 in. (15 to 30 cm) maximum.

Large vessels are usually furnished with a separate chemical distributor, which has two advantages. The chemical distributor can be designed for good distribution at the specific chemical flow rate desired. The second advantage is that, by locating the chemical distributor close to the resin bed, the volume of liquid in the freeboard area does not have to be displaced during regeneration and rinse steps. This reduces the rinse requirement. Some systems, such as mixed beds and certain types of countercurrently regenerated exchangers, have liquid collector distributors buried within the resin bed. These distributor collectors generally are designed in a similar fashion to the underdrain. There is significant hydraulic force against these distributors during the service cycle and potentially during the regeneration cycle. If the distributors are not firmly braced, they are prone to breakage.

Piping Design

Smaller systems, and in particular residential in-home softeners, generally have multiport valving systems. Multiport valves are inexpensive and fairly reliable up to about 2 in. (5 cm). Their main disadvantage is that they have a fixed cycle. Although several options are available, they are generally considered not suitable for most countercurrently regenerated designs. Larger systems usually employ a valve nest and a piping tree to deliver and collect the various flows.

Many piping systems materials are available; PVC is most commonly used for smaller systems and either plastic-lined steel or stainless steel for larger systems. The most important feature of the valves used is that they must be capable of tight shutoff over many cycles. The diaphragm-type valve has proven to be a good choice for this purpose, although it has a higher pressure drop and costs more than other types of valves. Two other commonly used valve types are ball valves and, in larger systems, butterfly valves.

Chemical Storage and Dilution

Some small systems simply use open-top tanks to mix and store the chemical solution at the required concentration before regeneration. Larger systems generally have some type of bulk chemical storage and in-line dilution stations. In these systems, concentrated chemical is pumped to a mixing tee where it is blended with dilution water before entering the exchange tank.

Smaller systems with dilution tanks generally require an operator to manually fill the tank, potentially exposing the operator to dangerous chemicals. Salt exchanges are not considered troublesome, but acid and base exchanges are considered hazardous because of the potential danger from acids and caustics.

Bulk chemical storage generally requires a containment structure in the area surrounding the dilution station where pressurized chemicals are used and some type of enclosure to prevent a worker from being accidentally sprayed should a rupture in the pipeline occur. For small systems, the eductor type of pump is ideal and can be used both for pumping and diluting the regenerant in one step. Larger systems generally use positive displacement pumps such as diaphragm, gear, and sometimes centrifugal pumps.

Metering pumps present problems from pulsation and are not particularly reliable. Gear pumps are far better in this respect, but are not free of maintenance problems. During regeneration, the resin bed changes size, and the pressure drop across that resin bed varies. It is harder to keep a constant flow rate with a centrifugal pump than with the other types of pumps.

Waste Collection and Disposal

In the past, wastewater from an ion exchange system was typically discharged to a sewer with no treatment or monitoring. Wastewater from ion exchange systems contains either concentrated salt solution or the concentrated acid and base solutions from the regenerant chemicals used. Overall ionic strength of the wastewater varies, but it is usually in excess of 10,000 mg/L and can be as high as 60,000 mg/L.

Most ion exchange systems today use a wastewater tank into which regeneration wastes are placed before treatment and disposal. The wastewater is then neutralized and monitored to ensure that it is neutral before being discharged. In some areas of the country, and for some types of wastewater, it is necessary to impound the spent regenerant portion of the waste and to haul it off-site for further treatment. Depending on the nature of the undesirable constituent and the location within the country, this cost of waste collection and disposal can actually exceed the operating cost of the ion exchange equipment itself. Disposal regulations are increasingly important concerns in the design of any ion exchange system.

Federal, state, and local regulations should be examined to determine what limitations are placed on the discharging of regenerant wastes and for specific objectionable substances such as barium, radium, and suspended solids.

The expected level of a specific contaminant in the waste can easily be estimated at the design stage. It takes about 80 gal/ft^3 (1,071 L/m^3) to regenerate an ion exchange resin, including the backwash, regeneration, and rinse cycles. The concentration ratio is found by dividing the throughput capacity in gallons per cubic foot of resin by the regeneration volume. Multiply the concentration ratio times the concentration of the contaminant in the source water level to determine the maximum level to be discharged.

For example, each cubic foot of softener resin requires about 80 gal for regeneration. Assume that for a specific installation the throughput is 2,400 gal for each cubic foot of resin during the service cycle. The concentration factor is 2,400 divided by 80 equals 30. The means that the hardness level in the regenerant waste will be 30 times as high as in the influent water.

Monitoring and Validating System Performance

The ion exchange process is predictable provided feed water conditions do not change and the condition of the resin bed remains reasonably constant, free of fouling and free of resin degradation. For this reason many ion exchange systems are designed with only throughput monitoring, and regeneration initiation is based on the throughput. In the case of home softeners, regeneration is usually based on time demand for an average expected usage.

However, where a potentially harmful contaminant is being removed by the process, more elaborate monitoring is required. Centralized treatment systems should have periodic monitoring of the objectionable constituent to verify effluent water quality from the ion exchange system remains within the desired limits. In some cases the possibility of an objectionable contaminant being dumped by the resin exists if the system is overrun past exhaustion. Treated water reservoirs can store enough water to significantly reduce overrunning for a limited time.

Certain ion exchange processes require the objectionable substance to be in a certain valence state, molecular configuration or physical form before it can be successfully removed by ion exchange, such as barium, silicate, arsenate, or selanate. In this case it is necessary to not only validate effluent quality but also to periodically monitor feed water quality to be certain the ions are in the state necessary for removal. If pretreatment equipment is provided to alter the ion into a form that can be removed, pretreatment equipment must also be monitored and validated.

Beyond these requirements, other parameters should be monitored depending on the size of the system and the intent of the user to operate the system in the most economical possible fashion. For instance, chemical concentration and dosage during regeneration is necessary to determine the cost of operation. This information is of limited value for any other purpose except for troubleshooting the system should a malperformance occur.

Resin Replacement

The usual rate of resin deterioration in clean water systems is slow and somewhat predictable. Nevertheless, the condition of the resin bed should be periodically monitored to verify resin condition. This monitoring helps determine when resin replacement is necessary and when maintenance and cleaning procedures should be performed before the bed deteriorates to the point where poor performance occurs.

It is also beneficial to periodically perform a complete water analysis of source and treated waters to verify that the overall water quality remains within desired limits. Because ion exchange resin beds are in some situations fertile breeding ground for bacteria, it is prudent to include occasional bacterial analysis as part of the validation of system performance.

Automation

Many ion exchange systems are best operated manually. However, there are advantages to having automatic regeneration in large treatment facilities where multiple tanks are used. Automation has an advantage over manual operation in that regenerations are performed in a predictable fashion time after time in the exact same way at the exact same time, reducing the possibility of human error. However, automatic regeneration increases the possibility of machine error.

In large, sophisticated automated systems with programmable logic control and computer data logging, the cost of automating a system can exceed the equipment cost itself. This is an important point to consider in planning an ion exchange system and in budgeting for the capital purchase expense. The more complicated control systems also require highly skilled technicians to properly maintain the automatic control hardware and software. Because the state of the art in computer automation is changing so rapidly, it is likely that any automation system purchased today will be obsolete within a few years. This can lead to significant future capital expenditures to upgrade obsolete hardware and software.

BOUNDARIES BETWEEN SUPPLIERS, CONSULTANTS, AND USERS

The process of creating an ion exchange system usually involves a resin supplier, an equipment supplier, a consultant or design engineering firm, mechanical and electrical contractors, and an end user or operator of the finished plant. Except for relatively small systems, it is unlikely that the role of these various suppliers and users is handled by a single company. These roles usually are handled by various companies who must cooperate to achieve a successful ion exchange plant.

The end user, usually the municipality or private owner, must begin the process of building the plant by collecting information about water quality, both feed water quality over time and the requirements for treated water. The end user must determine that the plant will be built, must economically justify it, and must secure capital for its construction. The end user then usually contracts with a design engineering consultant to create a specification for the finished ion exchange plant. The consultant then works with equipment suppliers and resin suppliers to determine the size and shape of the plant necessary to meet a certain objective. The finished design is generally put out to bid, and the consultant and end user select an equipment supplier.

The equipment supplier builds the equipment; the installation contractor installs the equipment; the end user and equipment supplier commission the plant; and the end user takes over the operation of the finished exchange plant. Traditional roles played by the various companies involved can lead to certain types of difficulties if the various parties do not communicate well. The bidding process, particularly if the selection is made solely on the basis of the lowest price, may lead to shoddy workmanship and poor-quality components.

Various types of partnering arrangements have been proven to provide a better finished product at an overall price that is little or no more expensive than the total cost provided by the traditional bid method of supply. Regardless of how the various roles are played, it is extremely important that the end user take an active interest in the design and construction of the ion exchange plant. The user who believes that by hiring a consultant his or her work is finished may be unpleasantly surprised when the finished product does not fit unexpressed needs. AWWA's manual *Construction Contract Administration, M47*, discusses common contracting issues, tradition and nontraditional approaches, and general issues such as partnering, communication, contract disputes, and payment procedures.

FUTURE TRENDS IN ION EXCHANGE TREATMENT

It has become common to see ion exchange used in conjunction with membrane technology, especially in high-salinity applications. In the future the role of ion exchange will likely shift from bulk removal of ions toward selective ion removal and polishing. Cation resins' use in water softening will continue as the largest single role of resins, including pretreatment for membrane processes. In some applications, ion exchange for bulk ion removal will diminish as membrane processes become more efficient. At the same time, ion exchange polishing of membrane-treated waters will continue to grow. Ion exchange will become more widely used for purification. Currently the areas of rapid growth in ion exchange are selective removal of contaminants such as nitrates, heavy metals, radioactive ions, and organic acids from drinking water supplies and for specific metal removal from wastewater streams. A recent trend in large-scale ion exchange systems is for service companies to build, own, and operate them.

There will be increasing reliance on the companies that provide exchange tank services. These services are ideally suited for point-of-use applications because they allow the regeneration of the ion exchange material in bulk at centralized facilities, including treating regenerant wastewater. Another recent trend is for owners to contract out operations and maintenance of their plants, eliminating manpower, training, and management costs. In many cases the outside service can perform equipment operations and maintenance at lower costs to the owners.

Exchange tank services are available for all flows, from point-of-use systems up to larger centralized treatment facilities. Advantages include the capital cost saved by not installing the treatment system and not having to regenerate and dispose of the wastewater produced by the regeneration process. Because of their economy of scale and experience in performing regenerations, the companies that offer these services are often able to charge a lower per-unit water cost than can be provided by a municipality or a homeowner. These services are more readily available in urban communities than in rural communities because they are transportation (location) sensitive. Some companies also perform monitoring and validation requirements of the ion exchange system on behalf of the end user.

BIBLIOGRAPHY

Arsenic Removal

Arsenic, Arsenite ($AsO2^-1$), Arsenate ($AsO2^-3$), Organic Arsenic Complexes. Recognized Treatment Techniques for Meeting the Primary Drinking Water Regulations for the Reduction of Arsenic ($AsO2^-1$, $AsO4^-3$, and Organic Arsenic Complexes) Using Point-of-Entry POU/POE Devices and Systems. Technical Application Bulletin. Lisle, Ill.: Water Quality Association.

Frank, Phyllis, and Dennis Clifford. *Arsenic (III) Oxidation and Removal from Drinking Water. Project Summary.* Cincinnati, Ohio: U.S. Environmental Protection Agency, February 1992.

Lead, Fluoride and Arsenic Selective Media. ResinTech Product Bulletin SIR-900. Cherry Hill, N.J.: Resin Tech, 1996.

Rosenblum, Eric, and Dennis Clifford. *The Equilibrium Arsenic Capacity of Activated Alumina. Project Summary.* Cincinnati, Ohio: U.S. Environmental Protection Agency, February 1984.

Rubel, F., and Williams. *Pilot Study of Fluoride and Arsenic Removal From Potable Water.* Cincinnati, Ohio: U.S. Environmental Protection Agency 600.2-80-100, August 1980.

Russo, R. V., and Bruce Saaski. "Arsenic: New EPA Regs To Trigger Extensive Remediation Efforts." *Water Conditioning and Purification.* 35(12):64, 1994.

Barium and Radium Removal

Clifford, Dennis. "Ion Exchange and Inorganic Adsorption." In *Water Quality and Treatment.* Edited by F. Pontius. New York: McGraw-Hill, 1990, p. 561.

Cole, Lucius, and Jack Cirrincione. *Radium Removal from Groundwater by Ion Exchange Resin.* Lisle, Ill.: Water Quality Association, 1987.

Determination of Radium Removal Efficiencies in Iowa Water Supply Treatment Processes. The U.S. Environmental Protection Agency Office of Radiation Programs, Technical Note, ORP/TAD-76-1, June 1976.

Myers, A. G., V. L. Snoeyink, and D. W. Snyder. "Removing Barium and Radium through Calcium Cation Exchange." *Journal AWWA,* 77(5): 69, 1985.

Snoeyink, V. L., C. Cairns-Chambers, and J. L. Pfeffer. "Strong-Acid Ion Exchange for Removing Barium, Radium, and Hardness." *Journal AWWA,* 79(8): 66, 1987.

Snoeyink, V. L., C. C. Chambers, C. K. Schmide, et al. *Removal of Barium and Radium from Groundwater.* EPA/600/M-86/021. Cincinnati, Ohio: Environmental Research Brief, February 1987.

Snyder, D. W., V. L. Snoeyink, and J. L. Pfeffer. "Weak Acid Ion Exchange for Removal of Barium, Radium and Hardness." *Journal AWWA* 78(9):98, 1986.

Subramonian, Suresh, D. Clifford, and W. Vijjeswarapu. "Evaluating Ion Exchange for Removing Radium from Groundwater." *Journal AWWA* 82(5), 1990.

Fluoride Removal

Bellen, Gordon, M. Anderson, and R. Gottler. *Defluoridation of Drinking Water in Small Communities.* Final Report. Ann Arbor, Mich.: National Sanitation Foundation Assessment Services, 1985.

Cole, Lucius, and J. Cirrincione. *Fluoride Radium Removal from Groundwater by Ion Exchange Resin.* Water Quality Association, 1987.

Johnson, W. J., and D. R. Taves. "Exposure to Excessive Fluoride during Hemodialysis." *Kidney International.* (5):451, 1974.

Rubel, Jr., Frederick. *Design Manual Removal of Fluoride from Drinking Water Supplies by Activated Alumina.* U.S. Environmental Protection Agency 600/2-84-134, August 1984.

Nitrate Removal

Clifford, Dennis, C. C. Lin, L. L. Horng, and J. V. Boegel. *Nitrate Removal from Drinking Water in Glendale, Arizona.* PB 87-129 284/AS. Springfield, Va: NTIS, 1987.

Subramonian, D. Clifford, and W. Vijjeswarapu. "Removal of Nitrate From Contaminated Water Supplies for Public Use." *Journal AWWA* (81):61, 1990.

Organics (Naturally Occurring) Removal

Black, B. D., G. W. Harrington, and P. C. Singer. "Reducing Cancer Risks by Improving Organic Carbon Removal." *Journal AWWA* 88(6):40, 1996.

Fu, P. L. K., and J. M. Symons. "Removal of Aquatic Organics by Anion Exchange Resins." *Proc. AWWA 1988 Annual Conference.* Denver, Colo.: American Water Works Association, 1988.

Gottlieb, Michael. "The Reversible Removal of Naturally Occurring Organics Using Ion Exchange Resins. Part 1." *Industrial Water Treatment* 27(3):41–52, 1995.

Gottlieb, Michael. "The Reversible Removal of Naturally Occurring Organics Using Resins Regenerated with Sodium Chloride. Part 2." *Ultrapure Water* 13(8):53–58, 1996.

"Operating Experiences with a New Organic Trap Resin." *International Water Conference Proceedings IWC95-1,* Pittsburgh, Pa., October 1995.

Organic Traps, Application Bulletin. Cherry Hill, N.J.: ResinTech, Inc., 1997.

The Role of Organics in Water Treatment Parts 1, 2 and 3. Amber-hi-lites, Rohm and Haas Company, Fluid Process Chemical Department, No. 169 (Part 3), Winter 1982.

Sinsabaugh, R. L., R. C. Hoehn, R. W. Knocke, and A. E. Linkins III. "Removal of Dissolved Organic Carbon by Coagulation with Iron Sulfate." *Journal AWWA* 78(5), 1986.

Symons, J. M., P. L. K. Fu, P. H. S. Kim. "The Use of Anion Exchange Resins for the Removal of Natural Organic Matter from Municipal Water." *Proceedings International Water Conference,* Pittsburgh, Pa., October 1992, p. 92.

Selenium Removal

Boegel, J. V., and D. Clifford. *Selenium Oxidation and Removal by Ion Exchange.* PB 86-171 428/AS. Springfield, Va.: NTIS, 1986.

Softening

Harrison, J. F. *Information Facts on Home Water Softening, Total Dissolved Solids, Chlorides, and Water.* CWS-V. Lisle, Ill.: Water Quality Association, 1991.

Uranium Removal

Clifford, Dennis, and Z. Zhang. "Modifying Ion Exchange for Combined Removal of Uranium and Radium." *Journal AWWA* 86(4):227, 1994.

Hanson, S. W., D. B. Wilson, N. N. Guanji, and S. W. Hathaway. *Removal of Uranium from Drinking Water by Ion Exchange and Chemical Clarification.* Project Summary. Cincinnati, Ohio: EPA/600/S2-87/076, 1987.

Jelinek, R. T., and T. J. Sorg. "Operating a Small Full-Scale Ion Exchange System for Uranium Removal." *Journal AWWA* 80(7):79, 1988.

Sorg, T. J. "Methods for Removing Uranium from Drinking Water." *Journal AWWA* 80(7):84, 1988.

Uranium: Recognized Treatment Techniques for Meeting the Primary Drinking Water Regulations for the Reduction of Uranium from Drinking Water Supplies Using Point-Of-Use/Point-Of-Entry Devices and Systems. Technical Application Bulletin. Lisle, Ill.: Water Quality Association, February 1992.

Zhang, Z., and D. Clifford. "Exhausting and Regenerating Resin for Uranium Removal." *Journal AWWA* 86(4):228, 1994.

Zeolites

Burn, P., D. K. Ploetz, A. K. Saha, D. C. Grant, and M. C. Skriba. "Design and Testing of Natural/Blended Zeolite Ion Exchange Columns at West Valley." *Adsorption and Ion Exchange,* AIChE Symposium Series, 259(83). 1983, pg 66.

Sherman, J. D. "Ion Exchange Separations with Molecular Sieve Zeolites, Linde Molecular Sieves." Ion Exchange Bulletin. Presented at the 83rd National Meeting of the American Institute of Chemical Engineers, Houston, Texas, March 20–24, 1977.

SORBPLUS Adsorbent Product Data Sheet. Warren, Penn.: Alcoa Industrial Chemicals, December 1992.

General

Anderson, R. E. "Estimation of Ion Exchange Process Limits by Selectivity Calculations." *Journal of Chromatography* 201, 1980.

Applebaum, S. B. *Demineralization by Ion Exchange in Water Treatment and Chemical Processing of Other Liquids.* New York: Academic Press, 1968.

Bellen, Gordon, M. Anderson, and R. Gottler. *Management of Point-of-Use Drinking Water Treatment Systems, Final Report.* Ann Arbor, Mich.: National Sanitation Foundation Assessment Services.

Calmon, Calvin, and H. Gold. *Ion Exchange for Pollution Control.* Vol. I. Boca Raton, Fla.: CRC Press, 1979.

Gottlieb, M. C. *Fundamentals of Ion Exchange.* Water Quality Association. *Construction Contract Administration, M47.1996.* Denver, Colo.: American Water Works Association, 1989.

Gottlieb, Michael. "Performance Projects for Anion Resins in Two Bed Demineralizers." International Water Conference, October 1989.

Gottlieb, Michael. "The Practical Aspects of Ion Exchange in the Service D.I. Industry." *Water Technology* February 1989.

Kunin, Robert. *Ion Exchange Resins.* Robert E. Krieger, Melbourn, Fla. 1985.

Owens, D. L. *Practical Principles of Ion Exchange Water Treatment.* Littleton, Colo.: Tall Oaks, 1985.

Pontius, F.W., editor. *Water Quality and Treatment: a Handbook of Community Water Supplies.* 4th ed. Denver, Colo.: American Water Works Association, 1971.

CHAPTER 13
MEMBRANE PROCESSES

A wide variety of membrane processes can be categorized according to driving force, membrane type and configuration, and removal capabilities. Membrane processes in the drinking water industry are used for desalting, softening, dissolved organics and color removal, particle removal, and other purposes. Although membrane technologies became commercially available more than 25 years ago, they are experiencing rapid development and improvements.

Membranes are increasingly cost-effective with better performance characteristics, and their applications continue to grow. For example, membrane filtration systems are now used to remove *Giardia, Cryptosporidium,* and other particles from surface water supplies and to treat backwash return waters from conventional water treatment plants.

This chapter is based on current state-of-the-art membrane system design criteria. Some of the design data will undoubtedly change as new developments occur. The reader should keep abreast of new developments in membranes and system components by attending training seminars, reviewing publications and ongoing research reports, and contacting manufacturers, consultants, and other professionals practicing in the field.

TYPES OF MEMBRANE PROCESSES

Membrane processes can be classified by the driving force used to promote the water treatment

- Pressure
- Electrical voltage
- Temperature
- Concentration gradient
- Combinations of more than one driving force

Pressure-driven and electrically driven membrane processes are the only commercially available and commonly used membrane processes for water treatment.

Pressure-Driven Membranes

The pressure-driven membrane processes are

- Reverse osmosis (RO)
- Nanofiltration (NF)
- Ultrafiltration (UF)
- Microfiltration (MF)

In these processes, pressurized feed water enters one or more pressure vessels containing membranes, called membrane modules. Membranes are permeable to water but not to substances that are removed. All membrane processes separate feed water into two streams. The permeate (for RO, NF, or UF) or filtrate (for MF) stream passes through the membrane barrier. The concentrate (or retentate) stream contains the substances removed from the feed water after being rejected by the membrane barrier.

Pressure-driven membrane processes can be designed for cross-flow or dead-end operating mode. In the cross-flow mode, the feed steam flows across the membrane surface and permeate (or filtrate) passes through the membrane tangential to the membrane surface. Cross-flow operation results in a continuously flowing waste stream. Sometimes a cross-flow system is designed with a concentrate recycle with a reject stream (feed-and-bleed mode). Many MF and some UF systems treating relatively low-turbidity waters are designed to operate in a dead-end flow pattern where the waste retentate stream is produced by an intermittent backwash.

Figure 13.1 shows the relative removal capabilities for pressure-driven processes and compares these processes with media filtration. MF and UF separate substances from feed water through a sieving action. Separation depends on membrane pore size and interaction with previously rejected material on the membrane surface. NF and RO separate solutes by diffusion through a thin, dense, permselective (or semipermeable) membrane barrier

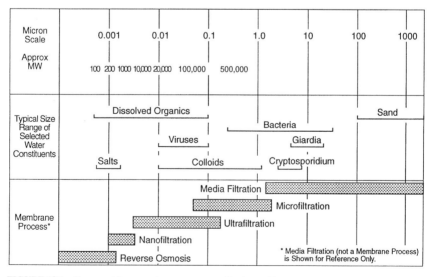

FIGURE 13.1 Pressure-driven membrane process application guide.

layer, as well as by sieving action. The required membrane feed pressure generally increases as removal capability increases. Table 13.1 presents typical feed pressures for pressure-driven membrane processes.

Electrical Voltage–Driven Membranes

The electrical voltage–driven membrane processes are:

- Electrodialysis (ED)
- Electrodialysis reversal (EDR)

Alternating anion and cation transfer ion exchange membranes in flat-sheet form are placed between positive and negative electrodes. Applying a voltage across the electrodes causes a direct current (DC) to flow, resulting in postively charged cations moving toward the negative electrode (cathode) and negatively charged anions moving toward the positive electrode (anode). This causes alternating compartments to become demineralized and the other compartments to become concentrated with ions.

A typical electrodialysis system has many anion- and cation-transfer membranes and spacers (membranes with spacers are called cell pairs) stacked vertically between electrodes. Feed water enters the stack under a pressure of about 50 psi (340 kPa), ions are removed, and the dilute (demineralized) stream leaves as product water. A typical design includes recycling the concentrate stream and discharging concentrate (blowdown) (feed-and-bleed mode) to waste. Electrodialysis does not remove electrically neutral substances such as silica or particulate matter because product water does not pass through a membrane barrier as it does in pressure-driven membrane processes.

Electrodialysis reversal is a variation of the electrodialysis process. The electrical polarity of the electrodes is reversed on a set frequency, typically about every 15 minutes, reversing the direction of ion movement to "electrically flush" the membranes for membrane scale and fouling control (Meller, 1984).

Selecting a Membrane Process

Membrane processes that satisfy treatment objectives at the lowest possible cost should be selected. The first step is to identify overall project goals and define the correct treatment

TABLE 13.1 Typical Feed Pressures for Pressure-Driven Membrane Processes

Membrane process	Typical feed pressure range	
	psi	kPa
Reverse osmosis		
Brackish water application		
Low pressure	125 to 300	860 to 2,070
Standard pressure	350 to 600	2,410 to 4,140
Seawater application	800 to 1,200	5,520 to 8,270
Nanofiltration	50 to 150	340 to 1,030
Ultrafiltration	20 to 75	140 to 520
Microfiltration	15 to 30	100 to 210

objectives. Source water quality should be compared with product water quality goals to determine the degree of removal required for various constituents. Removal may be by membrane process or associated pretreatment and posttreatment processes.

Sometimes small flat-sheet membranes with varying pore sizes are used with bench test equipment to characterize feed water in terms of solute fractions in various molecular weight size ranges.

Historical, current, and expected future water source quality data and product criteria should be considered.

The following factors should be considered in membrane selection:

- Source water characteristics and availability
- Pretreatment and posttreatment requirements
- Product water quality and quantity requirements and blending options (for example, split-treatment)
- Waste residuals disposal
- Capital and operation and maintenance (O&M) costs

Table 13.2 provides a starting point for a preliminary determination of which membrane processes, if any, apply to the treatment objectives. Membrane characterizations conducted during bench and pilot tests may be necessary to determine which process will provide the desired degree of removal (Bergman and Lozier, 1993).

Membrane Composition

Most membranes used in processes for municipal water treatment are prepared from synthetic organic polymers. Pressure-driven processes use either cellulosic or noncellulosic membranes. Cellulosic types include cellulose acetate, cellulose acetate blends, and cellulose triacetate. Noncellulosic types include polyamides, polyurea, polysulfone and sulfonated polysulfone, sulfonated polyfuran, polypiperazides, polyvinyl alcohol derivatives, acrylics, and other composites. Inorganic membranes, such as ceramic membranes with an alumina barrier layer, are available but are used to a lesser extent than are organic membranes.

ED and EDR membranes are essentially ion-exchange resins in flat-sheet form. These are synthetic polymers consisting of either cross-linked sulfonated copolymers of vinyl compounds (cation transfer type) or cross-linked copolymers of vinyl monomers with quaternary ammonium anion exchange groups (anion transfer type).

Membrane Configurations

Most pressure-driven membranes for municipal water treatment are arranged in spiral-wound and hollow fiber configurations and, to a lesser extent, tubular and plate-and-frame configuration.

Flat-sheet membranes for pressure-driven processes are most commonly assembled into a spiral-wound element (module) in which multiple membrane "leaves," each composed of two membrane sheets separated by a permeate carrier, are connected to a central permeate collector tube. A feed concentrate spacer is placed between each leaf and the leaves and spacers are rolled around the central permeate collection tube (Figure 13.2a). One or more spiral-wound elements are placed inside each pressure vessel in a series arrangement (Figure 13.2b).

RO hollow-fiber membrane modules (also called permeators) are commonly constructed by forming the hollow, fine fibers into a U-shape, "potting" the open ends in an

TABLE 13.2 Typical applications of membrane processes

Process	Application

Reverse osmosis and electrodialysis	Total dissolved solids reduction
	  Seawater desalting (RO favored)
	  Brackish water desalting (RO is typically more cost-effective than ED or EDR for greater than 3,000 mg/L TDS)
	  Brackish water desalting of high-silica waters (ED and EDR are favored)
	Inorganic ion removal
	  Fluoride
	  Nutrients (nitrate, nitrite, ammonium, phosphate)
	  Radionuclides (RO only)
	  Others listed under drinking water regulations
	Synthetic organic chemicals (SOCs) removal (RO only)

Nanofiltration	Hardness removal
	Organics removal
	  THMs, other DBP precursors
	  Pesticides (SOCs)
	  Color

Ultrafiltration	Organics removal (dependent on molecular weight) (sometimes powdered activated carbon is used with UF for improved organics removal)
	Particulate removal
	  Suspended solids
	  Turbidity
	  Bacteria
	  Viruses
	  Protozoan cysts
	  Colloids

Microfiltration	Particulate removal
	  Suspended solids
	  Some colloids
	  Bacteria
	  Some viruses (associated with particulate matter)
	  Protozoan cysts
	  Turbidity
	Inorganic chemical removal (after chemical precipitation or pH adjustment)
	  Phosphorus
	  Hardness
	  Metals

epoxy tube sheet, and placing the potted assembly inside a pressure vessel, as shown in Figure 13.3. The flow direction through RO hollow fibers is typically outside-in, with feed water on the outside of the fibers and permeate within the fibers' central bore. Some RO hollow-fiber membranes are available that can be located within standard spiral-wound pressure vessels. For UF membrane modules, larger-diameter hollow fibers are commonly potted at both ends and assembled into modules having a straight-through flow path for the feed-concentrate stream within the fibers and with the permeate passing on the outside of the fibers (inside-outside flow pattern), as shown in Figure 13.4.

Hollow-fiber MF often is configured in a straight-through flow pattern, but uses an outside-in flow with the filtrate within the fibers. One type of hollow-fiber MF places the

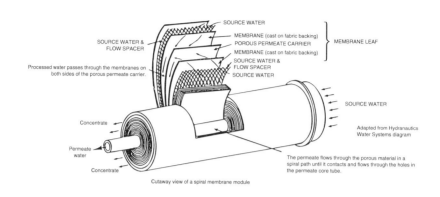
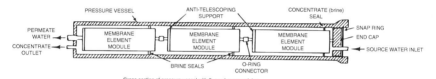

FIGURE 13.2 Typical spiral-wound RO membrane module for pressure-driven processes. *(Adapted from the U.S.A.I.D. Desalination Manual (Buros et al., 1980) and is used courtesy of the U.S. Agency for International Development.)*

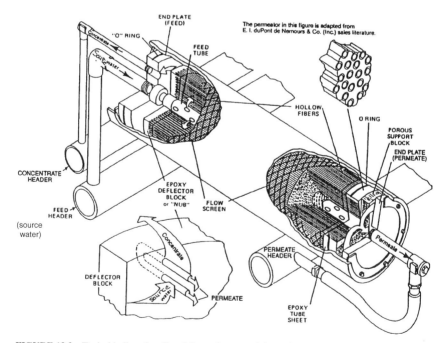

FIGURE 13.3 Typical hollow fine-fiber RO membrane module. *(Adapted from The U.S.A.I.D. Desalination Manual (Buros et al., 1980) and is used courtesy of the U.S. Agency for International Development.)*

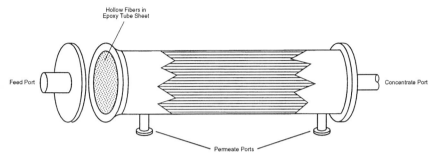

FIGURE 13.4 Representation of hollow-fiber UF module.

fibers directly in an open feed tank (not in a pressure vessel) and uses negative pressure on the filtrate side of the membrane (central bore of the fibers) as the driving force.

Electrically driven ED and EDR processes typically use flat-sheet membranes assembled into stacks of up to 500 membrane pairs with a single pair of electrodes, as shown in Figure 13.5. ED/EDR stacks can have multiple electrical and hydraulic stages.

MEMBRANE SYSTEM COMPONENTS AND DESIGN CONSIDERATIONS

Some of the more important considerations in membrane system design are the system components, feed water characteristics, and pretreatment that may be required.

Definitions

Applied pressure: feed pressure minus product pressure.

Demineralized stream pressure: dilute (product) stream pressure in an ED/EDR system.

Concentrate pressure (Pc): hydraulic pressure of the concentrate or retentate flow stream leaving the membrane modules or circulating in an ED/EDR membrane stack.

Delta-P or differential pressure (ΔP) for pressure-driven membrane systems: feed pressure minus concentrate pressure.

Delta-P or differential pressure (ΔP) for electrically driven membrane systems: pressure difference between the demineralized steam and the concentrate stream.

Feed pressure (Pf): hydraulic pressure of the feed water flow stream entering the membrane modules.

Flux (water flux): rate of product water flow through a pressure-driven membrane, commonly expressed in units of gallons per day per square foot of active membrane area (gal/day/ft^2) or meters per second (m/s).

Interstage pressure (Pi): in a multistage membrane system, concentrate pressure from one stage and feed pressure to the following stage.

Osmotic pressure (π): a natural pressure phenomenon exhibiting a force from a low-concentration stream (e.g., product) to a high-concentration stream (e.g., feed and concentrate). Osmotic pressure is related to the solution's ionic strength and must be

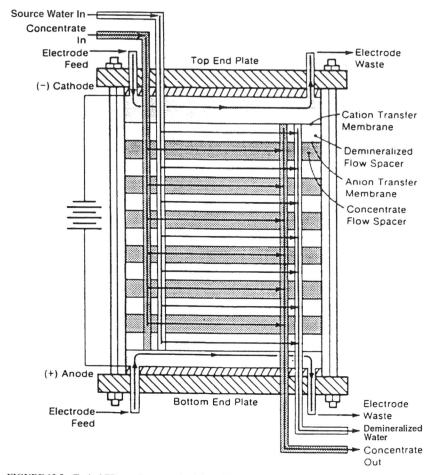

FIGURE 13.5 Typical ED membrane stack. *(Adapted from Electrodialysis-Electrodialysis Reversal Technology, Ionics, Inc., 1984.)*

offset by feed pressure. Osmotic pressure for saline water is about 10 psi (70 kPa) per 1,000 mg/L of total dissolved solids (TDS). Osmotic pressure can generally be ignored in UF and MF water treatment applications.

Transmembrane osmotic pressure (π_{tm}): feed stream (average feed/concentrate) osmotic pressure of a pressure-driven membrane minus permeate osmotic pressure. Mathematically expressed as:

$$\pi_{tm} = \frac{\pi_f + \pi_c}{2} - \pi_p$$

where π_{tm} = transmembrane osmotic pressure, psi
  π_f = feed osmotic pressure, psi
  π_c = concentrate osmotic pressure, psi
  π_p = permeate osmotic pressure, psi

Product pressure (P_p): hydraulic pressure of the permeate or filtrate flow stream at the outlet of the membrane modules.

Net driving pressure (NDP): the net driving force for a pressure-driven membrane system.

$$\text{NDP} = P_{tm} - \pi_{tm}$$

where NDP = net driving pressure, psi
P_{tm} = transmembrane (hydraulic) pressure differential, psi
π_{tm} = transmembrane osmotic pressure differential, psi

Transmembrane pressure (P_{tm}): feed stream (average feed/concentrate) pressure (cross-flow operating mode) or feed pressure (dead-end operating mode) minus the permeate (product) pressure. Mathematically expressed as:

$$P_{tm} = \frac{P_f + P_c}{2} - P_p = P_f - \frac{\Delta P}{2} - P_p$$

where P_{tm} = transmembrane pressure, psi
P_f = feed pressure, psi
P_c = concentrate pressure, psi
P_p = permeate pressure, psi
ΔP = feed/concentrate pressure differential $(P_f - P_c)$, psi

Recovery (Y): permeate or filtrate flow rate divided by the feed water flow rate, usually referred to as permeate or product water recovery for RO and NF and feed water recovery for UF and MF systems.

$$Y = (Q_p/Q_f) \times 100$$

where Y = recovery, percent
Q_p = product flow rate (volume)
Q_f = feed flow rate (volume)

Solute passage: solute passage is the fraction of solute present in the feed that remains in the permeate, typically expressed as a percentage. When the solute considered is TDS, it is usually called salt passage.

$$\text{SP} = (C_p/C_f) \times 100$$

where SP = solute (salt) passage, percent
C_p = solute (salt) concentration in the permeate, mg/L
C_f = solute (salt) concentration in the feed, mg/L

Solute rejection (removal): solute rejection or removal (or for TDS, salt rejection or removal) is the fraction of solute in membrane feed water that remains in the concentrate (retentate) stream and does not enter the product stream, expressed as a percent. For ED/EDR systems, *removal* is the term used instead of rejection.

$$\text{SR} = \frac{C_f - C_p}{C_f} \times 100 = 1 - \frac{C_p}{C_f} \times 100 = 100 - \text{SP}$$

where SR = solute (salt) rejection or removal, percent
SP = solute (salt) passage, percent
C_p = concentration of solute in permeate or filtrate stream, mg/L
C_f = concentration of solute in feed stream, mg/L

Membrane System Components

A simplified membrane system flow schematic is shown in Figure 13.6. Membrane systems typically include pretreatment, the membrane unit(s), product (permeate or filtrate) posttreatment, and possibly concentrate (or retentate) waste stream posttreatment. Membrane systems are commonly designed with multiple parallel process "trains" to give more flexibility in production output rates and to allow cost-effective incremental expansions. The flow schematic also shows that in some cases it is possible to bypass a portion of source or pretreated water around the membranes and blend it with the product to produce finished water, before disinfection (i.e., split treatment).

Membrane System Feed Water Characteristics

Membrane system design must consider source water quality and temperature. Design feed water composition should encompass current source water quality, anticipated changes in quality, and the effects of membrane pretreatment processes. Historical records are useful for determining a range of values and averages for specific water quality parameters and in estimating long-term trends. Seasonal variations for surface waters should also be considered.

Table 13.3 presents recommended items that should be analyzed for the feed water to a proposed pressure-driven membrane system plant.

The feed water flow rate required for a system depends on membrane system recovery. Maximum possible recovery of a membrane system is usually controlled by feed water quality. For RO, NF, and ED/EDR, maximum allowable recovery often depends on the concentration of sparingly soluble salts and, except for ED/EDR, silica in the source water supply.

Feed Water Pretreatment

Most membrane treatment systems require some pretreatment for source water. The type of pretreatment system required depends on feed water quality, membrane type, and design criteria for the membrane unit. Pretreatment may be used to:

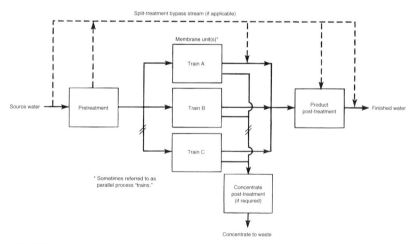

FIGURE 13.6 Typical membrane system flow schematic.

TABLE 13.3 Recommended feed water analyses for pressure-driven membrane systems

RO, NF, and ED systems	MF and UF systems
Temperature (field)	Temperature (field)
pH (field)	pH (field)
Alkalinity	Alkalinity
Hardness	Hardness
Conductivity (at 25° C)	Iron (total and dissolved)
Evaporative TDS (at 180° C)	Manganese (total and dissolved)
TDS by summation of major ions*	Turbidity
Calcium	Total suspended solids
Magnesium	Chlorine residual and other strong oxidants
Sodium	Total organic carbon
Potassium	Bacterial analysis
Barium	
Strontium	
Iron (total and dissolved)	
Manganese (total and dissolved)	
Bicarbonate*	
Carbonate*	
Hydroxide*	
Carbon dioxide*	
Sulfate	
Chloride	
Fluoride	
Nitrate and nitrite	
Phosphate (total)	
Silica (total)	
Silica (reactive)	
Ammonium	
Bromide	
Boron	
Turbidity	
Total suspended solids	
Silt density index (15 min) (field)	
Hydrogen sulfide	
Chlorine residual and other strong oxidants (field)	
Total organic carbon	
Radionuclides	
Bacterial analyses (total plate count)	

*Calculated value.

- Condition the feed water to allow membrane treatment to be effective; for example, using coagulants to create particles large enough to be removed by microfiltration
- Modify the feed water to prevent membrane plugging, fouling, and scaling; to maximize the time between cleanings; and to prolong membrane life

Design of specific components needed to pretreat membrane feed water is discussed in other chapters of this book. In general, surface waters require more extensive pretreatment than groundwaters because of the presence of significant levels of suspended solids and biological matter.

For example, Figure 13.7 shows typical groundwater and surface water pretreatment for RO and NF processes. In some situations, pretreatment can be relatively simple, such

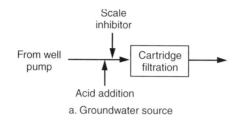

a. Groundwater source

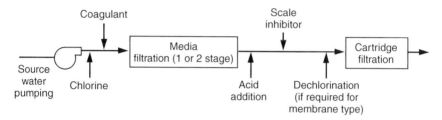

b. Low turbidity surface water source (includes seawater)

c. High turbidity surface water source

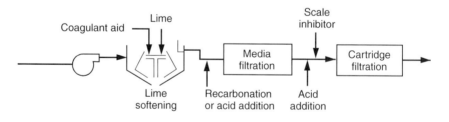

d. Excessive calcium water (surface or ground) source

FIGURE 13.7 Typical RO or NF pretreatment processes.

as adding chemicals, such as acid or scale inhibitor, and cartridge filtration. Other waters may require pretreatment as extensive as a full conventional water treatment plant with coagulation, flocculation, sedimentation, and filtration.

Suspended Solids Control. All membrane systems require control of suspended solids in the feed water to prevent plugging the membrane system. Pretreatment for UF and MF systems typically includes strainers or bag filters rated in the 100 to 500 μm size range.

Membrane manufacturers usually specify the maximum feed water suspended solids concentration or turbidity allowed for their particular products.

Feed water must be relatively free of suspended matter for RO, NF, and to a lesser degree, EDR systems. Small suspended particles (colloids) must be removed to the extent determined by the type of membrane. Improved suspended solids removal can significantly lower the frequency rate of required membrane cleaning.

Nearly all RO and NF membrane systems use pretreatment cartridge filters nominally rated at 1 to 25 μm, depending on the type of system. ED/EDR systems commonly use 20 to 25 μm cartridges. Where solids loading would result in frequent filter cartridge replacement, or where cartridge filters are desired for backup protection only (the preferred method), additional pretreatment, such as granular media filters with or without chemical addition, is used. For waters with high suspended solids loading (usually surface water sources), a coagulation-flocculation-sedimentation pretreatment process may be used.

The two most common indicators of feed water suspended solids content used today for RO, NF, and ED/EDR membrane systems are turbidity and silt density index (SDI), although the use of particle counters is increasing. SDI is determined from the rate of plugging of a 0.45 μm filter under a feed pressure of 30 psig (207 kPa) as described in ASTM D4189. ED/EDR, RO, and NF manufacturers usually specify maximum allowable turbidity or SDI limits. Typical turbidity and SDI limits for the various membrane desalting processes, depending on the particular membrane, are listed in Table 13.4.

Scaling Control. Scaling control is applicable to RO, NF, and ED/EDR. Design for all three processes must consider calcium carbonate and sulfate scaling control. RO and NF systems design must also consider the need for silica control. Because electrodialysis does not remove silica from the feed water and does not concentrate it in the concentrate flow channels, silica does not limit ED design recovery.

Depending on hydraulic recovery, the concentration of salt ions and silica in feed water can be increased during the treatment process by as much as tenfold. If concentrations exceed the solubility product of the compound at ambient conditions of temperature and ionic strength, scale can form within the modules, decreasing productivity and deteriorating permeate quality. More important, it can also cause failure of the membrane module. The sparingly soluble salts of concern in most waters are calcium carbonate ($CaCO_3$); the sulfate salts of calcium, barium, and strontium ($CaSO_4$, $BaSO_4$, and $SrSO_4$, respectively); and silica (SiO_2).

TABLE 13.4 Feed Water Turbidity and SDI Limits Recommended by Manufacturers for RO, NF, and ED/EDR Systems

	RO and NF		ED/EDR
	Spiral wound	Hollow fiber	
Maximum turbidity, ntu	1*	—	2 to 3
Maximum SDI (15 min)	3 to 5	3 to 4	—
Maximum SDI (5 min)	—	—	15

*Recommended turbidity less than 0.2 NTU.
Sources: Spiral-wound RO data based on product literature from FilmTec (Dow) Corp., Fluid Systems, Hydranautics, and TriSep. Hollow-fiber RO data based on product literature from the Du Pont Company. EDR data based on Ionics, Inc., Bulletin No. 121-E, *EDR—Electrodialysis Reversal,* March 1984.

Calcium Carbonate Control. The pH of calcium carbonate ($CaCO_3$) solubility can be estimated by:

$$pH_s = pCa + pAlk + K$$

where pH_s = solubility pH
 pCa = negative logarithm of calcium concentration
 $pAlk$ = negative logarithm of alkalinity (bicarbonate, HCO_3 concentration)
 K = constant related to ionic strength (and TDS) and temperature

The tendency to develop $CaCO_3$ scale during the treatment of fresh and brackish waters can be determined by calculating the Langelier Saturation Index (LSI) of the concentrate stream.

$$LSI = pH_c \text{ (of concentrate stream)} - pH_s$$

Figure 13.8 presents a nomograph that can be used to determine LSI. For seawater desalting, the Stiff and Davis Saturation Index (SDSI) is often used. Figure 13.9 shows graphs with which the SDSI K value can be determined.

Control for $CaCO_3$ scale can be achieved using the following methods:

- Acidifying to reduce pH and alkalinity
- Reducing calcium concentration by ion exchange or lime softening
- Adding a scale inhibitor chemical (antiscalant) to increase the apparent solubility of $CaCO_3$ in the concentrate stream
- Lowering the design recovery

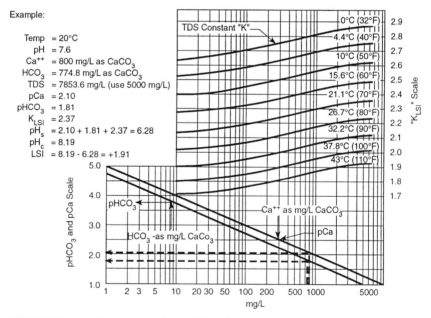

FIGURE 13.8 Langelier saturation index. *(Adapted from DuPont Permasep Products Engineering Manual.)*

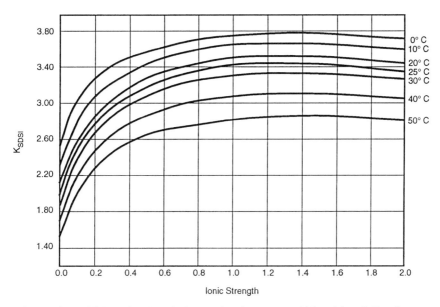

FIGURE 13.9 Stiff & Davis "K" vs. ionic strength and temperature. *(Adapted from DuPont Permasep Products Engineering Manual, 1993.)*

Feed water acidification is the most widely used method for controlling $CaCO_3$ scaling. Sulfuric acid is generally used because of its low cost and less hazardous (nonfuming) characteristics. When scaling from sparingly soluble sulfate salts is a concern and adding sulfate ions is undesirable, hydrochloric acid should be considered.

The need for acidification is lessened with the introduction of sodium hexametaphosphate (SHMP) or polyacrylic acid antiscalants. Polyacrylic acid antiscalants at dosages of less than 5 mg/L typically can provide scale control up to an LSI (or SDSI) value of about +2.0. Also, polyacrylic acid antiscalants are often used at plants employing noncellulosic membranes, and pH control is not required to minimize membrane hydrolysis—a concern with the use of cellulose products. When feed water LSI is initially high, softening may be required.

Sulfate Scale Control. Scaling potential of sulfate salts can be estimated by calculating the ion product for each salt in the concentrate stream and comparing it with the solubility product (K_{sp}) of the salt at the temperature of interest. Figures 13.10 and 13.11 present calcium sulfate ($CaSO_4$), strontium sulfate ($SrSO_4$), and barium sulfate ($BaSO_4$) solubility graphs.

The most widely used method of control for sulfate scale is the addition of SHMP or polyacrylic acid–type antiscalant. Polyacrylic acid antiscalants are commonly used for RO and NF applications because they permit a greater degree of supersaturation of critical ions in the concentrate stream than does SHMP.

Silica Control. The silica (SiO_2) scaling potential of the concentrate stream usually can be estimated by the following formula (Du Pont Company, 1992):

$$SiO_{2\,(max)}(mg/L) = SiO_{2\,(temp)}(mg/L) \times \text{pH correction factor}$$

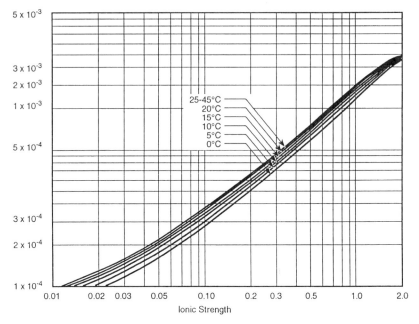

FIGURE 13.10 K_{sp} for $CaSO_4$ vs. ionic strength. *(Adapted from DuPont Permasep Products Engineering Manual, 1992.)*

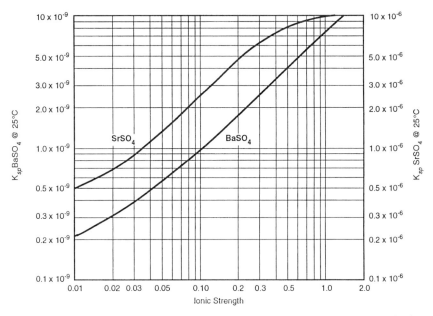

FIGURE 13.11 K_{sp} for $BaSO_4$ and $SrSO_4$ vs. ionic strength. *(Adapted for DuPont Permasep Products Engineering Manual, 1992.)*

where $SiO_{2\ (max)}$ = maximum concentrate stream silica, mg/L
$SiO_{2\ (temp)}$ = silica solubility concentration at pH 7.5 as a function of water temperature, mg/L
pH correction factor = silica solubility factor at various pH values

Concentrate silica as a function of temperature is shown in Figure 13.12, and concentrate stream pH correction factors are shown in Figure 13.13.

In some cases, RO systems have successfully operated with concentrate stream silica levels exceeding theoretical solubility limits, especially with some applications using new silica-specific antiscalants. Silica scaling has occurred in other systems at concentrations much less than theory would predict because of complexation and precipitation of silica with trivalent ions such as oxidized iron or aluminum. Where feed water silica concentrations approach solubility limits, it is essential that the presence of these metal species be minimized.

The following pretreatment methods can be used to control silica scale:

- Reducing hydraulic recovery to reduce concentrate stream silica concentrations
- High lime softening to reduce feed water silica concentrations
- Increasing the temperature of the feed water
- Increasing feed water pH to 8.5 or higher (taking into account the impact of increased pH on $CaCO_3$ scaling potential)
- Adding specific silica antiscalant chemical

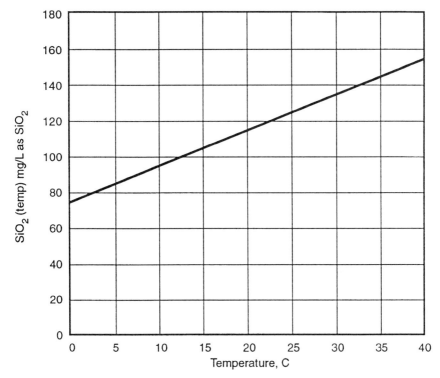

FIGURE 13.12 Effect of Temperature on SiO_2 Solubility. *(Adapted from DuPont Permasep Products Engineering Manual, 1992.)*

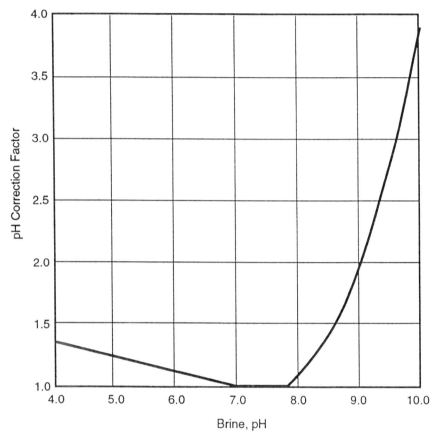

FIGURE 13.13 SiO$_2$ pH Correction Factor. (Adapted from DuPont Permasep Products Engineering Manual 1992).

Concentration of Sparingly Soluble Salts and Silica. Concentrations of sparingly soluble ions and silica (for RO and NF) in the concentrate stream can be estimated by dividing feed water concentrations of each by 1 minus the desired hydraulic recovery:

$$C_c = C_f \times 1/(1 - Y)$$

where C_c = concentration of solute in the concentrate stream, mg/L
C_f = concentration of solute in the feed water stream, mg/L
Y = hydraulic recovery (permeate flow rate divided by feed flow rate), expressed as a decimal

This expression overestimates bulk stream concentrate stream ion concentration because it assumes that membrane rejection of each ion is 100%, which is never the case. A more accurate method uses a formula taking into account the membrane's percent rejection value for each ion, if known. This formula is:

$$C_c = C_f \times [(1 - Y \times SP_c)/(1 - Y)]$$

where SP_c is the solute passage for the specific ion, expressed as a decimal.

With available computer programs for the projection of RO, NF, and EDR system performance, the need to manually calculate scaling potential has declined. It is more convenient to send the feed water quality analysis to the membrane manufacturer, system supplier, or membrane treatment consultants and request scaling predictions. The desired recovery and the type of membrane to be used are also variables in the analyses. Table 13.5 presents typical concentrate stream solubility criteria used in RO and NF design.

Microbial Control. Microbial growth must be controlled because it can foul membrane systems. In addition, certain types of cellulose acetate RO and NF membranes might be degraded by several species of bacteria. Chlorination is the usual method of disinfection, although disinfection by-product formation potential may preclude the use of free chlorine. Chloramines are also used for bacterial control in membrane systems where the membranes can tolerate the combined chlorine concentration.

Chlorine dosage rate must be closely controlled. Most types of cellulose acetate RO and NF membranes can tolerate up to 1.0 mg/L free chlorine. The various types of polyamide (PA) RO and NF membranes cannot tolerate chlorine or other strong oxidants. If chlorine is used, a dechlorination process, such as sodium bisulfite or sulfur dioxide addition, must be used immediately before the RO or NF system.

Chlorination-dechlorination is also used for ED systems because most ED membranes cannot tolerate chlorine. Some EDR membranes may have relatively long life with a continuous free chlorine exposure of up to 0.3 mg/L. Some MF and UF membranes can tolerate chlorine where other membranes cannot. It is important to contact membrane manufacturers to verify chlorine tolerance levels.

TABLE 13.5 Typical RO and NF Concentrate Stream Solubility Design Criteria Used by Some Manufacturers

	Concentrate stream criteria[a,b]		
Compound[c]	Typical (without scale inhibitor)	Typical (with scale inhibitor)	Maximum (with scale inhibitor)[c]
Calcium carbonate ($CaCO_3$)	$LSI^d < -0.2$	$LSI < +1.5^h$	$LSI < +2.3^h$
Calcium sulfate ($CaSO_4$)	$IP^e < 0.8 \times K_{sp}^f$	$IP < 1.5 \times K_{sp}$	$IP < 2 \times K_{sp}$
Strontium sulfate ($SrSO_4$)	$IP < 0.8 \times K_{sp}$	$IP < 8 \times K_{sp}$	$IP < 8 \times K_{sp}$
Barium sulfate ($BaSO_4$)	$IP < 0.8 \times K_{sp}$	$IP < 40 \times K_{sp}$	$IP < 60 \times K_{sp}$
Silica (SiO_2)[g]	$IP < 0.8 \times K_{sp}$	$IP < 1.5^g$	<200 mg/L[g]

[a] The membrane and antiscalant manufacturers should be contacted to confirm design criteria for the specific products for the specific application.
[b] In some cases, maximum calcium fluoride and calcium phosphate levels are specified.
[c] Assumes addition of an appropriate scale inhibitor and that, when shut down, the system is flushed within the specified time period, commonly less than 4 hours.
[d] LSI is the Langelier Saturation Index. Positive LSI values indicate potential for calcium carbonate precipitation. Negative LSI values indicate corrosive tendencies.
[e] IP is the ion product of the compound based on the concentrations in the concentrate stream.
[f] K_{sp} is the solubility product. If the ion product (IP) is greater than the K_{sp}, precipitation can occur. If the IP is less than the K_{sp}, precipitation usually does not occur.
[g] Presence of polyvalent metal oxides also affects criteria. Silica precipitation usually occurs from polymerization, which is a relatively slow process, but is significantly affected by other ions in the water.
[h] For brackish water greater than about 10,000 mg/L TDS and seawater applications, the Stiff and Davis Saturation Index (SDSI) is typically used instead of LSI. SDSI values typically are not greater than +0.5 for RO design.
Sources: Du Pont Company: Wilmington, Del. Hydranautics: San Diego, Calif. Dow–FilmTec: Midland, Mich. Fluid Systems: San Diego, Calif. TriSep: Goleta, Calif.

Granulated activated carbon (GAC) is sometimes used for dechlorination, but microbial matter released from GAC may cause problems in the membrane system. Ultraviolet light disinfection has also been used for some membrane systems, but the lack of a residual can result in regrowth problem. Many types of membranes can accept higher short-term concentrations of chlorine for cleaning or shock treatment. Biological control through periodic shock treatments with sodium bisulfite or other nonchlorine compounds is also used for membrane systems. While not in service, membranes are typically placed in storage solutions of various formulations to retard microbial growth.

Hydrogen Sulfide Control. Where hydrogen sulfide is present, such as in some well waters, no chlorine or exposure to air can be allowed; colloidal sulfur will form and foul an RO or NF membrane system. Because chemical cleaning is typically ineffective in removing sulfur from the membrane surface, membranes that become fouled are usually be replaced. When low levels of hydrogen sulfide exist in feed water, the system can be designed to be airtight and hydrogen sulfide can be removed in posttreatment degasifiers. Hydrogen sulfide greater than 0.1 to 0.3 mg/L can cause problems with ED/EDR systems and must be removed in pretreatment. RO and NF membranes can tolerate relatively high concentrations of hydrogen sulfide without damage.

Iron and Manganese Control. Iron and manganese may cause problems with RO, NF, and ED/EDR membrane systems. However, as long as these metals are kept in their reduced state and dissolved in water, they cause little problem with RO and NF systems. Because iron and manganese form precipitates in the presence of oxygen or at high pH, acid addition is often used for metal oxide control (in addition to its other uses previously discussed).

Sequestrants such as sodium hexametaphosphate are also commonly fed to inhibit iron and manganese deposition. Some types of polyacrylic acid antiscalants react with iron and manganese, causing fouling problems in downstream membranes, and should not be used. Proprietary antifoulants (scale inhibitor/dispersants) are also commonly used to control fouling and scale in the presence of iron and manganese.

Iron and manganese can foul ED/EDR membranes even when they are in the reduced state. In addition, manganese can plate out on the electrodes, decreasing their efficiency. For ED/EDR systems, iron removal is usually recommended if feed water iron concentration exceeds 0.3 mg/L or if the manganese level exceeds 0.1 mg/L.

If iron or manganese precipitates, or if either metal is present from corrosion products, ED/EDR, RO, and NF systems can be fouled. Appropriate materials should be used for construction of feed water supply or pretreatment systems.

Three pretreatment processes can be used for iron or manganese removal:

- Oxidation using air, chlorine, or another oxidant, followed by granular media filtration
- Oxidation using potassium permanganate, followed by a manganese greensand filter
- Cation ion exchange softeners

Residual oxidant levels must be monitored and controlled before entering the membrane system. If lime softening is used for scale control, it has the additional benefit of lowering iron and manganese levels in feed water.

Organics Control. Organic matter may be classified as either suspended or dissolved. Suspended (and colloidal) organic solids and microbial matter (previously discussed) adversely affect RO, NF, and ED systems.

Naturally occurring dissolved organics cause fouling in ED/EDR systems and generally, to a lesser degree, in RO or NF systems. In fact, some NF and RO systems are used to remove natural and synthetic dissolved organics and color. Membrane manufacturers are aware of many organic compounds that cause problems in their systems and can be contacted for specific information. There is no definitive correlation between the quantity of organics present, such as measured by total organic carbon (TOC) analyses, and performance decline. However, if TOC exceeds 10 mg/L in feed water, the potential effects of fouling usually warrant further investigations. Pilot testing is often used to determine the effects of dissolved organics on membrane systems.

Oil, greases, hydrocarbons, and various organic solvents and other chemicals can damage or foul RO, NF, and ED membranes and should not be allowed to enter the feed water supply. Conventional coagulation, flocculation, and sedimentation processes and lime softening can reduce the organic content of feed water before it enters membrane units. Activated carbon filters are sometimes used to remove organics or for dechlorination of feed water entering RO, NF, and ED systems, but problems caused by carbon fines or biological microorganisms released from carbon filters have occurred in some systems.

Control of pH. All membrane types have specified pH ranges in which they should operate. Many types of ED and polyamide RO and NF membranes can tolerate continuous exposure to feed water in the pH range of 4 to 10 and intermittent exposure to an even wider pH range. The various types of cellulose acetate RO and NF membranes normally require pH between approximately 4 and 6.5 to minimize degradation caused by hydrolysis. Many types of polysulphone and polypropylene MF and UF membranes tolerate a pH range of at least 2 to 13. Inorganic MF and UF membranes can generally accept feed water with nearly any pH.

Temperature Control. Water temperature significantly affects membrane systems in several ways:

- It alters membrane material characteristics and membrane life.
- Water viscosity and density affect membrane hydraulic performance and required membrane area.
- It changes solubility of sparingly soluble salts and silica, which limits the design recovery of membrane desalting processes.

Membrane systems must operate within the manufacturer's temperature guidelines to maximize membrane life. The various types of cellulose acetate membranes deteriorate from hydrolysis at increasing rates as temperatures rise. Flux decline caused by compaction of RO and NF membranes also is greater at higher temperatures.

The maximum temperature limit for the various brackish water cellulose acetate RO and NF membranes ranges from 35° to 40° C, depending on the membrane. The various types of polyamide RO membranes typically are rated at maximum temperatures ranging from 40° to 50° C, but their useful life can be significantly reduced at these elevated temperatures.

ED/EDR membranes commonly can operate at temperatures up to 45° C. Typically, polysulphone, polypropylene, and several other types of MF and UF membranes tolerate a temperature up to at least 45° C. Inorganic (ceramic) MF and UF membranes can tolerate temperatures well over 100° C.

Viscosity and density of water increase at colder temperatures, and for pressure-driven systems greater membrane area is required to produce a specified product flow at a given feed pressure. For example, assuming all other factors (including membrane area and feed pressure) remain constant, the permeate productivity of a pressure-driven membrane system is about 30% to 40% less at 15° C than it is at 30° C. For RO and NF systems, an approximation of permeate flow at any temperature relative to flow at 25° C, assuming all other factors are constant, is as follows:

$$Q_p = Q_{p(25°C)} \times 1.03^{(T-25)}$$

where Q_p = permeate flow at temperature T
$Q_{p(25°C)}$ = permeate flow at 25° C
T = water temperature in degrees C

For UF and MF systems, an approximation of permeate flow at any temperature relative to flow at 20° C is as follows:

$$Q_{pT} = \frac{Q_{p20°C}}{e^{-0.0239(T-20)}}$$

where Q_{pT} = permeate flow at temperature T
$Q_{p20°C}$ = permeate flow at 20° C
T = water temperature in degrees C
e = 2.71828

Typical membrane system design necessitates evaluation of performance at minimum and maximum temperatures assuming new (initial start-up) and "used" (often 1, 3, or even 5 years of operation) membranes.

A maximum design flux or feed pressure is assumed and the minimum temperature is used to determine the required membrane area for "used" membranes. Given the membrane area, the performance at maximum temperature is then evaluated to verify that product water quality goals are also met.

Temperature is one of the most important factors affecting ion removal of an ED/EDR system. Ion removal increases about 2% per degree centigrade temperature rise. A two-stage ED/EDR system can have TDS removal approximately 25% greater at 30° C than at 15° C.

Temperature affects membrane system performance not only directly by influencing product water flow rates or salt removal, but indirectly by affecting solubility of compounds, which can precipitate and foul the system. Some compounds, such as calcium carbonate, have a greater tendency to scale at higher temperatures. However, most compounds normally encountered in natural waters have better solubilities as the temperature rises.

MEMBRANE UNIT DESIGN

Once treatment objectives have been identified, design criteria can be developed for the membrane process. The design considerations are grouped for:

- RO and NF units
- UF and MF units
- ED and EDR units

Reverse Osmosis and Nanofiltration Unit Design

Because of the high solute rejection of RO and NF membranes, which concentrate inorganic ions among other constituents, there are many special considerations in the treatment unit design.

Design Equations. The basic behavior of permselective (semipermeable) membranes can be described by the following two diffusion model equations. Permeate flow through the membrane may be expressed as:

$$F_w = A \times (P_{tm} - \pi_{tm})$$

where F_w = water flux (g/cm^2-s)
 A = water permeability coefficient (g/cm^2-s-atm)
 P_{tm} = hydraulic pressure differential applied across the membrane (atm)
 π_{tm} = osmotic pressure differential across the membrane (atm)

The solute (or salt) flux through the membrane may be expressed as:

$$F_s = B \times (C_1 - C_2)$$

where F_s = solute (or salt) flux (g/cm^2-s)
 B = solute (or salt) permeability constant (cm/s)
 $C_1 - C_2$ = concentration gradient across the membrane (g/cm^3)

Water and solute permeability coefficients are characteristic of the particular membrane type.

Water flux depends on applied pressure, but solute flux does not. As pressure of membrane feed water increases, water flow through the membrane (water flux) increases, while solute flow remains essentially unchanged. Permeate quantity increases with increased applied pressure, as does the quality (decreased solute concentration).

Water flux decreases as the salinity of the feed increases because of increased osmotic pressure differential resulting from increased salinity. As increasing amounts of water pass through the membrane system, the salinity of the remaining feed water (concentrate) increases. Concentrate osmotic pressure increases, resulting in a lower water flux with increasing overall percent water recovery.

Finally, because salinity of the feed-concentrate stream increases with increasing permeate production from a given volume of feed, and the membrane rejects a fixed percentage of solute, product water quality decreases (higher concentration) with increasing recovery. Table 13.6 presents typical feed pressures and recoveries for RO and NF systems.

TABLE 13.6 Typical RO and NF System Feed Pressure and Recovery

Application (with example TDS)	Typical operating pressure (psi)	(kPa)	Typical product water recovery (percent)*
Reverse Osmosis			
Seawater (35,000 mg/L TDS)	800 to 1,200	5,520 to 8,270	30 to 45
Brackish water (5,000 mg/L TDS)	350 to 600	2,410 to 4,140	65 to 80
Brackish water (1,000 mg/L TDS)	125 to 300	860 to 2,070	70 to 85
Nanofiltration			
Fresh water (500 mg/L TDS)	50 to 150	340 to 1,030	80 to 90

*Maximum allowable recovery is site specific and depends on the feed water's scaling potential, membrane rejection and product water quality requirements, concentrate stream osmotic pressure, and the use of scale-inhibiting chemicals.
Source: Adapted from Bergman, Robert A. "Anatomy of Pressure-Driven Membrane Desalination Systems." *1993 Annual AWWA Conference Proceedings—Engineering and Operations.* Denver, Colo.: 1993. American Water Works Association.

Recovery Considerations As recovery increases, the following factors must be considered in designing an RO or NF system.

Scaling. The concentration factor and potential for scaling increase as recovery increases. The source feed water composition must be evaluated to estimate maximum operating recovery and the necessary pretreatment requirements (for example, pH adjustment or scale inhibitor addition).

Hydraulics. Optimal performance requires adhering to minimum concentrate and maximum feed flow conditions for membranes. Feed flow to the first element in a pressure vessel and concentrate flow from the last element in a pressure vessel must satisfy the manufacturer's stated requirements.

System design must provide adequate membrane concentrate flow. Concentrate staging of membranes and pressure vessels is typically used for recoveries greater than 50% to 60%. Two stages are commonly used for recoveries between 50% and 60% and between 75% and 85%; three stages are used for higher recoveries (up to about 90%). Some small systems are designed with concentrate recycle to produce flows above the minimum specified by the membrane manufacturer.

Source Water Use. The required volume of source feed water necessary to produce the same volume of permeate decreases as the recovery rate increases. Maximizing recovery rates minimizes both the source water requirement and the volume of concentrate generated.

Permeate Water Quality. Feed-concentrate average salinity increases as recovery increases. Because the flow of solutes through the membrane is a direct function of their concentration in the feed-concentrate stream, permeate quality decreases as recovery increases.

Solute Rejection and Solute Passage. The removal, rejection, or passage of solutes in a membrane system requires consideration of several variables.

Manufacturer's Specifications. RO and NF membranes are rated for nominal and minimum rejections based on a specific test condition. Each RO membrane manufacturer typically provides both design and minimum specifications relating to percent rejection for sodium chloride (NaCl). With NF modules, specifications are also given for selected divalent salts, for example, magnesium sulfate ($MgSO_4$), and possibly organics.

For example, low-pressure spiral-wound RO membrane elements are commonly rated at 225 psig (1,550 kPa) feed pressure; 25° C feed temperature; 8% to 15% recovery; 2,000 mg/L NaCl solution; and a pH of 5.7 to 7.0.

Seawater spiral-wound RO elements are typically rated at an 800 psig (5,520 kPa) feed pressure; a 25° C feed temperature; 7% to 10% recovery; a 32,000 to 35,000 mg/L NaCl solution; and a pH of 5.7 to 7.0.

Inorganic versus Organic Solutes. Both RO and NF membranes reject ionic and many nonvolatile organic solutes to a high degree. Composite membranes typically reject organic compounds better than do cellulose acetate or polyamide hollow fine-fiber membranes.

In general, volatile organic compounds are poorly rejected by all membrane types (less than 50%), although certain composite formulations have considerably higher rates. NF membranes reject multivalent ionic and many nonvolatile organic solutes to a high degree (90% or greater); monovalent solute rejection is much lower (75% or less) and is strongly dependent on the types of co-ions present in the feed and the feed pH.

Types of Membranes. Membrane composition may affect solute rejection over time. Composite membranes are generally stable and maintain their rejection properties over long periods of operation. Cellulosic membranes continuously undergo hydrolysis, which gradually diminishes their rejection properties. The extent of rejection loss depends on the rate of hydrolysis, but often the salt passage is assumed to double within 3 years of operation.

Increased temperature also increases the rate of cellulose acetate membrane hydrolysis. Cellulose acetate membrane module manufacturers typically recommend that feed pH be controlled between 5.0 and 6.2. The useful life of a membrane is also highly dependent on external (nonmembrane) factors associated with the application such as pretreatment operation, the type of cleaning chemicals, and frequency of cleaning.

Operating pH. Feed water pH may affect the rejection properties of composite membranes and the degree of rejection for certain ionic constituents. Composite membranes generally have an optimum pH at which rejection is maximized, although it is not always a published parameter. Rejection of certain ions such as fluoride and bicarbonate can vary with pH. As pH increases, fluoride and bicarbonate ion rejection also increase in many types of membranes.

Flux. For a system containing more than one stage, the flux value can be expressed as a system average or as individual values for each stage. The individual module flux rate within a system generally decreases from beginning to end. Lead modules operate at a higher flux rate than trailing modules because feed TDS (and osmotic pressure) is lowest and feed pressure highest at the lead end of the system.

System capacity is determined by the amount of membrane area provided (number of elements in the system) and the flux rate. The number of modules required can be calculated given an average membrane flux.

As flux rate increases, the loading rate of potential foulants at the membrane surface increases. Theoretically, there is an optimum flux rate for every water supply and membrane combination. In general, groundwaters allow higher design flux rates and surface water applications are restricted to lower design flux rates.

The water production capacity of a membrane module specified by the manufacturer, typically expressed in gallons per day per square foot (gal/day/ft^2), is based on a membrane flux achieved during laboratory testing at standard test conditions. Typically, manufacturer fluxes are 20 to 30 gal/day/ft^2 (0.034 to 0.051 m/h). Rarely, if ever, can this rate be achieved during system operation because of fouling constraints and the impact of recovery. Typical average system flux rates for groundwater are 15 to 18 gal/day/ft^2 (0.025 to 0.031 m/h) and for surface waters between 8 and 12 gal/day/ft^2 (0.014 and 0.020 m/h).

Low-pressure RO and NF systems are sometimes designed with flux balancing to better balance the flux rates of the various stages. This is an attempt to minimize fouling rates and extend the intervals between membrane cleanings, or to improve the overall hydraulics of the system. Where flux balancing is desired, either permeate backpressuring the first stages of the system or boosting the pressure of the interstage feeds can be used. These techniques lower transmembrane pressure of the first stages to limit their production, shifting a greater proportion of the production to the later stages.

Temperature. Water temperature is a factor that must be considered in all membrane systems.

The important effects of temperature on membrane system design are:

- Membrane operations using high-temperature waters require lower operating pressures to achieve a given flux, compared with operation on low-temperature waters.

- Membrane permeate quality degrades as water temperature increases and membrane flux is held constant (water flow through the membrane is constant and solute flow increases). Minimum temperature should be used to determine maximum anticipated feed pressure (and thus pump pressure and motor horsepower) at the design flux rate because the net driving pressure (NDP) is less at a higher temperature.

- If the design flux rate is held constant, maximum temperature dictates the worst permeate quality condition because solute passage is greater at higher temperatures.

All polymeric membranes have maximum operating temperatures. The use of high-temperature waters may exclude the use of cellulose acetate membrane, and if temperatures are too high, membrane feed water cooling will be required for all types of membranes.

Feed Pressure Requirement. The required feed pressure depends on the following:

- Membrane type
- Flux
- Recovery
- Osmotic pressure
- Temperature
- Permeate pressure
- Changes over time

All of the above points have been discussed previously except changes over time. To ensure that a system continues to produce the desired quantity and quality of permeate over the expected life of the membrane, the feed pump must provide sufficient membrane feed pressure throughout the design life of the system. The three main causes of time-related performance changes are membrane fouling and scaling, membrane compaction, and membrane degradation.

To illustrate the effect of these factors on water flow through the membrane, system productivity is typically plotted as a function of operating time (Figure 13.14), referred to as a flux decline curve. Flux decline is defined as the loss in system productivity, expressed as a percentage of initial productivity, that occurs with operating time, assuming constant feed water quality, pressure, temperature, and recovery. Flux decline causes an increase in required feed pressure. The principle causes of flux decline are membrane compaction and the effects of fouling and scaling that cannot be reversed by cleanings (referred to as irreversible fouling or scaling). Additional design pumping pressure, typically at least a 15% to 20% increase in available net driving pressure, must be included in the design to offset flux decline.

Membrane compaction occurs when the membrane compresses under the applied pressure of operation. The degree of compaction depends on the specific membrane type. This compression increases membrane resistance to water flow and increases NDP. A majority of compaction (and flux loss) occurs during the first 100 hours of operation. Membrane projections allow for decreases in productivity over a period of time, usually three to five years.

Fouling or scaling occurs when inorganic scales, suspended solids, organics, or biofilms collect on the membrane surface or the membrane module. Proper pretreatment system design should minimize membrane fouling. A membrane cleaning system should be included in the design to periodically clean and restore membrane performance.

Product (or permeate) pressure must be offset by feed water pressure to produce the proper net driving pressure. Feed pump requirements are minimized by designing the facility for minimum permeate backpressure. For some low-pressure RO and NF systems, a permeate backpressure valve is installed for the first stage to help balance flux rates throughout the system in an attempt to minimize fouling. Some membrane manufacturers specify maximum allowable permeate pressures for their products to prevent module damage.

Membrane Module Arrays and Staging. The arrangement of membrane modules used to achieve the desired process flow or optimum hydraulic configuration is called the membrane module array. Arrays are based on the number and location of pressure vessels (modules or permeators). For example, a 2:1 array would have two parallel pressure vessels feeding one. Depending on the quality of the feed water or permeate quality requirements,

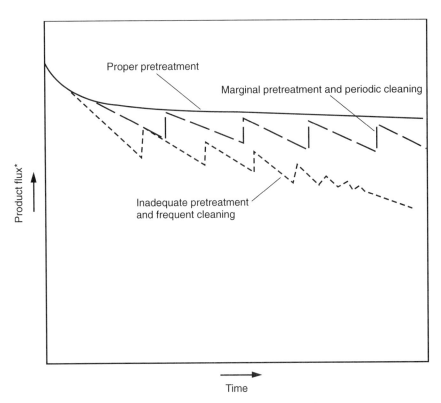

FIGURE 13.14 Membrane flux decline curves under different pretreatment conditions. *(Adapted from DuPont Permasep Products Engineering Manual, 1992.)*

RO and NF system arrays are either concentrate staged or permeate staged. Sometimes permeate staging is referred to as multiple "passes."

Concentrate-Staged Design. Standard 40 in. (102 cm) long spiral-wound RO or NF membrane elements have a maximum individual operating recovery of 8% to 15%. To achieve higher recoveries, elements are loaded in series into pressure vessels, with the concentrate flow from the first element becoming the feed flow to the second element, and so on.

In a typical large, spiral-wound system design, six or seven elements are loaded into each pressure vessel (referred to as a 6M or 7M vessel). Each 6M vessel in the design provides for an operating recovery of approximately 50%. In a concentrate-staged design (Figure 13.15) containing up to three stages, pressure vessels are arranged so that the concentrate flow from the first stage serves as the feed to the second stage, and the concentrate flow from the second stage serves as the feed to the third stage. In this flow configuration, concentrate staging maximizes system recovery.

To achieve 75% recovery, two parallel 6M vessels in the first stage can be arranged to feed a single vessel in the second stage (called a 2:1 array). If feed flow is 100 gpm (6.3 L/s) to the first stage using 6M vessels operating at 50% recovery, permeate flow will be 50 gpm (3.2 L/s) and the concentrate flow 25 gpm (1.6 L/s) from each per vessel, or a total of 50 gpm (3.2 L/s). This concentrate flow would then be fed into a single 6M vessel

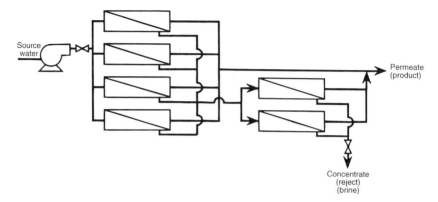

FIGURE 13.15 Concentrate staging (two-stage design shown).

in the second stage, where 50% recovery for this stage would yield a concentrate flow and a permeate flow of 25 gpm (1.6 L/s) each.

The total system produces 75 gpm (4.7 L/s) of permeate and 25 gpm (1.6 L/s) of concentrate. Commonly, two-stage and three-stage design using 6M vessels can yield recoveries of 75% and 85%, respectively. Designs using 7M vessels can achieve a recovery of 65% in one stage and 85% in two stages.

For hollow-fiber RO membrane modules (permeators), each pressure vessel contains a "bundle" of membrane fibers. Modules in each stage are placed in parallel with designed pressure loss in each module's concentrate outlet piping or tubing (before concentrate header piping). This pressure drop helps balance flows between each module. For one manufacturer (DuPont Company, 1992), a minimum pressure drop of 35 psi (240 kPa) is used for single-stage systems and the final stage of a multistage system. Typically, maximum recoveries up to 50%, 75%, and 90% are used for one-, two-, and three-stage hollow-fiber membrane systems, respectively.

Permeate-Staged Design. In applications in which the TDS of the feed water is too high to produce a permeate of sufficient quality with a single pass through the membrane, system design incorporates permeate staging (Figure 13.16). This design is common to RO

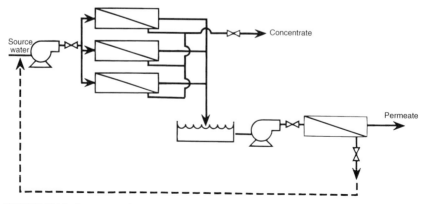

FIGURE 13.16 Permeate staging (multiple-pass design).

systems treating seawater with very high salinities (45,000 mg/L or greater) and when treating typical seawater salinity (35,000 mg/L), where permeate quality requirements are more stringent than typical drinking water standards. In this type of design, permeate from the first RO train (which can be concentrate staged) becomes feed water to the second RO train. The first-pass system uses seawater RO membranes, and the second-pass system uses brackish water membranes.

Normally, not all first stage permeate requires treatment in a second pass and some can bypass and be blended in finished water. This offers the advantage of reducing the size of the second pass. Concentrate from the second-pass system is generally recycled back to serve as feed water to the first-pass system.

Blending Membrane Permeate and Bypass Water. In cases in which feed water quality is such that membrane treatment provides a permeate with significantly better than needed finished water quality, it may be possible to blend the permeate with source or pretreated water. The advantages of blending are:

- Reduction of the required membrane capacity for a given finished water flow rate
- Reduced concentrate flow needing disposal for a given finished water flow rate
- Corresponding reduction in plant capital and O&M costs, including concentrate disposal costs
- Reduction in the amount of permeate posttreatment required for corrosion control

The opportunity for blending depends on the relative concentrations of critical constituents in both the membrane bypass and permeate relative to treated water goals and the composition of the membrane feed (bypass water).

The potential for blending can be determined using mass balance techniques for each product water quality constituent to determine the allowable blend ratios. Generally, the designer can easily determine the most critical constituents that need to be evaluated for a given application. The constituent requiring the permeate flow to be the highest percentage of the finished water flow is the limiting constituent and controls the overall allowable blending rate. The potential for blending can be calculated as follows:

$$Q_p/Q_{fin} = (C_{source} - C_{goal})/(C_{source} - C_p)$$

where Q_p = membrane permeate flow rate
Q_{fin} = finished water flow rate
C_{source} = concentration of solute in source (feed) water
C_{goal} = concentration of finished water (after blending) solute quality goal
C_p = concentration of solute in the permeate

The resulting bypass flow rate is:

$$Q_{byp} = Q_{fin} - Q_p$$

where Q_{byp} is the membrane bypass flow rate that is blended with the membrane permeate to form the finished water (before subsequent posttreatment processes).

The required source water flow rate can be calculated as follows:

$$Q_{source} = Q_{fin} \times [(BR/Y) + 1 - BR]$$

where Q_{source} = source water flow rate
BR = blend ratio = Q_p/Q_{fin}
Y = membrane system recovery = Q_p/Q_f
Q_{fin} = finished water flow rate

Using Membrane Computer Performance Projections for System Design. An RO or NF system is seldom designed by hand calculations. Most major membrane and system manufacturers, as well as some consultants, have developed computerized modeling programs that generate theoretical feed pressures and permeate quality at both initial and future operating conditions. However, designers must use caution because many projections are accurate for initial operating conditions but can be inaccurate in predicting future performance, particularly if the effects of site-specific feed water membrane foulants on membrane performance have not been determined. The following variables are required to predict system performance:

- Design feed water composition (including temperature)
- Process train capacity and desired recovery
- System array, number of elements per vessel, and membrane module type
- Fouling characteristics of feed water
- Membrane age (number of operating years)

Other Design Considerations. Depending on the type of membrane process and the specific application, the following additional items need to be considered.

Energy Recovery. For some applications, it is cost-effective to include energy recovery devices in the design. Such devices are often included when the potential for energy savings using a particular energy recovery device exceeds associated capital and maintenance costs for the device. Some commonly used energy-recovery devices are:

- Impulse turbine
- Integrated turbopump
- Turbocharger

RO and NF systems commonly include feed water pumps with adjustable (variable) frequency drives.

Automatic Flushing Systems. The membrane system design should include a means of automatically flushing membrane modules after planned or unplanned shutdown of the membrane train. The flushing system exists to remove from the modules concentrate that would otherwise remain if flushing were not performed. Flushing is particularly important when concentrate contains sparingly soluble constituents in excess of their theoretical solubility (supersaturated) that may precipitate and scale the modules. A scale inhibitor may be used to prevent this under normal operating conditions. However, in many cases the inhibitor will not permanently prevent scaling if concentrate remains in the modules for long periods of time. Flushing displaces the sparingly soluble components to eliminate this scaling potential.

Permeate is often used for flushing because it is generally the best-quality water (lowest in foulants and scalants) available on site, although in many cases membrane feed water is acceptable. Design should include provisions for storing a sufficient amount of permeate to flush the system and to displace concentrated water. If chlorine-intolerant composite membranes are used, only unchlorinated permeate should be stored and used for flushing. If cellulosic membrane modules are used, low-chlorine permeate is often preferred because of its ability to inhibit bacterial action during the downtime period. If multiple membrane trains are provided, a portion of total permeate flow can be diverted to flush an individual train and limit the amount of storage needed.

Permeate Drawback Tank. Hollow-fiber seawater RO systems typically have the permeate drawback tank piped such that, after feed pump shutdown, an immediate backflow of permeate water is available to flow back through the membranes by means of the natural osmotic pressure differential across the membranes. The water prevents potential damage from this "osmotic drawback."

Ultrafiltration and Microfiltration

MF and UF membrane water treatment system design should involve communication with manufacturers and others familiar with the products. Many systems have proprietary design features that the designer should know. MF and UF systems are relatively new to the municipal water treatment marketplace, and new products continue to become available. Facility design requirements can vary greatly between the commercially available MF and UF systems. The following design parameters, however, are applicable to most, if not all, membrane filtration systems.

Recovery. The recovery of a UF or MF system, typically called feed water recovery, is the system's final product volume over a given time period divided by the feed water flow volume. The quantity of source water or permeate/filtrate used for backwashing and flushing is considered in calculating feed water recovery. Typical recoveries for UF and MF systems range from 85% to greater than 95%. Overall recovery of a system can sometimes be improved by recycling backwash water after solids removal.

Solute Rejection. In MF and UF systems, solute or particle rejection is primarily a function of membrane type and pore-size rating. However, previously rejected substances remaining on the membrane surface also can act as a "dynamic membrane" and affect solute removal, especially for UF systems. Typical MF membrane pore-size ratings range from 0.1 to 0.5 μm. Commonly used UF membranes range from about 2,000 to 100,000 dalton molecular-weight-cutoff (MWCO).

Flux. The permeate or filtrate flux through UF and MF membranes depends significantly on transmembrane pressure and water temperature. Design flux rate used for a system is also strongly defined by membrane feed water quality because the rate of plugging and frequency of backwashing and cleaning are affected by the operating flux rate. Design flux rate for a specific application is often determined by bench or pilot testing. Typical flux rates for UF and MF membranes range from about 40 to 100 gal/day/ft^2 (0.068 to 0.170 m/h).

Temperature. Temperature affects the required driving pressure for a UF or MF system because of changes in feed water viscosity. For example, UF flux rate at 0° C can be as much as 35% less than at 20° C (Laine and Jacangelo, 1992).

Pumping Requirement. UF and MF systems may require up to three pumps: (1) feed pump, (2) recirculation pump, and (3) backwash pump.

The feed pump for a system is sized to meet the maximum required pressure and flow rate considering the membrane type, temperature, flux, piping and other pressure losses (separate from the transmembrane pressure), and the degree of fouling over time. Additionally, one major MF manufacturer with a proprietary gas backwash system uses an increased-capacity feed pump to provide the desired feed water flow rate for flushing out suspended solids removed from membranes by the gas backwash. Typical feed pressures for MF systems range from 15 to 40 psi (100 to 280 kPa) and UF systems range from about 20 to 75 psi (140 to 520 kPa).

If the design includes retentate stream recirculation (common in feed-and-bleed crossflow operating mode), recirculation pumps are sized to overcome pressure losses through the system while providing the desired recirculation flow rates, typically 3 to 6 times greater than the source water flow rate.

Backwash pumps, if required, are sized for the specific needs of the membrane system and commonly provide targeted flow to the filtrate or permeate side of the membranes at low pressure.

Membrane Module Arrays. MF and UF systems are typically designed using multiple parallel units. For large-capacity systems, multiple groups (each containing several membrane units) form modular arrays. Each array has common manifold piping to control service flows, backwashing, and cleaning-in-place (CIP).

Cross-Flow Velocity. A hollow-fiber membrane filtration system operating in an "inside-out" flow pattern often requires a specific minimum flow rate for sufficient cross-flow velocity to control fouling and to flush retentate from the fibers. Typical cross-flow velocities range up to 3 ft/s (0.9 m/s).

Backwashing. Depending on the specific membrane system, backwashing uses unchlorinated water, chlorinated water, or gas (e.g., air). Gas-backwashed systems first remove foulants from the membrane by transferring gas through the membrane from the filtrate to the feed side and then flushing the foulants to waste using feed flow. Backwash commonly is accomplished using relatively frequent (5 minutes to several hours) and short duration (3 to 180 seconds) backwashes, depending on the specific application and membrane system.

Cleaning System. MF and UF systems typically include chemical cleaning systems used for periodic major cleanings when routine backwashes are not adequate. Some membrane systems have manually initiated, automatically controlled chemical cleaning systems.

Membrane Integrity Tests. Most systems include instrumentation that continuously monitors filtrate quality and membrane performance. This instrumentation may include turbidity monitors, particle counters, and other devices. Some MF systems have components allowing periodic integrity testing using air, based on the bubble point theory. Some UF membrane systems can also use air for system integrity testing.

Electrodialysis and Electrodialysis Reversal

Electric current through an ED/EDR system can be calculated using Faraday's law (AWWA Electrodialysis and Electrodialysis Reversal, M38, 1995):

$$I = \frac{F \times F_d \times \Delta N}{e \times N}$$

where I = direct electrical current in amperes
F = Faraday's constant (96,500 amperes-seconds/equivalent/cell pair)
F_d = flow rate of demineralized stream through the membrane stack (L/s)
ΔN = change in normality of demineralized stream between the inlet and outlet of the membrane stack
e = current efficiency
N = number of cell pairs in the membrane stack

Electrical resistance can be calculated by Ohm's law:

$$R = E/I$$

where R = resistance (ohms)
 E = voltage (volts)
 I = current (amperes)

ED/EDR current efficiency is defined as follows:

$$e = \frac{F \times F_d \times \Delta N \times 100}{I \times N}$$

Recovery. Recovery of an ED/EDR system is often limited by the concentration of sparingly soluble salts, similar to RO and NF units. However, silica scaling is not a concern because silica is not concentrated (or removed) by ED/EDR. Pretreatment processes, such as those described previously for RO, can be used to control the concentration of sparingly soluble salts (for example, $CaCO_3$, $CaSO_4$, $SrSO_4$, $BaSO_4$) to allow higher recoveries. One major EDR manufacturer allows LSI values up to +2.2 and $CaSO_4$ concentrations up to 175% of saturation without chemical addition (Meller, 1984). Higher values are allowed by the manufacturer with appropriate antiscalant addition.

For EDR systems, recovery is lowered somewhat by discharge of concentrate off-spec water immediately after polarity reversal when the concentrate compartments become the dilute (demineralized) water compartments.

System recovery is also affected by electrical water transfer. Typically 0.5% of demineralized stream flow passes through to the concentrate stream per 1,000 mg/L TDS removed electrically (not hydraulically) along with ions (Meller, 1984).

Solute Removal. The following factors affect solute removal of an ED/EDR system.

Desalination Ratio. Salt removal in an ED/EDR system is directly proportional to current flow across the stack. Typically, TDS is concentrated about 3.0 to 3.6 times (i.e., the desalination ratio) in a single pass through the system (Wolfe, 1993). The desalination ratio varies with different membrane types and varying ions. For instance, the desalination ratio may be as high as 10 to 1 for nitrate and only 1.5 to 1 for sulfate using monovalent selective membranes (Wolfe, 1993).

Limiting Current Density. There is a limit to the rate that ions can be transferred through an ED/EDR membrane. Polarization occurs when too few ions are present to allow proper current flow and the resulting high electrical resistance causes water molecules to split. Typically, 70% of the limiting current density is used as the maximum design value.

Current Leakage. There is also a limit to the voltage that can be applied to an ED/EDR membrane stack. At that limit, excessive electric current leaks through an adjacent membrane to the concentrate stream, causing heat and resulting in damage. Maximum design voltage is typically 80% of the voltage where current leakage occurs.

Back Diffusion. If the concentrate is too concentrated, ions can diffuse (against the force of the DC potential) back to the demineralized stream. Typical design practice lowers membrane stack efficiency when the concentration of the concentrate exceeds 150% of the concentration in the demineralized stream.

Types of Membranes. A number of membrane products are available. Some membranes are designed to maximize specific ions, such as monovalent-specific membranes that do not concentrate calcium sulfate as much as conventional membrane. This may allow higher recovery operation, if calcium sulfate solubility is the limiting criteria.

Temperature. Temperature is one of the most important factors affecting ion removal. Ion removal increases about 2% per degree centigrade temperature rise.

Product Water Quality. Because electrical resistance increases as the demineralized stream becomes more and more dilute, ED/EDR systems often are designed to limit demineralized flow streams to a minimum of about 200 mg/L (although lower concentrations are possible).

Power Consumption. Power is required for pumping power and current flow across membrane stacks. The pumping power for a single-pass system can be estimated assuming a flow rate of about 1.3 times the feed flow rate (for dilute and concentrate) and the pressure loss through the stack at about 50 psi (340 kPa).

An estimate of the stack power requirement can be made from the following formula (Wolfe, 1993):

$$\text{Desalting kWh} / 1{,}000 \text{ gal} = \frac{F \times \Delta C \times 3.788}{1{,}000 \times \text{eq wt} \times e}$$

where F = Faraday constant, 26.8 amp-hr/equivalent
ΔC = ion concentration removal, mg/L
eq wt = equivalent weight for ion removed (58.5 g/eq for NaCl)
e = average current and rectifier efficiency (typically about 0.83)

Temperature and pH. Maximum water temperature limitation for ED/EDR stacks is generally about 45° C. Commonly used membranes can tolerate a pH range from 1 to 10.

Staging. ED/EDR systems have electrical stages and hydraulic stages, sometimes in a single stack. Multiple hydraulic stages within one stack are created using special interstage membranes designed to accommodate elevated pressure differentials. An electrical stage is created by using a pair of electrodes (anode and cathode).

Typically, 40% to 67% salt removal is possible in one hydraulic stage (American Water Works Association, 1995). Greater salt removals are accomplished using additional hydraulic stages in series. In an ED/EDR system, salt removal is directly proportional to electrical current and inversely to flow rate because of less available detention time. Because current flow can be regulated independently for each electrical stage, multiple hydraulic and electrical stages optimize hydraulic and electrical parameters.

Differential Pressure. Demineralized stream pressure is usually controlled to be about 0.5 to 1.0 psi (3 to 7 kPa) greater than concentrate stream pressure to ensure that any cross leakage moves toward the concentrate stream and does not lower product water quality (American Water Works Association, 1995). This differential pressure is maintained slightly positive.

Operating Mode. An ED/EDR membrane system can be designed for batch or continuous-flow operating mode. Municipal water treatment systems commonly use the continuous mode when the feed stream is demineralized and passes from the system as demineralized water. Some of the concentrate is recycled through the stack and some leaves as blowdown (feed-and-bleed design). Chemicals are sometimes added to the recycle stream for scale control.

Electrode Compartments. Electrodes are usually made of platinum-coated titanium. Chlorine gas, oxygen gas, and hydrogen ions are produced at the anode, creating an acidic condition. Hydrogen gas and hydroxide ions are generated at the cathode, raising the pH and increasing the conditions for scaling. Water from electrode compartments is usually mixed together and transferred to a degasifier before final disposal or possible recycle.

OTHER MEMBRANE PROCESS DESIGN CONSIDERATIONS

Other considerations necessary in membrane process design include the need for post-treatment, membrane cleaning, disposal of waste residuals, instrumentation and control, efficient O&M, and building design.

Posttreatment

Membrane product flow streams usually require some form of posttreatment before distribution. Posttreatment provides disinfection, corrosion control, and removal of dissolved gases and volatile compounds.

Posttreatment Disinfection. Most pathogens are removed by RO, NF, UF, and MF membranes, and microbiological quality of the permeate (or filtrate) is usually excellent. However, it is possible for microbes to pass through during the process when imperfections are present in the membrane system. One type of imperfection that could allow microbe passage is an O-ring leak or contamination before or during operation of the permeate side of the membrane module or permeate piping. In ED/EDR systems, the product water does not pass through a membrane barrier. Consequently, product water posttreatment usually includes disinfection, typically with chlorine.

Posttreatment disinfection design is similar to conventional water treatment plants and is described in Chapter 10.

Posttreatment for Corrosion Control. Corrosion control is not usually required for UF and MF processes. However, RO, ED/EDR, and to a lesser extent NF product streams are commonly corrosive because of the lack of calcium and alkalinity and the acidic pH. Posttreatment operation for corrosion control can include the following activities:

- Releasing carbon dioxide in a degasifier
- Adding caustic (with or without previous CO_2 addition), sodium bicarbonate, or sodium carbonate to increase bicarbonate alkalinity and pH
- Adding a corrosion inhibitor chemical
- Adding calcium chloride to increase calcium levels
- Adding lime to increase both calcium ion and alkalinity and pH (but this may cause unacceptable turbidity)
- A combination of these processes

Selecting the most appropriate posttreatment method for corrosion control should be site specific and depend heavily on membrane product water quality.

Removal of Gases and Volatile Organic Compounds. Membrane processes do not remove dissolved gases and, in general, provide poor removal of most volatile organic compounds (VOCs). Therefore these components must be removed during permeate posttreatment.

Hydrogen sulfide (if present in the source water) and carbon dioxide are usually the predominant gases present in membrane product water. Sulfide is present primarily in shallow groundwater source waters, and carbon dioxide can be present in groundwaters or generated by feed water acidification. Unless the permeate has sufficient alkalinity (which is generally not the case), carbon dioxide removal is not desirable because the gas can be used as a source of bicarbonate alkalinity for corrosion control. VOCs are often a problem when treating groundwaters contaminated by industrial processes.

The most common method of treatment is the use of a degasifier, commonly a packed tower with a blower. Tower design should be a function of the critical gas or VOC present in the product and the degree of removal required. When carbon dioxide and hydrogen sulfide are both present, sulfide removal usually controls the degasifier design.

If the pH of the permeate is greater than 6.5, hydrogen sulfide removal may require that the permeate be acidified before degasification. Acidification ensures that a large proportion of the sulfide is present in the gaseous form, in contrast to sulfide ions. Degasifier off-gases containing sulfides may also require scrubbing to minimize odor and corrosion problems and for safety reasons. Scrubbing is typically achieved by conveying the gas through a caustic or chlorine solution to convert the sulfide to an ionized form. Degasification for VOC removal may also require off-gas treatment with granular activated carbon (GAC), depending on state or local air quality regulations.

Membrane Cleaning System

With few exceptions, all membrane systems are subject to fouling by one or more source water components and therefore require periodic cleaning. Membrane cleaning is usually performed without removing membranes from the pressure vessels or the system (i.e., cleaned in place). A cleaning system is designed to prepare and recirculate chemical solutions through some or all membrane modules at low pressure.

The cleaning system can also be used to feed special membrane posttreatment chemicals (not to be confused with membrane system posttreatment) that sometimes are used to improve membrane performance. The cleaning system also serves to prepare and transfer membrane storage solutions, or preservatives, used to prevent microbial growth and in some cases to prevent freezing when the membrane system is shut down for extended periods, typically more than a week.

The cleaning system for an RO or NF system should be designed to accommodate all cleaning and membrane storage solutions expected to be used at the plant. Cleaning systems typically consist of the following basic components:

- Tanks with mixers for holding cleaning solutions.
- A pump providing sufficient head to circulate the solution from the tank through the membrane system and back to the tank.
- A cartridge filter to intercept any suspended solids present in the cleaning solution and prevent them from clogging the module flow passages.
- Piping and valves to transfer the solution to and from the RO train feed, permeate, and concentrate headers. The design should consider the use of removable piping spools or hoses to provide positive air-break in the cleaning system connection points. An air-break prevents chemical solutions from entering the product header piping and being transported to downstream product facilities when the membrane system is being put back into service.
- Instrumentation and control for proper system operation, including tank level, temperature switches, and a pH monitor.
- A cleaning solution heating device (not always required), such as an immersion heater, to increase solution temperature and improve the efficiency of cleaning, particularly when cleaning with detergents.
- A cleaning solution cooling device (not always required), such as a heat exchanger external to the cleaning solution tank or cooling coils inside the tank, if it is determined that the maximum solution temperature needs to be controlled. The need for cooling increases when the cleaning solution makeup water is at relatively high temperature or the heat energy expected to be transferred to the cleaning solution from the cleaning pump while recirculating is too great.

Typical design allows for separate cleaning of each stage of a multistage train. With large-capacity trains, individual sections of a stage can be designed to allow separate cleaning, minimizing cleaning tank volume and pumping requirements.

Residuals Disposal

The designer must consider the least expensive, acceptable method of disposing of wastes generated in both pretreatment processes and residuals resulting from membrane treatment processes.

Concentrate from membrane processes is typically considered a waste stream that must be disposed of in accordance with applicable local discharge regulations. Because these regulations vary considerably from one geographical location to another, it is not practical to provide more than general information on the proper methods of concentrate disposal for a certain treatment. Disposal alternatives are easier for MF and UF systems, which do not concentrate salts, than for RO, NF, and ED/EDR systems. Potential concentrate disposal options for RO, NF, and ED/EDR include (Mickley et al., 1993):

- Discharge to surface water
- Land application (e.g., irrigation, possibly after blending with other low-TDS waters)
- Injection wells
- Evaporation ponds
- Evaporators
- Wastewater collection system
- Wastewater treatment plant effluent

Spent cleaning and membrane storage solutions are typically disposed to a sanitary sewer system after pH neutralization. Often spent solutions are transferred to storage tanks with provisions for adding acid and caustic to neutralize low- and high-pH cleaning solutions. Neutralized spent solution is then transferred to the disposal location.

Instrumentation and Control

Instrumentation and control (I&C) facilities vary greatly depending on membrane process type, application, and degree of automatic monitoring and control desired. Some membrane systems are fully automatic for normal operation and are designed to be monitored remotely by means of modems connected to computer-based controls.

A minimum of I&C equipment is necessary for all systems to protect equipment from damage. For example, low-suction pressure switches or other devices automatically shut off the system on loss of adequate flow. It is also common to provide on-line instrumentation to measure feed and product water qualities.

If source water temperature and turbidity are relatively variable, the membrane system should be provided with instrumentation to continuously measure water temperature and feed water turbidity. Product water turbidimeters or particle counters usually monitor UF and MF systems. For RO, NF, and ED/EDR systems, conductivity monitors installed on the feed and product flow stream monitor salt removal performance. Where acid is fed for feed water pH adjustment, an on-line pH meter with high/low alarms is commonly used. ED/EDR systems typically provide for pH measurement of the feed, product, and concentrate flow streams.

Pressure-driven membrane systems typically have either feed and product or product and concentrate flowmeters. Pressure gauges indicate all critical pressures and pressure differentials, such as membrane feed, interstage, concentrate, permeate, and, unless determined solely by calculations, differential pressure across each stage.

ED/EDR systems typically have flowmeters for the dilute-in, concentrate-makeup, and electrode flow. Pressure gauges are usually provided to measure inlet and outlet pressures of the dilute and concentrate flow streams, as well as the pressure of the electrode flow stream of each membrane stack. Additionally, ED/EDR systems continuously monitor current flow and voltage; for EDR systems, measurements are needed for each electrical polarity.

Elapsed time meters provided on the power supply to the membrane system track accumulated run time. Sampling stations should be provided at all important process locations in the piping system for periodic sampling and analyses.

Most drinking water membrane systems designed today have programmable logic controller (PLC) based controls. These systems can be designed to control flows, pressures, and other parameters. Many RO and NF systems have controls to provide automatic flushing upon shutdown, and most MF and UF systems have automatically controlled backwash/backflush facilities. Many large RO and NF plants are designed with computer systems that compute normalized performance data for operation monitoring and determine when to perform membrane cleanings.

Operations and Maintenance (O&M) Design Issues

Membrane systems require proper O&M to perform according to design and to extend membrane life. Depending on the type of membrane system and application, O&M requirements vary greatly. As a minimum, operations staff should monitor and record all critical parameters, operation incidents (such as feed water quality upsets and shutdowns), and maintenance performed. The membrane plant designer should provide all needed instrumentation and representative sampling points for operations staff.

System design should consider adequate spacing and access for equipment maintenance, as well as devices for lifting and moving equipment. Proper storage facilities should be provided for membranes, chemicals, and other equipment replacements.

Membrane Process Building Design

Floor plans of membrane process buildings vary greatly depending on the type of membrane process, the site-specific conditions and design constraints, and the number and types of nonprocess areas included. Process areas include the following pieces of equipment, depending on the membrane process and application:

- Pretreatment filters, such as cartridge filters, basket strainers, and bag filters
- Chemical feed systems for pretreatment and posttreatment chemicals
- Membrane feed pumps
- Membrane treatment units
- Membrane cleaning and flushing systems
- Posttreatment equipment
- High service pumps
- Holding tanks (where applicable) such as feed and spent backwash water tanks for a package MF system
- Electrical switch gear and motor control center
- Emergency generator

Typical nonprocess areas commonly incorporated into building designs are as follows:

- Control room
- Laboratory
- Chemical storage area
- Maintenance shop
- Spare parts storage room
- Rest rooms and locker rooms
- Offices
- Mechanical room for heating, ventilation, and air conditioning equipment
- Lunch/break room
- Training room

Particular attention should be given to the size and location of doorways. In some cases removable wall panels should be provided to allow large equipment to be removed and installed.

Materials of Construction

It is particularly critical to use proper construction materials in membrane systems because corrosion products can foul membranes, resulting in increased cleaning frequency and shortened membrane life. In addition, process flow streams and chemicals used are commonly highly corrosive, and, for pressure-driven desalting systems, high pressures are often required.

Where pressures and temperatures allow, nonmetallic (PVC, CPVC, FRP, etc.) piping and valves are commonly used. Nonmetallic piping is typical throughout MF, UF, ED/EDR, and in some cases NF systems. For applications requiring higher pressures and temperatures, such as most RO systems, appropriately selected stainless steels are typical. For example, a 5,000 mg/L TDS brackish groundwater RO system commonly has the following components (other available material options can also be considered):

- Piping from supply well to pretreatment cartridge filter—PVC or FRP
- Cartridge filter housing—316L stainless steel or FRP
- Piping from cartridge filter to membrane feed pump suction connection—PVC or FRP
- Membrane feed pump discharge piping—316L stainless steel
- RO pressure vessels—FRP
- RO membrane unit interstage and concentrate lines (up to and including the concentrate control valve)—316L stainless steel
- Concentrate line downstream of the concentrate control valve—PVC or FRP
- Permeate piping—PVC or FRP

For lower-salinity applications, 304L stainless steel is common for high-pressure piping. For seawater RO system high-pressure lines, 316L stainless steel piping is considered minimum. Higher-alloy stainless steels, with a molybdenum content of 6% or more, are also commonly used. Alternatively, polyethylene-lined carbon steel piping has been used for seawater RO high-pressure piping, although piping failures have been reported when the lining was improperly manufactured or damaged after manufacturing.

Because many membrane systems require concentrated chemicals, storage and handling systems must be designed with operator safety in mind.

Process Reliability and Redundancy Issues

In designing membrane systems the same reliability and redundancy issues considered for conventional treatment facilities should be evaluated. Membrane treatment systems have been used for water treatment for over 30 years, and when properly manufactured, installed, operated, and maintained, the systems are very reliable.

Membrane treatment facilities commonly have multiple parallel process trains. Where required, additional membrane units can be installed to provide needed redundant capacity when one or more units are out of service for maintenance or membrane cleaning. In many cases, particularly where the membrane system provides expansion to an existing water supply system, only critical system components are provided with redundancy, rather than the entire membrane unit. This not only minimizes the financial investment, but also allows the membrane system to operate as continuously as possible. Because there is no complete standby membrane unit, the need to prepare and feed membrane storage solutions is minimized, as well as other advantages.

If a product water quality problem is detected in a membrane system, the specific membrane modules within the system having the problem can be identified for remedial action. With many membrane systems, the deficient membrane modules can be isolated and the rest of the system operated until there is a convenient time for maintenance.

Bench and Pilot Testing

Bench testing is often performed in conjunction with a feasibility study to determine which, if any, membrane process is applicable to treatment requirements. If a particular process has been selected, bench testing results can develop basic characterization data on the process. Bench testing is typically 1 day to 2 weeks long and uses a simple membrane characterization apparatus consisting of a feed pump; a pressure assembly containing a flat-sheet, hollow-fiber, or tubular membrane; and valves to set flow rates, pressures, and recovery. Typical operating data derived during bench testing include initial membrane feed pressure to produce a given flux and initial permeate quality (representative of low-recovery operation) at a given membrane flux.

Compared with bench testing, membrane pilot testing is lengthy, often taking 1 to 4 months of operation. Test duration is generally governed by the amount of time the proposed membrane and pretreatment systems must be operated to correctly quantify longer-term trends in membrane performance, such as flux decline rate and rate of change of solute transport.

Bench and pilot testing both have limitations that the designer must consider. For example, source water may not be representative of water quality over the life of the membrane system. The designer should consider extra allowances, or safety factors, for critical design parameters.

Designer and Vendor Interface Issues

Many membrane products are available for water treatment applications, and new products continue to enter the marketplace. It is important that the membrane facility designer contact membrane manufacturers and system suppliers to identify the best candidate membrane products for the application and to identify all critical design issues applicable to each product. This not only minimizes potential problems with the constructed facility and allows the best candidate membranes to be considered, but it also provides for the designer the benefit (directly or indirectly) of the latest computer design software to minimize design calculation time.

BIBLIOGRAPHY

"Alternative Water Supplies—Desalting and Reuse." *Proceedings of AWWA 1992 Annual Conference.* Vancouver, B.C. Denver Colo.: American Water Works Association, 1992.

American Water Works Association. *Water Quality and Treatment.* 4th ed. New York: McGraw-Hill, 1990.

American Water Works Association. *Proceedings 1991 Membrane Technology Conference.* Orlando, Fla. Denver, Colo.: American Water Works Association, 1991.

American Water Works Association. *Proceedings 1993 Membrane Technology Conference.* Baltimore, Md. Denver, Colo.: American Water Works Association, 1993.

American Water Works Association. *Electrodialysis and Electrodialysis Reversal, M38.* Denver, Colo.: American Water Works Association, 1995.

American Water Works Association Research Foundation, Lyonnaise des Eaux, and Water Research Commission of South Africa. *Water Treatment Membrane Processes.* New York: McGraw-Hill, 1996.

AWWA Water Desalting and Reuse Committee Report. "Membrane Desalting Technologies." *Journal AWWA* 81(11), 1989.

Applegate, L. E. "Membrane Separation Processes." *Chemical Engineering* (6):64, 1984.

Applegate, L. E. "Posttreatment of Reverse Osmosis Product Waters." *Journal AWWA* 78(5):59, 1986.

Baker, R. W., et al. *Membrane Separation Systems. A Research Needs Assessment.* Final Report by the Department of Energy Membrane Separation Systems Research Needs Assessment Group. Springfield, Va.: National Technical Information Service NTIS-PR-360, U.S. Department of Commerce, 1990.

Bergman, R. A. "Anatomy of Pressure-Driven Membrane Desalination Systems." *Proceedings AWWA 1993 Annual Conference—Engineering and Operations.* San Antonio, Tex. Denver Colo.: American Water Works Association, 1993.

Bergman, R. A., and Lozier, J. "Membrane Process Selection and the Use of Bench and Pilot Tests." *Proceedings of the Membrane Technology Conference.* Baltimore, Md. Denver, Colo.: American Water Works Association, 1993.

Buros, O. K., et al. *The U.S. Agency for International Development Desalination Manual.* Produced by CH2M Hill International, Denver, Colo.: 1980.

Du Pont Company. *PEM Permasep Products Engineering Manual.* Wilmington, Del.: Du Pont Company, 1992.

"Engineering Considerations for Design of Membrane Filtration Facilities for Drinking Water Supplies." *Proceedings of AWWA 1992 Annual Conference.* Vancouver, B.C. Denver, Colo.: American Water Works Association, 1992.

Jacangelo, J. G., N. L. Patania, and J. M. Laine. *Low Pressure Membrane Filtration for Particle Removal.* Denver, Colo.: AWWA Research Foundation, American Water Works Association, 1992.

James M. Montgomery, Consulting Engineers, Inc. *Water Treatment Principles and Design.* New York: John Wiley, 1985.

Katz, W. "Electrodialysis for Low TDS Waters." *Industrial Water Engineering.* June/July 1971.

Laine, J. M., and J. G. Jacangelo. "System Components and Process Design Considerations for Ultrafiltration of Untreated Drinking Water Supplies." *Proceedings AWWA 1992 Annual Conference—Engineering and Operations.* Vancouver, B.C. Denver, Colo.: American Water Works Association, 1992.

Meller, F. H., editor. *Electrodialysis (ED) and Electrodialysis Reversal (EDR) Technology.* Watertown, Mass: Ionics, Inc., 1984.

Mickley, Mike, R. Hamilton, L. Gallegos, and J. Truesdall. *Membrane Concentrate Disposal.* Denver, Colo.: AWWA Research Foundation and American Water Works Association, 1993.

Sourirajan, S. *Reverse Osmosis.* New York: Academic Press, 1970.

Sourirajan, S., and T. Matsuura. *Reverse Osmosis/Ultrafiltration Process Principles.* Ottawa, Ontario: Division of Chemistry, NRC Canada, 1985.

Taylor, J. S., S. J. Duranceau, W. M. Barrett, and J. F. Goigel. *Assessment of Potable Water Membrane Applications and Research Needs.* Denver, Colo.: AWWA Research Foundation and American Water Works Association, 1989.

U.S. Congress, Office of Technology Assessment. *Using Desalination Technologies for Water Treatment.* OTA-BP-O-46. Washington, D.C.: U.S. Government Printing Office, 1988.

U.S. Environmental Protection Agency. *ICR Manual for Bench- and Pilot-Scale Treatment Systems.* Cincinnati, Ohio: Technical Support Division, Office of Ground Water and Drinking Water, U.S. Environmental Protection Agency, 1996.

Wale, R. T., and P. E. Johnson. "Microfiltration Principles and Applications." *Proceedings AWWA 1993 Membrane Technology Conference.* Baltimore, Md. Denver, Colo.: American Water Works Association, 1993.

"Water Desalting: a Viable Technology for Water Supply." *Proceedings of AWWA 1993 Annual Conference.* San Antonio, Tex. Denver, Colo.: American Water Works Association, 1993.

"What Membranes Can Do for You?" Seminar S5. AWWA 1993 Annual Conference. San Antonio, Tex.

Wolfe, T. D. "Electrodialysis Design Approaches." *Proceedings AWWA 1993 Membrane Technology Conference.* Baltimore, Md. Denver, Colo.: American Water Works Association, 1993.

CHAPTER 14
ACTIVATED CARBON PROCESSES

Activated carbon is an adsorbent material that provides a surface on which ions or molecules in the liquid or gaseous phase can concentrate. In drinking water treatment, activated carbon adsorption is used to remove natural organic compounds, taste and odor compounds, and synthetic organic chemicals from source water.

CHARACTERISTICS OF ACTIVATED CARBON

Activated carbon has a random structure that is highly porous, with a broad range of pore sizes ranging from visible cracks and crevices down to molecular dimensions. Intermolecular attractions in the smallest pores create adsorption forces. These forces cause large and small molecules of dissolved contaminants to be condensed and precipitated from solution into the molecular scale pores. Activated carbon is an effective adsorbent because it provides a large surface area on which the contaminant chemicals can adhere.

Activated carbon is available in two different forms: powdered and granular. Powdered activated carbon (PAC) is added to water, mixed for a short period of time, and removed. Adsorption of molecules occurs while the PAC is in contact with the water. PAC is usually added early in the treatment process and then either settles out with the floc in the pretreatment basins or is removed from the filter beds during backwashing.

Granular activated carbon (GAC) is generally used in beds or tanks through which the water passes for treatment. As GAC is used for treatment, surfaces within the pores gradually become covered with chemical molecules until the carbon is no longer able to adsorb new molecules. At that point, the old, or spent, carbon must be replaced with new, virgin, or fresh, reactivated carbon. The capacity of spent GAC can be restored by thermal reactivation.

Adsorptive properties of GAC and PAC are fundamentally the same because they depend on pore size, the internal surface area of the pores, and surface properties independent of overall particle size. Each brand of commercially available PAC or GAC has properties making it most suitable for particular applications. Besides adsorptive capacity and selectivity in removal, these properties include the ability to withstand thermal reactivation and resistance to attrition losses during transport and handling.

The use of activated carbon in water treatment in the United States has been limited primarily to removing taste- and odor-causing compounds, pesticides, and other organic contaminants. In both Europe and the United States, PAC is typically fed as a powder using dry feed equipment or is batched as a slurry and fed with metering pumps or rotodip feeders. In conventional treatment plants, PAC is usually added to the plant influent, the filter influent, or both.

One method of using GAC is to install it as a partial or complete replacement for the conventional granular media in conventional gravity filters. In this case, GAC acts as a filtering medium and provides limited adsorbance as water passes through the filters. The other common type of installation is to pass the water through tanks filled with GAC as the last step in the treatment process.

Physical Properties of Activated Carbon

Surface area within porous carbon structure provides the capacity to adsorb dissolved organic materials such as natural organic matter (NOM), disinfection by-products (DBPs), or taste and odor compounds. Carbons used for adsorption in drinking water applications have a minimum surface area of 73 acres/lb (650 m^2/g), with typical surface areas on the order of 112 acres/lb (1,000 m^2/g). A series of tests developed for testing the suitability of various types and brands of activated carbon are detailed in American Water Works Association (AWWA) Standard B604, Granular Activated Carbon and B600 for Powdered Activated Carbon.

Iodine Number and Molasses Number. The quantity of small and large pore volumes in a sample of activated carbon is described by the iodine number and molasses number. Adsorption tests are also used to approximate the distribution of pores available for adsorption. Using standard reference adsorbates (such as iodine and a molasses solution) allows the activity characteristics of different carbons to be compared. Iodine's small molecular size can characterize the small pore volume of a carbon and its ability to adsorb contaminants small in molecular size. The iodine number is the mass of iodine adsorbed (in mg) from a 0.02 N bulk solution by 1g of carbon. The molasses number is a measurement comparing the color (optical density) of the filtrate from a standard activated carbon with the color of the filtrate from the carbon being investigated.

When activated carbon is in use, the iodine and molasses numbers decrease with time as adsorption occurs and available adsorption sites are filled. Some water treatment systems use iodine or molasses numbers to determine when to replace carbon. To use the iodine number as a surrogate measure (replacing actual bench-scale testing of the GAC with the water needing treatment), a correlation between the iodine number and the degree of carbon exhaustion should be developed for each specific adsorption application.

The AWWA standard indicates a performance requirement that the adsorptive capacity of the granular activated carbon, as measured by the iodine number, shall not be less than 500 mg/g carbon. The procedure for determining the iodine number of activated carbon is ASTM D4607.

Carbon Weight. The weight of carbon in air is the apparent density. For water treatment applications, carbon density is described by bulk density or as backwashed and drained. For a filter of known dimensions, density determines the weight (or mass) of GAC required to fill that filter. Typically, carbon density should be determined on an as-received basis, and calculations can then be made to correct for moisture content. The apparent-density test apparatus and the procedure for determining the apparent density are described in the AWWA standard.

Apparent density is important because, for new GAC systems, the initial quantity of activated carbon is typically specified either by volume or by weight. Volume (cubic feet or cubic meters) for first-time installation is specified as backwashed, drained, and in place.

After the initial installation, additional GAC needed to replace lost carbon may be specified by volume or by weight. The standard specifies that the apparent density of GAC shall be at least 0.25 g/cc.

Moisture Content. The AWWA standard indicates that moisture content of GAC shall not exceed 8% by weight as packaged or at the time of shipment in the case of bulk shipments.

Abrasion Resistance. Abrasion resistance is a property describing carbon durability. Activated carbon is exposed to abrasion during shipping, installation, backwashing, and regeneration. If the carbon is not durable enough, abrasion can generate undesirable fines or crushed carbon. Increased fines can result in increased head losses across the filter bed, increased loss of carbon, or degradation of water quality. Abrasion resistance is expressed in terms of the abrasion number; the greater the number, the more resistant the carbon is to abrasion.

Durability. Although the industry has not yet agreed on a standard test for predicting carbon durability, the AWWA standard suggests two tests. The stirring abrasion test is recommended for lignite-based GAC, and the Ro-Tap abrasion test is recommended for bituminous-based GAC. For either test, the retention of average-sized GAC shall be at least 70%.

Ash Content. Ash content reflects the purity of the carbon. In the United States, most activated carbons are manufactured from coal. Higher-quality coals, such as metallurgical-grade bituminous coals, produce carbons with ash contents of approximately 5% to 8%. Subbituminous coals produce carbons with ash contents of approximately 10% to 15%. Lower-grade coals, such as lignite, produce activated carbons with the highest ash content of approximately 20%. Ash found in these coal-based carbons can contain calcium, magnesium, iron, and silica. These constituents can form precipitates in areas with hard water supplies. According to AWWA Standard B604, water-soluble ash in GAC should not exceed 4%.

GAC Particle Size. The particle size of GAC used in a filter affects pressure drop, filtration abilities, requirements for backwash rate, and the rate at which adsorption equilibrium is reached. Smaller particle sizes increase the pressure drop across the carbon bed and necessitate lower backwash rates. However, smaller GAC particles reach equilibrium more rapidly than large particles because of the smaller (shorter) distance organics must diffuse to reach the center of the particle. The larger surface area–to–volume ratio of smaller GAC particles also allows adsorption equilibrium to be reached more quickly. Mesh size describes the range of particle sizes to be used in a filter.

The effective size of GAC is defined in the AWWA standard as the size opening through which only 10% of a sample of representative filter material will pass. For example, if the size distribution of the media grains has 10% finer than 0.600, the effective size of the GAC is 0.600 mm.

The uniformity coefficient is a ratio of the size opening that will pass 60% of a representative sample of the filter material to the size opening that will pass just 10% of the same sample. The maximum uniformity coefficient recommended by AWWA is 2.1 after backwashing and draining a GAC filter. If two types of GAC media have the same effective size, the one with a larger uniformity coefficient (i.e., 1.7 to 2.4) will have a greater number of fines in the upper layers of a stratified bed and larger particles in the lower layers. Typical uniformity coefficients for GAC produced in the United States are less than or equal to 1.9.

Manufacture of Activated Carbon

Activated carbon is made from organic materials having a high carbon content. A wood product such as sawdust, coconut shells, or wood may be used, or a coal product such as bituminous coal, lignite, or peat is converted to activated carbon by heating the material to between 300° and 1,000° C. The resulting carbon material provides large surface areas and, accordingly, a large number of adsorption sites.

DESIGN OF POWDERED ACTIVATED CARBON FACILITIES

PAC is used by surface water treatment plants either full time or as needed for taste and odor control or removal of organic chemicals.

Characteristics of PAC

PAC is made from a variety of materials, including wood, lignite, and coal. The apparent density of PAC ranges from 23 to 46 lb/ft^3 (0.36 to 0.74 g/cm^3) depending on the type of material and the manufacturing process. Iodine and molasses numbers are often used to characterize PAC. Specifications for some of the commercially available PACs are summarized in Table 14.1. The AWWA standard for PAC specifies a minimum iodine number of 500.

Smaller PAC particles adsorb organic compounds more rapidly than larger particles, so PAC is a very fine powder. Typically, 65% to 95% of commercially available PAC passes through a 325-mesh (44 μm) sieve.

Selecting a PAC Application Point

PAC is primarily used to control tastes and odors. The following points should be considered when selecting a PAC application point:

TABLE 14.1 Manufacturers' Specifications of Some Commercially Available PACs

Parameter	PAC 1[a]	PAC 2[b]	PAC 3[c]	PAC 4[d]	PAC 5[e]	PAC 6[f]
Iodine number—mg/g	800	1,199	600	900	1,000	550
Molasses decolorizing index	9	—	—	14	18	—
Moisture as packed—percent	5	3	5	10	10	4
Apparent density—g/cm^3	0.64	0.54	0.74	0.38	0.38	0.50
Ash content—percent	—	6	—	3 to 5	3 to 5	—
Passing 100 mesh—percent	99	—	99	95 to 100	95 to 100	99
Passing 200 mesh—percent	97	—	97	85 to 95	85 to 95	95
Passing 325 mesh—percent	90	98	90	65 to 85	65 to 85	90

[a] Aqua-Nuchar, Westvaco, Covington, W. Va.
[b] WPH, Calgon Corp., Pittsburgh, Pa.
[c] Aqua, Westvaco, Covington, W. Va.
[d] Nuchar S-A, Westvaco, Covington, W. Va.
[e] Nuchar SA-20, Westvaco, Covington, W.Va.
[f] Hydrodarco B, American Noril, Jacksonville, Fla.

- The contact time between PAC and organics in the source water is important and depends on the ability of the carbon to remain in suspension. Providing a minimum of 15 minutes of contact time is generally sufficient for most taste and odor compounds. Considerably longer contact times may be needed for 2-methylisoborneol (MIB) geosmin removal.
- The surfaces of PAC particles should not be coated with coagulants or other water treatment chemicals before the PAC has had adequate contact time with the source water.
- PAC should not be added concurrently with chlorine or potassium permanganate because PAC will adsorb these chemicals. The best location for PAC addition is usually at the head of the plant, either in the source water pipeline or in a basin dedicated to rapid or flash mixing chemicals. Applying PAC at the earliest point in the treatment process allows the longest contact time possible before the application of other chemicals. If PAC must be added later in the treatment process, dosages may be higher to account for shorter contact times and interference by other chemicals, such as coagulants and chlorine.

If PAC is added to the sedimentation basin effluent or filter influent, particular care must be taken in filter operation. Because of its small carbon particle size, PAC can pass through a filter and cause black water complaints from consumers. An approach used by some systems is to add carbon at two or more application points, with part of the carbon added to the source water and smaller doses added before filtration to remove remaining taste- and odor-causing compounds. Table 14.2 summarizes some of the important advantages and disadvantages of PAC addition at each of these points.

PAC dosage depends on the type and concentration of organic compounds present. Common dosages range from 2 to 20 mg/L for nominal taste and odor control, but doses can go as high as 100 mg/L to handle severe taste and odor episodes or spills of organic chemicals.

TABLE 14.2 Advantages and Disadvantages of Different PAC Application Points

Point of addition	Advantages	Disadvantages
Intake	Long contact time, good mixing	Some substances may adsorb that otherwise would be removed by coagulation, thus increasing the activated carbon usage rate
Slurry contactor preceding rapid mix	Excellent mixing for the design contact time; no interference by coagulants; additional contact time possible during flocculation and sedimentation	A new basin and mixer may have to be installed; some competition may occur from molecules that otherwise would be removed by coagulants
Rapid mix	Good mixing during rapid mix and flocculation; reasonable contact time	Possible reduction in rate of adsorption because of interference by coagulants; contact time may be too short for equilibrium to be reached for some contaminants; some competition may occur from molecules that otherwise would be removed by coagulation
Filter inlet	Efficient use of PAC	Possible loss of PAC through the filters and into the distribution system

Methods of Applying Powdered Carbon

Although some treatment plants feed PAC continuously at a low dosage, a far greater number of plants keep a supply available for use during emergencies, such as organic pollution, that may last only a day or so. Other plants may apply PAC for only a few weeks out of the year during periods of poor source water quality, such as during algae blooms. A major dilemma facing the designer is how to provide a PAC feed and storage system to meet the full range of feed rates. PAC system design must also deal with how best to provide for a chemical that is dirty, difficult to handle, and difficult to store for long periods, but is needed on a moment's notice.

The designer must first consider the specific needs of the owner and develop a PAC feed strategy. The following questions should be asked before beginning the design:

- How often, over the course of a typical year, is carbon needed?
- What is a reasonable range of PAC feed rates to meet historical source water quality characteristics?
- What is the worst case scenario for PAC feed?
- How quickly can a shipment of PAC be delivered to the plant site?

A PAC system must be ready when needed, quick to respond, and flexible over a wide range of needs.

Powdered Carbon Application Equipment

Powdered activated carbon can be fed as a powder using dry feed equipment or as a slurry using metering pumps or rotodip feeders. A dry feed system should be considered if PAC feed is infrequent and the maximum feed rate is less than a few hundred pounds per hour. The feed system usually includes a bag-loading hopper, an extension hopper, a dust collector, a dissolving tank or vortex mixer tank, and an eductor.

A much better way to feed PAC is to mix it with water to form a slurry, but if PAC is not required regularly, the slurry tank either sits empty or full. If the tank is full, the carbon and water mixture must be kept continuously stirred to keep the carbon in suspension. If the tank is kept empty with the intent of mixing slurry when it is needed, the taste-and-odor episode may be over before the slurry system is operable.

A slurry system (Figure 14.1) should be considered if PAC is frequently used and the maximum feed rate exceeds several hundred pounds per hour. The metering pump system usually includes a PAC slurry storage tank and can either be a hydraulic diaphragm pump or rotary volumetric feeder.

PAC is typically available in 50 lb (23 kg) bags or in bulk form from trucks or railroad cars. PAC is usually purchased in bags if dry feed is to be used and in bulk if it is to be fed as a slurry. If delivered in bulk form, carbon is removed from the tank car or truck using an eductor, or it is blown into the slurry tank directly from the delivery truck. The slurry can be formed by the eductor and then transferred to a storage tank, or slurry can be batched directly inside the tank. In either method, slurry should be batched at about 1 lb of PAC per 1 gal of water (0.1 kg/L). A batch meter should be provided on the water supply line to accurately measure the amount of water added to a specific load of carbon. The system illustrated in Figure 14.1 incorporates the following design features:

- A 30,000 gal (113,500 L) square concrete bulk storage tank
- A 30 hp vertical turbine mixer, with type 316 stainless steel shaft and blades

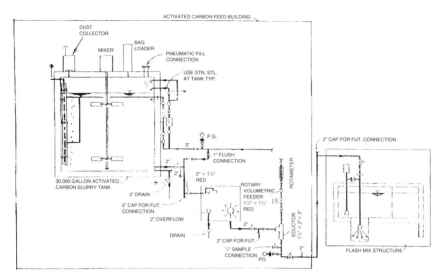

FIGURE 14.1 Activated carbon feeding system.

- A bag loader and dust collector
- An air bubbler and stilling well for slurry tank level measurement, with a water flush connection for periodic purging of the stilling well chamber
- A pneumatic fill connection directly to the bulk delivery truck
- A batch meter and flush connection, with gauges for monitoring supply pressure
- A single rotary volumetric feeder (because carbon is not fed constantly at this facility, only a single feeder has been provided)
- An eductor, equipped with a rotameter for measuring water supply
- A solenoid valve on the water supply line that keeps the water flowing for a period of time after the feeder is turned off to flush the discharge line
- Sample connections
- Cleanouts

The maximum plant flow rate is 90 mgd (340 ML per day). At maximum flow and an average dosage of 1.5 mg/L, the tank provides 26 days of storage. At a maximum feed rate of 10 mg/L, only 4 days of storage is available.

Plants using PAC on a regular basis should consider at least two slurry tanks so that a shipment of carbon can be placed into one tank before the other is empty. However, most applications can be satisfied by a single storage tank. A carbon slurry storage tank can be circular or square and made of steel or concrete. Tank storage capacity should be approximately 20% greater than the maximum carbon load delivered by railroad car or truck. A protective lining can be provided to inhibit corrosion, but unlined concrete tanks are usually sufficient. Mechanical mixers should be provided in the storage tank to keep the PAC slurry in suspension. All metal surfaces should be type 316 stainless steel.

Slurry can be pumped from the main storage tank to a day tank that holds the volume of slurry to be fed over a period of several hours, or slurry can be fed from the main storage tank directly to the chemical feeder. The day tank, if provided, should be plastic or steel with a corrosion-resistant lining. The tank must be equipped with a mixer to keep the slurry in suspension. Slurry should then flow by gravity to the volumetric feeder.

An eductor usually moves the slurry from the feeder to the application point. Piping should slope downgrade to the application point, with provisions for flushing any carbon that may settle out and clog the pipe. Slurry piping should be corrosion and erosion resistant, such as rubber, plastic, or stainless steel. Pump impellers and mixing blades in the slurry tank and day tank should also be of stainless steel or fiberglass. To avoid feed pipe clogging, a minimum flow velocity of 5 ft/s (1.5 m/s) should be maintained in the pipelines.

Wet activated carbon removes oxygen from the air, and as a result, slurry tanks or other enclosed spaces containing carbon may have seriously reduced oxygen levels. Personnel who must enter these spaces should use an oxygen meter to check the atmosphere and also have attached safety belts and another worker present to pull them from danger if necessary.

Common PAC System Operating problems

The most common operating problem with PAC is chemical handling. Most dry feed systems currently in use can be labor intensive. Dust is a major problem if a dry feed system is used, but the amount of dust allowed to become airborne can be minimized by good design practice.

A problem that is not uncommon is PAC passing through the filters and entering the distribution system, provoking complaints from consumers. Black water is usually caused by inadequate coagulation or sedimentation or high doses of PAC added just before the filters. The problem is usually eliminated if carbon is fed into the source water intake or into the rapid mix basin.

Careful attention must be paid to the interaction of PAC with other water treatment chemicals. Activated carbon chemically reduces substances such as free and combined chlorine, chlorine dioxide, ozone, and potassium permanganate, and the demand for these chemicals will be substantially increased. Mixing PAC with chlorine also reduces the adsorption capacity of the activated carbon for selected compounds. Competition between the two chemicals must be avoided. Adding PAC to a source water supersaturated with $CaCO_3$ or other precipitates, or after lime softening, may lead to particle coating of the PAC and a corresponding decrease in adsorption efficiency.

Carbon dust is potentially explosive. Explosion-proof motors should be used if contact with carbon dust is a possibility. Because PAC adsorbs organic compounds, including gases that could reduce its effectiveness, PAC storage must be carefully located. Carbon should be stored in a separate, climate-controlled storage area. The stock of bagged PAC should always be stored in a manner that allows stock to be rotated. Storing carbon outside on pallets under canvas or plastic sheets for long periods of time is not recommended.

PAC slurry solidifies if not mixed frequently or, preferably, continuously. Feed lines become clogged, as do the bags of filter dust collectors. Hydraulic eductor lines carrying slurry also become clogged if they are shut down without being thoroughly flushed. Mixers should be set on timers to allow scheduled, unattended operation.

If a slurry-feed system is used, carbon should be fed below the water surface in the slurry tank so that only a small amount of dust is produced during the loading period. A dust collector should be provided on top of the slurry tank to prevent any dust from escaping into the plant.

Some bag storage should be maintained to supplement a slurry-based system for use if sudden, extremely heavy use exceeds the slurry tank capacity. A bag loader can be provided on top of the slurry tank to allow the tank to be recharged with PAC without calling for a full tanker truck or to meet lower feed rates. Although not recommended except for worst case conditions, carbon bags can be dumped directly into the intake well or elsewhere at the head of the plant.

The PAC feed system should be located as close to the application point as possible. It is suggested that the system be periodically operated, even without a need for carbon feed. Working parts of the system should be checked to be sure the system is always ready for use. Periodic operation also ensures that feed lines are cleaned, carbon is mixed, and the system is flushed. The discharge piping for any feed device should include a flushing line, automatically set to flush the pipelines for a period of at least 60 seconds after carbon feed is stopped. Because moist PAC is highly corrosive, all metal parts that come in contact with carbon should be type 316 stainless steel.

DESIGN OF GRANULAR ACTIVATED CARBON FACILITIES

As illustrated in Figure 14.2, three basic options are available for locating a GAC treatment step for both new and existing water treatment plants:

- Prefiltration adsorption, ahead of the conventional filtration process (prefilter adsorber)
- Postfiltration adsorption, after the conventional filtration process (postfilter adsorber)
- Filtration/adsorption, combining water filtration and GAC filtration into a single process (filter-adsorber)

Under certain plant configurations and site-specific conditions, prefiltration adsorption may be feasible. However, this option has limited applications and benefits and therefore is not discussed further in this chapter.

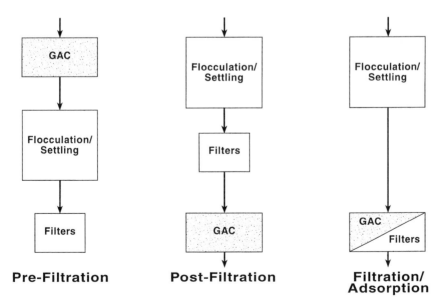

FIGURE 14.2 Basic options for locating GAC treatment.

The more conventional option for most applications is postfiltration adsorption in which the water receives complete treatment before GAC application. In this option, the highest-quality water is applied to the GAC system, the only objective being to remove dissolved organic compounds. As a result, the carbon usage rate is maximized and the frequency of regeneration is minimized, providing an efficient GAC design.

General Considerations

Major design considerations for either prefiltration or postfiltration GAC treatment include:

- Design flow rate
- Types and amounts of organic compounds to be removed
- Location relative to other treatment processes
- Type of GAC (size, depth, and specific gravity); many types of GAC are available, each with different filtration, adsorption, and backwash characteristics
- Filtration rate
- Empty bed contact time
- Number and size of contactors
- Type of contactors
- Gravity
- Pressure
- New or retrofit
- Upflow or downflow
- System hydraulics
- Backwashing requirements
- GAC delivery and receiving facilities
- GAC storage
- GAC transport for placement and removal

GAC Properties.   GAC particles are similar in size to filter media, typically ranging between 1.2 mm and 1.6 mm in diameter. PAC particles are smaller, typically less than 0.1 mm in diameter.

General properties of typical types of GAC produced in the United States are presented in Table 14.3. Some commonly manufactured mesh sizes available for GAC and their associated size ranges appear in Table 14.4.

Contact Time and Breakthrough.   Primary factors the designer must consider in determining the volume required for a GAC adsorber or contactor are breakthrough, empty bed contact time, and the design flow rate.

Breakthrough.   Breakthrough can be defined as the point when the concentration of a contaminant in the effluent adsorption unit exceeds the treatment requirement. Total organic carbon (TOC) is often used as the performance standard defining GAC exhaustion. As a rule, if the GAC filter effluent concentration is greater than the performance standard for three or more consecutive days, it is time to regenerate or replace the GAC.

Breakthrough depends on the quality of both the influent stream and the design of the carbon bed. As illustrated in Figure 14.3, the time from start-up to exhaustion depends on

TABLE 14.3 Properties of Granular Activated Carbon Produced in the United States*

	The Carborundum Company	Calgon Corporation	NICIT		Witco Chemical
Product name	GAC 40	Fitrasorb	Hydrodarco HD-1030	Hydrodarco 83 Plus	Witcarb 950
Base material	Western bituminous coal	Bituminous coal	Lignite	Bituminous coal	Petroleum coke
U.S. standard sieve size	12 × 40	12 × 40	10 × 30	8 × 30	12 × 40
Effective size, mm	0.6	0.8 to 0.9	0.8 to 0.9	0.8 to 0.9	0.8 to 0.9
Uniformity coefficient	≤1.9	≤1.9	≤1.7	≤1.9	≤1.7
Apparent (or vibrating feed) density, g/cm^3	0.47	0.4 to 0.5	0.40 to 0.50	0.47	0.46 to 0.53
Washed density, lb/ft^3	25	25	23.5	26	29 to 33
Iodine number, mg/g	1,050	1,050	600	900	1,050
Surface area, m^2/g	1,000 to 1,100	1,050	650	1,000	1,000
Available in:	60 lb bags, bulk	60 lb bags, bulk	40 lb bags, bulk	50 lb bags, bulk	50 lb bags, 200 lb drums, hopper trucks, tote bins

*"AWWA Standard for Granular Activated Carbon" (AWWA B604-74, *Journal AWWA* 66(11):672, 1974) provides minimum specifications on the properties of granular activated carbon used as an adsorption medium for treating drinking water.

TABLE 14.4 Commonly Manufactured GAC Mesh Sizes

Mesh size (U.S. standard sieve size)	Particle diameter range (mm)
8 × 16	1.18 to 2.36
8 × 20	0.850 to 2.36
8 × 30	0.600 to 2.36
10 × 30	0.600 to 2.00
12 × 40	0.425 to 1.70
14 × 40	0.425 to 1.40
20 × 40	0.425 to 0.85
20 × 50	0.300 to 0.85
Sand	0.38 to 0.65
Anthracite	0.45 to 1.6

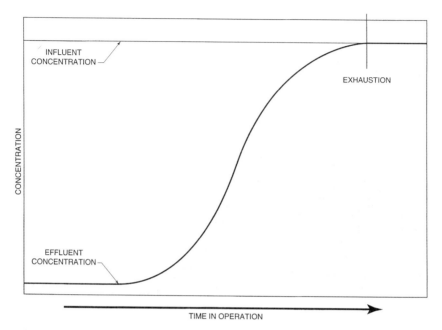

FIGURE 14.3 Breakthrough pattern for GAC filter.

the type and depth of carbon and specific solute characteristics. Breakthrough curves are important to the designer because they define the relationship between the physical and chemical parameters of the carbon system (e.g., flow rate, bed size, carbon exhaustion rate), the determination of the number of beds or columns, their arrangement (either series or parallel), and treatment plant effluent requirements.

Empty Bed Contact Time. Empty bed contact time (EBCT) is calculated as the volume of the empty bed (occupied by the GAC) divided by the volumetric flow rate of water through the carbon. Alternatively, EBCT can be defined as the depth of GAC divided by the linear velocity of water flowing through the carbon bed. It should be noted that EBCT is a false-residence time.

EBCT is used instead of detention time because of the ease of calculation. An actual detention time would have to account for the porosity of the bed, a variable that changes with carbon size and type. EBCT can be varied by changing the bed depth at constant flow or by changing flow with constant bed depth. Together, the design EBCT and the design flow rate define the amount of carbon to be contained in the adsorption units.

Longer EBCTs can delay breakthrough (to a point) and improve carbon usage rate; shorter EBCTs can expedite breakthrough. Thus the time of GAC operation between replacement or regeneration depends on the EBCT. For most water treatment applications, EBCTs range between 5 and 25 minutes. In addition, a factor having a greater influence on operating costs than EBCT is volume throughput, which is the number of bed volumes of water processed before the breakthrough concentration is reached.

Adsorber Volume and Bed Depth. After the EBCT has been determined, carbon depth can be selected. The adsorber design volume depends on bed volume and the amount of freeboard (excess vessel capacity beyond design operating levels). Freeboard can range up to about 50% for fixed and expanded bed systems. If bed expansion is unnecessary, a freeboard of 20% to 30% may be adequate to allow for proper bed expansion during backwashing.

No freeboard is needed for upflow pulsed beds. An economic evaluation is usually made of capital and operating costs to compare carbon columns of various depths.

Hydraulic Loading Rate. The surface loading rate for GAC filters is related to the design flow of a particular treatment plant. Surface loading rates are defined in the same manner as conventional granular media filters. The surface loading rate is the rate of a volume of water passing through a given area of GAC filter bed, usually expressed as cubic meters per square vector (m^3/m^2) or gallons per minute per square foot (gpm/ft^2). Surface loading rates for GAC filters range from 2 to 10 gpm/ft^2 (5 to 24 m/h), although rates of 2 to 6 gpm/ft^2 (5 to 15 m/h) are more commonly used as design criteria.

Surface loading should be kept large for compounds, where the mass transfer rate is controlled by the rate of transfer of the chemical from the bulk liquid to the interior pores of the GAC. Typically this is the case for highly adsorbable compounds (e.g., many SOCs). When mass transfer is controlled by the rate of adsorption (and transport) within the GAC particle, surface loading is not important. This is the case with most less-adsorbable compounds. Figure 14.4 illustrates the relationship between hydraulic loading and pressure drop for several brands of GAC.

Backwashing. A GAC filter bed is backwashed using the same general procedures typically used for backwashing conventional granular gravity filters. If GAC is installed as a sand filter replacement, a redesign of the backwash supply system, including the rate of flow control and washwater troughs height, is often necessary. This redesign is necessary because of the difference in particle density between GAC and sand—about 1.4 g/cm^3 and 2.65 g/cm^3, respectively. If GAC is used as a simple replacement for anthracite coal as a filter medium, the backwash system may be adequate because particle densities are nearly identical.

GAC particle size distribution and wetted density vary among different carbon brands and even among different deliveries of the same carbon. Appropriate backwash rates can be obtained from the manufacturer for each type of carbon (Figure 14.5). Backwash rates must be adjusted to account for specific media characteristics and for changes in backwash water temperature (Figure 14.6). Installing a surface wash or air scour system to assist with filter cleaning may be necessary to control mudball formation.

A good conservative design should allow for 75% to 100% expansion of light GAC media, but 50% is generally considered to be adequate. The design should provide for sufficient freeboard to reduce media losses during the backwashing cycle.

Carbon Usage Rate. The carbon usage rate (CUR) determines the rate at which carbon will be exhausted and how often the carbon must be replaced. The CUR essentially determines the size of the entire regeneration system. The CUR for GAC systems removing organic compounds may be determined using physical models or adsorption isotherm models. A pilot-scale test is often used to evaluate the complexity of multiple chemical interactions. A quicker and more economical method for evaluating GAC column performance is the rapid small-scale column test (RSSCT). Small systems can be designed to best simulate the performance of a full-scale system using dimensional analysis. Dimensionless parameters for the full-scale system and the small system are designed to be equal.

Carbon treatment effectiveness improves as contact time increases. The percentage of total carbon in a bed that is exhausted at breakthrough is greater in a deeper bed than in a shallower bed. At a point beyond the optimum bed depth, the additional adsorber volume provided acts primarily as a storage capacity for spent carbon. The actual selection of bed depth and corresponding adsorber volume also depends on

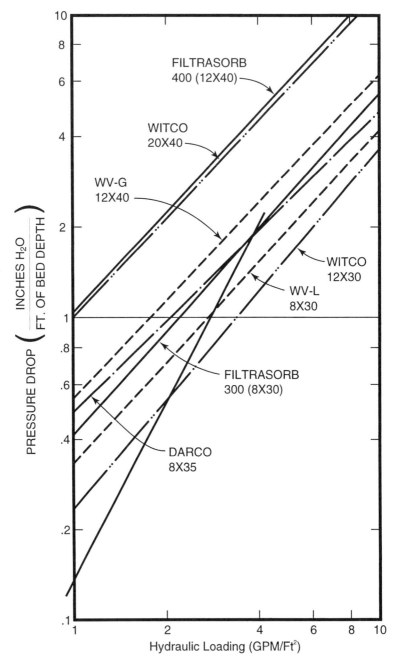

FIGURE 14.4 Carbon bed pressure drop vs. hydraulic loading.

ACTIVATED CARBON PROCESSES 391

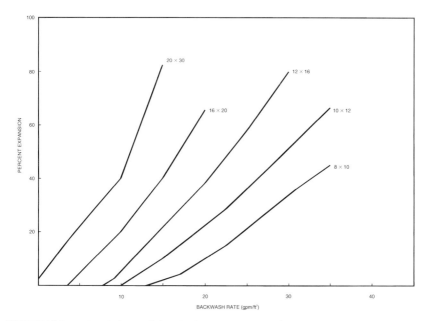

FIGURE 14.5 Backwash characteristics of various mesh sizes at 25° C.

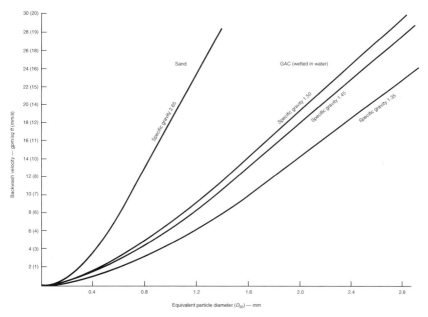

FIGURE 14.6 Determination of appropriate medium size and backwash rate for a water temperature of 68° F (20° C) *(backwash velocity = 1.3 × V_{mf} for D90)*.

reactivation frequency and contactor construction cost. The relationship between increased adsorber volume with reduced reactivation frequency can be compared against a reduced adsorber volume and increased reactivation frequency to determine the optimum characteristics for design. Together, CUR and EBCT have the greatest effect on capital and operating costs for GAC processes.

Design of GAC Adsorption Beds

For filtration/adsorption, either as a filter cap or full-depth GAC, major design considerations include:

- Filter run lengths
- Means for cleaning the GAC bed as a part of routine operations
- Life expectancy of the GAC
- Chlorine, polymer, and ozone feed points
- Size and depth of GAC as filter media
- Matched GAC and sand barrier layers during backwash (GAC cap)
- Head loss restrictions
- Backwash restrictions
- Flow rate restrictions—inlet and outlet piping and conduits

Adsorber Configuration. GAC adsorption units can have one of three basic configurations, with single or multiple adsorbers operated in series or in parallel:

- Downflow fixed bed
- Upflow fixed (packed) or expanded (moving) bed
- Pulsed bed

The most common GAC application in drinking water treatment is the downflow fixed bed in parallel operation. Downflow operation is appropriate when the carbon bed is to be used as a suspended solids filter as well as an adsorber. In this design, carbon contained in the adsorber remains stationary. Flow is divided equally to each contactor, and each contactor is sized for the design EBCT. Suspended solids are periodically removed by backwashing in a manner nearly identical to that used for sand and dual-media (sand and anthracite) filters. When the carbon adsorber is preceded by conventional filtration, downflow operation can sometimes be used with or without reduced backwashing.

Larger installations are likely to be gravity fed. Pressure adsorption units are cost-effective for smaller installations. Pressure flow can be used for either downflow or upflow beds. Pressure flow achieves higher hydraulic loadings than would be economically feasible with gravity flow. This higher loading reduces the required adsorber cross-sectional area. Pressure flow also permits operation at higher suspended solids concentrations with no backwashing or with less frequent backwashing than is possible with gravity flow. Pressure adsorbers can be less expensive to design and construct because they can be prefabricated and shipped to the site.

For many systems, the decision between pressure or gravity adsorbers is based solely on initial installed cost. Most existing and many new water treatment plants require that the total plant flow be pumped to postfiltration GAC facilities because of insufficient hydraulic head downstream of the existing or new conventional filters. Pressurized adsorbers (commonly called contactors) may be more cost-effective in these installations.

Upflow expanded (moving) beds are best suited for waters with high suspended solids concentrations when suspended solids are to be removed by subsequent processes (filtration). For high suspended solids concentrations, upflow beds may be preferred, because solids accumulation and corresponding head losses would be excessive in downflow adsorbers. For low suspended solids concentrations, upflow adsorbers can be considered, because the carbon bed is not needed as a solids filter.

In an upflow bed, the upward movement of water causes the carbon bed to expand slightly (approximately 10%). A higher CUR is expected for expanded beds because mixing of the carbon creates a longer mass transfer zone. Mixing may allow the release of carbon fines into the effluent flow.

A second type of moving bed, a pulsed bed adsorber, is periodically pulsed to discharge a portion of the exhausted carbon from the bottom of the bed while a portion of fresh replacement carbon is added at the top of the bed. The liquid and the carbon move in true countercurrent flow to each other.

GAC Facility Sizing. Perhaps the most important considerations facing the designer are choosing the proper EBCT and process flow configuration (adsorbers in series or parallel) to maximize GAC effectiveness. The type of regeneration system used for reactivating spent carbon is also an important process design element.

Separate GAC adsorbers after conventional filtration offers increased flexibility for handling GAC and for designing specific adsorption conditions. Considerably longer contact times can be obtained than are typical for the filter-adsorber.

Retrofitting Existing High-Rate Granular Media Filters

One option readily available for most water treatment facilities is to replace conventional filter media with GAC. Existing filter boxes can be converted to GAC filter-adsorbers simply by removing all or a portion of the granular media from the filter box and replacing it with GAC. Alternatively, an entirely new filter box, underdrain, and backwashing system for the GAC can be designed and constructed. In these applications, GAC is used for turbidity and solids removal, biological stabilization, and dissolved organics removal by adsorption.

Filter-adsorbers must balance the constraints of both adsorption and filtration. They typically require backwashing at about the same frequency as conventional filters. A filter-adsorber design can be easily installed as a retrofit for an existing conventional granular filter, but the designer must recognize and account for the fact that filter-adsorbers:

- Have shorter filter run times and must be backwashed more frequently than postfilter adsorbers, which in some applications do not require backwashing at all.
- May incur greater carbon losses than postfilter adsorbers because of more frequent backwashing.
- May cost more to operate than postfilter adsorbers because carbon usage is less efficient.

Washwater Trough Considerations. In retrofitting existing filters, modifications may be necessary to position the washwater collection troughs to allow for proper bed expansion to take place. Although some loss of GAC is expected over time as a result of the backwash process, the trough must be located at the correct elevation above the top of the filter bed to ensure that particle losses are not excessive. The height and spacing of washwater troughs is illustrated on Figure 14.7. Given the high cost of GAC as compared with anthracite, excessive and unnecessary GAC loss during backwash is costly.

GAC Support System. For support, GAC can be placed directly on sand or graded gravel. In either case, particular attention must be paid to the type of underdrain system used. Underdrain products are available that allow GAC to be placed directly on top of the underdrain. The most common types are the stainless steel or plastic wedge wire screen and block underdrains. In the block underdrain design, a permeable membrane is placed over the block underdrain. The membrane prevents GAC loss through the underdrains and provides a surface for the uniform application of filter washwater. A cross-sectional view of a wedge wire underdrain is shown on Figure 14.8.

The block underdrain with a porous membrane is shown in Figure 14.9. To convert a dual-media filter to a GAC postfilter adsorber, the process is a little more complex. Many

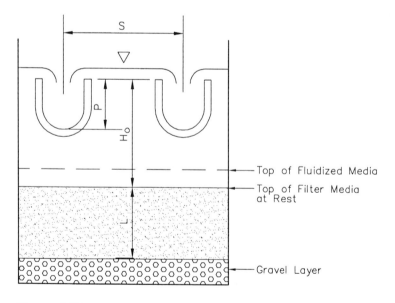

FIGURE 14.7 Design guidelines for filter washwater troughs.

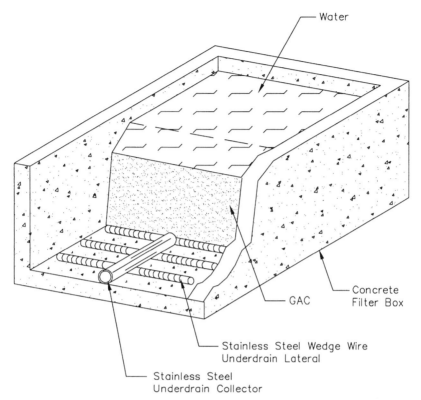

FIGURE 14.8 Wedge wire underdrain for GAC filters.

larger gravity dual-media filters—usually greater than 1,000 ft^2 (about 100 m^2—have center filter gullets that split the filter into two physically separate but hydraulically connected cells. The separation is designed to separate the total filter area into two manageable areas that can be backwashed separately, one cell at a time.

This design reduces the size of the backwash facilities (pumps and storage tanks can be half the size) and ensures that the rate of backwash flow is not disproportionate to the total instantaneous plant flow. The center filter gullet supplies source or settled water to the filter in the upper gullet, collects the filtered water, and supplies filter backwash water (reverse flow) to the underdrains in the lower gullet.

Many current dual-media filters were at one time sand filters. Converting sand media to dual media, to attain higher filtration rates without the need for physical plant expansion, has been found to be beneficial for many treatment plants.

However, in many of these applications, the filter box is barely deep enough to allow the media change to sand and anthracite coal. Excessive media losses, from losing anthracite over the backwash troughs of shallower filter boxes, are common. These filters typically have very shallow filter box depths relative to the depth that would be provided for either a new dual-media filter design or a GAC filter or postfilter adsorber design. Providing empty bed contact times for GAC filtration or adsorption presents physical demands that cannot be easily solved with this type of design.

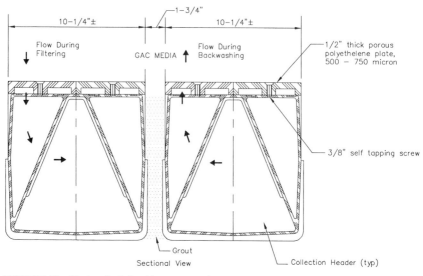

FIGURE 14.9 Block underdrain with porous membrane.

A simple method of converting dual-media gravity filters to GAC contactors is illustrated in Figure 14.10. The depth of GAC needed to provide the empty bed contact time can be accommodated with minor physical modifications. The filter can be retrofitted to provide downflow in one cell and upflow in the adjacent cell. As a result, the depth of GAC needed for EBCT can be provided with a shallow box. In many applications, converting existing dual-media filters to GAC, just as these same filters were once converted from sand to dual media, can be accomplished at a fraction of the cost of constructing entirely new contactors. When contemplating a conversion to GAC postfilter adsorbers, the value and cost benefit of retrofitting the existing filters should not be overlooked.

The conversion from sand or dual media to GAC can be accomplished as follows:

- Remove all filter media from the filter (sand or sand and anthracite coal and the support gravel) either physically or by eduction.
- Inspect the underdrain system.
- Increase the height of the backwash troughs if required for the filter media.
- Alternatively, replace existing underdrains with an underdrain system using a permeable membrane layer to achieve the depth needed for the new or deeper media bed.
- Check existing backwash supply facilities. Redesign may be necessary because GAC may require more or less backwash flow.
- Inspect the existing surface wash or air scour system. A system should be considered if none is currently in place.
- Train plant staff in new backwash procedures.

Capping Existing Slow Sand Filters

Slow sand filter operations can be improved by adding a GAC filter cap. Slow sand filtration is a process of filtering water through a sand layer at a very low flow rate. Backwashing is not

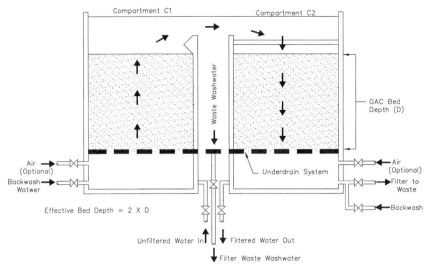

FIGURE 14.10 Conversion of split-cell filters to GAC contactors.

required. Instead, the filters are scraped after one to six months of operating time. Scraped sand is replaced with new sand to restore the sand filter to the desired depth. The depth is usually 24 to 40 in. (625 to 1,000 mm), supported by 12 to 18 in. (300 to 450 mm) of graded gravel. Underdrains typically consist of perforated pipes, porous concrete slabs, or channels formed by double layers of loose brick. Filter areas may be as large as 50,000 ft^2 (4,645 m^2). Water height above the sand surface is usually maintained between 3 and 5 ft (1 and 1.8 m).

Slow sand filter performance can be also be improved by adding a GAC sandwich layer, placed 4 to 6 in. (100 to 150 mm) below the top of the sand layer, as shown in Figure 14.11. The GAC layer is typically 3 to 8 in. thick (75 to 200 mm), depending on water quality conditions. Operation is similar to the normal sand filter. At the time of sand scraping, sand is removed down to the GAC level. Sand is then be added again if the GAC is determined to be active (not exhausted or spent). When the GAC layer is spent, it can be removed and reactivated, then returned to the sandwich filter for reuse.

Design of Postfiltration Adsorbers

Separate GAC adsorbers after conventional filtration offer increased flexibility for handling GAC and for designing specific adsorption conditions. Considerably longer contact times can be obtained than usual for the filter-adsorber.

Postfilter adsorbers typically use GAC with small effective sizes and large uniformity coefficients to promote rapid adsorption of organic compounds, better stratification, and reduction of overall carbon loss. These adsorbers can also provide an improved barrier against microbial penetration, are more compatible with other advanced treatment processes such as ozonation (or ozone and hydrogen peroxide), and use more of the adsorptive capacity of the carbon.

Adsorber Configurations. Various combinations of the three GAC contactor configurations and flow patterns can be made. Table 14.5 provides a summary of the most common configurations, and each is described below.

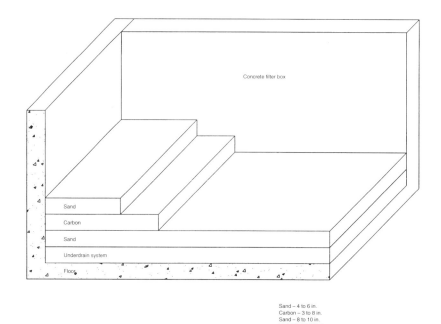

FIGURE 14.11 Media arrangement in a GAC sandwich filter.

TABLE 14.5 Guide to Selection of GAC System Configuration

Adsorption system configuration	Applicable if:
Single adsorber	Low carbon exhaustion rate (usage rate) occurs.
Fixed beds in series	High carbon exhaustion rate occurs. High effluent quality must be ensured at all times.
Fixed beds in parallel	High system pressure drop is minimized. Large total flow rate is required.
Moving beds, upflow	High carbon exhaustion rate occurs. High effluent quality is required. Either some carbon fines can be tolerated in the effluent, or carbon adsorption is followed by filtration.
Expanded beds, upflow	High suspended solids are present in the influent, and some suspended solids can be tolerated in the effluent.

Fixed Beds in Series. In this configuration, flow is downward through the carbon bed for each unit connected in series, with the EBCT divided between the number of adsorbers. The first in the series receives the highest contaminant loading and is thus the first to be exhausted. The last in the series receives the lightest contaminant loading and serves as a polishing step. When carbon is removed for reactivation, the first adsorber in the series is removed from service, with the next adsorber in line becoming the lead unit. This system can be constructed with an extra adsorber on standby to become the first adsorber when the lead adsorber is taken out of service. The configuration can be arranged and controlled so that the freshest GAC is always the last step in the series.

Capital cost considerations usually limit the number of adsorbers in series to four or fewer, because the increased cost of piping and valving counters the cost benefit of reduced carbon usage. Countercurrent flow provides highly efficient usage of the carbon by maximizing carbon exhaustion in the lead adsorber before it is removed for reactivation. Premature breakthrough of organics may also be reduced. Carbon usage rates under series operation can approach one-half the rate under parallel operation, reducing operating costs accordingly. The lead absorber can be backwashed to remove suspended solids that accumulate in the carbon bed.

Fixed Beds in Parallel. In a parallel configuration, each carbon bed receives essentially the same quantity and quality of flow. Start-up of individual units is staggered so that exhaustion of the carbon occurs sequentially. This allows removal of all carbon from each adsorber, one at a time, for reactivation. For systems operating at full design capacity, a spare adsorber can be provided to bring on line when an adsorber is taken out of service.

In this arrangement, the level of carbon exhaustion when an adsorber is discharged is not as high as that of fixed beds in series because no adsorber is operated completely to breakthrough. However, carbon use in a given adsorber can be increased by blending effluent from all adsorbers. One or more adsorbers can be run slightly beyond the breakthrough point while other adsorbers produce water with concentrations below the breakthrough point.

Because effluent from each of the units is blended, each unit can be operated until it is producing a water with an effluent concentration in excess of the treated water goal. Only the composite flow must meet the effluent quality goal. For example, if 10 adsorbers are used in parallel, each adsorber can process 10,000 bed volumes of throughput if the effluent TOC criterion is 50% of the influent, compared with 5,000 bed volumes if only a single contactor were used or if all contactors were operated in parallel but were replaced at the same time. This method of operation may be most appropriate for large plants.

Expanded Beds. Expanded beds permit removal of suspended solids by periodic bed expansion. In general, expanded beds can tolerate relatively high suspended solids while using finer carbon particle sizes without excessive head loss. Expanded beds are not appropriate for achieving very low effluent standards. In addition, carbon usage rates may be slightly higher for expanded beds compared with fixed beds. Expanded beds should not be used where downstream contamination by suspended solids or carbon fines passing through the bed would be a problem.

Pulsed Beds. A pulsed bed operates in an upflow mode, with water and carbon flow moving countercurrent. Pulsed bed adsorbers permit intermittent or continuous removal of spent carbon from the bottom of the bed while fresh carbon is added at the top without system shutdown. The chief advantage of this system is better carbon use because only thoroughly exhausted carbon is reactivated. In contrast with fixed beds, a pulsed bed cannot be completely exhausted. This prevents any contaminant breakthrough into the effluent that may cause effluent water quality standards to be exceeded.

The performance of pulsed beds is affected by suspended solids or biodegradable compounds that cause extensive biofilm growth on the activated carbon, so that the bed must be backwashed. Backwashing leads to mixing of the fresh activated carbon at the top of the bed with spent activated carbon deeper in the bed and destroys some of the beneficial countercurrent effect. Additionally, some activated carbon fines may be produced during upflow that may require removal by a subsequent process.

Another characteristic of the pulsed bed system is that a steady-state constant effluent concentration (assuming a constant influent concentration) is achieved. In fixed beds, effluent concentration gradually increases with time.

Adsorber Design. Various adsorption vessels can be considered by the designer. The design depends on the size of the plant, type of adsorber selected, mode of operation, and number of adsorber units. Adsorption units are typically cylindrical steel vessels (pressure or gravity flow) or cast-in-place concrete rectangular structures (gravity flow only). The type of vessel used depends on plant size and specific site constraints. Basic characteristics of typical adsorbers are presented in Table 14.6. The four fundamental adsorber types are:

- Open top steel, gravity flow
- Enclosed steel, gravity flow
- Enclosed steel, pressure flow
- Open top concrete, gravity flow

Treatment plants of 10 mgd (38 ML per day) or less usually select pressure steel vessels, although many designers prefer open-top beds to allow the operator to see the backwash. Larger plants usually have open-top concrete tanks. Other features that differentiate the individual adsorbers are the details of internal hardware, such as liquid distributors and collectors, carbon bed support methods, underdrains, and backwashing apparatus.

Adsorber sizing is based on flow rate, hydraulic loading, and EBCT. These variables determine the adsorber volume, depth, cross-sectional area, and the number of individual adsorber vessels. Once these quantities are known, adsorber design can proceed.

Circular Adsorbers. Circular adsorbers include both pressure and gravity flow steel vessels. Adsorbers in parallel or in series affect equipment layout because of the differing needs for piping and valving.

The selection of the number of adsorbers for any given design depends on vessel size and permissible hydraulic loading. Commercially available steel adsorbers have standard diameters. Shop-fabricated steel vessels are limited to a diameter of 12 ft (3.7 m) and a length of 60 ft (18.3 m) because of trucking transportation constraints. Larger vessels can be installed by shipping partial pieces (circular rings and end pieces) and then fabricating the complete vessel in the field. However, this approach is not likely to be cost competitive with cast-in-place concrete alternatives for any but the largest installations. American

TABLE 14.6 Basic Characteristics of Adsorbers

Adsorber type	Material	Volume* range, ft^3	Diameter†	Remarks
Steel, gravity flow	Lined carbon steel	6,000 to 20,000	20 to 30 ft	Field fabricated by welding or bolting preformed steel plate; mounted on concrete slab foundation
Steel, pressure flow	Lined carbon steel; stainless steel	2,000 to 50,000	Up to 12 ft	Shop fabricated; over-the-road transportation constraints limit size
Concrete, gravity flow	Standard reinforced	1,000 to 200,000	Usually rectangular	Field constructed; designs vary; 2:1 length-to-width ratio is common

*To convert ft^3 to m^3, multiply by 0.0283.
†To convert ft to m, multiply by 0.3048.

Society of Mechanical Engineers (ASME) standard diameters for unfired pressure vessels for shop-fabricated and field-fabricated steel vessels are given in Table 14.7.

Rectangular Adsorbers. Rectangular adsorbers are usually constructed of conventionally reinforced concrete. Multiple vessels are generally built into a single large concrete structure with individual treatment units sharing common walls, pipe galleries, and operating areas. The major design features subject to variation are the length-to-width ratio of the individual unit, the design of the influent and effluent channels and conduits, bed depth, and the type of underdrain system.

A typical length-to-width ratio is 2:1, similar to the ratio used for conventional gravity filtration. Influent and effluent conduits are usually oriented to the inside of the common pipe gallery to facilitate placement of isolation valves, meters, and instrumentation and control equipment. Bed depth is determined from other factors, but it is usually not less than 600 mm and can be as deep as 6,000 mm or deeper. A cross section of a typical open-top concrete, rectangular, gravity flow contactor is illustrated in Figure 14.12.

Design Considerations. GAC contactor design is similar to design for conventional gravity dual-media filtration. Hydraulically, the design must accommodate the maximum flow of the plant. GAC volume is determined based on design flow for the GAC facilities and the EBCT established for this flow. Because flow rates and organic concentrations often vary seasonally, careful analysis is needed to prevent oversizing the contactors.

The surface loading rate at maximum flow determines the surface area of the contactors. Typical design ranges for surface loading rates are 2 to 10 gpm/ft^2 (5 to 24 m/h) and for bed depths are 10 to 20 ft (3 to 6 m).

With bed volume and surface area established, the number and size of contactors can be determined. Design should allow for the full plant flow to be treated with one contactor out of service for maintenance or carbon replacement. As with the design of conventional filters, fewer large contactors are generally more cost-effective than constructing a greater number of smaller contactors. Whether or not operation of the units is staggered, the method of effluent blending may also affect the number of contactors.

The number of contactors is influenced by:

- Initial capital costs
- Long-term operating costs
- Ease of operation—generally the fewer the units, the easier to operate
- Rate at which backwash water must be applied because this is a function of contactor surface area
- Difficulties in providing uniform distribution of backwash water over a large area that may limit the effective maximum area of a single unit

TABLE 14.7 Standard Diameters for Circular Steel Vessels

Shop fabricated	Field fabricated
7'11"	15'6"
9'6"	21'6"
10'0"	29'9"
11'0"	38'8"
12'0"	55'0"

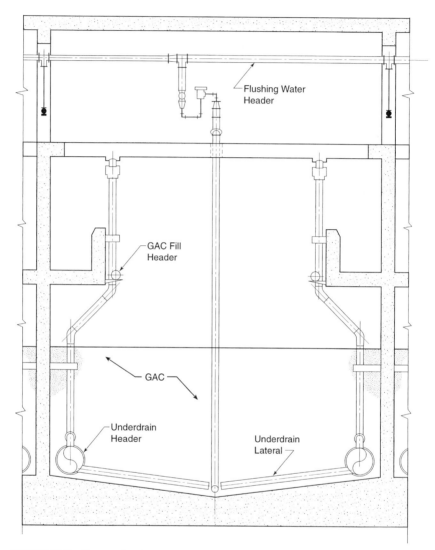

FIGURE 14.12 Cross section through a GAC contactor.

- Reduction in capacity with one unit out of service for backwash, regeneration, or maintenance
- Number and size of valves required
- Structural considerations

For purposes of determining the optimum number of contactors, estimates of capital costs should be developed for various contactor configurations. A plot of the relative capital cost as a function of the number of contactors aids in selecting the best configuration.

Perhaps the most important factor in the design of a GAC facility is selecting the type of underdrain system to be used. Underdrains must perform several important functions:

- Collect water at a uniform rate to ensure GAC retention in the contactor
- Prevent binding or plugging of the collection system
- Distribute water evenly over the entire contactor area
- Provide uniform fluidization of GAC at a low backwash rate during transfer of spent carbon out of the contactors so that carbon moves to the points of discharge

Common Operating Problems

Operating problems encountered when using activated carbon as a filter medium or adsorber are similar to those found with most rapid sand or dual-media filters. The most common operating problems that the designer should be aware of when preparing a design include:

- Controlling the carbon depth
- Properly backwashing the carbon, as carbon bed density and depth change during backwashing
- Removing carbon from the contactors and storage tanks
- Transporting the carbon
- Regenerating the carbon
- Maintaining a good-quality contactor influent
- Controlling contactor flow rate
- Controlling biological activity in the carbon adsorber

Controlling the Carbon Depth

A means must be provided for continuously monitoring and measuring the levels in all storage vessels, including contactor and storage tanks (virgin, spent, and regenerated carbon). The designer must recognize that slurry concentrations and flow rates change during the filling process. When carbon is placed in an adsorber, it typically has a density of approximately 90% of the apparent density for random-packed graded media. When a carbon column is first backwashed, density decreases from approximately 90% to approximately 83%, resulting in a permanent expansion of 8%.

An 8% expansion in a 24 in. (60 cm) filter is only 1.9 in. (4.8 cm) and does not present any major operating problems. However, an 8% expansion in a 15 ft (4.6 m) deep carbon bed entails a permanent expansion of 14 in. (35 cm). This can create a major problem when trying to closely control the surface elevation of a deep bed.

The problem of controlling surface elevation of a deep bed, graded media is complicated further by the fact that the bed also compacts during operation. The same 15 ft (4.6 m) column could see a compaction of 5 to 15 in. (13 to 38 cm). If surface wash sweeps are used, they must be placed to avoid this expansion and contraction of the bed.

Backwashing the Carbon Properly. The basic objectives of the backwashing process are to expand the bed for cleaning and to restratify the bed after washing to maximize the carbon usage rate. These simple objectives are not easily attained.

GAC density, wetted in water, is appreciably lower than sand, and the carbon bed can easily rise as a plug. The force required to lift any bed is the difference between the force of gravity and the buoyant force. For a carbon bed, an upward pressure of 3 psi (21 kPa) is sufficient to lift the bed. The backwash must be gradually applied, or the carbon bed could rise as a plug rather than expanding from the top downward. This problem is compounded if the entire carbon bed has been drained, because air in the bed prevents the upper bed from expanding when it is not submerged in water. Although only a small pressure is acting on the bed, the cross-sectional area is so large that a tremendous force can develop. This situation can result in damage to the carbon bed and tank internal equipment.

To prevent the carbon bed from rising as a plug, the backwash rate should be gradually increased to ensure that the bed expansion occurs from the top downward. Slowly ramping or stepping the backwash rate from approximately 3 to 4 gpm/ft^2 (8 to 10 m/h) to 14 to 15 gpm/ft^2 (34-37 m/h) should be sufficient to ensure proper bed expansion. Conversely, ramping the backwash rate down ensures proper bed stratification. Because of the potential damage that can result from a rising carbon plug and the difficulty of maintaining a constant bed depth due to expansion and compaction of the carbon, conventional surface sweeps are not recommended for deep bed carbon systems.

Removing Carbon from the Adsorbers. Because granular carbon flows as a slurry when diluted with transport water, it is most easily removed from a contactor in the form of a slurry. Dilution or transport water can be added a number of ways—through the underdrains or tangential to the vessel wall at the bottom of the wall. GAC slurry can be withdrawn through ports at the base of the wall or in the vessel side wall. Multiple ports are very efficient, but single ports are also effective.

Controlling Biological Activity. Biologically active carbon must be controlled to avoid undesirable affects. The larger surface area provided by GAC also provides an excellent attachment medium for microorganisms. Anaerobic conditions may develop, resulting in odor problems if the system is not kept aerobic. Chlorine application to GAC filter-adsorbers or postfilter adsorbers is not recommended because chlorine does not prevent growth, and it increases the concentration of adsorbed chloro-organics. Chlorine can make activated carbon more brittle because it destroys some of the activated carbon by chemical reduction. Brittle carbon is then susceptible to increased breakup during filter backwash.

Backwashing for filter-adsorbers and periodic physical replacement (for both filter-adsorbers and postfilter adsorbers) is necessary to control biological growth. Backwashing should not be excessive because of its possible detrimental effect on adsorption efficiency. Mixing may also take place during backwashing. As a result, GAC with adsorbed molecules near the top of the bed move deeper into the bed, leading to possible early breakthrough. The large uniformity coefficient of most commercial activated carbons promotes restratification after backwash. If the underdrain system does not properly distribute the washwater or if backwashing procedures do not aid restratification, substantial mixing of the activated carbon can occur with each backwash.

Biological activity can be beneficial for removing some compounds. Biological treatment reduces assimilable organic carbon (AOC) and other biodegradable compounds, resulting in a biologically stable water. The oxidation step (ozone or ozone with hydrogren peroxide) renders more of the AOC biodegradable, resulting in a greater biological activity on the GAC, rather than in the distribution system. Bacterial growth in a distribution system is undesirable because it can accelerate pipeline corrosion, produce tastes and odors, and increase the amount of disinfection needed to maintain a residual throughout the distribution system. Maximizing the production of AOC within the treatment plant itself, and then providing for a biological treatment step with GAC, can be an effective method for reducing the concentration of AOC delivered to the distribution system.

Any disinfectant residual applied to a biologically active GAC adsorber is quickly consumed, wasting chemicals and inhibiting biological activity on the GAC. Adding chlorine ahead of GAC produces chlorinated disinfection by-products that are adsorbed. Thermal regeneration oxidizes these by-products to dioxins and furans. For this reason, primary disinfection with chlorine should be provided after contact with GAC. The design engineer should also be aware that microbial growth on filter-adsorbers can cause increased rates of head loss buildup and shorter filter runs if growth is excessive and the backwash system is not able to control the growth. These possible conditions should be considered in the design.

Biological activity is lower for colder water (8° to 12° C) than for warmer water (25° to 35° C). As a result, biological oxidation cannot be counted on throughout the year if temperatures are low enough to affect biological activity.

The designer must also be aware that it is possible for activated carbon particles to migrate through and penetrate the filter-adsorber or postfilter adsorber underdrain system, providing a habitat for microbial growth that cannot be controlled by disinfection. Zooplankton and other undesirable organisms can also grow in GAC filters that are biologically active. Increased backwashing may be necessary (once every five days) to keep filter beds clean and free of these organisms before they have an opportunity to flourish.

As described previously, biologically active carbon must be controlled to avoid undesirable effects. Anaerobic conditions may develop, causing odor problems, if the system is not kept aerobic. This can occur if large concentrations of ammonia are allowed to enter the filter, insufficient dissolved oxygen is in the water, or the bed is removed from operation for an extended period of time. Proper operation should ensure that sufficient oxygen is present at all times.

Maintenance Requirements

In activated carbon adsorption, maintenance requirements fall into preventive and repair categories for the main components of the activated carbon system.

Adsorbers. Steel adsorbers are either lined with rubber, painted with epoxy, or constructed of type 316 stainless steel. The life of a rubber lining is a function of the frequency and extent of backwashing and the frequency and changeout of the activated carbon. Minimal maintenance should be involved, and it is expected that the lining will need to be replaced every 10 to 15 years. Underdrains should be inspected after each carbon replacement.

Transfer Equipment. Design of pipeline velocities should minimize the effects of erosion and corrosion. Lines should be thoroughly flushed to keep any residual carbon from plugging the lines and accelerating corrosion. The life of an eductor is generally a function of the amount of carbon transferred. A properly designed and operated eductor can handle 2 to 4 million pounds (0.9 to 1.8 million kilograms) of carbon before being replaced. Slurry pumps should be avoided because of the initial costs and the amount of subsequent maintenance required.

Carbon Losses. In a facility using GAC and on-site reactivating, there are three areas where carbon losses can occur:

- Within the adsorbers
- In the transport system
- In the reactivation furnaces and ancillary equipment

For a new shipment of GAC, fines are typically specified in the range of 4.0% maximum, although requiring 2.0% is common. The majority of initial carbon fines are backwashed out of the system within one to two days after the GAC is placed into service. During regular backwashing operations, abrasion of particles occurs. The mean particle diameter of typical coal-based GAC products is usually reduced by 0.1 to 0.2 mm during the on-line time of two to four years.

A loss of GAC bed depth of 0.2 to 1.0 in. (0.5 to 2.5 cm) per year is common. For a 24 in. (0.6 m) deep bed, this translates to 1% to 4% loss per year. Losses beyond these values indicate that there may be a problem with the underdrain system or that the applied backwash rate is excessive, resulting in the migration of carbon over the tops of the washwater troughs.

Medium for Biological Treatment. Both filter-adsorbers and postfilter adsorbers become biologically active when disinfection occurs after the GAC process. When ozone or an advanced oxidation process (AOP), such as ozonation and hydrogen peroxide, precedes the GAC treatment step, the GAC also becomes biologically active. Ozonation makes some nondegradable compounds biodegradable by breaking the compounds into smaller, more biodegradable products and thereby provides an additional means of organics removal.

Practical Design Suggestions

Carbon adsorption system design is not complex. In fact, most of the design elements and features are similar to those found in the design of a conventional filter system. However, there are several design considerations worth noting that should be taken into account when preparing a carbon adsorption design.

Design of Adsorber Vessels.

- For gravity downflow, open-top concrete adsorbers, consider dual cells to reduce the backwash rate. In the desire to limit the total number of contactors, the backwash flow rate required for the large filter area may be too great or take too much of the instantaneous plant flow for backwashing. Providing dual cells allows each filter cell to be backwashed separately, reducing instantaneous backwash flow rate by one-half. Dual cells reduce the size of the washwater system but increase the number of valves. In large systems, the number of valves to open, close, operate, and maintain is an important consideration.

- If a stainless steel wedge wire underdrain system is used, underdrain laterals should be at least 1 ft off of the vessel floor to facilitate carbon removal. Sloping the bottom from the back of the vessel to the front assists flushing.

- Underdrain systems with nozzles are not recommended, because they are subject to plugging. Systems with false bottoms or plenums in which carbon fines can accumulate also are not recommended. Where nozzles and false bottoms are used in combination, plugging of the nozzles may create uplift pressures during backwashing, leading to structural failure of the false bottoms. Allowing carbon fines to collect in the underdrain system is also undesirable because these fines can provide a habitat for the growth of microbes and other undesirable organisms.

- Provide a method for periodic evaluation of the carbon bed depths by placing a permanent reference mark such as a stainless steel plate, laminated plastic staff gauge,

or painted gauge marks directly on the vessel wall.
- For conventionally reinforced concrete structures in contact with carbon, a minimum concrete cover of 5 mm should be provided over reinforcing steel. Carbon is extremely corrosive and attacks reinforcing steel through cracks in concrete.
- Provide a washdown hose site on the vessel deck to facilitate cleaning.
- Provide air release valves on the backwash water supply piping to ensure that air has been completely purged.
- Provide adsorber-to-waste connections (similar to conventional filter-to-waste) to permit additional removal of GAC fines remaining after backwashing.

Design of Carbon Storage and Transport Facilities. Carbon replacement is a major expense associated with carbon adsorption systems. Because slurry storage and transport can cause major attrition losses, facilities must be designed with care. Carbon deliveries in excess of 9,000 kg (10 tons) usually justify an eductor/carbon slurry transport system. Design guidelines include the following:

- To minimize carbon-to-carbon abrasion, avoid air-assisted transfer of carbon.
- Transfer all carbon in a water slurry form, designed for a slurry of approximately 1 to 3 lb/gal (0.12 to 0.34 kg/L).
- Minimize the length of carbon slurry transfer pipes through optimum arrangement of structures and equipment.
- Lines delivering carbon to tanks should discharge below the water surface to minimize attrition and overflows from the dewatering screw. Tank overflow should be screened or the discharge diverted to a collection point where carbon can be recovered.
- Provide a metering device to measure carbon slurry flow rate in the piping system.
- Limit velocities in carbon slurry transfer lines to 3 to 5 ft/s (91 to 152 cm/s).
- Avoid the use of throttling valves on carbon slurry lines and on potable water dilution lines ahead of the point of carbon introduction.
- Use a minimum diameter of 50 mm type 316 stainless steel pipe for carbon slurry transfer lines and long sweep bends to minimize pipe erosion and improve hydraulics. Mitered bends should not be used for piping carbon slurry.
- Minimize the number of movements (valve changes, flow direction changes, pump starts) to effect a complete carbon transfer.
- Limit the speed of recessed impeller, rubber-lined, centrifugal carbon transfer pumps to 900 rpm or less.
- Design carbon storage bins to receive the entire wet contents of carbon transferred from a single carbon vessel, plus at least 20% for expansion.
- Install pressure gauges with diaphragm seals throughout the eduction system to monitor system performance.
- Use double-seated valves for in-line shutoff and isolation throughout, because the system may see carbon flow in both directions. Double-seated knife gate valves, plug valves, or ball valves, fabricated from materials suitable for carbon contact, can be used.
- Design the carbon slurry system to move slurry through the system continuously until all carbon has been flushed from piping. The process must not be allowed to stop until all carbon has been moved through the piping or carbon will settle and plug the lines. Provide cleanouts and flushing connections throughout the piping system.
- Provide a vacuum cleaning system for picking up loose carbon.

- Provide a method for automatically and accurately measuring carbon levels in storage bins and vessels. A backup manual system should be provided to track bed and bin levels. Depths can be monitored by placing a permanent reference mark such as a stainless steel plate, laminated plastic staff gauge, or painted gauge marks directly on the inside wall of each bed and bin.

Materials of Construction. Materials of construction for carbon adsorption plants vary. Epoxy-coated steel, type 316 stainless steel, concrete, or fiberglass are used for vessels; stainless steel, lined carbon steel, fiberglass, and plastic have been used for piping. Problems have been reported with plastic due to abrasion and scouring from the transport of carbon slurries.

Dry carbon and carbon in water slurries are not corrosive. Damp or wet carbon, however, is extremely corrosive. Precautions must be taken to ensure that only proper materials are used in contact with damp carbon. All metal in contact with damp carbon must be corrosion resistant (type 316 stainless steel) or noncorrosive (fiberglass). Tanks, vessels, and fittings may be fabricated of epoxy or rubber-coated steel, stainless steel (type 316), glass-reinforced polyester (fiberglass), or concrete. At existing plants, embedded metals can be field lined with rubber, glass-reinforced polyester, vinyl ester, or epoxy. Valves and instruments such as flowmeters and eductors must be protected from corrosion.

REGENERATION OF GRANULAR ACTIVATED CARBON

Facilities for on-site regeneration of GAC are expensive, and smaller water systems are generally designed to waste spent carbon and replace it with new. Larger systems must carefully look at the economics of installing regeneration equipment.

Regeneration Facilities

GAC is expensive compared with sand and anthracite filter media. As a result, it is often cost-effective to regenerate and reuse GAC. Two basic approaches to regenerating GAC are off-site regeneration and on-site regeneration.

The rate at which contaminants break through the carbon bed determines the size of the regeneration system. The primary design criterion for a reactivation system is the rate of carbon regeneration (mass per unit time, e.g., kg/hr). A complete regeneration or reactivation process typically consists of a furnace system, including a feed system, a drying or dewatering scheme, and a reactivation process. A simple regeneration system schematic is presented as Figure 14.13.

Typical carbons used in water treatment require regeneration from every 6 months to 5 years, depending on the application. On-site regeneration is generally not cost-effective unless the carbon exhaustion rate is over 910 kg per day. Current U.S. applications fall in the range of off-site regeneration (225 to 700 kg per day). For facilities using less than 225 kg per day, off-site disposal should be considered.

For off-site disposal applications, virgin carbon is first purchased from a carbon supplier. Once carbon becomes exhausted, it is transported in slurry form by gravity to a drainage collection tank where the supernatant is routed to the plant headworks for treatment. Recovered carbon is then manually packaged or conveyed to hopper trailers and

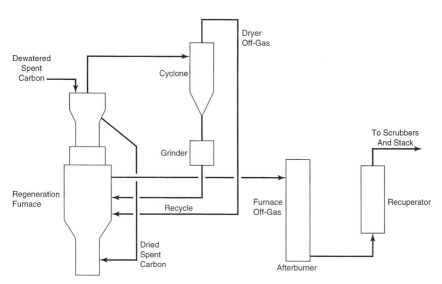

FIGURE 14.13 Activated carbon regeneration system schematic.

shipped to a landfill or an incineration facility, depending on the chemicals that have been adsorbed.

The off-site disposal concept is a simple solution for applications with small carbon exhaustion rates. Because toxic or hazardous materials are likely to be absorbed on the spent carbon, landfill disposal may not be a feasible long-term option if the disposed carbon is considered a hazardous waste. Fortunately, this is not often a concern for GAC used in drinking water treatment. Incineration may be necessary to satisfy environmental concerns regarding proper ultimate disposal.

Off-site regeneration is similar to off-site disposal, possessing many of the economic benefits of carbon reuse. However, the number of handling steps and resulting carbon attrition involved in off-site regeneration must be considered. In most cases, GAC manufacturers are capable of retrieving, transporting, and regenerating the spent GAC at their own facility on a contract basis, but most regeneration facilities combine every type of carbon they receive.

Custom regeneration is expensive. Before entering into an agreement with a carbon supplier for off-site regeneration, assurance should be given that only virgin GAC will be returned to the plant. Although it may not be possible to track and monitor, receiving regenerated activated carbon obtained from municipal or industrial waste treatment processes should be avoided.

Regeneration Alternatives

The three most common methods of GAC regeneration are steam, thermal, and chemical. Of the three methods, thermal regeneration is the most widely used. Chemical and steam regeneration techniques are used primarily in industrial systems designed to recover the adsorbate. Only thermal regeneration will be discussed in detail in this chapter.

Thermal Regeneration. Thermal regeneration is typically carried out in five steps:

1. Dewatering to 40% to 50% moisture content by draining. It is more efficient to dewater moisture than to evaporate extra water in the drying stage.
2. Drying at temperatures of up to 200° C.
3. Baking or pyrolysis of adsorbates at 500° to 700° C.
4. Activating carbon at temperatures above 700° C.
5. Quenching regenerated carbon.

Reactivation Rate. The primary consideration in designing a carbon reactivation system is the reactivation rate. This rate depends on carbon loading and carbon usage or exhaustion rate in the adsorbers. Carbon loading is the amount of organic matter removed from the water treated per unit quantity of carbon and is usually expressed in terms of milligrams per gram (mg/g) or pounds per pound (lb/lb) of organic materials to carbon. Carbon usage is often expressed as carbon dosage or kilograms of carbon per megaliters of water treated and is inversely related to carbon loading.

Carbon usage (exhaustion) rate is the mass of carbon exhausted per unit time. It is usually expressed as kilograms or pounds per day and is inversely related to reactivation frequency. Reactivation frequency measures the time between the reactivation of a given bed of carbon. The CUR is the minimum rate at which carbon would be reactivated in a continuous system with no intermediate storage. The designer should provide sufficient carbon storage between the adsorbers and reactivator so that the furnace is sized above the minimum reactivation value. Operation at a higher rate takes into account times when furnaces are out of service for maintenance or repair, while still satisfying the carbon usage rate of the adsorbers.

Types of Furnaces. Four types of regeneration furnaces may be used to remove adsorbed organics from activated carbon:

1. Electric infrared oven
2. Fluidized bed furnace
3. Multiple-hearth furnace
4. Rotary kiln

Typical parameters for the multiple-hearth, fluidized bed, and infrared furnaces are summarized in Table 14.8.

In selecting the type of furnace to be used, the following criteria should be considered:

- Capital cost
- Operating experience and service record
- Ability to change feed rates
- Individual control of the drying, pyrolysis, and activation zones
- Uniformity and quality of the regenerated carbon
- Design and operating parameters that may lead to potential problems such as explosions, fires, and dust
- Efficiency with respect to downtime during regeneration
- Downtime
- Energy requirements

TABLE 14.8 Typical Operating Parameters for Three Reactivation Systems

		Multi-hearth furnace	Fluidized bed furnace	Infrared furnace
Furnace loading:	(lb/ft²/day)	40 to 115	1,460 to 1,700	—*
	(kg/m²/day)	195 to 561	7,124 to 8,296	—*
Residence time (min)		120 to 240	10 to 20	25
Natural gas fuel (ft³/lb carbon)		3.0 to 5.5	1.5 to 2.2	—*
Steam (lb/lb carbon)		1.0 to 3.0	0.4 to 0.8	—*
Heat loss (%)		15	9	—*
Carbon losses (%)		5.7	2 to 5	5

*Data not available.

- Operating and maintenance costs
- Experience of system suppliers and availability of service

Electric Infrared Oven Furnaces. An infrared furnace consists of an insulated enclosure through which carbon is transported on a continuous metal conveyor belt. Heat for reactivation is supplied to the furnace by a number of electric infrared heating elements mounted in the top of the tunnel. Spent carbon is fed to the reactivator through an airlock controlling carbon feed rate and minimizing air intake. Carbon is spread evenly onto the belt conveyor and transported slowly through discrete chambers or zones of the furnace.

Chambers provide increasingly higher temperatures. Atmosphere and temperature can be controlled with this type of furnace. Carbon discharges into a conventional water-filled quench tank for cooling. Modules are made of a steel shell lined with thermal shock-resistant ceramic fiber insulation and include support rollers for the conveyor belt.

Fluidized Bed Furnaces. A fluidized bed furnace suspends carbon particles by an upward-flowing gas stream. Gas velocity is controlled so that the weight of particles in the bed is just balanced by the upward force of the gas. The use of a fluidized bed in this application offers the advantages of uniform temperatures within the bed and high heat and mass transfer rates. Steam must be injected to control the temperature in the furnace. A typical fluidized bed regeneration furnace is shown in Figure 14.14.

Dewatered carbon is dried in the first zone to a moisture content of about 1%. It then flows by gravity to the bottom zone. The lower zone is maintained at about 982° C. As the carbon temperature increases to the temperature of the zone, adsorbed organic materials are pyrolyzed and gasified. After leaving the furnace, carbon is cooled in a water-filled quench tank.

Air is injected into the space between the two zones that serves as a second combustion zone. This zone burns organic compounds and the hydrogen and carbon monoxide released during reactivation in the lower chamber. Part of the heat produced dries wet carbon in the upper bed, with the remainder recycled to the lower bed reactivation zone.

Multiple-Hearth furnaces. Figures 14.15 and 14.16 show cross sections of a typical multiple-hearth furnace. This type of furnace consists of a refractory lined steel shell containing five to eight circular hearths. Burners are generally located on the bottom hearth to provide heat necessary for reactivation. Burners can also be located on higher hearths to provide improved temperature profiles for increased performance and flexibility of operation.

Spent carbon enters the furnace through a dewatering screw conveyor onto the top hearth. A rotating center shaft with supporting arms and rabble blades moves carbon in a

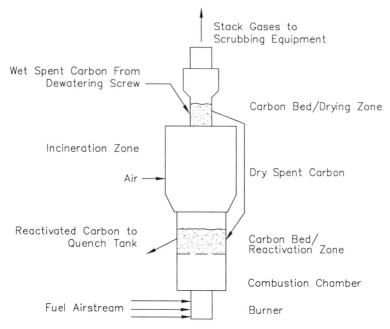

FIGURE 14.14 Fluidized bed regeneration furnace.

spiral path so that it drops from hearth to hearth and finally out of the bottom of the furnace into a quench tank for cooling. Residence time in the furnace is controlled by the rotation speed of the center shaft. The hollow rabble arms are cooled by ambient air blown through them. The atmosphere within the furnace is tightly controlled to prevent excessive carbon oxidation. Steam is usually added in the lower hearths for temperature control.

Rotary Kiln Furnaces. Multiple-hearth furnaces are the traditional method of granular activated carbon reactivation, although fluidized bed and infrared furnace installations are more common. Rotary kilns are more energy intensive and are becoming less competitive with other technologies.

Furnace Size. Furnace size is based on the rate at which carbon is charged per unit of furnace dimension and varies with the type of reactivation furnace chosen. For multiple-hearth furnaces, values are usually given in pounds of carbon per square foot of hearth per day (or kg/m²/day). For multiple-hearth furnace design, typical values are 40 to 115 lb/ft²/day (145 to 560 kg/m²/day). Values for fluidized bed furnace loadings range from 1,460 to 1,700 lb/ft²/day (7,125 to 8,300 kg/m²/day). These numbers are based on the cross-sectional area perpendicular to the flow of combustion air. In an infrared furnace, carbon flow depends on the width of the conveyor belt and residence time.

Regeneration By-Products

Activated carbon must be regenerated in a way that does not pollute the environment. For this reason, scrubbers and afterburners are used to minimize particulate and gaseous

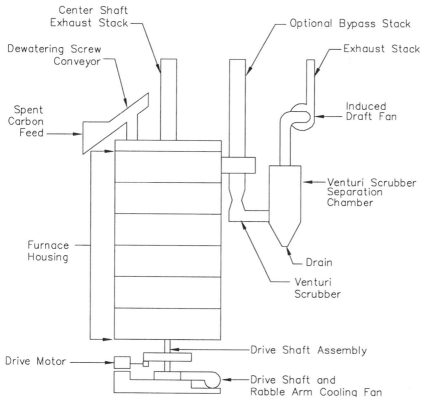

FIGURE 14.15 Typical multiple-hearth furnace regeneration system.

emissions, respectively. The afterburner oxidizes organic compounds, and the scrubber removes particulate matter and any soluble chemical species from the gas stream. Off-gas control is important in controlling dioxins and furans produced during regeneration. For these reasons, local air quality regulations should be considered early in the design process. A supplementary dust collector is used only where it is necessary to collect particulate matter not normally removed by scrubbers.

Transporting Carbon from the Quench Tank

Eductors typically transport carbon away from the quench tank. The quench tank is normally a small tank that receives hot regenerated carbon from the furnace, and the water level must be controlled to prevent hot carbon from being exposed to air. When a method other than an eductor is used to transport carbon, the volume of makeup water must be greatly increased. The increased volume of makeup water causes turbulence that keeps carbon in suspension and increases carbon losses. With an eductor, most of the transport and dilution water is supplied by the water treatment system.

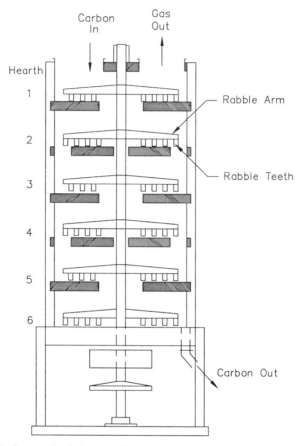

FIGURE 14.16 Cross-sectional view of multiple-hearth furnace.

BIBLIOGRAPHY

Adsorption of Pesticides by Powdered Activated Carbon. Denver, Colo.: AWWA Research Foundation, American Water Works Association, 1994.

Clark, R. M., and B. W. Lykins, Jr. *Granular Activated Carbon Design, Operation and Cost.* Boca Raton, Fla.: Lewis, undated.

Control of Organic Compounds with Powdered Activated Carbon. Denver, Colo.: AWWA Research Foundation, American Water Works Association, 1991.

Crittenden, J. C., D. W. Hand, H. Ariham, and B. W. Lykins, Jr. "Design Considerations for GAC Treatment of Organic Chemicals." *Journal AWWA* 79(1):74, 1987.

Culp, R. L., and R. M. Clark. "Granular Activated Carbon Installations." *Journal AWWA* 75(8):398, 1983.

"Design and Use of Granular Activated Carbon: Practical Aspects." *Proceedings Technology Transfer Conference.* May 9 and 10, 1989. Denver Colo.: American Water Works Association, 1989.

The Effect of Metals on Thermal Regeneration of Granular Activated Carbon. Denver, Colo.: AWWA Research Foundation, American Water Works Association, 1994.

GAC Filter Adsorbers. Denver, Colo.: AWWA Research Foundation, American Water Works Association, 1987.

Graese, S. L., V. L. Snoeyink, and R. G. Lee. "Granular Activated Carbon Filter-Adsorber Systems." *Journal AWWA* 79(12):64, 1987.

The Hazardous Potential of Activated Carbons Used in Water Treatment. Denver, Colo.: AWWA Research Foundation, American Water Works Association, 1994.

J.M. Montgomery Consulting Engineers. *Water Treatment Principles and Design.* New York: John Wiley and Sons, 1985.

Kawamura, Susumu. *Integrated Design of Water Treatment Facilities.* New York: John Wiley and Sons, 1991.

Knappe, D. R. U., V. L. Snoeyink, G. Dagois, and J. R. DeWolfe. "Effect of Calcium on Thermal Regeneration of GAC." *Journal AWWA* 84(8):73, 1992.

Najm, I. N., V. L. Snoeyink, B. W. Lykins, Jr., and J. Q. Adams. "Using Powdered Activated Carbon: a Critical Review." *Journal AWWA,* 83(1):65, 1991.

Optimization and Economic Evaluation of Granular Activated Carbon for Organic Removal. Denver, Colo.: AWWA Research Foundation, American Water Works Association, 1989.

Organics Removal by Granular Activated Carbon (GAC). Denver, Colo.: American Water Works Association, 1989.

Oxenford, J. L., and B. W. Lykins, Jr. "Conference Summary: Practical Aspects of the Design and Use of GAC." *Journal AWWA* 83(1):58, 1991.

Standardized Protocol for the Evaluation of GAC. Denver, Colo.: AWWA Research Foundation, American Water Works Association, 1992.

Technologies for Upgrading Existing or Designing New Drinking Water Treatment Facilities. Cincinnati, Ohio: U.S. Environmental Protection Agency, 1990.

Weisner, M. R., J. J. Rook, and F. Fiessinger. "Optimizing the Placement of GAC Filtration Units." *Journal AWWA* 79(12):39, 1987.

CHAPTER 15
CHEMICALS AND CHEMICAL HANDLING

A principal issue in modern water treatment plant design is the decision of what chemicals are to be used for treatment processes, how they are to be shipped and stored, and what type of chemical feed equipment should be used. This chapter summarizes alternatives and issues associated with gaseous, dry, and liquid chemicals and the associated equipment and piping necessary for their use. Other issues for the designer to consider involve the many regulations that must be met and safety considerations in planning the chemical feed systems.

Additional discussions of the theory of chemical treatment may be found in detail in *Water Quality and Treatment*. Details of specific chemicals commonly used in water treatment processes are provided in Appendix A.

RECEIVING AND STORING PROCESS CHEMICALS

Many alternatives are available for receiving and storing process chemicals, depending on plant size and location, treatment processes to be used, and many other factors.

Sizing Storage and Feed Systems

Sizing storage and feed systems begins with an investigation of dosage requirements for each chemical used. Chemical feed rates (lb/day, ft^3/h, gal/day, kg/day, etc.) can be computed from dosages (mg/L) and plant flow rates (million gallons per day, mgd, ML/day). For chemicals in dry form:

$$\text{Feed rate (lb/day)} = 8.34 \text{ lb/gal} \times \text{Dosage (mg/L)} \times \text{Flow (mgd)}$$

For chemicals in solution form:

$$\text{Feed rate (gal/h)} = \frac{8.34 \text{ lb/gal} \times \text{Dosage (mg/L)} \times \text{Flow (mgd)} \times 100\%}{24 \text{ (h)} \times \text{Concentration (\%)} \times \text{Density (lb/gal)}}$$

Chemical storage or inventory requirements can be computed from feed rates and the number of days of storage required.

Capacities and Feed Rates. Many different conditions must be evaluated to determine the range of feed rates used to select feed equipment capacities for each chemical. For an existing treatment plant, or where there is another plant using the same water source, historical records should be investigated to obtain peak- and minimum-hour dosages for the plant's chemicals. For a new plant without records, bench- and pilot-scale testing are required. Table 15.1 provides guidelines for chemical feed and storage capacities, but these guidelines should be integrated with consideration to regional and state standards, operating records of similar plants, and experience.

Minimum feed rate capacities and the feed equipment's capability to accurately feed at low rates are as important as maximum capacities. Once dosage and feed rates are calculated, these usually must be adjusted upward to account for chemical purity or percent availability for reaction.

Sizing the Inventory. Inventory size is based on factors that vary from plant to plant, even within the same system. First determine the maximum monthly feed rate of the chemical, accounting for dosage and plant flow rate. Next, the appropriate number of days of chemical storage should be established for each chemical, typically between 7 and 30 days. Major considerations include the general availability of the chemical, the location of chemical suppliers, reliability of suppliers during periods of shortage, normal delivery time, and delays in shipment that may be caused by weather conditions. Another consideration is the possibility of delays in receiving shipments because of transportation strikes or natural or human-caused emergencies. Some state regulatory agencies specify a minimum number of days for certain chemicals.

Once the number of days of storage for each chemical has been identified, it is possible to determine whether bulk deliveries are practical. In smaller plants using bulk delivery chemicals, the quantity stored may be dictated more by the size of the bulk shipping containers in use in the area than by the desired number of days' supply. Many regulators require minimum storage requirements equal to one and one-half truckloads of a bulk delivery.

Excess inventory is undesirable if the chemical has a limited shelf life. Chemicals such as soda ash and quicklime that possess either hygroscopic properties or are otherwise affected by moisture can become difficult to handle after several months in bulk storage unless precautions are taken to control the environment. Chemicals furnished in returnable containers, such as chlorine cylinders or ton containers, may be subject to demurrage charges unless these are waived by agreement in the purchase contract.

Delivery, Handling, and Storage of Chemicals

Modes of transporting chemicals include railroad tank cars, trucks, and, in rare instances, barges. The type of unloading facilities depends on whether chemicals arrive in bags, drums, or bulk; the form in which the chemical will be fed to the process; the type of carrier (rail or truck); and the location and type of storage silo, tank, or other storage facility.

TABLE 15.1 Suggested Chemical Feed and Storage Design Parameters

Design parameter	Dosage condition	Plant flow rate condition
Maximum feeder capacity	Peak hour	Peak hour
Minimum feeder capacity	Minimum hour	Minimum hour
On-site storage	Maximum month	Maximum month
Day tank volume	Maximum day	Maximum day

Receiving Shipments. Whether the chemical is liquid or solid, truck delivery is usually preferred because of its simplicity, maneuverability, and generally prompt and predictable delivery. In contrast to rail shipments, truck delivery makes more load-size options available while still preserving the economies of bulk purchasing. Modern air-slide and pneumatic unloading equipment on these vehicles permits quick and easy delivery of bulk loads below grade and to overhead silos.

Unloading platforms or docks must be provided in all but the smallest plants to accommodate truck deliveries of containerized chemicals, including bagged material, drums, small gas cylinders, and the like. Horizontal transport of nonbulk chemicals from the unloading dock is normally accomplished by hand or power trucks, conveyors, or a monorail system. Storage in areas above or below the unloading area requires installation of an inclined conveyor, dock leveler, hoist, or elevator designed for the maximum anticipated loads.

Receiving and Storing Pressurized, Liquefied Gases. Gases commonly used in water treatment are chlorine and ammonia for disinfection processes and carbon dioxide for stabilization and pH control after softening. Oxygen is sometimes used for the gas feed to ozonation generators. All these gases are shipped as pressurized liquids.

Chlorine. Chlorine used in the water industry is seldom stored in on-site receivers. Instead, the shipping container is used for storage. If chlorine is received at a treatment plant in a tank truck or barge, a sufficient pressure differential must be maintained between the shipping container and the storage container during transfer.

Gaseous chlorine is commercially available in containers of the following sizes:

- 150 lb cylinders
- 1-ton cylinders
- 15- to 17-ton tank trucks
- 16- to 90-ton railroad tank cars

Cylinders of 100 lb (45 kg) weight are also available, but are rarely used. The dimensions of the 150 lb (68 kg) and 1-ton (907 kg) cylinders are shown in Figure 15.1. Because chlorine tank trucks do not have standard sizes, dimensions should be obtained from local chlorine suppliers.

In all the containers listed above, liquid chlorine occupies a maximum of approximately 85% of the volume when the product is delivered. This 15% allowance provides room for liquid chlorine to expand if the cylinder becomes warm. No chlorine container should ever be directly heated. If the liquid were to become warm enough to expand and fill the entire container, tremendous hydrostatic pressure would result, and the container would rupture. As a safety precaution, cylinder outlet valves are equipped with a small fusible plug that melts at approximately 158° F (70° C) and releases some liquid chlorine to cool the cylinder before a more serious accident can occur.

Chlorine is most often fed by withdrawing gas from the top of the container, and the reduced pressure above the liquid then causes some of the liquid to evaporate, providing additional gas. The maximum withdrawal rate with this method is about 40 lb per day (18 kg per day) for a 150 lb cylinder and 400 lb per day (181 kg per day) for a 1-ton cylinder. Maximum continuous withdrawal rates for containers of other sizes can be approximated by comparing the surface area available to absorb the heat required to replace the heat of evaporation lost. If chlorine is being used in this manner, the containers should be maintained in an environment that can be heated to 65° F (18° C). The design-maximum 24 h withdrawal rate of a system designed to withdraw gas from the top of the container should not exceed the continuous withdrawal rate of the containers on-line.

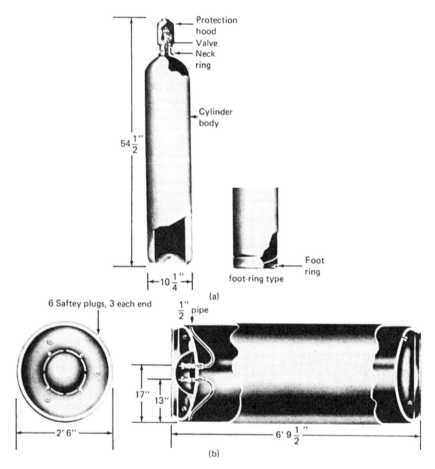

FIGURE 15.1 Dimensions of standard chlorine cylinders. (a) 150-lb cylinder. *(Courtesy of PPG Industries, Inc., Pittsburgh, Pa.)* (b) One-ton cylinder. *(Courtesy of the Chlorine Institute, Inc.)*

Bulk Shipment of Chlorine. Chlorine tank barges have either four or six tanks, each with a capacity of from 85 to 185 tons (77 to 168 metric tons), and must comply with Coast Guard regulations for piping, loading, and unloading arrangements. The valves are similar to those on tank cars, but the arrangement is not standard, and on some barges the valves are larger.

Tank trucks consist of cargo tanks for chlorine permanently attached to a motor vehicle. U.S. Department of Transportation (DOT) regulations are explicit as to tank and accessories design and maintenance, marking, and emergency handling. Only a few chlorine tank trucks are currently being operated in the United States. Rail car deliveries are practical only for the largest plants and are subject to the economic feasibility of constructing a siding from the nearest rail line.

Ton containers of chlorine are authorized for rail shipment, but only on a special tank car frame known as a multiunit tank car designed to hold 15 containers. The initial rail shipment of these containers is unloaded from the car for use at the consumer's plant. In subsequent shipments, the full containers are exchanged for empties, which are returned for refilling. No freight is charged on the return carload of empty containers because tank cars are entitled to free return movement.

It is impractical to ship less than 15 containers because the transportation fee is figured for 15 full containers at prevailing carload rates. Multiunit tank cars must be consigned for delivery and unloading on a private track. If a private track is not available, containers may be removed from the car frame on carrier tracks with previous written permission. Regulations provide that one or more ton containers may be transported on trucks or semitrailers under special conditions.

If logistics show that purchase of chlorine by tank truck or railroad tank car is feasible, stationary storage facilities should be considered. The user can purchase chlorine for a better price if the tank car or trailer is on-site only for the period of time it takes to unload it. On the other hand, the use of tank truck trailers and railroad cars for on-site storage is also common.

Stationary chlorine facilities should be designed in complete accordance with the recommendations of The Chlorine Institute as described in *Facilities and Operating Procedures for Chlorine Storage*. Provision must also be made for a weighing device, either a lever scale system or, more commonly, load cells.

An air padding system is recommended for unloading the tank car and removing gas from the tank before inspection. The air should be dried with a heat-reactivated, desiccant-type air dryer. Facilities must be provided to vent chlorine gas or chlorine-and-air mixtures from the storage tank to the consuming process or other disposal system. When transferring chlorine from the storage tank to the consuming process, air padding may be necessary. The procedure is essentially the same as that required for emptying tank cars.

Tank cars should be emptied through a suitable-metal flexible connection that accommodates the rise of the car as its springs decompress. Tank cars are almost invariably emptied by discharging liquid. Liquefied gases may be unloaded by their own vapor pressure. Cold weather usually decreases the unloading rate. Sometimes it is desirable to place an air pad over the chlorine vapor in the car to facilitate unloading. The air pad may be provided by the chlorine supplier or at the point of use. Weight-measuring devices are preferred to determine tank contents; gauge glasses should not be used. Adequate lighting, including auxiliary power sources, should be provided for night operations.

The location, design, maintenance, and operation of chlorine bulk storage tanks may be subject to local or state regulations and to insurance requirements. The number and capacity of storage tanks should be consistent with the size of shipments received and the rate of consumption. Receiving and unloading areas and safety precautions applicable to the handling of single-unit cars, cargo trucks, and barges are subject to DOT, Coast Guard, and other regulations.

Handling Chlorine Ton Containers and Cylinders. Cylinders should be stored upright and secured in a manner that permits ready access and removal. Ton containers should be stored horizontally, slightly elevated from ground or floor level, and blocked to prevent rolling. A convenient storage rack for ton containers can be constructed with rails or I-beams to support both ends of the containers.

Ton containers should not be stacked or racked unless special design provisions are made for easy access and removal. Storage areas should be clean, cool, well ventilated, and protected from corrosive vapors and continual dampness. Cylinders and ton containers should preferably be stored indoors, in a fire-resistant building, and away from heat sources, other compressed gases, and flammable substances. If containers are stored outdoors, the area should be shielded from direct sunlight and accumulations of rain and snow. If natural ventilation is inadequate, storage and use areas should be equipped with suitable mechanical ventilators.

All storage areas should be designed so that personnel can quickly escape in emergencies. If possible, two exits should be provided from each separate room or building in which chlorine and other gases are stored, handled, or used.

Various mechanical devices, such as skids, troughs, or upending cradles, should be provided to facilitate the safe unloading of chlorine cylinders. Specially designed cradles or carrying platforms are recommended if it is necessary to lift the cylinders by crane or derrick. Chains, lifting magnets, and rope slings that encircle cylinders are unsafe. Hand trucks are preferred for lateral movement.

Special provision should be made for lifting ton chlorine containers from a multiunit tank car or truck by means of hooks designed to fit on a chain sling or lifting clamp, in combination with a hoist or crane with at least a 2-ton (1,800 kg) capacity. Containers should be moved to point of storage or use by truck or by a crane and monorail system.

Ammonia. Anhydrous ammonia is most commonly shipped in 150 lb (68 kg) cylinders and 1-ton (907 kg) containers similar to those used for chlorine. The guidelines described for handling and storing chlorine containers also apply to ammonia. Ammonia and chlorine gases should not be stored or fed in the same room.

Carbon Dioxide and Oxygen. Carbon dioxide and oxygen gas can be manufactured on-site for use in water treatment processes. More commonly, it is delivered by tanker truck and pumped into refrigerated, pressurized, outdoor storage tanks on the plant site. The liquids are vaporized and sometimes heated to the desired temperature for process use. The storage tanks are equipped with fill lines having quick-connect couplings located for easy access by delivery trucks. Tanks and all piping segments capable of isolation must be provided with pressure relief safety valves.

Evaporators. Chlorine, ammonia, oxygen, and carbon dioxide are usually stored as liquid and must be evaporated to a gas for metering and flow control to the point of application. Depending on ambient temperatures and rates of withdrawal, some of these gases must be evaporated by warming in vaporizers.

If chlorine containers are 1 ton or larger and the withdrawal rates exceed the quantity available by direct evaporation, evaporators may be required. Evaporators are available in capacities of 4,000, 6,000, and 8,000 lb per day (1,800, 2,700, and 3,600 kg per day). When an evaporator is used, liquid is withdrawn from the bottom of the container and transported to the evaporator, where it is converted to gas. The most common type of evaporator uses an electric resistance heater in a hot-water bath surrounding a vessel in which the liquid is converted to gas.

Sometimes the bath is heated with steam or with a separated, recirculating hot-water system. Figure 15.2 shows a cross section of a typical evaporator. The heat of evaporation of chlorine is low, approximately 69 cal/g, compared with 540 cal/g for water. However, evaporators should be designed with extra capacity to ensure that the gas is superheated and does not recondense on the downstream side.

When an evaporator is being used beyond its capacity, misting occurs. A chlorine gas filter should be installed on the exit gas line from the evaporators to remove impurities in the chlorine that would be detrimental to the chlorinator. Evaporators should be equipped with an automatic shutoff valve to prevent any liquid from passing to the chlorinators.

When possible, all portions of the chlorine feed system that contain liquid chlorine should be designed and operated with all liquid in the system as a continuous medium. To shut down the evaporator, it is necessary only to close the effluent valve on the evaporator. No other valves between the evaporator effluent valve and the liquid chlorine container should be shut. Very long liquid chlorine lines should be avoided; if they are unavoidable, expansion chambers should be provided. It should be emphasized that liquid chlorine has a high temperature-expansion coefficient. Unless expansion is permitted, the temperature increase in trapped liquid will increase the pressure enough to rupture the pipes.

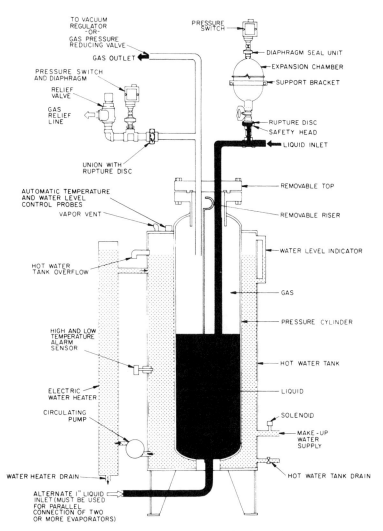

FIGURE 15.2 Cross section of typical evaporator. *(Courtesy of Wallace and Tiernan Division of Pennwalt Corp.)*

Ammonia feed systems are similar to those used for chlorination. Unlike chlorine and ammonia, carbon dioxide is usually vaporized and maintained with positive pressure during the metering and flow rate control process.

Delivery, Handling, and Storage of Dry Chemicals. Dry chemicals may be purchased and stored in bags or barrels if relatively small quantities are to be used. For larger use, economy of bulk delivery and storage must be considered.

Bulk Delivery by Truck or Rail Car. When large quantities of the product are consumed, bulk delivery provides not only economic benefits, but also protection against supply shortages and transportation difficulties. Obviously, the type of storage facilities provided vary with the chemical.

Flow is the key word in chemical handling of bulk solids. In general, the flow of material increases with particle size, uniformity, hardness, smoothness, absence of fines, and lower moisture content. These physical parameters can be controlled to some degree by purchase specification.

Solids unloading is usually accomplished with pneumatic equipment (blower or vacuum), air slides, or mechanical screw conveyors or bucket elevators. The latter equipment is satisfactory for unloading lumpy or coarse material where excessive dust is not a problem.

Pneumatic truck unloading of dry bulk chemicals is done through the user's pipe conveying system, consisting of a truck inlet panel, interconnecting piping to silos or storage bins, safety release valve, and dust collector mounted on top of the storage bin. Piping diameter is usually standardized at 4 in. (10 cm), with bends having a minimum radius of 4 ft (1.2 m).

Maximum length of piping between the inlet panel and storage depends on the nature of the material being transported. Pebble lime, for instance, may be blown as much as 100 ft (30 m) vertically, providing the total length of run does not exceed 150 ft (46 m). Lightweight powdery material can easily be transported up to 300 ft (90 m) over a combined vertical and horizontal distance. Pneumatic trucks equipped with air compressors are generally available in capacities of from 700 to 1,300 ft^3 (20 to 40 m^3).

Covered railroad hopper cars used for delivery of dry bulk chemicals may have capacities up to about 3,700 ft^3 (105 m^3). Some of these cars are constructed with two to four compartments, each provided with its own bottom discharge gate. Material is withdrawn through these gates positioned over undertrack hoppers. Air vibrators are used where necessary to facilitate the movement of fine, powdery material through the hopper.

A more widely used type of rail car permits a method of pneumatic unloading similar to that used on pneumatic trucks. Others employ air slide conveyors to move the material out to the truck hopper. Canvas connectors, or stockings, generally connect to the truck hopper that feeds the conveying system. The user must provide air compressors to pressurize the cars or the vibrators and air slides, because compressors are not part of the car equipment, as is the case with pneumatic trucks. Hopper cars may also be unloaded using a vacuum system consisting of suction pump, filter receiver, and discharge air lock or rotary gate at the top of the silo or storage bin.

Storage of Bulk Dry Chemicals. Once the desired material is delivered, proper bin or silo capability must be provided. Dry chemical flow in the bins must be maintained by installing vibrating or pulsating devices, live bin bottoms, and internal devices to control packing and arching. The height of bins or silos is usually approximately 2.5 to 4 times the diameter, with the discharge area as large as possible.

Hopper bottoms should slope at least 60 degrees from the horizontal; for storage of hydrated lime, an even greater slope is desirable. Offset hopper sections are commonly used, with the outlet on one side of the vertical bin axis. The distortion produced by the varying slope angles tends to prevent arching in the conveying section. The use of vibrators to maintain flow requires caution and consideration of the type of material being handled; the worst possible situation occurs when fine materials (such as hydrated lime) are overvibrated and packed. Such material can only be vibrated intermittently—for instance, by a 2- to 4-second pulse repeated several times a minute. By contrast, lumpy material such as pebble lime can be vibrated continuously during discharge. The interrupter unit used to control the vibrator must be interlocked electrically with the process feeder to allow vibration only during discharge.

Air jets and pulsating air pads are also commonly used to fluidize light materials such as hydrated lime. Numerous other devices are available, the most popular being the "live" bin bottom. These units operate continuously during discharge and use gyrating forces or upward-thrusting baffles within the hopper to eliminate bridging and rat-holing. Less sophisticated devices include double-ended cones supported centrally within the hopper, rotating chains or paddles, and horizontal rods run from wall to wall.

Quicklime and hydrated limes are abrasive but not corrosive, and steel or concrete bins and silos can be used for storage. It is imperative that the storage units be airtight as well as watertight to reduce the effect of air slaking; this includes relief valves, access hatches, dust collector mechanism, and so on, all of which are normally exposed to the weather. Bins and silos can be designed with rectangular, square, hexagonal, or circular cross sections; the first three make optimum use of plant space, but the circular silo is less susceptible to sidewall hangups, which tend to occur in corners of bins of other shapes. Regardless of the cross-sectional configuration of the vertical storage unit, the bottom is always designed with a hopper or conical base right up to the discharge gate. The design volume of any silo or bin should be based on the average bulk density of the chemical, with an allowance for 50% to 100% extra capacity beyond that required to accommodate a normal-size delivery.

Because there is a tendency for water of crystallization from alum to partially slake lime, it is necessary to avoid mixtures of alum and quicklime. In a closed container, this may lead to a violent explosion. For the same reason, equal care should be taken to avoid mixtures of ferric sulfate and lime.

Powdered activated carbon, a finely ground, low-density material, is capable of producing copious quantities of black dust with the least disturbance. As a result, large users of this material who can accept air slide truck or rail bulk shipments prefer to have the carbon unloaded directly into slurry tanks with a dust collector or scrubber on the vent. In this way, further handling of the powder into and out of storage is avoided, and dust problems are minimized. If dry storage is necessary or preferred, a vacuum transfer system minimizes dust. Suppliers should be consulted about trucks or rail cars that can accommodate vacuum hookups.

Storage of Materials in Bags and Drums. Areas used for dry chemical storage of bagged material, drums, and other chemical containers must be fireproof, dry, and well ventilated. Compressed gases in cylinders should be stored separately in an area provided with separate mechanical ventilation or exhaust fans.

Containment of Dry Chemical Dust. Any unloading or transfer of dry chemicals creates dust, especially when airflow equipment is used to unload the material. Operation of this type of equipment requires discharging dust-free conveying air to the atmosphere, a job for which a bag-type filter is best suited. Because these filters collect chemical dust that can be returned to the process, it is common practice to mount the collector on top of the silo or dry chemical feeder so vibration of the bags drops the chemical back into the original container.

Dust collector operation is remotely controlled from the unloading point or truck inlet panel. It is possible to have one dust collector serve more than one storage receiver provided the same chemical is being handled, but under no circumstances should a mixture of different chemical dusts be allowed to accumulate within one collector. A dust collector's size depends on the volume of air to be handled. Air blowers mounted on pneumatic trucks commonly have the capability of producing up to 750 ft^3/min (21 m^3/min) of air.

Bag filters work for most dust collection applications, but vortex scrubbers should be considered for powdered activated carbon slurry tanks. The unloading rate for carbon into a slurry tank potentially could overwhelm a bag filter and require excessive maintenance. Properly specified vortex scrubbers can provide adequate removal of carbon dust with relatively minimal maintenance. When mounted on top of the carbon slurry tank, vortex scrubbers operate with a water spray entering near the top and traveling countercurrently to the carbon-laden air entering from the bottom. Water and trapped carbon discharge directly into the slurry tank.

Delivery, Handling, and Storage of Liquid Chemicals. Some of the chemicals most commonly used in liquid form in water treatment are listed in Table 15.2. When computing storage and feed rate requirements, the effective density must be used. The effective density is determined by the bulk density, active component of the chemical molecule, and solution strength, as shown in the table.

Liquid chemicals, such as caustic soda, liquid alum, acids, or corrosion inhibitors, are generally delivered by tank truck. To facilitate safe unloading, it is the user's responsibility to provide appropriate fill-pipe connections, clearly labeled and equipped with protective caps. A concrete drip sump protected with a chemically resistant coating should be provided beneath all fill-pipe connections.

Local and state regulations may require secondary containment for the entire unloading area. Storage tank overflow pipes should terminate into the secondary containment. If the stored chemical has fuming or corrosive properties, the overflow must be sealed with a vapor check valve.

Vent pipes should terminate outdoors with screened elbows facing downward. Vent and fill lines must be run without traps. Overflow lines may interconnect sealed tanks containing the same material, but they should not be allowed to terminate in the open air unless they are inside the secondary containment area. Such a line would otherwise become the primary vent and permit a spill to occur in the event of an overfill. As with any chemical, but particularly with liquids, the user must be certain that the quantity ordered can be accommodated in storage at the time of delivery.

Accidental overflows are best prevented by providing dependable level indicators and alarms on all tanks. To further guard against overfilling a storage tank, a high-level audible alarm should be provided as part of the in-plant chemical tank-level indicating system. This alarm should be mounted outside at the unloading station to alert the vehicle operator. The alarm can be common to all tanks served by the particular unloading station.

Several level sensing systems are available with local or remote indicating and alarm capability. Pneumatic level sensors using bubble pipes are effective and accurate, but before they are used, it must first be determined that air bubbles will not deteriorate the chemical whose level is being measured.

TABLE 15.2 Liquid chemical effective densities

Chemical name	Formula	Specific gravity	Density of solution (lb/gal)	Total molecular weight	Active molecular weight	Active (%)	Typical solution strength (%)	Effective (lbs/gal)
Sodium hypochlorite	NaOCl	1.21	10.1	74.5	51.5	69.1	12.5	0.87
Fluorosilicic acid	H_2SiF_6	1.21	10.1	144.1	144.1	79.2	23.0	1.84
Alum	$Al_2(SO_4)_3$	1.34	11.2	342	342.0	100	50.0	5.6
Ferric chloride	$FeCl_3$	1.44	11.9	162.5	162.5	100	40.0	4.76
Phosphoric acid	H_3PO_4	1.57	13.1	98	95.0	96.9	75.0	9.52
Caustic soda	NaOH	1.54	12.8	40	40.0	100	50.0	6.4
Ammonia, anhydrous	NH_4OH	0.9	7.48	35	35.0	100	29.0	2.17

Types of Storage Tanks. Storage tanks for liquid chemicals are commonly fabricated from steel, stainless steel, fiberglass-reinforced plastic (FRP) (also referred to as reinforced thermoset plastic [RTP]), and various forms of polyethylene, such as high-density, cross-linked polyethylene (HDXLPE).

For very corrosive chemicals, such as ferric salts, sodium hypochlorite, and sodium hydroxide, the nonmetallic materials are normally most appropriate. Polyethylene tanks should be designed in accordance with ASTM D-1998. Whenever possible, nozzles should be constructed as integrally molded flanged outlets (IMFO) to avoid stress points and potential leak sites. Polyethylene is more appropriate for smaller day tanks.

The use of FRP for chemical service began in the mid-1950s. ANSI Standard RTP-1, *Reinforced Thermoset Plastic Corrosion Resistant Equipment,* should be the basis for design of FRP tanks having internal pressures less than 15 psi (103 kPa). FRP has greater structural strength and flexibility for installation of nozzles and other appurtenances in the side walls. A variety of polyesters, vinyl esters, and other materials are available for the resin or double laminate construction, depending on chemical resistance requirements.

The choice of tank material or lining, in addition to its chemical resistance, also depends on the concentration of chemical, pressure and temperature conditions, and tank dimensions. The tank manufacturer must be consulted about the types of materials to use for the proposed chemical. Improper material selections could lead to tank failure.

Open-top tanks can be used for stable, nonvolatile liquids, but their use is restricted. Buried tanks must be strapped to anchor blocks with sufficient mass to prevent flotation when the tank is empty and must meet U.S. Environmental Protection Agency (USEPA) leak prevention criteria. These criteria mandate the use of double-walled tanks with leak detection alarms. Buried tanks must also be equipped with manholes, vent lines, and other connections.

The majority of storage tanks are located in the lower or basement areas of buildings, where air is naturally cooler, or outdoors if winters are not too severe. As a result, heating and insulation may be required for those tanks that contain chemicals subject to crystallization. Installing insulation and thermostatically controlled hot-water or steam coils, or electric immersion heaters within or around the tank, ensures positive protection against slushing or freezing. If the problem is marginal or occurs only during severe seasonal weather, applying electrical heating tape on the tank exterior usually offers sufficient protection.

Spill Containment. Chemical storage and feed facilities require adequate provision for spill containment with consideration for worker safety; water quality; customer protection; local, state and federal regulations; and general environmental protection. Depending on the capacity of chemical storage facilities, the strictest regulations require secondary containment for chemical unloading areas, tertiary containment for chemical storage tank and feed equipment areas, and double-walled piping with leak sensors in the annular space for outside chemical piping runs.

Liquid chemical storage areas should have dike walls, or other impermeable structures, providing adequate secondary containment capacity based on the volume of the single largest storage vessel, a safety factor (e.g., 10%), and freeboard. The model facility provides common secondary containment for bulk and day tanks, pumping equipment, safety valves, and immediate piping associated with each chemical. Consequently, the structure can contain leaks from the most vulnerable areas, including storage tank fittings and pump connections, without the risk of mixing incompatible chemicals. For corrosive chemicals, secondary containment structures require a chemical resistant coating or liner to protect and prolong the structure's useful life.

Secondary containment areas generally should not include any floor drains or other floor or wall penetrations that could compromise the integrity of the containment structure. Design should provide access to critical isolation valves without requiring the operator to enter the containment structure, or motorized isolation valves with remote control or automatic interlocks. As an alternative, the facility can include a floor drain connected to a separate secondary containment vessel at a lower elevation.

The design also should provide a leak sensor in a sump or low point within the containment structure, local and remote alarms, isolation valve interlocks, and other instrumentation and controls necessary to enhance safety and ensure rapid response to emergency conditions.

Secondary containment provisions for liquid chemicals should extend to unloading areas with tank filling connections, as well as inside-outside chemical piping runs. Inside piping should run in troughs or channels that allow access for repairs. As a minimum, exposed inside piping should run along walls where a leak would not endanger personnel or other passersby. Outside piping or tubing should run inside conduit or concrete encasement allowing access for repairs. Some plastic pipe manufacturers provide integrated double-walled piping for secondary containment purposes.

Alum Storage Facilities. Alum is probably the most commonly used liquid chemical in water treatment plants. In liquid form [$Al_2(SO4)_3 \cdot 14H_2O$], it contains 48.5% alum by weight, the balance being water. As a liquid it has a light green to light yellow appearance. Specific gravity of the bulk liquid is 1.33.

A tank truckload contains between 3,300 and 5,500 gal (12,500 and 20,800 L), and a rail car contains 8,000 to 18,000 gal (30,300 to 68,100 L). At 48.5% strength, its freezing point is 2° F ($-17°$ C), but it can begin localized crystallizing at warmer temperatures. The freezing point is highly dependent on the percent strength. Both higher and lower concentrations have higher freezing temperatures, so alum is rarely handled at other concentrations. For example, at the slight increase to 50% strength the freezing point rises dramatically to 30° F ($-1°$ C).

In freezing climates, storage tanks should be installed indoors or in a heated enclosure to maintain liquid temperature between 45° and 60° F (7° and 16° C). Storage and feed lines exposed to low temperatures should also be insulated and equipped for flushing with water if crystallization should occur. Storage tanks should be sized for at least 7,500 gal (28,400 L) if alum is delivered by truck.

Storage tanks are most commonly constructed of FRP and polyethylene. Alternatively, rubber-lined steel and type 316 stainless steel can also be used. Piping can be schedule-80 polyvinylchloride (PVC) with socket end joints if liquid temperatures do not exceed 120° F (49° C). The temperature of liquid alum when delivered may occasionally exceed 120° F (49° C), so the unloading line for the tank should be chlorinated polyvinylchloride (CPVC), 3 in. (80 mm) diameter, as a precaution. Plug, needle-and-ball valves, 150 lb intregal flanges with PVC or type 316 stainless steel construction, and tetra fluoroethylene (TFE) or similar material sleeves or retaining rings are recommended. A schematic layout of an alum feed system is shown in Figure 15.3.

Caustic Soda Storage. Liquid caustic soda is delivered by rail car and tanker truck. Tanker trucks can haul up to 3,000 gal (11,400 L) per load. Storage tanks should be constructed of stainless steel, FRP, polyethylene, or steel lined with rubber or polypropylene. As with alum, schedule-80 PVC with socket end joints is the recommended piping material if liquid temperatures do not exceed 120° F (49° C). Caustic soda generates heat when mixed with water resulting in temperatures approaching this level, so use of CPVC piping should be considered for this application. Valves can be steel or ductile iron with Teflon components, rubber lined, or PVC.

The freezing point of caustic soda solution is highly dependent on the solution strength (Figure 15.4). Freezing points for solutions greater than 50% quickly elevate making it impractical to handle and store. Because 50% caustic soda begins to crystallize at approximately 54° F (12° C), storage tanks must be indoors or insulated and heated to avoid crystallization. A typical chemical feed schematic for caustic soda appears in Figure 15.5.

CHEMICALS AND CHEMICAL HANDLING

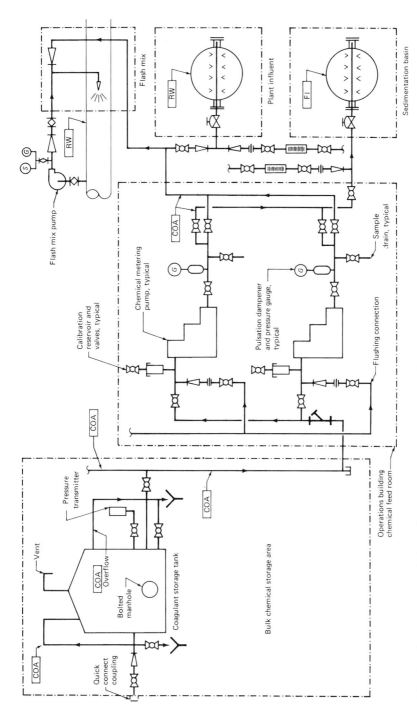

FIGURE 15.3 Coagulant alum/ferric chloride system. (*J. M. Montgomery Consulting Engineers.*)

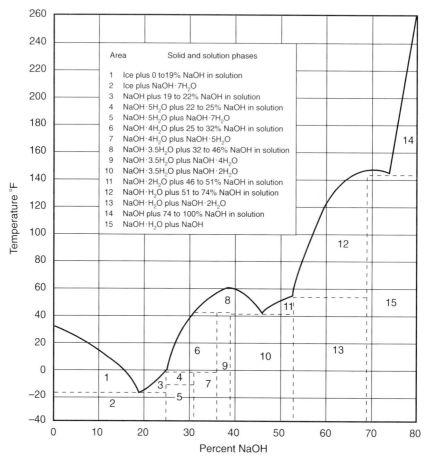

FIGURE 15.4 Freezing points of caustic soda solutions.

Sodium Hypochlorite Storage. Commercial sodium hypochlorite (NaOCl), or liquid bleach, is marketed in carboys and rubber-lined drums holding up to 50 gal (190 L) volume, and in trucks. Storage tanks should be constructed of FRP, polyethylene, or steel lined with rubber or polypropylene. As with alum, schedule-80 PVC with socket end joints is the recommended piping material if liquid temperatures do not exceed 120° F (49° C).

Valves may be a plug type made of steel (lined with PVC or polypropylene), or PVC diaphragm valves. Although PVC ball valves would not react with the chemical, hypochlorite releases small amounts of gas as it decomposes, and PVC ball valve failures due to gas buildup have been reported. A similar type of failure has occurred with stainless steel (SST) ball valves in hydrogen peroxide service. Any type of ball valve, especially 1 inch and larger sizes, may allow hypochlorite to weep through the seal and crystallize. To avoid these problems, a diaphragm valve constructed of composite material using a Teflon diaphragm can be used.

When hypochlorite is added to water, it hydrolyzes to form hypochlorous acid (HOCl), the same active ingredient that occurs when chlorine gas is used. The hypochlorite

CHEMICALS AND CHEMICAL HANDLING

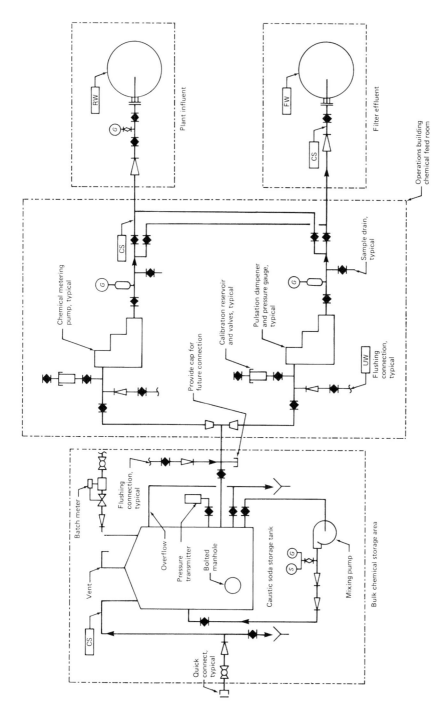

FIGURE 15.5 Caustic soda system. *(J. M. Montgomery Consulting Engineers.)*

reaction slightly increases the hydroxyl ions (pH increase) by the formation of sodium hydroxide, whereas the reaction of chlorine gas with water increases the hydrogen-ion concentration (pH decrease), forming hydrochloric acid. In most waters, these differences are not significant, but when high chlorine doses are used in poorly buffered waters, these effects should be considered. They can be evaluated by calculation or by simple laboratory tests.

In the commercial trade, the concentration of sodium hypochlorite solutions is usually expressed as a percentage. The "trade percent" is actually a measure of weight per unit volume, with 1% corresponding to a weight of 10 g of available chlorine per liter. Common household bleach, at a trade concentration of 5.25%, has approximately 5.25 g/100 mL or 52.5 g of available chlorine per liter. Hypochlorite available for municipal use usually has a trade concentration of 12.5% to 17%. These are approximate concentrations and should always be confirmed by laboratory procedures.

Because increasing the concentration of any salt lowers the freezing point of a solution, the freezing points of various solutions of sodium hypochlorite are a function of their concentrations, with the more dilute concentrations approaching the freezing point of pure water. Figure 15.6 shows the freezing temperature of hypochlorite solutions as a function of concentration in the concentration ranges normally experienced.

The chlorine concentration in hypochlorite solutions is adversely affected by high temperature, light, low pH, and the presence of certain heavy metal cations. Iron, copper, nickel, and cobalt are the most common problem-causing cations. The concentration of hypochlorite itself also has a major impact on hypochlorite degradation. Table 15.3 shows the half-life of a hypochlorite solution as a function of chlorine concentration and temperature.

When purchasing bulk sodium hypochlorite, purchasing specifications should be used, delineating the acceptable ranges for available chlorine (12.5% to 17%) and pH (11 to 11.2), as well as maximum contaminant limits for iron (2 mg/L) and copper (1 mg/L). Specifications should also require that shipments be free of sediment and other deleterious

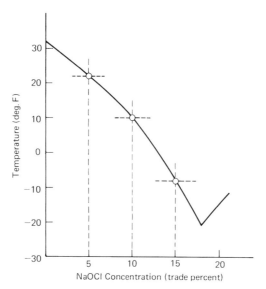

FIGURE 15.6 Freezing temperatures of hypochlorite solutions.

TABLE 15.3 Influence of Temperature and Strength on Decay of Hypochlorite Solutions

% Cl$_2$	Half-life (days)			
	100° C	60° C	25° C	15°C
10.0	0.079	3.5	220	800
5.0	0.25	13	790	5000
2.5	0.63	28	1800	—
0.5	2.5	100	6000	—

particulate material. On arrival, shipments should be analyzed for the concentration of chlorine, pH, and the concentration of metal contaminants.

Sodium hypochlorite storage must be carefully managed to limit degradation and the formation of chlorate. The rate of degradation is a function of storage time, temperature, and chlorine concentration, so storage time should preferably be limited to less than 28 days. The chemical may be delivered at temperatures up to 85° F (30° C), which must be considered in the system design.

Storage temperatures should not exceed about 85° F (30° C) because above that level, the rate of decomposition increases rapidly. Although storage in a cool, darkened area greatly limits the deterioration rate, most manufacturers recommend a maximum shelf life of 60 to 90 days.

All hypochlorite solutions are corrosive to some degree and can affect the skin, eyes, and other body tissues. Accordingly, rubber gloves, aprons, goggles, and similar suitable protective apparel should be provided for preparing and handling hypochlorite solutions. Areas of skin contact should be promptly flushed with large quantities of water. Every precaution should be observed to protect containers against physical damage, to prevent container breakage, and to minimize accidental splashing.

CHEMICAL FEED AND METERING SYSTEMS

The types of chemical feed equipment for the water treatment plant must be part of the decision of whether gaseous, dry, or liquid chemicals are to be used.

Gas Feed Equipment

Chlorine and ammonia gases are used in disinfection treatment processes and require careful selection of equipment because of the aggressiveness and dangers associated with the gases.

Chlorinators. A conventional chlorinator consists of:

- An inlet pressure-reducing valve
- A rotameter
- A metering control orifice
- A vacuum-differential regulating valve

A simple schematic is shown in Figure 15.7. The driving force for the system comes from the vacuum created by the chlorine injector. The chlorine gas flows to the chlorinator and is converted to a constant pressure (usually a mild vacuum for safety reasons) by the influent pressure-reducing valve. Present design practice locates the influent pressure-reducing valve as close as possible to the storage containers to minimize the amount of pressurized chlorine gas piping in the plant.

The chlorine then passes through the rotameter, where the flow rate is measured under conditions of constant pressure (and consequently constant density), and then through a metering or control orifice. A vacuum differential regulator is mounted across the control orifice so that a constant pressure differential (vacuum differential) is maintained to stabilize the flow for a particular setting on the control orifice. Flow through the control orifice can be adjusted by changing the opening on the orifice. The control orifice has a typical range of 20 to 1, and the vacuum differential regulator has a range of about 10 to 1. Thus the overall range of these devices combined is about 200 to 1. On the other hand, a typical rotameter has a range of about 20 to 1. Thus the chlorinator should be selected based on design capacities, and the rotameter installed should be appropriate for current demands.

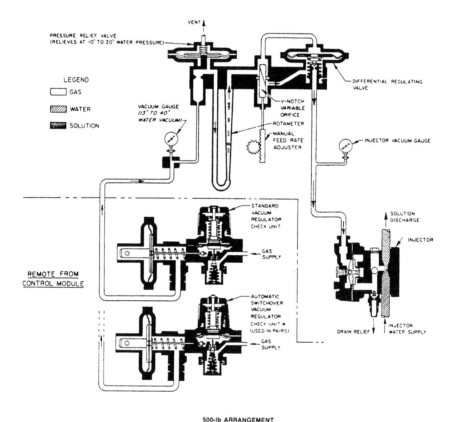

FIGURE 15.7 Flow diagram for conventional chlorinator. *(Courtesy of Wallace and Tiernan Division of Pennwalt Corp.)*

Chlorine Gas Piping. Between the chlorinator and the injector, the chlorine gas flows in a vacuum. Although the head loss of the gas flow is usually small, it is critical that the vacuum created by the injector be transmitted to the chlorinator without significant dissipation. As a consequence, the diameter of the chlorine vacuum lines should always be designed rather than arbitrarily selected. Lines should be sized to limit the total pressure drop over the pipe length to between 1.5 and 1.75 in. of mercury under maximum injector vacuum levels (22 to 23 in. of mercury).

Dry Chemical Feed Equipment.

Feeders for powders and granules may be classified as gravimetric (accuracy range 0.5% to 1% of set rate) and volumetric (accuracy range 1% to 5% of set range depending on material fed). Gravimetric feeders are preferred when accurate feeding of chemicals with varying bulk densities is important.

Loss-in-Weight Gravimetric Feeders. Gravimetric feeders use a feeder hopper suspended from scale levers, a material feed-control mechanism, and a scale beam with motorized counterpoise. The rate of weight loss of the hopper equals the weight loss equivalency of a traveling counterpoise when the feeder is in balance. If it does not, the scale beam deflects, and the feed mechanism increases or decreases the feed accordingly. Although these feeders are highly accurate, their capacity is usually less than 1,000 lb/h (453 kg/h). The total amount of material fed may be recorded or read directly off the weight beam at any time.

Belt-Type Gravimetric Feeders. These feeders are available in numerous forms and are usually designed to handle specific types of material. Weight belts can be of the pivoted type for heavy feed rates of 250 tons (227 metric tons) per hour and up, or the rigid belt type passing over a live or weigh-deck scale section, with a feed hopper at one end of the belt and a control gate to regulate the flow and depth of material placed on the belt. A scale counterpoise is adjusted to establish the desired belt loading, and the control gate is automatically repositioned in proportion to the error signal. Various gate-control systems are available depending on the material to be handled, response time desired, feed range, and capacity. Normally belt speed is varied to produce the desired flow of material. Total quantity of material fed can be read directly on a totalizer or similar device.

Volumetric Feeders. Although more than a dozen types of volumetric feeders are available, all operate on the principal of feed-rate control by volume instead of weight. Advantages of volumetric feeders include low initial cost, good overall performance at low feed rates, and acceptable accuracy for materials with stable density and uniformity. Disadvantages include unresponsiveness to density changes, fixed orifices or openings subject to clogging, and calibration by manual sampling (which must be done regularly).

The many types of volumetric feeders available permit a good choice based on capacity requirements and the nature of the material to be fed. The roll feeder forms a smooth ribbon of material of adjustable thickness and width; this feeder is unique in its ability to handle very low feed rates of fine ground materials such as hydrated lime. It cannot be used for coarse granular material.

The screw feeder is a popular unit employing rotating or reciprocating feed screws that can handle most dry chemicals. Most of these require hopper agitation or vibration to maintain screw loading. The range of feed is good (at least 20 to 1), with capacities up to 600 ft^3/h (17 m^3/h) using 6 in. (15 cm) helical screws. Minimum feeds on certain models are very low when using fine powdery material through a small-diameter screw.

On belt feeders, the material is deposited on a moving belt from an overhead hopper and passes beneath an adjustable vertical gate. The speed of the belt and the position of the gate establish the volume of material passing through the feeder. These are high-capacity feeders that can handle anything from powder to 1.5 in. lump materials at rates from 600 to 3,600 ft^3/h (17 to 100 m^3/h) depending on belt width. A schematic of a typical volumetric feeder system for potassium permanganate is shown in Figure 15.8.

Rotary paddle feeders consist of a paddle or series of compartments revolving within an enclosure that receives material from the hopper and releases it through a discharge chute as rotation proceeds. Feed rate is normally controlled by using a sliding gate or varying the speed of the paddle shaft. These feeders have a unique application in that they can deliver material into vacuum or pressure systems because they form an airlock. They are also commonly used to feed chemicals that tend to flow or gush out of control through a fixed orifice.

Vibrating feeders employ a vibrating mechanism attached to a slightly inclined feed trough. Flow is controlled by regulating the depth of material and the intensity of vibration. These feeders are used only on dry, nonhydroscopic, free-flowing materials. They are generally used in smaller installations, and their accuracy is acceptable as long as the material is of consistent quality and there are no large voltage fluctuations that would affect the amplitude of the vibrator.

An oscillating hopper feeder consists of a main hopper fitted with an oscillating apex section that discharges to a stationary tray or plate. Oscillation of the hopper pushes previously deposited material off the tray in one or more directions. Capacity is controlled by adjusting the depth of chemical deposited and regulating the length of stroke. Because these feeders can handle a variety of chemicals from powder to pebble lime, they are popular in smaller plants.

Virtually all dry feeders, gravimetric or volumetric, can be equipped to operate automatically in proportion to a flow or other process signal. The means for accomplishing this varies depending on the nature of the feeder's control mechanism. In its simplest form, time-duration control using a resettling time can provide proportional feed using a manually adjusted feeder, providing a flow-proportional pulse or contact signal is available from the flowmetering system.

Lime Slakers. Slaking means combining water with quicklime (CaO), in various proportions, to produce milk-of-lime, a lime slurry or viscous lime paste. Slakers operate at elevated temperatures, with or without auxiliary heaters, because of the exothermic reaction between CaO and water.

Aside from capacity, operational flexibility, and the desired concentration of the slaked lime product, the most appropriate method of and equipment for slaking quicklime depend on the characteristics and quality of the quicklime supply. Lime slakers come in two basic varieties: slurry (detention) and paste slakers. Elements common to both varieties include a quicklime (dry) feeder, a water flow control valve, temperature controls, a grit removal device, a dilution chamber, and a final reaction vessel. All slakers require an integral water vapor and dust collector to maintain a slight vacuum within the slaker and discharge clean air.

The difference between the two varieties relates to the temperature and consistency of the lime-water mixture as it passes through the slaker. A slurry (detention) slaker typically mixes lime and water at a weight ratio between 1:3 and 1:4. Paste slakers, and some slurry slakers with auxiliary heaters, mix the lime and water at a ratio of about 1:2. The slurry slaker uses a mechanical (typically an impeller) mixer and maintains the slurry level in the slaker by regulating the water flow rate. A paste slaker uses a pug mill type of agitator and regulates the water flow rate based on the torque imposed on the agitator. In this manner, the paste slaker adds only enough water to achieve and maintain the desired consistency.

CHEMICALS AND CHEMICAL HANDLING 437

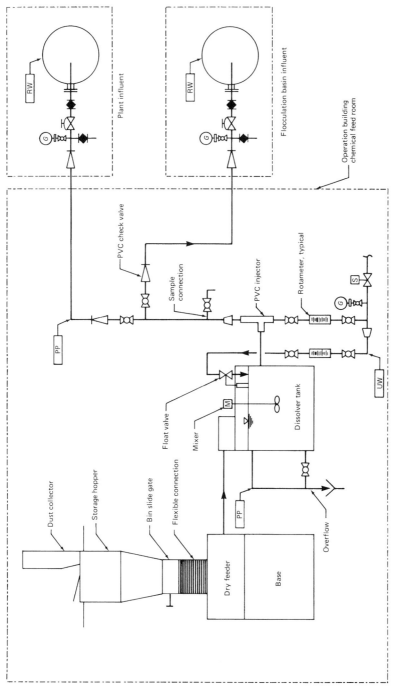

FIGURE 15.8 Potassium permanganate system. *(J. M. Montgomery Consulting Engineers.)*

Batch Mixing. In most water treatment facilities today, batch mixing typically applies only to potassium permanganate, polymer, and powdered activated carbon. Batch mixing of potassium permanganate in "saturation tanks" provides a constant strength solution for application to multiple points at variable rates via positive displacement pumping equipment. For continuous operation, the design should include at least two tanks to maintain consistent permanganate dosages while preparing the solution in one of the tanks.

Batch mixing can work in a similar fashion to prepare solutions from dry or emulsion polymers. Again, the design should include at least two tanks to maintain consistent polymer dosages while preparing and aging the solution in one of the tanks. Special automatic batch mixing equipment for dry polymers is available.

Activated Carbon. Powdered activated carbon can be stored in bins and extension hoppers to supply dry feeders. This type of installation is appropriate in smaller plants and in plants where carbon usage is intermittent and the carbon is received and stored in paper bags. Activated carbon is combustible and will burn when ignited. The ignition point of the activated carbons varies from 600° to 800° F (315° and 426° C). After ignition, activated carbon does not burn with a flame, but glows or smolders until all carbon material is oxidized. At least one U.S. utility has experienced spontaneous combustion in its bulk storage bins. The combustion was smothered by injecting carbon dioxide in the bottom of the bin and placing dry ice on the top of the carbon.

Storing carbon in paper bags presents a hazard in that the paper burns more rapidly. Bags of powdered carbon should be stacked in rows with aisles between so that each bag is accessible for removal in case of fire.

In the event of an activated carbon fire, the safest procedure, if possible, is to place the smoldering material in a metal container and haul it outside the building. A smoldering carbon fire may be extinguished by means of a very fine spray or mist of water from a hose or by a foam-type chemical extinguisher. Do *not* attempt to extinguish the carbon by a direct stream of water, because this causes the light, smoldering particles to fly into the air and spread the fire.

Installing an overhead sprinkler system in the storage and feeding rooms is a practical precautionary measure. Activated carbon should not be stored where it can come into contact with gasoline, mineral oils, or vegetable oils. These materials, when mixed with carbon, slowly oxidize until the ignition temperature is reached. Never mix or store carbon with such materials as chlorine, lime hypochlorites, sodium chlorite, or potassium permanganate. Such mixtures are known to be spontaneously combustible.

Activated carbon is an electrical conductor and should not be allowed to accumulate as dust near or on open electrical circuits. Some activated carbons are subject to deterioration in storage, so carbon storage areas should be relatively free of such air contaminants as sulfur dioxide, chlorine, hydrogen sulfide, and organic vapors. Normal safety equipment, such as protective clothing, respirators, neck cloths, gloves, and goggles, should be provided for workers handling powdered activated carbon.

Sodium Chlorite. Sodium chlorite is a dry, flaked salt, which, because of its powerful oxidizing nature, is shipped in steel drums bearing an Interstate Commerce Commision (ICC) "yellow" label classification. It is stable when sealed or in solution, but it is highly combustible in the presence of organic material. For this reason, the solution should not be allowed to dry out on floors, but should be hosed down with minimum splashing. Technical-grade sodium chlorite is an orange-colored flaked salt.

Sodium chlorite should be stored in an enclosed space specially prepared for the purpose, and removed from the storage room only as needed for immediate use. Empty containers should be returned to the storage room immediately after each use unless they are shipping containers, in which case they should be thoroughly flushed with water (to the sewer) as soon as they are empty and should be immediately disposed of well away from any building. Shipping containers should never be used for any other purpose after they are empty.

Calcium Hypochlorite.   Although calcium hypochlorite is a stable, nonflammable material that cannot be ignited, contact with heat, acids, or combustible, organic, or oxidizable materials may cause fire. It is readily soluble in water, varying from about 21.5 g/100 ML at 32° F (0° C) to 23.4 g/100 ML at 104° F (40° C). Tablet forms dissolve more slowly than granular materials and provide a fairly steady source of available chlorine over an 18 to 24 h period.

Granular forms usually are shipped in 35 or 100 lb (16 or 45 kg) drums, cartons containing 3.75lb (1.7 kg) resealable cans, or cases containing nine 5 lb (2.3 kg) resealable cans. Tablet forms are shipped in drums and in cases containing resealable plastic containers.

Because of its strong oxidizing powers and reactivity with organic materials, calcium hypochlorite should be segregated from other chemicals or materials with which it can react, or stored in a separate location. To minimize the loss in available chlorine content that occurs with elevated temperature, cool storage areas should be provided. Containers should be kept dry and located in a darkened area unless the containers themselves shield out excessive light. Their size should be consistent with use requirements. Stored containers should be arranged such that they can be easily moved from the storage area in the event of leaks.

Sodium Carbonate (Soda Ash).   Soda ash used to soften water is a grayish-white powder containing at least 98% sodium carbonate. It may be shipped in bulk, in bags, or in barrels. Soda ash is noncorrosive and may be stored in ordinary steel or concrete bins or silos and fed using a conventional chemical dry feeder. Its solution may be transmitted through conventional pipelines or troughs. Hazards associated with soda ash are primarily those of a chemical dust. Protective clothing and devices such as gloves, respirators, and goggles should be provided.

Sodium Chloride.   Sodium chloride has a tendency to absorb moisture and to cake under certain conditions. It should be protected from moisture and is best stored in concrete bins. Sodium chloride is highly soluble in water, may be readily made up to a desired concentration, and may be fed by a standard liquid chemical metering device. The solution may be transmitted through rubber or bronze lines. Exposure of the skin to large amounts of the dry salt would have a tendency to cause the skin to dehydrate. Protective clothing and devices such as gloves and face shields should be provided. Large users of brine solution prefer to receive delivered salt directly into a saturator tank and avoid the problems of dry storage.

Liquid Chemical Feed Equipment

Liquid chemical feed equipment includes various types of positive displacement pumps, centrifugal pumps, dipper-wheel feeders (discussed in the section on slurries), and eductors. The designer must select appropriate pumping equipment for the intended application and pump components compatible with the intended chemicals.

Positive displacement pumps handle a nearly constant rate of flow regardless of the back pressure they pump against, making them ideal for metering precise chemical flow rates. Several types of positive displacement pumps are manufactured; those most commonly used in water treatment plants are piston and diaphragm and progressive cavity.

Piston and Diaphragm Pumps. Piston and diaphragm pumps are accurate, their capacity range dependent on stroking speed and length of stroke. Diaphragms pump the liquid so that the mechanical components of the pump do not come into direct contact with corrosive chemicals. Diaphragms are set in oscillating motion by a piston either directly connected to the diaphragm or indirectly connected through a hydraulic fluid.

Flat-faced diaphragm and tubular diaphragm pumps come in different configurations, with some advantages and disadvantages to each, but all operate based on the same principle, with an elastomeric diaphragm, compression chamber, and inlet-outlet check valves. Diaphragm pumps can accommodate a wide range of capacities and operate against relatively high back pressures. These pumps are driven by motors or are solenoid driven.

Air-operated diaphragm pumps are also available, the discharge rate controlled by regulated air admission and exhaust to the power side of the diaphragm. Pump stroking speed should be limited to less than 120 strokes per minute. Flow rate adjustments are made by varying stroke length, speed, or both.

Although manufacturers sometimes claim the capability of high turndown ratios using a combination of stroke length and speed, pumps should be sized so that the turndown ratio does not need to exceed 10 to 1, to help ensure accuracy at low feed rates. For adjustable-speed drives, the turndown ratio should not be expected to exceed 5 to 1 in order to maintain stability at low speeds. In addition, a tachometer for feedback to the control system should be considered to help provide stable speeds.

Progressive Cavity Pumps. Progressive cavity pumps, with their unique rotor and stator elements, are capable of pumping thick pastes, gritty slurries, or viscous shear-sensitive fluids. These pumps can wear more rapidly than other pumps handling slurry under similar conditions unless care is exercised in selecting proper construction materials. Progressive cavity pumps are specialized service pumps with a relatively high initial cost; with preventive maintenance, they can be expected to give trouble-free service.

Other Pumps. Properly selected peristaltic pumps have some applications for liquid chemical feed applications, particularly at lower capacities. Operating similarly to peristaltic pumps, hose pumps can meet high-capacity liquid chemical feed requirements. Gear pumps operate similarly to other positive displacement pumps but use gear teeth as the motive elements. Gear pumps have particular applications for highly viscous liquid chemicals such as polymers.

Valves and Appurtenances. With the possible exception of peristaltic pumps, all liquid chemical feed pumping equipment requires certain valves and other accessories to provide a safe and reliable system. All positive-displacement pumps should have a suitable pressure-relief valve installed in the discharge piping to prevent overpressurizing piping due to inadvertently closed valves or other obstructions. In some cases these valves are integral with the pump. Calibration chambers for calibrating the discharge from metering pumps should be installed on the suction piping with valves and fittings to isolate it from the chemical supply tank so that measurements can be made with the pump operating against normal discharge conditions.

Back pressure valves are used to prevent gravity flow or siphoning to a low-pressure application point and to ensure consistent operation. A typical set point for back pressure valves in this application would be at least 30 psig. Even when chemical feed pumps normally discharge to a high-pressure application point, an accidental loss of pressure in the system could result in an overdose by gravity flow or siphoning action, which back pressure valves would mitigate.

Figure 15.9 shows a typical liquid chemical feed system, including isolation valves, strainers, and a calibration chamber on the suction side and a series of safety valves (pres-

CHEMICALS AND CHEMICAL HANDLING

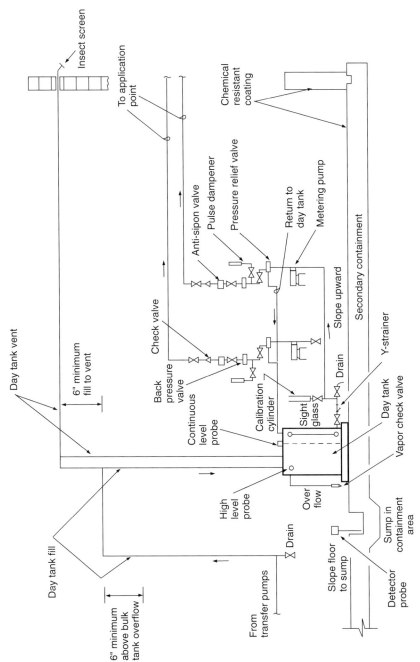

FIGURE 15.9 Typical arrangement for valves and apparatus.

sure relief, check, back pressure, antisiphon) along with isolation valves on the discharge side of the chemical feed pumps. Vents for calibration chambers should extend above storage tanks or whatever the maximum level of the liquid is in the piping system.

Many positive displacement pumps have integral check valves, and some are equipped with combination pressure-relief, priming, back pressure, and antisiphon valves. Even when a pump has its own combination of safety valves, the design should provide at least a redundant antisiphon valve to help protect the water supply from an overdose of chemical should a system upset occur.

Low pressures, which could encourage gases to come out of solution and form bubbles, should be avoided. This requires larger suction lines that are as short as possible. Suction piping should have a positive static pressure when possible. A standpipe (or calibration column) near the pump suction should be considered to divert bubbles that may be present in the suction line.

Positive displacement pumps typically require some type of pulsation dampener on the discharge (and sometimes suction) piping to reduce intermittent high pressures and partly equalize the flow of liquid chemical to the application point. Pulsation dampeners should be placed upstream of back pressure valves to work properly. Even with pulsation dampening, the piping pressure rating should be at least 150% of average working pressure.

Valves for isolation should be used often on chemical feed piping. Even long lines with no attachments or equipment should be equipped with isolation valves for maintenance and cleaning. Unions should be installed around pumps and valves to facilitate their removal.

Eductors. Eductors, also referred to as ejectors or jet pumps, are useful for introducing gases, solutions, or slurries into a diluted water solution followed by application to the process water. A stream of water passing through a venturi in the eductor creates a subatmospheric pressure, which draws the chemical into the eductor. Because eductors are incapable of flow rate control, the chemical must be metered in some fashion before entering the eductor. The chemical introduced at the eductor can be monitored by a flowmeter and its flow rate controlled by a valve in response to a reading from the meter. This practice should be used only when adequate suction pressure is available from the chemical supply source, to allow the control valve to accurately control the chemical entering the eductor.

Centrifugal Pumps. Centrifugal pumps are used primarily in chemical feed systems for chemical transfer from bulk tanks to day tanks. Centrifugal pumps for chemicals operate similarly to centrifugal pumps for water. Speed variation affects both discharge flow rate and head. For this reason, centrifugal pumps typically are not used to deliver chemical directly to the injection point, except in a recirculating system with rate control valves at the application points. Centrifugal pumps come with mechanical seals that contain the liquid chemical within the pump body, or with sealless magnetic drives that eliminate the need for mechanical seals. Centrifugal pumps also have important applications for supplying high-pressure water to eductors.

Feed Systems for Slurries. As used in water treatment, a slurry can be described as a suspension of a relatively insoluble chemical in water. Powdered activated carbon and lime (calcium hydroxide) are the most common slurries handled in water treatment. Most questions about these slurries deal mainly with handling and mixing. Once a slurry (carbon or lime) has been prepared, a pumping and piping system conveys the material to the point or points of application. Different techniques are required to handle carbon and lime slurries.

Carbon Slurries. Numerous tests have demonstrated that there is no loss of adsorptive activity when carbon is held in water as a slurry over long periods of time (up to one year). Mixing equipment used to wet and slurry activated carbon should operate with the agitator revolving at a minimum of 60 to 70 rpm. It is preferable that the agitator motor be dual speed, providing about 80 rpm for initial wetting and 40 rpm to maintain suspension. Agitators normally consist of two sets of stainless steel paddles, one set near the bottom of the tank and one set placed approximately 18 in. (0.46 m) from the top. With such an arrangement, activated carbon can be slurried almost as rapidly as it is discharged from the delivery vehicle. With the proper equipment, tests have shown that an air slide car containing 21 tons (19 metric tons) of carbon can be unloaded in less than one hour. Water treatment plants purchasing less than carload quantities of powdered activated carbon may also find a slurry system desirable.

Tank sizing is based on a slurry concentration of 1 lb/gal (2 gr/L). It is common practice to size the tank to have a total volume equal in gallons to the maximum load (in pounds) of carbon to be received plus 20% for freeboard.

Carbon slurry tanks are usually square concrete structures with a bitumastic or epoxy lining. If steel tanks are used, the surface must be cleaned to bare metal before lining. The mixer agitator shaft, impeller, assembly bolts, pump suction piping, and other such parts must be made of stainless steel or be rubber covered. Sufficient horsepower should be available to handle carbon slurry concentrations up to 1.5 lb/gal (180 g/L). A schematic drawing of a carbon slurry feed system is shown in Figure 15.10.

Mechanical failure of the mixing equipment renders the slurry system inoperative. For this reason, those installations that are critically dependent on uninterrupted carbon feed could be equipped with an air-agitation system for backup. Although not as efficient as mechanical mixing, adequate suspension can usually be maintained until mixer repairs are made. Obviously, large plants that use carbon routinely should consider the feasibility of installing duplicate slurry tanks, not only for mechanical standby but because it is almost impossible to maintain a constant-strength slurry during the recharge and wetting period after a new delivery of carbon.

Tank level readings are best accomplished using an air bubbler system with a stainless steel bubble pipe. Slurry is controlled by measuring makeup water to the tank through a water meter.

Carbon, unlike lime, does not react chemically when diluted in water, and therefore it does not cause scaling and salt precipitation problems. On the other hand, construction materials for piping, pumps, and valves must be resistant to the corrosive nature of powdered carbon slurries. Suitable materials for this service include type 316 stainless steel, rubber, silicon bronze, monel, Hastelloy C, Saran, and fiberglass-reinforced plastic.

On smaller installations, where feed control is achieved by regulating the flow of dry carbon through a dry feeder equipped with a slurry pot, the feed line carrying the carbon suspension from the machine should be of ample size to handle the volume of water required for dilution. If possible, the line should be installed with a continuous downgrade to the point of application. Provision should be made for cleaning out any carbon that may settle in the line and cause clogging.

Whenever feeding is stopped for any reason, the line should be flushed with clear water. In many smaller plants, a rubber hose is used for the entire feed line, so that stoppages may be easily eliminated by manipulation. If an ample volume of flushing water is not available through the slurry pot, dilution water should be added downstream.

Larger installations using agitated slurry storage of constant strength employ metering pumps for feed control. These pumps should discharge to a dilution box located at an appropriate elevation to permit gravity feed. Dilution water is added through a rotameter to maintain a velocity of at least 1.5 ft/s (46 cm/s) to prevent settling. The point of application should be above the water surface, and unavoidable high spots in the line must be vented.

Carbon slurry concentrations generated by dry feeder–slurry pot systems naturally vary, because dry carbon feed is variable. On the other hand, agitated slurry storage

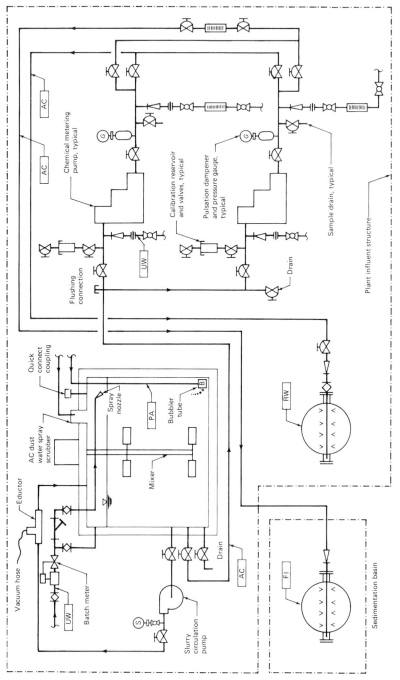

FIGURE 15.10 Activated carbon system. *(J. M. Montgomery Consulting Engineers.)*

systems are commonly batch mixed and maintained at 1 lb of carbon per gallon of water for convenience in computing feed rates. For this reason, additional dilution water should be added after the metering pump.

Lime Slurries. The concentration of lime slurry suspension is commonly expressed as pounds of CaO or $CaOH_2$ per gallon of water. It is also expressed as percent lime solids, although some confusion can arise unless the dry weight basis (oxide or hydroxide) is specified. Lime suspensions are commonly referred to in three categories of concentration:

- Paste (as prepared by paste-type slakers). A wet, lime, puttylike mass containing about 30% to 35% quicklime with sufficient body to exhibit some angle of repose.
- Slurry (as prepared by slurry-type slakers). A creamy suspension containing about 20% to 25% quicklime that may be poured or pumped.
- Milk-of-lime (ready for application). Waterlike consistency containing up to 1.5 lb/gal (180 g/L) hydrated lime. Milk-of-lime, the final diluted product from paste and slurry slakers and dry feeder slurry pots, is readily pumpable.

A typical waterworks milk-of-lime ready for application contains between 1 and 1.5 lb/gal (2 and 3.2 gr/L) of hydrated lime (10.7% to 15.2%). At 59° F (15° C) these suspensions exhibit specific gravity readings of 1.07 and 1.11, respectively. Lime concentration can readily be checked using a hygrometer and specific gravity tables.

The formation and handling of lime slurries involves both chemical and mechanical considerations. Lime slurries result from adding controlled amounts of water to either slaked quicklime or hydrated lime. A slurry of reasonable purity is not corrosive and is relatively easy to keep in suspension provided that the slurry has been stabilized. Stabilization occurs when all chemical reactions between the water used and the calcium hydroxide have been completed. This essentially is a softening reaction.

If slurries are transported before stabilization is complete, dissolved solids contained in the dilution water will precipitate on piping, valves, and pumps as a scale. A detention time of 15 minutes in an agitated slurry-storage tank is usually sufficient for maximum development of salt crystals, which tend to grow on themselves. By maintaining suspension, these crystallized scale-forming compounds can be pumped and transported to the process with a minimum of problems. Conversely, attempts to pump directly from hydrated lime-wetting tanks or from slaker sumps can produce intolerable maintenance problems.

Agitated detention or stabilization of lime slurry does not resolve all problems. Quicklime slaked on-site contains fine settleable impurities that can be highly abrasive. This material is usually too fine to remove by slaker grit-removal systems, so purchase specifications should emphasize or reward purity of product. Water quality for slaking and slurry dilution is also important, particularly with respect to the amount and type of dissolved solids present.

Water of medium or high hardness, or containing sulfates is particularly objectionable in that it is likely to interfere with the hydration process itself. In such cases it may be feasible to consider installing water-softening equipment of ample capacity to satisfy water supply needs for slaking and slurry dilution. In less severe cases, the addition of 5 to 10 mg/L of hexametaphosphate to the supply line sequesters the mineral hardness present when a 2- or 3-minute detention time is provided before reaching the slaker.

Once the quicklime has been slaked or the hydrate wetted to form slurry, it is essential that it be discharged as directly as possible into an agitated storage tank, where mixing and stabilization occur. Mixers must be designed to maintain the impurities and precipitated salts, as well as the lime, in suspension. As a general rule, about 0.5 to 1 hp (370 to 746 w) per 1,000 gal (3,785 L) of tank capacity is required for cylindrical tanks of less than 3,000 gal (11,400 L) capacity and slurry concentration not exceeding 1 lb/gal (120 g/L). Smaller tanks are usually equipped with propeller agitators; tanks larger than 2,000 gal (7,570 L) capacity resort to turbine-type mixers. Speeds vary from approximately 350 rpm for propeller mixers to 100 rpm or less on larger turbine units.

Because most slurry storage tanks are agitated continuously, care must be taken to position mixing impellers in the lower portion of the tank when slaker operations are automatically controlled by level controls within the tank. Under no circumstances should the impeller be located so that exposure above the liquid surface is possible. Tanks should be equipped with peripheral baffles set 90 or 120 degrees apart to control vortex formation. Baffles should terminate above the bottom and be attached to the sides with intermittent spaces to prevent lime buildup in corners. Level sensors for slaker control, probes or sonic devices, and alarms are usually required. Float-actuated controls are usually impractical, even with stilling wells, because of heavy solids buildup that results.

One of the principal problems encountered with small systems using gravity feed lines for unstabilized slurry is the periodic line clogging because of low velocity. As a result, a compromise solution to the problem is to design the line for easy cleanout, removal, or flushing. Other measures commonly used to convey lime slurry include:

- Polyphosphates in dilution water
- Flexible hose to allow breakup of scale
- Open troughs or channels
- Duplicate lines and passing chlorine solution through these lines on an alternating schedule

On larger installations it is preferable to employ pumps that circulate the slurry to the point or points of application and back to storage. A properly designed system has the pump and piping sized to produce a minimum velocity of 3 to 4 ft/s (0.9 to 1.2 m/s) in the return line. At each application point several methods can be used to control the flow of slurry to process:

- Control valve, manually operated or automatically modulated by an external controller analyzer
- Discharge into a local day tank equipped with agitator and metering pump; tank refills through an automatic valve actuated by level sensors

The recirculating line must be run as close as possible to the points of application to minimize any additional length of piping where optimum velocities cannot be maintained. Takeoffs that are periodically closed, as in refilling day tanks, must be tapped and valved off the top of the pipe to prevent settling in the static line behind the closed valve. Throttling valves in continuous service may be tapped from the side or bottom of the line. Flow stability through these diversion valves requires some back pressure in the recirculating line; this is normally accomplished by installing a throttling valve in the line as it returns to storage.

Although ball valves and plug-type valves perform reasonably well in open-close service, they are not recommended for throttling service. Similarly, diaphragm valves tend to collect compacted sludge behind their weirs. Pinch valves operate well in throttling service, provided they are properly sized and not subject to a vacuum. These valves are available with manual, electric, hydraulic, or pneumatic operators. Flow modulation is possible with each type of power operator.

Mild steel piping is satisfactory for most slurry lines where rigid piping is preferred. Tees and crosses should be used as elbows to facilitate cleaning. A freshwater flushing system should be installed to flush out piping, pumps, and valves when the system is shut down for any reason. Reduction of line diameter must not be so abrupt that it causes a violent hydraulic disturbance that can result in dewatering and compaction of the lime.

Pipeline designers usually use a coefficient of $C = 100$ for slurry lines carrying up to 3 lb/gal (594 g/L) of hydrated lime. Piping should not be excessively oversized initially to accommodate estimated future system capacity. The penalty for oversizing is usually increased maintenance problems during the early years. Dissolved solids tend to be precipitated out of the process water where lime slurry or any other alkaline substance is applied. For this reason, an air gap is preferable between the end of the feed line and the surface of the water being treated. If a submerged application cannot be avoided, some means must be provided for periodic cleaning of the end of the slurry pipe to break up the precipitated mass, which may eventually plug the feed line.

Slurry Pumps. Slurry pumps generally fall into two categories: centrifugal pumps and positive displacement, or controlled-volume, pumps. Centrifugal pumps are generally employed for low-head transfer or recirculating service. With proper selection of casing and impeller material and an appropriate shaft seal, satisfactory service can usually be attained at a reasonable cost. Replaceable liners and semiopen impellers are preferred. It is important that the pump design allow easy dismantling for cleanout and repair. Lime slurry requires the lowest speed of rotation (1,725 rpm or less) consistent with hydraulic requirements to control impeller plating. Using water-flushed seals on centrifugal lime slurry pumps is not recommended, because it usually results in localized scaling.

Controlled-volume pumps are typically used where metering or positive control of slurry flow is required, such as at the point of application to the process. Several types are available for slurry service. Piston-type pumps where the slurry is in direct contact with the cylinder walls are not recommended for slurry service because of uncontrolled wear and abrasion. Similarly, peristaltic or squeeze-type pumps are subject to wear and excessive tubing replacement and are not normally used for slurry pumping.

Dipper-Wheel Feeders. The rotating dipper feed has been a longtime favorite for feeding slurries where gravity feed is possible between the feeder and the point of application. The feeder consists of a tank in which the slurry level is maintained by a float valve (or overflow weir if gravity return to slurry storage is practical), a dipper wheel with variable-speed drive, and a totalizer to register wheel revolutions. The dipper wheel is usually divided into eight segments, or dippers, each containing about 500 ml of slurry liquid. As the wheel rotates, an agitator bar maintains the slurry in suspension. The inlet float valve to the tank must be routinely cleaned, particularly if it is connected to a pressurized slurry recirculating system. An overflow connection must be provided when a float valve is used; the selection of an appropriate discharge point for this overflow requires planning for each project. Discharging the overflow to the process is not recommended. The overflow line is not required where gravity return of excess slurry to storage is possible using an overflow weir instead of the float valve.

Flow Control Valves. Sensitive flow control valves are capable of being positioned or modulated by an external control signal. The valve functions to control the flow of chemical to the process in proportion to the signal output analyzer. For this type of feed system to operate satisfactorily, chemical must be supplied to the valve under reasonably constant head, free of suspended material that could clog the valve. Because this constitutes what is referred to as a closed-loop system, it is imperative that the response and sample detention time of the overall control loop be properly designed to prevent cyclic overfeeding and underfeeding.

Splitter Boxes. Custom-designed splitter boxes can provide a constant ratio of slurry discharge rates by gravity flow to multiple application points. For example, one splitter box with three weirs discharging to separate troughs can supply lime slurry to three different basins. The weir design, which can provide for adjustable weirs, dictates the flow split ratios.

Feed and Metering Systems for Polymers. Many treatment plants feed commercial-strength liquid polymer direct from shipping containers or storage tanks, or manually prepared dry polymer solutions from batch-mixing tanks. The relatively high chemical cost requires maximum activation and minimum waste of polymers in solution, and pre-engineered polymer feed equipment can often achieve both goals.

Several pre-engineered equipment packages automatically mix and dilute commercial-strength liquid polymers and deliver a diluted (and in some cases, aged) polymer solution to the point of application. These packages typically include a positive displacement pump, check, pressure-relief and back pressure valves, a dilution water flow control valve, an integral mixing chamber, and feed rate control instrumentation. Many of these equipment packages require the operator only to connect hoses to the inlet and outlet ports and the power plug to an electrical outlet.

Semiautomatic and fully automatic batch mixing and solution feed equipment is also available for dry polymers. Skid-mounted equipment packages typically include a dry polymer storage hopper, dispenser and conveyor (pneumatic or hydraulic), dust collector, mix tanks and agitator, an aging tank, a water flow control valve, positive displacement pumping equipment to deliver the polymer solution to the point of application, and all necessary instrumentation for the batching process and solution feed rate control.

Equipment manufacturers also provide hybrid equipment packages that can accommodate liquid or dry polymer stock. These have manual, semiautomatic, or fully automatic control features and in some cases include redundant mix tanks or storage tanks for aging and activation of the polymer solution. These equipment packages typically include solution transfer and feed pumps in addition to the stock liquid polymer pump or the dry polymer eductor.

Ancillary Feed Equipment. This section describes several equipment items necessary for a complete chemical feed system.

Transfer Pumps. Transfer pumps recirculate during dilution of chemical solutions in the storage tanks, move the chemicals from storage to day tanks, or, for some polymer batching systems, transfer solutions from a mixing tank to an aging tank. These pumps are generally the low-head centrifugal type, with a capacity several times greater than the maximum application rate of the process. Progressive cavity or controlled-volume units are sometimes used for transfer pumping where liquid polymers or other shear-sensitive liquids are handled.

Materials for fabrication of pump liners, body, impeller, and shaft depend on the corrosive and abrasive nature of the material pumped. In some instances, temperature can be a factor if nonmetallic parts are involved. When ordering a pump for chemical service, it is best to consult with the pump supplier and check with the chemical manufacturer. Total reliance on published tables of corrosion- or abrasion-resistant materials for various chemicals can be misleading.

In itself, speed is not a major design factor where true solutions are being handled. However, slurries, viscous liquids, and polymer solutions require special handling and use of low-rpm pumps. In no case should a centrifugal-type slurry pump operate in excess of 1,750 rpm. Liquid, undiluted polymers may be transferred using a 1,750-rpm pump; however, once dilution and aging have taken place, the product is subject to molecular shearing if it is transferred in a centrifugal pump operating at any speed. Screw-type progressive cavity pumps are recommended for this purpose.

Generally, pumps designed to use flushing water on shaft seals should not be used for chemical transfer. Chemical dilution, scaling in lime slurries, and the expense of furnishing a non–cross-connected seal water system are some of the problems encountered. Pumping of carbon slurry is one of the few exceptions.

Modern technology has made available a variety of dry-mechanical-shaft seals to handle most types of liquids and slurries. In addition, indirect magnetic drives are available for smaller pumps.

Day Tanks. Day tanks generally minimize the amount of hazardous chemicals that may be stored within a chemical feed area or minimize the amount of chemicals that will be lost if there is a rupture in the suction lines between the storage tank and the chemical feeders. The use of day tanks is primarily a safety issue and applies to hazardous chemicals such as acids or caustic soda.

In addition, day tanks locate stored chemicals closer to pulsating diaphragm metering pumps to minimize losses in suction piping, provide a near-constant suction head, and allow for proper operation of the metering pump. Locating day tanks close to pumps is useful when bulk storage tanks are relatively remote because of chemical delivery and unloading constraints. Day tanks are typically sized to provide 24 hours of chemical storage based on the maximum chemical volumes metered to the application points. Smaller day tanks can limit the volume of hazardous chemicals and subsequently prevent the area from receiving a more stringent hazardous classification under building and fire codes.

The filling mechanism between the bulk storage tank and the day tank should be carefully designed to prevent overfilling the day tank. Controls include automatic shutdown of transfer pumps or valves controlling the flow between bulk storage tanks and the day tank, and alarms on high levels in the day tanks. Visual and electronic level measurements are recommended, and secondary containment of day tanks should also be provided. Some systems require an operator to hold a safety switch to continue the transfer operation.

Day tanks can be mounted on scales to monitor feed rates and calibrate chemical metering equipment. Materials used for day tanks are generally the same materials used for bulk storage.

Piping and Conduits. Designing piping systems to convey chemicals must include provisions for redundancy, isolation, and maintenance. These provisions are essential when considering problems that can occur from deposits and precipitates when eductors and positive displacement pumps are used to convey slurries. Eductors and piping that handle slurries should be sized to maintain minimum velocities to minimize solids deposition, especially for powdered activated carbon slurries. Flushing provisions are desirable for systems with slurries or polymer solutions to evacuate chemicals in the piping and prevent deposition or solidification of chemical solids when the system is off for any length of time.

Suction lines between supply tanks and pumps must be kept as short as possible, and the entire system should be located close to the application point. Where possible, it is important to avoid a layout that permits siphoning by simultaneous existence of positive suction pressure and low or negative discharge pressure on the pump. Back pressure valves installed on the discharge line prevent this condition and can help maintain the accuracy of discharge on diaphragm metering pumps. Back pressure valves should not be used for slurry applications.

Piping system design for use with oscillating or reciprocating pumps must be carefully prepared to avoid cavitation in the pump chamber and vapor lock in the suction piping from low suction pressures below the net positive suction head (NPSH) required. These effects occur because of the inertia effect on the liquid column from the movement of the piston or diaphragm as it changes direction on either end of the stroke.

The NPSH required must be established for the chemical being conveyed and then compared with the available NPSH based on piping head losses calculated for both viscosity head loss and acceleration head loss. The higher of these two head losses is used to calculate available NPSH. Suction chambers located close to metering pumps can minimize head losses in suction piping due to pulsating flows. Discharge pressures based on piping head losses must also be determined using the higher of these two types of head losses.

Table 15.4 is a condensed listing of common water treatment chemicals and the corrosion resistance of various types of piping materials at temperatures up to 104° F (40° C)

TABLE 15.4 Piping Applications

Chemical	Iron or steel	Type 316 stainless	Type 304 stainless	Copper	PVC—type 1	Fiberglass reinforced polyester (FRP)	Polypropylene	Rubber tubing	Glass
						Piping material			
Activated carbon (slurry)	NR	X				X			
Alum	S	S	NR		S	X	X		X
Ammonia, aqua	S	X			X			X	
Calcium hydroxide (slurry)	S	X	X		X		X	X	
Calcium hypochlorite					X		X		X
Carbon dioxide (dry)	S	X	X	X	X	X	X	X	
Chlorinated copperas					X		X	X	
Chlorine (dry gas)	S			X	NR				
Chlorine solution	NR	NR	NR		S	X		X	
Chlorine dioxide (3% soln.)					X				X
Coagulant aids	Consult manufacturer—generally not corrosive								
Copper sulfate		X	X		S	X	X	X	
Dolomitic lime (slurry)	X	X	X		X		X	X	
Ferric chloride	NR	NR	NR	NR	S	X	X		X
Fluosilicic acid	NR	NR	NR		X		X		NR
Hydrochloric acid	NR	NR	NR	NR	X	X	X	X	
Potassium permanganate (2% soln.)	X	X			X	X	X		

CHEMICALS AND CHEMICAL HANDLING   **451**

TABLE 15.4 Piping Applications (*Continued*)

Chemical	Iron or steel	Type 316 stainless	Type 304 stainless	Copper	PVC—type 1	Fiberglass reinforced polyester (FRP)	Polypropylene	Rubber tubing	Glass
Sodium carbonate (soln.)	S				X	X	X	X	
Sodium chloride		X			X	X	X		X
Sodium chlorite					X	X	X		X
Sodium fluoride (1% to 5% soln.)		X			X	X	X	X	
Sodium hexametaphosphate (soln.)		X			X	X		X	
Sodium hydroxide (to 50% soln.)	X	X	X		X	X	X	X	
Sodium hypochlorite (to 16% soln.)					S		X	X	X
Sodium silicate	S	X	X		X	X	X	X	
Sodium silicofluoride		X			X			X	
Sulfur dioxide (dry gas)	X	X			X				
Sulfur dioxide (soln.)		X	X						X
Sulfuric acid (conc.)	S								
Sulfuric acid	NR				S	X	X	X	X

Key: S = Industrial standard or excellent for handling
      X = Suitable for handling
      NR = Not recommended

It is important to remember that temperature and operating pressures are equally important parameters in piping materials selection, particularly when using plastic materials. With plastic materials, temperatures in excess of 104° F (40° C) generally reduce the maximum safe working pressure rating of the pipe.

The designer should consult manufacturers' data for special plastics acceptable for temperatures up to 176° F (80° C). All chemical piping should be schedule-80 thickness, particularly if threaded joints are used. Double-walled pipe with an annular space between the carrier pipe and the containment pipe should be considered for aggressive chemicals such as acids or caustic soda. This design minimizes the release of aggressive and hazardous chemicals to the environment in the event of a leak. In many cases, a leak detection mechanism can be included to provide warning that a leak of the chemical into the annular space has occurred.

Unions or flanges and suitable isolation valves must be provided at each pump or feed-controlling device to permit removal for routine maintenance. Plastic pipe fittings sized 3 in. (80 mm) and smaller are typically the solvent welded type. Plastic pipe fittings 4 in. (100 mm) and larger are similar, except that pipe joints may use solvent-welded flanges instead of sleeves. Threaded pipe fittings are not generally recommended.

Valves used for open-close service should be the straight-through pattern, maintaining full-line size. Plug, ball, and diaphragm valves fabricated of the appropriate material are normally used for solutions free of suspended matter. Pinch valves are capable of handling slurries in both shutoff and modulating-control service. The relatively complicated internal parts of globe and gate valves would be exposed to chemical flow, so their use is not recommended.

It is good practice to support rigid plastic piping on hangers at intervals of 4 ft (1.2 m) or less. Piping must not be clamped by the hanger, because movement resulting from expansion and contraction should not be restricted. Flexible piping and rubber tubing are best supported by channel troughs or sections of steel piping supported on hangers or brackets. In these cases, gaps should be left in the support piping at bends and changes of direction to facilitate tubing installation and removal.

Where rigid metallic piping is used for the solution line, support hangers or brackets should be provided at intervals that prevent sagging for the size of pipe used. In any case, supports should be provided at least every 10 ft (3 m). Burying chemical piping, particularly under concrete slabs, is not desirable due to the possibility of undetected leaks. Piping trenches can be used to convey piping below grade while still allowing access to the pipe for inspection and repairs.

Rotameters. Rotameters indicate flow in conjunction with flow-control valves and can become part of a feed-control system. Certain types of rotameters can be equipped with rotor-position transmitters to permit remote recording of flow rates. Rotameters more generally provide a visual indication of flow, either of water or chemical solutions, to assist in manually setting flow rates for dilution or carrying water.

Visual rotameters, where the float position can be seen relative to a calibrated scale, are practical for clear liquids free of iron or other impurities that obscure scale calibrations. Where liquids do not allow visual indication, magnetically actuated indicators may be used. Branching chemical feed streams through more than two or three rotameters to different application points usually requires continuous manual flow rate adjustment. A more reliable method of feeding multiple application points is to use separate feed pumps dedicated to each application point.

Miscellaneous Feed Systems. Numerous other methods and devices, some of which date back to the nineteenth century, can be used to feed liquid chemicals or to feed dry chemicals that are dissolved or slurried. Many of these devices are still in operation at smaller plants throughout the country, in testimony to their simplicity and adequacy for the job at hand. Some of the more common chemical feed devices are described in the following paragraphs.

Dissolving Tanks. Dissolving tanks have a perforated wooden or metal basket filled with a weighed quantity of the chemical to be dissolved and hung in the upper part of a water-filled tank. Mechanical mixers can be installed to assist in dissolving the chemical. When all material is dissolved, the batch is ready for use, and it is applied by pumps or other flow-control devices. This method is popular for handling glassy polyphosphates in smaller installations and for other lumpy materials that dry feeders do not handle well.

Pot Feeders. Pot feeders are normally used with coarse chemicals, such as lump alum. The pot is charged manually through a pressure-tight cover. An orifice plate or gate valve in the pressure line to be treated generates a differential pressure that is tapped off the line and permitted to flow through the pot via a flow-control valve. In theory, the chemical dissolves and maintains a constant saturated solution in the pot. As the chemical is consumed through the pot, it is displaced and replenished in proportion to the flow. However, the system is subject to clogging with chemical impurities.

Gravity Orifice Feeders. Gravity orifice feeders generally consist of a constant-head supply tank and a fixed or adjustable orifice to deliver the desired rate of flow. These devices can be started and stopped remotely, but they do not lend themselves to slurry feeding or to proportional feed control.

Displacement or Decanting Feeders. A decanting feeder consists of a tank equipped with an overhead mechanism that lowers a displacement cylinder or decanting pipe arm into a prepared batch of chemical solution. The rate of feed flowing over the overflow weir or through the decanting pipe is related to the speed at which the cylinder or pipe is lowered. Mixers can be used if slurries are to be fed.

Utility Water. High-pressure, or utility, water has several uses in chemical feeding systems, including diluting chemical solution, flushing chemical feed systems, providing carrying water for neat chemicals, and washing down chemical feed areas. This water is generally obtained from plant-pumped finished water systems or from the distribution system served by the water treatment plant. Use of this water within the chemical feed area requires the precautions described below to maintain the drinking water status of the plant utility water system. It may be useful to install a flowmeter on this water supply to track in-plant water usage.

Dilution of Chemicals. Many chemicals, such as caustic soda and polymers, are delivered in concentrations inappropriate for direct use in unit processes. Utility water reduces the concentration of the chemical to an acceptable level before delivering the chemical to the application point. Polymers are unique in that they must be properly activated in the dilution water before reaching the application point. Many methods of activating polymer are available, but they all generally involve adding dilution water to neat polymer.

Because caustic soda has a high freezing point, it is diluted in many areas of the country to prevent freezing under normal seasonal temperatures. Care must be taken with dilution water used for caustic soda (or other liquids with a high pH) because of its tendency to precipitate calcium and the exothermic nature of the dilution reaction. It is recommended that dilution water used with high-pH solutions be softened before use.

The amount of dilution water used must be carefully calculated to know the concentration of the diluted chemical and to calculate the feed rate required for the desired dosage. In many cases a flowmeter for batch mixing is used on the dilution water to provide the exact quantity of water necessary for the desired dilution ratio.

Flushing Feeders and Chemical Feed Lines. For most chemicals, it is not desirable to allow a concentrated solution to remain for long periods of time in chemical metering equipment and chemical lines. For polymers, slurries, some coagulants, and chemicals that cause precipitates or crystals to form in the pumps and piping, a flushing water system is recommended. This system flushes chemical metering equipment and downstream chemical pipelines whenever chemical metering is disrupted for any length of time.

Control systems can be designed to automatically flush the feed equipment and piping whenever a feed system is turned off. Flushing time should be sufficient to completely evacuate the discharge pipeline of all chemicals to the application point.

Conveying Water. It is sometimes desirable to meter a chemical dose into a stream of water to convey the chemical to the application point. The amount of water used to carry the chemical is not necessarily important, but a flushing velocity of 2 to 5 ft/s (0.6 to 1.5 m/s) may be desirable. Chemical is metered in its neat form with the dose dependent on the plant flow rate and the amount of chemical fed, not the amount of carrying water used.

A typical example of this condition is the use of an eductor or injector (jet pump) to provide the motive force for moving the chemical from its metering device to the application point. Care must be taken if carrying water is to be used with alum because the hydrolyzing chemical reaction begins to occur as soon as the alum is added to water. In most cases, the carrying water is metered to determine the volume of water so that it can be kept to a minimum while still maintaining adequate velocities in the solution pipeline.

Carrying water is almost always used with gaseous chlorine. The carrying water, in passing through an injector, creates a vacuum that draws gaseous chlorine from the storage tanks through a vacuum regulator. Chlorine is not controlled through carrying water but by other means, such as controlling gas flow in the chlorinator. Carrying water provides the means of creating a vacuum and then mixing gaseous chlorine with water to create a chlorine solution. The quantity of carrying water for chlorine is dictated by the amount necessary to create a vacuum in the injector against the back pressure created in the downstream piping from the carrying water flow.

Conditioning and Heating Requirements. Whenever water is mixed with chemicals, a reaction occurs that may or may not create a reason for concern. In the case of ammonia or caustic soda, water should be softened before adding the chemical to prevent precipitation of calcium or magnesium carbonates and ultimately clogging of the line downstream of the addition point. Softened water should also be used for any chemical that significantly elevates solution pH or alters the chemicals' solubility. Commercial ion exchange equipment can generally soften utility water adequately for mixing and conveying most chemicals used in water treatment.

Some chemicals go into solution more readily or in higher concentrations in warmer water than colder water. These chemicals include dry polymers, ferric sulfate, sodium silicofluoride, soda ash, and some quicklime (slaking rather than dissolving). Blenders may combine hot water with utility water to achieve a preset temperature before mixing with the chemical. Care must be taken to adequately size hot water heaters in this application to ensure that an adequate quantity of hot water is available for all chemicals that need water with an elevated temperature. Hot water is also desirable for use in washing down polymer feed areas. Spilled polymers are more easily flushed with hot, rather than cold, washwater.

Cross-Connection Control. Whenever utility water is used in direct connection with chemical feed systems, cross-connections are a concern. If negative pressures are created in the plant water system, chemicals can be drawn back into the potable water system and contaminate the drinking water. Most health agencies require reduced pressure backflow prevention devices to be installed in all utility water systems that are connected to nonpotable water systems, such as chemical feed systems. A careful analysis will identify all potential cross-connections and ensure installation of acceptable backflow prevention devices between the potential cross-connection and the potable water system.

Diffusion of Chemicals at Application Points

It is critical to the efficient use of chemicals in treatment processes to adequately disperse the chemical throughout the flow. When a chemical is applied at turbine rapid mixers or static mixers, the chemical should be injected into the center of the flow stream immediately upstream of the mixing devices. The mixing devices then disperse the chemical evenly throughout the flow. However, in many cases a mixing device is not used, and the chemical must be dispersed evenly throughout the flow using other means such as diffusers and the hydraulics of the conduit or basin where the chemical is added.

With the exception of carbon dioxide and ozone, most chemicals are brought to the application point in the form of liquid solutions or slurries. Although these gases can be injection fed, carbon dioxide and ozone are usually direct fed as fine bubbles to the process water through relatively deep contact tanks. Typically, a counterflow (water flowing downward while the bubbles rise) is used with fine bubble diffusers to provide efficient transfer of the gas to the water.

For applying solution chemicals to both pipes and channels, the chemical is usually dispersed in the flow using diffuser pipes with multiple orifices. Figure 15-11 shows several of the more common diffusers used in dispersing chemicals. If chemicals are in slurry form, dispersion is usually accomplished using an open-ended pipe for pipe flow or a trough with notches or holes at the bottom for channel flows.

When disinfection is entirely by free chlorine, it is important to distribute the chlorine dose uniformly across the flow cross section. Traditionally, good designs have accomplished this with open-channel or pipeline diffusers. It can be shown, however, that truly complete mixing across the flow cross section only takes place some distance downstream from diffusers of this type. When free chlorine residuals are used, this may not be a critical problem.

When water contains significant amounts of ammonia (10% of the chlorine dose or more), initial mixing takes on new importance. Under these conditions there is evidence that chlorine is more effective during the first few seconds after its addition. With ammonia present, failure to mix the chlorine solution rapidly with all the water can result in a break-point reaction with part of the water and little or no reaction with the remainder. The break-point reaction is not reversible. Thus part of the water receives excess chlorine, and the remainder receives little or none. Consequently, good initial mixing becomes a much more important process step, and initial mixing devices should be designed to spread chlorine throughout the flow cross section in the shortest possible time with the least possible backmixing. This can be accomplished by in-line mixers, gas aspirator devices, a hydraulic jump, or a pump-diffuser system. Flash-mixing chambers such as those used for coagulants in a conventional water treatment plant are not satisfactory because they result in backmixing.

Careful attention must be given to the dispersion of scale-producing chemical solutions that can clog diffuser pipe orifices. These chemicals include metal salt coagulants, such as alum or ferric chloride, and alkali chemicals, such as caustic soda, soda ash, lime-saturated solutions, and ammonia solutions. For these types of chemicals, provisions must allow cleaning of the diffusers.

Redundant diffusers may also be included to allow one diffuser to be taken out of service for cleaning while the other diffuser remains in operation. A relatively unique self-cleaning design for ammonia solution diffusers uses short lengths of Teflon tubing at each orifice. These lengths of tubing move in the flow stream and can break off scale products that form at the end of the tubing where the solution contacts the main process water.

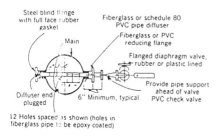
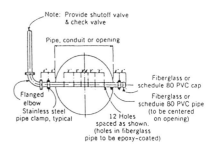
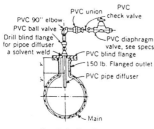
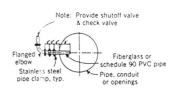
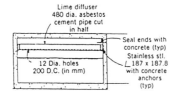
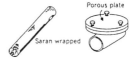

FIGURE 15.11 Types of chemical diffusers.

DESIGNING FOR SAFETY AND HAZARDOUS CONDITIONS

A general overview of safety requirements and suggested protective measures for handling various chemicals is presented in Table 15.5. Suggested protective measures shown in the table apply to the various chemicals used whether they are in dry, liquid, or gaseous state. Refer to the chemical supplier's most current Material Safety Data Sheet (MSDS) for up-to-date chemical safety requirements and protective measures. See below for a complete description of an MSDS. Another useful general reference for chemical safety, exposure limits, and incompatibilities is the *NIOSH Pocket Guide to Chemical Hazards* (National Institute of Occupational Safety and Health, 1994).

TABLE 15.5 Protective Measures for Water Treatment Chemicals

Chemical (D = dry; L = liquid; G = gas)	Positive ventilation	Protective clothing	Neck cloths	Gloves	Rubber boots	Rubber gloves	Goggles	Face shields	Rubber aprons	Respirator	Gas mask	Avoid skin contact	Safety shower and eye baths	General
Activated alumina (D)	■													Store away from gasoline, mineral or vegetable oils, calcium hypochlorite (HTH), lime, sodium chlorite, or potassium permanganate
Activated carbon Powder (D)	■	■	■	■										
Granulate (D)	■		■	■										
Alum sulfate (D)	■	■				■	■			■				Similar to other acid
Alum sulfate (L)						■		■	■	■		■	■	
Ammonium hydroxide (L)	■											■	■	Moist NH_3 reacts with many metals and alloys—liquid contact produces burns
Ammonium sulfate (D)	■	■	■	■								■	■	See alum sulfate above
Anhydrous ammonia (G)	■	■					■				■	■	■	Fire sprinklers and water hoses effective in removing gas
Bauxite (D)	■													
Bentonite (D)	■			■										
Calcium Carbonate (D)	■		■											
Calcium hypochlorite (D)		■		■		■				■		■		
Carbon dioxide (G)											■	■		
Chlorine (G)	■			■						■	■	■		Avoid contact with hydrogen or organic compounds or other flammable materials
Chlorine dioxide (G)		■					■			■		■	■	Solution is corrosive
Copper sulfate (D)		■	■		■	■				■		■	■	Very corrosive
Ferric chloride (D)		■	■		■	■				■		■		Very corrosive
Ferric sulfate (D)		■	■		■	■	■			■		■		
Ferrous sulfate (D)	■	■	■	■								■		
Ferrous sulfate (L)	■	■										■	■	

TABLE 15.5 Protective Measures for Water Treatment Chemicals (*Continued*)

Chemical (D = dry; L = liquid; G = gas)	Positive ventilation	Protective clothing	Neck cloths	Gloves	Rubber boots	Rubber gloves	Goggles	Face shields	Rubber aprons	Respirator	Gas mask	Avoid skin contact	Safety shower and eye baths	General
Fluorosilicic acid (L)	■	■	■			■		■	■	■		■	■	Have lime slurry on hand
Fluorspar (D)	■	■	■			■		■		■		■	■	Etches glass when moist
Hydrated lime (D)	■	■	■				■					■	■	Can burn eyes or skin
Hydrochloric acid (L)	■	■			■	■		■	■			■	■	
Iron-exchange resins (D)	■	■												Hydrogen cation resins acid
Ozone (G)	■									■				
Potassium permanganate (D)	■	■	■		■	■		■	■			■	■	Large quantities present fire hazard
Quicklime (D)	■			■			■					■		Can burn eyes or skin
Sodium aluminate (D)	■	■		■			■					■	■	
Sodium aluminate (L)	■	■		■								■	■	
Sodium bisulfate (D)	■	■		■			■	■				■	■	
Sodium carbonate (D)	■	■		■			■					■		
Sodium chloride (D)	■	■	■	■								■		
Sodium chlorite (D)	■	■		■			■			■		■	■	
Sodium fluoride (D)	■	■		■			■					■	■	
Sodium polyphosphate, glassy (D)	■	■		■								■		
Sodium hydroxide (D)	■	■		■			■					■	■	Can dehydrate skin
Sodium hydroxide (L)	■	■			■	■		■	■			■	■	Rinse any spills immediately with water
Sodium hypochlorite (L)	■	■				■		■	■			■	■	
Sodium silicate (D)	■	■		■								■		
Sodium fluorosilicate (D)	■	■		■			■					■	■	
Sodium sulfite (D)	■	■		■								■		
Sodium dioxide (G)	■									■	■	■		
Sulfuric acid (L)	■	■			■	■		■	■			■	■	

With all safety procedures, a written program should be developed and reviewed regularly with all personnel. This establishes what should be done in emergency situations before they occur. Every new employee who may be expected to use the equipment should be instructed in safety emergency procedures. Documented periodic review and training sessions are recommended. These may be mandatory under worker safety, transportation, or environmental regulations.

Federal, State, and Local Regulations and Codes

At present, regulatory requirements related to worker safety, public safety, and environmental protection are in a state of flux, as are many basic standards for drinking water safety. The use of many water treatment chemicals is subject to overlapping regulation by local, state, and federal statutes and codes. Many water treatment chemicals are classified as hazardous materials and regulated accordingly.

Early in the design process it is highly advisable to consult with appropriate agencies to determine which have input or enforcement authority over the project. These typically include local and state public health agencies, local and state environmental agencies, state or regional federal Occupational Safety and Health Administration (OSHA) offices, and local fire marshals. Federal OSHA's jurisdiction is the safety and health of private as opposed to public works, sector workers and employees, and related workplace conditions; however, some states have adopted and enforce federal OSHA regulations for municipal and other governmental employees. Local fire departments are concerned with acute effects of a chemical release both on-site and off-site. Failure to address concerns of these authorities early in the treatment plant design could lead to costly changes later in the design process.

Some states have their own OSHAs authorized to enforce state regulations instead of federal OSHA regulations. These state regulations may be more restrictive than the federal OSHA requirements cited in this chapter. Sources of assistance in identifying pertinent agencies include staff environmental and safety personnel available to the design team; state-specific compliance guides developed by state agencies, chambers of commerce, or private publishers; and outside consultants.

Disclosure requirements may be necessary under the Emergency Planning and Community Right-to-Know Act (EPCRA), also known as Title III of the Superfund Amendments and Reauthorization Act (SARA) of 1986 (42 USC 11,001). Sections of EPCRA require facilities with chemicals above the thresholds given in Table 15.6 to report to their State Emergency Response Commission (SERC) and coordinate with their appropriate Local Emergency Planning Committee (LEPC). This can include providing an annual copy of their chemical inventory to the SERC, the LEPC, or the local fire department. Construction of a new facility exceeding these amounts will require notification.

Transportation, loading, and unloading of hazardous materials is also regulated by the U.S. Department of Transportation (49 CFR 171 through 180).

An additional subcategory of hazardous materials mandated by the Clean Air Act Amendments of 1990 [Section 112(r), 42 USC 741(r)] receives additional regulatory requirements because of their particularly acute ability to injure workers, the public, or the environment. Table 15.7 lists typical water treatment chemicals so regulated. All of these laws stem from the 1984 toxic cloud in Bhopal, India, which killed or injured thousands of individuals, and from subsequent petroleum refinery and chemical transportation accidents.

Storage or use of chemicals above these threshold quantities may be subject to OSHA's current Process Safety Management (PSM) regulation (29 CFR 1910.119), to the USEPA Risk Management Plan (RMP) regulation (40 CFR 68), or both.

TABLE 15.6 EPCRA Threshold Planning Quantities

Chemical	Threshold planning quantity (lb)
Chlorine	100
Chlorine dioxide	Not listed
Anhydrous ammonia	500
Aqua ammonia	Not listed
Hydrogen peroxide (conc. 52% or greater)	1,000
Sulfuric acid	1,000
Ozone	100

TABLE 15.7 Selected Accidental Release Program Regulatory Thresholds

Chemical	USEPA threshold quantity (proposed RMP) (lb)	OSHA threshold quantity (PSM) (lb)
Chlorine	2,500	1,500
Chlorine dioxide	1,000	1,000
Anhydrous ammonia	10,000	10,000
Aqua ammonia	20,000 (conc. 20% or greater)	15,000 (conc. 44% or greater)
Hydrogen peroxide	Not listed	75,000 (conc. 52% or greater)
Sulfuric acid	Not listed	Not listed
Ozone	Not listed	100

OSHA requires evaluation and minimization of risks to workers (within the facility itself, or "inside the fence") through safety programs; USEPA scrutinizes public health and environmental effects (beyond the facility boundaries, or "outside the fence") using air-dispersion modeling studies. Both regulations share an identically defined requirement for a detailed failure and release evaluation of the entire system containing the listed chemical. This evaluation is called a Process Hazard Analysis (PHA).

Because of the potential costs and effects of these regulations, in 1995 the American Water Works Association (AWWA) funded development of a water industry–specific guidance document. This document should be available in 1997. In the interim, general technical guidance publications are available from the American Institute of Chemical Engineers.

If a facility is required to comply with PSM or RMP, the design team should schedule project time and funds to conduct a PHA on the design. A PHA can be accomplished using one of several alternative techniques and normally requires a group approach with both engineering design and facility operations expertise. The results and recommendations of the PHA can directly and indirectly affect plant design and operations, equipment options, treatment chemical management, and even community relations.

Before starting up the process equipment for the first time a formal Pre–start-up Safety Review is also required to ensure that all required design elements, training, and management systems are in place. Subsequent changes to process equipment must follow a documented Management of Change procedure.

NSF Listings and Certifications

Drinking water standards commonly cite the publications of NSF International (NSF—formerly the National Sanitation Foundation). NSF develops voluntary standards for drinking water system components and drinking water chemicals. These can include chemicals added to the treatment process, piping, lubricants in water treatment equipment, and so on. Purchased process chemicals may have to be certified as meeting NSF maximum contaminant levels. NSF standards may also be adopted by reference in state drinking water regulations. If new chemicals are added to the facility water treatment process, finding certified suppliers may become an issue.

Material Safety Data Sheets

All chemicals sold should be accompanied by a Material Safety Data Sheet (MSDS). Use of MSDSs are mandated by OSHA as part of the Hazard Communication Standard (29 CFR 1910.1200). ANSI standard Z400.1 details the recommended contents, as shown below. The MSDS contains important basic information for the design team (Table 15.8). Current copies should be obtained from potential suppliers.

Confined Spaces

A "confined space" is an area large enough for an employee to enter that has limited or restricted means of exit and is not designed for occupancy (29 CFR 1910.146). Examples include tanks, silos, storage bins, vaults, and pits. They are regulated by OSHA because:

TABLE 15-8 Required Contents of Material Safety Data Sheets (ANSI Z400.1)

1. Chemical product/company product identification
2. Composition/information on ingredients
3. Hazard identification
4. First aid measures
5. Firefighting measures
6. Accidental release measures
7. Handling and storage
8. Exposure controls/personal protection
9. Physical and chemical properties
10. Stability and reactivity
11. Toxicological information
12. Ecological information
13. Disposal considerations
14. Transport information
15. Regulatory information
16. Other information

1. They have the potential to create an oxygen-deficient or hazardous atmosphere.
2. A worker could become engulfed or suffocate in the storage material.
3. The space has converging walls that could trap a worker.

Because of the extensive management program mandated for confined spaces, the design team should work with the operations staff to identify and minimize confined spaces in the plant. This will significantly improve the facility's long-term ease of operability.

Safety Equipment

Safety equipment may be used by many people and must be easily cleaned and disinfected. A regular inspection and maintenance program will ensure that all protective equipment is kept clean and in good repair. Protective clothing must include foot protection to prevent injury from falling objects. This is especially important in receiving and transferring inventory. In general, regulatory agencies prefer engineering controls on potential health and safety dangers and use of personal equipment when such controls are not feasible. Engineering controls include enclosure or confinement of the operation, general and local ventilation, and substitution of less toxic materials.

Protective Equipment and Clothing. Protective clothing used in handling chemicals should be impervious and should cover exposed areas of the body, including arms and legs. Where gloves are used, they must be appropriate to the exposure. They may be standard work gloves of heavy canvas or leather construction or rubber or synthetic gloves impervious to various liquid chemicals. Gloves should generally protect the forearms as well as the hands.

Hard hats are normally required where workers may be subject to injury from falling or flying objects and must meet ANSI Standard Z89.1. Safety shoes should conform to the ANSI National Standard for Safety Toe Footwear, Z41.1.

Eye protection is generally provided by protective goggles fitted to each individual worker. Where skin exposure is a concern, the face should be protected by an 8 to 10 in. (20 to 25 cm) high face shield that normally covers the full face. Personnel performing emergency response duties may require higher levels of chemical protection as required by OSHA (29 CFR 1910.120).

Respirators and Masks. Respirators or gas masks should be provided, as indicated in Table 15.5 or by the product's MSDS. Respirators may be of the particulate filter type, commonly referred to as filter respirators. These should be properly fitted. They are generally used for short, intermittent, or occasional dust exposures. Respirators must be approved for protection against the specific type of dust that may be encountered. Respirators should be selected in accordance with American National Standard Practices for Respiratory Protection, Z88.2. Normally, respirators are used as a protection from particulate matter, dust, or mist. Breathing air can be supplied to the airline-type respirator from cylinders or air compressors. Oxygen must never be used with airline-type respirators.

For gas or gas and particulate matter not immediately dangerous to life, the airline-type respirator is normally satisfactory. This can be a hose mask without a blower, or it can be a chemical-cartridge respirator provided with a special filter for the specific contaminant present in the atmosphere.

For gas or gas and particulate matter that is immediately dangerous to life, a self-contained respirator or gas mask should be used. A hose mask with blower or a gas mask with a special filter can also be used. In unknown conditions, self-contained breathing apparatus and the pressure-demand-mode positive pressure should be employed.

Personnel using respirators should be certain that they experience minimum face mask leakage and that the respirator is fitted properly. A semiquantitative fit test should be set up annually for each user of a nonpowered particulate respirator. Previous medical clearance or periodic medical review may also be necessary.

Wearing contact lenses with a respirator in a contaminated atmosphere is not advisable. Individuals wearing corrective glasses may find that they cannot achieve a proper seal with a gas mask or full face mask. In such cases, corrective lenses or lens inserts should be provided in full face pieces and should be clearly identified. Facial hair can also prevent a proper seal and may require a "clean-shaven" policy.

Respirator use is governed by OSHA regulation 29 CFR 1910.134, Respirator Protection. Requirements such as a monitoring program, a buddy system, and confined space restrictions may apply.

Safety Showers and Eye Washes. Where indicated for specific chemicals, suitable facilities for quick drenching or flushing eyes through the use of an eye or total body wash must be provided. Both facilities must use potable water from a reliable source and be located within the work area when a person may be exposed to injurious or corrosive chemicals. The water supply pipe to these devices can be equipped with a flow switch that activates an alarm when used. Safety showers installed outdoors must be freeze-proof.

Safety Screens. Safety screens are a simple engineering control that can prevent worker injuries resulting from accidental contact with process equipment and chemicals. Appropriate locations should be identified during design through a combined effort of the design team and the facility operations staff. General guidelines can be found in OSHA 29 CFR 1910.23.

Protection from Toxic Gases

As previously discussed, typical gaseous treatment chemicals such as chlorine, anhydrous ammonia, and ozone are classified as hazardous or toxic under worker safety, public health, environmental, and transportation regulations. They are also governed by regional model building and fire codes.

Effective water treatment chemicals can be powerful and dangerous materials and must have the full respect and attention of the designer. Treatment chemicals that can become airborne, such as chlorine gas, are of particular concern. Chlorine is classified by the U.S. Department of Transportation as a toxic and corrosive gas. It is also a strong oxidizer and can support combustion, including steel combustion. When chlorine gas is inhaled or contacts mucous membranes, such as the eye or lung, it reacts to form acids. At sufficient concentrations, these acids can maim or kill the victim by causing severe damage resembling a burn to body tissue. Direct contact with liquid chlorine can also cause freeze burns.

Stored under pressure in the liquid phase, chlorine liquid evaporates to fill 460 times its original volume. Consequently, even a small indoor release of chlorine liquid can flood an enclosed room with vapor and overcome the occupants. Chlorine gas is heavier than air and tends to remain concentrated at ground level. Consequently, a large outdoor release of liquid or gaseous chlorine can be life-threatening miles downwind from the leak. It is extremely important for the safety of plant staff and neighbors that designers follow recognized codes and practices for chlorine systems. Where design varies from standard practice it must be rigorously analyzed for system reliability and integrity.

Building and Fire Code Requirements

Three regional organizations publish model building and fire codes, as well as the National Fire Protection Association (NFPA), publisher of model fire codes and standards. Each organization takes a different approach to hazardous materials control. A code does not have the force of law until it is adopted as written or with local modifications by an enforcement body, such as a city, county, or state fire marshal.

It is consequently difficult to generalize about code requirements. Codes may be specific and used to enforce requirements, such as containment and treatment systems for toxic gases. At the beginning of the design process, it is necessary to determine which agencies enforce building and fire code requirements and what version of the respective codes have been adopted.

Ventilation Requirements

Positive ventilation indicated for specific chemicals is essential to remove gases or dust that may develop from materials handling. Proper ventilation design can also avoid some areas becoming designated as "confined spaces." The contaminated area must be isolated from the rest of the plant, and the exhaust systems must be run for a sufficient period of time to remove all contaminated air before that area is reopened.

The rate of exhaust must be sufficient to promptly clear the laden air from the contaminated area. Doors to the area should be flanged and sealed tightly when closed. The construction, installation, inspection, and maintenance of the exhaust system should conform to requirements in American Society of Heating, Refrigeration, and Air Conditioning (ASHRAE) and *Recommended Standards for Water Works* (1992), together with state and local codes governing design and operation of exhaust systems.

The static pressure in exhaust ducts leading from the ventilation equipment should be periodically checked to ensure continuing satisfactory operation. If there is an appreciable change in pressure drop, the system should be cleaned.

It is normally good procedure in handling dust-laden air to have the ventilation equipment discharged through dust-collecting equipment. This equipment must be set up so that accumulated dust can be removed without contaminating other working areas.

Heating, ventilation, and air conditioning (HVAC) of rooms containing equipment regulated under PSM or RMP, such as chlorination equipment, must be carefully designed. In the event of a leak it may be desirable to automatically shut down the room ventilation to contain the leak. It may also be desirable to automatically shut down air conditioning systems in adjacent buildings to prevent fugitive vapors from being drawn inside. Either way, the HVAC system for chlorine storage and feed rooms should be separated from the HVAC systems serving other areas of the same building.

Containment and Treatment Systems

Containment and treatment systems are design approaches intended to prevent the release of process chemicals that could cause employee, public, or environmental injury. These can be active, meaning that they have a human, mechanical, or other energy source, or passive, meaning that they do not have an energy source. Passive systems can include dikes, catch basins, and drains for liquids and enclosures for both liquids and gases. Containment for liquids is discussed in the section on secondary containment for liquids. Passive systems are often preferred by regulatory agencies because there is less opportunity for failure during emergency operation. Active systems include chemical scrubbers that neutralize a released chemical with another chemical and water sprays that can partially lower an airborne chemical.

No national consensus approach exists at present specifically addressing containment and treatment systems for water treatment plants. One regional model fire code promotes containment and neutralization of specified chemicals, although opinions differ on the value of such systems. General references on accidental chemical release prevention are available. When considering neutralization systems, the designer must balance risks introduced by storage of the neutralizing agent.

The chlorine scrubbing system, or scrubber, represents existing technology for controlling accidental or routine releases of chlorine gas. A scrubber is basically a process equipment unit that collects chlorine vapor and mixes it with alkaline solution (caustic) to remove the chlorine. The main reaction product is hypochlorite. The most common chemical used for scrubbing is sodium hydroxide (caustic soda), although others such as sodium carbonate and calcium hydroxide may be used. Scrubbers are often associated with neutralizing accidental releases, but they are also used for neutralizing routine process releases.

Proper sizing and design of the scrubber are important. Significant heat is generated by the caustic-chlorine reaction, so materials must be chosen for heat and chemical resistance as well as for strength.

Scrubbing systems are a tested and proven technology for chlorine storage vessels up to one ton in size, but they have not been reliably tested beyond this range. Nevertheless, scrubbers and enclosures have been built for 90-ton chlorine railcars, although at this time performance of these larger systems has not been verified by a full-container release.

Scrubbers require storage of significant quantities of caustic, and neutralization capacity is also critical. If the neutralizing caustic becomes saturated, chlorine gas, along with steam and oxygen, may evolve, causing foaming, which could become a significant emergency situation. Caustic storage imposes its own chemical management requirements, such as secondary containment, as well as additional handling risks for facility personnel. Caustic can also react with some metals to form hydrogen, creating a potential fire or explosion hazard.

Scrubber designs include venturi systems, horizontal sprayers, and packed media and usually recirculate caustic solution. Efficiency can be improved by increasing the caustic/gas interface ratio. Proper attention must be paid to container enclosure design, such as the storage room. The structure must provide for any expected pressurization, as well as wind, snow, and seismic loadings. Ventilation must also be sized for optimum air transfer between the enclosure and the scrubber unit, including any recirculation requirements.

Enclosing a leaking chlorine container generally limits options for direct access by response personnel because current chemical protective clothing is not designed for the concentration of chlorine that can be reached in containment rooms. For this reason, accidental release scrubbers generally should be sized to neutralize the complete contents of the largest container. Local fire codes may specify performance parameters, such as the permissible concentration level of chlorine in scrubber exhaust.

Reliability is essential for accidental-release scrubber systems. Appropriate chlorine sensors, alarms, and control systems are needed. The designer should also consider using redundant systems and conservative structural safety factors. For example, scrubbers may be designed for either dry or wet chlorine gas. For emergency purposes it is more conservative to assume the gas is wet and select materials of construction for the more corrosive operating environment.

Fire and Explosion Prevention

Although generally rare, fire and explosion are potential risks at treatment facilities. For example, chlorine supports combustion in steel at 483° F (251° C). Fuels such as gasoline and propane may be present at support operations. Chemicals used to adjust pH levels can react with each other. It is important that plant design be carefully evaluated to avoid the potential for mixing incompatible materials during transportation, storage, and use.

The National Fire Protection Association (NFPA) develops model standards to address fire and explosion issues. Among these is the joint ANSI-NFPA publication *NFPA 101— Life Safety Code,* which discusses fire design considerations for a variety of building types.

Once facility personnel are familiar with emergency plans, it is advisable to conduct an orientation session for the local fire suppression agency. Joint facility and fire personnel drills may also be useful.

RECENT TRENDS IN CHEMICAL HANDLING AND USE

New trends in chemical use at water treatment plants include increased use of sodium hypochlorite, new safety requirements for gaseous chlorine installations, new chemicals for water treatment, and increased use of bulk chemicals.

Sodium Hypochlorite Systems

Chlorine has been successfully and safely used at municipal water and wastewater treatment facilities for many years. In recent years, however, utilities have been considering the use of hypochlorite rather than chlorine gas because of an increased emphasis on safety where large amounts of gas are stored in or transported through urban areas.

If cost were the only criterion for selecting the chemical form, gas would be chosen rather than sodium hypochlorite. Hypochlorite costs two to four times the equivalent quantity of gaseous chlorine. Local code and regulatory requirements for employee training and safety equipment for gaseous chlorine in some communities, however, cause hypochlorite and gaseous chlorine to be more comparable in total chlorination costs. For this reason, many large and small utilities have converted from gas to hypochlorite, and more are expected to do the same.

Chlorine Gas Leak Containment and Treatment Systems

Another reason for the recent emphasis on safety issues associated with chlorine handling has been brought about by code officials and public leaders concerned about protecting those who operate the facilities, those who reside and work in the surrounding area, and those who respond in case of an emergency. Heightened awareness of safety issues and the desire to provide as safe a system as possible have made it necessary to consider emergency chlorine scrubbers and other available safety measures in any new or rehabilitated facility that handles chlorine or other hazardous or toxic compressed gases.

Providing the impetus for this trend are the Uniform Fire Code, Superfund Amendment and Reauthorization Act (SARA Title III), Clean Air Act Amendments (CAAA), and regional and local building codes. OSHA regulations for process safety management of highly hazardous chemicals apply in those states where OSHA has jurisdiction over employees operating and maintaining municipal treatment plants and in all states over employees of privately owned plants.

New Water Treatment Chemicals

Water treatment chemicals that are relatively new include liquid oxygen, polyaluminum chloride, phosphoric acid, and several forms of phosphates, hydrogen peroxide, and polymers.

Liquid oxygen is used for generating ozone and results in more efficient production than an air source. It is conveniently delivered and stored on site. Polyaluminum chloride is a primary coagulant, and its chief advantage is a residual, which generates a smaller volume of sludge that is easier to dewater than some other coagulants.

The use of phosphoric acid and other phosphates is steadily increasing to help control internal corrosion in the distribution system piping. Designers should be aware that some phosphate products lower in phosphates accomplish the same purposes, and chemical selection should factor into environmental considerations. Hydrogen peroxide is sometimes used together with ozone for oxidation purposes. Although polymers have been used in water treatment for years, the increased interest in enhanced coagulation and maximizing particle removal through filters will probably increase the use of polymers as coagulant, flocculent, and filter aids.

Bulk Storage and Handling

In the past, bulk storage and handling facilities were often limited to plants consuming large quantities of chemicals. Now smaller plants use bulk facilities to avoid potentially hazardous unloading and handling procedures and to simplify labor-intensive demands of handling containers and bags of chemicals. Before deciding on bulk service, plants using small quantities of chemicals should determine whether a long shelf life (which may result with large, bulk storage tanks) would affect the quality of stored chemical. If excessive shelf life is a concern, it may be practical to use smaller storage tanks and order partial truckloads of chemicals, although the delivery cost would be higher.

Acknowledgements

The following manufacturers contributed information on chemical properties and characteristics: Air Products and Chemicals, Inc.; Albright & Wilson Industrial Chemicals Group; American Cyanamid Co.; American Norit Company, Inc.; Calgon Carbon Corp.; Delta Chemicals, Inc.; Dow Chemicals U.S.A.; General Chemical Corp.; HVC, Inc.; International Dioxide, Inc.; Interox America; Mayo Chemical Company, Inc.; Monsanto Company; Occidental Chemical; Olin Corporation; PVS Chemicals, Inc.; and Sternson Water Treatment Chemicals.

BIBLIOGRAPHY

American Institute of Chemical Engineers. *Guidelines for Hazard Evaluation Procedures.* New York: Center for Chemical Process Safety, 2nd Ed. 1992.

ANSI/NSF. *Standard 60. Drinking Water Chemicals—Health Effects.* NSF International Arlington VA 1994.

ASCE and AWWA. *Water Treatment Plant Design.* 2nd ed. New York: McGraw-Hill, 1990.

AWWA Standards *B200* through *B703,* which cover the topics of softening, disinfection chemicals, coagulation, scale and corrosion control, taste and odor control, and prophylaxis. Denver, Colo.: American Water Works Association.

AWWA. *Water Quality and Treatment.* 4th ed. New York: McGraw-Hill, 1990.

The Chlorine Institute. *The Chlorine Manual,* 5th ed. Washington, D.C.: The Chlorine Institute, 1986.

The Chlorine Institute. *Chlorine Scrubbing Systems.* 1st ed. Pamphlet 89. Washington, D.C.: The Chlorine Institute, 1991.

The Chlorine Institute. *Facilities and Operating Procedures for Chlorine Storage.* Pamphlet 5. Washington, D.C.: The Chlorine Institute.

Kawamura, Susumu. *Integrated Design of Water Treatment Facilities.* New York: John Wiley and Sons, 1991.

NAS Chemicals Codex. 2nd ed. Washington D.C.: National Academy of Science, 1985.

National Fire Protection Association. *NFPA 101—Life Safety Code,* Quincy, Mass.: National Fire Protection Association, 1985.

National Lime Association. *Lime Handling, Application, and Storage.* Bulletin 213. Arlington, VA.: National Lime Association, 1990.

National Institute of Occupational Safety and Health. *NIOSH Pocket Guide to Chemical Hazards.* New York: Nostrand Reinhold, 1994.

Recommended Standards for Water Works. Report of the Great Lakes–Upper Mississippi River Board of State Sanitary Engineers. Albany, N.Y.: Health Education Services, 1992.

Uniform Fire Code. Article 80. Whittier, Calif.: International Conference of Building Officials, 1994.

CHAPTER 16
HYDRAULICS

The calculation of a water treatment plant hydraulic grade line involves much more than the simple summation of head losses through the processes. Coincident with determining the hydraulic profile, obviously needed to prevent spillage over the channel and tank walls, are other important hydraulic considerations. One important consideration is the necessity for equal distribution of flow among the various unit processes. There is also the need to prevent negative pressures where air is admitted into the plant's pipelines causing unexpected head losses.

HYDRAULIC DESIGN

Good hydraulic design is a foundation of a well-designed water treatment plant. However, although hydraulics is a common element in the education of every environmental engineer, the nuances of plant hydraulics is not. This chapter provides advice on the types of head losses that need to be calculated, reference sources for critical head loss data, and information on equal flow distribution.

Preliminary Hydraulic Design

The first step in the hydraulic design of a new treatment plant is preparing the preliminary hydraulic grade line. The preliminary hydraulic grade line is not by definition the definitive solution, but rather serves to expose areas of concern in developing the plant design.

Establishing Design Criteria. The criteria used in preparing the hydraulic design depend on such factors as the type of treatment system to be used, the design flow of the plant, plant layout, and any special criteria and limitations of the site or equipment that may affect the hydraulics of the total system.

Type of Treatment Plant. The type of plant that is basic to the public water supply field is the conventional rapid sand filter treatment plant. There are, however, many varieties of water treatment plants, including packaged treatment with pressure filters, ozonation treatment, air stripping, reverse osmosis membrane process, and water softening ion exchange. The conventional rapid sand filter plant is the focus of discussion for this chapter.

A conventional plant consists of rapid mixing, flocculation, clarification, and granular media filters. Pretreatment basins are sometimes installed before the mixing process.

Filter effluent is usually stored in a clearwell that provides both equalization of the incoming and outgoing plant flow and also detention time for the disinfectant before the finished water is pumped to the distribution system. The conventional plant is sometimes modified where the level of treatment allows the exclusion of the clarification step. Such a plant is referred to as a direct filtration plant.

Coordinating with the Project Design Team. It is essential to a well-designed plant that the members of the team designing the various plant components meet regularly to coordinate the interaction of the design functions. The initial meeting is important to set a firm foundation from which the design can proceed; for example, at the initial meeting the hydraulic designer receives required information. It is presumed for the purposes of this chapter that hydraulic design is a separate function; however, the project team may have members that provide dual design functions for a project.

Essential information to be determined by the hydraulic designer before beginning hydraulic computations for a water treatment plant are maximum and average daily design flows, water level at plant inlet and outlet, number of units out of service during maximum flow, and the required filtering head and required freeboard at tanks and channels. Design average flow and maximum hydraulic flow must be established by the project team before hydraulic calculations proceed.

Maximum hydraulic flow can be slightly higher than maximum day demand of the distribution system to accomodate filter backwash water requirements when adequate system storage is provided. Design flow can be the peak hour demand of the distribution system at the extreme, where no storage is provided. Any value between the maximum day and peak hour may be the design maximum flow.

Design Flow and Available Plant Hydraulic Head. Maximum plant flow must be known to compute the maximum hydraulic grade line (HGL). This, in turn, is necessary to design wall elevations to contain water levels within tanks and channels. Average flow is needed to determine the average flow HGL, which sets average depths and volumes in the plant's unit processes. Inlet and outlet water levels are needed to determine available plant hydraulic head.

The outlet level, usually the maximum clearwell level, sets the start of the HGL and, hence, the elevation of the treatment plant units. The outlet level must be adjusted during and even after the calculation of the initial HGL to reconcile the physical profile of the plant with the surrounding plant site.

The hydraulic designer must also be provided with the anticipated head needed for the filter to trap material. This head is referred to as the available filter head (AFH) and is the difference in head between clean filter media and dirty filter media. To fail to account for the head loss of a dirty filter in addition to that of the clean filter causes excessive backwashing and leads to failure of the plant to provide the design volume of treated water. The required AFH is usually based on the results of filter runs determined during pilot studies. Pilot studies are recommended unless good data are available from an existing plant treating the same or similar source water.

Filter box elevation is typically set so that the top of the filter media is slightly above the clean water hydraulic grade line elevation (HGLE) at the maximum flow condition (with the required number of units out of service). The clean water HGLE may be up to 1 ft (0.3 m) below the media surface. This minimizes the required depth of the filter box and prevents excessive negative pressures within filter media as the maximum amount of solids is trapped. Setting the filter box too high could cause excessive negative pressures as the media become dirty. This in turn causes air to come out of solution and leads to hydraulic binding of the media with air bubbles (Steel, 1960).

The freeboard required to safely contain flow within plant facilities during the maximum flow condition must be discussed with the project design team before the initial HGL is calculated. Alternatively, the freeboard can be adjusted to the appropriate value as a result of the initial HGL calculation. For basins and channels with calm water surfaces, a freeboard of about 12 in. (25 cm) is considered adequate. Basins with turbulent water surfaces, such as rapid mix basins, require a higher freeboard because of splashing. Ozonation basins require a freeboard of 2 to 3 ft (0.6 to 1 m) or more to allow draw-off of the ozone gas.

Plant Layout. Before the preliminary or initial plant HGL can proceed, a scaled layout of plant facilities must be provided by the project team. The hydraulic design engineer can then use this layout to determine the length of connecting pipes and channels between the various plant facilities. The path of original connecting pipes and channels can be provided by the project team or by the hydraulic designer. The final setting of the pipe and channel routes is often a result of the give-and-take of the project team and the hydraulic designer as the plant's design develops.

Special Criteria and Limitations. When operating at maximum design flow, a number of plant process units could be out of service. To provide for such an eventuality, the hydraulic designer must obtain from the project team the number of units that can reasonably be expected to be out of service during maximum flow.

Although it is crucial that the plant be designed to be hydraulically adequate to meet the design year conditions, it is also important to plan for stages of expansion beyond the design year. For some treatment plant layouts, future staging will govern the size of pipes and channels.

Minimum velocity is not usually a concern in itself when dealing with relatively clean water through a treatment plant. A minimum velocity must be maintained in source water lines to prevent solids from settling in the piping. Some regulatory agencies, however, may establish a maximum velocity limit for all piping. The limit does not usually come into play as a critical hydraulic factor because it is the available hydraulic head that sets the allowable velocities. That is, because high velocities go hand in hand with high head losses, hydraulic design naturally progresses toward reasonable velocities.

The exceptions are the permissible velocities for flocculated water. Floc particles must not be severely agitated or they may shear, causing flocculation problems and inhibiting clarification. To prevent shearing of floc, flow velocity must be kept low from the outlet of the flocculation basins, where floc has formed, to the inlet of the clarification basins, where floc is removed.

Actual maximum velocity that floc particles can tolerate depends on the type of floc generated and its strength. Maintaining velocities at approximately 0.5 ft/s (15 cm/s) or less prevents shearing of the floc particles under most conditions. Maximum velocity should correlate with maximum flow condition through channels, pipes, or gates. For direct filtration plants where clarification is not used, flocculated water is directly applied to filters, and floc shear may not be a concern.

Determining the Preliminary Hydraulic Profile. Preparing a preliminary HGL is usually required only for the worst case condition—the maximum flow condition (hydraulic capacity) with the prescribed number of units out of service. All remaining HGLs can be prepared after resolving all issues and problems and after calculation of the final HGL to the maximum flow condition.

Before proceeding with the preliminary HGL, the location of the process units and the routes of the connecting pipes and channels must be provided, as well as the downstream control water surface elevation. Downstream control water elevation is needed because all treatment plant HGL evaluations must proceed from a hydraulic control point and in an upstream direction.

Ongoing discussions with the project team can often resolve problem issues as they arise. Other issues will be resolved after the initial HGL is completed. The problems that arise are often interconnected, with the solution to one problem affecting others. For this reason, all areas of concern must be reviewed and eventually resolved by the project design team. For example, simply raising the walls of process units to contain the flow may affect the type of pump selected, or raising walls to accommodate the HGL may be too costly or may not be a viable alternative because of incompatibility with the ground profile.

Areas of concern must be fully documented for eventual resolution. Documentation may take the forms of a list of notes or a comprehensive design memorandum. The objective is to compile a list of concerns as the hydraulic analysis proceeds, not letting minor problems delay the completion of the HGL. Minor problems can be resolved after the HGL is completed.

Of course, any fatal flaws in the design must be brought to the attention of the project design team as soon as possible for resolution. The most common problem issues are related to plant head limitations that require a tight HGL and flow distribution where the flow stream is not equally split to process units. Flow distribution is an especially insidious problem because geometry alone, even with symmetry of the supply channels and pipes, does not ensure equal flow splitting to the process units. Instead, flow distribution requires deliberate design and often additional head loss.

Presentation and Adjustment of the Preliminary HGL. Once the preliminary HGL is calculated and concerns are highlighted, the resulting problems must be addressed and resolved by cooperation between the hydraulic design engineer and project design team. The hydraulic design engineer must present the results of the HGL to the project team. The presentation and related meetings should be an open discussion in which problem solving is required from all parties.

Effects on project scheduling and on each of the engineering functions must be considered. The most difficult issues are in areas where multiple engineering functions have valid considerations that conflict. Creative thinking and team cooperation must be used to resolve these issues. All issues that are resolved during the discussions with the project design team must then be incorporated into the hydraulic design and the HGL adjusted accordingly.

Intermediate and Final HGL Design

After modifications resulting from determining the preliminary HGL are incorporated into the plant design, the next round of HGL calculations is then performed. The result is the intermediate HGL. This level of HGL may be sufficient for the final HGL if no additional issues or problems are uncovered.

For large, complex treatment plants, another round or two of hydraulic analyses may be required to resolve all hydraulic issues. As for the preliminary HGL, presentation of results to the project design team and ensuing discussions are required to resolve problems. All modifications decided on are incorporated into the design, and the final HGL is calculated.

Once the final HGL is completed for maximum flow condition with units out of service, other HGLs can be calculated. Other HGLs typically required are the maximum flow condition with all treatment units operating and the average HGL with all units operating. The minimum HGL is not normally required but can be done at this time if desired.

After completion of the required HGLs, the hydraulic analyses should be fully documented in a hydraulic design report and presented to the project design team. All major changes and issues should be summarized in the report. The report should contain all piping and channel sizes and elevations, as well as HGL elevations throughout the treatment plant. All HGLs are presented on the hydraulic profile sheet of the design drawings. Figure 16.1 presents an example of a typical hydraulic grade line profile.

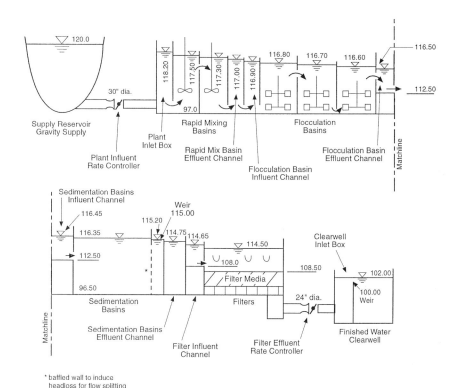

FIGURE 16.1 Typical hydraulic gradeline-maximum day flow (50MGD) conventional rapid sand filter water treatment plant.

HEAD LOSS TYPES AND CALCULATIONS

Head losses through piping, valves, filters, channels, gates, and weirs all have an important bearing on determining the plant hydraulic grade line.

Pipelines

The hydraulics of pipelines and the corresponding head loss determinations are the most understood of all head loss calculations by engineers. Many references (Daugherty, Franzini, and Finnemore, 1985; WEF, 1991) adequately present the topic of pipeline losses. In general, the Hazen-Williams equation is traditionally used for pressurized pipes and the Mannings equations for free surface pipe and channel flow. The Darcy-Weisbach equation is the most theoretically correct of the pipeline head loss equations, but because it requires multiple steps, it is not commonly employed in the design of water treatment plants or in the environmental field as a whole. For this reason, the Darcy-Weisbach equation is not presented here but can be found in other references (Daugherty, Franzini, and Finnemore, 1985; WEF, 1991). The respective head loss equations are as follows. The Hazen-Williams equation:

$$\mathrm{HL} = \left(\frac{V}{K \times C \times R^{0.63}}\right)^{1.85} \times L$$

where   HL = head loss (ft) (m)
V = velocity (ft/s) (m/s)
C = head loss coefficient
R = hydraulic radius, A/P (ft) (m)
A = pipe area (ft^2) (m^2)
P = pipe perimeter (ft) (m)
L = pipe length (ft) (m)
K = 1.318, English units (0.85, metric units)

Mannings equation:

$$\mathrm{HL} = \left(\frac{V \times n}{K \times R^{2/3}}\right)^{2} \times L$$

where   HL = head loss (ft) (m)
V = velocity (ft/s) (m/s)
n = head loss coefficient
R = hydraulic radius, A/P (ft) (m)
A = pipe area (ft^2) (m^2)
P = pipe perimeter (ft) (m)
L = pipe length (ft) (m)
K = 1.49, English units (1.0, metric units)

Typical head loss coefficients for the most commonly used materials employed in treatment plant design are $C = 120$ for cement-lined ductile iron pipe and $n = 0.013$ for concrete and channels. For other channel and pipe lining materials, the designer should consult the manufacturer's literature.

Valves and Fittings

Valve and fitting head loss calculations are familiar to most engineers in the environmental engineering field. Although some engineers use the equivalent pipe length method, the method that is more representative of the theory behind fitting and valve head loss computations is based on the use of K values according to the following equation:

$$\mathrm{HL} = (K \times V^2/2g)$$

where   HL = head loss (ft) (m)
K = loss coefficient
V = velocity (ft/s) (m/s)
g = gravitational acceleration constant (32.2 ft/s/s) (9.81 m/s/s)

K values to determine fitting and valve head loss more closely represent the cause of the head loss—turbulence from flow contraction or expansion. K values can be thought of as related to coefficients of contraction, C, based on the following equation:

$$K = 1/C^2$$

The K value method also has the advantage of wider availability of values. The main confusion in using the K value or equivalent pipe length method to determine head loss is the wide variability of values listed in the various references. It should, however, be kept in mind that any errors resulting from overstated or understated K values rarely cause failure in the hydraulic design. Although it is not critical to use the "correct" value, whatever that may be, it is critical that a reasonable head loss estimate be made for each valve and fitting.

Many references (Daugherty, Franzini, and Finnemore, 1985; Heald, 1988; King, 1963; Miller, 1990) list K values for various valves and fittings. The most commonly used K values are:

Fitting type	K
90-Degree elbows	0.2 to 0.3 (where $K = 0.2$ for 12 in. diameter pipes or greater)
45-Degree elbows	0.1 to 0.2 (where $K = 0.1$ for 12 in. diameter pipes or greater)
Pipe entrance with tank	0.5 (flush entrance) to 1.0 (re-entrant)
Pipe exit	1.0
Gate valve (full open)	0.1
90-Degree miter bend	1.2

The greatest difficulty usually comes in selecting the proper K values for complex fittings such as tees with combining or dividing flow. Miller (1990) is an excellent source of K values for tees.

The most accurate information on head loss through valves can be obtained from valve manufacturers. The manufacturers commonly represent the head loss coefficient as a Cv value where this value is defined as the flow in gallons per minute associated with a 1 psi pressure drop through the valve. Using the fitting head loss K value in the equation $K = 1/C^2$, the Cv value can be transformed into a K value to determine head loss on the same basis used for fittings.

Of special concern are valves in rate controlling systems. The K value used must correspond to the value when the valve is about 60 to 70 degrees open to allow for good hydraulic control characteristics. For valves used for isolation purposes only, the full open K value is appropriate.

Filter Media

The major concentration of head loss in a conventional water treatment plant using rapid sand filters occurs at the filtering media and the filter box effluent piping. This location typically accounts for over 10 ft (3 m) of head loss through dirty filter media and effluent piping, including the rate controller. The clean media head loss is affected by the types of media installed in the filter box. Pilot studies commonly done in conjunction with the design of the treatment plant not only provide important data on the expected head loss through the media selected, but also information on required filter run time and head loss to be expected for the dirty filter. The difference between dirty filter head loss and clean filter loss is the AFH. Typical values of AFH can run from an extreme low value of about 5 ft (1.5 m) to a high value of about 7 to 10 ft (2.1 to 3 m).

The types of media used in filters are garnet sand, silica sand, granulate anthracite coal, and activated carbon. Head loss through sand depends on characteristics of the sand, such as porosity and gradation of media particles. A rough initial estimate for head loss through sand is 1 ft (0.3 m) per 1 ft (0.3 m) of sand layer thickness at 6 gpm/ft^2 (0.004 m^3/sec/m^2).

Using established equations (Fair, Geyer, and Okun, 1968), head loss can be calculated knowing the physical properties of the media. To calculate filter media head loss, either physical properties of the sand media or head loss characteristics of the sand media must be obtained from suppliers. Head losses of proprietary media such as activated carbon also are best obtained from the supplier.

Sand and other types of filtering media are supported by gravel or other layers of coarse materials such as garnet. As with media, the physical characteristics of the material or head loss characteristics of the coarse support material must be obtained from the material's supplier for head loss calculation. Head loss through the coarse support material is much less than head loss through the media. For example, head loss through coarse gravel material is often a nominal value of about 0.1 ft (0.03 m) at nominal flow rates.

Filter bottoms are designed primarily to distribute the much higher backwash flows. Therefore the head losses they cause during filtering operation are so low as to be negligible. However, it is best to check head loss characteristics with the underdrain system manufacturer.

Filter Piping

Much of the head loss at the filters is caused by the many fittings and valves in the piping. Head loss though the effluent piping immediately adjacent to the filter box may be as high as 2 to 4 ft (0.6 to 1.2 m) at maximum flow conditions. Piping size is driven by the limited space available and the allowable head loss. Figure 16.2 show a typical piping arrangement at the outlet of a rapid sand filter box.

The principal feature of effluent piping is the rate controller that maintains a constant level in the filter as treatment plant flows fluctuate and as filters become dirty with trapped material. The typical rate controller is composed of a flowmeter followed by a downstream butterfly valve. The rate controller must be sized to allow travel of the butterfly valve disc over a reasonable range of operation to provide good hydraulic control characteristics. If

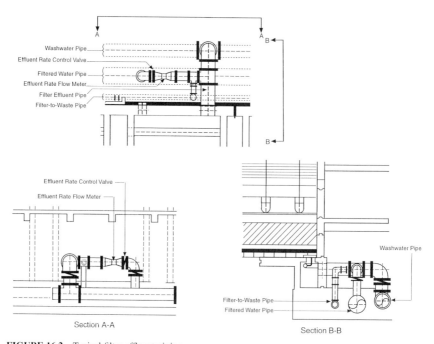

FIGURE 16.2 Typical filter effluent piping.

possible, the butterfly disk should be sized to limit operation between 20 and 70 degrees. For operation outside these limits, advice of a control valve and instrumentation specialist should be obtained. The rate controller is often sized smaller than the adjacent pipe size.

The flowmeter component of the rate controller must be sized to allow flow measurement over the range of flows to be conveyed. The flowmeter used in some rate controllers is a shortened variation of the venturi tube called a flow tube. For adequate flow measurement at low flow, a flow tube must be sized to permit at least 0.5 in. (1.3 cm) of head differential to be developed at extreme low flow.

The flow tube size must also accommodate measurement at maximum flow without excessive head loss. Maximum head differential is limited to about 5 ft (1.5 m). Actual head loss through the flowmeter depends on the size and type of flowmeter and is usually a small percentage of the head differential. Information on appropriate limits and head loss should be obtained from the flowmeter manufacturer.

Channels and Troughs

The hydraulics of channels and troughs is governed by the hydraulics of open channel flow. References are available (Chow, 1959; Henderson, 1966) that discuss in detail the methods used to calculate backwater profiles or head loss through open channels or free-surface flowing pipes. Commercial computer software programs are available to calculate head loss through open channels.

There often are channels in a treatment plant that collect flow as the flow proceeds downstream, such as the effluent collector channel of a number of parallel sedimentation basins or the filter trough when backwashing is occurring. This type of open channel flow is referred to as spatially varied flow and must be analyzed using spatially varied flow methodology (Chow, 1959) or computer software programs using such techniques.

Gates and Ports

Many gates and ports are used in a water treatment plant design. At the inlet of unit processes such as sedimentation basins, gate openings through the inlet wall not only allow flow into the basin but also distribute flow across the basin width through multiple gates or ports. The required head loss though ports can be high.

Determining head loss for gates and ports depends on the degree of contraction and expansion of the flow through the opening. The more contraction and expansion that occurs, the higher the head loss will be. The degree of contraction and expansion depends on the submergence of the opening, the thickness of the wall opening, and the number of sides causing the contraction. A submerged gate causes contractions at four sides; an unsubmerged gate causes contraction at two sides.

Hydraulic handbooks (Davis and Sorensen, 1969) provide contraction coefficients based on these parameters. Having obtained the appropriate contraction coefficient, head losses through submerged gates and ports can be calculated using equations previously presented for fittings and valves. For unsubmerged gates and ports, the equations still apply, with velocity based on the depth through the opening relative to its downstream side.

Weirs

Weirs are employed at the outlet of process basins to maintain a relatively constant level within the basins. Many hydraulic textbooks and handbooks (Davis and Sorensen,

1969; Henderson, 1966; King and Brater, 1963; WEF, 1991) present the theory behind the operation of weirs, as well as the coefficients that correspond to the types of weirs available. The most commonly used weirs within a treatment plant are sharp, broad-crested, and V-notch weirs. The basic form of the equation for sharp and broad-crested weirs is:

$$\text{Hw} = \left(\frac{Q}{C \times L} \right)^{2/3}$$

where Hw = weir head (ft) (m)
  Q = flow (ft³/s) (m³/s)
  C = weir coefficient (in either English or metric)
  L = weir length (ft) (m)

For sharp-crested weirs at the effluent end of relatively deep basins, the weir coefficient is approximately 3.27 (1.8, metric). For broad-crested weirs, however, the weir coefficient (C) depends on the height of weir head and the thickness of the weir wall (King and Brater, 1963).

For short-length weirs where flows at the ends of the weir contract and reduce the effective hydraulic length of the weir, the weir length (L) must be adjusted accordingly. Effective weir length is determined by subtracting a weir length equal to 0.1 times the weir head for each end where contracted flow occurs.

The 90-degree V-notch weir is the most commonly used type of weir. It is described by the following equation:

$$\text{Hw} = \left(\frac{Q}{C \times N} \right)^{(1/2.48)}$$

where Hw = weir head (ft) (m)
  Q = flow (ft³/s) (m³/s)
  N = number of notches
  C = 2.48, English units (1.34, metric units)

Venting of the weir nappe (the area just below the weir) must be done to prevent fluctuation of the weir head. The V-notch weir is naturally vented; but for sharp- and broad-crested weirs, care must be given to the venting of the weir nappe. For weirs with contracted ends, venting is ensured. For weirs where the flow is not contracted, small pipe vents should be installed.

The above weir equations presume that the weir is free flowing (i.e., not submerged by the downstream water level). For submerged weirs, the effect on weir head can be determined from equations and empirical data presented in hydraulic handbooks (King, and Brater, 1963). For weirs submerged above their critical depth, a good approximation of the change in head across the weir can be obtained by multiplying the velocity head relative to the downstream side times a K value of 1.5. This K value presumes that head loss across a submerged weir can be treated as an abrupt contraction and expansion.

HYDRAULIC DESIGN HINTS

This section presents a number of design hints that the hydraulic designer should be aware of in designing water filtration plants.

Use of Computer Programs

Proprietary hydraulic computer programs can be used to calculate the HGL through water treatment plants, but most of the programs are available only to the personnel of the engineering consulting firm that developed the program. Undoubtedly, many more hydraulics programs, some more reliable than others, will be developed in the future.

Unfortunately, computer programs may be incapable of capturing the nuances of a design that lead to its success or failure. Therefore, no matter how good the program is, the most important component in the calculation of any hydraulic profile is the user. The use of any computer program to calculate HGL must be accompanied by an engineer familiar with treatment plant hydraulics. In short, use computer programs with caution and always check their output for reasonableness.

Available Filter Head

As discussed in preceding sections, AFH is the difference between the dirty filter and clean filter head loss at maximum design flow. The reason that AFH is mentioned again here is to reiterate its importance. An uncommon but serious design error is to account only for the clean water head loss through the filter. Without accounting for the head loss caused by the filter as it becomes dirty, the entire design and operation of the water treatment plant becomes severely compromised.

Flow Distribution

To minimize chemical usage and to maximize water treatment, the distribution of flow to and across each unit process must be addressed. Symmetry alone does not ensure proper flow distribution because locally generated velocity currents often adversely effect the presumed flow distribution attributed to symmetry. Also, flow distribution can be adversely affected by uneven setting of weir elevations. Flow distribution must be maintained either by rate controllers, as is done at filters, or by inducing head loss at points of the desired distribution.

Distribution of flow by rate controllers to unit processes other than filters is not often desirable because of equipment cost and additional maintenance required. Where feasible, flow distribution should be obtained by inducing head loss. The principle behind flow splitting is to provide a high enough head loss at points where equal flow distribution is required so that the head differential acting against equal flow distribution is made relatively small in comparison. An inlet header channel or pipe that feeds flow to a number of parallel process units is usually a location where equal flow distribution is required. The following equation (Camp and Graber, 1968; Fair, Geyer, and Okun, 1968) describes flow distribution along an inlet header to parallel units fed by gates or ports:

$$m = (1 - H/\text{HL})^{(1/2)}$$

where m = the ratio between the lowest and highest flow
H = head differential along the channel between the first and last port (ft) (m)
HL = head loss across the port (ft) (m)

The head differential, H, is based on the maximum velocity head and the friction loss in the channel or pipe. For a conduit where flow is taken off uniformly along its length, friction loss is one-third of the head loss that would occur for a conduit conveying the entire flow along its length, with no flow taken off. The following equation allows calculation of the differential head, *H,* along a conduit with uniform discharge of flow along its length:

$$H = V_o^2/2g \times (1 - hl/3)$$

where H = head differential along the inlet header channel (ft) (m)
V_o = velocity head at upstream end of channel (ft/s) (m/s)
g = acceleration constant (32.2 ft/s/s) (9.81 m/s/s)
hl = frictional head loss along the inlet header channel (ft) (m)

For conduits that approach uniform discharge of flow along their length to a reasonable degree, the one-third friction loss presumption still yields a good approximation of head loss. Frictional head loss along the inlet header channel is usually computed using the Mannings equation. For a conduits where only a few takeoff ports are located, the presumption of uniform discharge of flow along its length is not valid. For such cases, manual analysis using detailed open channel flow techniques is required. Flow distribution is discussed in further detail in several references (Camp and Graber, 1968; Fair, Geyer, and Okun, 1968; James Montgomery, 1985).

Another item related to distribution of flow is the placement of perforated baffle walls within sedimentation basins. Baffle walls are installed to prevent local velocity currents within the basin that could short-circuit flow and upset formation of the sludge blanket. Both occurrences would adversely affect settled water quality. The location of perforated baffle walls is empirical and is therefore based on past successful designs or established design standards (James Montgomery, 1985).

Where head loss cannot be tolerated at the inlet to sedimentation basins to distribute flow, multiported baffle walls can be installed at the downstream end of the basin to affect the required head loss. This arrangement avoids disruption of the floc.

Filter Effluent Piping

To minimize the head loss through filter piping, the designer should be particularly careful in laying out multiple 90-degree bends, butterfly valve alignment, and conditions that may allow air entrainment.

Compound 90-Degree Bends. The layout of effluent piping must avoid using combined 90-degree bends in series and at right angles unless they are spaced sufficiently apart. Reference material (Padmanabhan and Nystrom, 1985) has presented research data indicating that closely spaced 90-degree combined bends cause swirling flow leading to head losses much higher than losses normally expected with two 90-degree bends. The reference concludes that typical spacing required for combined 90-degree bends should be about 6 pipe diameters.

Butterfly Valve Alignment. When a butterfly valve is placed near a bend or tee where the flow turns, the disk of the valve should be aligned in the same plane as the turning flow to offer minimum resistance to the turning flow. This minimizes additional head loss that would result from the disk encroaching into the turning flow stream.

Avoid Air Entrainment. Air entrainment in filter effluent piping can be prevented by ensuring that piping, especially the packing boxes for valve stems, is not subjected to negative pressure under any flow condition. To prevent this from happening often requires that a minimum HGL elevation be maintained downstream from effluent piping. To buoy up the HGL elevation, a weir is sometimes installed at an effluent box immediately at filter effluent discharge piping or at an inlet box at the downstream clearwell.

Unless a pressure disk diaphragm is installed at the low-pressure tap of a flow tube, the low-pressure tap should be installed below the minimum HGL elevation to always ensure positive pressure. Proper calculation of the HGL elevation at the low-pressure tap requires that the velocity head at the throat of the meter be subtracted from the associated energy grade line elevation.

Rapid Mixing Basins

Conventional rapid mixers usually add energy, or at least do not deplete energy, from the flow stream, so no head loss is typically presumed for action of the mixers themselves. Other types of mixers, such as in-line units and jet mixers, are used in many plants, and these units may cause some head loss. The manufacturer of these units should be contacted for information on the head loss that they will cause.

If conventional mixing basins are used, head loss associated with the basins is caused by their respective inlet and outlet openings. Because the configuration and opening sizes affect the mixing process, the layout and design of mixing basins should be reviewed by the mixing equipment manufacturer.

ANCILLARY HYDRAULIC DESIGN

In addition to establishing the hydraulic grade line through the treatment plant, a number of individual pump and piping systems must be considered and designed by the hydraulic designer.

Filter Backwash System

The filter backwash system must be hydraulically designed to ensure proper backwashing. Required backwashing flow rates must be obtained from the project team before the hydraulic analysis can proceed. Some systems pump backwash water, and other systems are designed for complete gravity operation using a storage tank that furnishes water at the required level. Both types require a slow opening and closing valve to prevent surging in the filter that could cause media washout or failure of the filter bottom.

Whether the filter backwash is a gravity or pumped system, backwash piping proceeds from the supply to the filter backwash system. The control valve should be designed to operate within a range of 20 to 70 degrees open. Filter bottoms and supply pipe should be designed to allow enough distribution of flow over the filter bottom to achieve adequate washing of the media while at the same time preventing media mounding and washout. Filter bottoms must be selected with good flow distribution characteristics, and the supply pipe and inlet gullet preceding the filter bottom must be configured to enhance the distribution of flow through the filter bottoms.

Information on head loss through filter bottoms must be obtained from the manufacturer. Head loss through the media depends on the degree of expansion required. For fully expanded beds, head loss through the media bed is equal to the weight of media (Fair, Geyer, and Okun, 1968; Steel, 1960) and can be calculated according to the following equation:

$$HL = (s - 1)(1 - f) \times l$$

where: HL = head loss though media (ft) (m)
s = specific gravity of media
f = porosity of media
l = media depth (ft) (m)

Head loss through the supporting gravel depends on the gravel's physical characteristics such as effective size and porosity. Knowing its characteristics, head loss can be calculated with equations describing flow through media (Fair, Geyer, and Okun, 1968). One reference (Steel, 1960) states that head loss can be approximated at 0.1 ft per 1 ft (0.1 m/m) of gravel depth at a backwash vertical rise rate of 12 in./min (30 cm/min). Head loss is directly proportional to gravel depth and backwash rise rate. This head loss presumes a typical support bed of coarse gravel. Where support beds include other materials such as garnet, the material's supplier must be consulted to obtain its physical characteristics or, if known, its head loss characteristics.

Filter Waste Discharge

Dirty backwash water must be directed to waste. The waste stream is conducted from filter troughs to a waste gate that is opened to direct flow to a water reuse basin or to a waste lagoon or sewer. If directed to a reuse basin, the decanted reuse water is pumped to the head of the treatment plant, and the residuals are conveyed to a waste lagoon or sewer. Residuals may be conveyed by gravity or may require pumping. If filter waste discharge is conveyed directly to a waste lagoon or sewer, the system may be gravity driven or may require pumping. For any of these cases, the filter waste discharge system must be hydraulically designed.

Water that initially passes through a filter after backwashing is often initially diverted to the filter inlet channel for reuse or to waste. If directed to the filter inlet channel, in-line pumping is required. If directed to waste, the system may be gravity driven or require pumping. In either case, the filter-to-waste system must be hydraulically designed. A significant hydraulic concern is the potential of "shocking" the filter in the process of switching from filter-to-waste and back to filter effluent. Special modulating valves should be considered to perform these functions gradually without disrupting the filter.

Pumps

A variety of pumps and pumping systems are needed for the operation of water treatment plants. Most treatment plants require pumping the source water to the treatment plant, as well as pumping treated water to the transmission and distribution system. The types of pumps employed for these services are horizontal split case centrifugal and vertical turbine pumps.

Pumping systems within the treatment plant include filter backwash, filter waste discharge, and filter-to-waste systems. Horizontal split case centrifugal and vertical turbine pumps are usually found in backwash systems. The smaller ancillary filter pumping systems are often serviced by end suction pumps. All pumping systems require hydraulic analyses for proper design and operation. Some of the references that amply cover pump design are Karassik et al. (1976) and Sanks et al. (1989).

BIBLIOGRAPHY

Camp, T. R., and S. D. Graber. "Dispersion Conduits." *Journal Sanitary Engineering Division, ASCE,* 94(2):31, 1968.

Chow, V. T. *Open-Channel Hydraulics.* New York: McGraw-Hill, 1959.

Daugherty, R. L., J. B. Franzini, and E. J. Finnemore. *Fluid Mechanics with Engineering Applications.* 8th ed. New York: McGraw-Hill, 1985.

Davis, C. V., and K. E. Sorensen. *Handbook of Applied Hydraulics.* 3rd ed., New York: McGraw-Hill, 1969.

Fair, G. M., C. G. Geyer, and D. A. Okun. *Water and Wastewater Engineering.* Vol. 2. *Water Purification and Wastewater Treatment and Disposal.* New York: John Wiley and Sons, 1968.

Heald, C. C. *Cameron Hydraulic Data.* 17th ed. Woodcliff Lake, N.J.: Ingersoll-Rand 1988.

Henderson, F. M. *Open Channel Flow.* New York: Macmillan, 1966.

James M. Montgomery Consulting Engineers, Inc. *Water Treatment Principles and Design.* New York: John Wiley and Sons, 1985.

Karassik, I. J., W. C. Krutzsch, W. H. Fraser, and J. P. Messina. *Pump Handbook.* New York: McGraw-Hill, 1976.

King, H. W., and E. F. Brater. *Handbook of Hydraulics.* 5th ed. New York: McGraw-Hill, 1963.

Miller, D. S. *Internal Flow Systems.* 2nd ed. Houston: Gulf, 1990.

Padmanabhan, M., and J. B. Nystrom. *International Conference on the Hydraulics of Pumping Stations.* September 17–19, Manchester, England. Cranfield, Bedford, England: British Hydromechanics Research Association, 1985.

Sanks, R. L., et al. *Pumping Station Design.* Boston, Mass.: Butterworth, 1989.

Steel, E. W. *Water Supply and Sewerage.* New York: McGraw-Hill, 1960.

WEF Manual of Practice No. 8. *Design of Municipal Wastewater Treatment Plants.* Vol. I. Joint Task Force of the Water Environment Federation and American Society of Civil Engineers. Brattleboro, Vt.: Book Press, 1991.

CHAPTER 17
PROCESS RESIDUALS

Historically, process wastes from most water treatment plants were disposed of by discharge into natural waterways. It is estimated that until 1971, more than 90% of water treatment plants were discharging wastes to streams or lakes. Since that time, dramatic changes in regulations, land availability, and environmental sensitivity have caused essentially all treatment plant wastes to be recycled or treated in some manner before disposal.

TYPES OF PROCESS RESIDUALS

The first wastes to receive attention for treatment were sludges from metal hydroxide coagulation plants, lime softening facilities, and iron and manganese removal operations. Growth in the use of new processes such as granular activated carbon, membrane filtration, and ion exchange processes have created new types of wastes that require unique handling and disposal techniques. The types of wastes produced from different types of treatment processes are illustrated in Figure 17.1.

Another trend has focused on treatment of liquid wastes from water treatment processes. At plants using on-site treatment, the recovered liquid portion of the waste has generally been returned to an early segment of the treatment plant as a means of water conservation and the easiest method of treatment. Growing concerns about the level of cyst contamination in this portion of the waste stream have resulted in additional design considerations for handling liquid wastes.

The general steps (Figure 17.2) in designing residuals handling, recovery, and disposal are:

- Determining the types, characteristics, quality, and quantity of waste flows
- Evaluating treatment and disposal options that are available
- Reviewing the regulations and restrictions affecting each disposal method
- Reviewing treatment changes that could reduce the quantities of wastes, including economies and advantages of recycling some of the wastes
- Reviewing the economics, advantages, and disadvantages of all alternatives
- Selecting the best treatment or disposal alternatives

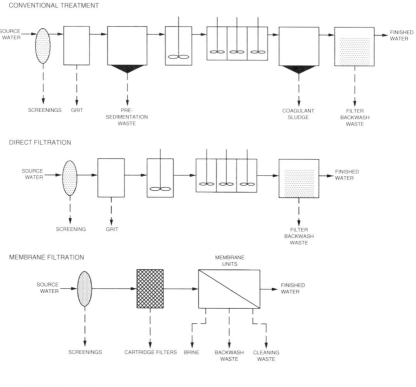

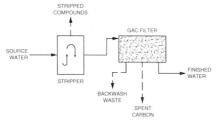

FIGURE 17.1 Water treatment plant wastes.

Sludge Types, Quantities, and Characteristics

Wastes from water treatment plants have traditionally been referred to as sludge or solids; however, there are also nonsolid wastes that must be treated or controlled, and the broader term *residuals* is becoming more common. The term *waste* itself may be "politically incorrect" because it fails to recognize potential beneficial reuse of some of the by-products of treatment processes.

Sludge generated from water treatment processes includes suspended solids removed from source water and chemical precipitates created by the treatment process. The various types of sludge resulting from water treatment processes generally fall into the following categories:

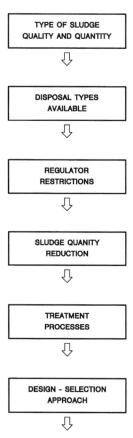

FIGURE 17.2 Steps in development of a sludge handling system.

- Solids in filter backwash water
- Aluminum or iron coagulant sludges
- Iron and manganese precipitates
- Softening plant sludges

In deciding what treatment or disposal methods are appropriate for each type of sludge, consideration must be given to the:

- Type of solids
- Quantities of sludge being generated
- Sludge characteristics that may influence sludge dewaterability

Filter Backwash Water Sludge. Filter backwash water can be troublesome to handle. It has low solids content even after thickening, it usually dewaters poorly, and most of the solids are difficult to gravity separate without using coagulant aids. Filter backwash water deserves careful consideration not only with regard to solids disposal, but because of potential recovery of a significant volume of water.

Filter backwash water typically represents 2% to 5% of total water processed. The quantity of solids depends on filter efficiency and the amount of solids applied to the filter, but the concentration generally varies from 50 to 400 mg/L. The amount of solids applied to the filter depends on the pretreatment provided.

Where sedimentation precedes filtration, typical suspended solids of water entering the filters ranges from 4 to 10 mg/L and the backwash contains a solids loading of about 35 to 85 lb/mil gal (4 to 10 kg/ML). Attainable sedimentation-basin effluent turbidity can be less than 1 turbidity unit (ntu); however, typical sedimentation effluent turbidities are 2 to 6 ntu. In the direct filtration process, flocculated water is applied directly to the filter. As a result, solids loading is a function of coagulant dosage and source water turbidity.

Filter backwash solids are typically difficult to separate from liquid. Washwater recovery ponds sized to hold backwash water for 24 hours or more may recover up to 80% of solids with the use of polymers or other coagulant aids. Reclaimed water is then usually reprocessed through the treatment plant.

Coagulant Sludges. Aluminum and iron coagulants result in inorganic sludges containing compounds such as aluminum hydroxide and ferric hydroxide along with clay, silts, and organic and inorganic matter precipitated by the coagulant. The nature of sludge produced is highly variable, depending on source water quality. Seasonal variations in source water also affect such characteristics of the sludge as its thickening density and dewaterability.

The characteristics of coagulant sludges also vary with the proportion of material removed from the water. High-turbidity waters usually result in sludges that are more concentrated and less difficult to dewater; low-turbidity waters present a more difficult sludge processing problem. In general, settled iron sludges have a higher solids concentration than alum sludges, and the addition of polymer or lime increases the solids concentration of both. Coagulant sludges are essentially biologically inert, having low biodegradable organic content and a near-neutral pH. Iron and alum coagulant sludges may be characterized as follows:

Solids content	Sludge character
0% to 5%	Liquid
8% to 12%	Spongy, semisolid
18% to 25%	Soft clay
40% to 50%	Stiff clay

Alum Sludge. Alum sludge is gelatinous with poor compactability. It generally concentrates to 0.5% to 2.0% (5,000 to 20,000 mg/L) in sedimentation basins. When filter alum $[Al_2(SO_4) \cdot 14H_2O]$ is added to water, it forms aluminum hydroxide $[Al(OH)_3]$. For every mg/L of alum (as 17.1% Al_2O_3) added, 0.44 mg/L solids is formed. The quantity of aluminum escaping the filters and appearing in the finished water depends on the pH of the water.

Determining optimum coagulant dosage to minimize sludge quantities should be developed from a series of jar or pilot filter tests. Repeat tests and sludge quantity analyses should be run at least seasonally to account for variations in water quality.

Suspended matter in the source water is usually reported in nephelometric turbidity units, but there is no absolute correlation between turbidity units and total (dry weight) suspended solids (TSS). However, based on observed values where both parameters have been measured, the ratio of TSS to ntu normally varies from 1.0 to 2.0, and can often be as high as 10.0. Two examples on estimating solids residue from alum coagulation are shown below.

Example 1. Alum sludge formed in treatment of a low-turbidity source water:

Source water turbidity	10 ntu
Alum dose	30 mg/L
Aluminum hydroxide sludge	30 mg/L $\times$ 8.34 $\times$ 0.44 = 110 lb/mil gal (13 kg/ML)
Raw water solids	10 ntu $\times$ 1.5(assumed) $\times$ 8.34 = 125 lb/mil gal (15 kg/ML)
Total solids	110 + 125 = 235 lb/mil gal (28 kg/ML)

Example 2. Alum sludge formed in treatment of a high-turbidity source water:

Source water turbidity	150 ntu
Alum dose	60 mg/L
Aluminum hydroxide sludge	60 mg/L $\times$ 8.34 $\times$ 0.44 = 220 lb/mil gal (26 kg/ML)
Source water solids	150 ntu $\times$ 1.5(assumed) $\times$ 8.34 = 1,876 lb/mil gal (225 kg/ML)
Total solids	2,096 lb/mil gal (251 kg/ML)

Iron Salts Coagulation Sludge. Iron coagulants used in water treatment include ferric sulfate [$Fe_2(SO_4)_3$], ferrous sulfate ($FeSO_4 \cdot 7H_2O$), and ferric chloride ($FeCl_3$). The precipitate formed is ferric hydroxide, $Fe(OH)_3$. Like alum sludges, ferric hydroxide is hydrophilic and thickens poorly. The amount of sludge formed can be estimated as 2.9 mg/L of solids formed for every mg/L of iron added. The solids may also be estimated as shown in the following example.

Example 3. Iron sludge formed in the treatment of a low-turbidity source water:

Source water turbidity	10 ntu
Ferric sulfate dose (as iron)	15 mg/L
Ferric hydroxide sludge	15 $\times$ 2.9 $\times$ 8.34 = 363 lb/mil gal (43 kg/ML)
Source water solids	10 ntu $\times$ 1.5(assumed) $\times$ 8.34 = 125 lb/mil gal (15 kg/ML)
Total	488 lb/mil gal (59 kg/ML)

Iron and Manganese Precipitates. Water treatment to remove iron and manganese consists of first oxidizing soluble iron and manganese by means of aeration or by adding a chemical such as permanganate, chlorine, or ozone. Precipitates formed are principally ferric hydroxide, ferric carbonate, or manganese dioxide. These precipitates are then removed in sedimentation or filtration processes. The sludge produced is inert and typically red or black.

For each milligram per liter of iron or manganese in solution, 1.5 to 2 mg/L of sludge production may be anticipated. However, because concentrations of iron and manganese found in most natural waters are typically low, overall iron and manganese sludge volume is generally much lower than the volume of coagulation or softening sludge.

Iron and manganese sludge removed from filters by backwashing generally settles sufficiently in 2 hours to allow decanting and recycling of backwash water to the head of the water plant. Concentration of the remaining sludge varies considerably, with typical values of 10% to 30% of the total backwash water volume.

Softening-Plant Sludge. Water softening with lime or soda ash produces sludge containing precipitates such as calcium carbonate, calcium sulfate, magnesium hydroxide, silica, iron oxides, aluminum oxides, and unreacted lime. Coagulated organic and inorganic substances typically constitute a small fraction of the sludge mass. When highly turbid waters are to be softened, turbidity is normally removed by coagulation before softening.

Softening sludge is relatively inert and stable, and biologically inert as a result of the high pH caused by unspent lime and high alkalinity. Softening sludge is normally easier to concentrate than coagulant sludges. The solids content of lime sludges typically ranges between 2% and 15%.

The quantity of lime-softening sludge produced at water treatment plants varies greatly depending on the hardness of the source water, the source water chemistry, and the desired finished water quality. The volume of softening sludge produced is typically much greater than that produced by coagulation processes. Solids in lime-softening sludge normally concentrate in the sedimentation basin to 10%.

Lime-softening sludges are primarily calcium carbonate with varying amounts of other constituents. The dewaterability of the sludge varies with the concentration of magnesium hydroxide that has been captured, ranging from a few percent to as much as 30%. Sludges low in magnesium hydroxide may be dewatered to cakes having 60% solids, whereas cake solids may be as low as 20% to 25% with higher magnesium hydroxide concentrations. Magnesium hydroxide solids are gelatinous and similar in nature to aluminum and iron coagulant solids. The calcium carbonate solid is more discrete and crystalline and more readily dewatered. The character of lime sludges at varying moisture contents is generalized as follows:

Solids content	Sludge character
0% to 10%	Liquid
25% to 35%	Viscous liquid
40% to 50%	Semisolid, toothpaste consistency
60% to 70%	Crumbly cake

Lime sludge cakes in the 50% to 65% moisture content range are generally sticky and difficult to discharge cleanly from dump trucks.

Brine Wastes

Methods of disposal of wastes from membrane and ion exchange processes are growing in importance as these processes are increasingly used for drinking water treatment. At the same time, governmental regulations on acceptable methods of disposal are becoming more strict.

Membrane Process Waste. Membrane processes used in water treatment include microfiltration (MF), ultrafiltration (UF), nanofiltration (NF), electrodialysis reversal (EDR), and reverse osmosis (RO). These processes involve separation of particles or ions from source water by passing the flow through a semipermeable membrane. Waste from a membrane treatment unit consists of constituents that do not pass through the membrane and are termed *reject, concentrate,* or *brine.* The volume of the reject can be calculated using the following formula:

$$Q_c = Q_f \times (1 - R)$$

where Q_c = reject flow
Q_f = feed water flow
R = recovery rate

Recovery rate depends on various factors, including the source water quantity, fouling potential, feed rate, operating pressure, and type of membrane. Typical recovery rates are as follows (Mickey, 1993):

Membrane process	Typical system recovery rates (%)
Microfiltration	90 to 95
Ultrafiltration	90 to 95
Nanofiltration	75 to 90
Electrodialysis reversal	70 to 90
Brackish water reverse osmosis	60 to 85
Seawater reverse osmosis	20 to 50

The quantity of the reject water is highly dependent on the type of membrane process and source water quality. Reject water from large pore processes has water quality comparable to backwash waste from a conventional water treatment process. RO and EDR are typically applied to remove total dissolved solids (TDS), and the concentration of specific ions or salts is relatively high in the reject water. The relative concentration of ions in a reject water may be determined using the following equation:

$$I_c = I_f \times \text{CF}$$

where I_c = ion concentration in the reject water
I_f = ion concentration in the feed water
CF = concentration factor $1/(1 - R)r$
r = fractional rejection rate for an ion

If an ion is completely rejected by a membrane, the fractional rejection rate (r) is 1.0 and the following equation applies:

$$\text{CF} = 1/(1 - R)$$

For example, a membrane process with an 80% recovery rate would concentrate an ion in the reject water five times the concentration in the source water.

One of the advantages of membrane processes is the decreased use of chemicals such as coagulants. Chemicals added before the membrane process generally include acid, antiscalant, and, in some cases, low concentrations of disinfectant. These chemicals generally pass through larger pore size membranes.

Ion Exchange Process Waste. Brines are also produced as a waste from ion exchange processes. Ion exchange has primarily been used as a softening process, but other applications include removal of specific ions such as nitrate and barium. Waste flow from the ion exchange process is produced during backwashing, regeneration, and rinsing. The quantity of brine waste is generally between 1.5% and 10% of the water treated by the process unit (ASCE/AWWA/EPA, 1997).

Ion exchange waste brine typically has high TDS and low suspended solids. The concentration depends on source water hardness, target ion concentration, regeneration rate, rinsing procedure, and resin capacity. Typical ranges of values are shown in Table 17.1.

Cationic ion exchange resins are usually regenerated with an acid solution. The low pH backwash water may require neutralization before disposal depending on the disposal method. Similarly, anionic ion exchange wastes may require neutralization because they are regenerated with a basic solution of salt.

PROCESS RESIDUAL DISPOSAL METHODS

Disposal of treatment plant wastes includes removal of the wastes from the treatment plant or reuse of some of the residuals.

Sludge Disposal Alternatives

Six methods of disposal are typically considered for process wastes:

- Discharge to a natural waterway
- Discharge to a sanitary sewer system
- Discharge to permanent lagoons
- Burial in a landfill
- Land application
- Reuse of all or a portion of the wastes

Each of these disposal methods involves different regulatory requirements and may require varying levels of pretreatment before disposal.

Discharge to a Natural Waterway. Although discharge to natural waterways has traditionally been the predominant form of both handling and discharging process wastes, requirements of the National Pollution Discharge Elimination System (NPDES) have essen-

TABLE 17.1 Typical Chemical Constituents of Ion Exchange Wastes

Constituent	Concentration range (mg/L)
TDS	15,000 to 35,000
Ca^{2+}	3,000 to 6,000
Mg^{2+}	1,000 to 2,000
Hardness (as $CaCO_3$)	11,600 to 23,000
Na^+	2,000 to 5,000
Cr	9,000 to 22,000

Source: Mickey, 1993.

tially terminated all direct discharges. The primary concern has been the introduction of pollutants into the aquatic environment. The impact of aluminum toxicity from alum coagulant wastes on aquatic biota has been a chief concern. There is also concern about the impact on wildlife and the environment of wastes containing high levels of solids, total dissolved solids, pH, various trace metals, nitrates, and chlorine, as well as the potential for creating excessive flow rates. Each of these items may be included in the NPDES permit if a plant is granted permission to discharge wastes to a waterway.

Discharge to a Sanitary Sewer. The practice of disposing of water treatment plant solids to sanitary sewers has become increasingly common. The economies of scale provided by treating water and wastewater treatment solids together is attractive, and dilution of the inorganic sludges with organic sludges makes the resulting sludge more acceptable for land disposal.

Discharge of sludge to sanitary sewers must be coordinated with the sewer authority operation and maintenance department and wastewater treatment plant authorities. The impact of both the chemical nature and the volume of the sludge on the wastewater facility needs particular consideration. Additional solids loading needs to be assessed, because it affects the waste treatment plant solids handling capacity. The additional liquid and solids load also increases operational and maintenance costs to some extent. Discharges to a sewer should be monitored and controlled to minimize the possibility of large quantities of relatively inert sludge filling the digesters and upsetting the wastewater treatment process.

Managers of some wastewater utilities are often concerned that water treatment plant solids will adversely affect their treatment processes. However, experience has shown that controlled addition of water treatment plant sludge does not cause disruption, and there may even be benefits to the waste treatment operation as a result of enhanced sedimentation.

Culp and Wilson (1979) studied the effect of adding alum sludge to an activated sludge wastewater treatment facility. They reported no identifiable benefit or detriment to the treatment process or to the anaerobic digester. The increase in wastewater sludge quantities was reported to be in proportion to added water treatment solids.

Discharge of lime sludge to sanitary sewers should be considered carefully because sludge may produce encrustations on weirs, channels, and piping. Because the volume of softening sludge disposed of is also typically large, it may be more than the waste treatment plant facilities can handle.

Discharge to a Lagoon. If adequate land is available, dilute sludge may be diverted directly to lagoons where coagulant solids concentrate to 6% to 10% over time. If a water treatment plant operating at an average rate of 1 mgd (3.8 ML per day) produces 2,000 lb/mil gal (240 kg/ML) of solids, 25 to 50 acre-ft (3,080 to 6,160 m^3) of storage capacity is required for every decade of operation.

Softening sludge, on the other hand, can be expected to concentrate to 20% to 30%, and may attain a 50% concentration over a period of years. For a plant with a 1 mgd (3.8 ML per day) treatment rate that produces 2,000 lb/mil gal (240 kg/ML) solids, about 10 acre-ft (1,230 m^3) of storage capacity is required every decade. This option may be attractive for small treatment plants, but it is impractical for most larger treatment plants.

Lagoon storage of water treatment plant solids may be an attractive alternative for the short term. However, the lagoon will eventually be filled. Reuse of the land may require varying degrees of reclamation of the lagoon.

Burial in a Landfill. Water treatment plant wastes disposed of in a sanitary landfill must first have solids concentrated to a semisolid or cake form. When properly dewatered, sludge may be disposed of using sludge-only trenches or area fill techniques. Alternatively, sludge may be codisposed with refuse. Because of new environmental laws, the number of acceptable landfills has been greatly reduced and the costs of opening and closing landfills is much higher than it was just a few years ago. Most landfill operators probably will strongly resist accepting additional water treatment plant sludge.

Land Application. Direct application of water treatment plant sludge to land surface has been attempted at a number of locations. The problem is that alum sludge solids tend to clog soil pores and prevent seed germination. However, it has been found that breaking up the crust somewhat mitigates this problem. Lime sludge has often been promoted as a soil additive for certain soil types, such as clays. It has been claimed that it stabilizes clay soils (from shrinking or swelling) and increases the pH of acidic soils.

Disposal of Brine

Brine disposal options for an RO or EDR process plant include discharge to natural waterways, evaporation, or discharge to an injection well.

Discharge to freshwater generally requires a permit from the state and is limited by the TDS concentration of the brine in comparison with the freshwater flow rate. Coastal water treatment systems have an advantage in that there is less objection to discharging the brine to the ocean. There is also a potential for discharge to brackish water marshes (Boyle Engineering, 1994).

In areas where surface discharge is not possible, evaporation ponds have been used. However, groundwater protection laws in most areas now require lining the pond system to prevent contamination of groundwater. Industrial processors have used several mechanical forms of evaporation, including brine concentrators and crystallizers, but these technologies are generally considered cost prohibitive for water treatment facility use.

Pumping water treatment brine wastes into injection wells is a possible option if wastes do not contain any hazardous constituents. Injection wells are now closely controlled by state and federal regulations, and the permit to allow injection includes showing that contaminants will not migrate to or adversely affect, an underground source of drinking water.

Residuals Disposal Regulatory Requirements

This section provides a general discussion of federal regulations affecting disposal of water treatment plant wastes. However, each disposal situation may face different requirements because of the wide variation in state and local regulations.

Most federal regulations do not address water treatment discharges specifically. Broad categories of wastes that are regulated include solid, hazardous, and radioactive waste. Requirements for each should be reviewed to determine applicability for water treatment plant residuals. Most water treatment plant wastes do not fall into either the hazardous or the radioactive categories.

Regulation of Discharges to Natural Waterways. The 1972 Clean Water Act established the NPDES, which can affect water treatment process waste discharges into waterways. However, the application of this program has differed throughout the United States. Some states ban water treatment plant discharges completely; other states have set discharge limits based on total suspended solids and pH. There are some regions where direct discharges may be allowed for filter backwash waste and overflow from residuals handling processes such as lagoons and thickeners.

The Clean Water Act originally set procedures and prepared guidelines for discharges from water treatment plants, but the guidelines were never implemented. Guidelines identified best practical control technologies (BPT) for different types of water treatment, and the requirements were established in terms of total suspended solids and pH. The approximate discharge limit suggested for total suspended solids was between 30 and 60 mg/L; this limit was based on the assumption that the discharge was 2% of the stream flow. Because these guidelines were not fully adopted, it has fallen to U.S. Environmental Protection Agency (USEPA) regional offices and individual states to set specific limits for discharges.

Regulation of Discharge into Sewer Systems. Discharging water treatment residuals to a sanitary sewer system is generally covered by the Clean Water Act, which requires that each utility with a wastewater treatment plant capacity over 5 mgd (19 ML per day) must develop a pretreatment program. The program is to establish minimum pretreatment standards for all industrial customers. State and local agencies may develop more restrictive standards that may limit flows and contaminant levels in water treatment discharges into sewerage systems.

Waste treatment plants that accept water treatment plant wastes must be sure there are no contaminants in the waste that will pass through the treatment systems and cause the wastewater plant to exceed one of the monitoring parameters of their NPDES discharge permit.

Regulation of Permanent Lagoons and Landfills. Subtitle D of the Resource Conservation and Recovery Act (RCRA) applies to the management of nonhazardous solid waste. The USEPA sets the regulatory direction and provides technical assistance. Planning, regulation, and implementation of solid waste programs are generally carried out by state and local agencies.

Regulations include location restrictions, facility design criteria, operation criteria, groundwater monitoring requirements, corrective action requirements, financial assurance requirements, and closure and postclosure care requirements. Lagoons and water plant residual landfills usually must meet these requirements under state direction.

Regulation of Land Application. Regulations for land application of water treatment residuals are usually found at the state and local levels. Some states specifically regulate the land application of alum coagulation residuals (ASCE/AWWA/EPA, 1997).

Federal regulations for land application are typically not issues because of the following:

- Standards for the Use or Disposal of Sewage Residuals, 40 CFR 503. Water treatment residuals are specifically excluded from these criteria.
- Criteria for Classification of Solid Waste Disposal Facilities and Practices, 40 CFR 507. Pollutant concentrations listed in these rules are typically higher than contaminant levels found in water treatment solids.

Regulation of Deep Well Injection. Underground Injection Control (UIC) rules were developed by the USEPA after the 1980 amendments to the Safe Drinking Water Act. The regulations are compiled in 40 CFR 122, 144, and 146. Some states have accepted primary enforcement responsibility (primacy) for the program and have developed their own injection well requirements. Other states have relinquished regulatory control to the USEPA.

Well injection is defined by federal regulations to include bored, driven, drilled, or dug wells. Injection wells are further separated into classes I through V. Water treatment waste disposal injection wells are generally described as class I wells. In this category, nonhazardous industrial or municipal wastes may be injected beneath the lowest formation containing a source of drinking water, and must be at least 0.25 mile (402 m) from the nearest drinking water well. There are also a variety of additional siting, permitting, and monitoring requirements placed on a permit for an injection well.

DESIGN CONSIDERATIONS AND CRITERIA

Methods for processing and disposing of residuals from a water treatment plant should be investigated early in the planning phase. Sludge disposal may represent a substantial portion of the investment and operating costs of providing treated water and may influence the source water selection and the method of treatment. When evaluating design alternatives for processing and disposing of waste solids, the following items should be considered:

- Sludge handling and transport
- Sludge dewatering techniques
- Sludge disposal
- Recovery and reuse of coagulants or lime

Minimizing Sludge Quantities

One of the first considerations for a designer beginning to design a new water treatment plant is how sludge handling and disposal costs can be kept to a minimum. There are two general methods of reduction: reducing sludge production through process modifications and recovering spent coagulants and lime for reuse.

One method of reducing the quantity of sedimentation wastes is by using polymers to enhance the performance of coagulants such as alum or ferric chloride. For some source water, the ratio of alum dose to polymer dose required for effective coagulation is approximately 50:1; this results in a sludge production ratio of aluminum hydroxide to polymer of 7:1.19, realizing a significant reduction of sludge volume.

In lime-softening plants where a significant fraction of hardness is attributed to magnesium, split-flow lime softening can reduce total sludge production compared with excess lime softening. A high pH is necessary for magnesium hardness removal. In addition, most of the calcium hardness must be removed before magnesium hardness will precipitate.

In the split-flow process, the lime dose required to treat the entire flow is added in the first softening basin. Magnesium and calcium hardness precipitate in the primary softening basin as a result of the elevated pH. Unreacted lime enters the recarbonation basin or second softening basin where it removes additional calcium hardness from the bypassed source water. Total sludge production is usually lower because less lime is used than in excess lime softening.

If more than one water source is available, possible differences in residuals created from the treatment of different qualities of source water should be considered. For example, selecting softer water may result in smaller sludge quantities.

Where softening of a water supply is required, alternative processes can be selected. For instance, the softening plant can be operated to selectively remove only the calcium fraction total hardness. However, the resulting higher magnesium concentrations may reduce the life of water heaters installed on the water system. Ion exchange softening may also be considered if there is a feasible method of disposing of the waste brine.

Recovery of Coagulants

Recovery of coagulants for reuse has been examined as a means of helping resolve the waste disposal problem of water treatment plant solids. Recalcination of spent lime is a proven technology in many locations, but coagulants recovery and reuse has been more elusive. Recent technology in alum recovery shows promise, and research efforts are being made to recover and reuse spent iron coagulants.

Alum Recovery. The traditional scheme for alum recovery consists of thickening, reducing pH, and separating residual precipitates from the dissolved aluminum decant. The recovered alum solution is decanted from the separation stage and reused. Aluminum recoveries of 60% to 80% have been reported at pH levels of 3.0. However, some sludge may require pH values as low as 1.0.

When 1.9 lb (0.9 kg) of sulfuric acid is reacted with 1 lb (0.5 kg) of aluminum hydroxide, 1.9 lb (0.9 kg) of alum is formed. This acidic alum recovery process was eagerly pursued in several Japanese water treatment plants, but concern about the potential buildup of heavy metals in the recovered alum stopped further use of the process in 1972. Metals such as chromium and iron can be converted to a soluble form during acidification. Other metals and impurities may also be present in the sulfuric acid, and these impurities may become concentrated in the recovered alum. Further detractions for the process included the expense and critical operating control required.

Another process that may be considered for alum recovery is the liquid ion exchange process (Westerhoff and Cornwell, 1978). Thickened alum sludge is acidified to a low pH (about 2.0), separated from remaining precipitates in a sedimentation tank, and subjected to liquid ion exchange. The aluminum is extracted into a liquid carrier that is immiscible in water and separated. Aluminum in the separated liquid carrier is stripped with sulfuric acid, and the carrier liquid is recycled. The extractant is selective for aluminum, creating no buildup of impurities or heavy metals. A 95% recovery of alum is achievable under laboratory conditions.

If sodium aluminate proves to be a good coagulant for a specific water, the alkaline method of alum recovery may be applicable. Aluminum can be redissolved from alum sludge by raising the pH to 12 to 12.5 with sodium hydroxide. This converts the aluminum hydroxide to sodium aluminate. Aluminum recoveries of 90% to 95% are reported.

Iron Coagulant Recovery. Recovering iron coagulants involves acidification of the ferric hydroxide and a recovery technique similar to that described for the acidic alum recovery process. The pH must be reduced to 1.5 to 2.0 to attain 60% to 70% recovery of iron. Because of the expense and the poor dewatering characteristics of the sludge, there has been little interest in this process.

Lime Recovery. A coagulant recovery technique for lime-softening plants has also been developed (Black and Thompson, 1975). It is based on a combination of water softening and conventional coagulation procedures that can be applied to all types of water. Magnesium carbonate is used as the coagulant, with lime added to precipitate magnesium hydroxide as the active coagulant. The resulting sludge is composed of $CaCO_3$, $Mg(OH)_2$, and the turbidity removed from the source water. Sludge is carbonated by injecting CO_2 gas, which selectively dissolves the $Mg(OH)_2$. Carbonated sludge is filtered, with the magnesium recovered as soluble magnesium bicarbonate in the filtrate. The magnesium bicarbonate coagulant is then recycled to the point where chemical is added to the source water. At that point, it is precipitated as $Mg(OH)_2$, and a new cycle is initiated. Filter cake produced in the separation step contains $CaCO_3$ and the turbidity removed from the source water.

The process may be expanded to recover lime. The filter cake ($CaCO_3$ plus turbidity particles) is slurried and processed in a flotation unit to separate turbidity particles from the $CaCO_3$. The purified $CaCO_3$ can then be dewatered and recalcined to quicklime.

Residuals Handling and Transportation

Methods for transporting water treatment plant solids are similar to other materials handling methods. Sludge in liquid form may be pumped through pipelines or trucked. Alternatively, sludge in cake form may be trucked or barged.

Dilute concentrations of coagulant sludges flow by gravity or may be pumped using centrifugal pumps with nonclog impellers. Gravity sewers should be designed to maintain nonsettling velocities at minimum flows. Lime sludge or thickened alum sludge (8% to 15%) should be pumped with a positive displacement pump.

Precautions are required to protect pumps and pipeline materials from corrosion and abrasion. Pipeline scaling has not been reported to be a severe problem, because the chemicals are stable. Abrasion of pump impellers is common with all water treatment plant sludges. In some instances pneumatic ejectors are used to transport sludges to avoid abrasion problems with pump elements. Conveyor belts or screw conveyors are commonly used for thickened lime sludge (30% or greater) and thickened coagulant sludges (15% or greater).

Because trucking sludges to remote sites often compresses solids from vibration and release of free water, the trucks should be watertight. Lime sludge commonly is sticky and adheres to the truck container, making it difficult to dump. Special surfaces on the truck dump body should be considered to reduce adherence.

Sludge Treatment Techniques

Economics, regulatory requirements, and various other factors may result in the need for utilities to provide treatment of residuals before transport and disposal. The purpose of these processes is usually to reduce the quantity of solids. Figure 17.3 shows treatment categories that generally handle coagulant sludges, softening sludges, and filter backwash waste. The three primary categories of residuals treatment processes are thickening, dewatering, and drying.

These categories are typically delineated according to the ranges of solids concentration of the cakes produced as the end product of each process. Following are the normal ranges of concentration of alum or metal hydroxide sludge by various processes:

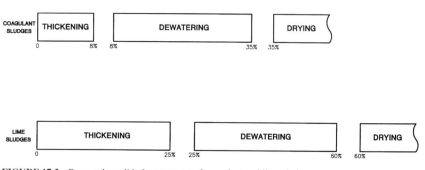

FIGURE 17.3 Percent dry solids for treatment of coagulant and lime sludges.

Process	Solids concentration
Thickening	≤ 8%
Dewatering	8% to 35%
Drying	> 35%

The processes generally achieve a higher percent solids concentration for lime sludge than for a metal hydroxide sludge. For example, a centrifuge may dewater a metal hydroxide sludge to 15% while lime sludge is dewatered to 50%. Following are the normal ranges of concentration of lime-softening water treatment plant sludge by various processes:

Lime sludge process	Solids concentration
Thickening	≤ 25%
Dewatering	25% to 60%
Drying	> 60%

Thickening Processes. Thickening is typically the first step toward reducing the quantity of water treatment residuals. The products of the process are a low solids return flow and a thickened solids by-product flow. Return flow is generally returned to the water treatment processes, termed *used water* recovery. The thickened solids flow is either further thickened or transported off site for disposal.

Thickening has traditionally been used at water treatment plants in tandem with a dewatering or drying step. The function of thickening is to reduce land area requirements or mechanical needs of dewatering. There are three types of thickening:

- Gravity
- Flotation
- Mechanical

Gravity Thickening. Gravity thickening is the most predominantly used technique at water treatment plants. It reduces sludge bulk, provides a more consistent feed material, and reduces the size of subsequent dewatering units. Polymer is often added to increase particle size and reduce solids carryover in reclaimed water. A section of a typical gravity thickener is shown in Figure 17.4, and a picture of a gravity thickener is shown in Figure 17.5.

Sedimentation basin sludge is usually less than 1% solids when drawn off and can often be thickened to 2% solids. Aluminum and iron hydroxides may be conditioned with the aid of polymers to provide improved thickening. Typical design parameters reported for alum sludge thickening are 100 to 200 gpd/ft^2 (240 to 490 m/h) when conditioned with polymers. Alum sludges mixed with clay or lime have exhibited thickened concentrations ranging from 3% to 9% at higher overflow rates than sludges without clay or lime.

Lime sludge thickening provides more concentrated solids and a more consistent feed material for dewatering units. Solids loading levels from 60 to 200 lb of solids per square foot (290 to 980 kg/m^2) of thickener surface area per day are common. High magnesium hydroxide concentrations reduce sludge dewaterability and thus reduce the density of thickened sludge.

Flotation Thickening. Flotation thickening has been used with a significant amount of success in the wastewater industry. Dissolved air flotation and recycle flotation are the two forms of flotation thickening.

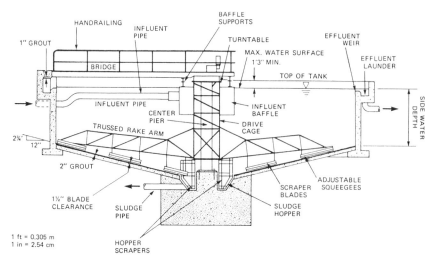

FIGURE 17.4 Gravity thickener section. *(Source: U.S. EPA Process Design Manual: Sludge Treatment and Disposal, 1979.)*

FIGURE 17.5 Gravity thickener.

In the dissolved air flotation thickening process, air is added at pressures in excess of atmospheric pressure either to the incoming residual stream or to a separate liquid stream. When pressure is reduced and turbulence is created, air in excess of that required for saturation at atmospheric pressure leaves the solution as very small bubbles of 50 to 100 μm in diameter. Bubbles adhere to the suspended particles or become enmeshed in the solids matrix. Because the average density of the solids-air aggregate is less than that of water, the agglomerate floats to the surface. Water drains from the material that floats to the surface, called float; the float is continuously removed by skimmers.

Recycle flotation is most often used to concentrate metallic hydroxide residuals. In this type of system, a portion of the clarified liquor, or an alternate source containing only minimal suspended matter, is pressurized. Once saturated with air, it is combined and mixed with the unthickened residual stream before it is released to the flotation chamber. The advantage of this system is that it minimizes high shear conditions, an extremely important factor when dealing with flocculent-type residuals (metallic hydroxides).

Coagulant residuals can be thickened in flotation thickening systems to a concentration of about 2,000 to 3,000 mg/L using a solids flux rate of 10 to 30 lb/day/ft^2 (50 to 150 kg/day/m^2). This is higher than can be achieved by simply settling flocculated surface water in a clarifier, but less than can be achieved by gravity thickening. Flotation units are usually best operated with a continuous feed. Optimum operation requires adding polymer or other conditioning agent at a dosage of about 2 kg per dry ton (2 kg/907 kg dry).

Mechanical Thickening. In situations where it is not possible to obtain adequate thickening of sludge using gravity or flotation thickening, mechanical thickening equipment may be used. The most commonly used type is gravity belt thickeners, but other mechanical thickening processes may also be considered.

Gravity Belt Thickeners. A gravity belt thickener concentrates solids by letting gravity pull the free water through a moving porous belt. Free water passing through the belt is called filtrate. Figure 17.6 illustrates a typical gravity belt thickener. A polymer solution is normally injected and mixed into the solids in a chamber that feeds the gravity thickening zone. Fixed or adjustable plows guide the solids as water drains through the moving belt. Thickened solids are then discharged from the end of the thickener as the belt reaches the end roller. A scraper or an adjustable ramp is used to assist removal of the solids to a sludge hopper. A belt wash station then cleans the belt before it rotates back to the front roller.

Gravity belt thickeners have maximum hydraulic and solids loading rates that vary depending on the manufacturer's design and the character of the sludge to be thickened. Actual loading rate is usually determined by pilot or bench testing. Concentration of from 2.5% to 4.5% solids of metal hydroxide sludges can usually be achieved.

Advantages of a gravity belt thickener include simple design, low operating cost, limited operator attention, minimal chemical conditioning, and suitability for rapidly settling sludges such as lime. Disadvantages include two waste streams, filtrate and belt washwater, labor required for operation and maintenance, and usually chemical conditioning of the sludge.

Gravity belt thickeners are usually equipped with 1, 2, or 3 m wide belts, with 2 m most common. Depending on the manufacturer's design, equipment is furnished with a belt tracking system that may be hydraulic, pneumatic, or mechanical. Before purchasing a gravity belt thickener, plant operations and maintenance staff should be consulted to see if they have a preferred design.

Provisions must be made to furnish equipment with a washwater supply of about 45 to 60 gpm (3 to 4 L/s) at the pressure specified by the manufacturer, and also for receiving and disposing of the spent washwater. Other design considerations are the necessity of furnishing a polymer dosing system, splash control, and providing building humidity and ventilation control.

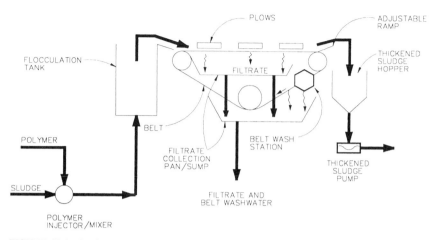

FIGURE 17.6 Gravity belt thickener schematic. *(Source: WEF Manual of Practice No. 8, Design of Wastewater Treatment Plants.)*

Other Mechanical Thickening Processes. Gravity belt thickening is the most common type of mechanical thickening. However, other mechanical devices have been used or studied for sludge thickening at water treatment plants, and these processes are discussed later in this chapter.

Natural Dewatering Processes. The division between concentrating (or thickening) and dewatering has traditionally been vague. Concentrating has generally been defined as raising the solids concentration of a liquid stream, and dewatering has been defined as the separation of liquid from a solid.

Natural dewatering refers to those methods of sludge dewatering that remove moisture either by natural evaporation, gravity, or induced drainage. Most air-drying systems were originally developed for dewatering wastewater treatment sludge.

Air-drying processes are less complex, are easier to operate, and require less energy to operate than mechanical systems. However, they are not often used because they require a large land area, the operation depends on climatic conditions, and they are labor intensive. The effectiveness of air-drying processes is directly related to weather conditions, type of sludge, conditioning chemicals, and materials of construction for the drying bed. Sizing requirements for a natural dewatering process are shown in Figure 17.7.

These processes may include evaporation and percolation for unlined lagoons and beds or underflow for drying beds with underdrain or vacuum systems. Understanding these mechanisms is important in sizing the process.

Typical loading criteria for lagoons and drying beds are not well documented but coagulant sludges are typically applied to sand drying beds at between 2 and 5 lb/ft² (1.0 to 2.4 g/m²). The loading rate is a function of both the application depth and the applied solids concentration as shown in the following equation:

$$\text{SLR} = D_i/12 \times 62.4 \times DS_i/100$$

where SLR = solids loading rate
 D_i = the applied depth in inches
 DS_i = the applied solids concentration in %

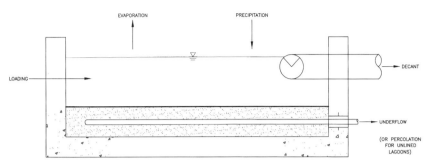

FIGURE 17.7 Natural dewatering sizing parameters.

The required time to achieve a desired solids concentration on the bed can be obtained from the following equation:

$$T = \frac{[D_i - (D_i \times VU) - (D_i \times VD)] - (D_i \times DS_i/DS_f)}{E}$$

where T = evaporation time in months
VU = % volume reduction through underflow/100
VD = % volume reduction through decanting/100
DS_f = finished solids concentration in %
E = seasonal net pan evaporation rate in in./month

One study has shown that this equation is a conservative estimate of the drying time (Vandermeyden, 1993). Design involves setting up a monthly balance of the loadings on each drying bed. This balance will show the anticipated cycling of the beds in and out of service.

One factor impacting the design of dewatering basins is local groundwater protection regulations. The regulations may require lining of a dewatering basin to protect an underlying aquifer. The state of Arizona, for instance, requires that a liner with a minimum equivalent permeability of 10^{-6} cm^{-1} be provided beneath waste lagoons. Lining increases the cost of the waste basin by both the liner installation cost and also additional area and volume required due to the loss of the percolation function.

Drying Beds. Drying beds usually consist of a sand underdrain with gravel and perforated pipe. In dry climates, shaped, shallow, earthen basins without underdrains are used that rely only on evaporation to separate solids from water. With either type of drying bed, sludge storage facilities must also be provided for periods when climatic conditions prevent effective dewatering.

Some designs incorporate additional drying beds that hold sludge until the right season. The size of drying beds should be based on the effective number of uses that may be made of each bed and the depth of sludge that can be applied to the bed. The required area can be estimated using the formula

$$A = \frac{V}{N \times D \times 7.5}$$

where A = drying bed area, ft^2
N = number of uses of beds each year
D = depth of sludge to be applied, ft
V = annual volume of sludge for disposal, gal

The number of times that the beds may be used depends on the drying time and the time required to remove solids and prepare the bed for the next application. The bed is usually considered dewatered when sludge can be removed by earth-moving equipment (such as a front-end loader) and does not retain large quantities of sand. Alum sludges generally attain solids concentrations of 15% to 30%, and lime-softening sludges attain 50% to 70% solids content.

Alum sludges require from 3 to 4 days to drain, but polymers may accelerate this to 1.5 to 3 days. These are optimal times and do not reflect realistic field conditions. Both field tests and a detailed study of climatic variations are required to apply this option. Bed uses usually range from 1 to 20 per year, depending on climate. In northern locations, drying beds are sometimes designed for one use per year, partially to take full advantage of the natural freezing.

The depth at which sludge may be applied ranges from 8 to 30 in. (20 to 75 cm) for coagulant sludge and from 12 to 48 in. (30 to 122 cm) for lime sludge. Greater sludge depths require proportionally longer drying times. For example, alum sludge at Kirksville, Missouri, required 20 h per percent solids concentration for an 8 in. (20 cm) application, and 60 h per percent solids concentration for a 16 in. (40 cm) application. To obtain a dewatered cake on the bed with a finished thickness suitable for removal with a front-end loader, at least 16 to 24 in. (40 to 60 cm) of sludge should be applied. For example, for a 1 mgd (44 L/s) average treated water quantity, 2,000 lb (900 kg) of sludge per million gallons (3.8 ML) treated, and 20 bed uses per year, a 2% concentration sludge applied at a 16 in. (40 cm) depth requires:

$$A = \frac{4{,}357{,}000}{20 \times 1.33 \times 7.5} = 22{,}000 \text{ ft}^2 \, (2{,}044 \text{ m}^2)$$

Freeze-Assisted Sand Beds. Alum residuals have a jellylike consistency that makes them extremely difficult to dewater. By freezing and then thawing the sludge, the bound water is released from the cells, changing the consistency to a more easily dewatered granular type of material. Freezing alum residuals changes both the structure of the residuals slurry and the characteristics of the solids themselves. In effect, solids tend to be compressed into large discrete conglomerates surrounded by frozen water. When they thaw, drainage occurs instantaneously through the large pores and channels created by the frozen water. Cracks in the frozen mass also act as conduits to carry off the melt water.

Freezing can be done mechanically or naturally. Because of the high cost associated with mechanical systems, natural systems are most common. The optimum effects of both the freezing and thawing portions of the cycle can be obtained by exposing solids on uncovered beds. Water may drain during thawing at a faster rate and produce a greater volume when compared with applying the same unconditioned solids to a conventional sand bed.

The critical operational requirement is to ensure complete freezing of the solids layer before the next layer is applied. Hand probing with a small pick or axe usually helps make this determination.

Solar Drying Beds. Until recently, paved drying beds were constructed with an asphalt or concrete pavement on top of a porous gravel subbase. Unpaved areas, constructed as sand drains, were placed around the perimeter or along the center of the bed to collect drainage water. The main advantage of this approach was the ability to use relatively heavy equipment for solids removal. However, experience has shown that pavement inhibits drainage, so the total bed area must be greater than that of conventional sand beds to achieve the same results in the same period.

Recent improvements to the paved bed process include a tractor-mounted horizontal auger or other device to regularly mix and aerate the sludge. Mixing and aeration break up the surface crust that inhibits evaporation, allowing more rapid dewatering than conventional sand beds. Although underdrain beds are still used in some locations, the most cost-effective approach in suitable climates is to construct a low-cost, impermeable paved bed and depend on decantation of supernatant and auger/aeration mixing for evaporation to reach the necessary dewatering level. A solar drying bed is pictured in Figure 17.8.

Vacuum-Assisted Drying Beds. In vacuum-assisted drying, a vacuum is applied to the underside of rigid, porous media plates on which chemically conditioned sludge has been placed. The vacuum draws free water through the plates and essentially all sludge solids are retained on top, forming a cake of fairly uniform thickness. Solids can be concentrated to between 11% and 17%, depending on the type of solids and the kind and amount of conditioning agents used.

One problem encountered with this system involves improper conditioning of the sludge. The wrong type of polymer, ineffective mixing of polymer and solids slurry, and incorrect dosage result in poor performance of the bed. In addition, overdosing polymer may lead to progressive plate clogging and the need for special cleaning procedures to regain plate permeability.

Plate cleaning is critically important. If not performed regularly and properly, media plates will clog and the beds will not perform as expected. The special cleaning measures then required are costly and time consuming.

Wedgewire Beds. The wedgewire, or wedgewater, process is physically similar to the vacuum-assisted drying beds. The medium in this case consists of a septum with wedge-shaped slots about 0.01 in. (0.25 mm) wide. This septum supports the sludge cake and allows drainage through the slots. Through a controlled drainage process, a small hydrostatic suction is exerted on the bed, removing water from the sludge.

Lagoons. Lagoons are one of the oldest processes used to handle water treatment residuals. Lagoons can be used for storage, thickening, dewatering, or drying. In some instances, lagoons have been used for final disposal of residuals.

The lagoon process involves discharging residuals into a large hole in the ground, with the anticipation that the solids will be retained there for a long period of time. Solids eventually settle to the bottom, and liquid can be decanted from various points and levels

FIGURE 17.8 Solar drying bed.

in the lagoon. Evaporation may also be used in the separation process if the residuals are to be retained in the lagoon for a long period of time.

The traditional lagoon consists of either an earthen berm built on the ground surface or a large basin excavated in the ground. Various types of systems are installed into lagoons to decant the supernatant and ultimately drain the lagoon. State and local regulations have become more stringent with regard to preventing pollution of groundwater and may affect the design of water treatment residual lagoons. Liners using materials such as high-density polyethylene (HDPE), leachate collection systems, and monitoring wells are becoming common features of lagoon designs. Lagoon depth typically varies from 4 to 20 ft. The surface area of lagoons ranges from 0.5 to 15 acres (AWWARF, 1987). A typical section for a lined lagoon is shown in Figure 17.9.

The effectiveness of lagoons in concentrating solids typically depends on the method of operation. Operating lagoons at full water depth without further air drying of solids typically results in a solids concentration of 6% to 10% for metal hydroxide solids when solids are retained in the lagoon for one to three months. Solids concentrations of 20% to 30% may be achieved for lime sludges under the same conditions. Some facilities achieve solids concentrations above 50% by stopping influent to the lagoon and allowing drying through evaporation. This process may require over a year of holding solids in the dewatering lagoon.

The lagoon process may incorporate certain modifications similar to sand or solar drying bed systems. Using a freeze-thaw process for lagoons is a common approach in northern climates.

Mechanical Dewatering. Mechanical equipment used for dewatering water treatment plant sludge includes filter presses, belt presses, centrifuges, and vacuum filtration.

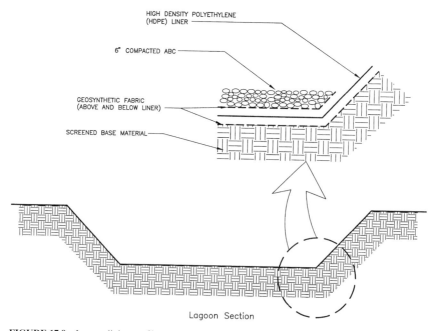

FIGURE 17.9 Lagoon lining profile.

Filter Presses. Pressure filters are typically divided into two categories: plate and frame presses and diaphragm presses. Plate and frame presses are also called fixed-volume presses and are the older category of units. Presses use a series of rectangular or circular frames with a filter cloth spanning the edges of the frame. A hydraulic drive locks the plates together at the beginning of a cycle, and solids are pumped into each chamber formed by the presses.

Pressures within each chamber range between 100 and 225 psi (690 and 1,550 kPa). Two pumps are sometimes used in this procedure: one with a relatively high pumping rate to fill the chambers and one lower-rate pump to maintain pressure in the chambers after the fill is completed. At the end of the cycle, the plates are disengaged, allowing the dewatered solids to drop from the unit. Some units are termed *recessed chamber filter presses,* which generally denotes the method of mounting the cloth on the frame.

Diaphragm filters began to be used at water treatment plants during the 1980s; *variable volume presses* is another term for these units. Operation is an enhanced version of the traditional plate and frame unit. After the fill and filter steps are completed, flexible diaphragms along the plate are expanded with compressed air or water to further compress solids in each chamber. This action is shown in Figure 17.10. Diaphragm filters generally provide higher filter cake solids and decreased cycle times compared with the traditional plate and frame press. Filtered water from the press, termed *filtrate,* flows to the end of the units for drainage away from the dewatering equipment.

Chemicals used to condition sludge in filter presses include lime, ferric compounds, and polymers. Precoat materials have also been used to minimize the potential for blinding of the filter media. It is becoming more common for filter presses to use polymer as the sole conditioning chemical.

The primary operating issue with the filter press is the accumulation of solids on the filter media after each cycle. High-pressure rinse water is incorporated into the operation of most models as a sequence at the end of each cycle. Some manufacturers also recommended periodic washing of the filters with an acid-based cleaning fluid. A plate and frame press is shown in Figure 17.11.

Although the filter press is applicable for dewatering lime-softening sludges, it is primarily used for metal hydroxide coagulant sludges. Filter presses are often used for difficult sludges because the batch operation can maintain solids under pressure for extended periods of time until the desired consistency is attained. With alum or iron salt sludges, either polymer conditioning, lime conditioning, or diatomaceous earth precoat is often required. Fly ash may also be appropriate. Cake solids contents of 20% to 50% result with pressing cycle times of 2.5 to 22 h, with approximately 8 h being average. Filter cloth life is usually about 12 to 18 months.

Filters can be provided in a variety of materials, weaves, and air permeability. The material should be selected based on durability, cake release, blinding, and chemical resistance. Selection is usually guided by either pilot testing or experience with similar solids streams.

Layout and profile of a filter press system are shown in Figure 17.12 with recommended clearances around the unit and head space above the filter for crane access. A traveling bridge crane is recommended for assistance in removing plates and drivers from the device.

Advantages of pressure filter dewatering include the following:

- Generally highest cake solids of any mechanical dewatering process
- High-quality filtrate
- Good mechanical reliability
- Adaptable to varying influent solids conditions

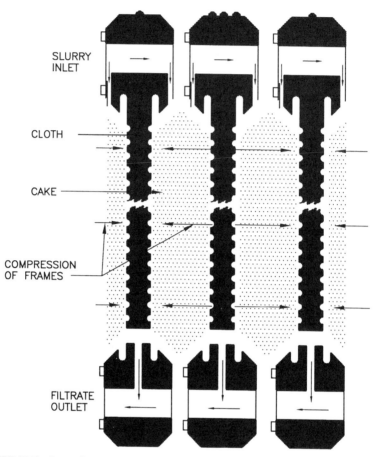

FIGURE 17.10 Dewatering mechanism of plate and frame filter press. *(Source: EIMCO Process Equipment Co.)*

Disadvantages include relatively high construction and operating costs and the problem of some sludges blinding the media.

Belt Presses. More than 10 different belt filter presses are manufactured at present. Although each of these units uses a unique design, the three schematic stages of operation in common are:

- Chemical conditioning zone. Solids entering the press are flocculated with a chemical additive.
- Gravity drainage zone. Flocculated solids flow onto a porous traveling belt where free water drains from the solids by gravity.
- Compression zone. Solids are compacted between two belts.

The last stage is sometimes differentiated into a low-pressure zone and a high-pressure zone. Solids are initially wedged together as belts come together in the low-pressure zone and a cake is formed at this point that withstands the shear forces of the high-pressure

FIGURE 17.11 Filter press.

zone. In the next stage, forces are exerted by the upper and lower belts passing through a series of decreasing-diameter rollers to further dewater the sludge. Chemical conditioning for most belt filter press systems consists of polymer injection.

The cake has a tendency to collect on belts, creating a long-term effect of blocking the porous openings. Solids are removed in one of two ways. On some models, cake discharge blades scrape solids from the belt. On other designs, a spray wash system rinses both belts after the cake has dropped from the belts.

Provisions for disposal or recycling of both filtrate and washwater are usually called for in regulations.

Belt presses used to dewater coagulant sludges generally produce cake solids of 15% to 20% when aided by optimum polymer dosage.

The belt filter press involves a large number of key components critical for proper operation. Research should be conducted to determine the right configuration for a given application. The two belts for the unit can be constructed of various materials, including rayon and nylon, with nylon being more appropriate for high-pH applications. Belts can be obtained in either seamed or seamless configuration. Seamed belts are more prone to wear at the joint; seamless belts avoid this wear but have a higher installation cost.

Rollers on belt filters are generally subject to failure caused by the high compression forces imposed on them. Rollers must be corrosion resistant, and some manufacturers use perforated rollers to improve drainage. Roller bearings should be self-aligning, and should be sealed appropriately against exposure to drainage and washdown. They should also have an L-10 life of 300,000 hours (the number of hours of operation at which 10% of the bearings could be expected to fail). Accessibility of replacement parts should be reviewed for each manufacturer's design.

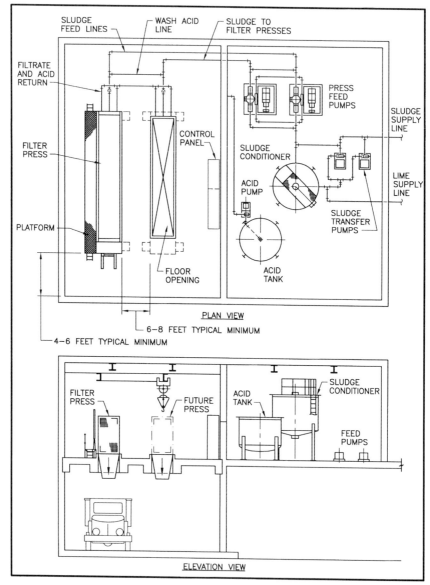

FIGURE 17.12 Typical filter press layout. *(Sources: AWWARF Handbook of Water Treatment Plant Waste Management and WEF Manual of Practice No. 8, Design of Municipal Wastewater Treatment Plants.)*

Filter belts are prone to tearing by abrasives in the incoming solids stream, but in most plants all abrasives have been removed by the treatment processes. However, there remains the possibility for sticks to blow into sedimentation or thickener basins. If this is a possibility, it may be desirable to add grinders to the suction side of the system feed pumps.

Layout and profile of a filter press system are shown in Figure 17.13. An elevated walkway may be desired to access the top of the unit.

Some of the principal advantages of using a filter press system are:

- Low energy consumption compared with other mechanical dewatering techniques
- The most experience of available mechanical dewatering processes
- Price competitive from large number of manufacturers

Principal disadvantages are that they generally require a large amount of operator attention, and odor control is more difficult than for other mechanical dewatering techniques.

Centrifuges. A centrifuge performs solids separation by applying a centrifugal force to the contents of a spinning bowl. The centrifugal force applied in municipal sludge dewatering is typically over 1,500 times that of gravity.

Several types of centrifuges have been used for solids separation in industry. The solid bowl centrifuge has developed into the principal unit used for large-scale municipal water treatment residuals dewatering. It has also been referred to as the scroll or decanter centrifuge. A schematic view of a solid bowl centrifuge is shown in Figure 17.14. The centrifuge uses two rotating elements: the bowl and the scroll. The bowl provides the solids separating force; the scroll moves the solids toward discharge from the unit. The bowl and the scroll operate at different speeds.

Solid bowl centrifuges operate continuously rather than in a batch mode. Liquid separated from the solids is termed *centrate*. Inside the spinning bowl the liquid level is a constant distance from the rotating axis. Liquid level is controlled by weirs or dams at the end of the machine.

The centrifuge shown in the figure is called a countercurrent centrifuge. In this design the solids and the centrate are discharged from opposite ends of the machine. An alternative design approach, called a cocurrent centrifuge, diverts centrate into a series of conduits that travel parallel to the direction of the solids discharge. Manufacturers publish the relative advantages and disadvantages of each machine type. A centrifuge facility design must either predetermine a centrifuge type or be flexible enough to handle alternate solids and centrate discharge points. Centrifuges are rarely operated without a conditioning chemical.

Centrifuge loading is generally defined in terms of hydraulic loading (gpm or L/s) and solids loading (lb/day or kg/day). Hydraulic capacity is generally defined by the centrifuge supplier. Increasing hydraulic loads affects the ability of the unit to separate solids. Using a hydraulic loading higher than the unit capacity generally leads to a decrease in centrate quality. Higher hydraulic loads may also require a higher chemical conditioning dosage.

Using a bowl centrifuge to dewater alum sludge produces widely varying results. Some installations report that as little as 1 to 2 lb (0.5 to 1 kg) of polymer per ton of solids yields 98% recovery with a 30% cake. In other locations with low-turbidity water, as much as 4 lb (1.8 kg) of polymer per ton of solids may be required to produce a 15% cake. When the bowl centrifuge is used to dewater lime-softening sludge, it yields a cake solids concentration ranging from 33% to 70%.

The bowl centrifuge may be used to selectively separate magnesium hydroxide from calcium carbonate or to produce a cake. Separating magnesium hydroxide precipitate from calcium carbonate precipitate is desirable when waste lime sludge is to be recalcined. However, separation is difficult to achieve with many waters. Magnesium hydroxide may also be recovered by carbonation of the calcium carbonate/magnesium hydroxide sludge. This process solubilizes magnesium, resulting in a high magnesium content in the centrate. Separation of the magnesium and calcium precipitates with a centrifuge results in 60% to 75% of the magnesium in the centrate. With the recarbonation/magnesium carbonate

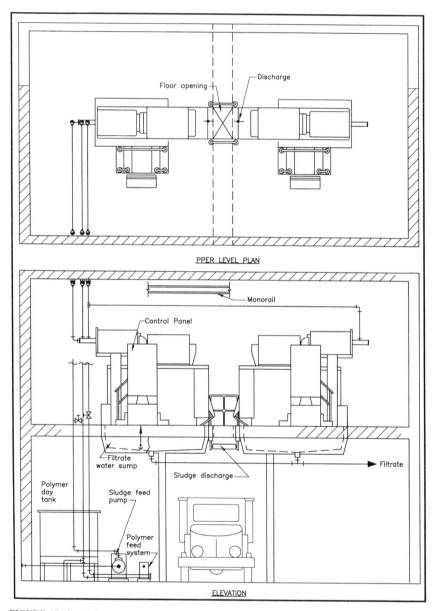

FIGURE 17.13 Typical belt filter press installation. *(Sources: AWWARF Handbook of Water Treatment Plant Waste Management and WEF Manual of Practice No. 8, Design of Municipal Wastewater Treatment Plants.)*

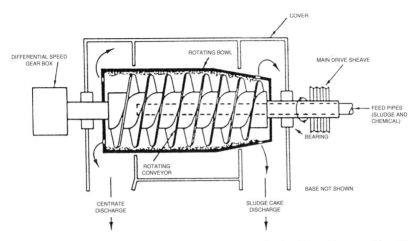

FIGURE 17.14 Centrifuge Schematic. *(Source: AWWARF Handbook of Water Treatment Plant Waste Management.)*

recovery process, magnesium recovery of 80% in soft water and nearly 100% in hard water is achievable.

Specifications for centrifuges should be tailored to protect the unit from corrosion, abrasion, and wear. Areas prone to abrasion include the bowl wall and the scroll blades. Stainless steel or ceramic liners should be considered for the inside of the bowl. Hardened materials or replaceable units may be considered for certain centrifuge components such as the scroll blades. Research on each type of centrifuge should include anticipated life of components, ease of parts replacement, and availability of replacement parts.

Layout and profile of a centrifuge system are shown in Figure 17.15, which shows recommended clearances around the units. A typical design issue is the removal of the centrifuge cover for access to the adjustable weirs or dams. It is standard to provide some lifting mechanism to remove the centrifuge cover, and the designer should consider installing a traveling bridge crane to move the cover out of the way while centrifuge adjustments are being made. Figure 17.16 shows a centrifuge installation.

Advantages of centrifuges include:

- Continuous feed operation
- Minimal operator attention required compared with other mechanical dewatering techniques
- Units have an excellent safety record
- Produce a cake with relatively high percent solids
- Maintenance requirements relatively low compared with other mechanical dewatering techniques

Vacuum Filtration. A vacuum filter consists of a horizontal rotating drum that rotates while partially submerged in a basin. The drum is partitioned into several zones. These zones are shown in Figure 17.17. Vacuum is applied across the submerged cake formation zone, causing solids to adhere to a filter medium extended across the drum face. The drum then rotates through the drying zone while a vacuum is applied to continuously draw liquid from the sludge cake through the filter medium. The vacuum is shut off as the drum enters the discharge zone, where a scraper removes the dewatered cake from the filter medium.

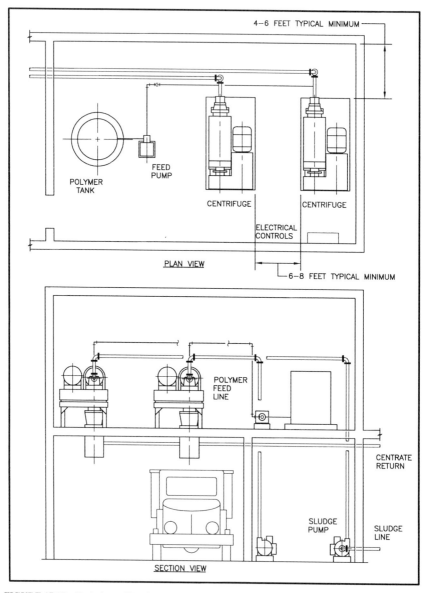

FIGURE 17.15 Typical centrifuge layout. *(Sources: AWWARF Handbook of Water Treatment Plant Waste Management and WEF Manual of Practice No. 8, Design of Municipal Wastewater Treatment Plants.)*

FIGURE 17.16 Centrifuge system.

Because of the growing use of other mechanical dewatering methods, vacuum filtration is becoming less popular, particularly with metal hydroxide sludges, because of low yields. However, the system continues to be a viable option for lime-softening wastes.

Vacuum filtration of metal hydroxide coagulant sludges is often ineffective. The dilute solids (even if thickened), high compressibility, and resistance water flow through the sludge result in low yields and poor recoveries of hydroxide sludges. When chemically conditioned with polymers, and when large concentrations of inert solids are present, vacuum filters can dewater alum sludges to 20% solids. A diatomaceous earth precoat on the vacuum filter has also been suggested as a means of attaining a 20% cake (including precoat solids).

Vacuum filters used to dewater lime-softening sludges yield 10 to 20 $lb/ft^2/h$ (50 to 100 $kg/m^2/h$), typical for sludges with high magnesium hydroxide content. Yields of 40 to 90 $lb/ft^2/h$ (195 to 440 $kg/m^2/h$) have been reported for sludges with low magnesium hydroxide content. Although cake solids content is typically 40% to 50%, some applications attain as high as 70% solids content. Supplemental chemical conditioners are not used, and solids recovery ranging from 96% to 99% can be expected.

Drying. Drying water treatment plant residuals has historically revolved around the issue of economics—how to reduce disposal transportation costs through reducing solids volume. Drying to solids concentrations greater than 35% is becoming a regulatory issue in many areas. For instance, the state of California requires that solids concentration of a water treatment plant waste be at least 50% before disposal in a landfill. Similar to the dewatering process, the drying process may be carried out either through open air means or through mechanical devices.

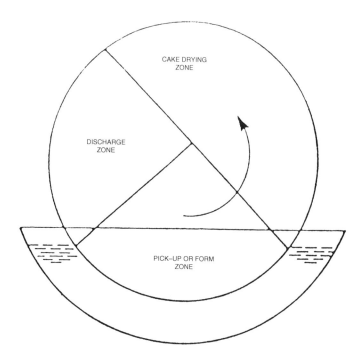

A. OPERATING ZONES

FIGURE 17.17 Vacuum filter. *(Source: AWWARF Handbook of Water Treatment Plant Waste Management.)*

Open-Air Drying Any of the solar drying or lagoon processes presented in the discussion on dewatering may be used for sludge drying. Drying depends only on evaporation so it may require years to achieve the desired solids concentrations. One method of accelerating the drying is to use a tractor to periodically furrow and mix the solids to increase exposure to sun and air.

Mechanical Drying. Of the mechanical dewatering methods discussed above, only the filter press has shown the ability to consistently produce solids concentrations greater than 35%. However, the ability of this process to achieve solids concentrations greater than 50% is not proven.

Thermal processes used for drying wastewater solids include the Best process, the Carver Grainfield process, and various forms of incineration. However, these processes are more applicable to a solid with a high organic content. There is no conclusive evidence available to evaluate the effectiveness of mixing wastewater and water solids together before feeding into one of these thermal processes.

Auxiliary Solids Handling Systems. The preceding sections of this chapter have focused on the individual processes for waste treatment. There are two additional elements that are common components of any waste handling system: transport and chemical conditioning.

Handling and Transport. For purposes of considering handling alternatives, process residuals can be divided into two classes: low-concentration wastes, having less than 15% solids; and high-concentration wastes, with over 15% solids.

Low-Concentration Wastes. Sludge in liquid form may be pumped through pipelines or transported in tank trucks. Dilute concentrations of coagulant sludges may flow by gravity or may have to be pumped using centrifugal pumps with nonclog impellers. Pumps and pipelines should be protected from both corrosion and abrasion. Abrasion of pump impellers is a particularly common problem when pumping sludge. Some treatment plants use pneumatic ejectors to transport sludge to avoid pump abrasion problems.

When sludges contain from 4% to 15% dry solids, positive displacement pumps are generally used. The most common type is the progressive cavity pump with a coil-shaped metallic rotor in a synthetic stader chamber. Important details in the installation are to provide pressure relief and flushing connections. If the pump is to be used to feed a mechanical dewatering device such as a belt filter press, the pump will need to be equipped with a variable-speed drive.

High-Concentration Wastes. Cake solids are typically trucked away from the treatment plant site. Three traditional methods for truck loading are:

- A truck drive-through directly under the dewatering equipment
- A series of conveyors from the dewatering equipment to the truck loading dock
- Use of special cake-handling pumps

Designing a truck driveway requires previous knowledge of the truck type and dimensions. The unloading system shown in Figure 17.18 uses a protected chute to reduce spillage of material around the truck. The truck unloading pad should be sloped to facilitate cleanup of spilled material. A truck wash-down area may also be necessary, either at the plant site or at the dump location.

FIGURE 17.18 Truck unloading system.

Conveyor belts or screw conveyors are commonly used to transport thickened lime sludge (30% or greater) and thickened coagulant sludge (15% or greater). Conveyor manufacturers should be consulted about anticipated sludge quality before selecting equipment. The rise rate of the conveyor is dictated not only by building space, but by the angle of repose of the particular sludge. Special belts may be specified to minimize backflow along the belt. Provisions for cleaning the belts include flexible scrapers with adjustable tension and rinse water sprays. Mobile conveyors may also be provided to enhance the flexibility of an installation. Figure 17.19 shows a mobile conveyor used to lift dried solids outside a building.

Positive displacement pumps are available that can handle caked sludge. One manufacturer uses a piston-type pump fed by a screw conveyor.

Trucks used for transporting sludge should be watertight because free water usually results from the solids compression caused by vibration. Lime sludge is sticky and adheres to the truck body, so special surfaces in the container shell should be considered to reduce adherence.

Chemical Addition. Sludge thickening and dewatering often require a chemical for conditioning. Polymers are most commonly used for this purpose. Designing polymer feed systems is complicated by the wide variety of polymer types and brands available. Bench and pilot testing are almost always used to determine the best type of chemical and dosage. Dewatering equipment suppliers may also provide information about chemicals that work best in their equipment.

If any sidestream from the dewatering process is recycled into the treatment process, only chemicals meeting the standards of the National Sanitation Foundation for addition to drinking water can be used. Polymers can be used in dry or solution form. Solution-based polymers can be made up of varying degrees of water, oil, and polymer. Some feed

FIGURE 17.19 Conveyor system.

equipment, such as the unit shown in Figure 17.20, are equipped to feed either liquid or dry chemicals.

The method of activating the polymer is an important consideration in plant design. The traditional approach has been to provide a batch tank for aging the diluted polymer. Several equipment suppliers now have devices that activate polymers through the addition of energy, and they have been found to be successful with certain types of polymers.

Polymer system sizing is difficult because of the wide range of possible feed rates. Information required includes pounds of neat polymer per hour, dilution to be used, quantity of dilution water required per hour, and quantity of diluted polymer used per hour.

Recycling/Disposal of Liquid Wastes. As Figure 17.21 illustrates, each type of waste treatment produces a liquid sidestream such as decant or filtrate. These may fit into either of two definitions:

- Recoverable water—water that may be recirculated to the main water treatment process
- Nonrecoverable water—water that may not be recirculated to the plant and must be disposed of through an alternative means, such as to a sanitary sewer or by evaporation

It is generally desirable to recover as much water as possible, both for water conservation and to minimize the cost of alternative disposal. Some sidestreams are automatically defined as nonrecoverable. This category includes filtrate from a mechanical dewatering process where a nonapproved polymer has been used, but most sidestreams have the potential for recovery.

The main concern regarding recycling recoverable water is its impact on treatment plant finished water quality. The parameters of concern are:

FIGURE 17.20 Polymer feed equipment.

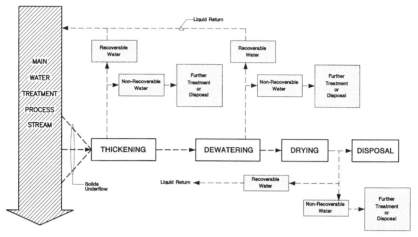

FIGURE 17.21 Concentration and side streams.

- Increased concentration of *Giardia* and *Cryptosporidium* cysts
- Increased concentration of assimilable and total organic carbon
- Increase in total trihalomethanes and trihalomethane precursors
- Increase in turbidity and particle counts
- Increase in aluminum concentration

Recent concern has focused on the potentials for cyst concentration and filter breakthrough because of recycling recovered water. This concern places a greater emphasis on the effectiveness of backwash waste treatment such as thickening.

A recent study in Salt Lake City evaluated the effectiveness of backwash waste treatment in the removal of *Giardia*- and *Cryptosporidium*-size particles. Tests for the removal of 2 to 5 micron size particles through settling at overflow rates between 160 and 550 gpd/ft^2 (6,500 to 22,400 L per day/m^2) indicated particle removal rates greater than 90%. It was also shown that addition of an anionic polymer before thickening significantly improved particle removal (Ashcroft et al, 1997).

RESIDUALS HANDLING CONSIDERATIONS

Among the decisions that must be made by the water treatment plant designer regarding residuals disposal are selecting the disposal method and equipment best suited for the plant, methods of transporting residuals, and the final disposition of liquid and solid wastes.

Residual Handling Process Performance

The residual handling process for a treatment plant is selected after a thorough comparison of available treatment and disposal alternatives.

Comparison of Thickening Processes. Factors that affect performance of the various types of thickening processes include the type of solids, their characteristics, influent concentration of the solids to the process, climatological conditions for open-air processes, variations in influent flow rates, and type and dosage of chemical conditioner.

This section provides typical ranges of values for percent solids that each thickening and dewatering process may generate. The comparison includes solids loading on the units and the solids concentration of the various types of residuals often experienced in these processes.

A comparison of thickening processes is shown in Table 17.2, and a summary of dewatering process performance provided by *Water Treatment Plant Waste Management* (AWWARF, 1987) is shown in Table 17.3.

Definition of Residuals Processing Requirements. In order to identify the ideal combination of unit processes necessary to form a preliminary residuals processing alternative, it is imperative that specific requirements of the residual processing system be defined. To define specific requirements of a system, the following fundamental information must be obtained (or generated if it does not already exist) (ASCE/AWWA/EPA, 1997):

- Residuals disposal limitations
- Quantity and quality of residual sources
- Resource recovery potential limitations
- Residuals mass balance

TABLE 17.2 Comparison of Thickening Processes

Type of process	Residual	Percent solids
Gravity	Carbonate	1 to 30
Gravity	Hydroxide	1 to 3
Flotation	Hydroxide	0.2 to 0.3
Gravity Belt	Hydroxide	2.5 to 4.5

TABLE 17.3 Range of Cake Solid Concentrations Obtainable

| | Percent solids concentration | |
Process	Lime sludge	Coagulant sludge
Gravity thickening	15 to 30	3 to 4
Scroll centrifuge	55 to 65	10 to 15
Belt filter press	—	10 to 15
Vacuum filter	45 to 65	—
Pressure filter	55 to 70	35 to 45
Sand drying beds	50	20 to 25
Storage lagoons	50 to 60	7 to 15

Disposal Limitations. In most cases the ultimate method of residuals disposal determines limitations, which in turn define process design requirements. The percent solids content of the residual is considered the primary criterion to define acceptable limits of a disposal option. Landfill and land application disposal options can, of course, be accomplished with a wide range of solids content, but both disposal options have different equipment requirements at the low and high solids limits of their ranges. The following tasks will help establish disposal limitations that a water utility must consider for further evaluation of residual process alternatives:

1. Identify the disposal limitations of available residual disposal alternatives. The range of normal residual solids content of the six most common methods used in the water industry are:

   - Land application: <1% to 15% solids
   - Landfilling (codisposal): >15% to 25% solids
   - Landfilling (monofill): >25% solids
   - Discharge to sewer: <1% to 8% solids
   - Direct stream discharge: <1% to 8% solids
   - Residuals reuse: <1% to >25% solids

2. Identify external disposal costs, if they apply. Disposal by landfilling in a codisposal municipal landfill can be associated with a per-unit-weight tipping fee levied by the landfill authority. Discharges to a sewer also have a user charge that can be assessed based on flow rate units or the combination flow rate and suspended solids concentrations. In some instances residuals reuse requires the water utility to pay a fee to the user to cover additional costs of special handling, or the difference in production costs associated with accommodating the sludge over a more traditional material.

Quantity and Quality of Residuals. Many water utilities have not closely monitored the quantity and quality of the treatment plant residuals. This deficiency of data is typical and requires additional effort and time to collect the following information:

1. Inventory residual sources. Typical residual sources associated with conventional water treatment plants are:

   - Headworks bar screens
   - Grit basins
   - Presedimentation basins
   - Sedimentation basins
   - Filter backwash
   - Filter to waste

2. Quantify flow rates and quantities of residual produced. Flow rate data collected for each residuals source should include the following:

   - Maximum, minimum, and average flow rate
   - Maximum, minimum, and average volume at each release
   - Frequency of events
   - Average event duration

- Maximum number of events per day
- Potential for concurrent events

3. Determine the quality of each residual source. Parameters that should be analyzed are:

   - Aluminum and iron
   - pH
   - Total dissolved solids
   - Total organic carbon
   - Trihalomethane formation potential
   - Trace metals
   - Toxic characteristic leachate procedure (TCLP)

Resource Recovery Potential. Although standard water conservation strategies support the recovery of reusable residuals, consideration must also be given to potential risks of continuous recycling of infectious organisms. Typical recycling of untreated residual streams such as backwash and filter-to-waste residuals back to the plant influent can result in accumulation of organisms in treatment processes. In light of these concerns, the value of water continues to increase with time, and recovery potential for reusable residual streams should be evaluated using the following procedure:

1. Assess the value of recovered water. Essentially the value of recovered water depends on the value of source water and the relative abundance of supply. Establish whether the cost of recovery is justified.
2. Identify residual sources that are easily recovered. Backwash and filter-to-waste residual streams generally constitute the largest volume of residuals generated with the lowest percent solids concentration. Treating these residual streams usually is justified for the value of the water that is recovered. Other residual streams should be evaluated closely with regard to their potential impact on finished water quality.
3. Identify residual sources unsuitable for recovery. The most difficult residual streams to justify recovery of are the sidestreams of mechanical dewatering processes. They usually contain many components that are undesirable to return to the plant.

Residuals Mass Balance. The most valuable exercise in evaluating residuals generated at a water plant is the development of a mass balance. A mass balance is a theoretical accounting of the water and solids content that enter the plant, solids that are added to the water, the water and solids content that leave the plant in the finished water, and the residual streams. This assessment must include the assessment of the effects of normal variations in source water quality and the associated chemical feed rates necessary to treat the water on the residual production rates. The objective of the mass balance is to specifically define the following parameters:

1. Establish maximum, minimum, and average solids production rates for the source water and each residuals sidestream in terms of percent solids, flow rates, and total solids produced per million gallons of water treated.
2. Define the allowable return residuals stream limits in terms of hydraulic and percent solids limits (e.g., maximum backwash returned to the plant influent should not exceed 5% of operating production capacity).
3. Define the allowable solids content in the finished water.

4. Define the allowable flow rate and solids content for available residuals disposal options.
5. Based on the allowable return limits, effluent limits, and disposal limits for each residual stream, determine a required residuals processing efficiency for each residual stream. The objective of this exercise is to identify potential solids content objectives for each residual stream process.

Residual Handling Process Design Issues

Methods used in handling water treatment plant residuals depend on the character of wastes produced, the quantities of materials to be produced, and type and size of treatment equipment to be used.

Estimating Solids Production. Solids mass balances are prepared to develop estimates of hydraulic and solids loading on residuals handling unit processes. Solids production from a water treatment plant may be estimated based on source water suspended solids removal and quantity of process chemical added. The following formulas may be used in developing solids loadings (AWWARF, 1987):

$$S = Q(0.44\ Al + SS + P)(8.34\ \text{lb/mil gal/mg/L})\ \text{for an aluminum sulfate coagulant sludge}$$

$$S = Q(2.9\ Fe + SS + P)(8.34\ \text{lb/mil gal/mg/L})\ \text{for a ferric chloride coagulant sludge}$$

where S = solids produced, pounds dry solids per day
Q = water production, million gallons
Al = alum dosage, mg/L (alum as 17.1% solution)
SS = suspended solids concentration, mg/L
Fe = ferric chloride dosage, mg/L (dosage as Fe)
P = polymer dosage, mg/L

A similar equation has been developed for solids produced from a lime-softening plant:

$$S = Q(2.0\ Ca + 2.6\ Mg)(8.34\ \text{lb/mil gal/mg/L})$$

where Ca = calcium hardness removed as $CaCO_3$
Mg = magnesium hardness removed as $CaCO_3$

Note that factors (0.44Al, 2.9Fe, P) should be added when metal hydroxide coagulants or polymers are used in a lime-softening facility.

A series of issues must be addressed in estimating solids production for the preliminary design of a facility:

1. Historical trends of source water suspended solids loading must be developed to determine the average day, maximum day, and peak hour loadings. These values are typically used to evaluate the following design issues:
   a. Average day: used in analyzing annual solids disposal fees, land availability for open air dewatering and drying processes.
   b. Maximum day: used to size thickening, dewatering, and drying processes.
   c. Peak hour: used for the sizing of piping and pumping.
2. If equipment such as centrifuges or belt filter presses are sized based on maximum day loadings, the residuals handling systems must be designed with the capacity to absorb the difference between peak hour and maximum day loading. This may be

done by incorporating equalization basins into the design. An alternative approach is to consider the ability of a given process to handle short-term spikes in the solids loading rate.
3. Most water treatment facilities record solids loadings in terms of turbidity (ntu) rather than suspended solids. A sampling program should be established to define a statistical correlation between turbidity and suspended solids.
4. Estimating source water suspended solids loadings is inherently more difficult for new water treatment plants as opposed to the design of residuals handling facilities for existing plants. The following options are available for estimating loadings for a new plant:
   a. Use historical records for a water treatment plant with similar processes and a comparable water source.
   b. Develop a water quality database through a program of sampling, jar testing, filtration, and analysis. Care must be taken with this program in that a short-term program represents conditions for only one period in time and may not represent the long-term water quality for the site.

Determining Equipment Sizing. Water treatment plant residuals handling systems vary from simple to complex. The mass balance provides the starting point for the design process. The next step is generating a schematic based on refined flow rates and individual equipment loadings.

Issues that must be determined at this preliminary stage in design are the number of anticipated shifts, coordination with main plant operations, and provision for redundant process trains. Projected equipment suppliers also play a key role in this design phase as specific equipment capabilities and requirements are identified.

The open-air dewatering and drying system is one of the most common forms of residuals handling processes. Although the process approach and system equipment are relatively simple, the actual design depends on local climatological conditions, regulatory requirements, and projected use.

Contingency Planning. Contingency planning design issues that should be incorporated into a water treatment plant residuals handling facility are listed in Table 17.4, along with some potential design resolutions.

Backup systems and storage spaces are key contingency planning items often incorporated into residuals handling facility designs. For example, for a facility that uses

TABLE 17.4 Residual Handling Facility Contingency Planning Issues

Potential problems	Potential resolutions
Equipment breakdowns	Provide redundant units and piping. Allow for solids storage during repair. Base design on single shift; second or third shift can be used on emergency basis.
Solids disposal disruption	Plan for a secondary disposal approach. Allow for solids storage during disruption.
Solids production greater than expected	Provide for safety factors in process design. Allow for solids storage during peak periods. Base design on single shift; second or third shift can be used on emergency basis.

centrifuges as the primary mode of dewatering, solar drying beds may also included in the facility for the following purposes:

- To handle peak solids loadings the centrifuges may not be able to effectively treat
- To serve as a dewatering process in the event of a centrifuge failure
- To provide a dewatered solids storage area in case off-site solids hauling is disrupted

Residuals Disposal Unit Process Selection

The decision to install or modify solids handling processes at a water treatment plant can result from a variety of forces ranging from state or local regulation of waste stream discharge to minimizing operating costs. In any event, the decision is accompanied by a selection procedure to identify the residuals handling processes appropriate for the plant.

Selecting residuals handling processes is a more complicated task than selecting traditional liquid stream water treatment processes for the following reasons: there is much less existing operating experience with residuals handling processes to use as a basis of comparison, and residuals handling processes are more difficult to test with a procedure such as a jar test or a pilot filter because of the interdependence of solids handling with other treatment processes.

The following selection criteria are typically used in screening and selecting residuals handling processes:

- Discharge limitations and the effective operating range of the residuals handling process
- Similar operating experience with unit process
- Bench and pilot scale testing of unit processes
- Environmental impact of unit process
- Economics

Environmental Impacts. Environmental impacts have become a standard criterion in developing desired alternatives for any public facility. However, environmental criteria can be frustrating as a basis for comparison because of their subjectivity. The environmental impact of a water treatment plant residuals handling process varies greatly depending on whether the viewpoint is that of an operator, a neighbor, or a migrating egret. However, environmental criteria are typically part of the selection process. Some environmental criteria are described below:

- Effectiveness in meeting discharge requirements. The ultimate criterion for selecting a process or a combination of processes is whether the process will reliably and consistently meet the regulatory discharge requirements. This depends on many factors, including the sizing of the application and process loading conditions.
- The effect on groundwater quality. Any process with the possibility of contaminating groundwater is a possible culprit in this category. Unlined lagoons and drying beds are typically of greatest concern.
- Creation of noise. This criterion is a concern to plant employees and site neighbors. Noise is a typical concern for thickening and dewatering equipment using mechanical equipment. Open air processes may also involve noise in the form of front-end loaders or other devices used to move and remove solids. Off-site transport of solids is another

solids concern. Mitigation techniques include supplying hearing safety equipment for workers, enclosing the operation in acoustically sealed buildings, and specifying operating hours for front-end loaders and other vehicles.
- Creation of odors. Many operators emphatically state that there are no odors associated with water treatment residuals, and in most cases these individuals are correct. However, the possibility always exists for odors with open-air processes and sometimes even with mechanical systems. The traditional technique for odor mitigation is an operational change in the loading and removal of solids from process units. Chemicals such as caustic and chlorine may have to be added to stabilize and destroy odor-causing organisms in the sludge. Odor control units may also be required for systems installed indoors.
- Energy use. Energy use is a function of both direct usage by equipment and the cost of removing solids from the site. Mechanical dewatering techniques involve the highest energy consumption, although there is a wide variation among different processes. Different applications of processes can also affect the percent solids concentration of the residual before disposal. A higher concentration translates to a lower volume for disposal. This lower volume then relates to less transportation energy.
- Attraction of insects and other pests. Insects are a potential concern with open-air processes. Submerged lagoons or gravity thickeners may serve as breeding grounds for mosquitoes and other flying insects. Flies and gnats may be a concern with drying residuals in the open air.
- Impact on the neighborhood. This broad category encompasses many other environmental concerns that result in the lowering of local property values.
- Creation of air pollution. Air pollution is a more typical concern in evaluating wastewater processes. Concerns about stripping volatile contaminants to the air with water treatment systems are minimal. Possible concerns may arise from the potential use of gas- or diesel-driven engines with mechanical equipment.
- Space requirements. Many plant sites do not have the benefit of a large property area or the possibility of expansion onto adjacent vacant land. Limited space typically drives process selection away from open-air processes to mechanical processes.
- Employee and public safety. Mechanical handling processes typically involve the highest safety concern of potential employee accidents. All processes involve some potential hazard, ranging from suffering heatstroke while working in a solar drying bed to drowning in a thickener.

One issue to note in comparing environmental criteria is the interrelationship of these criteria with both construction and operations and maintenance costs. For example, groundwater quality concerns associated with lagoons can be resolved by installing a relatively impermeable liner beneath the lagoons. Thus the liner mitigates the environmental concern at a significant increase in construction cost. This should be considered in the analysis of different residuals handling approaches.

Bench Tests. Bench and pilot tests serve two functions in selecting residuals handling processes: they provide additional data for the sizing of the full-scale equipment, and they provide an indication of the process performance (in other words, what percent solids can be achieved).

Bench and pilot testing should be considered after the number of alternative residual handling schemes have been screened to a few options requiring further study. Testing may not be needed if there is similar experience with process at a comparable facility, such as a plant of similar size with approximately the same source water quality and unit processes.

The following bench scale tests are commonly used for thickening and dewatering:

- Thickening: settling tests, flotation test, capillary suction time test
- Dewatering: time-to-filter test, filter leaf test, capillary suction time test, settling tests

Bench testing does not provide any direct correlation to the performance of a given process. Bench tests are typically used in conjunction with full-scale operation to develop correlation, for instance, between the results of capillary suction time tests and the performance of centrifuges. The application of the bench test is then to analyze the effect of different sludge conditioners. Bench tests may have application to selecting a residuals handling process if the bench tests can be correlated to process performance based on previous testing. Equipment suppliers should be asked for the availability of past bench test comparisons.

Pilot Testing. Pilot testing involves the use of test equipment similar enough in size to the full-scale operating equipment that testing results can be directly compared between pilot test and full-scale operation. This testing may involve constructing a scaled-down version of a process basin such as gravity thickener or a sand drying bed. Suppliers of mechanical thickening and dewatering equipment typically have test units available for rent or loan.

The use of bench and pilot tests has many pitfalls. Testing must be as representative as possible of real-world conditions. Testing procedure must simulate all processes involved with an alternative. This is a particular concern with thickening and dewatering, where incoming solids must represent the solids concentrated and transported from the previous process. Even when the bench and pilot testing successfully model process performance, the results may not reveal operational problems that are part of the process.

The duration of bench and pilot testing is of particular concern. Year-round operation may involve significant changes in source water quality that may greatly affect the operation of the residuals handling process. One week's worth of pilot testing may not reflect this impact. It is generally not feasible to perform year-round pilot testing. The use of short-term pilot testing to correlate results with bench tests may allow long-term testing on a lesser scale.

Analysis of pilot testing results should consider the limitations of the test setup and protocol. If only limited testing was performed, a lower emphasis should be placed on the testing results.

Additional Information Sources

Regulatory restrictions have placed an increased focus on the design of water treatment plant residuals handling facilities in the 1980s and 1990s. This design field is complicated by regulations, site neighbor concerns, and a growing number of technical options for treatment and disposal. The water treatment plant designer also has a limited number of reference materials to utilize.

Two texts are particularly suggested for a more detailed coverage of this subject. The handbook *Water Treatment Plant Waste Management* (AWWARF, 1987) provides a comprehensive overview of residuals management issues. It also presents a series of cost curves that are useful in the preliminary screening of treatment and disposal alternatives.

The design procedures presented in this chapter originate from *Technology Transfer Handbook: Management of Water Treatment Plant Residuals* (ASCE/AWWA/EPA, 1997). This text provides a broad background into the regulatory requirements and solids characteristics associated with residuals management. This test provides additional data on cost estimating residuals handling systems.

BIBLIOGRAPHY

American Water Works Association. *Water Quality and Treatment.* 3rd ed. New York: McGraw-Hill, 1971.

American Water Works Association. *Water Treatment Plant Waste Committee Report.* Unpublished data, December 1971.

American Water Works Association Committee. "Water Treatment Plant Sludges—an Update of the State of the Art." *Journal AWWA* 70(9):498, 1978.

American Water Works Association Research Foundation. *Disposal of Wastes from Water Treatment Plants.* Washington D.C.: Federal Water Pollution Control Administration, publication PB 186157. 1969.

ASCE/AWWA/EPA. *Technology Transfer Handbook: Management of Water Treatment Plant Residuals.* New York: American Society of Civil Engineers, 1997.

Ashcroft, Craig, et al. *Modifications to Existing Water Recovery Facilities for Enhanced Removal of Giardia and Cryptosporidium.* Proceedings WEF and AWWA Joint Residual Conference. Alexandria, Virg.: Water Environment Federation, 1997.

AWWA, ASCE, CSSE. *Water Treatment Plant Design.* New York: McGraw-Hill, 1969.

AWWA Committee Report. "Water Treatment Plant Sludges—an Update of the State of the Art: Part 2." *Journal AWWA* 70(10):548, 1978.

AWWARF. *Water Treatment Plant Waste Management.* Denver, Colo.: American Water Works Assn., 1987.

Benjes, H. H., Jr. "Treatment of Overflows from Sanitary Sewers." Presented at the 9th Texas WPCA conference, Houston, Tex., July 1970.

Bishop, S. L. "Alternate Processes for Treatment of Water Plant Wastes." *Journal AWWA* 70(9):503, 1978.

Black, A. P., and C. G. Thompson. *Plant Scale Studies of the Magnesium Carbonate Water Treatment Process.* Publication EPA-660/2-75-006. Cincinnati, Ohio: U.S. Environmental Protection Agency, 1975.

Black and Veatch, Engineers-Architects. *Report on Water Treatment Plant Waste Disposal.* Wichita, Kan.: Black and Veatch, 1969.

Burris, M. A., et al. "Coagulation Sludge Disposal Studies for a Surface Water Supply." *Journal AWWA* 68(5):247, 1976.

Cleasby, J. "Iron and Manganese Removal—a Case Study." *Journal AWWA* 67(3):147, 1975.

Cornwell, D. A. "An Overview of Liquid Ion Exchange with Emphasis on Alum Recovery." *Journal AWWA* 71(12):741, 1979.

Cornwell, D. A., and R. M. Lemunyon. "Feasibility Studies on Liquid Ion Exchange for Alum Recovery from Water Treatment Plant Sludges." *Journal AWWA* 72(6):64, 1980.

Cornwell, D. A., and J. A. Susan. "Characteristics of Acid Treated Alum Sludges." *Journal AWWA* 71(10):604, 1979.

Culp, R. L., and W. I. Wilson. "Is Alum Sludge Advantageous in Wastewater Treatment?" *Water and Waste Engineering* 16(79):16, 1979.

Dean, J. B. "Disposal of Wastes from Filter Plants and Coagulation Basins." *Journal AWWA* 45(11):1229, 1953.

Dlouhy, P. E., and A. P. Hager. "Vacuum Filtration Solves Problems of Water Softening Sludge." *Water and Waste Engineering* 5(68), 1968.

Foster, W. S. "Get the Water Out of Alum Sludge." *American City and County* 90(75), 1975.

Gruinger, R. M. "Disposal of Waste Alum Sludge from Water Treatment Plants." *Journal WPCF* 47(75):535, 1975.

Gumerman, R. C., R. L. Culp, and S. P. Hansen. *Estimating Water Treatment Plant Costs.* Publication EPA 600/2-79-162. U.S. Environmental Protection Agency, 1979.

Hagstrom, L. G., and N. A. Mignone. "Centrifugal Sludge Dewatering Systems Can Handle Alum Sludge." *Water and Sewer Works* 125(5):54, 1978.

J.M. Montgomery, Consulting Engineers, Inc. Water *Treatment Principles and Design.* New York: John Wiley and Sons, 1985.

Knocke, R. W. "Thickening and Conditioning of Chemical Sludges." *Proceedings ASCE Environmental Engineering Conference,* San Francisco, Calif., July 1978.

Krasaukas, J. W. "Review of Sludge Disposal Practices." *Journal AWWA* 61(5):225, 1969.

Lawrence, Charles. "Lime Soda Sludge Recirculation Experiments at Vandenberg Air Force Base." *Journal AWWA* 55(2):177, 1963.

Mickey, M. *Membrane Concentrate Disposal Issues.* Proceedings of the 4th Joint WEF and AWWA Conference on Biosolids and Residuals Management. Alexandria, Virg.: Water Environment Federation, 1993.

Pigeon, P. E., et al. "Recovery and Reuse of Iron Coagulants in Water Treatment." *Journal AWWA* 10(7):397, 1978.

Reynolds, T. D. *Unit Operations and Processes in Environmental Engineering.* Belmont, Calif.: Wadsworth, 1982.

Singer, P. C. "Softener Sludge Disposal—What's Best." *Water and Waste Engineering* 11(12):25, 1974.

Taflin, C. O., et al. "Minneapolis Keeps on Trucking." *Water and Waste Engineering* 12(5):24, 1975.

U.S. Environmental Protection Agency. *Process Design Manual: Sludge Treatment and Disposal.* Cincinnati, Ohio: U.S. Environmental Protection Agency, 1979.

Vandermeyden, C. *Design and Operation of Nonmechanical Dewatering Systems.* Proceedings of the 4th Joint WEF and AWWA Conference on Biosolids and Residuals Management. Alexandria, Virg.: Water Environment Federation, 1993.

"Water Plant Waste Treatment." *American City and County* 94(3), 1979.

WEF/ASCE. *Design of Municipal Wastewater Treatment Plants.* Alexandria, Virg.: Water Environment Federation, 1992.

Westerhoff, G. P., and D. A. Cornwell. "A New Approach to Alum Recovery." *Journal AWWA* 70(12):709, 1978.

Wilhelm, J. H., and C. E. Silverblatt. "Freeze Treatment of Alum Sludge." *Journal AWWA* 68(6):312, 1976.

Williamson, J. Jr. "Something New in Sewage Treatment." *Water and Sewage Works* 96(49):159, 1949.

CHAPTER 18
ARCHITECTURAL DESIGN

Successful treatment plant design must address an array of issues, many outside of the traditional area of water quality engineering. The facility must fulfill its primary objective, which is to provide a quality product, but it is also important how the facility is perceived by those who work in it, visit it, live near it, and see it every day. It is often the architectural and landscape design forms, the visual portions of the project, that mold public perception.

Careful consideration must be given during the planning and design process in order to address the needs of plant personnel for a well thought out, conveniently arranged, pleasant working environment. These features will engender positive employee response and provide an efficient and functional workforce.

Construction systems and materials must also be selected that will combine to create a facility that is cost-effective, as well as resource and energy efficient.

THE ROLE OF THE ARCHITECT IN WATER TREATMENT PLANT DESIGN

The design team for a water treatment plant design project generally includes an architect. In a team, the whole is more than the some of its parts. Associated with the benefits inherent in the early involvement of the architect is the additional benefit of team synergy. Team synergy occurs when professionals focusing on the same problem, but approaching it from different perspectives, offer solutions that spark additional ideas that could be realized only with the cooperation of all team members. It is this kind of complementary and cooperative effort that will produce the best work and that should be cultivated. Only when all design professionals collaborate does the owner truly receive the highest-quality service. It is through such collaborations that the best water treatment engineering is combined with the highest quality building design concepts.

Aesthetics and Contexturalism

The architect is trained to view the buildings and environment of a project to establish an aesthetic approach for the design of the facility. The architect will often point out the need to "visually relate" or "respect the scale of" the proposed buildings with the communities that are adjacent to the plant. Even where the location of the facility is relatively isolated, the design should be visually pleasing, because, as major works of substantial expense, they are reflections of community pride and achievement. The architect will balance these considerations against available design and construction budgets and suggest construction materials appropriate to each building. The architect is also trained to propose the proper design of spaces for "people functions" and relate the various plant functions to each other and to the site.

The architect should preferably begin meeting with the design team during the time of facility planning and will need to determine factors such as the number of personnel, scope of laboratory services, and extent of on-site maintenance that is being planned. The architect should particularly be involved in coordinating the placement of external building elements that affect the architectural design, such as louvers, light fixtures, and stack penetrations. If these decisions are to be made with the participation of the owner, it is generally best to meet as a team with the owner so that all perspectives are considered when important decisions are made.

Codes and Standards

It is the responsibility of the architect to investigate, understand, and implement applicable provisions of building construction codes, design standards, and regulations such as the American National Standards Institute, National Fire Protection Association, and Americans with Disabilities Act. Many trade organization standards define the level of acceptable workmanship and materials for construction systems. A list of the sources of many design standards has been included at the end of this chapter.

State and local building codes often reference some of these standards by name, and, in such cases, the standard becomes part of the construction code. If a particular standard is not referenced, it may mean that the local code has different or more specific requirements, and those requirements must be followed. If a construction code is mute on a particular topic and there is no higher governing code, national standards can be used as the basis for prudent design.

Standards specific to water treatment process design should remain the responsibility of the engineers, but these standards often impose construction and fire protection requirements of which the architect will need to be made aware. Codes and national standards represent minimum standards, and the design team should always consider improvements that supersede code requirements if they will improve safety, lower operating and maintenance costs, or provide the owner with a more functional design.

It is important to determine at the beginning of the design process what the owner's code compliance responsibilities are. Some large municipal clients may be able to work with state and local planning, zoning, and building code officials in shaping code requirements to the project needs. The design team should also involve the local code enforcement officer early in the project. Valuable insight on local requirements can often be provided to the design team, which expedites the entire design process. Knowing exactly what is required may also save on costly design changes that would be caused if uncovered toward the completion of design or, worse, during construction.

The architect must also be familiar with the requirements of local architectural review boards, planning and zoning departments, and any possible effects of other governing authorities on the design of the project.

Environmental Design

One important duty of the architect is to attempt to minimize the impact that a new facility has on the natural environment. The architect typically has little control over site selection and process considerations but can play a critical role in overcoming negative public perception of a new project. This can often be accomplished through inventive architectural design and site planning. The public must also be shown that it is necessary to see the total environmental benefit of the project, not just the facilities alone.

Such phrases as "green architecture" and "sustainable architecture" are now making their way into design vocabulary to indicate resource- and energy-efficient buildings. This philosophy seeks to take advantage of regional construction opportunities and methodologies, site orientation, natural daylighting, and resource-efficient recycled materials in order to save energy, conserve resources, and improve the quality of the indoor environment.

Sealing of Documents

Many states now require that certain building types must involve the services of a licensed architect, and the completed contract documents for applicable portions of the work must bear the architect's seal and signature. In order to provide this, most jurisdictions require that the work must be done under the architect's personal direction. Many states restrict the corporate practice of architecture regardless of an individual licensee's status. The issue of who may seal what types of documents should be established at the outset of a project.

Consideration of the Site

Architectural form is greatly affected by regional differences in area topography, climate, and individual site details. Plant site design should include adequate space for visitor, employee, and delivery parking that is accessed from a clearly defined traffic pattern that splits off traffic according to its type and destination. It is preferable that employee parking be isolated from the main entrance and be near an employee entrance that is close to lockers and other personnel facilities. This will reduce traffic near administrative and public areas.

If possible, administrative areas should be nearest the main entrance, not only for the convenience of visitors, but for control of visitor access. The public areas must comply with the requirements of the Americans with Disabilities Act, and, because compliance will require additional space, this should be considered early in the site planning process.

Landscaping offers a powerful mechanism with which to shape a site and goes far beyond obvious aesthetic benefits. Careful use of deciduous and evergreen plant material can be used to bring clarity to the service road network by highlighting directional approaches while concealing less desirable or confusing views. Landscape material also gives scale to structures by integrating them with the ground plane and by softening visually hard edges and long expanses of blank walls. Plant material can make important contributions to energy conservation. Deciduous plant material allows winter sun to warm the building while providing shade from the heat of summer sun, thus reducing the energy consumption of structures.

In colder climates, evergreen windbreaks and earth berms can be used to buffer buildings from winter winds. A similar approach can be used in warm climates to channel breezes, which will provide natural cooling.

Architectural Programming.

The client often does not fully know, or has not thought about, many of the factors that will affect internal architectural design and will need to be guided by the consultant. Some of the questions that can be asked are:

- What will be the staffing of the plant?
- What special administration or billing needs are anticipated?
- What accommodations should be provided for visitors?
- How much and what kind of laboratory analysis will be performed at the facility?
- What degree of employee training is anticipated?
- How many offices and conference rooms should be provided?
- How extensive should the maintenance facilities be?
- Should vehicle storage be provided?

The answers to these and other questions will uncover important planning considerations, which then can be discussed in more detail.

It is also important to know if the client is composed of more than one bureau or responsible entity. Does your immediate client contact need to seek input from other departments in order to determine planning choices for the plant? Once the full scope of facility programming is understood and agreed on, it is important to establish how each programmatic element interrelates.

The information gathered from the client can next be presented as "adjacency and circulation diagrams" that show the relationship between the various work and storage areas in the building. On more complex projects, it may be necessary to present secondary relationships between spaces, which may affect their locations within buildings. These areas can then also be related to site orientation, topography, views, and other important external planning criteria because these factors will also influence the location and architecture of the buildings. The information that has been gathered will also determine how compact or open the plan of buildings should be.

Planning for Persons with Physical Disabilities

The Americans with Disabilities Act (ADA) of 1990 represents landmark civil rights legislation that is the culmination of a series of related laws that began with the Civil Rights Act of 1964. The law essentially requires that the rights of individuals with life-limiting handicaps be afforded the same opportunities of access to places and jobs that able-bodied individuals enjoy. Although the law itself is complicated and is still being interpreted by the courts, the important point for designers to focus on is the need to provide adequate access for the public and employees to certain spaces within the project. The designers should remember to check with both the requirements of the state building code and the federal ADA accessibility guidelines to determine the requirements for the project, because the more restrictive of the two standards will apply.

In general, it is the owner who is responsible for complying with the law, so it is important that the engineer and architect review requirements with the owner. It is also important to have the interpretation arrived at by the design firm and owner in writing. In any case, it may be desirable for the owner to hire qualified physically handicapped individuals at some time in the future, and it is much less expensive to accomodate the necessary facilities initially than to add them later. Typical accommodations include elements such as ramps, elevators, accessible toilets, doors with adequate latch side clearances, hallways wide enough for wheelchairs to pass, and special design guidelines for stairs, elevators, and various equipment.

FACILITIES DESIGN

The architectural design of water treatment plant facilities generally involves consideration of administrative areas, facilities for the plant staff, laboratory spaces, and maintenance facilities.

Administrative Facilities

The design of administrative facilities must include consideration not only of offices, but areas that are available to the public and areas for conferences and training.

Lobby and Reception Area. Most plants will need a lobby or reception area for use on occasions when the public visits the plant. Heightened public interest in both public health and environmental issues is creating increasing requests for visits by groups of children, civic groups, and individuals, especially in larger facilities. These activities require more attention to public spaces and may increase the required lobby or reception area considerably beyond that which would otherwise be necessary.

Office Area. Office space for a treatment plant may vary from only a superintendent's or operator's desk to many offices for a large number of employees. Any planning for offices must include space for plant records. These records include contract documents, drawings, operations and maintenance manuals, and operating records, including recorded charts and logs. In larger treatment plants, it may be advantageous to provide a records storage room or technical library and training room dedicated to this function. It is not recommended that a lunch room serve as a library and training facility.

Incorporating a technical library as part of a public room or conference room often will work well. If a room large enough to serve as a lecture room is provided, it may be equipped with a projection screen, visual aid panels, a lectern, and demonstration tables. Consideration should also be given to providing an assistive listening system if an audio system is installed in the space in response to the requirements of ADA accessibility guidelines. Figure 18.1 shows a plan for a water treatment plant administration building in which lobby space is used to provide a small public exhibit area where visitors can learn how the plant functions and view some typical equipment used in the processing of the water.

Personnel Facilities

The design of new water treatment plants gives increased attention to making the facilities comfortable for employees, and in particular, consideration must be given to providing facilities for women as well as men.

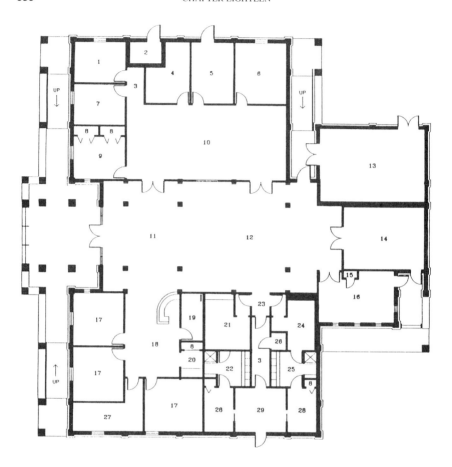

FIGURE 18.1 Administration building.

Planning for Change. Employment at water treatment facilities offers opportunities for job descriptions to be filled by either male or female applicants, and because of this, it is important to consider planning approaches that permit flexibility in accommodating the shifting proportions of male and female workers. Facilities that must be made available to workers of either sex include water closets, lavatories, lockers, shower facilities, drying areas, and laundry rooms. Unfortunately, current practice is to simply double the facilities that are planned in order to accomodate the shifting number of male and female employees. Even though most of one of the modules may not be used at present, the situation could change in the future. One alternative is to locate relatively fixed facilities such as toilet rooms and showers at opposite ends of the area that will be dedicated to lockers. The area between them could be allowed to shift in use through the relocation of partition walls if required in the future. In smaller plants it may be possible to make only a reasonable assumption on future male/female employment.

Locker Rooms. Many water treatment plants are designed to assign two or sometimes three lockers to each employee. A smaller locker is used for street clothes, and a larger locker is provided for work clothes. Some employees also require an equipment locker for assigned tools, boots, and rainwear.

A food locker for each employee is also provided in some treatment plants. These lockers are usually located in the lunchroom and are used to store personal food items. Handicap access to locker rooms is required by ADA. The type of locking mechanisms to use on lockers should be discussed with the client. Figure 18.2 shows a modular personnel area for a small treatment plant. The layout accommodates a staff of nine males and nine females and provides each employee with three lockers.

Lunchroom. Providing a separate lunchroom facility in a treatment plant contributes to the hygiene and health of plant employees by providing a central location for food storage and preparation, by fostering proper wash-up procedures, and by discouraging consumption of food in process areas. Providing a separate lunchroom space also serves to minimize disruption of process and administrative activities while employees are eating.

The lunchroom space should preferably be located near an accessible route, away from administrative activities and conference rooms, and near the toilet and locker facilities so employees will find it easy to wash before eating. Modular seating and tables should be provided so that they can be reconfigured for various activities, and durable kitchen equipment should be installed for employee use. The room should be open and light filled and, preferably, with good views out to the building exterior or process areas of the plant.

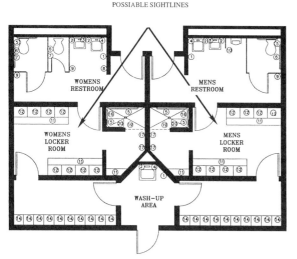

FIGURE 18.2 Locker room, toilet, and shower room plan.

The architect should also acknowledge that a growing number of states and local governments are instituting regulations governing smoking in areas that are not private offices, and this needs to be addressed in the treatment plant design. Some plants provide special smoking areas with separate ventilation systems; otherwise employees may have to leave the building to smoke. Signs prohibiting or limiting smoking as required by law or directed by the client should be furnished as part of the building design.

Laboratories

Since passage of the Safe Drinking Water Act, water treatment regulations, quality standards, and monitoring requirements have become increasingly strict, which, in turn, has caused the laboratory requirements for a plant to continuously escalate to perform more and increasingly sophisticated analyses.

Laboratory Planning Standards. In terms of total regulations, laboratories are the most regulated portion of the water treatment plant. They are usually regulated by the state's Department of Health and are regularly audited for strict compliance. They also must conform to Occupational Safety and Health Administration (OSHA) standards and may occasionally be audited by OSHA. Laboratories are also regulated by many states for compliance with clean air standards. Many cities and local jurisdictions have also developed their own regulations governing laboratories.

The National Fire Prevention Association (NFPA) Standard 45, *Fire Protection for Laboratories Using Chemicals,* and other standards referenced as part of this document are used by most local fire marshals and inspectors as a basis for their own standards. There are also other NFPA standards that most states or local authorities refer to in determining compliance. Finally there are the Environmental Protection Agency Testing Methodology Standards, which describe required testing, equipment, and facilities that laboratory personnel must use in order to maintain their laboratory certification. The designer must be thoroughly familiar with all applicable regulations and standards before beginning design.

Many larger laboratories must be separated into organics, inorganics, and bacteriological testing areas. These distinct areas must be kept environmentally isolated from each other, or cross-contamination of test results can occur. Even the use of the same laboratory coat by a worker who performs tests in both the organics and inorganics testing areas of the lab can affect results, because the solvents used in one area are the very chemicals being tested for in another.

Although these areas must be separated, it is important for the safety of laboratory personnel that visual connection be provided between the laboratory rooms by the use of glazed safety windows in walls and large vision panels in doors where they are allowed by the building code.

Safety must be a prime consideration in the laboratory layout. First aid kits, fire extinguishers, fire blankets, eye washes, and emergency showers are usually required by codes and governing authorities. Even if they are not mandated by code, they should be provided as a matter of good practice.

If possible, fume hoods should be located farthest from room entry doors, which will allow employees to exit away from an accident that occurs in a fume hood. At the same time, it also permits fresh air to be introduced at the door side of the room so that occupants will be moving toward the source of fresh air as they exit the room. Contaminants that have temporarily spilled into the room will then eventually be exhausted through the fume hood. If the laboratory is using sophisticated equipment, such as gas chromatographs and atomic absorption units, it is wise to include an automatic fire suppression system to protect equipment from damage by fire.

Laboratory chemicals should be stored in approved cabinets and vented to fume hoods or in an approved manner. Solvents and flammables should be stored in specially designed rooms complying with NFPA 30, *Flammable and Combustible Liquids Code* (or other standard referenced by local governing authorities). In some situations, flammable liquids may be stored in individual cabinets in compliance with NFPA 30.

Treatment plant laboratories should be provided with their own independent air supply and exhaust systems. Many codes require this, but in the interest of safety, it should never be overlooked.

Estimating the Laboratory Requirements. The principal factors in estimating needs for the treatment plant laboratory are the type of testing to be done, the quantity of tests to be run each day, and the type of equipment that will be used within the foreseeable future. Establishing the requirements for these will help determine some of the following items that must be furnished:

- Number and type of fume hoods
- Size of chemical storage rooms
- Types and quantities of gas cylinder to be stored
- Size of lab personnel support areas
- Quantity and types of bench space and laboratory furniture
- Electrical, gas, vacuum, and compressed air systems that will be required
- HVAC requirements for the laboratories
- Need for a deionized water system

Understanding these parameters will also help clarify the choice of finish materials for walls, ceilings, and floors and establish the importance of various safety issues as they affect the choice of partitions, doors, and glazing materials. It is best, if possible, to interview the owner's laboratory staff and learn firsthand the kind of testing and equipment that must be provided. Establishing this at the early phases of a project will lead to far fewer changes and a much happier client when the job is completed.

A number of factors affect the laboratory staffing requirements, such as:

- Type of laboratory equipment to be used
- Extent of automatic monitoring and recording instruments
- Amount of work that will be outsourced to other laboratories
- Level of training of laboratory personnel

The staffing needs should be projected as far into the future as possible. There should be enough built-in flexibility in the design of the laboratory to accommodate a reasonable amount of changes in operation caused by regulatory changes, increase in the size of the plant or water system, and more extensive testing due to technological changes.

Estimating the Necessary Floor Area Once the annual number of person-hours has been estimated, the number of persons involved can be determined by considering the actual on-the-job hours per person. A guideline provided by OSHA suggests 2.5 ft (0.75 m) of fume hood per person working in the laboratory. Other sources suggest 4 ft (1.2 m) of workbench per person. However, planning rules of thumb need to be checked once additional staffing and equipment information is available.

A methodology has been developed to derive laboratory area requirements from known bench space requirements. A completely efficient room would provide bench space along all four walls without any interruptions for circulation or openings in walls. An extremely large room can generally be laid out to be the most efficient. Smaller rooms are less efficient because the same amount of space dedicated to circulation and openings makes up a larger percentage of the total area available.

Some rules of thumb use 200 to 300 ft^2 (19 to 29 m^2) per staff member to determine the area of the laboratory, with bench-top working surface assumed to be 30% to 40% of the room area. Sinks should be provided at locations that are convenient but out of main circulation paths, each serving about 20 ft of bench-top work area. Sinks should not be adjacent to instruments.

A modular layout of electrical outlets should be established along the tops of all work surfaces in the laboratory. Some types of bench-top equipment must be hardwired, and some equipment also requires power conditioning. If some equipment must be furnished with an uninterruptable power source, space for battery units may have to be considered as part of the laboratory design.

Most laboratory areas should also be furnished with compressed air, burner gas, and vacuum outlets. The compressors and pumps necessary for these systems should be located outside of the lab to avoid vibration. Most laboratories also require quantities of deionized water. Systems for producing relatively small amounts of deionized water can be located within the lab, or central systems can be provided for labs using larger quantities.

Equipment and supplies in addition to those required by the state should be selected based on the frequency of tests, the sophistication of the unit processes, cost-effectiveness, and the desire to achieve optimum plant efficiency. The selection of major equipment items can be made from the equipment requirements suggested in *Standard Methods for the Examination of Water and Wastewater* and AWWA Manual M12, *Simplified Procedures for Water Examination*. Figure 18.3 shows a typical laboratory layout for a small plant.

Laboratory Operation Considerations. If the laboratory under design is to perform analyses from other treatment plants, it will need to be provided with a way of receiving and properly storing samples that are brought in, as well as an orderly arrangement for tracking and preparing the samples for analysis. This is usually done by providing a separate sample receiving room with refrigerators, bench tops, and fume hoods, as well as a generously sized storeroom. Larger laboratories now use a computer program designated LIMS (laboratory information management system) to record, track, and report on the analyses of samples that are processed by the laboratory.

Laboratory lighting must receive particularly careful consideration. Glare can render computer screens unreadable, and extreme variations of light and shadow can be annoying and disorienting. Opening large areas of the laboratory to northerly light will provide good, uniform lighting during daylight hours, but in cold climates it will cause unnecessary energy loss and possibly even discomfort to personnel. Methods of overhead skylighting with shade control or a south-facing clerestory with light shelves to distribute light along the plane of the ceiling should be considered as part of a well-designed laboratory.

Laboratory balances are highly sensitive to vibration, so balance tables are commonly placed against a bearing wall, next to a column, or at a location that is least likely to pick up vibrations from the heavy equipment operating in the plant. If there is no placement within the laboratory that is satisfactory, various types of vibration-isolated floor construction are available and should be considered.

Adequate space should be provided between back-to-back work areas. The minimum distance is usually considered to be 4 ft (1.2 m), and 5 ft (1.5 m) is generally better. The requirements of fume hood dimensions will usually determine clear ceiling heights in laboratory spaces and cause them to be higher than code-required minimums.

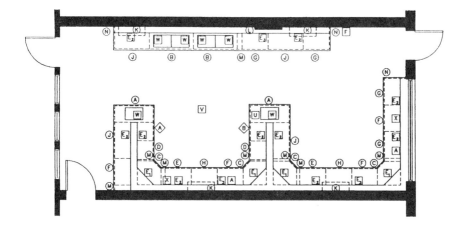

FIGURE 18.3 Process laboratory.

Maintenance Facilities

When designing a new water treatment plant, careful consideration should be given to providing the necessary repair shops for the types of work that will be performed by the plant staff, storage rooms for tools and spare parts, and basic tools for the shops.

Repair Shops. The size, number, and function of repair shops are, to some degree, related to the size of the plant. It is also important to determine early in the design stage whether or not the plant will serve as a maintenance or materials receiving hub for other plants. Some clients choose to use outside facilities at other treatment plants or have access to a central maintenance complex off-site, and this will be equally important to establish at an early point in the project design.

Some large plants have the maintenance functions located in a separate building to isolate noise, to allow for free movement in handling large equipment, and for truck delivery.

In addition, maintenance facilities often store costly equipment, and security is improved by the provision of a centralized location. Wherever the maintenance shop is located, there should be provision for truck delivery of large equipment. Doors should be made wide enough and high enough to allow equipment to pass in and out of the shop area.

Where plant size and anticipated maintenance workloads warrant, a separate electrical repair shop and a paint shop may be provided. Electrical repair shops require a high degree of cleanliness. Paint solvents are flammable, and most building codes will require strict standards of safety for these spaces.

The type and complexity of work that will occur in the maintenance shop must be determined early in the design. This will, in turn, dictate the type and size of repair machinery that should be provided. Clearances required around shop equipment must also be considered so that long, heavy, or bulky items can be handled and safely worked on. It is important to check all shop design layouts for safe clearance around each piece of shop equipment.

Adequate bench space is necessary in all shops. Benches should be sturdy and selected for heavy-duty service, allowing rough work to occur on top surfaces selected of stain- and scratch-resistant material. Drawers should be installed in workbenches to hold small, frequently used parts and hand tools. Tool storage shelving and storage systems should be selected to permit easy retrieval of tools near the point of work.

Attention should be given to bringing generous amounts of controlled natural lighting into the space, supplemented with task lighting at the equipment. The environment of the shops should be buoyant but not distractive. This can often be achieved by the use of high ceilings and clerestory or sawtooth monitors to bring light deeply into spaces. The use of skylights is less desirable because, as the sun moves across the sky, it will be difficult to control glare and hot spots in the shop and could, during certain times of the day, make the use of some areas impossible or at least unsafe.

One item commonly overlooked in the design of repair shops is the need to use water and to dispose of wastewater. Testing of solution tanks for feed machines, the injectors of chlorinators, or electric mixers commonly results in spilled water. It is valuable for a shop to have a drainage area in a corner or an end of the room, set off by a 4 in. high concrete curb, for use in wet testing equipment. A floor drain should also be installed at the low point of the shop floor so that the entire room can be hosed down if desired. If possible, it is also useful for a shop to have an area that can be used for solvent washing or steam cleaning equipment. Figure 18.4 shows a typical layout for a small maintenance shop.

Maintenance Shop Tools. A wide variety of tools are required for a water treatment plant maintenance shop. Heavy tools typically include drill presses of various types, bench and pedestal grinders, hydraulic presses, a portable hoist, a portable pump, pipe threaders, milling machines, an arbor press, welding rigs, lathes, band saws, and cutoff saws. Light tools should include those for plumbing, automotive, electrical, painting, carpentry, sheet metal, and masonry work. Also included should be grease guns, oilers, and trouble lamps.

For housekeeping, mops, pails, brooms, and vacuum cleaners are needed and will require dedicated storage areas. For outside maintenance, wheelbarrows, rakes, shovels, hoses, lawn sprinklers, and a power lawn mower should be provided and given adequate storage space. In cold climates, snow removal equipment is needed. Safety equipment should include first aid equipment, a resuscitator, and portable breathing equipment.

Storage, Spare Parts, and Security. Spare parts should generally be kept in a separate, secure storage area that can be locked for inventory control. The room should include a section of drawers, bins, cabinets, racks, a large amount of shelf space, and some peg racks

ARCHITECTURAL DESIGN

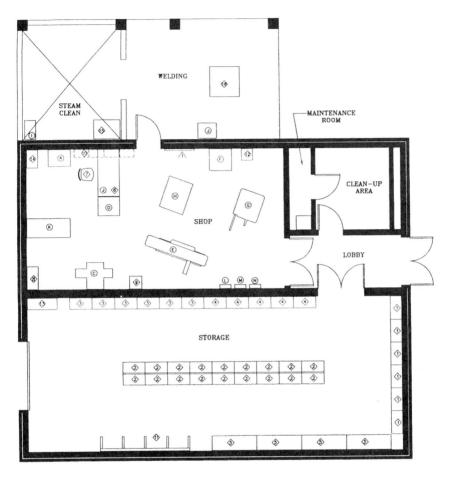

FIGURE 18.4 Shop and storage building.

on which roll material such as gasket stock, screen wire, and hardware cloth may be stored. For ease of inventory control, the spare parts storage area should preferably be located near the maintenance shop.

CONSTRUCTION ALTERNATIVES AND BUILDING MATERIAL SELECTION

There are a great number of alternative building materials available to water treatment plant designers today. The selection of which one will be best for a particular situation must be made on a consensus of the design team and client of what will be durable, economical, and aesthetically pleasing.

Structures over Process Units

The decision of whether to place superstructures over process units generally depends on the climate. In areas where there is minimal danger of freezing, structures such as flocculation and sedimentation basins can be left uncovered, which greatly reduces the capital costs. In addition, the visual impact of the plant is easiest to handle when the area and height of structures are kept to a minimum. In very hot climates it is often best to cover equipment such as filter operating consoles, both to protect the equipment and to make operations more comfortable for plant personnel.

Pre-Engineered Structures

For the smallest and simplest facilities, pre-engineered, or even premanufactured, structures may offer a viable option for reducing capital costs because less detailing of construction systems will be required. Pre-engineered structures are primarily used for warehouses and as enclosures for manufacturing facilities, but they have been used for covering some water treatment facilities.

They are best used when the function and floor plan of the building are straightforward and there is no need to support heavy overhead pipelines, monorails, or cranes. Some manufacturers of pre-engineered structures can accommodate some limited support of overhead loads by altering the design of the structure. These buildings are generally designed to be economical, so they offer little opportunity to accommodate unforeseen future additional loadings.

When selecting a pre-engineered building to house chemical facilities, care must be exercised that there are no adverse effects of the chemicals on the wall panels and structural supports. A review of local building codes should also be made to determine whether there are any restrictions on noncombustible, unprotected structures, based on building area, height, and occupancy type.

Planning for the use of pre-engineered buildings must start early, and it is best to consider only standard systems available from the manufacturers. The architect should keep in mind that, although these buildings are pre-engineered, compliance with building code requirements still must be ensured by the project designer, so a careful and complete building code study should be made before full commitment is made to this direction as the most cost-effective building alternative.

It will often be found that the use of an engineered concrete masonry building will be no more expensive when all costs are evaluated, and perhaps will be more durable and long-lived than a metal building. This, of course, must be determined individually for each project because there are many factors to be considered. Although pre-engineered structures can be erected quickly after they arrive on-site, the project engineer must remember that there is a fabrication lead time that must be considered.

Building Panel Systems

The use of panelized building modules will speed the enclosure of structures and will usually permit erection to continue regardless of weather conditions. There are many panel system options to chose from. The architect may suggest the use of precast concrete, insulated steel or aluminum (either field or factory assembled), translucent fiberglass, or even preassembled panels of thin masonry or ceramic tile applied to strong-backs and metal decking. Advancements in structural silicones and acrylic additives make these preassembled panels relatively lightweight and long lasting. A number of exterior insulation and stucco wall systems are now available that share the advantages of fast erection. When building codes allow their use, and the detailing, specification, and installation supervision of these systems are done carefully, they can be highly effective for construction of water treatment plant facilities.

Environmental Considerations in Design

In all designs of water treatment plant facilities, no matter what the size or budget of the project is, there is an opportunity to build in a sustainable manner. Creating healthy indoor environments requires careful selection of durable, easily maintained, and stable materials that do not continue to off-gas over their life span.

In a world of high energy costs the architect should focus on how well the materials that are selected insulate from hot or cold, as well as how much energy was consumed in the manufacture, transportation, and erection of the materials—their so-called embodied energy. Many governing jurisdictions have adopted one or another of the model energy conservation codes, and all designers should address the requirements. Where does our structure get its energy? Is it renewable? Can we obtain some of the energy we will need to power, light, heat, and cool our structures from alternative energy sources? Some states offer incentive programs through one or another of their agencies or through the National Energy Conservation Act of 1992 and the Department of Energy, and the designer needs to find out if monetary or other incentives will benefit the project.

Materials that close the recycling loop or that contribute to environmental cleanup should be explored. Many construction materials are now being offered that use all or some recycled materials in their manufacture. Some use currently unrecycled wastes and turn them into usable products. The designer should appraise these. Some are undoubtedly using the current interest in the environment for gain while others truly are assisting environmental cleanups by recycling waste material.

Insulating glass with wavelength-selective indium oxide and silver-coated polyester interlayers that reject unwanted heat in summer and retain indoor heat in winter are available. These units can be purchased with inert gas-filled (argon or krypton) "dead-air" spaces, which further improves their resistance to heat flow, as will systems of insulated edge sealants available from some glass manufacturers. The designer should never fail to specify insulated doors and thermal-break windows, which offer significant improvements over their counterparts. Weather stripping and seals save tremendous energy, whether in a hot or cold climate.

Included in the list of available standards listed at the end of this chapter are some sources where designers can learn about sustainable design. Contacting even just a few will provide a wealth of environmental building knowledge.

Selection of Interior Finishes

In general, the selection of interior finishes in process spaces should be based on durability and low maintenance, whereas in administrative and office environments, more attention must be given to providing a pleasant working environment. Considerations also include any possible effects on indoor air quality by avoiding materials that tend to off-gas or absorb pollutants and give them off over time, or are difficult to fully clean such as carpeting. The designer should refrain from using any phenol formaldehyde adhesives and minimize the use of laminates within occupied spaces. Zero-VOC paints are now available. Materials that require frequent cleaning or strong cleaning solutions to maintain should be avoided.

Materials such as ceramic tile or vitreous glazed structural tile make excellent choices for many types of spaces found at the plant. They are available in many styles, and the installation system should be as is suggested in publications provided by the Tile Council of America. Other recommended selections include regular or thin-set terrazzo for floors and gypsum board containing recycled gypsum and recycled paper for walls in areas not subject to dampness. For areas subject to dampness, such as shower and locker rooms, cement board or concrete block is usually used as a base for ceramic tile, or glazed structural tile may be used as both wall and finish in one system.

Areas such as basements, boiler rooms, and pipe galleries are generally not given a special finish unless the walls are concrete block, in which case a high-durability paint can be applied to facilitate maintenance and improve light distribution.

Selection of Colors. The American National Standards Institute, in ANSI A13.1, *Safety Color Code,* and in ANSI Z535.1, *Scheme for the Identification of Piping Systems,* has tried to standardize the identification of piping using color coding. Many water systems have their own standards, which have served them well for years, and the designer needs to be aware of this. There are no restrictions on color schemes except that care should be used in the laboratory. The determination of small concentrations of minerals is often done colorimetrically, and light reflected from variously colored areas in the laboratory may interfere with these tests. Laboratories should be painted in subdued colors such as off-white, light gray, or light blue.

Selection of Surface Coatings. Steel surfaces at a water treatment plant that particularly require protective coating include clarifier mechanisms, structural members, storage tanks, and piping supports. Most governing jurisdictions have maximum allowable volatile organic content for paint, and most manufacturers have responded with low-VOC paint systems. Any paint in contact with potable water must be approved in conformance with National Sanitation Foundation International standard 60.

The durability of an applied coating is generally only as good as the preparation that allows it to bond to the steel or material being coated. For this the recommendations of both the manufacturers and the Steel Structures Painting Council should be adhered to. Some surfaces, such as galvanized steel and fiberglass, are generally difficult surfaces on which to obtain adequate adhesion, and on these surfaces it is best to specify a light 60- to 80-mesh sandpaper finish.

Painting can be avoided by the selection of materials that do not require protection, such as stainless steel or ceramic tile, but there are many locations throughout the plant where it is not economically feasible to use corrosion-resistant materials. In these cases, there is no way to avoid the use of paint. It is generally best for future maintenance if the designer selects as few painting systems as possible for the plant. Paint systems that can be applied to a broad variety of substrates and still offer excellent wearability and longevity are preferred. Materials such as cured epoxies and polyurethane enamels are state of the art in current plant painting technology and offer many variations to meet specific needs.

DESIGN STANDARD PROMULGATING ORGANIZATIONS AND ORGANIZATIONS OFFERING DESIGN RECOMMENDATIONS

Aluminum Association
900 19th St., NW, Suite 300
Washington, DC 20006

American Architectural Manufacturers Association
1540 E. Dundee Rd., Suite 310
Palatine, IL 60067

American Concrete Institute
P.O. Box 19150
Detroit, MI 48219

American Forest and Paper Association
2nd Floor, 1250 Connecticut Ave., NW
Washington, DC 20036

American Hardboard Association
1210 W. Northwest Highway
Palatine, IL 60067

The American Institute of Architects
Center for the Environment
AIA Environmental Resource Guide
1735 New York Ave., NW
Washington, DC 20006

American Industrial Hygiene Association
2700 Prosperity Ave., Suite 250
Fairfax, VA 22031

American Institute of Steel Construction
One East Wacker Dr., Suite 3100
Chicago, IL 60601-2001

American Iron and Steel Institute
1101 17th St., NW
Washington, DC 20036-4700

American Institute of Timber Construction
7012 S. Revere Parkway, #140
Englewood, CO 80112

American Lumber Standards Committee
P.O. Box 210
Germantown, MD 20875

American National Standards Institute
11 West 42nd St., 13th Floor
New York, NY 10036

American Plywood Association
P.O. Box 11700
Tacoma, WA 98411

Asphalt Roofing Manufacturers Association
6000 Executive Dr., Suite 201
Rockville, MD 20852-3803

Adhesive and Sealant Council
1627 K St., NW, Suite 1000
Washington, DC 20006-1707

American Society of Heating, Refrigerating and Air-Conditioning Engineers
1791 Tullie Circle, NE
Atlanta, GA 30329

American Society for Testing and Materials
1916 Race St.
Philadelphia, PA 19103-1187

Architectural Woodwork Institute
P.O. Box 1550
13924 Braddock Rd., #100
Centerville, VA 22020

American Wood Preservers' Association
P.O. Box 286
Woodstock, MD 21163-0286

Builders' Hardware Manufacturers Association
355 Lexington Ave., 17th Floor
New York, NY 10017

Brick Institute of America
11490 Commerce Park Dr.
Reston, VA 22091

Center for Maximum Potential Building Systems, Incorporated
8604 FM 969
Austin, TX 78724

Center for Resourceful Building Technology
P.O. Box 3413
Missoula, MT 59806

Consumer Product Safety Commission
5401 Westbard Ave.
Bethesda, MD 20207

Carpet and Rug Institute
P.O. Box 2048
Dalton, GA 30722

Ceramic Tile Institute of America
12061 S. Jefferson Blvd.
Culver City, CA 90230

Door and Hardware Institute
14170 Newbrook Dr.
Chantilly, VA 22021-2223

Decorative Laminate Products Association
13924 Braddock Rd.
Centreville, VA 22020

EIFS Industry Manufacturers Association
2759 State Road 580, Suite 112
Clearwater, FL 34621

Flat Glass Marketing Association
White Lakes Professional Bldg.
3310 S.W. Harrison St.
Topeka, KS 66611-2279

Factory Mutual Systems
1151 Boston-Providence Turnpike
P.O. Box 9102
Norwood, MA 02062

Federal Specification (from GSA)
Specifications Unit (WFSIS)
7th St. and D St., SW
Washington, DC 20407

Gypsum Association
810 First St., NE, Suite 510
Washington, DC 20002

Hardwood Manufacturers Association
400 Penn Center Blvd.
Pittsburgh, PA 15235

Indiana Limestone Institute of America
Stone City Bank Building, Suite 400
Bedford, IN 47421

Marble Institute of America
33505 State St.
Farmington, MI 48335

National Association of Architectural Metal Manufacturers
600 S. Federal St., Suite 400
Chicago, IL 60605

National Building Granite Quarries Association
P.O. Box 482
Barre, VT 05641

National Center for Appropriate Technology
P.O. Box 3838
Butte, MT 59702

National Concrete Masonry Association
2302 Horse Pen Rd.
Herndon. VA 22071-3406

National Fire Protection Association
One Batterymarch Park
P.O. Box 9101
Quincy, MA 02269-9101

National Recycling Coalition
1718 M St., NW, Suite 294
Washington, DC 20036

National Renewable Energy Laboratory
1617 Cole Blvd.
Golden, CO 80401

National Roofing Contractors Association
10255 W. Higgins Rd., Suite 600
Rosemont, IL 60018-5607

National Sanitation Foundation
3475 Plymouth Rd.
P.O. Box 130140
Ann Arbor, MI 48113-0140

National Terrazzo and Mosaic Association
3166 Des Plaines Ave., Suite 132
Des Plaines, IL 60018

Occupational Safety and Health Administration
200 Constitution Ave., NW
Washington, DC 20210

Resilient Floor Covering Institute
966 Hungerford Dr., Suite 12-B
Rockville, MD 20805

Steel Door Institute
30200 Detroit Rd.
Cleveland, OH 44145

Sheet Metal and Air Conditioning Contractors National Association
4201 Lafayette Center Dr.
Chantilly, VA 22021

Southern Pine Inspection Bureau
4709 Scenic Highway
Pensacola, FL 32504

Single Ply Roofing Institute
20 Walnut St.
Wellesley Hills, MA 02181

Steel Structures Painting Council
4516 Henry St.
Pittsburgh, PA 15213

Tile Council of America
P.O. Box 326
Princeton, NJ 08542-0326

Underwriters Laboratories
333 Pfingsten Rd.
Northbrook, IL 60062

West Coast Lumber Inspection Bureau
P.O. Box 23145
Portland, OR 97281

Western Wood Products Association
Yeon Building
522 SW 5th Ave.
Portland, OR 97204-2122

U.S. Department of Commerce
14th St. and Constitution Ave., NW
Washington, DC 20230

U.S. Green Building Council
1615 L St. NW, Suite 1200
Washington, DC 20036

CHAPTER 19
STRUCTURAL DESIGN

The structural design of a modern water treatment plant is a dynamic, interactive process that requires specialized expertise. Because most structural engineers are primarily familiar with the design of conventional buildings, this chapter focuses on the design features that are unique to treatment plant structures.

THE DESIGN PROCESS

The engineer responsible for the design of structures at a water treatment plant must be intimately familiar with the general principles and codes governing structural design. The designer should work with the project team during preliminary design to establish the configurations of the facilities based on economy, structural efficiency, and performance criteria. The design should then be developed and detailed in a manner that accounts for the special considerations required for a water treatment plant.

Responsibility of the Structural Engineer

Historically, the civil, process, architectural, and structural designs of a water treatment plant were prepared by the same individual and shown on the same set of drawings. Although this comprehensive approach resulted in a well-coordinated design and an efficient use of construction documents, the additional complexities of a modern treatment plant make this approach impractical.

Engineering, like most technical professions, has advanced to the point where no one engineer can adequately address all of the considerations associated with the design of a large water treatment plant. Today's design team, therefore, requires an engineer who specializes in structures to support the project team in developing structural drawings and specifications depicting the design of the structures.

The structural drawings should include all of the information critical to the structural design, including dimensions, details, and materials that form the basis of the design. The final construction drawings should be stamped and signed by the professional engineer responsible for the structural design.

Interaction with the Design Team

The engineer who leads the design of a water treatment plant establishes the basic process and plant arrangements. The structural engineer should provide input early in the design process to optimize the efficiency of the structures. The civil and structural engineers and architect should work closely together in establishing the distance between structures, the materials of construction, and the foundation lines. Each party should appreciate the fundamental principles governing the other's design.

The mechanical engineer, who also supports the project team, is usually responsible for the design of plumbing, HVAC, and special mechanical systems. The detailed mechanical design usually begins after the basic civil and structural arrangements have been set. The structural and mechanical engineers coordinate the equipment support, the locations of pipe penetrations, and the reactions and support of the piping systems.

The electrical engineer usually designs the power, lighting, and instrumentation and control systems. This design generally follows the process design and often parallels the detailed structural design. The electrical and structural engineers coordinate the support of electrical equipment and the locations of electrical conduits.

Use of Codes and Standards

The specific codes that govern structural design should be established early in the design process. Most of the structures at a water treatment plant do not conform to traditional building codes because their function and behavior differ widely from those of conventional buildings. Consequently, the structural engineer must often base the design of treatment plant structures on standards or reference documents used for nonbuilding structures.

Although many water treatment facilities are exempt from local building codes, the structural design should conform to nationally recognized codes and standards and exceed the minimum requirements of local building and safety codes. Local building codes should also be used to establish basic design criteria such as wind and earthquake loadings.

STRUCTURAL DESIGN CONSIDERATIONS

Among the many basic considerations that the structural engineer must consider before any design can begin are the types of materials that will be used, how the facilities are laid out, and what local site conditions will influence the design.

Material Selection

The materials selected for use in the design of a treatment plant are influenced by many factors, including the size and shapes of the structures, the relative costs of materials, soil and other local conditions, architectural design, and environmental conditions.

Concrete. Concrete is extensively used for water-containing and water-conveying structures because it is generally durable and resistant to corrosion. The type of cement specified should take into account the composition of the source and treated water, the treatment chemicals, and the exposure of the site to atmospheric pollutants and seawater. The arrangement of reinforcing, jointing, and crack control should be thoroughly addressed by the structural engineer. Concrete alone should not be considered a watertight material, and the owner should be informed about the amount of cracking to be expected.

Steel. Steel is often used in the construction of plant superstructures, tanks, platforms, and walkways. The high strength and modulus of elasticity of carbon steel make it an efficient and economical construction material. Primary steel shapes and plates can be made corrosion resistant by applying protective coating systems. Structural steel is well controlled by national standards, which should be specified in the construction documents.

Masonry. Many water treatment plant buildings and partition walls are constructed with masonry such as concrete block, brick, and stone. Masonry materials are durable and have good corrosion resistance. Masonry structures should be designed in accordance with local building codes, using materials that are available in the region.

Aluminum. Aluminum is often used for walkways, grating, and handrails in areas exposed to a mildly corrosive environment. The structural components should be designed and specified to nationally recognized standards. Connections should be welded in the shop, and dissimilar materials should be separated. The engineer must check and limit deflections in aluminum structures during design.

Stainless Steel. Stainless steel is used in water treatment plant construction at locations where corrosion resistance is critical and where the design requires a high-strength material. Stainless steel is available only in a limited number of shapes and sizes and is significantly more expensive than steel and aluminum. When possible, stainless steel components should be welded in the shop rather than on the job.

Fiberglass. Fiberglass is often used for fabrication of equipment and parts that will be located in a highly corrosive environment. The designer must consider the deflection and fire resistance of fiberglass, and the material and performance requirements must be specified in detail in the contract documents.

Configuration of the Structures

The structural arrangements should be defined during preliminary design, based on process requirements and efficient use of structural elements.

Layout of Structures. The configuration of a structure often establishes its relative economy and long-term performance. The project team generally proposes a conceptual layout based on functional requirements. The structural engineer should then review the proposed arrangements and coordinate with the project engineer to economize the cost of the structures. Once set, the structural layout should not be changed during detailed design.

The structural engineer needs to first review the conceptual layout for horizontal and vertical symmetry. The configuration should be simple, and discontinuities that might cause a buildup of stress should be avoided. Units or components should be single, continuous structures, or control joints should be provided between components with different performance characteristics. If asymmetrical structures are unavoidable and cannot be separated into balanced units, additional strengthening should be provided at the discontinuities.

The stiffness of each structural element needs to be considered individually. Where incompatibilities exist, structural elements should be isolated from each other. The cost-effectiveness of vertical offsets needs to be estimated and scrutinized.

The designer must use good judgment when varying the thickness of cast-in-place concrete members. The relative cost of additional concrete must be compared with the added costs of reinforcing bar splices and forming if the thickness of a structure is varied.

Wherever possible, foundation lines should remain relatively consistent. The construction costs of excavation and soil preparation for foundations must be evaluated, along with the ability of the soil to support the imposed loads.

Once the preliminary design is substantially complete, the structural arrangements should be reviewed by a senior structural engineer who was not involved in the preliminary design. This review will primarily focus on cost-effectiveness and overall structural integrity. Establishing the proper configuration of the structures during preliminary design will minimize initial construction cost and reduce long-term operating costs for the owner.

Efficiency of Structures. Many design firms establish general layout criteria for conceptual design. By developing economical member size-to-span ratios, each project can use cost-effective structural components. Because structures at a water treatment plant are modest in size, a highly refined design of individual members is seldom justified.

Interaction between Structures. When establishing the basic geometry of structures, the designer must consider the relative rigidity between structural components. Common walls separating structures of different types must be carefully considered. Double-walled construction may be appropriate where structural isolation is required and where all process systems can be readily isolated between structures. A constructibility review should be performed on all structures located immediately adjacent to each other.

The interaction between individual building components should also be considered during preliminary design. A diaphragm's ability to deform must be evaluated against its boundary conditions. The architectural, mechanical, and electrical systems must be compatible with the drift of a flexible superstructure, and the reactions from intermediate platforms and walkways need to be applied to the building's superstructure during final design. Breaks should be provided in catwalks and ladders spanning independently moving structures.

The different types of equipment in water treatment plants often create dynamic forces on a structure. Oscillating equipment should be located as low as possible in the structure, and its effects on individual structural components must be analyzed. Dynamic or heavy equipment should preferably be supported on separate foundations. Equipment base isolators may also be used.

In seismic areas, if equipment such as pump motors, conveyors, blowers, and hoists must be located on upper floors, the supporting structure must be rigid and the equipment rigidly attached or adequately guyed. The structure and equipment must respond as a unit with provisions to resist amplification of seismic loadings by the building structure.

Jointing of Structures. Where necessary, structures should be subdivided by joints. Control joints can be either expansion or contraction type and are generally used to isolate structures or structural elements from each other.

An expansion joint permits relative expansion and contraction between elements. The use of shear transfer devices or waterstops at these joints depends on the degree of isolation and watertightness required.

A contraction joint permits relative contraction, but not expansion, between concrete elements. It often serves as an intentionally weakened plane between concrete elements to encourage shrinkage cracking at the joint instead of at random locations. Contraction joints are typically detailed to transfer shear across the joint. Waterstops may or may not be required at contraction joints depending on the need to prevent the passage of water. Reinforcing is not usually continued across control joints, but the designer may choose to continue a small percentage of reinforcing for shear transfer.

Construction joints are normally used to facilitate construction and not to isolate elements from one another. They are normally used to bond two adjacent concrete elements together as though the joint did not exist. Such joints are not usually considered in the design calculations, although special detailing at construction joints should be indicated on the drawings. Construction joints are not contraction joints unless so detailed.

Concrete joints that affect the design and performance of a structure should be clearly indicated on the drawings and described in the specifications. Ambiguous joint requirements on the drawings and in the specification will create problems in the field that may eventually affect the performance of the structure.

Alternating or delaying pours between joints or installing delayed pour strips can accommodate some initial shrinkage. However, shrinkage within concrete will continue at a decreasing rate for several years after placement. Minimum shrinkage reinforcement should be in accordance with American Concrete Institute (ACI) Standard 350 and should be determined by the spacing of the control joints, not construction joints.

Waterstops are specified wherever it is necessary to inhibit the passage of moisture across a control or construction joint. Waterstops should be included in walls and slabs of concrete structures containing treated water and in walls where one side is wet and the other side is dry. Under normal circumstances, waterstops are not required in divider walls between reservoir or basin chambers.

Public Access Considerations. Some structures at a water treatment plant are exempt from building code requirements pertaining to public access. This should not deter the project team from using good judgment when establishing the access in and out of a structure. Buildings that are regularly occupied should meet building code requirements for access. Provisions of the Americans with Disabilities Act (ADA) should also be satisfied to the degree desired by the owner and required by the governing authorities. Structures that are seldom occupied should meet OSHA standards and should include provisions for safe access and ready evacuation. Caution should be used when designing structures with only one means of egress.

Environmental Considerations

Depending on where a water treatment plant is to be located, a number of environmental considerations might be important in the design.

Freezing Considerations. Structures built in locations subject to freezing should be designed to meet the local building code requirements pertaining to minimum depth of foundations. In addition, a geotechnical report should identify a minimum depth of foundation based on site-specific soil characteristics. Basins that are constructed with slabs on grade must be considered carefully for the potential of frost heave if the basin may at any time be completely emptied during the winter. Unless the soil is resistant to frost heave, the structure must be surcharged to resist upward forces. Pile foundations exposed to frost heave should be designed to resist uplift forces or should be isolated through the frost zone.

Structural Watertightness. A unique consideration for structures at a water treatment plant is that many of the structures must be watertight. This criterion influences much of the structural design. The design load factors, shrinkage reinforcing, and reinforcing cover should comply with ACI Standard 350.

Almost every concrete structure is subject to some shrinkage cracking, a phenomenon that the owner should be made aware of. The designer's challenge is to control and minimize shrinkage cracking consistent with the use, cost, and importance of the structure. Shrinkage cracking can usually be effectively controlled by proper location and installation of joints and shrinkage reinforcing. The percentages of shrinkage reinforcement should vary with the spacing of control joints as identified in ACI Standard 350. It should also be kept in mind that the effects of concrete quality and curing, as well as member restraints (intentional or otherwise), on shrinkage cracking can be significant.

The special load factors for liquid containment structures indicated in ACI Standard 350 serve to reduce service load tensile stresses in the reinforcement and to limit the width of shrinkage and flexural cracking. Minimizing crack width in watertight structures is important for both leak prevention and protection of the reinforcing steel. The load factors should be used for members in direct contact with liquid, members directly above large areas of free water, and members exposed to high humidity or potentially corrosive environments. The structural design memorandum for the project should clearly indicate where the additional load factors will be used. The reinforcing selected for concrete structures should be checked for crack control in accordance with ACI Standard 318. Reinforcing for liquid containment structures should have a Z-value less than that specified in ACI Standard 350.

Dampproofing should be applied to concrete walls exposed to water or backfill, if the opposite side is above grade or inside of a room. The specified dampproofing should be compatible with the liquid to which it will be exposed.

Corrosion Protection. The choice of protective coating systems for corrosion control of structural members depends on the degree of exposure to corrosive environments. Concrete members in water treatment plants rarely need protective coatings, but epoxy-coated reinforcing may be specified for concrete members exposed to high concentrations of chlorides. Steel components in a water treatment plant can usually be adequately protected by application of an epoxy coating. The specified epoxy system must be compatible with the structural steel primer.

Aluminum, stainless steel, and fiberglass are often used in highly corrosive environments. However, even these materials may require a protective coating system suitable for particularly corrosive conditions.

Foundation Considerations

There are often many deep foundations located adjacent to one another when constructing a water treatment plant. The ability to construct these structures and the sequence of construction must be carefully considered by the structural designer. Shoring systems are usually required for buried structures and individual components constructed below grade. Buried pipes adjacent to or beneath a structure also must be given special consideration when establishing the bearing capacity the foundations.

The Geotechnical Report. A soils report stamped and signed by a professional geotechnical engineer should be prepared for the site of a new water treatment plant or for an expansion or modification to an existing plant. Plant site conditions are particularly significant in the design of concrete basins and vaults because these massive structures require stable bearing. The report should include the following information:

- Level of groundwater
- Active, passive, and at-rest soil pressures

- Allowable bearing pressures
- Minimum depth of foundations
- Overexcavation and backfill recommendations
- Corrosivity readings
- Estimates of settlement
- Recommendations for appropriate foundation systems
- Soil information necessary to construct the plant
- For areas of seismic activity, site-specific seismic information

In addition, the report should identify the potential presence of chemically reactive soils or shrinking and swelling clays and provide recommendations on how to mitigate their effects. After the plans are partially completed, the geotechnical engineer should review the final foundation design for compliance with the report's recommendations.

Use of Mat Foundations. Mat foundations are often used in the design of liquid-containing structures and to minimize differential settlement of large structures. Where a mat is constructed on fill, the fill materials should be of uniform thickness and compacted to a degree that will ensure comparatively high soil shear strength, little future consolidation, and no differential settlement.

A mat foundation may be subject to uplift forces from external water pressures acting on the mat. Structures that extend below the groundwater table should be checked for a minimum factor of safety against buoyancy of 1.1. The groundwater level assumed in the calculations should be based on the design flood level, which typically has a recurrence interval of 100 years or more. No live load or backfill friction should be assumed to be acting on the structure. The stiffness of the structure should be considered when relying on the full structure dead weight. Long-span slabs, in particular, may be too flexible to mobilize the dead weight of interior walls.

Buoyant forces causing flotation can be controlled in several ways:

1. By providing a positive tie-down mechanism
2. By tying the foundation to supporting piles designed to resist uplift
3. By increasing the weight of the structure by using mass concrete, adding to the overburden on top of the tank, or keeping the tank at least partially filled
4. By providing a positive drainage system to lower the groundwater table in the vicinity of the structure

Pressure-reducing valves (PRVs) are sometimes used to relieve ground water pressure but should be used only to alleviate a temporary high water condition, because continuous flows through the valve will eventually cause it to clog. In addition, the soil surrounding the valve must be carefully selected to establish a drainage path for the groundwater and to maintain the valve operation.

Use of Spread Footings. The geotechnical engineer should specify the allowable bearing pressure and the maximum settlement potential for spread footings. The fact that spread footings have a higher potential for differential settlement than mat foundations should be considered in the structural design. Because of the multiple foundation depths normally required in the design of a water treatment plant, the designer should verify the feasibility of locating spread foundations adjacent to deeper foundations and, where necessary, extend the footing down to native soils or design the higher footing for bearing on compacted fill.

Use of Deep Foundations. Some plant sites may require the use of pile or pier foundations. Water treatment plants are typically located in low-lying areas and on alluvial plains. It is usually necessary in these locations to support structures on deep foundations or to otherwise stabilize the underlying soil. Deep foundations will significantly add to the cost of a treatment plant and should be used with discretion. Basins or structures constructed on different foundation systems should be separated by a flexible joint.

Piles installed to resist seismic-induced bending stresses must be provided with ductility. The tops of piles should be embedded in the structure as deeply as feasible in order to transfer forces. Individual pile caps and piers of every building should be interconnected by ties. The effects of driving new piles on the support of foundations at an existing treatment plant should be evaluated before selecting pile foundations.

Lateral Soil Pressures. The structural engineer should carefully consider the ability of a structure to deflect before selecting the applicable lateral soil pressure. Walls must be free to rotate or significantly deflect before active soil pressures can develop. A basin wall laterally restrained by a top slab or side walls will be subject to at-rest soil pressures. A structure must undergo substantial movement before passive pressures can develop.

Construction below the Water Table. The groundwater elevation should be carefully considered when setting the depth of structures. Dewatering a site for construction is usually relatively costly. In addition, the long-term risks of uneven settlement, flotation, and water leakage when a building extends below the water table should be a significant consideration in the structural design.

Seismic Design

The design of treatment plants in seismic areas requires careful attention to detail. Geotechnical investigation is a logical first step in determining foundation conditions and structural stability issues. The next step should be to properly identify all the pertinent seismic forces and to design the structures accordingly. Anchorage of the components and equipment also warrants careful consideration so that a structural system may survive a sizable seismic event with minimal impact.

Danger of Soil Liquefaction. Depending on the soil profile, liquefaction of soil can occur in a high-seismic region. Liquefaction is the loss of shear strength or shearing resistance in loose, saturated, cohesionless soils as an aftermath of rapid shaking and reconsolidation of ground caused by moderate to strong earthquakes. Liquefaction may cause lateral movement of soil masses above liquefied soil layers along a downward slope or unrestrained surface. Surface fissures and sand boils may occur when liquefied material attempts to migrate upward, pushing through shallow overlaying strata of unliquifiable material near the ground level.

In addition to causing lateral movement or spreading, liquefaction may result in extensive, unanticipated ground settlement. Consequently, it is essential to consult a geotechnical engineer experienced in facilities built on soils that have the potential to liquefy.

If a treatment plant must be built over a stratum with liquefaction or densification, the designer may alter the soil characteristics or design the structure to overcome earthquake effects. Depending on the local topography, an effective drainage system may be constructed to prevent liquefaction. Sensitive clays may be removed and replaced with stable fill material, or structures may be supported on piles extending through the sensitive clay layer. Studies have shown that the costs of soil stabilization for a complete facility are high; therefore proper siting can be critical to mitigating seismic damage.

STRUCTURAL DESIGN   561

Special Tank Design Considerations. Another potential problem associated with earthquakes is failure of tank foundations. One possible reason is the increased localized loading caused by the tank overturning moment. The earthquake motion may cause the soil structure to liquefy, lose shear strength, or simply consolidate (settle), depending on the soil conditions. This may allow the tank to tip or settle unevenly, causing the tank shell or roof to buckle and sometimes fail. Proper design includes provisions for friction or other restraint of surface basins and vaults.

Seismic-induced forces must also be properly identified and applied to the structure and its components. If a structure is to contain liquid, the sloshing effects of the liquid must be taken into account by using accepted and recognized analytical methods in designing the containment structure. The analysis may involve the effective mass method described in American Water Works Association (AWWA) Standard D. The designer may also refer to Haroun and Housner (1980) for a complete, detailed analysis of earthquake-induced loadings on water-holding basins.

Another consideration in the design of tanks and basins located in seismic areas is that adequate freeboard must be provided so that the contents will not spill out and flood the immediate area or damage the roof of the structure.

Equipment Design Considerations. All of the essential components of a water treatment plant should be designed according to seismic risk analysis. The process units that are critical to the plant's serviceability after an earthquake are identified first. The unit processes that are not essential to short-term continuation of service are then identified and the acceptable duration of their downtime established. Nonessential facilities might include carbon regeneration, lime regeneration, sludge collection, sludge disposal, and similar functions that are not directly related to producing safe treated water for short periods.

Anchorage of all components and equipment to the structure must be designed for the appropriate seismic forces as required by the governing building code and standards. Piping connections should be flexible enough to allow for differential movements of adjacent components. Special care must be taken in locating equipment, storage tanks, and feed lines containing hazardous materials.

Most regulatory agencies require that chlorination facilities be located in a separate, well-ventilated room, away from the rest of the treatment facility. Chlorine feeders should also be separated from chlorine storage rooms. Similar precautions should be taken with other hazardous materials. All gas bottles and containers of hazardous materials must be restrained in appropriate racks. Chlorine cylinder scales must be equipped with hold-down devices so that the cylinders will move with the scales or will be restrained.

Vertical end suction pumps, often used for solids handling, may be tall, slim structures, so their bases and anchorages must be designed to resist overturning. Horizontal pumps and their motors should be mounted on a single structure so that they respond together. This concept also applies to a battery of pumps connected to a single header. Pumps and piping must always be separated by a flexible connection to allow them to respond independently to earthquake movement.

Heavy electrical equipment must be anchored to base pads or buildings. Precision equipment, such as residual analyzers, recorders, indicators, electronic instrumentation, switch gear, equipment instrumentation, and communications systems, should be mounted rigidly to avoid amplification of seismic acceleration.

Anchor bolt embedments should be used in preference to expansion anchors where continuous dynamic loads are encountered (e.g., rotating equipment). Expansion anchors are acceptable for static loads and infrequent dynamic loads such as earthquakes.

Seismic design factors for basins are usually determined as specified in Chapter 9 of ACI Standard 318 rather than Chapter 21, because the basin is not a building frame system. The load factor for hydrostatic forces is typically reduced for the seismic case as allowed by ACI Standard 318, because the amount of water is clearly defined.

Loading Criteria

The various types of structures used in a water treatment plant must each receive careful consideration of the loading under all possible conditions.

Types of Structural Loads. Unlike buildings, tanks and basins must continually resist both vertical and lateral loads while in service. Lateral loads can be separated into external loads typically resulting from the surrounding soil and internal loads resulting from the contents of the structure.

External soil loads consist of either at-rest or active soil pressure, compaction pressures, surcharge from trucks or adjacent structures, and groundwater. Internal service loads include the hydrostatic pressure exerted by the contained water. As a minimum, the structure should be analyzed for the hydrostatic loads resulting from an overflow condition. A load commonly overlooked is the tensile forces in slabs and walls resulting from the internal pressures on the structure.

Vertical loads can include soil overburden, equipment and construction loads working above the structure, and roof or floor dead and live loads. The roof slab of a tank or basin may also be subject to vacuum pressures created by the treatment process.

Load Factors and Load Combinations. Ultimate load factors and load combinations for concrete structures are defined in ACI Standards 318 and 350. The increased load factors for flexure, tension, and shear applicable to tanks and basins are intended to reduce the stress in the reinforcing to a level comparable with the working stress design used in the past. In addition, load factors and load combinations from the local building code should be reviewed and used for design if more severe than the national standards.

DESIGN OF BASINS, VAULTS, LARGE CONDUITS, AND CHANNELS

The basins, vaults, large conduits, and channels used in water treatment plants are special structures that must be designed to withstand the worst case loading that may be imposed on them.

Loading Considerations

Many treatment facility structures are rigid and box shaped, often buried in the ground, and sometimes without a superstructure. These types of structures are unique in many respects, and their design is normally based on a few simplifying assumptions. The walls of the structures are designed as plates to loads derived from the U.S. Bureau of Reclamation tables or from Portland Cement Association (PCA) tables. Walls with aspect ratios greater than 2:1 are generally designed as one-way slabs. The designer should be aware that the moment in the center of the plate may be underestimated because of moment redistribution at the corners of the walls and that the walls intersecting the plate must be designed for the same moment as determined for the end conditions of the plate.

The footings of a structure may also be subject to an increase in positive moment at the center of the span because of rotation of the footing on elastic soil. This phenomenon applies to both strip footings and mats. A strip foundation is normally used for basins that are particularly wide. A mat foundation is commonly used for vaults, conduits, and channels or where needed to resist groundwater uplift or to distribute bearing pressures over a wider area.

The stiffness and strength of the support should be evaluated for adequacy. For example, a basin wall could appear to be braced by several floor slabs of connecting channels; however, the slabs may not be adequately restrained to function as a brace.

Support conditions for unique structures should also be thoroughly examined for variations. Normally, the roof slab of a conduit or channel is supported by the two sidewalls, except where the channel turns and the support conditions change.

The structural design should separately consider the internal and external loads acting on treatment plant structures. Lateral soil pressures should never be used to counteract internal hydrostatic forces, because the backfill is likely to be excavated at some time when the plant must remain in operation. The design of environmental structures should consider at least these factors:

- Settlement
- Earth-retaining forces
- Flotation
- Appurtenant items such as baffles, miners, sludge collectors, water troughs, and filter surface washers
- In high-seismic areas, the internal effects of sloshing

Design Recommendations

Specific design recommendations for tanks, vaults, and similar structures are as follows:

- Specify the appropriate type of backfill used behind tank walls during construction. A noncohesive soil is normally used, because it is easier to attain a high compaction density with minimum effort.
- Use flexible connections between storage facilities and all inlet and outlet pipelines. These connections or joints can be mechanical, restrained expansion, rubber, ball-and-socket, or gimbals-restrained bellows-type couplings.
- Design walls serving as shear walls or diaphragms with expansion and construction joints keyed to carry the shear forces. To maintain watertightness at joints where some movement is anticipated, use flexible joints with waterstops.
- Design the top slab of buried structures for skip loading of the overburden soil to accommodate construction. If the overburden soils above a structure are subject to truck traffic and are less than 2 ft (0.6 m) thick, the top slab should be designed for the actual truck loading applied directly to the slab as required by the American Association of State Highway and Transportation Officials (AASHTO) *Standard Specification for Highway Bridges,* in addition to the overburden soil.

BUILDINGS AND SUPERSTRUCTURES

The performance requirements for structures at a water treatment plant are unique in many respects. While the design of these structures must comply with the local building codes, it must also account for the particular service conditions the facilities are subject to. Crack control, corrosion resistance, and unusual structural geometry all play an important role in the design of water treatment plant structures.

Design Criteria

Control buildings, buildings housing equipment, and other superstructures for treatment facilities should generally be designed in accordance with the local building codes. The design approach used for these types of buildings is similar to that used for conventional buildings, with a few exceptions.

Phantom loads are often added to floor and roof members of treatment plant buildings to simulate small, miscellaneous mechanical and electrical loads imposed either as part of the initial facility or by future modifications. The building may also be subject to unique live loads not covered by the building codes. The designer should consult American Society of Civil Engineers (ASCE) Standard 7 for a comprehensive listing of live loads that should be applied to industrial structures such as those found at a water treatment plant.

In high-seismic areas, a dynamic analysis will be required for structures with plan or vertical irregularities. The design engineer should particularly keep in mind that the seismic design criteria prescribed in the building code are considered minimum requirements to ensure life safety. The important structures at a water treatment plant should be designed to remain functional during and after an earthquake.

Analysis and Design of Buildings

The structural framing system for a building should be simple and symmetrical. Vertical and plan irregularities should be avoided when at all possible to minimize construction costs and to improve long-term performance of the structure.

The selection of the structural framing system should be based on functional requirements and construction costs. In general, the gravity framing system and the lateral load-resisting system are analyzed separately. For concrete framing, the gravity system is usually analyzed and designed as a string of continuous beams supported by hinges or columns, depending on the restraint condition. The effects of patterned live loads also should be taken into account during design. For steel framing, the gravity system is typically analyzed and designed as a series of simply supported members.

In buildings with rigid diaphragms, the lateral forces should be distributed to each frame based on their relative rigidity. For a building with flexible diaphragms, the lateral forces are distributed to each frame based on their tributary width. Whether a diaphragm is rigid or flexible is based on its relative rigidity as compared with the vertical shear-resisting element. For simple symmetrical buildings or asymmetrical buildings with flexible diaphragms, only a two-dimensional analysis is required for the lateral design.

Other Design Considerations

The fire resistance rating of a structure should be determined early in the design and taken into account when selecting the framing system. Many buildings in a treatment plant are not entirely enclosed or air-conditioned. This exposes structural components to corrosive environments that may affect the fireproofing or coating systems. The design details must address special concerns and issues in water treatment plants, such as the inability of protective coatings to protect the top flange of a beam after welding the metal deck to the beam.

The potential effects of temperature change and shrinkage should also be considered in the design of large superstructures. Many industrial-type buildings at treatment plants are subject to large temperature fluctuations caused by the equipment they house.

The structural framing system and the building enclosure should be compatible. A building cladding system is often initially selected based on aesthetic considerations and durability. The preferred system may or may not be compatible with the framing system chosen solely for structural efficiency. The structural engineer and the architect should work together to identify the optimal combination of building systems and adjust their individual designs accordingly.

The building enclosure and other attachments to the structure must be detailed to accommodate the anticipated vertical and horizontal movements. If masonry walls are used to enclose a building with a moment framing system, expansion joints must be provided in the masonry walls near the corners to allow for building movement. The structural framing system should also be compatible with the building cladding and partition details, equipment limitations, and the HVAC and piping systems it supports.

BIBLIOGRAPHY

American Association of State Highway and Transportation Officials. *Standard Specification for Highway Bridges.* Washington, D.C.: American Association of State Highway and Transportation Officials, 1996.

American Concrete Institute. *Environmental Engineering Concrete Structures, ACI 350-89.* Detroit, Mich.: American Concrete Institute, 1989.

American Concrete Institute. *Building Code Requirements for Structural Concrete and Commentary, ACI 318-95.* Detroit, Mich.: American Concrete Institute, 1995.

American Society of Civil Engineers. *Minimum Design Loads for Buildings and Other Structures, ASCE 7-93.* New York,: American Society of Civil Engineers, 1993.

American Water Works Association. *Standards for Welded Steel Tanks for Water Storage, ANSI/AWWA D-84.* Denver, Colo.: American Water Works Association, 1984.

Haroun, M. A., and G. W. Housner. *Dynamic Analysis of Liquid Storage Tanks.* Report EERL 80-04. Pasadena, Calif.: California Institute of Technology, Earthquake Engineering Research Laboratory, 1980.

Portland Cement Association. *Information Sheet # I500303 Rectangular Concrete Tank.* Skokie, Ill: Portland Cement Association, 1969 (Revised 1981).

W. T. Moody, U.S. Bureau of Reclamation. *Moments and Reactions for Rectangular Plates.* Engineering Monograph No. 27. Denver, Colo.: Bureau of Reclamation, 1963 (Reprinted 1990).

CHAPTER 20
PROCESS INSTRUMENTATION AND CONTROLS

Many large water treatment plants have been manually operated for years and have produced high-quality water with little on-line electronic instrumentation except flow recorders and chlorine residual analyzers. At these plants, staff gauges indicate levels, and gauges indicate pump suction and discharge pressures. Each plant operator takes pride in water production and has an opinion of how long a filter should be backwashed and how to adjust chlorine dosages. Why, then, is there a need to add expensive electronic instrumentation, automatic controllers, and computer systems? Some of the reasons are:

- The Safe Drinking Water Act has mandated strict control and verification of treated water quality.
- Economics dictates more efficient use of chemicals and energy.
- Labor is increasingly expensive.
- Plant managers require fast response to changing source water quality and emergency situations and the ability to alter water production rates.
- Better-trained staff are available.
- Water quality is closely scrutinized by a well-informed public.

In addition, more stringent water treatment plant effluent standards require tighter control of water treatment processes such as coagulation, sedimentation, and filtration, so these process must generally be under automatic, continuous control.

Optimal control requires accurate signals from process-variable transmitters. Accurate signals require high-quality electronic instrumentation installed with attention to detail. Accurate reporting of process parameters in orderly and timely fashion now requires a computer system for a treatment plant of almost any size.

The basic need to produce a better product (improved water quality) drives the need to engineer, operate, and maintain more complex instrumentation, controls, and computer systems. The advent of the microprocessor has yielded improvements only dreamed of a decade ago.

PURPOSE OF INSTRUMENTATION AND CONTROLS

Optimal water treatment can be made possible by a coordinated application of hydraulic, process, and equipment monitoring, coupled with hydraulic and process controls, as illustrated in Figure 20.1.

Hydraulic Monitoring

Just as assembly lines move products through a factory, water treatment requires accurate control of plant hydraulics to move the water from the plant source water inlet to the treated water outlet. The rate of plant production is adjusted to meet the demands for treated water.

In most water treatment plants, a single flowmeter measures the total plant flow rate and other meters measure flow at various stages of the process. In addition, individual flowmeters are used on each filter if constant-rate filter flow control is employed; smaller flowmeters monitor chemical and injection water feed rates.

Level transmitters are often used to monitor the level in the filter inlet channel to match filter flow rate to the plant inflow rate. Level is also commonly monitored in spent filter

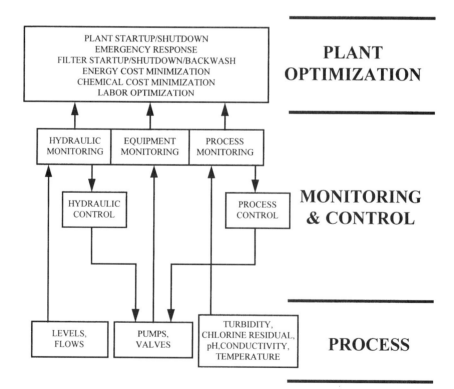

FIGURE 20.1 Monitoring and control for water treatment.

backwash recovery ponds and backwash storage tanks for filter backwash control interlocking. Level switches are used to start and stop source water, filter backwash, and filtered water pumps. Level control loops can control adjustable-speed pumps.

Pressure switches are used for plant air compressor controls and to detect discharge overpressure in positive displacement chemical metering pump applications.

Process Monitoring

Many new devices have become available at reasonable cost within the last few years that can monitor the operation of various treatment processes. Examples are chlorine residual monitors to ensure that the correct residual is present in the plant effluent, on-line particle counters to correlate filter performance with relative size distribution of suspended solids in filtered effluent, and turbidimeters to control backwash and optimize filter operation.

Equipment Monitoring

Most water treatment plants have a graphic panel or computer system displaying the status of major pieces of equipment, such as valves, pumps, air compressors, and blowers. Larger and specialized motors are candidates for protective devices such as vibration switches, phase loss detectors, and phase reversal detectors. Plant electrical service can be monitored for overvoltage, undervoltage, phase loss, phase reversal, real and reactive power, and circuit breaker status.

Hydraulic Controls

Water flow must be controlled to meet water production needs. Most of the controls in a water treatment plant are designed to move and direct the flow of water. Water plants use continuous control for most operations, such as maintaining, raising, and lowering plant flow rates. By contrast, during filter backwashing, valves, pumps, and blowers are sequenced in a batch control operation to treat a confined volume of water.

Process Controls

Treated water must be free of solids and properly disinfected. Chemical is added by means of continuous control, with the chemical feed rate varied in proportion to plant flow rate. The most common process controls are used for adding coagulants, pH control chemicals, and disinfectants.

Water Production Controls

In a small treatment plant, plant start-up, shutdown, and water production adjustments can be accomplished by manually turning equipment on and off, adjusting chemical feeders, and opening and closing valves. In a medium to large plant, automatic controls can make these operations relatively simple, with minimal staff. Automated filter backwash scheduling can minimize plant flow disturbances as filters are placed in and taken out of service.

Emergency Response

Earthquakes, line breaks, bomb threats, source contamination, and major fires require rapid adjustments in plant production. Centralized control is essential to reduce the time required to make changes in water production during crisis situations. Centralized automatic controls can reduce the response time even further.

Minimization of Energy, Chemical, and Labor Costs

Plant operating costs can be substantially reduced by using controllers that select pumps in a sequence that consumes the least energy. In a similar manner, backwash turbidimeters can reduce energy and washwater consumption by terminating backwash cycles when washwater light transmittance has reached a preset limit. Control systems that automatically place filters and pumps in service reduce the workload on operators, allowing them time to handle other duties.

Other Benefits of Instrumentation

Automated controls help ensure consistent quality of water treatment. Computers automatically log and report information and continuously provide the operator with data, as well as providing an alarm for equipment malfunction. In other words, the operator's time can be spent optimizing the process, rather than recording data onto a clipboard form.

TYPES OF INSTRUMENTATION FOR WATER TREATMENT

American Water Works Association (AWWA) Manual M2, *Automation and Instrumentation,* contains detailed descriptions and illustrations of instrumentation commonly used in water treatment plants. The following is intended to supplement the material contained in that publication and to address specific water treatment application issues for major classes of instruments.

Flowmeters

There are a wide variety of meters available for monitoring the flow of water and chemicals in a water treatment facility. Each has advantages and disadvantages to be considered such as cost, reliability, accuracy, and required maintenance.

Venturi Flowmeters. Venturi flow elements equipped with differential pressure (DP) transmitters are the most common type of flowmeter used in water treatment plants. Venturi meters first constrict flow and then allow velocity recovery of the flow stream, and in doing so are relatively tolerant of flow profile anomalies produced by upstream piping bends.

In filter control applications, repeatability is usually more important than absolute accuracy, and venturi meters are often close-coupled to the downstream filter flow control valve. Long-form venturi meters also have the lowest nonrecoverable head loss of all DP meters, making them ideal for gravity filter applications where flow would be restricted by higher piping head losses.

Orifice Plates. Orifice plates consist of a steel plate with a carefully sized hole that is mounted between pipe flanges to restrict flow. They are inexpensive but not very accurate and have higher nonrecoverable head loss. Orifice plates are useful in applications where pumped energy is not being added to the process.

Pitot Tubes. Pitot tubes require more straight upstream and downstream piping than venturis and orifice plates. Averaging pitot tubes have multiple pressure-sensing ports distributed across the pipe profile and thus require less straight pipe for metering runs than single-port types. They are still extremely sensitive to flow disturbances. Where accuracy is required, at least 10 to 20 diameters of straight pipe upstream and 5 diameters downstream are recommended by the manufacturers for averaging pitot tubes. Additional pipe diameters of straight pipe flow are required downstream of pumps and control valves, or in metering runs with bends in more than one plane.

Averaging pitot tubes can be installed in water lines up to 6 ft (1,500 mm) in diameter and can operate at line velocities up to about 15 ft/s (460 cm/s), if opposite end supports are provided for the sensor tube. Line velocities must be reduced to about half maximum value during insertion and retraction of the sensor tube. An averaging pitot tube with integral flow transmitter is shown in Figure 20.2.

Differential Pressure Transmitters for Flow Measurement. All DP flowmeters are "square-law" devices; that is, flow is proportional to the square root of the differential pressure. Transmitters develop a 4 to 20 milliampere direct current (mADC) signal converted by an electronic device to produce a readout of the flow rate. Modern DP transmitters are microprocessor based, have high accuracy over a wide range of flows, and have built-in square-root extraction with no loss in accuracy.

Typical accuracy of a DP transmitter is ±0.1% of full scale, including nonlinearity, hysteresis, and repeatability. A flowmeter element producing 100 in. (2,540 mm) of water column differential pressure at 100 mgd (380 ML per day) full scale has a transmitter error of ±0.1 in. (2.5 mm). This corresponds to an error of ±0.1 % of actual transmitted flow rate.

The same meter measuring 10 mgd (38 Ml per day) produces only 1 in. (25 mm) of DP. The transmitter produces the same DP error, which corresponds to an error of 10% of actual transmitted flow rate. A 10% error is significant if the flow signal is used for billing

FIGURE 20.2 Averaging pitot tube with integral flow transmitter. *(Photo courtesy of Dieterich Standard.)*

purposes or for pacing chemical addition, but it becomes less significant if the meter is used only for flow balancing.

Good engineering practice limits use of DP-type meters to a maximum of 5-to-1 rangeability. Higher rangeability is possible using parallel DP transmitters with split ranges, although calibration time and parts count are doubled. One manufacturer of averaging pitot tubes has recently introduced a specially characterized single-DP transmitter with claimed accuracy of ±1% of actual transmitted flowrate over a 10-to-1 flow range.

Electromagnetic Flowmeters. Electromagnetic flowmeters measure flow rate based on the velocity of the fluid through a known cross-sectional area of a pipe. Magnetic coils induce electric potential in the flowing fluid, which is proportional to fluid velocity. Potential is detected by electrodes in contact with the fluid.

Spoolpiece magnetic flowmeters are nonintrusive, have ranges up to 40 to 1 with little or no head loss, are highly accurate, and require metering runs of only four diameters upstream and two diameters downstream of the electrode plane for most situations. In small pipe sizes, the straight run piping requirements may be satisfied within the flow tube itself. Additional pipe diameters are required downstream of pumps and control valves. Magnetic flowmeters are also the most expensive meters available. They are heavy, and the spoolpieces are not field repairable. A cross-sectional view of a spoolpiece-style electromagnetic flowmeter is shown in Figure 20.3.

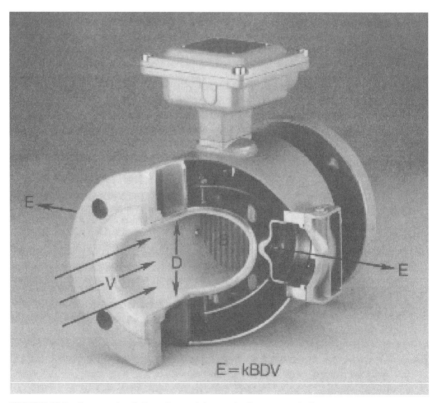

FIGURE 20.3 Cross-sectional view of a spoolpiece-style electromagnetic flowmeter. *(Source: Rosemount, Inc.)*

Magnetic flowmeters are also available with "hot-tap" style elements for installation on existing pipes. Probe meters that mount through the pipe wall are useful for smaller pipes but require longer straight metering runs than spoolpiece magmeters. Single-element magnetic probes must extend far enough into an existing pipe to be away from the viscous boundary layer of the flow profile. The magnetic sphere of influence is only a few inches around the tip of the probe, limiting accuracy in large pipes.

Multielement insertion magmeters install across a pipe diameter and provide better flow averaging, slightly higher than that of an averaging pitot tube. Multielement probes are available in lengths up to 60 in. (1,500 mm). The probe is supported only from the insertion end, so line velocity must not exceed 10 ft/s (300 cm/s). Where accuracy is required, at least 10 to 15 diameters of straight pipe upstream and 5 diameters downstream are recommended for multielement probe-style magmeters. Additional pipe diameters are required downstream of pumps and control valves and in metering runs with bends in more than one plane. A multielement insertion magnetic flowmeter is illustrated in Figure 20.4.

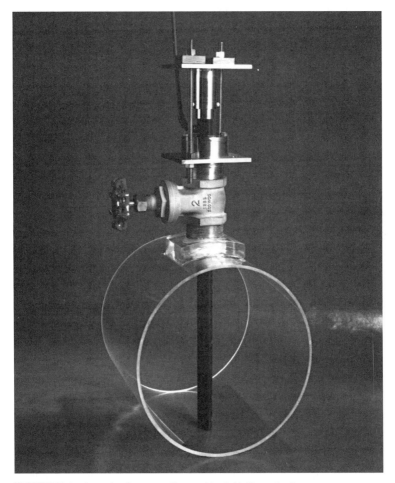

FIGURE 20.4 Averaging flowmeter. *(Source: Marsh-McBirney, Inc.)*

Ultrasonic Flowmeters. Ultrasonic transit-time flowmeters are nonintrusive and are available with clamp-on transducers that transmit acoustic energy through the walls of existing pipes. A pair of staggered transducers on opposite sides of the pipe measure differences in signal transit time upstream and downstream, and fluid velocity is inferred from this difference.

Accuracy is limited by installed piping configuration, because the computed velocity is an average taken over only one narrow path. Multipath transit-time ultrasonic meters are used on larger pipe to improve accuracy. Dual-path, dual-beam meters are available that use two transducer pairs mounted orthogonally. The signal from each pair is transmitted through the pipe wall, reflected off the opposite interior pipe wall, and detected on the originating side. Because the transducers are offset along the pipe length, this effectively yields four paths.

Ultrasonic transit-time flowmeters should not be used immediately downstream of pump manifolds because of swirl and entrained air bubbles induced by the pumps. Where accuracy is required, at least 10 to 20 diameters of straight pipe upstream and 5 diameters downstream are recommended for single-path ultrasonic meters. For multipath ultrasonic flowmeters, these requirements are approximately halved. Where flow measurement is downstream of pumps, add 30 to 50 diameters to allow air bubbles to dissipate. Additional pipe diameters are also required downstream of control valves.

Clamp-on Doppler meters are also available that operate on the principle of measuring sonic reflections from suspended particles. These are not suitable for clean water unless air bubbles are present, and accuracy is not as high as that provided by transit-time meters.

Vortex-Shedding Flowmeters. Flags wave in a breeze because of the vortices alternately shed from the sides of the flagpole. In the same manner, vortex meters infer fluid velocity from the rate at which vortices are shed behind a "bluff body." This bluff body extends all the way across the pipe diameter for line-size meters and just past the center for insertion meters. Most vortex meters measure minute deflections in the rear portion of the bluff body with strain gage sensors. Some meters sense the vortices themselves using ultrasound.

Spoolpiece meters are available up to 12 in. (300 mm) in diameter. Insertion vortex flowmeters can be installed in any line size. Accuracy decreases as pipe diameter increases because flow velocity is measured only at one point on the flow profile.

Vortex meters have wide rangeability, moderate head losses, and relatively high cost. As with other flowmeters that measure across only one pipe diameter, installed accuracy depends on the piping configuration. Where accuracy is required, at least 10 to 20 diameters of straight pipe upstream and 5 diameters downstream are recommended for spoolpiece vortex flowmeters. For insertion vortex flowmeters, these requirements increase with pipe diameter. Additional pipe diameters are required downstream of pumps and control valves and in metering runs with bends in more than one plane.

Weirs and Flumes. Weirs and flumes are open-channel flowmeters that infer flow rate from the fluid surface level near a constriction. Weirs are dams with rectangular, V-notch, trapezoidal, or other geometrically shaped openings. Flumes are specially fabricated channels that shape the flow through a throat. Water level upstream of weirs and flumes is measured using floats, ultrasound, bubbler tubes, submerged hydrostatic sensors, or capacitance probes. The mathematical conversion from level to flow rate is nonlinear and is generally handled by signal transmitter electronics.

Weirs and flumes are inherently averaging flow elements, tolerant of upstream and downstream velocity profile anomalies. Accuracies are generally lower than those with most closed-conduit flowmeters.

Propeller and Turbine Meters. Propeller meters are electromechanical devices converting the rotary motion of a spiral-bladed propeller to pulses or electrical current proportional to fluid velocity. Both spoolpiece and insertion styles are available. Propeller meters are inherently sensitive to swirl, so extreme caution must be exercised to avoid piping configurations that induce swirl. Spoolpiece propeller meters have integral flow-straightening vanes to mitigate the effects of swirl.

Typical accuracy of propeller and turbine meters is ±2% of rate over the flow ranges of 6 to 1 and 8 to 1. Head losses are less than 1 psi (7 kPa) for line sizes over 10 in. (250 mm). Insertion-style propeller meters are available for pipes up to 10 ft (3 m) in diameter.

Turbine flowmeters are similar to propeller meters, but the center of the rotating element is bullet shaped, forcing the fluid to flow through the narrow annular space between the pipe wall and the rotating element. This increases fluid velocity for a given flow, increasing the rangeability of the meter. Typical accuracies are ±1.5% of rate over the flow ranges of 40 to 1 and 100 to 1. Head losses are several times higher than those of propeller meters. Turbine meters are generally limited to line sizes under 16 in. (400 mm) and are roughly three times the cost of propeller meters. Where accuracy is required, at least 8 to 10 diameters of straight pipe upstream and 5 diameters downstream are recommended for spoolpiece propeller meters. Insertion-style propeller meters require straightening vanes to be installed in the piping if swirl is present. Additional pipe diameters are required downstream of pumps and control valves. Compound turbine meters have two differently sized flow elements in separate chambers, yielding wide rangeability.

Flowmetering Piping Requirements. As a rule, flowmeters exhibit the accuracies described on the manufacturers' data sheets only when installed in ideal piping configurations. Even magnetic flowmeters that average the flow profile over a large cross-sectional pipe area are sensitive to severe swirl. The best piping configuration to have upstream is a long run of straight pipe, preceded by a 90-degree elbow.

The worst devices to have upstream are a three-plane pipe bend, a partially open control valve, or a pump. These generate asymmetrical flow profiles and swirl. Swirl is easily induced in a moving fluid and can be carried with little attenuation for 20 to 30 pipe diameters. The flowmeters most sensitive to swirl are propeller-type meters, which read high if the swirl is in the same direction as the normal propeller rotation and low for opposing swirl.

Unfortunately, long, straight flowmeter runs are seldom available at a water treatment plant. Flow disturbance mitigation measures such as straightening vanes and tube bundles can be installed upstream of the meter. These devices should not be installed immediately downstream of pipe fittings because vanes or bundles collimate the flow disturbances caused by the fitting and simply carry them up to the flowmeter. Most line-size propeller meters have built-in straightening vanes, but additional flow conditioning may be required for severe swirl conditions.

Table 20.1 illustrates how piping configuration can affect clean water flowmeter selection for common types of meters. The left column indicates the length of straight pipe, measured in pipe diameters, directly upstream of the flowmeter. The other three columns describe the types and size ranges of flowmeters usually recommended if piping further upstream contains bends in the same plane, or bends in two or three separate planes. A three-plane bend resembling a corkscrew represents the worst case, generating severe swirl conditions. As noted, additional pipe diameters are required downstream of pumps and control valves.

Level Switches and Transmitters

The liquid level control equipment used in water treatment systems ranges from simple float switches to electronic sensors.

TABLE 20.1 Effects of Piping Configuration on the Selection of Clean Water Flowmeters

Clean water flowmeter selection guide based on piping configuration (planes) and upstream straight pipe diameters available (<6" = Applicable to meters less than 6" diameter)

Diameters	One plane	Two planes	Three planes
5	Spoolpiece magmeter Single-point probe mag <6" Averaging pitot tube <6" Orthogonal pitots <12" Multimag <12" Single-path ultrasonic <6" Multipath ultrasonic <12" Spoolpiece vortex <8" Propeller/turbine with vanes	Spoolpiece magmeter Orthogonal pitots <6" Multipath ultrasonic <6" Spoolpiece vortex <4"	Spoolpiece magmeter
10	Spoolpiece magmeter Venturi Single-point probe mag <12" Averaging pitot tube <12" Orthogonal pitots <24" Multimag <18" Single-path ultrasonic <12" Multipath ultrasonic <24" Spoolpiece vortex <12" Propeller/turbine with vanes	Spoolpiece magmeter Single-point probe mag <6" Averaging pitot tube <6" Orthogonal pitots <12" Multimag <12" Single-path ultrasonic <6" Multipath ultrasonic <12" Spoolpiece vortex <8" Propeller/turbine with vanes	Spoolpiece magmeter Orthogonal pitots <6" Multipath ultrasonic <6" Spoolpiece vortex <4"
15	Spoolpiece magmeter Venturi Single-point probe mag <18" Insertion vortex <18" Averaging pitot tube <24" Orthogonal pitots <36" Multimag <30" Single-path ultrasonic <24" Multipath ultrasonic <36" Spoolpiece vortex Propeller/turbine	Spoolpiece magmeter Venturi Single-point probe mag <12" Averaging pitot tube <12" Orthogonal pitots <24" Multimag <18" Single-path ultrasonic <12" Multipath ultrasonic <24" Spoolpiece vortex <12" Propeller/turbine with vanes	Spoolpiece magmeter Single-point probe mag <6" Averaging pitot tube <6" Orthogonal pitots <12" Multimag <12" Single-path ultrasonic <6" Multipath ultrasonic <12" Spoolpiece vortex <8" Propeller/turbine with vanes
20	Spoolpiece magmeter Venturi Single-point probe mag <24" Insertion vortex <24" Averaging pitot tube <36" Orthogonal pitots <60" Multimag <48" Single-path ultrasonic <36" Multipath ultrasonic <60" Spoolpiece vortex Propeller/turbine	Spoolpiece magmeter Venturi Single-point probe mag <18" Insertion vortex <18" Averaging pitot tube <24" Orthogonal pitots <36" Multimag <30" Single-path ultrasonic <24" Multipath ultrasonic <36" Spoolpiece vortex Propeller/turbine	Spoolpiece magmeter Venturi Single-point probe mag <12" Averaging pitot tube <12" Orthogonal pitots <24" Multimag <18" Single-path ultrasonic <12" Multipath ultrasonic <24" Spoolpiece vortex <12" Propeller/turbine with vanes

Float-Type Level Switches. Float-type level switches are cable-suspended rubber or plastic buoyant spheres containing a tilt-switch. Where turbulence is not present, cables can be tethered from the top of a pump wetwell, basin, or reservoir. In some models, a weight is molded into the cable near the float to hold the cable taut. As a falling liquid level approaches the cable weight, the float tilts down and the switch trips.

Where turbulence is present, as in a pumping wetwell, the cable should be clamped to a vertical mounting rail on the sidewall of the wetwell. In water storage reservoirs, cable-suspended level transmitters should be installed away from inlet and outlet piping. Most float switches contain mercury, and although the mercury is encased in a sealed switch compartment, some authorities feel that these float switches should not be used in drinking water for fear of contamination. For this situation, float switches that do not contain mercury are available.

Conductance Probe Level Switches. Conductance probe level switches are cable-suspended, plastic-shrouded electrodes connected to a sensitive electronic switching circuit. This circuit limits the electrical energy to the probe electrodes to reduce the possibility of personnel receiving shocks when accidentally coming into contact with the probes. A reference electrode is required for nonmetallic tanks, such as underground concrete reservoirs. This electrode must be submerged below all measuring electrodes.

The weight of the electrodes holds the cables taut. The electrode voltage is limited to less than 12 volts AC, and the current is limited to less than 30 milliamperes. The electronics module can be located up to several hundred feet from the electrodes. A typical application is for reservoir high- and low-level alarm monitoring. Electrode materials must be resistant to chlorinated water.

Submersible Hydrostatic Reservoir Level Transmitters. In situations where top access to a tank or reservoir is available, the level can be sensed by cable-suspended, submersible pressure-sensing elements that measure the hydrostatic level of the fluid above the sensor. Some manufacturers place the signal processing and transmitting electronics at the top end of the cable; others place the electronics in a sealed enclosure at the sensor (Figure 20.5).

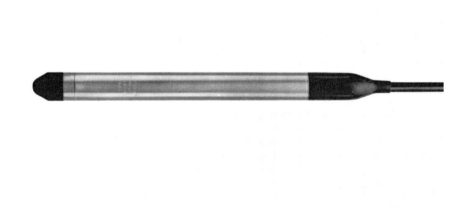

FIGURE 20.5 Cable-suspended submersible level transmitter with self-contained sealed electronics. *(Source: Druck, Inc.)*

Variations in atmospheric (barometric) pressure affect the hydrostatic reading, so a DP sensor is used with the reference side connected to a capillary vent tube, which is an integral part of the cable. If this vent tube is open to atmosphere at the top of the cable, condensation eventually fills part of the tube, causing measurement errors. Some manufacturers connect a dessicant cartridge to the vent tube to dry the air. This cartridge has an expected life of a year or two and is easily forgotten and not replaced.

The preferred design uses an elastic bladder that flexes minutely with variations in atmospheric pressure at the top of the vent tube, while preserving a factory-dried volume of air in the tube. Level ranges are available from 0 to 15 psig (0 to 103 kPa) to 0 to 300 psig (0 to 2,068 kPa), depending on the manufacturer's design. Typical accuracy is ±0.3% of span. One manufacturer provides cable lengths up to 1,000 ft (305 m) between the submersible transmitter and the electronics/breather bag enclosure.

Cable-suspended level transmitters used in large water reservoirs should be installed away from inlet and outlet piping. Care must be taken to avoid folding the cable/vent tubing because the minimum cable bending radius is about 8 in. (20 cm).

Continuous Purge Bubbler Level Transmitters. Continuous purge bubbler level transmitters measure hydrostatic level based on the pneumatic backpressure on a tube inserted into the fluid. A plastic sensing tube is clamped to the side of a basin or reservoir, or a weighted hose is dropped into the reservoir through a roof hatch. Two small-diameter (typically 0.25 in. (6mm) pneumatic impulse lines are then run from the top of the sensing tube (or hose). Air is fed from a compressed air supply at a minute, constant flow through one of the impulse lines to bubble air out of the bottom of the sensing tube. A gauge pressure transmitter measures the static backpressure on the second impulse line. A single impulse line can be used, but the pressure drop in the impulse line due to airflow results in level measurement errors. The sensing line should be protected from sunlight and other sources of heat that may affect the sensing line pressure.

The accuracy of the bubbler system is comparable with that of the cable-suspended submersible level transmitters described above. Bubbler sensing tubes are virtually maintenance free in source or treated water. However, dual compressors are required to ensure a reliable air source. Bubbler compressors, purge regulators, gauges, and component enclosures make bubbler systems more costly, require more room, and need more maintenance than submersible level transmitters.

Periodic Purge Bubbler Level Transmitters. This bubbler system employs an inverted chamber open at the bottom to trap air at the end of the bubbler tube. Without purging, air in the chamber dissolves gradually into the surrounding water, causing the water level to rise with no sensed change in backpressure. The unit is designed to have the tube and chamber charged with compressed air every hour or so to prevent inaccurate readings. The system is simpler mechanically than a continuous purge type, because the compressor is smaller and does not require an accumulator tank or a purge regulator. The controls are more complex, however, because the transmitted level signal must be held during the purge cycle. Integrated compressor/transmitter/controllers are available to simplify periodic purge system installation.

Ultrasonic Level Transmitters. Ultrasonic level sensing is noncontacting and uses transit time of ultrasonic pulses bounced off a liquid or solid surface to infer the distance to the surface. Available range spans are between 2 in. (5 cm) and 100 ft (30 m). Ultrasonic level transmitters are usually microprocessor based and have built-in signal characterization for computing volumes, or flows, through weirs or flumes. Transmitters incorporate logic to filter out effects of irregularly shaped vessels and agitator blades. Typical accuracy is ±0.1% of span.

Ultrasonic transducers should not be mounted directly in the top center of a vessel because beam reflections at nonvertical angles bounce off the sidewalls and are received at the transducer as time-delayed false echoes. Allowance must be made for a sensing "dead zone" between the maximum surface elevation and the transducer. This zone can vary from roughly 12 in. (0.3 m) for a 15 ft (4.6 m) span to 36 in. (0.9 m) for a 100 ft (30 m) span. A transducer mounting well may be required to provide proper distance between the transducer and the maximum surface elevation.

Ultrasonic level transmitters are more expensive than contacting-type transmitters and should be reserved for applications such as chemical storage tanks with top access, where contact with the fluid is not desirable.

Radio Frequency Capacitance Level Transmitters. Liquid level can be measured by changes in capacitance measured at a partially submerged coated probe or steel cable. This method uses radio frequency excitation between the probe and a grounded vessel, such as the water in a metal storage tank, to infer height of the fluid surface. Accuracies are typically $\pm 1\%$ of span. This corresponds to fairly significant errors of ± 3.6 in. (9.1 cm) in a 30 ft (9 m) deep reservoir and ± 24 in. (61 cm) in a 200 ft (61 m) deep water well. Capacitance-type level transmitters are usually microprocessor based and have built-in signal characterization for computing volumes, or flows, through weirs or flumes.

Differential Pressure Transmitters for Level Measurement in Closed Vessels. Differential pressure transmitters can be used to measure the hydrostatic level in closed vessels, such as hydropneumatic pressure tanks and surge tanks. Two pressure taps are required: a "wet leg" tap at the bottom of the vessel and a "dry leg" tap at the top. Static pressure within the vessel is measured equally on both legs of the DP measurement cell, so its effects on the measurement are canceled out. As described, DP transmitters are microprocessor based, and typical accuracy is $\pm 0.1\%$ of full scale, including nonlinearity, hysteresis, and repeatability.

Pressure Switches and Transmitters

Pressure switches used in water systems range from simple off-and-on applications to electronic equipment that monitors and reacts to a wide range of pressure differentials.

Bourdon Tube Switches and Transmitters. Bourdon tubes are hollow brass or stainless steel curved chambers that flex as internal pressure is applied. This flexure is linked to an indicating dial and to an electronic linear variable differential transformer (LVDT) sensor for signal transmission. Accuracy is typically $\pm 0.5\%$ of full scale, plus stability effects of $\pm 0.25\%$ of full scale for six months. Allowable overpressure is typically 130% of full scale. Bourdon tubes are not recommended on pumped treated water pipeline applications because of potential overpressure damage from pipeline water hammer.

A Bourdon tube can also be used to actuate either a mercury or snap-action switch. Accuracy is typically $\pm 0.5\%$ of full scale. Allowable overpressure is typically 130% of full scale.

Capacitive Pressure Transmitters. Capacitive pressure transmitters sense pressure variations as minute deflections of as little as 0.001 in. (0.025 mm). Deflections change the capacitance of a solid-state circuit connected for signal transmission. Accuracy is typically $\pm 0.2\%$ of full scale, and allowable overpressure is typically 2,000 psi (13,800 kPa).

Capacitive pressure transmitters are microprocessor based and are available with highway-addressable remote transducer (HART) digital communications capability. HART is a low-speed (1,200 bits per second), frequency shift–keyed (FSK) digital data stream, superimposed on the two current loop signal wires from the transmitter. This data can be used to remotely set the range, zero, span, damping, and other parameters, using a handheld programmer or computer connected anywhere on the current loop. Several pressure transmitter vendors have their own proprietary data communications protocol.

These "smart" transmitters offer two advantages. One is that a water system can purchase one spare transmitter that can then be reranged as needed to replace transmitters with different pressure ranges. The other is that calibration of a pressure transmitter can be accomplished using a handheld programming unit, rather than a screwdriver. The pressure at the transmitter can be easily bled to zero for zero calibration, but a full-scale pressure source is not always available. If the pressure being measured is known, the signal output of the transmitter can be forced to the correct value with the programmer. Given the high accuracy and linearity of these transmitters, this two-point calibration is often sufficient. A capacitive pressure transmitter is illustrated in Figure 20.6.

FIGURE 20.6 Smart pressure transmitter. *(Source: Rosemount, Inc.)*

Pressure Transmitter Accessories. Pressure and differential pressure transmitters should have shutoff valves and sediment traps installed as close as possible to the process piping. A sediment trap can be made from a vertical pipe nipple extending down from a tee or cross, with a cap or blowdown valve at the bottom. For highly corrosive applications, an isolation diaphragm must be installed between the process liquid and the transmitter.

Transmitter calibration is usually accomplished by connecting a nitrogen bottle and deadweight tester, or a hand-operated compressor, to the transmitter impulse (sensing) tubing. Valving should be provided to isolate the process pressure, relieve impulse line pressure for zero calibration, and connect to the calibration pressure source. Zeroing manifolds are one-piece multivalve accessories designed to accomplish these tasks for DP transmitters.

Temperature Switches and Transmitters

Temperature switches are typically filled-system bulb type. The switch is isolated from the process temperature by a filled capillary tube. For equipment monitoring applications, a brass or stainless steel bulb is mounted inside a thermowell. Typical accuracy is $\pm 2.5\%$ of full scale.

Temperature transmitters for water applications usually employ either resistance temperature detector (RTD) or thermocouple (TC) sensors. RTDs have a slightly higher accuracy but are slightly slower responding than TC sensors. Both types of sensors are installed in probes mounted directly to the transmitters. A thermowell (tapered metal sleeve screwed into a fitting welded to the process piping) allows the probe to be removed without shutting down the process. Thermowells add to the response time of the sensor because of their inherent thermal inertia. Water temperatures in tanks and pipelines do not vary rapidly, so thermowells are used for ease of maintenance

Temperature transmitters are microprocessor based and are available with HART communications capability. Accuracy is typically $\pm 0.2\%$ of full scale. Smart transmitters allow a relatively simple two-point calibration based on an ice bath and the known temperature of the water.

Analytical Instrumentation

Highly accurate bench-type laboratory instrumentation is used at water treatment plants to verify compliance with safe drinking water regulations. Critical water quality parameters can also be monitored on-line 24 hours a day for automated reporting and process control.

Chlorine Analyzers and Detectors. Chlorine gas dissolves in water to form hypochlorous acid and hydrochloric acid. The presence of hypochlorous acid and hypochlorite ion in solution is termed *free available chlorine*. If ammonia is present in the liquid being chlorinated, a reaction can take place that results in the formation of chloramines, termed *combined available chlorine*. Because there is usually limited ammonia in clean water, free chlorine analyzers are used for most applications.

Chlorine residual measurement can be affected by water temperature, pH, and total dissolved solids. Sample treatment before chlorine residual analysis generally includes a wye-type strainer for removing solids, and some analyzers require the addition of a reagent upstream of the analyzer. Several different measurement methods are used for chlorine residual, including colorimetric, amperometric, and polarographic. Amperometric and polarographic analyzers are similar in operation and use two dissimilar metals held in a solution or electrolyte. A voltage is applied to the two metals. The amount of current produced is proportional to the amount of chlorine in the sample. Colorimetric analyzers involve the reaction of the sample with a reagent that results in a color change proportional to chlorine concentration.

Amperometric Wet-Chemistry Chlorine Residual Analyzers. The wet chemistry chlorine residual analyzer is an amperometric device that includes pretreatment or conditioning of the sample. A pH buffering reagent and potassium iodide reagent are typically added to the sample before analysis. The conditioned sample is then analyzed using an electrode assembly, with the current developed in proportion to the chlorine concentration. One drawback to this type of analyzer is that the reagent is stored in bottles that require periodic replenishing. This analyzer can provide good repeatability and accuracy, because the samples are conditioned before the analysis. The analyzer is sensitive to changes in total dissolved solids (TDS), but this is not typically a problem in clean water, with normally low or relatively stable TDS.

Membrane-Type Chlorine Residual Analyzers. The membrane-type chlorine residual analyzer is an amperometric device designed not to require sample preconditioning for most applications. The sensor consists of electrodes mounted in an electrolyte solution, with the electrolyte separated from the sample by a permeable membrane. Chlorine molecules permeate the membrane and enter the electrolyte. The resulting change in chlorine concentration at the electrolyte causes a change in the electrical potential between the electrodes.

The potential depends on temperature, so most analyzers are provided with temperature compensation. Although these sensors can be mounted directly in the process flow stream, varying process head conditions can affect the transfer rate across the membrane and could cause variations in the chlorine residual readings. Flow-through assemblies in sample lines are recommended because they can maintain constant head conditions on the membrane.

Colorimetric Chlorine Residual Analyzers. Colorimetric analyzers use a reagent that reacts with either free or combined chlorine to produce a color change. The intensity of the color is proportional to chlorine concentration, and an optical sensor assembly determines the light absorbence of the sample as compared with an untreated sample. The wavelength of the light source is selected to match the characteristics of the reagent.

Drawbacks of this type of analyzer include requirement of reagent solution, interference effects of high turbidity or sample color, and slow response time. Benefits include generally low cost and simple calibration. These analyzers have typical accuracies of ±5% of range and sensitivity down to 0.5 mg/L of chlorine. They are mostly used for filtered water applications where samples are not highly colored.

Chlorine Leak Detectors. Chlorine gas leak detectors are required in facilities where gaseous chlorine is handled or is likely to become confined in a closed space. They are typically either reduction-oxidation or galvanic cell–based devices. Leak detectors are provided with low-level warning and high-level alarm contacts. Some detectors have a refillable electrolyte electrochemical sensor or an amperometric membrane sensor in a sealed assembly. Reagents are not required for the membrane-type sensor, but maintenance consists of replacement of the sensor assembly approximately every two years.

Turbidity Detectors. The measurement of the concentration of turbidity is a common tool for monitoring plant operation. Turbidity is often measured as the source water enters the plant, at various points in the treatment process, at the effluent of each filter, in the filter backwash water, and at the final plant effluent.

Surface Scatter Turbidimeters. Surface scatter turbidimeters consist of a light source and receiving photocell detector. Light is reflected off of the surface of the sample and measured by detector cells positioned at incident angles to the surface and light source. Detectors are typically oriented directly above the sample to measure the amount of light reflected at a 90-degree angle to the sample. Any particles or turbidity-causing components affect the amount of reflected light detected.

The surface scatter analyzer has a higher sensitivity than the transmittance type and has the advantage that the sensor and light source are not in contact with the sample. Where entrained air is present in the sample, bubbles scatter light, as do particles. For this reason, a bubble trap should be mounted in the sample line ahead of the analyzer. Surface scatter units are typically accurate to within 5% to 10% of reading in the 0.1 to 5,000 nephelometric turbidity units (ntu) range. Surface scatter turbidimeters are recommended for source water and filter influent.

Transmittance Turbidimeters. Transmittance-type detectors are used in clean water applications. A light source directed through the sample and a photocell detector measures the amount of light not absorbed or scattered by the suspended solids in the sample. Transmittance-type detectors are designed for use in the 0.1 to 100 ntu range. Bubble traps are recommended and are built into some models.

Accuracy is typically within 2% to 5% of reading in the 0.1 to 100 ntu range. Drawbacks include maintenance required to clean sensors. Transmittance turbidimeters are recommended for filter effluent and plant effluent applications.

Special purpose backwash turbidimeters are available for monitoring filter washwater clarity. These provide readings in 0% to 100% transmittance, not directly equated to ntu readings. The submersible sensor is mounted on or near the washwater trough. The amount of washwater required to clean a filter can be minimized by automatically controlling the wash cycle based on the water's transmittance value.

Particle Counters. Particle counters measure the size of particles in water by means of a laser light source directed through the sample and a photoelectric detector that measures the amount of transmitted light. The amount of light blockage is equated to particle size. Most analyzers are equipped with from one to eight sets of sources and receivers that can be calibrated to user-defined particle sizes.

Particle counters have a sensitivity to suspended particles in the range of 1 to 2,500 μm. The sensors are typically calibrated over a smaller measurement range, with the ranges selected to match the specific size particle of interest. For clean water applications particle counters may be calibrated to detect the following particle distribution: 0 to 1, 1 to 2, 2 to 5, 5 to 10, 10 to 15, 15 to 20, 20 to 40 μm. This way the analyzers provide information on a range of particle quantities of several size distributions. *Cryptosporidium* and *Giardia* are normally found in the 4 to 6 μm range.

Because particle counters determine the size and quantity of several particle size ranges, they produce a large quantity of data. As a result, transmitters can produce several 4 to 20 mA signals. Particle counters are now available with communications ports that allow the analyzer output data to be transferred digitally to computers. A custom software driver may be required to accomplish the data transfer.

Particle counters can detect particle concentrations that are not detectable with turbidimeters. For this reason, particle counters can be used in filtration applications where the particles of interest are smaller than can be detected using turbidimeters. Detection of particles in filter effluent can be used to initiate backwash and for filter-to-waste applications. Particle counters can also be used to help adjust chemical dosage rates because relatively small concentrations of suspended particles can significantly increase chemical dosage requirements.

Other Analyzers and Detectors. Many new automatic analyzers are now available that can continuously monitor some parameters, relieving the operator of the tedious task of sampling and running analyses, as well as providing better control of plant operations.

Fluoride Residual Measurement. Continuous on-line fluoride analyzers are based on amperometric techniques. On-line analyzers typically consist of a sample pump, sample conditioning, a constant head sample flow assembly, and a selective ion electrode. The electrical potential developed is in proportion to the fluoride concentration.

Sample conditioning normally includes adding a buffering reagent. The reagent is stored in bottles that require replenishing on a periodic basis. Typical reagent storage volume is adequate for 20 days of on-line operation. Sample temperature is also controlled for improved accuracy.

Measurement of pH. A glass membrane electrode and a reference electrode are used to measure pH. When submerged in the process fluid, an electrical potential develops between the sensor and the reference electrode. This potential is dependent on the free $[H^+]$ ion concentration.

Sensors are available in several mounting configurations, submersible style for open-channel applications and insertion type for mounting in pipelines. A common configuration consists of the insertion-type pH probe mounted in a sample side stream along with other instruments such as chlorine residual and temperature. Analyzers are available either as 120 volts alternating current (VAC) powered devices or 4 to 20 mA loop powered devices.

Streaming Current Detectors. Streaming current detectors are primarily used for on-line determination of dosage rates for coagulants. Traditional tests for determining coagulant dosing rates are usually analytical procedures such as jar tests.

Streaming current detectors measure the net effective charge of suspended particles in the sample. Samples are usually collected downstream of the coagulant dosing point, so the analysis determines the measurement on the unreacted particles only. The resulting signal can be used for feedback control of the coagulant dose.

Conductivity Sensors. Conductivity detectors are used to determine specific conductance of a solution. The analyzer is an electrical device that determines the ability of the solution to carry an electrical current. The two most common conductivity analyzers are the electrode type and the electrodeless type. The electrode type has two electrodes suspended in the sample to pass current through the sample. The electrodeless type uses induction coils linked by the sample solution. The amount of induction through the solution is affected by solution conductivity. Conductivity differs from pH in that all ions in solution affect the conductivity, but pH is affected only by hydrogen ions.

The concentration of ions in the sample solution is the main factor that affects the conductivity measurement, but temperature and ion type can also affect the conductivity determination. Solutions with few suspended solids or with relatively low conductivity usually require the electrode-type analyzer. Solutions with high suspended solids, high conductivity, or high corrosivity usually require the electrodeless type.

Sample Requirements for Analytical Measurement. Process time lag is one of the most difficult control problems to solve with analytical instrumentation. Control loops that rely on analytical signals, such as chlorine residual, operate best when the sample is analyzed as quickly as possible. If the analyzer is located some distance away from the process, by-passed sample pumping loops should be used. This arrangement pumps a relatively large volume through the sample piping to keep line velocities high. Most of the sample is by-passed back to the process by means of a drain, and a small portion is diverted to the analyzer through a wye strainer.

Analyzer process time lag is also affected by changing plant flow rate. A sample measured 30 seconds downstream of the disinfection point at maximum plant flow may be measured five minutes later at low flows. If the plant is subject to extremely wide variations in flow rate, it may be necessary to add a special process sample point for low-flow conditions.

CONTROL SYSTEM DESIGN CONSIDERATIONS

As water treatment plant controls become more complicated, many small details must be considered by the design engineer to ensure that the system is accurate and trouble free.

Signal Errors and Mitigation Measures

Signal errors introduced by process and electrical noise can render the best plant monitoring and control system ineffective. Proper installation of process transmitters and signal loops is essential for providing clean, stable signals.

Signal Errors from Process Noise. Modern process transmitters incorporate filtering and averaging to smooth their signal output. However, care must be exercised to install primary devices and signal transmitters to mitigate the effects of process noise. For example, turbidimeters and ultrasonic transit-time flowmeters are sensitive to entrained air bubbles. Sample color also affects readings of some turbidimeters. Because pressure transmitters measure minute movements of diaphragms, they should not be mounted directly on vibrating equipment. Vibration also affects many vortex flowmeters. Propeller flowmeters mounted too close to control valves or bends read erratically, and little can be done to smooth the electronic signal output without reducing accuracy and destabilizing flow control loops.

Electronic signals from process transmitters are usually 4 to 20 mADC for analog process variable loops, such as flow, pressure, and level. As the process variable changes, the current changes in proportion. Values less than 4 mADC ("live zero") are considered failed. Ground loops and electrical noise affect these signals.

Signal Errors from Ground Loops. A ground loop is an unintentional connection between two points in a signal loop by means of a ground path. Ideally, signal outputs from process variable transmitters and signal inputs at the other end of the loop are electrically isolated from ground. For example, the common wire from a 24 volts direct current (VDC) loop power supply is often grounded. The power supply may feed a pressure transmitter and a digital panel meter with a grounded 120 VAC power supply at the other end of the loop. A difference in earth potential between the two ends of the loop of only 10 mV will cause a ground current of 1 mA to flow in a loop with 10 ohms (Ω) resistance. That represents a 6% error in the 16 mADC transmitter span.

Ground loops can be avoided by using signal isolators—electronic amplifiers with no direct connections between signal input and output lines. Isolators can also be used as signal repeaters for driving multiple loads from one signal source.

Signal Errors from Electrical Noise. It is false economy to specify an accurate signal transmitter and allow electrical noise to couple into signal wiring. Any energized conductor of electricity is surrounded by an electric field. If the conductor is carrying current, a magnetic field is also present. Signal wiring is influenced by the electromagnetic fields produced by adjacent power wiring because the power and signal conductors become parallel plates of a capacitor and primary and secondary windings of a transformer.

If power conductors are carrying direct current, as from a battery or DC power supply, energy is transferred to adjacent signal wiring only when there are changes in power wiring voltage or current. If power conductors are carrying alternating current, as from a 120 VAC power panel, energy is transferred continuously to adjacent signal wiring. Table 20.2 is based on Institute of Electrical and Electronic Engineers (IEEE) Standard 518 and expanded by the author of this chapter. It provides general guidelines for separation of power and signal wiring conductors, assuming signal circuits are in galvanized rigid steel conduits.

Signal loops should be wired in shielded, twisted pair cables. Twisting conductors has the effect of canceling voltage generated by the transformer action of nearby magnetic fields, and the foil shielding acts as a barrier to electric fields.

TABLE 20.2 Recommended Separation Between Signal and Power Conduits

Power circuit AC or DC voltage	Maximum power circuit current (amps)	Minimum clearance between signal and power conduits, in. (mm)
0 to 125	Less than 30	2 (50)
0 to 125	30 to 100	4 (100)
0 to 125	100 or more	6 (150)
101 to 240	Less than 100	6 (150)
101 to 240	100 or more	6 (150)
Over 240	Less than 100	12 (300)
Over 240	100 or more	12 (300)

Types of Control Systems for Water Treatment Operations

Water treatment plants often use a combination of manual, semiautomatic, and fully automatic controls. For example, flocculator drives can be manually activated when a sedimentation basin is in service. Filter backwash sequencing may be manually initiated, with operator intervention required to advance through subsequent steps. Automatic controls for water treatment can be simple or complex, depending on the application.

Open Loop Control. The simplest and most common automatic control mode for water treatment does not make use of controllers at all. For instance, chemical addition at a fixed parts per million dosage is a linear relationship between chemical flow in pounds per day and plant flow in million pounds per day. This yields excellent results if water quality is constant. This flow-pacing scheme is an example of feedforward control.

Feedforward-only control is a special case of open loop control, where the control system does not monitor the effects of process changes due to control actions.

Two-Position/On-Off Control. Two-position is the simplest and most common automatic feedback control mode. As an example, a chlorine evaporator heater thermostat operates in two-position, or "bang-bang," control. The heater is either on or off. The thermal inertia of the heated chamber integrates, or smoothes, the rate of temperature rise and fall, resulting in temperature ramping up and down within the control band of the thermostat. Fixed-speed pumping controls operate similarly using high- and low-level float switches in a wetwell.

Proportional Control. A simple and elegant proportional feedback control system is found in every toilet tank. A ball float positions the water inlet valve proportional to the deviation, or "error," in tank level from the desired level "set point." The long lever arm on the ball float provides control system "gain," ensuring that a small excursion in tank level results in a large change in valve position.

An electronic equivalent of the above example may be found in a source water pump station containing variable-speed pumps. In this case the idea is to dispose of the water in the wetwell as fast as it enters, which means the level must not be allowed to change significantly. If the wetwell is sufficiently deep, the level may be allowed to vary over several feet.

The control system senses level as a continuously varying analog signal (usually 4 to 20 mADC) and an electronic controller compares this actual level (known as the process variable) with a desired minimum level set point stored in the controller. As the level rises above this minimum level, the controller generates an increasing output value (usually 4 to 20 mADC), which increases the speed of the pump. The difference between the minimum and maximum speed levels is known as the proportional band. A narrow proportional band requires a high controller gain, measured in percent pump speed per inch of wetwell level.

High gains often result in level oscillations, also known as hunting, when pump speed continually fluctuates. Extremely high gains result in limit cycling, where the pump speed commands are either minimum or maximum (the same as if this were two-position control). The wetwell level controller described above is direct acting, where increasing level results in increasing controller output. By comparison, the toilet tank controller is reverse acting.

Proportional Plus Integral Control. Proportional-only control requires some error to be present for a nonzero controller output. For example, if the set point for the above wetwell level controller is 6.0 ft (1.8 m), the level may reach 6.5 ft (2 m) before the pump is commanded to run at full speed. The proportional gain could be increased to reduce this constant error (referred to as droop), but control may become unstable. What is needed is another component to reset or raise the controller output when there is no deviation of the process variable from the set point. In proportional plus integral (P + I) control, this added component is the product of the error and the duration of the error, or

$$\text{Control output} = (\text{Proportional gain}) \times (\text{Error}) + (\text{Integral gain}) \\ \times (\text{Accumulated error} \times \text{Error duration})$$

The integral, or reset, term in this equation, like savings account interest, is affected by both the amount and the time held. A small control error, if present for a long enough period, results in a significant increase in controller output. The rate at which the controller integral term increases is the integral gain (analogous to savings interest rate). The integral term ramps up and down as the process variable exceeds or falls below the set point, like the charge on a capacitor.

When the error reaches zero, the proportional term is zero, but the integral term remains at its last value. In the above variable-speed pump control example, this means the pump runs at whatever speed is required to maintain the level at exactly 6.0 ft (1.8 m). Integral control generally allows system stability to be achieved with higher proportional gains than with proportional-only control. "Reset windup" can occur when the controlled device has been taken out of service, so the accumulated error times the duration eventually ramps the controller output to maximum. When the device is returned to service, the controller has to "unwind" the output back down using the proportional term. Digital control systems usually monitor the status of the controlled equipment, so a simple solution is to program the controller to force the integral term to minimum if the equipment is off.

Integral-Only Control. Integral, or floating, control is seldom used alone because it is slower than P + I control. It is also unstable if the process contains integrating elements. In the above wetwell level control example, the goal is to match pump speed to wetwell inflow based on level, which is the integral of net inflow over time. In this case integral-only control would be slow and could induce controller limit cycling. A step increase in flow would not be sensed until the level had risen for a sustained period. The resulting delays in pump speed changes would first cause underpumping, followed by overpumping, with a long period of oscillation.

Proportional-Speed Floating Control. Another integrating process element is a motorized valve, often used for filter effluent flow control. The valve actuator responds to open and close commands and stops when neither command is present. The valve travel speed is fixed by motor speed and gearing. For large errors, proportional control response is hindered by the ramping "process lag" of the relatively slow actuator. For small errors, the valve moves too far before its travel can be stopped. Reset (integral) control is not required, because the valve holds its last position when the error is zero. To minimize valve hunting for small controller errors, the duration of the open and close commands can be made proportional to the error. For small errors, the valve may be jogged only 0.25 second every 10 seconds. For large errors, the valve may be moved 8 seconds every 10 seconds.

Proportional Plus Integral Plus Derivative Control. Proportional plus integral plus derivative (PID) control has a derivative control term added to adjust controller output in proportion to the rate of change of the controlled process variable. If the process variable changes rapidly, controller output is spiked by an amount proportional to the change. This action is useful in compensating for integrating-type process lags, such as wetwell level in the above example. Derivative action is commonly used in temperature control applications where thermal inertia is a significant process lag. PID control is also known as three-mode control.

Model-Based Control. One of the most difficult control problems is process dead time, which prevents the process variable from being measured in sufficient time to make feedback control stable. For example, additional chlorine must be fed to an open reservoir to compensate for solar decay, which varies with the solar intensity and transport time in the reservoir. A mathematical model of solar decay can be included in a programmable chlorination controller that also measures solar radiation intensity, reservoir flow rate, and level.

Adaptive Gain Control. An automatically controlled process does not always have the same response to changes in controller output. A digital controller can be programmed to monitor the effect produced by the last control change on the process and to adjust the controller gain accordingly. For example, when gravity filters are clean, their flow control valves are only partially open, and a small valve movement produces a large flow change. The controller adjusts the duration of motorized filter effluent flow control valve commands, depending on the flow change produced during the last control interval.

Nonlinear Control. An automatically controlled process may have a highly nonlinear response to changes in controller output. A good example is a pH control loop, where the titration curve for the process is steeper on both sides of the neutralization pH. A controller with reduced gain near the neutralization point is more stable than a fixed-gain controller.

Fuzzy Logic Control. This comically named control method is capable of safely landing a helicopter with one rotor lost. Fuzzy logic uses fuzzy set theory, which tests overlapping logical conditions for degrees of membership and makes appropriate control decisions. Fuzzy logic emulates the way humans think, in shades of gray rather than in black and white.

If a car is parked halfway across spaces 2 and 3, you do not have to change your normal habit of parking in slot 1. If the car is parked 90% in slot 2 and 10% in slot 1, you still drive into slot 1, but slowly. If the car is parked 80% in slot 2 and 20% in slot 1, you stop and consider another slot. Your brain is overlapping the footprint area of your car with the available parking slot area left, applying rules concerning door-swing allowances, and deciding how much pressure to exert on the brake pedal.

Potential applications of fuzzy logic to water treatment include:
- Optimal treated water pump selection based on system delivery pressure, pump flow rate, pump speed, pump efficiency data, pump head-flow data, and rules of thumb contributed by plant operators
- Optimized chemical addition based on water quality data and rules of thumb contributed by plant operators

Water Movement Control

This section illustrates how the types of controls described in the previous section move water through water treatment plants.

Pressure Controls. Treated water pumps can be either manually controlled or automatically controlled, based on transmission or distribution system pressure or reservoir level. Multiple pumps can be automatically staged based on failure of the running pumps to maintain a certain pressure control range. Equally sized variable-speed pumps are usually all run at the same speed, because the pumps supply a common discharge header, and the discharge pressure reduces the flow produced by slower pumps. Pump speed is usually controlled using P + I or PID control. Pumps supplying a pressurized pipeline can be stopped based on sustained periods of reduced speed or flow rate.

Flow Controls. To stabilize coagulation, flocculation, and clarification processes, plant flow rate is generally adjusted only periodically. The plant operator issues commands to adjust inlet valves, gates, or pumps to set a new plant inflow rate.

Filters equipped with rate-of-flow controllers adjust automatically as plant inflow rate changes. A commonly used method is to transmit a level signal from the filter influent channel to a P + I level controller. The controller output adjusts the flow set points of the individual filter flow rate controllers. As plant flow rate increases, the filter influent channel level increases above the level controller set point, increasing the level controller output. This raises the filter flow controller set points, so the filter flow control valves open to dispose of more water. The filter influent channel level drops, eventually reducing the level error to zero, and the level controller integral term maintains the increased flow control set point. The level controller can be proportional-only control if the channel level can be allowed to rise a few inches at maximum plant flow.

If integral control is used, reset windup in the level controller must be prevented by suitable interlocks to force the flow to zero on plant shutdown and to the minimum flow set point on plant start up. The highest value of the level controller flow rate must always be limited to the maximum allowable filter bed flow rate.

The linking of one controller's output to another controller's set point input is known as cascade control. Figure 20.7 shows a cascade filter flow control block diagram.

Filter Backwash Controls. The operation of filters eventually progresses to the point where backwashing is required to clean the filter media. Backwashing can be initiated using any combination of the following:

- High head loss
- Elapsed run time
- High effluent turbidity
- High effluent particle count
- Predetermined schedule
- Operator judgment

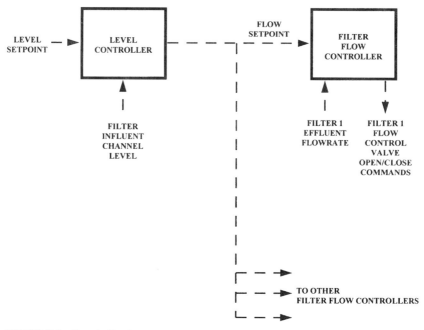

FIGURE 20.7 Cascade filter flow control block diagram.

High filter effluent turbidity can be caused by improper coagulant feed or improper filter aid feed, so turbidity is usually combined with other filter parameters to initiate backwash. Just like a batch of cookies is made in incremental steps using a predetermined procedure, once initiated, filter backwashing is primarily a batch process. As discussed in Chapter 8, several backwashing techniques are available.

A typical sequence proceeds as follows. First, a clean filter is placed in service. If none is available and the filters already on-line cannot handle the increased flow, plant inflow must be reduced or backwashing deferred. When a filter is backwashed, the dirty filter's inlet valve is closed, the filter level is allowed to drop, and the effluent valve closes. The drain valve opens, the air blower or surface wash is started, the air inlet valve slowly opens, backwash pumping begins, the backwash inlet valve slowly opens, the air wash or surface wash valve closes, the backwash pump flow ramps to low rate and holds, then high rate, and then low rate. (Backwash rates and ramping values are highly variable.) The surface wash valve (if used) closes. Later the drain valve closes to refill the filter with clean backwash water. Then the backwash valve closes and the pump stops. The filter is then clean and off-line, with all valves closed.

In many plants this procedure is manually controlled, usually from a control console on the filter deck where the operator can watch the process. Each operator has an idea of how long each step should take and why the process should not be automated. The first step in filter control system upgrading is to involve the operators in workshops to discuss their needs and preferences. As the first filter controls are placed in service, most operators ask when the rest will be done as they realize more time is available to perform other duties.

Given consistent washwater quality, a filter usually cleans repeatably in a fixed length of time. However, backwash turbidimeters can be added to the automatic controls to limit the amount of washwater pumped.

Chemical Addition Control

This section illustrates how the types of controls described above control and monitor the addition of chemicals used in water treatment processes.

Chlorination Controls. The most commonly used disinfectant for water treatment is chlorine. Where plant flow rate is constant and influent water quality is consistent, manual control of chlorination can sometimes be satisfactorily used. Most chlorination controls have automatic feedforward plant flow pacing.

Where water quality is not constant, automatic chlorine residual feedback control is required. Chlorine residual feedback control alone is not recommended, because residual analyzers can drift out of calibration if not carefully maintained. The use of both feedforward and feedback elements is known as compound control.

Flow feedforward control alone does not require a controller. The 4 to 20 mADC plant flow signal can be connected to multiple chlorinators through suitable ratio stations and signal-splitting isolating amplifiers. Each chlorinator in service will deliver from minimum to maximum chlorine as plant flow rate varies from minimum to maximum. Chlorination of plant effluent should be controlled using a flowmeter that measures flow at the point of chlorine addition as close as possible to the plant effluent. Plant inflow and outflow can differ, primarily because of backwash and plant water usage.

Most of the chlorination control-related problems for water treatment plants involve improper feedback control. For a feedback controller to maintain stable control, the measurement of a process change made as a result of control action must occur in a timely manner. If chlorine residual changes are only detected every five minutes, the controller output will ramp up and down in 10-minute cycles, as the error alternates polarity. The integral time must be increased to reduce the amplitude of these swings, which has the net effect of detuning the controller to the point where it cannot respond quickly to changes in chlorine demand.

Feedback delay, or lag time, includes three elements. The first is hydraulic transport lag within the chlorine contact tank, between the point of chlorine addition and the analyzer sample piping inlet. This time is inversely proportional to the plant flow rate. The sample piping inlet has to be far enough away from the point of chlorine addition to allow mixing of chlorine with water at the lowest plant flows. The second element is the fixed transport lag in the sample piping between the contact tank and the analyzer. The third element is the reaction time of the analyzer, which should be less than 10 seconds for moderate changes in residual.

Even if the analyzer is located directly adjacent to the contact tank, the sample should be pumped in a loop to the analyzer, with most of the sample bypassed back to the contact tank (Figure 20.8). This keeps the sample "fresh" for the analyzer, which can usually accept only small sample flows through its measurement cell. Ideally, the total lag time at lowest plant flows should not exceed three to five minutes. This may require the use of two or more sampling points in the contact tank.

Control of pH. It is often necessary to adjust pH to obtain optimum removal of inorganic contaminants, such as minerals and heavy metals, and for proper coagulation in low-alkalinity waters. Lime or caustic soda is usually added to raise the pH of the water. In that pH control is highly nonlinear, adaptive gain nonlinear control algorithms (step-by-step mathematical procedures) are necessary for optimal control.

Coagulation, Flocculation, and Filtration Aid Control. Coagulants are chemicals added to water to help suspended particles agglomerate during the flocculation process so they can settle out in the clarification process. Coagulant and coagulant aid chemicals are

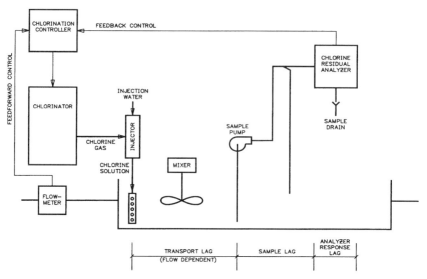

FIGURE 20.8 Compound chlorination control loop with time lags.

generally fed by positive displacement chemical metering pumps, generally under feedforward, flow-paced, open loop control. No closed loop controller is required.

DESIGN OF WATER TREATMENT PLANT CONTROL SYSTEMS

Both hardware and software available for water treatment plant automation are changing rapidly. Following is a review of the equipment and systems that the design engineer should consider in designing a modern control system.

Automated Control System Hardware

This section describes control system hardware options for implementing the water treatment plant applications described above.

Programmable Logic Controllers. Programmable logic controllers (PLCs) are industrial-grade, special-purpose microcomputers with input/output (I/O) subsystems for monitoring and controlling processes and equipment. PLCs were originally developed for factory applications where they were used for resequencing control activities when changes were made to production lines. PLCs are excellent choices for water treatment plant applications, where multiple PLCs are networked on high-speed data links to computer workstations.

The principal PLC programming language is relay ladder logic (RLL). RLL uses horizontal "rungs" as graphical representations of logical combinations of inputs or internally stored data that activate outputs. A typical rung may represent the following logical operation: "IF contact 1 is closed AND either contact 2 OR contact 3 is closed THEN output 1 is on."

Over the years analog function blocks, such as analog input averaging and PID control, have been added. Although the actual program takes up a relatively small amount of PLC memory space, RLL documentation for a typical filter backwash PLC program is about an inch thick, filled with cross-references between rungs at opposite ends of the program. PLC manufacturers provide higher-level programming tools that generate RLL code executed by the PLC but are easier for the programmer to use.

PLCs have typically been black boxes, tucked away inside control panels and closets, silently executing control logic. The current trend is to add local operator interface display panels to PLCs so that operators can observe process parameters, monitor progress of backwash sequencing, and call up alarm screens without walking to the plant control room.

PLCs generally require proprietary I/O subsystems and require specific I/O for a particular class of PLC. PLC systems are rack mounted (Figure 20.9), with a separate part number for each modular component, including cables and power supplies.

Remote Terminal Units. Like PLCs, remote terminal units (RTUs) are industrial-grade special-purpose microcomputers with I/O subsystems for monitoring and controlling processes and equipment. The differences between PLCs and RTUs diminish as products evolve. However, some general distinctions can still be made. Although a serial communications adapter can be added to a PLC for modem communications, RTUs were originally designed to be installed at remote sites, linked to a central station host by telephone or radio. RTUs tend to have somewhat more sophisticated communication capabilities, such as dual-communications port data access from two other sites, serial data store-and-forward, and select/check/operate control protocol.

FIGURE 20.9 Rack-mounted programmable controller. *(Source: G. E. Fanuc.)*

RTUs are generally a single circuit board with cable-connected I/O boards. PLCs generally have rack-mounted central processing unit (CPU) and I/O cards. RTUs are seldom programmed in ladder logic; they are generally programmed in C, BASIC, or proprietary programming languages. RTU manufacturers provide higher-level programming tools, which generate C or other code executed by the RTU.

RTUs are sold more as systems, rather than modules. They can often accommodate I/O from several different vendors and can use the same communication protocols as industry-standard PLCs. Many RTUs have built-in operator interface LCD displays and can accommodate external displays. RTUs generally lack the high-speed data highway communications capabilities of PLCs. A panel-mounted RTU is shown in Figure 20.10.

Distributed versus Centralized Control Systems Architecture. The first computerized control systems for water treatment plants relied on a central computer system for all monitoring and control logic, as well as for alarm and report generation and historical data archiving. Then PLCs became available, with higher control reliability than most computers. The PLC subsystem could handle control and data acquisition and leave the alarming, reporting, and data archiving to the computer system. The first PLC CPUs were installed in control rooms, with I/O subsystems located at various process areas.

FIGURE 20.10 Panel-mounted RTU. *(Source: Digitronics Sixnet.)*

Newer PLCs and RTUs are small, and the CPUs and I/O can be installed at various process areas within a treatment plant to split the data acquisition and control load. This trend toward decentralization will continue as microprocessors shrink in size and cost and grow in functionality. A properly designed decentralized process monitoring and control system increases fault tolerance, because failure of one subsystem affects only one process subarea, such as one filter or pump station.

Distributed Control Systems. Although PLCs and RTUs can be arranged in a distributed architecture, networked together, and supervised by a computer system, a more advanced approach is to distribute the functions of the supervisory computer system into the controllers themselves. True distributed control systems (DCSs) are a tightly integrated software and hardware solution to plant control. Control strategies can be rapidly developed on a DCS through the use of graphical configuration tools. New control logic can be tested and debugged using live plant data without disrupting the process, and then loaded into the DCS unit at the process.

DCSs are proprietary in both hardware and software, although many vendors now use desktop PCs as operator workstations. Some features available in general purpose computer process control software packages are not found in DCS systems. Examples are automated alarm paging and dialout to off-site facilities, and autoanswer voice-response status and alarm reporting.

Control Systems Design Planning

The following are guidelines for designing a control system to best use the features of modern water treatment plant monitoring and control systems.

Location of Operator Interfaces and Control Centers. A large plant should be designed so that the smallest possible staff can operate it easily and efficiently. Small plants often require more design features than larger ones, because there may be only one plant operator on duty. Some plants may be unattended on nights and weekends.

Operators should be able to respond quickly to process upsets, prompted by the plant control system, with recommended courses of action. The operator should be able to observe changing process trends as indication of potential operational or maintenance problems. If the plant has a small staff, a control system data highway should be routed through each major plant process area for operator interface workstations, even if no controllers are installed yet. This allows the operator to acknowledge alarms, access process information, and take control actions while away from the control room.

Another option is for the operator to carry a notebook-size computer when making rounds, connected by means of spread-spectrum radio to the data highway. If the plant is geographically spread out, control rooms can be provided at major process areas—any room with a tabletop to support a telephone, CRT, keyboard, and printer. This becomes a base of operations if an operator is dedicated to that process area, or a convenient location for a lone roving operator to obtain printed alarm logs and shift reports.

If there is no convenient location for a process area control room, design should specify process controllers with local operator interface (LOI) displays. These LOIs do not necessarily require full graphic displays, but they should have an alphanumeric display area large enough to display all local process variables, such as valve positions, flows, levels, and automatic commands being issued. The LOIs should also have access to plantwide data from the local controller.

Password protection should be included in the LOI for critical control actions. A water treatment plant is often the location of a supervisory control and data acquisition

(SCADA) system for source water supply or treated water distribution. The SCADA system can easily be integrated with the plant control system. Control room space should be allocated to accommodate SCADA equipment, operations staff, and mobile radio and telephone consoles.

Local Control Panels. Sooner or later any automatic control system will fail, possibly at an inconvenient time, such as the middle of an automatic backwash sequence. If the controlled process equipment is not easily accessible for manual operation and the process is critical to plant operation, local control panels may be required. These panels act as conveniently located control interfaces between process equipment (motorized valves, pumps, sensors, etc.) and automatic controllers, which are often mounted inside the panels.

Local-remote switches transfer control from the PLC, RTU, or DCS outputs to the local control switches. A common example is a local backwash control panel on a filter deck, where the operator has full view of the filtering process.

The alternative approach is to not have local panels at all, and require the operators to climb ladders in the filter pipe gallery to operate valve handwheels until the automatic controller is repaired. The logic of this approach is that 99.9% of the time the automatic control system is operational, and local control panels add considerable cost to the system. This approach is viable only if operators are continually retrained in manual operation of the plant. Automatic controls may operate perfectly for years, and operators may forget where ladders are and which valves must be operated. Most operators prefer the fallback manual control switch approach.

Control panel layouts require substantial thought by the designer to ensure that adequate space is provided for equipment, for separation of power from signal wiring, and for the panel itself in the process area. Panel layouts should be drawn to scale (Figure 20.11).

Unattended Plant Operation. A good test of a water treatment plant control system design is whether the plant operator feels comfortable leaving the control room for an extended period of time. Many small plants are attended only by roving operators or are unattended nights and weekends. Control systems designed for fully or partially unattended operation should incorporate automatic alarm paging for the operator and may require additional process and equipment monitoring.

Control Systems Software

This section describes control system software tools available to operations and engineering staff to optimize water treatment operations.

Expert Systems. Expert systems are software tools designed to apply rules taken from a knowledge base to current data in order to recommend appropriate actions to be taken by an operator. The knowledge base is a database comprised of logical rules, ideally based on years of water treatment experience by plant operators. The operator is prompted for manual data entry in a question-and-answer format, similar to the way a doctor diagnoses a patient. Recommended actions can be assigned degrees of probability. For example, a plant operator having difficulty with floc stability could enter current operating conditions into a computer system, such as plant flow rate, number of clarifiers in service, pH, conductivity, turbidity, and temperature. If the expert system software is capable of monitoring real-time data, manual data entry is minimized. The expert system computer could query the operator for special conditions such as rate of change of plant flow or water quality parameters, following knowledge base logic and rules of thumb, and then suggest the best steps

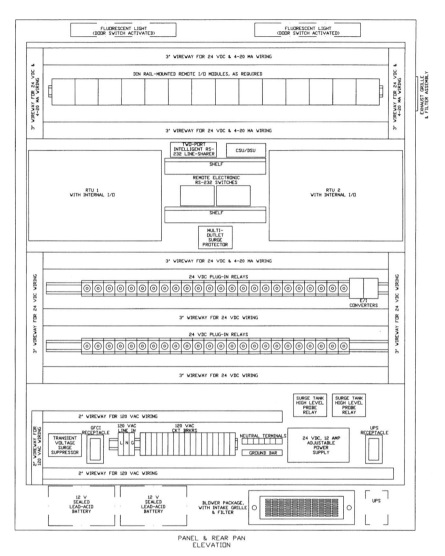

FIGURE 20.11 Pump station control panel layout, drawn to scale.

for corrective action. Expert systems can accommodate subjective parameters such as clarity, color, foaminess, and odor.

Neural Networks. Expert systems are not commonly used in water treatment. One possible reason is that knowledge base development requires considerable time, effort, and careful organization. Another reason is the time required for the operator to interact with the computer to obtain a diagnosis. In the years ahead, better tools will be available, probably using fuzzy logic in a neural network to make recommendations to operators in real time, before the process is adversely affected. Neural networks are software applications that model the human brain's thinking processes. "Input neurons" pass data through a parallel network of "hidden neurons" to a series of "output neurons."

There must be sufficient input neurons for all input variables (flow, pH, conductivity, clarifier status, etc.) that may affect the outputs (such as increase floc aid feed, add a clarifier). The network is "trained" based on dozens of sets of data elements representing input conditions and corresponding output conditions. The training process adjusts the connection weights among neurons. After many trials, neural pathways leading to the correct answer are strengthened, and incorrect pathways are weakened. Once trained, the network can generalize the appropriate output for new sets of input data.

Neural networks are best at solving problems that involve association, comparison, prediction, and pattern recognition. Alarm filtering is a potential neural network application. A dozen alarms occurring simultaneously could be distilled to a single message, such as "Backwash Control Panel Power Failure."

Information Flow to Advanced Computer Applications. The real-time database in a treatment plant's process monitoring computer can be shared with other computerized applications. For instance, water quality data can be shared with a laboratory information management system and equipment run times accessed from a maintenance management system. Combinations of water quality and equipment status data can be linked to an expert system.

Database compatibility can be ensured if the process monitoring computer system uses a structured query language (SQL) database format. Although computer databases vary in structure, most have SQL data exchange capability.

Computer-Aided Operations and Maintenance. Even the best instrumentation and control systems require specialized knowledge and routine maintenance for optimal operation. Vendor manuals provide equipment-specific operations and maintenance (O&M) information, including application software descriptions. This includes detailed descriptions for system components, such as calibration of sensors. Detailed descriptions of interrelationships among these components are best provided by the system design engineer.

Vendor and engineer-provided O&M manuals have traditionally consisted of a shelfful of seldom-referred-to, multiple-inch-thick binders stored in a closet. The same technology used to create interactive CD-ROM encyclopedias can now be applied to O&M manual production.

Interactive computerized O&M offers several benefits. First, descriptive text, photographs, digitally scanned material, computer-generated design drawings, and even video clips and animations can be indexed for random access with a few clicks of a mouse. Second, this information can be arranged in a hierarchical, layered structure, in which the user is presented with overview material first, such as block diagrams or aerial views of the plant. Plant process area-specific information is linked to computer screen targets. Keywords of text descriptions can be linked to additional text, photos, and other images.

A third benefit is that the document can be expanded and structurally modified as the plant expands. With proper training, this effort can be undertaken by water utility staff. Multimedia authoring software is maturing, and prices of computers, scanners, and digitizers are falling, making the development of computerized O&M manuals easier.

As an additional benefit, the computerization of O&M information is a valuable training tool for new operators to familiarize them with control system structure and operation. Training can also be enhanced with scored question-and-answer sessions.

Configuring a Process Monitoring and Control System

Computerized process monitoring and control systems are general purpose tools, not designed specifically for water treatment. This section describes the process of configuring control system software to acquire data and perform control actions required for a water treatment plant application.

Database Development. The software for a water treatment plant computer system assigns every monitored point a tag number, typically with a prefix such as "P" for pressure or "F" for flow, plus a number denoting the process area within the plant, and an instrument loop number. Each analog point also has an alphanumeric description (e.g., "Filter 1 Flowrate"), range limits (e.g., 0 to 2500), engineering units (e.g., gpm), alarm limits, and other data organized into database "fields."

The database includes records for all analog (continuously varying) inputs, analog outputs, discrete (on/off) inputs and outputs, digital data register points from process controllers, and computer-generated data points. Configuring the database is tedious and time consuming, but much of it is repetitive. For example, the database for Filter 2 is usually identical to that for Filter 1, except the area prefix is different. To take advantage of this, some software packages allow database development to be done in a spreadsheet format, with copying and editing capabilities.

Cathode-Ray Tube Screen Drawing and Linking. In more dvanced systems, the operator can interact with process computers using process schematic graphics animated with real-time data. As pumps turn on, their cathode-ray tube (CRT) schematic symbol color changes and can flash an alarm. Parameters such as flows, levels, and pressures display in small windows next to process schematic symbols for flowmeters, tanks, and pipe manifolds. These graphics can be highly detailed; CRT screen sizes of 21 in. (53 cm) and resolutions of 1,280 dots horizontal and 1,024 vertical are becoming commonplace.

CRT graphics consist of static portions containing piping, tanks, and structures and dynamic portions, such as pump and valve symbols that change color for various equipment states. Each dynamic element must be linked to the points database described above. These elements are drawn and linked as objects during the editing stage. Linked objects can be copied, rotated, scaled, and otherwise edited from process to process. CRT screen development is straightforward enough that water plant operators often compose their own process graphic screens.

Log and Report Configuration. Most computerized process monitoring and control systems use a spreadsheet format for composing logs and reports. Like the graphics described above, logs and reports contain both static and dynamic portions. The dynamic portions include database values and spreadsheet-calculated values, such as maximums, minimums, and averages.

Auxiliary Control Systems Equipment

This section describes ancillary subsystems required for the support of water treatment plant monitoring and control systems.

Uninterruptible Power Systems. Commercial power failures during critical process operations such as filter backwashing may result in undesirable process conditions, such as draining filter media dry. Momentary power failure and restoration causes RTUs, PLCs, and other types of programmable process controllers to reinitialize their programs. This can create a major problem if the equipment being controlled fails in an unsafe state when power is lost. Controller software must be designed to ensure that the process recovers in proper states when power returns.

Power upsets also cause computer systems to reboot, and momentary transients may cause unpredictable software operation, often referred to as glitches. Uninterruptible power systems (UPSs) can also protect equipment against harmonics and power line disturbances produced by adjustable frequency drives, and undervoltage and transient conditions resulting from starting large motors or other equipment.

UPS systems supply continuous alternating current power to the loads they serve and should be sized to handle valve actuators for critical processes, alarm panels, critical instrument loops, process controllers, control panels, computer systems, and operator interface stations. The systems are sized in kilovolt-ampere AC power (KVA) and battery operation duration in minutes. Both ratings affect physical size and cost. A power distribution panel should be included to supply all loads, and a bypass switch is recommended to allow commercial power feed directly to the load if the UPS electronics fail. Adequate room space must be allowed in the plant design for the UPS, battery cabinets, and bypass switch, with adequate working clearance around each.

Power Line Transient Voltage Surge Suppression. All electronic instrumentation is sensitive to power line disturbances. Power line transient voltages are spikes in the voltage and current waveforms whose amplitude may exceed twice normal values. The energy contained in these transients is proportional to the amplitude (volts) and duration (microseconds) and must be filtered by components sized for the anticipated worst case transient energy.

A transient voltage surge suppression (TVSS) unit is recommended for the power feeder serving a UPS system. The TVSS helps protect the UPS electronics and loads when the UPS is bypassed. Any instrumentation and control equipment not fed by the UPS should be powered from a TVSS-protected power feeder.

Communications Equipment. Space should be set aside adjacent to the main control room for fiberoptic or coaxial data highway cable entrances and equipment rack. Non-fiberoptic data highway cables should be fitted with properly grounded communication line TVSS modules. Source and treated water SCADA systems, if present, often require radio equipment, usually rack mounted in an adjacent room.

Instrumentation and Control Design Drawings

Once the instrumentation and control (I&C) system has been defined, technical drawings and specifications must be prepared to allow the owner of the plant to accurately visualize the implementation of the I&C system and to allow contractors to prepare bids for construction. This section describes some drawing types that should be included with I&C bid packages.

Process and Instrumentation Diagrams. Process and instrumentation diagrams (P&IDs) include schematic representations of piping, valving, and other process-related equipment, with all associated instrumentation and control components. P&IDs should be developed as early as possible and should be continually refined as design progresses. P&IDs are used to define the types and quantities of I&C devices, as well as the control system architecture.

Process equipment is generally drawn at the bottom of the sheet with heavy lines. Instrumentation and control components are shown as circles, or bubbles, and signal and control loop wiring is shown as dashed lines. P&ID drawing format should use Instrument Society of America (ISA) symbol and tag numbering standards.

Figure 20.12 shows part of an automated filter backwash control system P&ID. Instrument tag numbers all carry a prefix of "98," indicating their area location within the plant. Diamonds represent control interlocks, shown in more detail on elementary control diagrams for the control panels, designated by "FC" for "Filter Control Console" in this example. The locations of panel-mounted indicators and control devices are similarly designated.

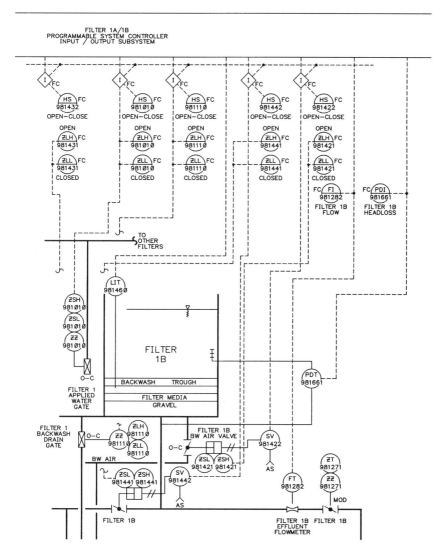

FIGURE 20.12 Excerpt from automated filter backwash control P&ID.

Interconnection Diagrams. Interconnection diagrams show the conduit and wiring linking major control system components, such as control panels, sensors, and valve actuators. Signal and control wiring connections are shown as single lines, grouped and labeled by logical functions. Multiple wiring groups may share the same conduit but are labeled individually.

Interconnection diagrams define the separation of power, control, and signal wiring into different conduits and show all inputs and outputs for each I&C component.

Instrument Loop Diagrams. Instrument loop diagrams are detailed multiline representations of unique signal and control loops within a plant. Usually only one or two loops are shown on a drawing sheet. Drawing format should use ISA loop diagram format and ISA symbol and tag numbering standards. Each wire, cable, and termination is shown, marked with signal polarity and function. Isolators and signal converters are included, as well as AC and DC power connections to instruments, and final control elements are shown.

Elementary Control Diagrams. Elementary control diagrams are "ladder" format electrical schematics for motor starter controls and control panels. Power supply to the control circuit is represented by vertical rails at the left and right sides of each diagram. The diagram shows power flow from left to right, through relay and switch contacts, to energize motor contactor and relay coils, indicator lights, and solenoids.

Elementary diagrams usually show how control logic is wired to operate on/off control devices in various control modes. They can also depict power supply and signal connections to analog control devices.

Specifications

Equally important to instrumentation and control design drawings are detailed design specifications. Typical specifications conform to the Construction Specifications Institute (CSI) division, section, and part formats. Instrumentation and controls are usually specified as Division 17. Typical sections are:

- Section 17,000: instrumentation and controls—general (project description, quality, submittal, testing, training, and documentation requirements)
- Section 17,100: process monitoring and control system (control room hardware and software requirements)
- Section 17,200: field instrumentation (data sheets and installation requirements for all instruments)
- Section 17,300: control panels (sizing, fabrication, wiring, and control diagram requirements)
- Section 17,400: programmable logic controllers (PLC hardware, software, testing, training, and documentation requirements)

Each of the above sections is divided into three parts: part one, general requirements; part two, products; and part three, execution. At the end of the I&C specifications are the most job-specific and time-consuming sections to produce:

- Section 17,901: control strategies (detailed monitoring/control descriptions for each control loop)
- Section 17,902: input/output schedules (listing of all PLC/RTU signals, plus computed points, such as flow totals)

- Section 17,903: instrument list (list of all field instruments and signal conversion devices)

The I/O schedules and instrument list are generally prepared using spreadsheet or database generation software on a personal computer. The remainder of the specifications are prepared with standard word processing software.

A Look to the Future

Increasingly powerful computer system hardware and software are overcoming limitations of previous generations of monitoring and control systems. PLC and RTU vendors are starting to implement versatile and graphical programming environments, such as the international IEC 1131 standard. The next five years will spawn at least one new generation of process controllers, display systems, computers, operating system software, and process monitoring and control software.

Trends will be toward more open controls architecture, in which controllers from several vendors can be linked on a single data network. Operator interface displays will be located conveniently throughout the plant, tied to any point on a network.

PLCs and RTUs will no longer exist as stand-alone systems. They will be replaced by distributed control modules built into control panel terminal blocks, communicating with digital sensors on an instrument network. Global monitoring and control logic will reside in local operator interfaces, supervised by control room computers. Treatment plant data networks will be fiberoptic or spread-spectrum radio.

Sadly, there will probably be little progress toward easy configuration of water treatment operations. There are only a few ways to backwash a filter and operate a pump station, yet who has seen menu-driven configuration software with basic options for basic filter backwash sequencing? Distributed control applications software is becoming more modular, graphic, and object oriented, but it is still basically a "tool kit" requiring water treatment applications to be built from scratch or from system suppliers' libraries of programs for similar jobs.

BIBLIOGRAPHY

Alstadt, R. J. "Use of Computers by Water Plant Operators." *Proceedings 1992 AWWA Computer Conference.* Denver, Colo.: American Water Works Association, 1992.

American Water Works Association. *Automation and Instrumentation.* Manual M2. Denver, Colo.: American Water Works Association, 1983.

American Water Works Association. *Instrumentation and Computer Integration of Water Utility Operations.* Denver, Colo.: American Water Works Association, 1993.

Cho, C. H. *Measurement and Control of Liquid Level.* Research Triangle Park, N.C.: Instrument Society of America, 1982.

Coggan, D. A. *Fundamentals of Industrial Control.* Research Triangle Park, N.C.: Instrument Society of America, 1992.

Considine, D. M. *Process/Industrial Instruments and Controls Handbook.* 4th ed. New York: McGraw-Hill, 1993.

DeLaura, T. J. "Building a Migration Plan for Computer Systems." *Proceedings 1992 AWWA Computer Conference.* Denver, Colo.: American Water Works Association, 1992.

Gaushell, D. J. "Expert System Development for Water Treatment Plant Backwash." *Proceedings 1992 AWWA Computer Conference.* Denver, Colo.: American Water Works Association, 1992.

Gillam, D. R. *Industrial Pressure Measurement*. Research Triangle Park, N.C.: Instrument Society of America, 1982.

Gillam, D. R. *Industrial Level Measurement*. Research Triangle Park, N.C.: Instrument Society of America, 1984.

Hughes, T. A. *Measurement and Control Basics*. Research Triangle Park, N.C.: Instrument Society of America, 1988.

IEEE Guide for the Installation of Electrical Equipment to Minimize Electrical Noise Inputs to Controllers from External Sources, Standard 518. New York: Institute of Electrical and Electronics Engineers, 1977.

Instrument Society of America. *Standard S51.5, Process Instrumentation Terminology*. Research Triangle Park, N.C.: Instrument Society of America, 1979.

Instrument Society of America. *Standard S5.4, Instrument Loop Diagrams*. Research Triangle Park, N.C.: Instrument Society of America, 1979.

Instrument Society of America. *Standard S5.1, Instrument Symbols and Identification*. Research Triangle Park, N.C.: Instrument Society of America, 1984.

Liptak, B. J. *Instrument Engineer's Handbook*. Radnor, Penn.: Chilton, 1969.

Mollenkamp, R. A. *Introduction to Automatic Process Control*. Research Triangle Park, N.C.: Instrument Society of America, 1984.

Murrill, P. W. *Fundamentals of Process Control Theory*. 2nd ed. Research Triangle Park, N.C.: Instrument Society of America, 1991.

Spitzer, D. W., editor. *Flow Measurement*. Research Triangle Park, N.C.: Instrument Society of America, 1991.

Water Environment Federation. *Manual of Practice 21, Instrumentation in Wastewater Treatment Facilities*. Alexandria, Va.: Water Environment Federation, 1993.

CHAPTER 21
ELECTRICAL SYSTEMS

The primary objectives for a water treatment plant electrical system are safety and reliability. The standard for safety adopted in virtually every locality is National Fire Protection Association (NFPA) Standard 70, the National Electrical Code (NEC), which states that an electrical system that complies with the code and that is properly maintained "will result in an installation essentially free from hazard but not necessarily efficient, convenient, or adequate for good service and future expansion of electrical use." The purpose of this chapter is to focus on the elements of electrical design that are not covered by the NEC. The system must be safe, thus code compliance is a must. The system also needs to be efficient, convenient, and expandable. Of particular importance for water plants, it must be reliable.

Any degree of reliability can be obtained if cost is no object. However, cost is important, so the electrical system must be a compromise between the competing objectives of maximum reliability and minimum cost. Water treatment plants have a fairly typical complement of loads: namely two large blocks of load for the source water and finished water pumping, and one or more smaller blocks of load for filtration, chemical addition, administration, and other functions. With this load complement in mind, we will look at specific design configurations and their effect on reliability and cost.

ELECTRICAL SYSTEM RELIABILITY

Electrical systems, particularly power distribution systems, are extremely reliable. Equipment is usually designed for a 20-year life, but it is not uncommon to find equipment that has been in service for 30 or 40 years. Fault-protective devices are expected to sit for years, often unattended and forgotten, waiting for something you hope never happens: a failure of the electrical system. When a failure occurs, the protective devices must react to clear the fault and prevent escalation of the damage.

Although systems and equipment are reliable, anything, of course, can fail. To evaluate reliability, the general rule is to look at the consequences of a failure of any one item. There is always the possibility of two major pieces of equipment failing at the same time, but few electrical systems cannot be completely disabled by two simultaneous failures at critical locations. In general, if the equipment reliability is high, the probability of multiple simultaneous failures is low. An electrical system designed to function with multiple simultaneous failures is extremely costly.

So, given the failure of any one item, what happens? How much of the plant is down? How long can the system get by without it? How long will it take to repair the failure?

If the water system can get by without power for a longer time than it takes to repair a problem, spending money for redundancy for that item is questionable. On the other hand, if repair will take longer than the system can stand, redundancy is definitely needed.

When looking at repair time, it is not necessary to look at the time required to repair something to its installed condition. Emergency or temporary fixes are often possible. For example, if a large substation or distribution transformer fails, it will typically take months to get a new unit. Before spending money on a redundant transformer, a check should be made with the power company to see what stock of transformers they maintain. Some power companies stock transformers in sizes up to 2,500 kVA in their standard voltage ratios, and they may be willing to loan one to the water plant if needed in an emergency. Theirs may be an outdoor, oil-filled unit where yours is an indoor, dry type, but you can still borrow theirs, place it outdoors, and temporarily cable into the building, thus reducing the repair time from four months to four hours.

Power Supply Reliability

In considering possible failures, the first place to look is at the source of power. How reliable is the utility supply? If an existing plant is being expanded, the answer is probably already known. If a new plant is being designed, a check should be made with the power utility to see if they have historical data on their system that tells the frequency of outages, the average length of outages, and the maximum length of an outage.

If the worst case outage is tolerable, the power utility's supply may be deemed satisfactory. Unfortunately, this is not always the case. If the treatment plant location is served by the utility subtransmission or transmission system (incoming voltage higher than 34.5 kV), it is probably reliable. More often, though, supply will be available from the utility distribution system (typically 15 to 25 kV) that is less reliable. The designer must keep in mind that a failure of the power supply normally disables the whole plant, and, unless there is sufficient storage in the water distribution system to last through the time the plant is down, some form of redundancy in the supply is needed.

U.S. Environmental Protection Agency (USEPA) guidelines for wastewater plants require that vital components be supplied by two sources of power. These may be two utility services from separate substations or a single utility supply and a generator. Although not mandatory for water plants, these are good guidelines to follow. In addition, individual states may have specific guidelines or requirements for water plants.

The starting point for the design engineer is to find out what the power company can offer. It is also necessary to ask if they charge extra for service arrangements above or beyond their standard. The utility may say they can bring in two lines to the plant, but it is necessary to find out where they originate. If both lines originate from the same substation, loss of that substation still leaves the plant without power. If they are at any point suspended from the same pole, there is also a possibility of them both being taken out by the same falling branch or auto accident.

In the same manner, if there are two separate feed lines but they both supply a single transformer for the treatment plant, loss of that transformer again leaves the plant without power. If it is possible to obtain two services that do not have a common point that could disable the power supply, it is necessary to stipulate that they must both be live at all times. This is so the plant operator can accomplish any switchover without having to call the power company. The power company may balk at this, their concern being that the two services could inadvertently be tied together, which could create a major problem on their system. They will probably require key interlocks, protective relays, or other features to prevent interconnection of the two sources.

If a sufficiently reliable service from the power company is not available or is prohibitively expensive, the alternative is to install an on-site generator. Generators are expensive, both to purchase and to operate. They also require regular use and maintenance to dependably provide standby power. For these reasons, generators are usually less desirable than a second service from the power company. If an on-site generator is installed, the power company may have rate incentives for operating the generator at their peak power demand times.

To size a generator, it is necessary to decide how many source water and finished water pumps will need to be operated for the duration of a power outage. If pumps are to be operated, a water plant needs to operate almost everything that requires power, or the few things that can be done without are of little consequence in terms of load. Therefore the generator size is usually the power required to operate essential pumps plus the power for the rest of the plant.

Power Distribution Reliability

Once a suitable power supply is decided on, the next step is to distribute the power to the plant equipment. Several system designs provide increasing levels of reliability and cost. For each water plant, the power system consists of four loads, two large loads for the source water and finished water pumping and two smaller loads designated filter/chemical and administrative/laboratory. For a typical plant, the two large loads are supplied at primary voltage (4,160 volts) and the two small loads have transformers for supply at a lower, or secondary, voltage of 480 volts.

A one-line diagram is used to depict the main elements of a power distribution system. These include the circuit breakers or switches, motor starters, transformers, cable interconnections, and significant loads. Figure 21.1 is a list of the symbols generally used for the one-line diagram.

Radial Electrical Systems. The simplest and least expensive power distribution system is the radial system (Figure 21.2). The radial system provides no redundancy, thus it is the least reliable of all system configurations. A failure at any point in the system interrupts power to everything downstream of that point. For the example in the figure, if the conductors between a source water pump starter and pump motor fail, that pump is without power. However, if the conductors between the main switch gear and the source water pump station fail, the entire source water pump station is without power. The higher up in the system the more the impact to the plant in the event of a failure.

Primary Selective Systems. The term *selective* means the ability to isolate a fault and switch to select a different path to provide power. A primary selective system provides this ability for some or all parts of the primary system. Stated another way, we have redundancy for some of the upstream portion of the system where a failure causes the greatest disruption of power.

Figure 21.3 shows a loop feed system, which is an example of a primary selective system. The various blocks of loads are connected to the main switch gear in a loop, starting from the supply, continuing each to the next, and then back to the supply.

In normal operation the loop is "broken" somewhere near the middle. In the figure, the break is at the switch designated N.O. (normally open) at the filter/chemical area. Now, on occurrence of a fault between the main switch gear and the source water pump station, the left feeder breaker in the main switch gear must trip to clear the fault. At this point both the source water pump station and the filter/chemical area are without power, but once the location of the fault is determined the loop can be "reconfigured." In this example, this is

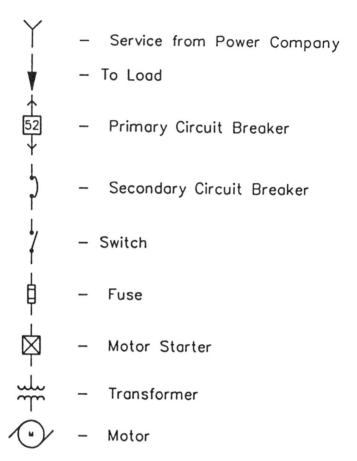

FIGURE 21.1 One-line diagram symbols.

accomplished by opening the left main breaker at the source water pump station to isolate the fault. Then the N.O. switch at the filter/chemical area is closed to restore power to all loads.

Comparing the loop system with the radial system, notice that the loop has five primary conductor runs between the main switch gear and the load areas, whereas the radial system has four. The one additional conductor run provides backup for the other four. Thus, based on conductor runs only, a slight increase in cost yields an appreciable increase in reliability. It should be noted, however, that the actual cost difference depends on the location of the loads, or the total length of conductors. The loop system also requires that each conductor run be sized for the total load, and additional switching is necessary.

Secondary Selective Systems. For both the radial and loop systems discussed above, a failure of the transformer supplying the filter/chemical loads results in a loss of these loads until the transformer can be repaired or replaced. Secondary selective systems merely ex-

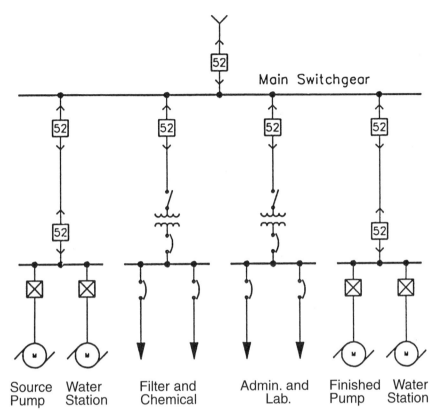

FIGURE 21.2 Radial distribution system.

tend redundancy downstream in the distribution system to provide an alternate path for power to secondary loads. Redundancy can be taken down as far as desired, but the cost increases exponentially the farther you go. This is because the distribution system is naturally branching out as you go down, from the entire plant, to areas, to groups of loads, to individual loads.

A popular configuration where very high reliability is desired is the "double-ended" or "main-tie-main" arrangement. It starts with two sources of power, two transformers if transformation is required, and two main breakers, each of which supplies half the loads. With this, no one failure can disable more than half the loads. A tie breaker is then added to allow either source to supply any load. Like the loop system, one more item is added to provide backup for several items. Figure 21.4 shows a double-ended configuration for the main switch gear and one secondary load area. In normal operation, the tie breakers are open and each transformer supplies half the secondary load. If a transformer fails, the associated main is opened and the tie breaker is closed. The other transformer then supplies the total load.

Full redundancy requires that the power train components (service conductors, transformers, and main breakers) be sized for 100% of the total load, even though they normally only carry 50% of the load. The tie breaker can be sized for 50% of the load because that

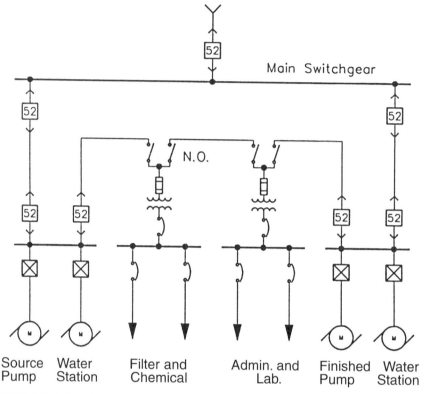

FIGURE 21.3 Loop feed system.

is all it will carry, but in practice the tie is sized the same as the mains so that it can double as a spare if a main breaker fails. The cost of full redundancy is usually more than twice that of the equivalent items in a radial system. The cost can sometimes be reduced by sizing components at a little less than full redundancy.

For example, perhaps the only time the plant actually operates at full load is for the hottest few weeks in the middle of August, so the transformers might be sized at 70% of the total load. Any other time, the 70% transformer would be sufficient to carry all loads. So if one transformer fails in the middle of August, the plant would have to either curtail some load or allow the other transformer to operate overloaded. Within limits, transformers can be overloaded, albeit with some decrease in the expected life. Because the only time overload operation would occur is on a transformer failure that happened to occur in mid-August, this might be considered an acceptable compromise to save 30% on transformer costs.

The design of a typical water plant might also use a combination of power distribution arrangements. A double-ended arrangement can be used for the main switch gear and the source and finished water pumping, and a loop feed system for the remainder of the plant loads. Large load–handling equipment generally takes longer to fix if something fails, so more redundancy is included. The loop feed system is an economical means to supply a number of small loads.

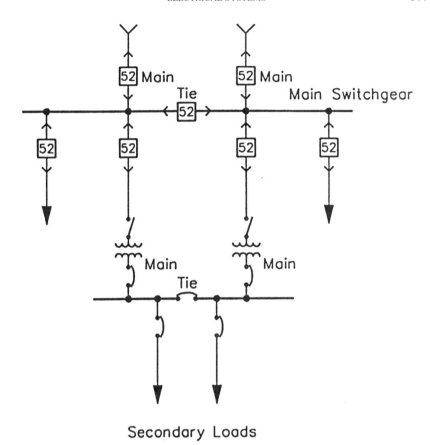

FIGURE 21.4 Double-ended configuration.

As another example, if a plant has 24 hours or more of source water storage, just about anything can be repaired or patched in this period of time, so consideration should be given to using a radial feed to the source water pump station.

ELEMENTS OF ELECTRICAL SYSTEM DESIGN

Reliability is second only to safety in a water plant electrical system. The power supply and power distribution system determine the overall reliability. However, there are other elements that make the electrical system efficient, convenient, expandable, and so on.

Voltage Selection

The most economical design results when the voltage is matched to the load. The larger the load, the higher the voltage should be. To select a service voltage, the designer should start by looking at the large loads, the pumps. For a single motor, including starter and conductors, the most economical voltages are:

- For motors up to 500 hp—480 V
- For motors 500 to 5,000 hp—5 kV (i.e., 2,400 or 4,160 volts)
- For motors larger than 5,000 hp—15 kV (i.e., 12,470, 13,200, or 13,800 volts)

For an entire plant, the approximate total horsepowers for different voltages are:

- Up to 2,500 hp—480 V
- From 2,500 to 10,000 hp—5 kV
- Above 10,000 hp—15 kV

It is generally desirable to have the service voltage match the most economical motor voltage for the large pumps. By so doing, there is no need to buy transformers for these loads. However, this does not work in all cases. For a small plant, 480 V will work fine for both the service voltage and the motor voltage. For a medium-size plant, 5 kV will probably be best for the service, but 480 V will still be the most economical for the motors. For a still larger plant 5 kV works fine for both, and so on.

In cases where the plant size is on the borderline, it is generally acceptable to stretch the various ranges. For example, the plant size may dictate a 5 kV service, but the largest pumps are only 300 hp. The designer can go ahead and use 5 kV motors, because the savings in the cost of the transformer can be applied to the somewhat higher motor and starter price.

When selecting the voltage to use, the horsepower values should include any future expansion contemplated. For any uncertainty, it is preferable to err on the high side when selecting the voltage. Captive transformers offer a good solution when dealing with large motor loads that may grow still larger in the future. A captive transformer is an arrangement where a breaker feeds (and protects) both a transformer and a motor as a unit. The transformer secondary voltage is matched to the motor size. On future plant upgrades the voltage may be raised or the transformer may be eliminated, but the primary distribution system does not change except for the relay settings.

Finally, voltage selected for a new plant should be coordinated with the power company to make sure the voltage that has been selected is standard with them. Also, the power company may offer incentives if you take service at a voltage higher than you might otherwise want, because this will save them the cost of transformation.

Equipment Selection

Electrical equipment that must be selected in the design of a water treatment plant includes fault protective devices, transformers, and various types of lighting.

Fault Protective Devices. Fault protective devices include fuses, protective relays, and circuit breakers. There are two types of circuit breakers: power circuit breakers and molded case circuit breakers. These devices all have the job of clearing a fault on the electrical system should one occur. To do this, they must have an interrupting rating that is greater than or equal to the maximum fault current that can occur.

Fuses have the advantage of high reliability and high interrupting ratings at relatively low cost. On the downside, fuses must be replaced when the fuse blows. Another problem is that there is the possibility of blowing one fuse in a three-phase circuit while leaving the other two fuses active, a situation called single phasing, which can quickly burn out a three-phase motor if the motor is not shut down immediately.

Circuit breakers can be reset and are not subject to single phasing, but their cost, particularly for high interrupting ratings, is higher. Power circuit breakers have the advantage of being maintainable, or repairable if they break. Molded case breakers, for the most part, are throwaway devices. As always there are trade-offs; power circuit breakers require periodic maintenance whereas fuses and molded case breakers do not.

Protective Device Settings. It is extremely important that protective devices be correctly set. Fuses must be selected for the correct amperage rating, and power circuit breakers must be correctly adjusted. The process of selecting settings is a coordination study, which involves a compromise between protection and coordination.

The best protection of equipment results when the nearest protective devices react as quickly as possible when they sense a fault. The best coordination results when the upstream protective devices do not react so quickly because it gives a downstream device a chance to clear the fault. Power circuit breakers have a definite advantage for coordination in that they can be set to not trip for a short time on any fault current up to their rating. Fuses and molded case breakers will react virtually instantaneously at high fault currents. This is inherent in their design, and there is no way to set them otherwise.

Transformers. Although many types of transformers are available, they can generally be categorized by the method used for insulation and cooling, the two most common being oil and air. Air (or dry) units are used almost exclusively in small sizes. Dry transformers are nonflammable and can be used indoors or outdoors.

In large sizes, oil-filled transformers offer a cost savings because oil is a more efficient cooling fluid than air. Oil-filled units are flammable and therefore are mostly used outdoors. To use them indoors requires a vault, which does away with the cost advantage. If an oil-filled unit is selected, the designer must be sure that it is situated such that no oil can find its way into the water supply in the event of a rupture of the transformer tank.

Lighting. Other than administrative spaces, which may warrant some aesthetic consideration, plant lighting should be functional and utilitarian. Industrial lighting fixtures are categorized by application and light source. Applications include indoor, outdoor, wet location, corrosion resistant (for example, where chlorine gas may be present), and others. Light sources include incandescent, fluorescent, and high-intensity discharge (HID).

Fluorescent fixtures do well in spaces that are intermittently occupied because they provide instant illumination when they are switched on. Although HID sources are highly efficient, they require time to warm up when first turned on. Therefore they are the best choice for areas occupied most of the time, where the lights can be left on for long periods of time or continuously. Incandescent lights are almost never used in a water treatment plant except for specialty lighting. For example, incandescent spotlights are often used over filters because they can penetrate the water surface down to the filter medium, allowing the operator to observe filter operation.

In considering lighting, the designer must not forget the plant exterior. Task lighting should be provided for equipment that requires nighttime operation or maintenance. General site illumination of at least 0.5 to 1 foot-candle (ft-c) should be provided to allow workers to safely see their way around the site and for security. Site lighting is usually controlled by a photocell or time clock to operate automatically.

An efficient way to light a site is by high-mast lighting. This consists of a group of high-wattage fixtures mounted on tall poles with a lowering mechanism so that lamp replacement can be done on the ground. It takes only a few poles to provide lighting over a large plant site. One caution, however, is that they should not be used close to a residential area because they will be bothersome to neighbors.

The use of extension lights around a water plant presents a definite hazard, so the design engineer should make a special effort to provide lighting that will minimize the need for extension lighting.

Lightning and Surge Protection

One of the advantages of the pneumatic control systems that were commonly used in water treatment plants until a few years ago is that they are not affected by any electrical disturbances. The solid-state, microprocessor-based controls and communication systems in use today are vulnerable to electrical surges. Water treatment plants present more problems than, say, an office building.

For an office building, a system of lightning rods on the roof can be connected to a good grounding system, which will provide adequate protection for the building envelope against a direct lightning hit. The electrical and phone services can also be easily protected against outside power surges, and the electrical equipment in the building is generally isolated. A water treatment plant, in contrast, consists of multiple buildings scattered over the site with numerous interconnections for power, process signals, and communications, so adequate protection involves several steps.

Lightning Rods. Lightning rods, also called air terminals, when connected to a good grounding system, provide the treatment plant's first line of defense against a direct lightning strike. They channel the current from the lightning through a defined path between air and earth. The alternative is to let the current find its own way, possibly through your computer network. Air terminals work best when located high. A water storage tank located at a treatment plant will provide an umbrella to shelter a significant portion of the adjoining plant.

Power System Arresters. Power system arresters protect against surges traveling on the power system. The best protection results by using a tiered approach, with the first protection at the power company main service panel. This is designed to absorb a large amount of energy and will take the brunt of a lightning hit. Branch circuit arresters can then be installed further down in the distribution system to provide additional protection.

Signal System Arresters. Signal system arresters are a continuation of the tiered approach. The devices provide protection at still lower voltages, thus providing still better protection. It is not uncommon to provide protection for all signal and communications terminations, particularly those that originate outside or in another building.

For the best in signal and communications protection, fiberoptics should be used where available. The price premium is not that great compared with the consequences of losing communications or data. Prices for fiberoptics continue to fall as technology advances, and problems that existed in the past, such as reliable terminations, have largely been overcome.

Surge Capacitors. The various types of arresters discussed above seek to lower the magnitude of a surge. However, motors, which abound in water treatment plants, are vulnerable not only to the magnitude of a surge, but also the surge voltage rate of rise. Surge capacitors are installed to lower the rate of rise. As a minimum, each motor control center should be equipped with a surge arrester or surge capacitor.

Power Factor Correction

Power factor is defined as the cosine of the angle between the voltage and the current in an electrical system. If the electrical load is resistive, the voltage and current are in phase, the angle is zero, and the power factor is at its maximum value of 1.0. Most types of electrical equipment (motors, ballasted lights, etc.) are a combination of resistive and reactive loads, which results in a power factor of less than 1.0. The consequence of low power factor is that the system must carry higher currents than would otherwise be required to supply the same load at a higher power factor.

Some power companies offer rate incentives for improved power factor, and some charge penalties for low power factor. The typical water plant, with pumps being the bulk of the load, will not achieve even 90% without some type of correction. Poor power factor results from motors and numerous other devices that consume reactive power, or VARs (volts ampere reactive). To improve power factor, something must be used that will supply VARs, or at least not consume as much. The two most common methods of correction are use of synchronous motors and installation of power factor capacitors.

Use of Synchronous Motors. Induction motors are generally more economical than synchronous motors. However, in high-horsepower (about 3,000 hp and up) and slow-speed (600 rpm or less) ratings, the total costs of the initial purchase plus long-term operation usually favors synchronous motors. These ratings are applicable to large plants, but the designer of a medium-size plant may still want to consider synchronous motors for power factor improvement.

Standard synchronous motors are available either rated unity power factor (they neither supply nor consume VARs) and 80% power factor (they supply VARs). When installing equipment that supplies VARs, they should never supply more than is consumed by other equipment in the plant. A host of evils, too numerous to mention, result if the power factor is overcorrected. If a plant should have overcorrection with a synchronous motor, it can be adjusted by lowering the field excitation.

Installation of Capacitors. Capacitors supply VARs in fixed amounts based on their rating. They cannot be adjusted, so to avoid overcorrecting they are usually switched with the load. When a load needing VARs (an induction motor) is turned on, an appropriately sized capacitor is turned on with it. It is not practical to equip every motor with a capacitor, nor is it usually necessary.

Depending on plant size, by providing capacitors for motors of 25 hp and above, the plant will usually achieve a power factor in the low- to mid-90% range. If additional improvement is needed, VAR controllers are available. These consist of several steps of capacitors that are automatically switched on and off in response to the overall plant power factor.

Motor Starting and Control

A common lament in water plants is "Whenever I start my large high-service pump, lights dim all over the plant." It is well known that motors draw more current when starting than when running. This causes the power system voltage to drop, and depending on the amount of drop, it can have various objectionable results. Some effects of voltage drop are as follows:

Percent drop	Result
3	Lighting flicker
10	The operating minimum for many solid-state devices such as variable speed drives
15	The minimum pickup for contactors and relays
20	The minimum operating voltage for fluorescent lamps
30	Operating motors may stall and contactors and relays drop out

Depending on other equipment in the plant, a 15% drop is usually tolerable. Some plants may be able to stand a 25% drop. At 30% or more, the power system begins to topple like a house of cards.

The voltage drop is related to the system fault current. Whereas a higher fault current results in higher protective device costs, as discussed previously, it helps by lowering the voltage drop.

The following formula gives the approximate voltage drop during motor starting:

$$VD = 100 \times IM / (IM + IS)$$

where VD = voltage drop in percent
IM = motor starting current
IS = system fault current

The motor starting current (IM) is approximately six times the motor full load running current if the motor is started with full voltage. To lower the voltage drop it is necessary to either lower the motor starting current or raise the system fault current. The system fault current is generally set by the power company and the configuration of the distribution system, so little can be done about it, but things can be done to change the motor starting current.

Reduced Voltage Starters. Full voltage starters are the least expensive and are used if the voltage drop is not excessive when motors are started. There is sometimes a concern about mechanical stress with full voltage starting, but this is almost never a problem with a direct coupled motor. Some designers also believe that reduced voltage starting will lower motor heating during starting (thus permitting more frequent starting) or that it will lower demand charges from the power company. These are both false.

The amount of motor heating and energy consumption during starting depends primarily on the final speed and the inertia of the pump, not the motor current or voltage. Unless the power company mandates a starting limitation, reduced voltage starters are usually needed only if the voltage drop with full voltage starting will be excessive.

Common reduced voltage starter types are autotransformers, wye-delta starters, and solid-state starters. Resistor and reactor types are seen occasionally, but these extract a penalty in the amount of torque the motor can deliver per amp of starting current.

Autotransformers. Autotransformers offer a selection of adjustments, or taps, of 80%, 65%, and 50% of full voltage for starting. Starting voltage is reduced to the percentage indicated by the tap value and starting current to the square of that value. For example, on an 80% tap, the starting current will be 64% of normal (i.e., 384% instead of 600%).

The starting dynamics require an 80% tap to accelerate the motor to the speed at which it develops maximum torque. The speed at maximum torque is the point where the current declines rapidly from the 600% starting value to the 100% running value. Stated another way, on the 50% and 65% taps you get a jolt of current (and corresponding voltage drop) when the starter makes the transition from starting to running. This jolt will approach the 600% full voltage value. It will not last as long as with full voltage starting, but it can still cause problems due to voltage drop.

These starters offer flexible reduction in starting current and a fairly simple design, and they require no special motor.

Wye-Delta Starters. A wye-delta starter is designed to connect the motor in a wye configuration during starting and then change to delta configuration after the motor has accelerated. The starting voltage is thus reduced by $1/\sqrt{3}$ (e.g., 277 volts for a 480-volt motor) during starting. In this way, the starting current is reduced to about 200% of running current compared with about 600% for full voltage starting. Wye-delta starters offer substantial reduction in starting current with a fairly simple design. On the downside, there are no adjustments available and it cannot be used on all motors. The motor must be delta connected with six leads out. The transition current can also be a problem.

Solid-State Starters. Solid-state starters apply a brief period of higher voltage to overcome breakaway torque, and then slowly raise (ramp) the voltage up from zero to full voltage over an adjustable time period. These starters provide flexible adjustment of starting conditions and do not have the potential transition current problem. The drawback is that their design is considerably more complex than wye-delta starters and autotransformer.

Adjustable-Speed Drives. Constant-speed motors should be used whenever possible because they provide the lowest cost and highest reliability. Adjustable-speed drives are considerably more complex and cost more both initially and for maintenance. However, hydraulic requirements sometimes dictate that they be used. Variable-speed drives is a vast topic that will only be touched on here.

Slip Drives. Slip drives include variable voltage, eddy current clutch, and wound rotor with secondary resistors or liquid rheostat. They were once popular but are rarely used anymore because of their poor efficiency. A slip drive can achieve an efficiency no better than the operating speed. For example, if you operate at 70% speed, the efficiency will be something less than 70%.

Variable-Frequency Drives. Variable-frequency drives (VFDs) have become the most popular type, and for good reason. They deliver efficiencies of about 95% or higher across the entire speed range. Although still a complex piece of equipment, the reliability of VFDs has improved greatly from their early years, when they were almost impossible to keep running.

They also deliver the ultimate in reduced voltage starting. They can bring the motor up to full speed and not exceed the full load running amps. VFDs operate with mostly standard squirrel cage induction motors. One problem is that a VFD can cause higher than normal motor heating and voltage stress; because of this it is advisable to use motors with lower than normal temperature rise or motors specifically designed for use with VFDs.

Wound Rotor Regenerative, Synchronous, and Direct Current Starters. Wound rotor regenerative, synchronous, and direct current motor starters offer high efficiency and are somewhat simpler than VFDs, but all require special motors. Because of the motor requirements, they are seldom used in water systems except in special cases.

ELECTRICAL SYSTEM TESTING AND MAINTENANCE

At the start of this chapter, it was noted that electrical systems and equipment are extremely reliable. This, unfortunately, sometimes leads to maintenance being neglected. Proper maintenance can head off many equipment failures. If an electrical fault does occur, proper maintenance will ensure that protective devices operate correctly to minimize the disruption and damage.

One of the duties of the design engineer is to make the owner and operating staff of a new facility aware of their continuing maintenance obligations to keep the electrical system operating efficiently and reliably. Below are some general maintenance guidelines. Whenever possible, the manufacturer's recommendations for maintenance of the particular equipment should be secured and included in the plant O&M documents.

Wire and Cable Testing

High-potential testing can be used to check conductor insulation. Unfortunately, the test may cause the cable to fail. This is considered acceptable, because the test can be scheduled at a convenient time, whereas failures cannot. This type of test is better suited to industrial plants, where it can be scheduled during a plant shutdown, but it is difficult to schedule for a water plant, which must operate continuously. High-potential testing may be worthwhile for an older plant that has experienced some cable failures.

The vast majority of wire and cable failures occur at the terminations because of heating caused by a loose connection. An excellent way to spot this problem is by infrared (IR) testing, which uses a camera with film sensitive to light in the infrared region. The picture colors, or shades, give an accurate indication of temperature. IR testing requires some specialized knowledge, so it is usually best to contract with a testing firm to perform the work. In addition to poor cable connections, the test can detect abnormal temperatures in switchgear bussing and other equipment. It is recommended that a water plant have IR testing once per year, or at least every three years.

Generator Maintenance

In addition to engine maintenance, generators should be operated at least once per month. Periodic operation, or exercise, is the only way to be confident that the generator will be available when it is needed. Time clocks can be incorporated into the generator transfer equipment to automatically exercise the unit.

Particular attention should be paid to the starting battery. Modern chargers are available that not only charge the battery, but also perform diagnostics and initiate an alarm if the battery is nearly out of life.

Switch Gear and Fault Protective Device Maintenance

As noted, molded case circuit breakers and fuses do not require maintenance. Power circuit breakers do require maintenance, and maintenance should be performed once a year. As molded case breakers, fuses, and power circuit breakers become older, it becomes increasingly important to maintain an adequate inventory of spare units and parts, both because they are more liable to fail and because it becomes harder to obtain replacements.

Medium-voltage vacuum switch gear is well accepted in the market, but there is no ready means to check for loss of vacuum. A vacuum circuit breaker with no vacuum cannot clear a fault. However, the vacuum bottle will usually contain the arc long enough for a backup breaker to react. The damage resulting from a breaker failing to clear is usually less with vacuum equipment than with older air-magnetic equipment.

Relay Testing

While performing annual switch gear maintenance, it is important to check the protective relays and trip devices. Simply pressing a test button to see if a device works normally performs only a partial test. A more positive method is called injection testing. With this, a high current is injected, usually through the current transformers. This is as close as possible to a "live" test of the relay and breaker operation.

Transformer Maintenance

Dry transformers do not require maintenance except that they should be periodically cleaned or vacuumed to remove any dust and spider webs from inside the units. Oil-filled transformers should routinely be checked to make sure the oil level is being maintained at the proper point. The transformer oil should also be periodically checked for dielectric strength. The recommended interval for dielectric testing is typically one year, although few plants test this often, if at all. Depending on how critical the transformer is, testing at three- to five-year intervals would seem to be a reasonable compromise.

BIBLIOGRAPHY

Broggis, M. C., and E. G. Potthoff. "Pump Motor Starting Problems." *Water and Sewage Works*, April 1956. (Reprinted by General Electric, Construction Works Technical Library, 1063.1.)

Construction Grants (C-85). Washington, D.C.: U.S. Environmental Protection Agency, 1973.

Design Criteria for Mechanical, Electric, and Fluid System and Component Reliability, Supplement to Federal Guidelines for Design, Operation, and Maintenance of Waste Water Treatment Facilities. Washington, DC.: Office of Water Program Operations, U.S. Environmental Protection Agency, 1973.

Fink, D. G., and J. M. Carroll. *Standard Handbook for Electrical Engineers*. 10th ed. New York: McGraw-Hill, 1969.

National Electrical Code 1993. ANSI/NFPA 70-1992. Quincy, Mass.: National Fire Protection Association, 1996.

CHAPTER 22
DESIGN RELIABILITY FEATURES

A reliable public water supply is essential to the health and welfare of the customers being served. Reliability provisions in the design of water treatment plants can ensure the production of an adequate quantity of high-quality water at all times. Selecting reliability provisions during design may be dictated by:

- Regulatory requirements
- Potential consequences of loss of part or all of the plant for a period of time
- Consequences if treated water does not meet drinking water standards

The design of water treatment facilities should be based on the premise that failure of any single plant component must not prevent the plant from operating at design flow or from meeting drinking water standards. Where source water quality varies, sufficient operational flexibility should be included in the design to handle a variety of water quality problems. Reserve or redundant capacity should be provided in each unit process so that process efficiency can be maintained when a single treatment unit within the process must be removed from service.

RELIABILITY AND REDUNDANCY CONCEPTS

Reliability and redundancy, although related to each other in plant design, have different meanings:

- Reliability refers to the inherent dependability of a piece of equipment, a unit process, or the overall treatment process train in achieving the design objective.
- Redundancy refers to the provision of any or all of the following:

1. Standby equipment or unit process
2. Having more than one unit in the process train that performs the same or a similar function as another unit in the same process train
3. Additional (or reserve) process capacity
4. Flexibility, enabling the operator to perform the same function using a different arrangement of treatment plant units

Reliability of Multiple Treatment Units

The reliability of multiple treatment units can be expressed using the formula

$$R_s = 1 - (1 - r)^m$$

where R_s = system reliability
 r = unit reliability (i.e., probability of being on-line)
 m = number of components in parallel

If, for example, a single rapid mix basin has been determined to be 90% reliable, it would have a reliability of 0.9. Using the formula it can be computed that, by providing two full-size rapid mix basins in parallel, the reliability of the unit process is increased to 0.99, or 99%, and provision of three full-size rapid mix basins exponentially increases the process reliability to 99.9%.

Providing two half-capacity basins might be considered more reliable than a single full-sized basin. One full-sized basin would have a reliability of 0.9, as indicated above, with a 0.1 probability of failure. Two half-size basins would have a reliability of only 0.81 to provide 100% of the design detention time but would have a reliability of 0.99 to provide at least 50% of the design detention time.

Two half-size basins would not be as reliable as providing a second full-size basin. A second full-size basin would provide sufficient redundancy to sustain process efficiency should an entire basin be lost for any reason.

Reserve Process Capacity

Some types of equipment can be designed to operate in excess of normal capacity for short periods of time in the event of an emergency. An example is a plant having three settling basins normally designated to handle 2 mgd (8 ML per day) flow, but each is sized for a maximum surface overflow rate equivalent to 3 mgd (11 ML per day). The facilities could still provide a capacity of 6 mgd (23 ML per day) if one basin must be out of service. In effect, this is the same as providing a standby process unit having the capacity of one-half of design flow.

One or More Processes Perform the Same Function

Overall plant reliability may also be improved when more than one process in the system is able to perform a given function. As an example, both sedimentation and filtration are capable of removing particulates. During normal operations, the function of removing particulates is enhanced by providing sedimentation ahead of filtration. If some sedimentation capacity is lost for a short period of time in an emergency, the filters would be able to accept the additional loading, although at a loss in operating efficiency.

Flexibility of Process Units

Overall treatment plant reliability is also influenced by the degree of flexibility of operation inherent in a given design. Flexibility can have a dramatic effect on a plant's capability to produce high-quality finished water. For example, having the flexibility to add potassium permanganate, ozone, chlorine dioxide, or powdered activated carbon to the treatment process at the onset of a taste-and-odor problem increases the capability to eliminate the problem in the finished water. Similarly, having the flexibility to bypass a unit or a malfunctioning mechanical component enhances system reliability.

Reliability obtained through the use of redundant units, conservative sizing, or functional unit arrangement presumes the use of the same treatment processes. Reliability obtained through flexibility does not. A flexible system offers operators a set of choices from which they may select the processes best suited to the needs of any particular circumstance.

Beyond this, flexibility can also offer an important element of redundancy. Redundancy does not always have to be exactly the same kind of units, but can be economically obtained with no sacrifice of reliability through multipurpose equipment. As an example, backup washwater supply for an elevated storage tank system may be achieved economically through appropriate piping and valves interconnected with high-service pumps.

Overall Water Supply System Reliability

Overall reliability for a public water supply system having multiple sources of supply or treatment affects design decisions about each individual component of the system. One example is a water system that has significant interconnection to another public water supply system or that has several days of in-system treated water storage. In this situation, the water system may be considered to have sufficient overall reliability that backup power may not be necessary for most water treatment functions.

Another example is a system with several different supplies of source waters, such as numerous wells, each with treatment before the connection with the distribution system. This water system would have a much higher degree of reliability than a system relying on a few high-production wells, all feeding source water to a single treatment plant.

Failure Analysis

The science of failure analysis has progressed significantly in fields of microelectronics and engineering specialties such as off-shore drilling rig evaluation. It is not within the scope of this chapter to go into detail about this subject except to explain that the goals of the use of probability and failure analysis are to avoid overdesign so as to save costs while maintaining reasonable reliability. Several references at the end of this chapter provide additional information on this subject.

DESIGN CONCEPTS FOR SYSTEM RELIABILITY

Many components and systems can affect the reliability of a water system. Some principal components that should be considered during water system design are:

- Unit processes
- Electrical equipment
- Mechanical equipment
- Hydraulic equipment
- Instrumentation and computer hardware
- Software error (i.e., human programming error)

- Human error (operational judgment and action error)
- Natural disasters
- Human-caused disasters

Unit Process Configuration

Process reliability features to consider during water system design include both individual process and entire process trains.

Reliability of Individual Unit Processes. Unit processes combining two or more process functions are inherently less reliable than single-purpose unit processes. For example, flocculation-clarifiers suffer from a deficiency known as the A-B syndrome. As illustrated in Figure 22.1, if unit A must be removed from service because of malfunction of the flocculation equipment, and during this time unit B has a malfunction in the clarifier (for instance, the sludge collector jams), both units could conceivably be out of service at the same time. If the flocculation and clarification steps are built as separate units, only one flocculator and one clarifier would be out of service under the same conditions.

Reliability of a Process Train. Another example of the A-B syndrome exists in a plant with two production trains, each with several unit processes, as shown in Figure 22.2. The production trains operate in parallel and are totally independent of one another. If the first unit in train A malfunctions and the second unit malfunctions in train B, the entire plant may be out of service.

An alternative design that is much more reliable is illustrated in Figure 22.3. With parallel trains interconnected between each unit process, any malfunctioning units can be bypassed. This minimizes the impact of the failure of any single unit on plant performance and nearly eliminates the possibility of both trains being out of service simultaneously.

Gravity Flow versus Pumping

Perhaps nothing in the world is as reliable as gravity. At the outset of design of a water treatment plant, careful consideration should be given to the use of gravity flow wherever possible to minimize the amount of in-plant pumping. When evaluating alternative loca-

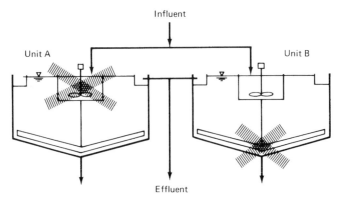

FIGURE 22.1 A–B syndrome—malfunction of two interrelated unit processes.

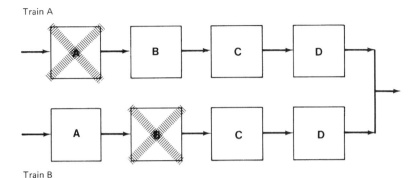

FIGURE 22.2 A–B syndrome—malfunction of two independent production trains.

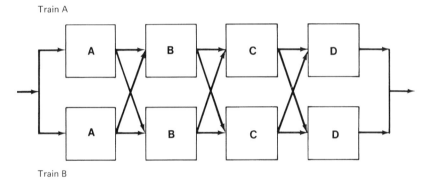

FIGURE 22.3 Interconnected parallel production trains minimize impact of single unit failure.

tions for a new plant or alternative process layouts, priority should be given to locations or layouts in which gravity flow can occur through the unit processes most critical to producing finished water.

If a new impoundment is to be built, every effort should be made to locate the plant so that the hydraulic grade line from the impoundment allows gravity flow completely through the plant. This practice eliminates a major source of potential unreliability due to failure of power, pumps, and other equipment.

If influent pumping cannot be avoided, one alternative approach is to use a source water storage basin upstream of the water plant to hold a significant amount of source water pumped from the source water source. Gravity flow from this point increases reliability by making the plant less susceptible to shutdown during a power failure.

In a similar manner, if a treatment plant is furnished with an elevated backwash water storage tank, filters can still be backwashed for a period of time if there is backwash lift pump malfunction or a power failure.

Chemical feed equipment should also be located, if possible, to feed by gravity. Chemicals such as alum, lime, soda ash, and activated carbon are often better controlled and fed through gravity systems, even though this requires elevated chemical storage facilities. Not all chemicals can be gravity fed, however. Chlorine gas, for instance, must usually be dissolved in water before feed to the process, and the solution process requires pumping of the injector water.

Individual Component System Reliability

The reliability of mechanical and electrical equipment is influenced by several factors:

- Equipment should be designed and specified with proper fault alarms and protective devices to alert operators to problems. Examples of alarms that can be incorporated into a motor include motor winding high temperature, water in the oil seal, high level of vibration, low voltage, and phase loss.
- The manufacturer's experience with design of a piece of equipment is crucial. It is generally unwise to purchase a completely new model without strong justification.
- Ambient operating environment should be reviewed, such as unusually wet or corrosive conditions.
- The availability of spare parts and repair services should be confirmed before equipment purchase.

The design of controls systems continues to make the transition away from analog controls and hardwired relays. As this trend continues, solid-state variable-speed drives (VSDs), industrial-grade computers known as programmable logic controllers (PLCs), and other digital controls will continue to become more reliable.

In the past, some applications of PLCs were plagued by reliability problems primarily caused by power system low voltage and power spikes, which have an adverse effect on solid-state electronics. Now, with high-quality power readily provided from uninterruptible power supplies, PLCs and other solid-state controls are more reliable.

Instrumentation reliability is influenced by several factors:

- Proper installation techniques
- Proper power supply and grounding
- An ambient environment that is not adverse
- Proper software
- Proper training for users and maintenance personnel
- Equipment that is readily available and repairable

One of the most important ways to achieve instrumentation reliability is to provide proper training for operations and maintenance staff. Repairs can often be made by maintenance personnel using parts in stock.

When large water systems are operated with computer controls, reliability can be increased by controlling individual processes, such as filtration with several PLCs linked by a data highway. As an example, 40 filters at a plant in Wylie, Texas, were retrofitted with eight PLCs, each controlling five filters, with a ninth PLC provided in the plant control room for monitoring and alarm functions.

RELIABILITY DESIGN PRINCIPLES AND PRACTICES

Among the more important reliability considerations that must be made by the design engineer are the possibilities of various types of disasters, as well as the reliability of chemical supplies, plant utilities, and plant processes.

Disaster Considerations

Disasters that may occur and disrupt the operation of a water system are generally divided into two classes: natural disasters and human-caused disasters.

Natural Disasters. In planning for the location, design, and operation of reliable plant facilities, careful consideration should be given to the potential for natural disasters such as tornadoes, hurricanes, earthquakes, fires, and floods. These considerations may affect both site selection and design of the treatment plant.

The effects of flooding can be minimized by choosing an appropriate elevation for plant facilities. If possible, the location should be outside the 100-year flood boundary. If this is not possible, special design features may have to be added, such as protective walls, gasketed watertight doors, and other measures to keep water out of critical locations within the plant if flooding should occur.

The disastrous effects of earthquakes can be counteracted to some extent by choosing structural design criteria appropriate for the plant location. Some types of soils susceptible to liquefaction within high-seismic hazard zones may have to be avoided altogether for water supply facilities construction. The AWWA handbook *Minimizing Earthquake Damage* provides information on assessing water system and system component seismic vulnerability in earthquake-prone regions.

Human-Caused Disasters. Consideration of the potential of human-caused disasters can also affect plant design. Such disasters can originate either from outside the plant or from within. Examples include strikes, civil disorders, sabotage, vandalism, explosions, fires, and chemical leaks.

Plant vulnerability to outside disasters can usually be incorporated into the plant design to some extent. Controlling access to the plant and providing plant security can alleviate the potential for acts of sabotage or vandalism. And the impact of strikes and civil disorders, such as truck blockades, can be reduced by providing for adequate storage of chemicals and other materials on the plant site.

On-site generation of some critical chemicals, such as chlorine, could also be considered to reduce a plant's dependence on truck or rail shipments. Although a supply of sodium chloride would still be necessary, maintaining a large inventory of sodium chloride is simpler and safer than storing large quantities of chlorine in liquid or gaseous form.

Vulnerability to explosions and fires within the plant can be reduced by establishing proper design safeguards with respect to handling hazardous materials. A thorough safety program established by the utility's management to protect operations and maintenance personnel is a critical element in preventing in-plant accidents.

At the beginning of the design process, a vulnerability analysis of the plant and the site should be performed. Its results can provide critical direction for the designers and the utility and can reduce the exposure of a plant to natural and human-caused disasters. The AWWA manual *Emergency Planning for Water Utility Management,* M19, is an excellent guide to assessing the need for many of these contingencies.

Source Water Supply Reliability

A primary concern in system design is the adequacy of the source water supply to provide the required quantity of water to the plant. Available flow or yield of a river, impoundment, or well field must be sufficient to deliver the design flow of the plant, plus estimated losses through treatment process. Some utilities have two or more sources of supply, for example, a well field in addition to river intakes. This redundancy not only improves reliability, but also offers the possibility of improved performance in colder climates where surface water becomes more difficult to treat as the temperature drops.

Another feature that can add to the reliability of the supply is multiple ports on intake structures, or multiple screens on a submerged well screen intake. With a river supply, multiple ports can reduce the problem of an intake structure becoming clogged with ice or debris. Multiple ports on an intake in an impounding reservoir make water of different quality available, improving the capability for producing treated water of the best possible quality.

Reliability of the source water supply can also be improved by off-stream source water storage reservoirs. If the water source is a river supply, off-stream storage with several hours' capacity allows a plant to reduce its exposure to chemical spills or toxic discharges upstream from the intake. A surface water supply of rapidly varying water quality may be more reliably treated if a source water storage reservoir is used upstream of treatment. This feature also allows the surface water intake to be taken out of service for maintenance for a short period of time.

Process Reliability

Because a complete discussion of individual process and equipment reliability characteristics requires more space than is available in this text, the following discussion provides a general overview of the subject.

Although the inherent reliability characteristics of a particular unit process are important, they are only one consideration in the selection process. As the number of separate component units provided to accomplish a specific function decreases, reliability becomes increasingly important. When only two units of a particular process are provided, reliability is the prime consideration.

Many states require a minimum of two component units for each unit process that is primarily responsible for meeting drinking water standards. For a surface supply, such processes might include rapid mix, flocculation, clarification, and filtration. Processes not normally required to have at least two components for reliability purposes could include aeration for carbon dioxide removal, waste washwater handling facilities, and sludge processing. Other examples of processes not absolutely essential for producing safe water include chemical feeds for fluoridation and for corrosion and scale control.

Minimum requirements for selected treatment processes from *Recommended Standards for Water Works* (commonly known as the 10-State Standards) are listed in Table 22.1.

The exclusion of some equipment and processes from this list is not meant to minimize the need for adequate reliability. Among the parts essential to operations are flowmeters, valves, chemical feed equipment, pumps, instrumentation and controls, liquid-level controls, piping, turbidity and other process-monitoring equipment, and the plant service water system. Fundamental to the design of water treatment facilities should be the philosophy that failure of any single structure, piece of mechanical or electrical equipment, element of pipe, or valve should not put the entire facility out of service.

An example of the importance of a single plant component to reliability is illustrated by the August 31, 1975, incident in Trenton, New Jersey. The inability to close a single cone valve led to a series of events that flooded the high-lift pump station. As a result, 211,000 customers were without drinking water and many businesses were closed for a 10-day period. The monetary loss was estimated at between $5 million

DESIGN RELIABILITY FEATURES

TABLE 22.1 Recommended Minimum Number of Process Units

Rapid mix	Two units
Flocculation	Two units
Clarification	Two units
Filtration*	Two units
Chlorinators	Two units
Coagulant feed	Two units

*Where only two filter units are provided, each should be capable of meeting the plant design capacity (normally the projected maximum daily demand) at the approved filtration rate. Where more than two filter units are provided, the filters should be capable of meeting the plant design capacity at the approved filtration rate with one filter removed from service.

and $10 million. The importance of each component of a plant facility to system reliability cannot be overstated.

Reliability of Plant Utilities

Water treatment plant reliability is critically affected by the reliability of its power source. Without power, all mechanical and electrical equipment cease to function. The Institute of Electrical and Electronics Engineers (IEEE) *Standard 446* contains recommendations and standards for selecting emergency and standby power systems. Most states prefer that water treatment facilities be served by two separate, independent sources of power. If two independent power feeds are not readily or economically available, standby power, such as generators powered by gasoline, natural gas, or diesel engine, may be necessary to ensure the plant's ability to operate during a power outage.

In addition to the power supply itself, dual-transformer substations, duplicate primary feeders, and other duplicate components may also be necessary to maintain the integrity of the power supply system. A standby source of power is of little value if failure of a plant's only primary substation transformer prevents the power from being delivered to the equipment.

Standby power supplies must be sized for the starting voltage dip and current surge of starting large motors. Switch gear and breaker low-voltage trip settings should generally be set at about 85% of full voltage to avoid unexpected motor trips during an already tense situation during a power outage. For a natural gas-fueled engine generator or pump drive, the worst case, lowest gas pressure in the public utility's main and the actual size and capacity of the gas main must be considered to specify the fuel feed system for standby engines.

If standby power generation is considered for only critical portions of a plant, the following areas are essential:

- To permit critical processes such as disinfection to continue
- To permit heating of areas where there is potential for freezing
- To permit ventilation of hazardous areas
- To provide egress illumination for safe exit

As an alternative to standby power generation for all equipment, engine-driven or dual-drive (electric motor and engine drive) high-service pumps are usually the most energy efficient and lowest in initial cost. If a plant is designed so that water flows through and is filtered by gravity flow, the need for on-site generated power is minimal except for high-service pumping.

If a facility has no staff on site, its components and systems need to be even more reliable than for a staffed facility. One question usually debated during design is whether to design equipment and controls to automatically restart after power failure.

Another design decision that must be made is the means of switching to an alternate source or to the emergency power source. Automatic, rapid-transfer switches are more costly and more complex. At a continuously staffed facility, if a power outage of 15 to 30 minutes is acceptable, only a manual transfer switch may be necessary. If an outage of several minutes is acceptable, motorized primary system switching may be acceptable. Rapid switching (15 seconds) requires an automatic transfer switch.

Some loads such as computers and PLC control systems may require power at all times, and uninterruptible power supplies (UPS) should be furnished for them.

Remote monitoring and control systems (SCADA) can make it easier to operate remote facilities during an emergency. For instance, if there is a power failure at the treatment plant but power is still on at a remote pumping station, the control system allows the remote station to continue operation. This is an example of designing to make a system easier to operate and therefore more reliable.

See Chapters 20 and 21 for discussions of reliability design of instrumentation and electrical systems.

Reliability of Chemical Storage

An adequate inventory of chemicals maintained on the plant site is critical to reliability. The inventory should be sufficient to ensure the production of water at the design flow for a reasonable period of time. In some instances, plant access may dictate inventory needs (e.g., the plant's only access road is located in a flood-prone area). In others, the inventory may be dictated by the length of a truckers' strike.

The location of the nearest source of the chemical may also affect storage requirements. Factors that should be considered in selecting both chemical feed equipment and chemical suppliers to gain maximum reliability are proximity of the company, their length of experience in business, and the extent and reputation of their service organization.

On-site generation of a chemical might be considered for reliability. For example, the use of ozone for disinfection or taste and odor control does not depend on delivery of any chemicals to the plant. Except for power and air, the generation process is self-contained. The designer must also keep in mind that an adequate supply of chemical on the plant site is of little benefit if a single conveyor, transfer pump, feeder, or other component can prevent the chemical from being applied.

Process Performance Reliability

Reliability of process design for a water plant can be best ensured by thorough, well-planned pilot work and by thorough, up-front source water quality analysis. This is especially true when source water quality is problematic and variable or finished water goals are especially demanding.

OPERATIONS AND MAINTENANCE FOR PLANT RELIABILITY

Equipment reliability can be optimized when equipment is installed to be safe and easy to operate and maintain and when good drawings, instructions, and training are provided to the plant staff.

Design for Ease of Maintenance and Operation

There are some basic rules to follow to design for ease of maintenance. Practically all equipment has to be moved or removed at some time for maintenance or repair. Proper lifting provisions and openings or hatches for access should be provided for any heavy or awkward pieces of equipment.

Working room around equipment and protection of outdoor working areas make it easier for operations and maintenance (O&M) staff to maintain equipment. Equipment systems are accordingly more reliable because they are more carefully maintained. Heat tracing and insulation of water-filled piping located out of doors contributes to reliability. In locations that experience extremes of cold weather, covers over walkways and basins may be considered necessary for reliable operation and maintenance.

Up-to-Date Drawings

As budgetary pressures force many systems to limit hiring levels, it becomes increasingly important to have good records of equipment O&M and staff training to maintain equipment reliability. Up-to-date, as-built drawings contribute greatly to the reliability of a water treatment plant. One trend in the future will be on-line equipment data sheets, parts lists, and O&M manuals, all accessible through operator interface terminals throughout a treatment plant. If computerized records are desired by an operations staff, they should be taken into consideration during the design stage.

Specifying Computerized Maintenance Systems

It is now generally accepted that computerized maintenance systems, if rigorously followed, improve treatment plant reliability. The design engineer can assist the O&M staff in establishing a computerized maintenance management system by requiring the contractor to submit complete information, such as manufacturer's recommendations for maintenance frequencies. An organized computerized maintenance management system moves plant maintenance out of the mode of responding to breakdowns to performing most maintenance before breakdowns occur.

Proper Training for O&M Staff

Proper O&M is crucial to plant reliability, and proper training of operators will prevent many process failures. Giving adequate attention to maintenance helps prevent equipment malfunction. Inadequate maintenance guarantees malfunction.

Safe Design

A safety program is a necessity at each water treatment facility. The safety practices of plant personnel are critical not only to their health, but also to plant operation. Accidents that disable plant personnel often also cause damage to machinery, equipment, or materials. The AWWA manual *Safety Practices for Water Utilities,* M3, offers information on developing a safety program that should be expected to improve reliability.

BIBLIOGRAPHY

American Water Works Association. *Safety Practices for Water Utilities.* M3. Denver, Colo.: American Water Works Association, 1990.

American Water Works Association. *Emergency Planning for Water Utility Management.* M19. Denver, Colo.: American Water Works Association, 1994.

American Water Works Association. *Minimizing Earthquake Damage.* Denver, Colo.: American Water Works Association, 1994.

Anton, W. F., R. M. Polivka, and L. Harrington. "Seismic Vulnerability Assessment of Seattle Water Department's Water System Facilities." *Proceedings of the 3rd U.S. Conference on Lifeline Earthquake Engineering.* August 22–23, 1991, Los Angeles, Calif. Reston, Va.: American Society of Civil Engineers, 1991, pp. 375–384.

Burazer, Jane, and S. N. Foellmi. "Tucson Water Plant Meets Community Expectations and Tough Quality Limits." *Water Engineering and Management* 140(8):20, 1993.

Cervi, Brian. "The General Electric Genius I/O and Increasing Reliability in Water and Sewage Control." *Proceedings of the 39th Annual Convention of the Western Canada Water and Wastewater Association.* October 21–23, 1987. Saskatoon, Saskatchewan: Western Canada Water and Wastewater Association, pp. 49–61.

Egleston, P. C., Jr., M. P. Steinert, and K. P. Hammer. "Plant Improvements and Reliability Are Aided by Computerized Maintenance." *Proceedings of the 1992 Industrial Power Conference.* March 29–31, New Orleans, La. New York: American Society of Mechanical Engineers, 1992, pp. 25–38.

Great Lakes–Upper Mississippi River Board of State Public Health and Environmental Managers. *Recommended Standards for Water Works.* Albany, N.Y.: Health Education Services, 1992.

Hamann, C. L., and G. Suhr. "Reliability and Redundancy: Dual Protection for Water Treatment Plants." *Journal AWWA* 72(4):182, 1980.

Hill, Bruce. "Water/Wastewater Facilities Get Better with Electronics." *Water Engineering and Management,* 135(6):34, 1988.

Institute of Electrical and Electronic Engineers. *IEEE Recommended Practice for Emergency and Standby Power Systems for Industrial and Commercial Applications (IEEE Orange Book).* ANSI/IEEE Standard 446. New York: Institute of Electrical and Electronic Engineers, 1987.

Jenny, J. A. "Effect of Partial Failure Modes on Reliability Analysis." *IEEE Transactions on Reliability R-18* 4(11):175, Institute of Electrical and Electronic Engineers, 1969.

Kennedy, J. B., and A. M. Neville. "Probability, Probability Distributions, and Expectation." In *Basic Statistical Methods for Engineers and Scientists.* 3rd ed. New York: Harper and Row, 1986.

Krause, T. L., and K. L. Zacharias. "Support Systems." In *Design of Municipal Wastewater Treatment Plants.* Water Environment Federation (MOP 8) and American Society of Civil Engineers (Manual no. 76). WEF: Washington, D.C.: Water Environment Federation, 1992, pp. 278–292.

Krishna, G. T. K. "Non-Sanitary Engineering Features Can Help Ensure a Plant's Success." *Water Engineering and Management* 134(10):32, 1987.

Male, J. W., and T. M. Walski. *Water Distribution Systems.* Boca Raton, Fla.: Lewis Publishers, 1990, pp. 81–82.

Moorman, Patrick. "Water Treatment Plant Implements Automatic Filter Backwash." *Proceedings AWWA Seminar on Computers in the Water Industry.* April 14–17, Houston, Texas. Denver, Colo.: American Water Works Association, 1991, pp. 863–865.

Pennsylvania Department of Environmental Resources. *Public Water Supply Manual and Revisions.* Harrisburg, Pa.: Division of Water Supplies, 1990.

"Risk Analysis Offers New Engineering Tool." *Automotive Engineering* 86(7):40, 1978.

Shorney, F. L., W. P. Grobmyer, and G. Taylor. "Zero Discharge Water Treatment Design Has High Reliability." *Power Engineering* 89(10):50, 1985.

U.S. Environmental Protection Agency. *Design Criteria for Mechanical, Electrical and Fluid System and Component Reliability of Wastewater Treatment Facilities.* EPA 430-99-74-001. Washington, D.C.: U.S. Government Printing Office, 1974.

CHAPTER 23
SITE SELECTION AND PLANT ARRANGEMENT

Siting water treatment plants is becoming more complicated as public awareness and environmental activism increase. From the public's perspective, water treatment plants are often perceived in the same light as wastewater treatment plants, solid waste transfer stations, incinerators, and compost facilities. They are often categorized as a LULU—locally unwanted land use, or locally unacceptable land use.

WATER TREATMENT PLANT SITE SELECTION

In this chapter, site selection and plant arrangement are portrayed as sequential tasks: first find the site, then determine the best plant arrangement for that site. It should be recognized, however, that the two steps are not strictly sequential and are closely interrelated. One must have a preliminary plant layout before the site selection process can begin. Without a preliminary plant layout, it would be difficult to evaluate potential sites with regard to the shape of the land parcel, site topography, aesthetic effects, and so on.

Having a preliminary plant layout is also an important requirement for any public participation efforts undertaken during the site selection process. Artist renderings and other graphics that show how the proposed plant will look are an important tool for educating the public about the project. The preliminary plant layout should be developed with due consideration to likely public opposition, because the plant layout can greatly affect public acceptance of the site selection. Feedback from the public often results in changes to the preliminary plant arrangement to address real or perceived concerns. Public concerns about the plant appearance and layout must be addressed during the site-selection phase. Otherwise sites that might otherwise be acceptable to the public will be dismissed.

Site-Selection Process

It is easy for those involved in planning a new water treatment plant to misjudge the reaction of the public. Both the owner and the design engineer tend to focus on technical issues. From their perspective, the primary function of the plant is to produce a finished water that is economical, reliable, and safe. Therefore sites that are conducive to these goals are favored.

The general public, on the other hand, takes for granted that drinking water should be economical, reliable, and safe. Aesthetic considerations are the foremost concern of the neighboring communities. By getting the public involved at an early stage, the owner gains an appreciation of the issues of most concern to the neighboring communities. The owner can then take those concerns into consideration in the site-selection process, as well as undertake measures to allay the concerns of the residents.

Multiple Treatment Plant Sites. Generally speaking, a water treatment plant and all its ancillary facilities should be located on one site. If a suitable site cannot be found to accommodate the entire treatment facility, consideration can then be given to locating the primary water treatment processes at one location and putting the sludge treatment facility at another location. However, this approach should be used only as a last resort, because pumping or hauling sludge over long distances is troublesome and expensive.

The cost of constructing and operating one large water treatment plant is normally less than that of building and operating two or more smaller plants. However, for cities with a very large service area and multiple water sources located far apart, it may be appropriate to provide multiple water treatment plants to save pumping costs and for increased reliability.

When designing a new water treatment plant for a city that already has one or more existing plants, there is little opportunity to eliminate facilities at the new plant by sharing operations with existing plants. Some ancillary facilities such as administrative areas, meeting rooms, and spare parts storage can be centrally located, but in general each plant must have its own full complement of treatment units and support facilities.

Factors that Influence the Site-Selection Process. One of the first things the project manager should do when beginning the design of a new treatment plant is make a list of factors that may influence the siting effort, noting the relative level of importance of each factor with regard to the particular site selection at hand. Table 23.1 is a list of factors that can influence the site-selection process. Some of these are within the designer's control, but most are not. Understanding these factors before the site-selection process begins in earnest is important because it allows the designer to anticipate potential roadblocks and plan countermeasures. For example, if an election is near, political considerations may have a big influence, and efforts to predict and deal with the political aspects would take on a high priority.

Selecting a Specific Site. Given all the technical, regulatory, financial, and political factors impinging on the site selection, identifying the optimum site is a complex and difficult process. The key to a successful siting is having a sound, methodical plan for carrying out the selection process. Table 23.2 shows a generic plan for performing a site-selection process. The specific details of each step should be adapted to suit the particular needs of each project.

Public Participation. An essential element of the site-selection process is public participation. The extent of public participation necessary depends on the type of project and location, but it is rare that the siting of a water treatment plant can be undertaken with no involvement by the general public. Furthermore, for municipal water treatment plants, the active involvement of public officials and general citizens should be viewed as a desirable component of proper governmental decision making.

In past years, the public was more trusting of the judgment of public officials and engineers with regard to the need for and design of capital improvements. Water utility staff and their consulting engineer typically assessed the need for a new water treatment plant, evaluated the most appropriate type of plant, and selected the location before going public. In recent years, the trend has been toward public distrust of government, environmen-

TABLE 23.1 Factors That Influence the Site-Selection Process

Factor	Characteristics that complicate the process
Project size and complexity	Large facility Multiple components Technically complex
Level of impact (real or perceived)	Major environmental impacts Close to residential areas Human health concerns Unfavorable press
Political agenda	Election year Involvement by different political bodies More than one community involved
Client experience	No previous siting experience Unsuccessful previous experience
Project costs	High costs
Regulatory drivers	No specific regulatory siting requirements No consent decree Major permitting effort required
Role of public	High level of public interest
Geographic base	Large geographic area Multiple communities

Source: Adapted from Camp Dresser & McKee Inc. *Guidelines for Site Selection of "Locally Unwanted Land Uses."* Internal document by The Engineering Center, 1993.

tal activism, and moves to limit government spending. Water utilities have not escaped this growing public skepticism. In the current political climate, there is a greater need for public participation in site selection and plant design.

Public participation involves two components: one-way public education and two-way public involvement in decision making. The goal of public education is to provide information on the water treatment plant and the planning process to a wide audience. Public education efforts may include direct mail, news releases in print media, the use of broadcast media, and public meetings.

Public involvement in the planning process typically involves public-opinion surveys, public meetings and public hearings, and citizen advisory committees or task forces. Surveys and questionnaires are not suited for complex issues and generally have limited value for plant site selection. Public meetings and hearings provide an opportunity to present information to a large group of interested citizens and to obtain general input and opinions from citizens. Public meetings are often legally required. However, they have limited effectiveness in decision making because of the size and format of the meetings.

Advisory committees and task forces composed of well-informed citizens who represent the community are better able to address detailed decision making. Care must be taken to ensure that such committees are representative of the community. For relatively noncontroversial issues, public interest and participation are generally low, and it is likely that the voluntary participants in advisory committees have some special interest and do not represent the community as a whole. However, as long as the decisions of the advisory committee are made public and there is some mechanism for the general public to intervene if it is perceived that the advisory committee is not properly representing the community, little else can be done. Also, the involvement of

TABLE 23.2 General Steps of Site-Selection Process

Step	Description
1. Develop work plan	The work plan defines the project and goals, technical approach, public involvement, task outlines, schedule, budget, etc.
2. Identify project components	Identify the elements of the project that must be sited. Prepare preliminary design information on these components such as size, layout, etc.
3. Identify players and their roles	Identify the various neighborhoods, political groups, politicians, government agencies, and special interest groups who have an interest in the project.
4. Public participation	Involve the public in the selection process to the extent appropriate for the type of project. Public participation involves both public education and public involvement.
5. Develop siting criteria	Select and define the siting criteria that impact the project.
6. Identify candidate sites	Using broad criteria, identify candidate sites that meet minimum project requirements.
7. Coarse screening	Coarse screening eliminates sites that do not meet a few critical constraints. This step shortlists the sites worthy of further study.
8. Detailed multi-attribute analysis	Apply the siting criteria to the remaining sites and evaluate the sites using qualitative or quantitative analysis methods.
9. Select preferred site	Based on the ranking of sites generated by step 8 and the input of the owner and public, select the preferred site.
10. Mitigation and compensation	Apply mitigation (measures to reduce the project impact to an acceptable level) and compensation (extraordinary measures beyond mitigation negotiated with the public).

citizens with special interests is appropriate from the viewpoint that the consent of these citizens is crucial to the success of the project, even though their opinions may not represent the entire population.

Becker (1993) conducted a survey of water utilities that have had experience with public advisory committees and offered the following advice about using such committees:

- Be open and honest with the committee.
- Clearly define goals and objectives at the beginning.
- Make it clear what decision-making power, if any, the committee has.
- Invite opponents to join the committee.
- Prepare an agenda.
- Seek professional assistance, if necessary, to coordinate the group.
- Do not be afraid of criticism.
- Be prepared to compromise.
- Establish a time frame.
- Expect the process to take longer than anticipated.

Siting Criteria. In the site-selection process outlined in Table 23.2, step 5 is to develop a list of siting criteria on which the candidate sites will be evaluated. This list should be long enough to include all important factors, considering the area and communities involved, but not so long that it makes the evaluation unwieldy.

A comprehensive list of potential siting criteria is shown in Table 23.3. In general, a criterion will be relevant to a particular siting decision if (1) it is an important consideration to at least one of the stakeholders, and (2) if this characteristic varies between candidate sites. For example, in an urban area with no tourism to speak of, impact on scenic areas might not be a significant concern to public officials or to citizens and therefore would not be included as a siting criterion. On the other hand, in an area with many picturesque locales and a thriving tourist industry, impact on scenic areas is obviously a major concern. If, however, all the candidate plant sites would have an equal impact in terms of aesthetic intrusion, impact on scenic areas is not a distinguishing criterion and would not be included as a siting criterion, although it would be relevant with regard to mitigation and compensation measures.

It is usually desirable to involve the public in choosing relevant siting criteria. If a citizen task force is formed, one of its responsibilities could be to identify the siting criteria to be used in the site-selection process. In addition, city and state agency regulatory requirements may dictate that specific criteria be included in the siting analysis.

Size Requirements for a Plant Site

The size of the site required for construction of a new water treatment plant generally depends on the size of the structures, the allowance for future expansion, and the need for a buffer zone around the structures.

Determining Minimum Site Area. A key siting criterion is the minimum site area required to accommodate the necessary plant facilities. It is difficult to generalize on the site area because many factors influence the minimum area required. Table 23.4 lists the main factors that determine how much area is required and includes generalizations about plant characteristics that necessitate more land.

The site areas for 21 conventional water treatment plants varying in capacity from 8 to 160 mgd (30 to 600 ML per day) are shown in Figure 23.1. The area shown includes the land required for the basic process facilities and buildings, including rapid mix, flocculation, and sedimentation basins; filters; clearwell; high-service pump station; operations/administration building; access roads; and parking. Not included is the area required for facilities such as source water reservoirs, finished water reservoirs, lagoons, and sludge drying beds that are not always required. The areas shown do not apply to "nonconventional" water treatment plants such as membrane plants.

According to Kawamura (1991), the minimum required site area for the basic process facilities of a conventional water treatment plant is:

$$A \geq Q^{0.6}$$

where A = area in acres
   Q = ultimate plant capacity in mgd

This equation has been plotted in Figure 23.1, and it is apparent that Kawamura's formula is a good approximation for the required area for conventional water treatment plants. The minimum site area required for a new water treatment plant using conventional treatment can be approximated by taking the area from the figure for the basic process facilities, but extra area must then be added for the following facilities that may be required:

TABLE 23.3 Checklist of Siting Criteria

Size requirements
  Area of the site
  Space for future expansion
  Shape of the land parcel
Environmental issues
  Land use
    Compatibility with surrounding land uses
    Compatibility with surrounding zoning
  Aesthetics and visual effects
    Impact on the existing character
    Visibility and prominence from key locations
    Impact of night lighting on neighbors
    Natural screening
  Noise level objections
    Construction noise
    Operational noise
    Traffic to and from plant
  Impact on air quality
    Plant odors
    Accidental releases of toxic gas
  Hazardous chemical considerations
    Impact of chemical spills and releases
    Transportation route to plant
  Traffic and transportation issues
    Construction traffic
    Operational traffic
    Accidents and safety
  Archaeological and historical issues
    On-site archaeological and paleontological resources
    On-site historical and architectural resources
    Potential incompatibility with off-site resources
  Ecological issues
    Effect on terrestrial habitat
    Effect on aquatic habitat
    Presence of rare or endangered species
    Presence of wetlands
  Natural resource issues
    Loss of trees
    Impact on scenic areas
    Loss of agriculture, pasture, fisheries
    Impact on mineral and energy resources
Technical issues
  Elevation of the site
    Pumping considerations
  Convenience of location
    Proximity to source water
    Proximity to service area
    Proximity to existing transmission mains
  Transportation and site access considerations
    Proximity to major highways
    Highway restrictions
    Proximity to railroads
  Utility services availability
    Proximity to power source
    Availability of second power source
    Proximity to existing sewer
    Capacity of existing services
  Disposal of residuals
    Options for waste disposal
Topography and soil conditions
  Site topography
    Slope constraints
    Grading of roads
    Erosion protection
  Subsurface conditions
    Difficulty of excavation
    Low bearing capacity soil
    Excessive settlement potential
    Expansive soil conditions
    Depth to water table
  Site drainage considerations
    Susceptibility to flooding
    Effects on existing drainage pattern
    Effects on amount of site runoff
    Effects on surface water quality
    Effects on groundwater quality
    Need for storm water detention
  Seismic activity potential
    Proximity of earthquake faults
Institutional issues
  Land availability
    Existing land ownership
    Land acquisition issues
    Easement acquisition for source water pipeline
    Easement acquisition for finished water pipelines
  Political issues
    Political agenda
    Public opposition
    Regulatory issues
    Special agency or public policy goals
    Permits and approvals
Financial considerations
  Costs that will be involved
    Land costs
    Construction costs
    Mitigation costs
    Compensation costs
    Operating costs
    Opportunity costs

TABLE 23.4 Factors that Influence Required Site Area

Factor	Characteristics that tend to require less land	Characteristics that tend to require more land
Plant capacity	Small capacity	Large capacity
Plant facilities		
Type of plant	Membrane treatment	Conventional treatment
Type of construction	Prefabricated package plants	Custom-designed, in-ground plants
Sedimentation process	High-rate tube or plate settlers Reactor clarifiers	Horizontal flow basins
Filtration process	High-rate gravity filters Pressure filtration	Slow sand filters
Waste washwater handling	Waste to sewer Settling basin with plate settlers	Recycle equalization basin Conventional settling basin Lagoons
Sludge dewatering	Mechanical dewatering	Gravity thickening Sludge drying beds Vacuum drying beds
Other on-site facilities		Low-service pump station Source water storage Presedimentation basins Finished water storage High-service pump station
Plant layout		
Structural configuration	Common-wall construction	Separate structures
Site topography	Flat site	Steep irregular site
Buffering and screening	No buffer No screening	Buffer required Screening required

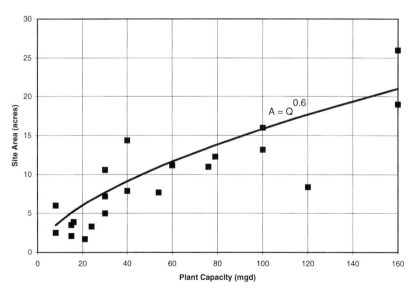

FIGURE 23.1 Minimum site area for conventional treatment plants.

$A = Q^{0.6}$

- Other supplemental facilities such as reservoirs, lagoons, and sludge disposal facilities
- Future expansion
- Buffer zones
- Other special requirements

Providing for Future Expansion. Although new water treatment plant facilities are typically designed for a 50-year time frame, the plant site is likely to be used for water treatment purposes for centuries. Water treatment plants are rarely relocated because of the huge investments in land, treatment basins, reservoirs, source water mains, and finished water transmission mains. When water treatment plant equipment and structures reach the end of their useful life, they are typically rehabilitated, upgraded, or rebuilt at the same location. It is important then to provide ample room at the plant site for future expansion, not only to accommodate the projected increase in treatment capacity, but also, if possible, for expansion beyond the immediate planning period and for unanticipated plant improvements, such as additional treatment processes required to meet future regulations.

Need for a Buffer Area. In localities where aesthetic considerations are an important issue, it may be necessary to provide a buffer zone between the plant facilities and neighboring residences, recreational areas, and other sensitive surroundings. This buffer could substantially increase the required site area.

If visual intrusion is the primary concern, the buffer zone should provide a screen of undisturbed natural vegetation, new landscaping, or earth berms to hide the plant facilities. Such screening will also control noise intrusion into surrounding areas. Odors are generally not a problem at water treatment plants, but if sludge drying beds or other related processes are likely to generate objectionable odors, a buffer zone may be required to distance the source of the odors from neighboring residences.

Environmental Issues Impacting Site Selection

Environmental impact is discussed in detail in Chapter 24, but aspects that pertain to site selection are reviewed below.

Aesthetic Issues. In most instances, the issue that generates the strongest public concern over a new water treatment plant is the aesthetic impact on the neighborhood. Residents are often concerned that the facility will have the appearance of an industrial complex and that their property values will be reduced. Indeed, many existing water treatment plants do look like an industrial complex with exposed tanks, pumps, piping, maintenance vehicles, and other equipment visible to passersby. Attempts to locate a water treatment plant in a sensitive area are likely to engender strong public resistance, despite assurances about aesthetic compatibility, because of the public's general skepticism of public officials.

Aesthetic concerns are not limited to the appearance of the plant during the day. Water treatment plants must be well lighted at night, both for security reasons and to enable plant operations to continue 24 hours a day. This lighting can have a negative impact on neighbors, particularly near residential areas.

There are two ways to approach the issue of aesthetic impact. One is to treat water treatment plants as an undesirable land use and seek neighbors who will be least impacted and least resisting to the intrusion. The other is to use mitigation measures such as innovative architectural treatments, landscaping, and screening to improve the aesthetics and make the plant more acceptable. The former approach restricts feasible plant sites to industrial and similarly zoned areas or to remote areas. The latter approach results in high costs for constructing and maintaining the architectural and landscaping treatments necessary to make the plant acceptable to a larger portion of the community.

Noise Level Concerns. It is difficult to evaluate the impact of noise in a community because of the different perceptions of how an objectionably noisy environment is defined. People respond differently to the same sound under different circumstances. Noise levels that are considered acceptable in a work environment may be deemed unacceptable in a residential setting. A noise that would not be bothersome during the day may be tolerated less at night. For a water treatment plant located on the edge of a small lake, even a low noise level may be offensive to residents on the opposite shoreline, because of the ability of sound to carry across water.

There are three chief origins of noise at a water treatment plant: construction site, routine operational, and traffic noise. In general, residents find construction noise the least offensive of the three, because they know it is temporary. Nevertheless, construction noise may affect the surrounding population for a year or more and may be quite offensive. Because most construction noise is generated by engines such as earth-moving equipment, cranes, trucks, and compressors, the greatest potential abatement measure is to ensure that all engines have suitable exhaust mufflers, intake silencers, and engine enclosures. Appropriate construction hours should also be established to minimize noise generation during evenings and weekends.

The main sources of routine operational noise at water treatment plants are pumps, generators, compressors, blowers, and personnel paging systems. Suitable noise abatement techniques include:

- Using low-noise emission equipment
- Enclosing the equipment in a sound-attenuating enclosure
- Erecting a sound wall, berm, or heavy landscaping between the source of noise and the neighbors

Traffic noise is discussed under a later section on traffic and transportation issues.

Air Quality Concerns. The Clean Air Act (CAA) sets limits on the emission of hazardous air pollutants. Volatile organic chemicals (VOCs), radionuclides, and other hazardous substances that may be emitted from water treatment plants are included under CAA requirements. The preliminary design should address the direct and indirect impact of the water treatment plant on air quality and recommend measures to mitigate any adverse effects.

Water treatment plants typically do not generate strong odors. However, citizens familiar with odors from wastewater treatment plants and solid waste facilities may have fears about potential odors. The most common source of objectionable odors at a water treatment plant is residuals handling.

Sludge drying beds and other sludge dewatering processes may generate odors. Sedimentation basins that are dewatered can also generate strong odors if the basins are not cleaned promptly and the sludge is allowed to go septic. Aeration towers can potentially cause odors in plants treating groundwater and some surface waters. Chemical storage and feed systems should not generate odors if they are operated properly.

Hazardous Chemicals Concerns. Water treatment plants use a number of hazardous chemicals. The transportation of these chemicals to the plant and the bulk storage of chemicals at the plant present a risk to the neighborhood, as well as to the plant staff. Accidental release of gaseous chemicals such as chlorine presents a significant health risk to adjacent communities. Safety concerns should influence both site selection and mitigation measures. (See Chapter 15.)

Traffic and Transportation Issues. The generation of truck traffic, both during construction and during normal operation, is an issue that can bring about a great deal of public opposition. Community concerns include public safety, disruption of neighborhood activities, and aesthetic matters. The degree to which the extra truck traffic impacts residential areas is an important siting criterion.

Mitigation measures that can be taken to reduce the impact of both construction traffic and routine delivery trucks must be carefully considered. One method is to designate truck routes that bypass residential streets and neighborhood commercial centers. Another method is to establish service schedules to avoid rush hours and the presence of schoolchildren on the streets.

Traffic noise from trucks traveling to, from, and around the plant, both during and after construction, can be minimized by the following measures:

- Identify the route to the plant with the least impact on the community.
- Locate the plant entrance where it will result in the least traffic noise.
- Avoid steep roads within the plant site.
- Erect sound walls, berms, or heavy landscaping adjacent to critical roads.

Archaeological and Historical Site Considerations. Sites with structures or objects that are listed or eligible for listing on the National Register of Historical Places should be avoided if at all possible. The National Historic Preservation Act (NHPA) requires that federal agencies consider the impact of any federal capital works on sites with historical, architectural, archeological, or cultural significance. Although the NHPA does not preclude development on historic sites if procedural requirements are met, public reaction and citizen lawsuits can delay or block projects from proceeding.

Ecological Issues. The importance of ecological issues has greatly increased in recent years. The presence of any rare or endangered species on or near a candidate site is a red flag. In fact, many would consider this to be an exclusionary criterion. That is, if a rare or endangered species is discovered at a site, that site is eliminated from further consideration.

Experts familiar with the Endangered Species Act (ESA) and other pertinent regulations should be consulted to assess candidate sites in terms of ecological issues and to assist with permitting once a final site has been selected.

Floodplain and Wetland Considerations. Floodplains and wetlands are also red flags for water treatment plant sites. Drainage and flooding are discussed later in this chapter. Wetlands are discussed in Chapter 24.

Technical Issues Impacting Site Selection

Technical issues that must be considered in site selection include the elevation of the site; the proximity to the water source, distribution system, and utility service; and transportation access.

Elevation and Pumping Considerations. The elevation of a plant site directly affects the pumping arrangement and pumping energy costs. Energy conservation is an important consideration because energy costs are a substantial portion of plant operating costs. Siting a water treatment plant at a low elevation results in a low head at the low-service pump station, but a high head at the high-service pump station. Conversely, a relatively high site results in a high head at the low-service pump station and a lower head at the high-service pumps.

For a plant located below a dam storage, it may be possible to gravity flow into the plant and eliminate the low-service pump station entirely. Similarly, if the plant is located at the highest elevation in the region, it may be feasible to eliminate the high-service pump station and gravity feed to the distribution system. The engineer must analyze the variables such as low-service pump size, diameter of source water pipeline, high-service pump size, diameter of finished water transmission mains, service area elevations, future service area limits, and power costs to determine the optimum configuration for the particular circumstances.

The most common situation is to locate the water treatment plant near the source water and to use low-head pumps to lift the source water to the plant and high-head pumps to deliver the finished water to the service area. One advantage of this approach is that less energy is expended in pumping water consumed by in-plant needs, such as for backwashing filters.

Proximity to the Water Source and Distribution System. The location of the water treatment plant in relation to the source water and the consumers has important implications for operating the source water pipeline, the plant itself, and the finished water transmission mains. Generally speaking, it is best to locate the water treatment plant near the source water, near the center of the distribution system, or somewhere directly between the two, to avoid duplicating the source water line and finished water lines. Whether it is better to locate the plant near the source water or near the distribution system can only be determined after a proper assessment of the advantages and disadvantages of the two approaches for the particular project circumstances.

If the plant is located close to the source water, there will be a short source water pipeline and a long finished water transmission main. Having a short source water pipeline simplifies pipeline maintenance operations. Source water pipelines typically require more maintenance than distribution system pipelines because of problems such as slime growth, silt deposits, and zebra mussels. Regular pigging or chlorination is often necessary to maintain design carrying capacity. Another advantage is that the long finished water transmission main can be used to provide service to customers along the pipeline route, or, if there are no customers drawing from the transmission main, the residence time in the transmission main can be used for primary disinfection credit.

A disadvantage of having a long finished water transmission main is the difficulty of maintaining a disinfectant residual. Long pipeline residence times may result in loss of disinfectant residual or excessive formation of disinfectant by-products. Also, if the water source is a remote location, siting the treatment plant at the water source may present difficulties in obtaining utility services, attracting plant staff, and getting deliveries.

Transportation and Site Access Considerations. The vast majority of water treatment plants receive regular chemical deliveries by truck, so an important site selection criterion is the proximity to major highways. Sites that are accessible only through residential streets are undesirable. If streets to the site have any restrictions on the type of traffic or cargo type, that is another concern. The site must be accessible to trucks hauling hazardous chemicals, sludge, and whatever other cargo types are anticipated to and from the plant.

A plant must be accessible 365 days a year under all weather conditions. Therefore the plant access road should have no low-level river crossings or stretches subject to flooding. In areas subject to severe winter conditions, access roads to the site and inside the site should have limited grades and gentle curves to allow vehicular access during icy conditions and to facilitate snow removal.

Sites with available service from a railroad spur offer the option of receiving chemicals by rail. The lower unit cost of bulk deliveries by rail is an important consideration for plants with high chemical usage. Sites bordering lakes or rivers may also have the option of receiving chemical deliveries by barge.

Proximity to Utility Services. Water treatment plants must have a source of electrical power, telephone communications, and sewerage service. It is also desirable to have a supply of natural gas. Dual electrical power sources are normally required for reliability, or an on-site backup generator will be required.

It should be noted that some or all of the cost of installing utility service lines from the nearest source to the plant site may be absorbed by the utility company. In this case, the cost of the utility connection incurred by the utility does not enter into the site selection analysis. If, however, the utility recovers the cost of the connection by way of a surcharge on the unit price of the service, ultimately the owner pays for the utility connection and the cost should be reflected in the site analysis.

In assessing the proximity of utility services, the required capacity of the service must be considered. Water treatment plants can have large power requirements, particularly plants with ozonation systems and large high-service pumps. Similarly, plants that discharge sludge to the sewer system can release large volumes of waste, and it may be difficult to locate a nearby sewer with enough available capacity to accept the additional flow.

Topography and Soil Condition Considerations

Important considerations in site selection are the topograpy of the land, soil conditons, site drainage, and potential for seismic activity.

Site Topography. Ideally, the site selected for a water treatment plant should have a uniform, moderate slope of 2% to 8%. This provides enough grade for surface drainage and to accommodate the hydraulic profile through the plant. A site steeper than 10% will probably require extensive cut and fill. Also, if process basins are aligned down the slope rather than across the slope, the vertical arrangement of the process basins will be steeper than required by the hydraulic profile. The excess drop between process units represents wasted pumping.

Sites with long slopes, steep grades, and highly erodible soils present special problems with erosion. The design of site drainage, landscaping, and other aspects of the civil design will have to be tailored to minimize erosion. During construction, special provisions will be necessary to prevent discharge of silt-laden soil to surrounding areas.

Sites with steep slopes also have the problem of potential landslides, particularly in view of the fact that leaking process basins or yard piping can saturate surrounding soils and increase the tendency for slope failures. Where soil conditions and topography are conducive to landslides, mitigation measures such as horizontal drainage, retaining walls, buttress fills, and tiebacks can be used. The cost of these measures should be factored into the site selection.

Subsurface Conditions. Subsurface conditions have a significant impact on the construction cost. Bedrock has an excellent bearing capacity, but if the bedrock is shallow, rock excavation will greatly inflate construction costs. On the other hand, soils that are easy to excavate often have a low bearing capacity, necessitating the use of expensive foundation types such as piles. A high groundwater condition also increases construction costs because of the need for dewatering the soil during construction, waterproofing to keep basement areas dry, and additional design provisions to resist flotation of empty basins.

Local knowledge of subsurface conditions should be used to estimate construction cost differences resulting from differing subsurface conditions. These costs should be included in the evaluation of alternative sites. Where cost differences arising from differing subsurface conditions are a major factor in the site selection, it may be advisable to have borings drilled at short-listed sites to gain a better understanding of subsurface conditions and refine estimated costs.

Site Drainage and Flooding Considerations. Sites that receive storm-water runoff from a large upstream catchment require a more extensive site drainage system than those where only a small amount of storm water drains onto the site. In jurisdictions where storm-water detention systems are required, evaluating differences in site drainage should include an assessment of the relative costs of providing the necessary detention basins and treatment system.

Many water treatment plants are located adjacent to a lake or river, because this provides close proximity to the source water. The disadvantage of siting a plant near a river is the risk of flooding. Protective measures such as earth levees can fail, leaving the community without a drinking water supply for an extended period. Sites well above the 100-year flood level offer increased reliability, as well as avoiding the cost of mitigation measures.

In addition to problems that flooding poses for site access and plant operation, any proposed construction in a floodplain triggers an array of permitting and regulatory requirements involving the U.S. Environmental Protection Agency (USEPA), the Federal Emergency Management Agency (FEMA), the U.S. Army Corps of Engineers, and other federal and state agencies. If construction in a floodplain cannot be avoided, adequate time should be included in the project schedule to deal with regulatory and permitting issues.

Potential for Seismic Activity. Damage from seismic activity results from ground shaking, ground displacement, slope failure, liquefaction of soil, lateral soil spreading, and forces due to liquid movement in the tanks (sloshing). Although all sites under consideration are likely to fall within the same seismic region, the topography and soil conditions at some sites may be more vulnerable to earthquake activity than others. Sites at which seismically induced slope movement or liquefaction of subsurface soil materials are likely to threaten essential structures should be eliminated from consideration. Sites on or near an active fault should also be avoided, because the large movements that can occur at faults cannot be accommodated in the design of structures or piping.

Institutional Issues to Consider in Site Selection

In addition to the many technical issues that must be considered in selecting a site, there are also many institutional reasons why a piece of land may or may not be available for a plant site.

Land Acquisition and Ownership. The number and type of owners of the proposed site can have a major impact on the difficulty of acquiring the land. The ideal case is where one individual owns the whole parcel and is amenable to sale of the property. Multiple ownership makes acquisition of the land more difficult and time consuming, particularly when parts of the land are used for residential housing or similar uses.

The assessment of land acquisition issues should include not only the plant site itself, but also the acquisition of easements for the source water pipeline and finished water pipelines. The ideal situation would be where pipelines can be installed within an existing utility easement, power company easement, railroad tract, or similar corridor. The least desirable situation is having to construct pipelines through a densely populated area.

Political Agenda. Local politics is always an important factor in the site-selection process. It is important to understand the political climate and local agendas. There are two approaches to dealing with potential political problems. One is to involve the pertinent officials, groups, and agencies in the site-selection process to gain their acceptance of the project goals and the selection procedures and to obtain their support. The other is to recognize the political reality of the situation and account for the political influences in the decision making. Both approaches should be used.

Permits and Approvals. If the owner of the water treatment plant is a city or county, it is generally preferable to choose a site within the owner's area of jurisdiction. This makes it easier to obtain permits and approvals (or to waive permitting requirements altogether) when dealing with one's own departments.

Financial Considerations in Selecting a Site

At the site selection stage, the prime objective is to evaluate the relative merits of the candidate sites to identify the optimum one. Detailed cost estimates for budgeting purposes are not normally required at this stage because project details are too incomplete to estimate costs with any degree of certainty.

It is usually sufficient to use comparative costs of alternative sites, with common costs being intentionally omitted. For example, equipment costs can be neglected as long as the same process equipment will be used at all of the sites. On the other hand, structural costs should be included if subsurface conditions or the structural configuration varies from site to site. The engineer should clearly denote in project documentation that the costs are for comparative purposes only and do not include all project components so that readers do not develop an unrealistic impression of the total project cost.

Capital Costs and Operating Costs. Examples of capital costs that vary between sites include land cost, new source water pipeline to the site, new finished water pipelines, pump stations, excavation and structural foundations, civil site work, utility connections, and mitigation measures to address concerns of neighbors. Operating costs relevant to the siting decision include energy costs for pumping and residuals disposal costs.

Taxes to Other Municipalities. If an agency purchases land and constructs a treatment plant in another municipality, that agency may be subject to local property taxes, even if that agency is itself a municipality. Such property taxes can be a significant annual cost. The plant owner should investigate options for reducing or eliminating the tax burden, such as offering payment instead of taxes. Compensation could be nonmonetary, such as an agreement to provide drinking water to city buildings.

Opportunity Costs. An opportunity cost is a potential benefit that is lost or sacrificed when the course of action makes it necessary to give up a competing course of action. Opportunity costs are relevant to the siting decision and should be included, even though there is no outlay associated with the opportunity cost. Potential future benefits foregone by selecting the site should be considered where such benefits differ between sites.

An example of an opportunity cost is property tax. Construction of a water treatment plant on a site means that land is no longer available for other uses. Suppose a city is evaluating various sites within the city limits and one of the candidate sites is located in an area with high property taxes. If that site is selected, the city will lose the income it would have derived had the land been used for commercial purposes. Therefore the loss of future taxes (or more specifically, the loss of future taxes in relation to the other candidate sites) should be considered.

Sunk Costs. Sunk costs are those costs that have already been incurred, and they should have no relevance to the decision making. They cannot be avoided, regardless of which course of action is taken. For example, if candidate sites include a tract the city owns, the purchase price of that land would be irrelevant to the siting decision. Of course, the fact that the land is already owned would favor that site, because acquisition cost would be zero, and the other sites would all have an acquisition cost. However, the original purchase cost, be it small or large, should have no bearing on the decision.

It is common for decision makers to erroneously include sunk costs in the decision process. Using the example above, if the city had paid an exorbitant sum for the land, there would likely be pressure by some officials to use that particular tract for the plant site to avoid "wasting" the previous investment. However, if the owned tract has certain undesirable features, a proper analysis neglecting the sunk costs may well determine that the correct course of action is to build the plant elsewhere.

Other sunk costs that should not enter into siting decisions are the cost of existing intake structures and pump stations, pipelines, reservoirs, and all other previous capital investments in the water supply system.

Net Present Value Approach. Capital costs generally are incurred at the beginning of a project's life when the facility is initially constructed. Operating costs are ongoing, periodic costs. Both are relevant to the siting decision. The most common method of comparing two or more alternatives with differing capital and operating costs is the net present value approach.

Each of the capital and annual costs is converted to an equivalent present value. These present values are added to yield the net present value, which can then be compared with other alternatives. Net present value represents the amount of money that, if invested now for a definite period, would provide the sums necessary for constructing the project and for keeping it in operation for that period.

The present value of a single capital cost is:

$$P = \frac{F}{(1+i)^n}$$

where P = present value
F = future value (i.e., estimated capital cost) at year n
i = annual interest rate
n = number of years

For the initial construction cost, normally $n = 0$ and $P = F$.

The present value of a series of equal costs (i.e., an annuity) is:

$$P = A \times \frac{(1+i)^n - 1}{i(1+i)^n}$$

where P = present value
A = periodic (annual) cost
i = annual interest rate
n = number of years

Texts on engineering economics or managerial accounting (Garrison, 1988) should be referenced for more information on net present worth calculations.

Inflation Effects on Net Present Value. A common error made in capital investment analyses is to make improper adjustments for inflation. It is common to adjust the interest rate for inflationary effects. Although current inflation is in the 3% range, over the last 20 years inflation has averaged almost 6%. Many engineers believe it is appropriate to account for inflation in long-term capital planning. However, the proper way to account for the effects of inflation is widely misunderstood, even in business texts (Garrison, 1988; Hanke, Carver, and Bugg, 1975).

Inflation can be handled in either of two ways. First, "real" dollars (also termed *uninflated* or *constant* dollars) and "real" (inflation-free) interest rates can be used in calculating net present worth. Cash flows are stated in uninflated (now) dollars. For example, if this year's labor cost was $100,000 and inflation is 5%, the projected real annual cash flow for future years would be $100,000, $100,000, $100,000, $100,000, etc.

Alternatively, "actual" (inflated) dollars and "combined" (inflated, market, or stated) interest rates can be used. Cash flows are stated in inflated (then) dollars. For the example above, the projected actual annual cash flow would be $100,000, $105,000, $110,250, $115,762, etc.

Much of the confusion occurs because the real interest rate is not the market interest rate published by financial institutions. Indeed, the real interest rate is a fictitious interest rate, frequently referred to in business texts but seldom clearly explained. Market interest rates (especially long-term treasuries such as municipal bonds) are combined interest rates because they have inflationary effects already built in. The real interest rate can be calculated using the following formula:

$$i = \frac{(i_c - f)}{(1 + f)} \cong i_c - f$$

where i = real (inflation-free) interest rate
 i_c = combined (market) interest rate
 f = inflation rate

It should be noted that it makes no difference whether the effects of inflation are included in capital investment analyses, as long as there is consistent treatment of the cash flows and interest rate. Net present value and the analysis outcome will be the same whether or not inflation is accounted for (Garrison, 1988; Canada and White, 1980).

Although this may seem surprising at first, on reflection it makes sense. If cash flows and interest rate both include the effects of inflation, inflationary effects cancel themselves out. The net present value is the same as if inflation had been ignored. If the reader needs more information, Canada and White (1980) provide a comprehensive and clear description of how to take inflation into account in capital project evaluations.

Although either of the two approaches described above provide the same result, it is generally considered preferable to use real dollars and real interest rates, because we intuitively think of future costs in terms of today's dollars. The USEPA requires that quantitative evaluations submitted in support of projects funded or controlled by the USEPA be performed without consideration of inflation. Remember, however, that in order to perform an inflation-free analysis, the market (combined) interest rate should not be used and the real interest rate must be computed using the above equation.

Some texts address the situation where the inflation movements of some component costs are projected to be different from the inflation of other component costs. For example, the escalation rate for energy costs is sometimes assumed to be different from the overall inflation rate. In this case, it is necessary to account for inflation effects. Goodman (1984) presents equations to handle differing inflation effects. However, given all the various assumptions and uncertainties inherent in long-term capital investment analyses, it is hard to justify the added complexity of such approaches.

Mitigation and Compensation. The last step in the site-selection process listed earlier in Table 23.2 was to apply mitigation and compensation measures. These are approaches for increasing the regulatory and public acceptance of projects.

Mitigation refers to specific measures taken to reduce or eliminate undesirable impacts of the project. For example, if the proposed plant site is close to a scenic lake, the owner may agree to undertake extensive landscaping to conceal all structures and facilities from the view of residents. Mitigation measures may be required by regulatory agencies as a condition of permitting, or they may be imposed by public pressure.

Compensation is an incentive or trade-off intended to make the project and its impacts more acceptable to the community. For example, a city may offer to build a neighborhood park. Building a park does not lessen the impact of building a new plant in the community, but it makes the construction of the plant more palatable to the community. Compensation measures are not normally imposed by regulatory agencies. They are typically used to gain political and public acceptance of a project once mitigation possibilities have been exhausted.

Compensation should be used with caution, because it may be perceived that the owner is trying to buy the decision. Compensation is typically reserved for negotiations with the host neighborhood once the site-selection process is near the end. Estimated costs of any mitigation and compensation measures anticipated at each site should be factored into the site evaluation process.

Site Evaluation and Selection Methodology

After all technical, institutional, and financial issues regarding available sites have been compiled, it is then necessary to perform an evaluation and screening to determine the best site.

General Considerations. The general site-selection process described earlier in Table 23.2 includes a four-stage evaluation methodology to select the preferred site:

1. Identify the candidate sites.
2. Perform coarse screening.
3. Conduct detailed multiattribute analysis.
4. Select the preferred site.

Such a staged approach is appropriate for a siting problem involving a large area and many candidate sites. If only a small area and few potential sites are involved, an abbreviated evaluation process could be used. Identifying candidate sites is characterized as an evaluation step because some minimum site requirements must be used to assess suitability.

In identifying candidate sites, the goal is to search the entire region under consideration and locate all potential sites. Only a few broad siting criteria should be used because the emphasis at this stage is to identify all possibilities. Long-time residents with a detailed knowledge of the local area may be able to identify candidate sites using an informal review of land-use maps, aerial photographs, and other reference materials.

This approach may be adequate where the area under consideration is relatively small and where political concerns are slight. For more complex siting decisions, a formal sieving analysis as discussed below is more comprehensive and objective.

Sieving Techniques. A rigorous technique for identifying candidate sites is to perform a sieving analysis using multiple overlay mapping (Noble, 1992), also called land suitability analysis (Lane, 1983). Using a few broad siting criteria, land suitability maps are prepared for the entire region under consideration. For each criterion, a map is prepared that indicates unsuitable areas. Individual maps are then overlayed to form a composite land suitability map. Areas on the composite map that have no shading meet all selected siting criteria.

In its simplest form, overlay mapping is performed using manual techniques. Transparent maps are superimposed on top of each other, and the resulting visual composite image is used. Alternatively, overlay mapping is an ideal application for computers using graphical information systems (GIS) software. It is increasingly common for cities to maintain their mappable data in GIS. Depending on the size of the GIS, data such as land use, zoning, ownership, wetlands, soil types, and sewer and water line locations may be included. If a computerized GIS has been developed, it enables the siting team to quickly and easily apply siting criteria over a large area.

The advantage of using an overlay technique, rather than selecting a few candidate sites based on local knowledge and judgment, is that all potential sites are identified. Also, the technique is objective and easy to understand, important considerations in obtaining support for controversial siting decisions. The overlay technique is also an effective tool for communicating and explaining the site-selection analysis to public officials and the public. GIS, in particular, is an excellent presentation tool.

An example of appropriate siting criteria for identifying candidate sites for a 50 mgd (190 ML per day) plant might be:

- The site must be contained within the municipal boundary.
- The site must be located no more than 5 mi (8 km) from the source water.
- The site should be located no more than 1 mi (1.6 km) from a major highway.
- The area of the site must be greater than 10 acres (4 ha).
- There must be no wetlands on the site.

These criteria are simple enough to be applied using a typical GIS.

Coarse Screening. Coarse screening reduces the number of candidate sites identified in the previous step to a manageable number. Because the amount of effort required to evaluate all candidate sites using all relevant siting criteria would be excessive in most cases, a few simple siting criteria are used for coarse screening. If the number of candidate sites identified in the previous step is relatively small, this intermediate step may not be required.

An example of coarse screening criteria for a 50 mgd (190 ML per day) plant might be:

- The narrowest dimension must not be less than 500 ft (152 m).
- There is no evidence of rare or endangered species at the site.
- There is no evidence of historic resources at the site.
- The lowest point on site is not less than 5 ft (1.5 m) above the 100-year flood elevation.
- The site slope is not not greater than 15%.

These criteria probably could not be evaluated using GIS methods as in the previous step. Candidate sites would be reviewed individually and those not meeting the additional criteria eliminated from further consideration. The coarse screening criteria should be ones that can be assessed relatively quickly. Screening criteria tend to be "fatal flaws"—either the site passes or it does not.

If only one or two sites pass the coarse screening, it may then be necessary to "loosen up" the screening criteria so that fewer sites are culled and a greater number remain for further investigation. Ideally, coarse screening should leave three to five sites remaining for detailed evaluation. If fewer sites remain, either the screening criteria were excessively stringent or the region has few suitable sites and the choice of site will probably be obvious. If eight or more sites remain, additional screening criteria should be used.

Multiattribute Analysis Techniques. The sites that pass the coarse screening should be rated in terms of the remaining siting criteria. Ratings normally include both monetary and nonmonetary attributes. Evaluating monetary criteria requires that estimates be made of capital costs and operating costs, which is why screening is used to limit the number of sites reviewed in detail. Examples of the criteria examined at this stage include differences in pumping requirements, foundation costs, and proximity of existing services.

The most basic analysis approach is to tabulate the pros and cons of each of the sites still under consideration and to make a reasoned judgment as to the best one. This approach is commonly used because, although engineers are skilled at determining the monetary and nonmonetary attributes of each site, they are often not knowledgeable in the use of multiattribute analysis techniques to interpret resulting ratings. In instances where one site has the preponderance of favorable ratings and clearly dominates all others, a more sophisticated technique is not necessary. However, where two or more sites have a similar overall desirability, more advanced techniques are warranted.

The simplest quantitative technique is a weighted score system. Each site is scored as follows:

$$S_A = W_1 \times R_{A,1} + W_2 \times R_{A,2} + W_3 \times R_{A,3} + W_4 \times R_{A,4} + \cdots$$

where S_A = total score for site A
W_1 = weight of first siting criterion
$R_{A,1}$ = rating of first siting criterion for site A

Ratings are assigned as a number on a common scale, for example 0 to 10, where 10 is best, 5 is fair, 7.5 is good, and so on.

This method can handle both qualitative and quantitative siting criteria. For qualitative criteria—for example, impact on scenic areas—judgment is used to arrive at a rating. For quantitative criteria—for example, foundation costs—costs are first estimated for each site and then these costs are converted to a rating as follows:

$$R_{i,j} = \frac{P_{\max,j} - P_{i,j}}{P_{\max,j}} \times 10$$

where $R_{i,j}$ = rating for criterion j at site i (i.e., a rating of 0 to 10)
$P_{i,j}$ = estimated cost (net present value) of criteria j at site i
$P_{\max,j}$ = estimated cost (net present value) of criteria j for the worst case

By using the above equation, the site with the highest cost will be assigned a rating of 0 (worst), any site with no cost for the criteria will get a rating of 10 (best), and sites with intermediate costs will be distributed linearly between 0 and 10.

Although this approach appears straightforward and easy to apply, a major difficulty arises in assigning criteria weights. The various stakeholders involved in the site selection have different perspectives and are unlikely to share the same opinion on which criteria are most important. Techniques for handling criteria weighting are discussed in the following section.

More sophisticated analysis approaches such as matrix methods, utility function techniques, and linear programming methods are available. Given the degree of uncertainty and judgment inherent in plant siting decisions, the simple weighted score system described previously is usually adequate. However, if political or other considerations dictate that a more rigorous analysis be used, advanced decision science texts such as those by Knowles (1989), Samson (1988), or Lang and Merino (1993) should be consulted.

Criteria Weighting. Criteria weights are determined on the basis of judgment, and different stakeholders often have different opinions on which criteria are most important. The owner will likely weight cost and reliability criteria highly; environmental groups will rate environmental issues above all others; and local activist groups will focus on local impacts. Assuming a task force consisting of various representatives of stakeholder groups has been formed, one approach is for the members of the task force to discuss the various siting criteria and try to come to a consensus on weightings.

If that does not work, another approach is to average individual weightings of stakeholders. Each member is asked to score each criterion using a common scale, and then all scores are averaged to produce an aggregate weight for the criterion. Other, more advanced, techniques for developing criteria weights, such as forced choice methods, are outside the scope of this text but are described in advanced decision-making texts.

Choosing Appropriate Siting Criteria. The discussion so far has assumed that appropriate siting criteria were identified before the screening and evaluation steps. Siting criteria should be carefully selected to ensure proper evaluations.

As a first step, the siting criteria identified as impacting the siting decision should include all regulatory-imposed requirements. Relevant federal, state, and local regulations must be adhered to unless it is expected that variances will be obtained. In the next step, criteria that do not discriminate among sites should not be included, even though they may be important to the facility design. For example, seismic activity has important implications for plant design, but if all sites have equal seismic activity, there is no value in including this criterion in the analysis.

The final siting criteria should be independent of each other, so that double-counting does not occur. For instance, if there is a concern that operational noise may be offensive to nearby residences, the criterion could be "operational noise" or "compatibility with surrounding land uses," but not both. To assign a poor rating for both criteria would be two strikes for the one problem. The listing of potential siting criteria presented previously in Table 23.3 includes both general and specific issues. The intent is for the designer to select the particular criteria that best suit the siting decision, but to avoid including multiple criteria describing the same problem.

Analysis of Risk and Uncertainty. An inherent characteristic of the planning process is uncertainty: uncertainty of future water demands, uncertainty of existing conditions, uncertainty of construction costs, uncertainty of future interest rates, and so on. These uncertainties have important implications for site selection. For example, if subsurface conditions are uncertain (which is normally the case), then whether a particular site is assumed to have favorable or unfavorable subsurface conditions may have a major bearing on the evaluation of that site.

At the simplest level, engineers deal with uncertainty by including contingencies for unexpected costs. Another approach is to perform sensitivity analyses to determine if the outcome of the evaluation is affected by whether optimistic or pessimistic assumptions are used. Advanced risk analysis or probabilistic analysis techniques are appropriate for complex evaluations under conditions of uncertainty, distinct from the deterministic analysis methods described earlier, which assume that uncertainty can be ignored.

With risk analysis, instead of making a single assumption about an outcome (for example, foundation costs will be $200,000), the planner describes potential outcomes in terms of probability distributions—for example, 75% probability that foundation costs will be greater than $150,000, 50% probability that foundation costs will be greater than $200,000, and 25% probability that foundation costs will be greater than $250,000. The details of how to perform sensitivity analyses and risk analyses are beyond the scope of this text. Texts on management science and decision analysis should be consulted for descriptions of how to apply these techniques if the nature of the project warrants such an approach.

Goodman (1984) offers the following methods of dealing with risk and uncertainty:

- Collect more detailed data to reduce uncertainty.
- Use evaluation criteria with better-known characteristics.
- Use more refined analytical techniques.
- Reduce the irreversible or irretrievable commitment of resources.
- Increase safety margins in design.
- Perform a sensitivity analysis.

ARRANGEMENT OF WATER TREATMENT PLANT FACILITIES

Once the plant site has been selected, the next step is to determine the arrangement of the treatment process units on the site. By this stage, specific process elements and support facilities will have been selected and sized. Careful consideration must be given to properly laying out the process units, buildings, and roads. A poor plant layout can have a negative impact on the plant treatment effectiveness, maintenance operations, construction and operating costs, ease of future plant expansion, and plant appearance.

The final plant layout evolves from the combined efforts of several interested groups. The key people typically involved in plant layout decisions are:

- Consulting engineers, who make recommendations on mechanical, structural, civil, and electrical issues
- The architect, who advises on the layout and design of buildings
- The plant superintendent and operators, who indicate operational preferences
- The owner's staff, who provide general input
- Public citizens (through advisory committees and similar avenues), who provide input on aesthetic issues
- State regulatory agencies, who review the construction plans to ensure that the layout complies with all minimum requirements for public water supply systems
- Environmental regulators, who review plans with respect to impacts on sensitive environmental features (such as wetlands)

Factors Impacting Plant Arrangement

Each site has unique characteristics that impact the plant arrangement. In addition, preferences of the owner's staff have a strong influence on the layout. Table 23.5 summarizes the major factors influencing plant layout and requiring careful consideration by the engineer. These factors are discussed in more detail in the remainder of this chapter.

Basic Layout Types

Most nonpackage water treatment plants are custom designed to suit specific requirements of the process train, site, and owner. Every layout is different, but there are three basic

TABLE 23.5 Major Factors Affecting Plant Layout

Site topography
Hydraulic profile
Degree of redundancy desired
Degree of flexibility desired
Climatic conditions
Architectural considerations
Operations and maintenance procedures
Provisions for future expansion of the plant
Provisions for future additional treatment processes
Types of chemicals and chemical feed systems
Types of sludge handling processes

---

patterns into which all plant layouts tend to fall: the linear layout, campus layout, and compact layout (Figure 23.2). Each layout type has advantages and disadvantages, as discussed below.

Linear Layout. The linear layout was for a long time the favored layout type. Water treatment plants were commonly arranged in essentially a linear fashion, unless site space limitations precluded it. The first edition of this text, published in 1969, advised that "with very few exceptions, a linear construction of a plant is preferable if the plant must be enlarged in the future."

One appeal of the linear layout is conceptual simplicity. The first impulse for laying out process basins on an open site is to arrange them in sequential order. The first process unit is located at the highest point on the site, and subsequent processes are arranged in order down the slope. This approach simplifies conveying water between process basins. Successive basins are connected by short pipes or channels.

Alternatively, basins can be constructed with a common end wall, in which gated ports are installed. The other advantage normally associated with the linear layout is ease of expansion. Plant capacity can be increased by simply constructing additional process trains adjacent to existing ones.

A major disadvantage of the linear layout is the unfavorable location of chemical feed points and mechanical equipment. These areas require a lot of operator attention, and the linear plant layout results in two centers of attention: one at either end of the plant. Water treatment chemicals are generally fed in the region near the rapid mix basins (e.g., predisinfectants, coagulants, flocculants, and pH control chemicals) and near the filters (e.g., filter aids, postdisinfectants, fluoride, and pH control chemicals). With a linear plant layout using conventional horizontal-flow sedimentation basins, rapid mix basins and filters are widely separated because of the flocculation and sedimentation basins. Consequently, either the chemical storage and feed facilities must be located at one end of the plant with long runs of chemical feed piping to serve the other end, or separate chemical feed facilities have to be provided.

Major mechanical equipment associated with process units also tends to be clustered near rapid mix basins and filters. Screening equipment, rapid mixers, flocculator drives, and sedimentation basin sludge collector drives are typically all located in the vicinity of the rapid mix basins. The plant control room and any transfer pumps and high-service pumps are often located near the filters at the outlet end of the plant.

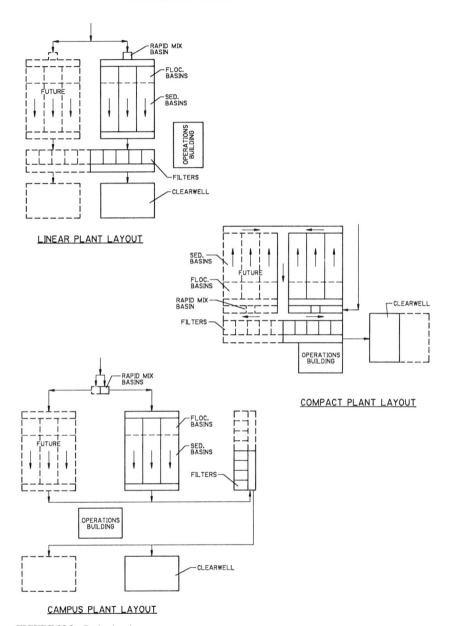

FIGURE 23.2 Basic plant layout types.

As for the segregated chemical facilities, this separation of mechanical equipment is undesirable from an operations viewpoint. It is not uncommon for large, linear water treatment plants to have two crews of operators—one based at the head of the plant to handle the chemicals and mechanical equipment in that area, and the other group operating the filters and chemical feed systems at the other end of the plant. Plant operation is much more convenient if the major mechanical equipment and chemical feed points are clustered at one location.

Campus Layout. The campus layout is so named because plant facilities are spread out over the site and resemble a university campus. Buildings and basins tend to be separate from one another and are each surrounded by a clear area. The campus arrangement is appropriate where physical separation of structures is desirable, for example, in areas with uneven subsidence or in areas subject to earthquakes. The main advantages of the campus layout are:

- Reduced structural complexity
- Better resistance to earthquake movement
- Better accommodation of differential settlement
- Allows vehicular access (e.g., trucks and cranes) to all sides of process basins
- Relatively simple to make modifications to connecting piping

The disadvantages are:

- Less convenient access within the plant because of longer travel distance
- Necessary to have buried utilities and chemical feed pipes between basins
- Entire plant takes up more room
- Long piping connecting the units results in higher head loss
- Construction cost is higher because it does not allow common wall construction

Compact Layout. The primary goal of the compact layout, also called cluster layout (Kawamura, 1991) and integrated layout (Corbin et al., 1992), is to consolidate areas requiring frequent operator attention. The compact plant layout, shown schematically in Figure 23.2, accomplishes the clustering of chemical feed points and mechanical equipment by locating filters and clearwells near the head of the plant. As discussed previously, the large separation of chemical application points and mechanical equipment inherent in the linear plant layout occurs because of the relatively long length of the sedimentation basins.

By moving filters from the outlet end of the sedimentation basins to a location near the rapid mix basins and routing the settled water to the filters using a channel or pipe, facilities needing frequent operator attention are clustered together. Figure 23.3 shows an example of a compact plant layout. Filters are positioned adjacent to the rapid mix basins, locating the motor-operated equipment for the rapid mix basins, flocculation basins, sedimentation basins, and filters in close proximity to the operations building.

Another characteristic of the compact layout is that the major process units and plant facilities are normally integrated into a single, compact structure with common wall construction, rather than having separate structures as in the campus layout. In this compact arrangement, rapid mix basins, flocculation basins, sedimentation basins, filters, clearwell, chemical facilities, operations building, and administration building may all be grouped into a single complex.

Enclosed galleries and tunnels can be constructed between process units at the lower level. Such tunnels allow covered operator access between processes and provide a convenient location for routing piping and electrical cables. Incorporating administration offices and operations facilities into a single building immediately adjacent to the filters means that maintenance facilities are central to the items of major equipment.

SITE SELECTION AND PLANT ARRANGEMENT   659

FIGURE 23.3   Aerial view of compact water treatment plant, Eagle Mountain WTP, Fort Worth, Texas.

The plant shown in Figure 23.3 has the bulk chemical storage area adjacent to the operations building. Chemical feed equipment is located in the lower level of the operations building. This arrangement provides for short suction lines between the bulk tanks and the day tanks and metering pumps. Chemical discharge pipes to the various application points are also relatively short, because all chemical application points are clustered in the same general location as the mechanical equipment. Unlike the linear plant layout, there should be no need for running any chemical feed pipes the full length of the sedimentation basin.

The advantages and disadvantages of the compact layout are summarized in Table 23.6. One of the disadvantages of the compact layout is that the settled water must be transferred from the outlet end of the sedimentation basin back to the filters located near the inlet end of the sedimentation basin. One way to do this is to construct a settled water return channel from the outlet end of the sedimentation basins back down to the filters. This channel can be made with an intermediate floor to provide an access tunnel running the full length of the sedimentation basin underneath the settled water channel. Such a tunnel provides a convenient place to route both chemical feed piping to the process basins and electrical conduits to the motor-operated equipment. It also provides a covered means of access to the far end of the sedimentation basins.

An alternative to constructing a settled water return channel is to use a two-tray sedimentation basin. In this arrangement, the sedimentation basin has an intermediate suspended floor, with flow in one direction on the lower half and returning in the opposite direction in the upper half. The plan area required for the sedimentation basins can be reduced significantly.

The theoretical settling area of the two-tray basin is twice the plan area. Because the turbulence generated at the 180-degree turn reduces settling efficiency, a plan area somewhat

TABLE 23.6 Advantages and Disadvantages of a Compact Plant Layout

Advantages	Disadvantages
Mechanical	
Major items of equipment are consolidated into one central area. Chemical feed lines are short.	Removal of equipment for off-site maintenance may be more difficult because of restricted access.
Operations	
Short walking distance to all equipment. Mechanical equipment is close to maintenance facilities. Access via covered galleries and tunnels. Required number of plant staff is reduced.	Careful equipment layout is required to ensure adequate space for maintenance activities.
Structural	
Reduced excavation volumes. Reduced concrete quantities because of common wall construction.	Larger slab penetrations because electrical conduits are concentrated in a smaller area. Extra provisions required for concrete shrinkage and expansion and for soil movement.
Electrical	
Reduced lengths of electrical and instrumentation conduits. Smaller cable size because of shorter cables. Cables can be installed in cost-efficient cable trays instead of in buried conduits. Cables routed indoors are more accessible.	Careful planning is critical because of the concentration of cables into a smaller area. Greater potential for conflicts between disciplines because other services (e.g., plumbing, HVAC) are also crowded into a smaller area.
Hydraulic	
Reduced hydraulic losses caused by less connecting piping between basins.	Long channel or pipe may be required to convey settled water to filters.
Architectural	
May be easier to develop an attractive architectural theme for a single mass than for several separate structures.	Mixed use of building (operations and administration) complicates building design.
General	
Less land area required. Fewer internal access roads and less paving. Less underground piping.	More difficult to comply with fire code requirements for access/egress because of the increased number of hallways, stairways, etc. Increased need for safety precautions arising from close proximity of bulk chemical storage to administration building.
Environmental	
Impacts are more concentrated. Generally more opportunities for buffering.	Impacts cover more area.

Source: Adapted from Corbin, D. J., et al.: *Journal AWWA* 84(8):36, 1992.

larger than half that of a conventional sedimentation basin is required. It is sometimes stated that two-tray sedimentation basins are economical only when available land is limited. However, engineers should not be too quick to dismiss them because the additional cost of the intermediate tray is offset by cost savings arising from the reduced volume of excavation, reduced length of walkways and handrails, reduced wall thickness and reinforcement (because walls are supported at midheight by the intermediate floor), and the deletion of the settled water return channel.

Another frequent criticism of compact plant layouts is the difficulty of making plant expansions. The integrated nature of the compact layout means that modifications are more difficult and disruptive than for a linear or campus layout. However, if the initial design includes adequate provisions for future modifications and expansion, this is not a major limitation.

Using a compact layout typically results in a lower capital cost than using campus layout because of savings in excavation and backfill, structural concrete, process piping, chemical piping, and electrical and instrumentation conduits. According to Ferguson (1991), savings for a 54 mgd (204 ML per day) plant with compact layout compared with a campus design is $560,000, or less than 5% of the total construction cost.

Example Plant Layout. Details of the 35 mgd (130 ML per day) Henrico County Water District water treatment plant in Virginia are shown in Fig. 23.4. This plant uses the compact layout approach with all facilities combined into one complex, including the processes of preozonation, rapid mixing, flocculation, sedimentation, intermediate ozonation, and filtration.

Operational Issues to Consider in Plant Arrangement

Some of the most important considerations in planning a plant arrangement are for the features that will make the plant reliable and convenient for operations and maintenance.

Need for Redundancy. The issue of plant redundancy is discussed in detail in Chapter 22. The impact of plant redundancy requirements on the plant layout is reviewed here.

Redundancy can be provided by adding standby units or by adding additional treatment capacity. Typically, redundancy for equipment items such as pumps is obtained by installing standby units. Treatment basins are generally oversized so that when one basin is out of service, the remaining basins can handle the flow. The size (i.e., loading rate) and number of process basins should take into account redundancy requirements. Also, the piping or channels connecting process basins should include sufficient valves and gates to allow some basins to be removed from service while the other basins are kept in operation.

Maintaining Flexibility of Operations. Operational flexibility is needed to handle a variety of operating conditions. Sufficient flexibility must be built into the plant arrangement to handle all likely variations in plant flow rate, source water quality, temperature conditions, and treatment goals. Flexibility is not the same as reliability, although flexibility can improve overall plant reliability. For example, having cross-linked interconnections between successive process basins improves both flexibility and reliability. Consider a water treatment plant with two process trains comprising rapid mix, flocculation, sedimentation, and filtration. If the trains have linked interconnections after each process, this offers the flexibility of taking one flocculation basin out of service at low flows to avoid having an excessively long period of flocculation, while not affecting rapid mixing or sedimentation. In addition, the interconnections enhance reliability, because if one flocculator fails the inoperative basin can be bypassed, rather than taking the complete train out of service.

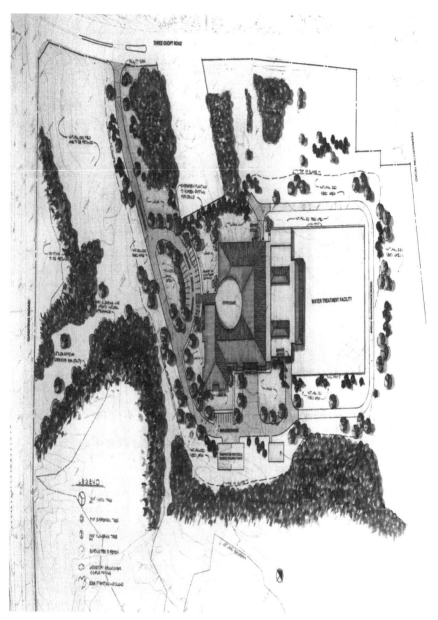

FIGURE 23.4 Henrico County water treatment plant site plan.

Providing for Convenience of Routine Operations. Plant layout can have a dramatic impact on the convenience of routine operations. A compact layout that centralizes facilities needing operator attention can significantly reduce the staffing requirements compared with a poorly arranged layout that separates operator work areas over a large area.

The daily duties of plant operators include:

- Monitoring equipment (looking for signs of developing problems such as leakage, vibration, unusual noises, overheating)
- Monitoring treatment processes (observing floc formation, water levels)
- Controlling treatment processes (filter backwashing, sludge drawoff)
- Routine sampling and testing

To perform these routine tasks, the operator has to walk around the plant to the various locations of mechanical equipment, process basins, sample points, and so on. If the plant is arranged for convenient access, operators are more likely to regularly perform these routine tasks. If the equipment and plant processes are spread over a wide area, operators are discouraged from making frequent trips to perform the required monitoring and control tasks, particularly during inclement weather.

Providing Access for Equipment Maintenance. Convenient access to all equipment is necessary for proper maintenance. The layout of basins and buildings should be designed with careful consideration of all necessary maintenance activities. Refer to Chapter 25 for a detailed discussion of operations and maintenance considerations.

A special concern in compact plant layouts is how to handle media replacement in filters and GAC contactors. Where the campus layout typically has room for vacuum trucks to drive up to one side of the filters or contactors, the process units in compact layouts are often difficult to access. Appropriate provisions for removing and replacing media, such as installing a permanent suction header to each of the filters, should be provided if direct access is impossible.

Providing Access for Disabled Persons. The Americans with Disabilities Act (ADA) sets out provisions for access by disabled persons. These access requirements may apply both to the general public (visitors to the plant) and to plant employees. State and local officials knowledgeable of the ADA requirements should be consulted to determine what portions of the plant must comply with the ADA and what access provisions are required.

Meeting the ADA requirements can have a significant impact on the plant layout. In order to provide wheelchair access to the upper level of aboveground process basins, it may be necessary to install long ramps or provide elevators. A vertical rise of several feet requires an unobstructed total ramp length of well over 100 ft (30 m). If such ramps are not planned at an early stage of the design, it may be difficult to add them later. It is common practice to provide short containment walls around storage areas housing liquid chemical bulk tanks, with stair access over the containment wall. If wheelchair access into the chemical storage area is deemed necessary, the dimensions of the storage area may have to be significantly increased to accommodate a ramp.

Control of Noise. Aside from reducing the impact of noise on adjacent properties as discussed earlier, measures to control noise within the plant must be undertaken to avoid occupational health problems for plant employees. The Occupational Safety and Health Administration (OSHA) defines maximum noise levels for working areas. The best way to deal with noise is to isolate noisy equipment from routine work areas. Available measures to reduce the sound level for plant personnel include (U.S. Environmental Protection Agency, 1976):

- Locate noise-generating equipment as far away as possible from routine work areas.
- Erect sound walls, berms, and heavy landscaping in the area surrounding outside noise areas.
- Install sound-attenuating enclosures around pumps, blowers, compressors, generators, ejectors, and other types of equipment that operate at high speeds.
- Install properly designed silencers and mufflers on equipment to reduce noise.
- Install isolation fittings in piping and equipment foundations to reduce the transmission of sound through rigid piping and structures.
- Whenever possible, select equipment that meets OSHA requirements for 8 hours of continuous sound exposure.
- Provide personal noise protection devices for all employees where other measures are not adequate.

Design of Internal Roads and Parking. Site layout should include permanent roadway access to all points in the plant where deliveries are made or where materials are loaded onto trucks for transport off site, as well as adequate parking space for employees, visitors, and city vehicles.

Design criteria for internal roads generally provides for an adequate turning radius for semitrailer chemical delivery trucks. In addition, the local fire department should be consulted with regard to specific requirements for fire lanes and equipment turnarounds. A pavement width of 25 ft (7.6 m) is appropriate for two-way traffic; a pavement width of 18 ft (5.5 m) is adequate for low-traffic roadways.

Parking should be adequate for the needs of both plant employees and anticipated visitors. Visitor parking should be near the main entrance, both for the convenience of visitors and to control visitor access. If tour groups are expected, allowance must be made for bus parking.

In areas with cold climates, roadway design should be appropriate for snow conditions. Alignment, gradient, and pavement design should be suitable for operation of snow removal equipment, and snow-dumping areas should be provided to eliminate windrows that catch drifting snow.

Providing for Chemical Delivery. The internal road layout must provide adequate space for chemical delivery and handling. Unloading bulk chemicals can take an hour or more, and the delivery trucks should not block access to other parts of the plant. Spill containment must also be provided at locations where liquid chemicals are unloaded, because the unloading operation presents a significant risk of spillage. Suitable materials handling facilities should also be included at the chemical delivery area. A monorail or hoist is highly desirable for unloading and transporting drums, carboys, and other large containers. A truck dock and forklift can also be used to load and unload pallets and drums. If chlorine is delivered in ton cylinders, a monorail is essential to lift and move containers to and from the storage area.

Providing for Plant Visitors. Most water treatment plants are publicly owned and generally allow or encourage visits by schoolchildren and public groups. Conference rooms may also be used by outside groups for meetings. The layout of the internal roadways and parking areas, location of the administration building, security measures, and signage should all be designed to ensure that visitors do not inadvertently end up at hazardous locations in the plant.

The designer should keep in mind that visitors who are not familiar with plant layout can easily become confused about where to go if roads and parking areas are not clearly designated. The best approach is for the administration building to have an obvious main entrance and for it to be located close to the front gate.

Providing for Site Security. Site security is important for two reasons: (1) for the safety of trespassers, and (2) to protect against vandalism and sabotage. Because children are particularly attracted to tanks and unfamiliar processes, keeping neighborhood children off the property is necessary both to safeguard against injuries to trespassers and to avoid the possibility of lawsuits brought about if a child should be injured.

Protection against vandalism and sabotage is required primarily to avoid loss of the drinking water supply resulting from equipment damage, and also to prevent deliberate contamination of the drinking water supply. Security measures that should be considered include perimeter fencing, motorized gates, security lighting, closed-circuit video cameras, and telephone communications with the front gate.

Technical Issues to Consider in Plant Arrangement

Technical issues provide some of the most important considerations in plant arrangement.

Impact of the Hydraulic Profile. One of the early steps in developing a site layout is to calculate the preliminary hydraulic profile through the proposed process basins. (See Chapter 16 for information on computing the hydraulic profile.) Once the hydraulic profile has been established, the top and bottom elevations of the main process basins can be determined. These basin elevations and how they relate to site topography are important considerations in locating the basins on the site.

The more treatment processes that are required, the greater the total head loss across the plant. If intermediate pumping is to be avoided, a site with adequate slope for flow to occur by gravity is needed. If necessary, upstream processes can be raised out of the ground to generate the necessary hydraulic head, but this increases construction cost, makes maintenance operations less convenient, and has a negative impact on the external aesthetics of the plant.

Optimum placement of structures relative to ground level depends on the site topography, soil conditions, and hydraulic profile, but as a general rule, it is desirable for plant treatment units to be approximately half underground and half aboveground. When process basins are located largely underground, the excavation costs, buoyancy problems, and depth of yard piping all increase. Where basins are mostly aboveground, access onto the basins becomes inconvenient, connecting piping becomes shallow, and basins become much more aesthetically obtrusive.

It should be emphasized that, at plant sites with a steep grade, it is wasteful to provide large elevation drops between basins. In instances where process basins are arranged in a linear layout down a steep grade, there may be several feet or more of water elevation difference from one basin to the next. The total drop in water surface across the plant can exceed 100 ft (30 m) in some cases.

The usual rationalization is that, because the site has plenty of grade, the large drop between basins does not matter. However, every foot that water falls through the plant represents another foot of pumping head, either at the low-service pump station or at the high-service pump station. Therefore the hydraulic profile through the plant should be kept as flat as possible, even if the site has a steep grade, to keep pumping energy costs as low as possible.

At least one plant process basin should include provision for overflow discharge to a watercourse. The overflow weir may be anywhere between the rapid mix basin and the filters, as long as the hydraulics are designed to ensure that the overflow weir will be the first place to overtop under both high-flow and low-flow conditions. The overflow weir and the grading of the ground away from the weir should be designed to accommodate the worst case scenario of the plant at maximum source water inflow with the filters completely stopped.

In addition to providing an emergency overflow onto the ground, it is desirable to provide overflows between adjacent treatment units. For example, the dividing walls between adjacent sedimentation basins or filters can be designed with the top elevation lower than surrounding walkways. This way if there is a valve operator malfunction or similar local failure that results in a rising water level in one unit, it can overflow into an adjacent basin rather than cause a spill onto the plant grounds.

Conveyance of Water Between Process Units. A common problem in water treatment design is improper hydraulic balance between parallel treatment units. It has been reported that over 70% of all existing plants have a problem with unequal flow split between basins (Kawamura, 1991). Ideally, basin layout should provide balancing of flows between parallel treatment basins without the need for automatic control devices such as throttling valves or adjustable weirs. In addition, connecting piping or channels should include bypass facilities for operational flexibility.

Many old plants use pipes or channels to convey water from flocculation basins to sedimentation basins. When water velocity in these pipes and channels is relatively high (greater than 3 ft/s [90 cm/s]), the turbulence created at entrances, exits, and bends tends to break up the floc. When water velocity is maintained too low (less than 0.5 ft/s [15 cm/s]), large floc may have a tendency to settle out. Wherever possible, connecting pipes and channels should be eliminated and flocculated water transferred directly through a common-wall baffle. If integrated basins are not feasible, open channels should be used to allow easy access for cleaning settled floc.

Single Application Points for Chemicals. It is highly desirable to provide a single chemical application point both before the pretreatment process units and before the filters. The application point should be a single pipe or channel through which all flow passes, as opposed to parallel conveyance systems. A single chemical stream can therefore be applied at that one location. This simplifies the chemical feed equipment considerably, because it is not necessary to use multiple metering pumps or flow-splitting devices in the chemical feed lines. Also, it eliminates problems of uneven chemical dosing due to uneven flow split of the chemical streams or of the plant flow.

One caveat is that the diffuser at the single chemical application point should be designed to disperse the chemicals uniformly into the flow. Alternatively, there should be sufficient turbulence and mixing in the piping or channel downstream of the application point to ensure a homogeneous blend before the flow divides between the downstream units. Otherwise, the downstream units receive water with varying chemical concentrations and treatment is degraded.

Separation of Filtered Water and Nonfiltered Water. State regulations for the design of water supply systems generally prohibit the separation of filtered and nonfiltered water by a single wall, because a crack or leak in the wall could result in contamination of filtered water. Nonfiltered water includes settled water channels feeding to the filters, all water in the filter box above the bottom of the media, and waste washwater channels and conduits.

Complying with this requirement sometimes becomes a challenge in compact plant layout design. It is not as severe a constraint in linear and campus layouts where common-wall construction is not used to the same degree. In instances where a compact layout configuration would result in common-wall construction between filtered and nonfiltered water, one option is to construct two walls (or two floors if vertical separation is the concern) with a narrow drainage space between the two walls.

Finished Water Storage. Water treatment plants located close to the source water and remote from the service area should generally be designed to provide a minimum quantity of finished water storage at the plant. It is preferable to locate finished water storage (either ground storage or elevated reservoirs) near the center of the service area. A clearwell is normally all that is needed at the plant to serve as a wet well for the high-service pumps. It also often serves as the backwash water source.

Treatment plants located within the service area and with sufficiently high elevation are usually designed with finished water storage reservoirs at the plant site. In that case, additional land area is necessary to accommodate the reservoirs. There are a number of different reservoir types: ground storage or elevated, buried or exposed, square or round, concrete or steel.

Plants designed to use an elevated tank for supplying filter backwash water should have the tank located as close as possible to the filters, because large-diameter piping is typically required to accommodate the required backwash flow. The volume of the backwash storage tank should be sufficient to hold at least two full backwashes, and the height of the tank should be sufficient to produce the maximum design backwash rate at the minimum usable tank level.

Source Water Storage. Source water storage may be needed in instances where the source water delivery system is undependable, or if the daily source water delivery is constrained by the capacity of the delivery system or by contractual conditions. However, it is generally more desirable to use finished water storage to address such problems rather than to construct source water storage.

Source water storage serves only to provide a safeguard against interruptions or restrictions to source water delivery. The same storage volume provided as finished water storage serves to maintain finished water delivery not only when source water supply is restricted, but also when water demand is extreme or when water treatment capacity is constrained. Source water storage can be in open basins, so its storage is significantly cheaper than covered finished water storage. In cases where source water delivery concerns necessitate an inordinately large storage volume, the best option may be a combination of source water and finished water storage.

Providing for Lagoons. If the residuals handling system includes sludge lagoons or drying beds, they should be located where they cause the least aesthetic concern. Lagoons and drying beds are generally feasible only when the site has a large expanse of flat land. See Chapter 17 for details on sludge handling methods.

Location of Hazardous Chemicals. Careful consideration should be given to the storage location of hazardous chemicals. As a rule, chemical storage should be close to the chemical feed equipment to reduce the length of suction piping. However, hazardous chemicals should not be close to administrative offices. The resourceful use of screen walls and careful location of windows and air intakes can enable the designer to keep hazardous chemicals close to the operations area and yet isolate the danger from administrative personnel. The requirement to provide containment in the event of a massive chemical spill also influences the choice of where to locate hazardous chemicals.

If chemical storage tanks are visible from the street, serious consideration should be given to constructing a wall or berm to screen the tanks from view, particularly for the case of pressure vessels. Screens improve aesthetics and protect the tanks from drive-by vandalism.

Providing for Flooding. The importance of protecting treatment plant facilities from flooding is obvious. Aside from flooding caused by rainfall and runoff, the designer should address flooding caused by internal sources. Careful consideration must particularly be given to the possibility of flooding of the filter pipe gallery and basement level from a pipe rupture, or overtopping the filters or clearwell.

Although these events are infrequent, serious flooding of water plants has happened enough times in the past that the possibility should not be ignored. If site grading permits, the filter pipe gallery can be designed so that it will be free draining. At flat sites where the gallery is substantially below ground level, the amount of mechanical and electrical equipment that could be flooded in the gallery should be kept to a minimum.

Climate Considerations. Cold climates require special measures that should be included in the plant design. Filters and other process basins must be covered, because cold weather can damage process equipment. It should also be kept in mind that cold weather damage of equipment can result from viscosity changes in lubricants, condensation freezing, and ice incrustation.

Snow mostly affects access around the plant site. The effects of drifting snow can be minimized by the use of trees, shrubs, and snow fences. Site design should also attempt to locate major roads parallel with the normal wind direction, and not locate roads directly upwind or downwind of large obstructions (WEF and ASCE, 1991). Structures should also be oriented, if possible, so that frequently used doors are on the sides of the building toward the upwind end.

Strong winds can induce circulating currents in open basins and interfere with settling, and long, shallow sedimentation basins are particularly prone to wind effects. The effects of strong wind can be counteracted by covering the basins with roofs, installing floating covers, constructing windbreaks beside the basins, or installing submerged baffles to impede circulating currents. Wind effects on long sedimentation basins are minimized by locating basins with their longitudinal axis aligned with the direction of the prevailing wind (AWWA, 1990; James M. Montgomery, Consulting Engineers, 1985).

Architectural Considerations in Plant Design

After the many operational and technical issues have determined the basic arrangement of a new plant, there should be architectural consideration of the design.

Building Orientation. Careful placement of water treatment plant buildings can greatly improve plant appearance. Buildings can be located to hide unattractive views of the facility. (See Chapter 18 for further information on architectural matters.) Site conditions and climate should also be considered in laying out the buildings on the site. Building design and orientation should be planned to minimize heat loss in the winter and heat gain in the summer and to use the potential for natural lighting.

Landscaping. Landscaping should be designed to be low maintenance. An attempt should be made to preserve as much existing vegetation as possible and to use native plantings, because they are more hardy than imports. Trees and shrubs should not be planted near open treatment basins to prevent leaves falling into the basins.

Areas to be mowed should have a slope of no more than three horizontal to one vertical. Steeper slopes should be planted with ground cover that does not need mowing and prevents erosion. Alternatively, steep slopes can be lined with a concrete slab, masonry pavers, or other suitable slope-protection system.

Provision for Future Changes in Plant Design

Provisions should be allowed in a plant design for both expansion and process changes.

Capacity Expansion. It is important that the initial plant layout allow for future capacity expansion. Many old water treatment plants have been expanded three, four, or more times. In some instances, plants have been enlarged far beyond the expectations of the original designer, and the resulting hodgepodge of process units results in poor hydraulic conditions, scattered chemical application points, and considerable operator inconvenience.

The long-range master plan for the plant site should allow space for future expansion up to and beyond the projected long-term treatment needs. Leaving space for expansion beyond the projected needs is prudent because "long-term" demand projections typically look ahead 20 to 50 years, but most water treatment plants will be in use 100 years after original construction.

Provisions for future expansion should include not only adequate space for future basins, but also provisions to aid in the construction of future expansions. For example, blind flanges on the end of piping runs, knock-out walls, reinforcing bar extensions, and similar provisions can be easily provided during original construction and will be of great assistance during future expansion. Channels to be extended should be constructed with stop plank grooves in the channel walls so that the plant can be operated during construction of the expansion.

Future Process Additions. Aside from future process basins required to expand plant capacity, plant layout (including hydraulic design) should allow for future additional treatment processes required by new regulations. As water quality regulations become more stringent, many plants are finding it necessary to add treatment processes such as GAC contactors and disinfectant contact basins.

In addition to regulatory changes, the emergence of new or improved treatment processes sometime causes plants to install new processes to support or replace existing ones. It is much easier and less costly to add new processes when the original designer makes allowances for such a possibility. This is especially true for compact plant layouts.

Support Facilities to Consider in Plant Design

Support facilities are those facilities that are necessary to keep the water treatment processes in operation or to accommodate the staff who operate and maintain the plant, but are not directly involved in the treatment of water. Some of the principle facilities of this type are the maintenance workshop and spare parts storage, administrative offices and meeting rooms, lunch room and locker rooms, and laboratory. Issues concerning the location of such support facilities within the plant site are reviewed below. See Chapter 18 for detailed information on the architectural design of plant support facilities.

Maintenance Facilities. Maintenance facilities should be located near the most equipment-intensive area. It is desirable to provide a monorail or bridge crane in the maintenance area for lifting and moving heavy items. The monorail or bridge crane should travel out to a loading dock to facilitate pickup or delivery of parts and equipment. It is also desirable for the monorail to extend into the pump rooms, pipe galleries, and other such areas containing heavy equipment to simplify movement of equipment into the maintenance area.

Laboratory Facilities. The water treatment plant laboratory is used for process control, testing for regulatory agencies, collection of historical data, and cost control. In many instances, the laboratory is also used for testing distribution system samples and to perform other chemical, bacteriological, and physical testing for the utility.

The laboratory should be at a convenient, central location because operating personnel will be using the laboratory many times each day. Laboratory equipment such as analytical balances are sensitive to vibration, and the laboratory should not be located near large pumps or other heavy operating equipment. Also, heating and air-conditioning systems for the laboratory should be independent of the main plant systems to prevent process area gases and fumes from being circulated into the laboratory.

The size and layout of plant laboratories vary considerably. Depending on the type of analyses to be performed on site, the laboratory could be quite small to relatively large. An alternative approach used in some large plants is to have two laboratories on the site: a small one located central to plant operations for conducting routine plant process control tests, and a full-service laboratory for complex analytical work and outside testing services located elsewhere on the site.

BIBLIOGRAPHY

AWWA/ASCE/CSSE. *Water Quality and Treatment.* 4th ed. New York: McGraw-Hill, 1990.

Becker, Joan. "Survey Says Water Utility Advisory Councils Are a Success." *Journal AWWA* 85(11):58, 1993.

Camp Dresser & McKee Inc. *Guidelines for Site Selection of "Locally Unwanted Land Uses."* Internal document by The Engineering Center, 1993.

Canada, J. R., and J. A. White. *Capital Investment Decision Analysis for Management and Engineering.* Englewood Cliffs, N.J.: Prentice-Hall, 1980.

Corbin, D. J., R. D. Monk, C. J. Hoffman, and S. F. Crumb, Jr. "Compact Treatment Plant Layout." *Journal AWWA* 84(8):36, 1992.

Ferguson, K. G. "Compact Water Treatment Plant Design." In *Proceedings of AWWA 1991 Annual Conference.* Denver, Colo.: American Water Works Association, 1991.

Garrison, R. H. *Managerial Accounting—Concepts for Planning, Control, Decision Making.* Plano, Tex: Business Publications, 1988.

Goodman, A. S. *Principles of Water Resources Planning.* Englewood Cliffs, N.J.: Prentice-Hall, 1984.

Great Lakes–Upper Mississippi River Board of State Public Health and Environmental Managers. *Recommended Standards for Water Works.* Albany, NY: Health Education Services, 1992.

Hanke, S. H., P. H. Carver, and P. Bugg. "Project Evaluation during Inflation." *Water Resources Research* 11(4), 1975, p. 511.

Hudson, H. E., Jr. *Water Clarification Processes—Practical Design and Evaluation.* New York: Van Nostrand Reinhold, 1981.

James M. Montgomery, Consulting Engineers. *Water Treatment—Principles and Design.* New York: John Wiley and Sons, 1985.

Kawamura, Susumu. *Integrated Design of Water Treatment Facilities.* New York: John Wiley and Sons, 1991.

Knowles, T. W. *Management Science—Building and Using Models.* Homewood, Ill.: Irwin, 1989.

Lane, W. N. "Community and Regional Planning." In *Standard Handbook for Civil Engineers.* 3rd ed. Edited by F. S. Merritt. New York: McGraw-Hill, 1983.

Lang, H. J., and D. Merino. *The Selection Process for Capital Projects.* New York: John Wiley and Sons, 1993.

Noble, George. *Siting Landfills and Other LULUs.* Lancaster, Pa.: Technomic Publishing, 1992.

Samson, D. *Managerial Decision Analysis.* Homewood, Ill.: Irwin, 1988.

U.S. Environmental Protection Agency. *Direct Environmental Factors at Municipal Wastewater Treatment Works.* EPA 430/9-76-003. Washington, D.C.: U.S. Government Printing Office, 1976.

Water Environment Federation and American Society of Civil Engineers. *Design of Municipal Wastewater Treatment Plants.* WEF Manual of Practice No. 8, ASCE Manual and Report on Engineering Practice No. 76. Washington, D.C.: Water Environment Federation, 1991.

CHAPTER 24
ENVIRONMENTAL IMPACT AND PROJECT PERMITTING

Environmental impacts must be considered at several junctures in the planning and design of a water treatment plant project. First, as described in Chapter 23, environmental issues play an important role in site selection. Once the site is selected and design commences, plant and site design must incorporate environmental controls that will minimize the impacts to the human and natural environment during plant construction and operation.

This chapter presents an overview of the environmental issues associated with plant construction and operation, describes general permitting requirements, and discusses the integration of environmental review into permitting, design, construction, and operation. This framework will enable the design engineer to be cognizant of the basic environmental review steps and to incorporate them in the planning and design phases of the project, before construction begins. Table 24.1 is a list of possible environmental review steps in a typical project.

ENVIRONMENTAL ISSUES ASSOCIATED WITH PLANT CONSTRUCTION

The environmental issues associated with construction of water treatment plants are similar to those of any medium- or large-sized construction project. The site features, neighboring land uses and ecological conditions, specific facility components and design, and level of community interest are all factors in determining the nature and extent of environmental impact review. The issues can range from minor considerations of land disturbance to more significant environmental impacts caused by truck traffic or major alterations to sensitive ecological areas.

The environmental impacts of constructing water treatment plants are typically of greater potential magnitude than the environmental impacts of operations. Construction impacts are characterized by higher amounts of trucks, noise, and dust than they will be later when the plant is operating. However, construction impacts are of shorter duration.

Operational impacts normally involve long-term issues related to design, daily operational performance, and maintenance of the facility—attributes that generally do not have the potential to significantly impact surrounding areas if a plant is well designed, well constructed, and properly maintained and operated. A general comparison of construction versus operational environmental issues is given in Table 24.2.

TABLE 24.1 Typical Environmental Review Steps

Project phase	Environmental review
Planning	
Site selection	Environmental siting criteria consideration of resource protection areas
Public hearings	Preliminary impact assessments
Initial approvals	State or municipal level environmental impact regulations
Design	
Technical design	Development of design criteria
Permits and approvals	Impact assessments and development of mitigation measures
Bidding	
Bid documents	Incorporate permit conditions and performance standards into bid documents
Construction	
Facility construction	Environmental monitoring and reporting
Operations	
Facility operations	Environmental monitoring and reporting Maintaining permit compliance

TABLE 24.2 Comparison of Construction and Operational Environmental Issues

Attribute	Construction	Operations
Study area	Site and adjacent areas Truck routes Noise Site compatibility	Site and adjacent areas Area of groundwater influence Area of surface water influence Delivery truck routes
Frequency	Daytime hours	24 hours
Duration	Temporary 6 months to 2 years	Long term 20 to 30 years
Magnitude	Ongoing disturbances for duration	Occasional disturbances
Mitigation	Timing Scheduling Physical barriers Equipment location	Performance standards Design features
Special concerns	Traffic safety Fuel spills	Chemical releases Solids disposal

Before construction, the project designer must analyze the potential environmental impacts associated with a proposed change in land use of the site and general project compatibility with surrounding land uses. This is particularly important when construction is proposed on a previously undisturbed site or within a nonindustrial area.

Determining the Major Environmental Issues

The process of identifying and evaluating issues of concern consists of four general steps:

1. *Evaluate site conditions and features.* This would include areas of particular environmental sensitivity such as wetlands or floodplains. The evaluation should include review of state and local regulations and ordinances that protect natural resources and adjacent land uses, because they provide excellent guides for identifying the types of site features that need to be considered before construction begins. These features should then be located on a site plan.
2. *Identify all major project components.* This might also include some off-site activities. Typical items are access and egress, staging areas, utility connections, pumping stations, source water intakes, groundwater well installations, dredging, dewatering areas, and solids disposal areas.
3. *Determine the plant arrangement.* The arrangement of the major plant components on the site must be initially determined.
4. *Develop a general construction plan.* The plan should include the basic construction phases and the approximate timing of the various construction activities, a description of the sequencing of those activities to determine which are concurrent activities, and the magnitude and duration of construction activities.

From the above steps, the environmental issues associated with construction can be readily identified. These are the issues that will most likely be the subject of permit application filings, environmental impact evaluations, and public concern about the proposed facilities.

The type of facility, size and location of the facility, and the site conditions are all important variables affecting the real environmental issues associated with facility construction. Table 24.3 shows some of the potential environmental impacts by construction phase and facility component.

There are also numerous potential environmental impacts associated with the selected site. For example, if the site has wetland areas, is within a flood zone, or has particular historic or archeological significance, those issues could be extremely important.

Environmental Issues Associated with Plant Construction

The principal environmental issues that must be considered during construction are site preparation, facility construction, and landscaping.

Site Preparation. Although site preparation is not always distinct from facility construction, it generally includes access road and staging area preparation, site clearing and grubbing, preliminary excavation, and placement of fill.

Environmental impacts associated with staging areas are generally temporary, although careful site selection is required to avoid sensitive areas. Access roads may be temporary or long term, depending on whether the road is for use during construction only or for facility access during operations. In either case, sensitive siting is important. In determining locations for access roads and staging areas, consideration should be given to disturbing as few natural features as possible. Consideration must also be given to the proximity of sensitive residential activities such as homes, schools, and hospitals to ensure that traffic conflicts and disruption from noise and dust are minimized during construction.

TABLE 24.3 Potential Environmental Impacts by Construction Phase or Aspect

Construction Phase	Land use	Traffic	Air quality	Noise	Ecology	Aesthetics	Vibration	Public safety	Water quality	Wetlands
Site preparation										
Access roads and staging areas	X	X	X	X	X					
Site clearing and grubbing	X	X	X	X	X					
Excavation			X	X		X	X			
Placement of fill		X	X	X						
Facility construction										
Materials delivery	X	X	X	X				X		
Dewatering systems		X							X	
Placing concrete				X						
Intake or groundwater wells					X				X	X
Landscaping					X	X				X

A certain amount of clearing and grubbing are needed on most undeveloped land. Site design should take into account any unique natural features that can be preserved, such as especially large or rare trees. There may also be vegetation or features that can be preserved to act as a natural aesthetic buffer between the plant and public view after construction. Potential impacts resulting from clearing and grubbing include construction traffic, equipment noise and emissions, dust, aesthetics, and ecological disruption.

The extent of excavation required depends on site-specific conditions. Generally, the more excavation required, the greater the anticipated impact in terms of construction traffic, noise, dust generation, and vibration, particularly if rock drilling or blasting is required. The duration of the site preparation construction is also an important consideration and is often a trade-off with the magnitude of impact. A shorter construction schedule on a particular site, for example, may result in greater daily noise and traffic impacts than a longer schedule.

Facility Construction. Site preparation and facility construction both involve site disruption and on-site activities that may pose conflicts with neighboring land uses. The greatest environmental impacts associated with on-site facility construction are generally air quality (dust) and noise caused by construction traffic. The installation of water mains to connect with the source water source and to connections with the water distribution system will have additional off-site impacts.

If a new intake is to be installed, it will undoubtedly impact a watercourse and will require additional piping. The method of intake installation and its location are critical factors in determining the extent and magnitude of environmental impacts on both aquatic species and downstream users. Environmental regulators will generally restrict construction activities in a river to times of low flow and will require that the river be restored to its original bottom contours to minimize ecological impacts.

Landscaping. Landscaping is often a mitigation measure, although it can have impacts of its own, depending on the amount required and its compatibility with the area aesthetics. Planting ornamental trees in a natural setting, for example, may have a negative aesthetic impact rather than a positive impact. However, in general, landscaping provides an opportunity to restore and even enhance the natural attributes of the site, as well as provide visual screening from neighbors.

ENVIRONMENTAL ISSUES ASSOCIATED WITH PLANT OPERATIONS

There are numerous environmental issues associated with plant operations, primarily related to the various waste streams and by-products of water treatment plant processes, but also to the impacts on the supply source as a result of water withdrawal.

The type of treatment process that will be used will obviously affect the type and extent of the plant's operational environmental impacts. In particular, the chemicals that will be used in the treatment process and the waste stream characteristics of that process are significant in determining the environmental impact. For example, the waste streams associated with filtration processes may contain higher levels of metals than some other processes, which could be of concern if the resulting residuals are applied to agricultural land.

Water Withdrawals

Whether water withdrawal will be from a groundwater supply or a surface water supply may have significant environmental ramifications. The cone of influence surrounding a well may change the hydrology and, as a result, the ecology of the surrounding area. Surface water withdrawals have the potential to affect downstream ecology and uses and to entrap fish.

Surface Water Withdrawals. Some of the environmental impacts that may be caused by new or increased withdrawals from a surface water supply include:

- Change in water temperature and quality characteristics as a result of river or reservoir depth changes
- Change in pollutant dilution as a result of flow changes in a river
- Ecological effects related to depth changes and reduction in flow caused by river withdrawal
- For river withdrawals in particular, entrainment of fish eggs and larvae as they are drawn into the intake, and impingement of animals and debris on the intake structure
- Potential conflicts with recreational or other human uses of the water body

Through proper design and operation, a balance can be achieved between the need to withdraw water and the goal of minimizing impacts on the water body. For example, selecting a proper type of screen and minimizing the flow velocity around an intake will help prevent entrapment of fish. Placing easy-to-read signs upstream of a river intake and buoys to mark a channel passage can help mitigate potential conflicts with boaters.

Groundwater Withdrawals. Well withdrawals can also affect surrounding ecology and human activities. Examples include:

- Movement of the interface between saltwater and freshwater in the aquifer in coastal or near-coastal areas
- Reduction of groundwater available for nearby domestic and agricultural well systems
- A reduction in the aesthetics and recreational value of surface water bodies in the area
- Increased activity of sinkholes
- Changes in the composition and health of terrestrial and aquatic ecosystems.

The type and extent of the drawdown impacts that will occur around a well vary widely, depending on many site-specific factors. Conducting pump tests and modeling the likely drawdown of groundwater levels at proposed withdrawal rates can help predict the magnitude of the change and determine what the optimal pumping rate will be, while minimizing impacts to the surrounding area.

Chemical Delivery, Handling, and Storage

Some of the chemicals used in water treatment are commonly found in household products; for instance, sodium hypochlorite is bleach and aqueous ammonium hydroxide is household ammonia. These chemicals, when properly handled and stored, are safe. However, because the chemicals are often concentrated and used in bulk at water treatment plants, delivery, handling, and storage are particularly critical and are sometimes the subject of public concern. Use of chlorine gas, for example, often generates public concern because of the inhalation dangers if there should be a leak.

A number of criteria for addressing public health concerns are described in the *Guiding Principles for Chemical Accident Prevention, Preparedness, and Response* published by the Organization for Economic Cooperation and Development (OECD, 1992). These guiding principles establish "general guidance for the safe planning, construction, management, operation and review of safety performance of hazardous installations in order to prevent accidents involving hazardous substances." Although the principles define hazardous installations as fixed plants or sites, many are also applicable to transporting hazardous substances. These guiding principles are used as the basis for much of the following discussion.

Proximity of Plant to Populous Areas. The guiding principles suggest separating people from hazardous substances, which can be achieved in a number of ways. One obvious method is to site the facility away from high population densities and particularly from very sensitive persons, such as children and the elderly. However, other siting constraints may preclude this. Furthermore, this method may not adequately address transportation of chemicals and worker safety. Therefore an emergency response plan should, as required, address the areas likely to require evacuation or protection in the event of a chemical spill.

Transportation. Safe delivery of chemicals to the plant is as critical as safe chemical storage and use at the plant. Delivery modes and routes are important factors to consider in transporting hazardous materials. To minimize public health risks, the ideal transportation route for hazardous substances is a large, little-traveled roadway in good condition or a rail line with few roadway crossings that passes through undeveloped land. If such a route is not available, a route should be selected that will be as short as possible and will avoid accident-prone areas and high-density populations.

Engineering and Design Controls. Once chemicals have been delivered to the plant, a number of safety measures are recommended to help prevent and contain spills. Preventive measures begin with designing the facility to reduce the possibility of releases. Good engineering and design controls can range from such straightforward methods as storing reactive chemicals separately to more advanced pollution control technologies, such as air scrubbers.

Typical engineering controls should include installation of sprinkler systems, construction of storage spaces with fire-resistant concrete, and providing separate containment areas for each chemical. Providing enclosed areas with proper ventilation for unloading chemicals is another design control that should be incorporated in the facility.

Residuals Treatment and Disposal

Until recently, the primary focus of the regulatory agencies and water treatment plant owners and operators was on providing high-quality water to meet federal and state drinking water standards. Little attention was paid to the treatment and final disposal of water treatment plant waste sludges, or residuals, which were typically discharged directly into a nearby water body or to a local wastewater treatment facility. Both of these practices, however, are receiving increased attention from the environmental and regulatory communities.

Potential localized effects of stream discharges include increased suspended solids and turbidity, creation of sludge deposits, and increased iron and aluminum concentrations. If water treatment residuals are discharged to a wastewater treatment plant, there may be an adverse impact on the plant's ability to meet discharge permit limits. As a result, alternative disposal methods are now being increasingly considered, including spreading residuals on agricultural or forest land, and landfill disposal or use as landfill cover material after thickening and dewatering.

Before land application or landfill disposal will be allowed by state or local authorities, it must be shown that groundwater will not be adversely affected. The toxic characteristic leaching procedure (TCLP) is commonly used to simulate the climatic leaching action expected to occur in a landfill. The TCLP identifies 8 metals and 25 organic compounds (pesticides, herbicides, etc.) with the potential to leach into groundwater.

With respect to land application, concern has also been expressed regarding the lack of organic value and high metals content of the residuals, and the potential effect on crops and groundwater. However, lime-softening or lime-treated sludges have been shown to have positive effects on agricultural and vegetative uses of land (Robinson and Witko, 1991). The general rule in evaluating land application alternatives is to consider the ultimate land use and the residuals quality. Residuals with a high metals content, for example, will probably not be appropriate for application on agricultural land.

A 20% solids content is generally required for landfill disposal of residuals. In addition to mechanical dewatering, drying of residuals in on-site lagoons is still a common practice to achieve the appropriate solids content for landfill disposal. When suitably dry, the residuals can be scraped out of the lagoons and transported to a landfill for final disposal. But the practice of drying in lagoons also has environmental concerns, including odors (described as musty, or earthy) and land requirements for drying. In addition, the overall quality of the residuals is also an important consideration in the feasibility of landfill disposal because the characteristics of the residuals will determine whether a sanitary landfill, hazardous waste disposal site, or low-level radioactive waste disposal site is most appropriate.

For use as landfill cover material, residuals could be dried to about 60% solids and applied directly, or dried to about 40% and mixed with soil. The residuals must be shown to be nonhazardous by a TCLP test and must have no free liquids. Some landfill operators and state regulators will not accept residuals for landfill cover because they are concerned that the material will be gelatinous when wet and dusty when dry.

Although there are many considerations in determining the most appropriate method of residuals treatment and disposal, environmental issues are gaining more significance as there is better understanding of the impacts on the receiving water body, treatment plant, landfill, and agricultural land. It appears likely that direct discharge to receiving streams, wastewater plants, and landfill disposal will become increasingly disfavored, and land application and other beneficial disposal alternatives will gain greater acceptance.

Wastewater Treatment and Disposal

Typical sources of liquid waste discharges from water treatment plants include filter backwash water, supernatant from sludge dewatering, emergency bypass water, water from washing equipment and flushing pipelines, and employee sewage. Disposal of employee sewage must be to an on-site septic system or to a nearby sewer. Water from equipment washing and pipeline flushing can be similarly disposed of, although in some cases it is discharged to a nearby water course.

Filter backwash water, supernatant, and bypass water are also often discharged to receiving waters, but in many cases these wastewaters are chlorinated, which has raised concern because of the potential toxicity effects to aquatic biota and potential health hazards. The impacts on the receiving water are primarily dependent on the concentration of chlorine residuals, which is in turn governed by the combined effects of the mixing process with the receiving waters and the time-dependent decay of the chlorine (Stanley and Smith, 1991).

There are two potential environmental hazards associated with chlorine residual discharges. First is toxicity to humans through downstream drinking water intakes. However, this concern appears to be minimal because the residual chlorine concentration is much less than the U.S. Environmental Protection Agency (USEPA) ambient water quality criteria for chlorine (Stanley and Smith, 1991). A second hazard is the potential impact of the chlorinated discharge on aquatic biota. Chlorine toxicological studies have shown that the effects on aquatic life are determined by many physical, chemical, and biological factors, including pH, alkalinity, hardness, temperature, species sensitivity to chlorine, and the speciation of the chlorine (Turner and Chu, 1983). These factors were all considered by USEPA in establishing limits for allowable total residual chlorine in receiving waters to be met by water treatment plants. As a result, any chlorinated discharges will require a permit from USEPA.

Storm Water Management

Even after construction is completed, proper management of storm water is necessary. Appropriate controls must be incorporated into the plant design to ensure that predischarge runoff rates from the site are maintained. If, for example, a new plant is constructed on a previously undeveloped site, the increased rate of runoff could be significant as a result of the increase in impervious surfaces. Typical storm water control measures include drainage channels and detention or retention basins.

Air Emissions

Water treatment processes can create air emissions and odors, which must be considered during plant design. Examples of emission sources at a treatment plant include ozonation, chlorination, emergency power generation, space heating, and solids handling.

During ozonation, some ozone off-gas is released. Because ozone is a strong irritant and will harm vegetation, the off-gas must be treated by means of a catalytic or thermal ozone destruction unit. These units are effective in reducing ozone emissions to below national or state ambient air quality standards. It is also important to protect plant workers from ozone exposure. Preventive measures include maintaining the ozone contact tanks under negative pressure to prevent leaks, and placing ozone monitors in working spaces designed to shut down the entire system if a leak is detected.

Power for normal plant operations is generally provided by the local electric utility; however, many plants have their own diesel generators to supply electricity for use during commercial power failure. Diesel generator exhaust is a source of carbon monoxide, nitrogen oxides, sulfur dioxide, and particulate matter, so the release of these pollutants may pose an environmental concern if in large enough quantities. Similarly, the plant space heat boilers may release pollutants that may be considered objectionable.

Although not as noxious as wastewater residuals, residuals from the water treatment processes may be a source of odors, in particular, the organic matter removed from the treated water. If the residuals are discharged directly to a sewer with no open-air storage or drying, odors are usually not an issue. However, if the residuals are stored and dried on site, off-site odors could be a concern.

If source water is of high quality and the resulting residuals have a low organic content, odors are generally not a problem. In most cases, the odors are of an earthy or musty nature and cannot be detected beyond 100 yd (90 m) of the residuals under the worst conditions. A survey recently conducted for Camp Dresser & McKee (1995) showed no significant odor problems reported by any of the respondents. Some of the responses included:

- In Englewood, Colorado, residuals are collected in large thickening lagoons and then are mechanically dewatered and stockpiled on the site for as long as two years. Some of the residuals may have been in the lagoon for up to 15 years. No odor complaints have been received from any of the adjacent neighbors.
- A facility in upper Michigan discharges alum residuals several times a year to a freeze-thaw lagoon. The residuals are excavated and disposed of each spring. In the past 20 years, no complaints have been received from residents living less than 200 yd (180 m) away. The plant manager did report, however, that on a hot windy day, the odor may be detectable 100 yd (90 m) away.
- A plant in Weymouth, Massachusetts, recently cleaned six to eight feet of residuals out of five lagoons that had not been cleaned in 30 years. The residuals were excavated by dredging and dried on site. After drying, the residuals were hauled to a landfill in Connecticut. No odors were reported from this operation, which took place in summer.

PROJECT PERMITTING

Numerous federal, state, and local environmental laws, regulations, and policies govern water treatment plant construction activities and operations. The focus of this discussion is on federal permits and approvals because there is a great deal of variability between the permitting processes in individual states and municipalities. It is important to note, however, that there is also some variation in implementation of federal requirements from state to state. It is therefore important that the designer carefully review the state permitting process for each particular project. Table 24.4 summarizes the most pertinent federal environmental regulations and specific triggering activities.

TABLE 24.4 Typical Permits, Regulations, and Triggering Activities

Federal permit and regulations	Responsible agency/authority	Triggering activity
Title III of Superfund Amendment and Reauthorization Act	USEPA	Tracking of hazardous chemicals and release reporting
Approval under CFR Title 49	U.S. Dept. of Transportation	Classification of hazardous materials; packaging and transport
Resource Conservation and Recovery Act (RCRA) of 1976 and Comprehensive Environmental Response, Compensation, and Liability Act (CERCLA) of 1980	USEPA	Review triggered by hazardous characteristics of residuals
Approval under 40 CFR 257	USEPA	Land application of residuals
Chemical accident prevention provisions under 40 CFR Part 68	USEPA	Potential accidental airborne releases of substances that pose a public health environmental threat
Title V Operating Permit Program under 40 CFR Part 70	USEPA	Air emissions exceeding certain thresholds
National Environmental Policy Act (NEPA) review to determine whether EIS is needed (42 USC 554321 et seq)	USEPA	Any federal action (including grants or permits); need for EIS depends on the magnitude of perceived impacts
National Pollutant Discharge Elimination System Permit under Section 402 of the Clean Water Act (40 CFR 122-124)	USEPA	Discharges of storm water, treated wastewater flows, or other process flows into surface waters or wetlands; storm water discharge from construction sites >5 acres in size
Section 404 of the Clean Water Act	U.S. Army Corps of Engineers	Discharges of dredged or fill material into U.S. waters (including wetlands); jurisdiction includes incidental discharges associated with excavation activities
Section 10 of the Rivers and Harbors Act of 1899 (33 CFR 322)	U.S. Army Corps of Engineers	Structures or work in or affecting navigable U.S. waters (navigable waters are generally defined as U.S waters that are subject to the ebb and flow of the tide shoreward to the mean high water mark or are used or have been used in the past to transport interstate or foreign commerce)

Federal Environmental Regulations

Federal environmental regulations can generally be categorized as affecting either plant construction or plant operations.

Construction Permitting. Construction permit requirements are primarily dictated by the site and surrounding area characteristics. Work in wetlands, for example, will generally require a permit from the U.S. Army Corps of Engineers. However, the type and extent of activity within that wetland will also determine permit jurisdiction, permit review procedures, and the relative ease or difficulty in obtaining the permit.

National Environmental Policy Act. Review under the National Environmental Policy Act (NEPA) is triggered by any federal grant or permit. If, for example, the Army Corps of Engineers determines through their review of a permit application that a proposed project will have significant environmental effects, an environmental impact statement (EIS) under NEPA may be required. Generally, an EIS is required only for large, environmentally disruptive projects such as construction of a water supply lagoon that may impact a protected endangered species. By minimizing environmental impacts up front in water treatment plant site selection and preliminary design, the potential need for an EIS is reduced.

NPDES General Permit. A permit under the National Pollution Discharge Elimination System (NPDES) will be required for storm water discharge during plant construction if more than five acres are affected by construction activities and the discharge will occur to a water body or wetland. The general permit application must be submitted to USEPA at least 48 hours before the start of construction, and a storm water pollution prevention plan must be prepared and kept on file. When construction is completed, a notice of termination must be sent to USEPA.

U.S. Army Corps of Engineers Permits. Under Section 404 of the Clean Water Act, a permit may be required from the Corps of Engineers for any filling of wetlands or water bodies. For structures or work affecting navigable waters, such as construction of a source water intake in a navigable river, a permit under Section 10 of the Rivers and Harbors Act would be required. The precise permitting requirements vary between Corps districts and between states; therefore it is important to understand the particular requirements within the state where the construction is to take place. Programmatic general permits or nationwide permits apply in some states, which allows certain activities involving minimal wetlands or water body disturbance to proceed without, or with limited, Corps involvement.

Operations Permitting. Federal programs regulating chemical delivery and handling, residuals disposal, storm water discharge, and air emissions are briefly summarized below.

Chemical Delivery and Handling There is potential for regulation of chemical delivery and handling by at least two federal agencies and programs. Title III of the Superfund Amendment and Reauthorization Act (SARA) of 1986 (also referred to as the Emergency Planning and Community Right-to-Know Act, or EPCRA) deals with the tracking of hazardous chemicals and release reporting. EPCRA notification allows local fire departments, local emergency planning committees, and state emergency response commissions to be informed about the existence of large quantities of hazardous materials at facilities (including water treatment plants) in their community or state. The information required to be reported is used to facilitate emergency response planning at both the local and state level. The goal of the reporting system is to inform and prepare communities hosting these facilities for potential hazards and emergency situations that could occur if there was an accidental release of hazardous chemicals.

More recent federal programs include the Occupational Safety and Health Administration (OSHA) Process Safety Management (PSM) program and the USEPA Risk Management Program (RMP). Although similar in many aspects, the PSM program deals with worker safety, and the RMP program focuses on potential impacts to the public and environment.

Transportation of hazardous materials is regulated by the U.S. Department of Transportation under CFR Title 49. Title 49, Parts 171 through 180, provides for the classification of hazardous materials and for the marking, labeling, and placarding (for bulk shipments) of hazardous materials packages and containers. These regulations also contain requirements for communications, emergency response, and training of persons who handle or transport hazardous materials. Other parts of the regulations cover the specific requirements for each transportation mode (rail, air, water, and public highway).

Residuals Disposal Regulations. The regulation of residuals disposal depends on the ultimate disposal method, and the ability to comply with applicable regulations may determine which disposal method is the most feasible. If residuals are to be discharged to a water course, the Clean Water Act (CWA) is applicable. Under the CWA, an NPDES permit would be required from USEPA for any discharges of sludge or semisolids to navigable waters. However, the need for compliance with water quality standards would prohibit these discharges in most cases. NPDES permit limits are determined by the appropriate water quality standards and criteria, which are based on protecting human and aquatic life.

Landfill disposal of residuals is potentially regulated under the Resource Conservation and Recovery Act (RCRA) of 1976 and the Comprehensive Environmental Response, Compensation, and Liability Act (CERCLA) of 1980 if the residuals are considered to be hazardous. Under the Toxicity Characteristics Leaching Procedure (TCLP), it is possible for water treatment plant residuals to exceed the threshold limits and therefore to be considered hazardous (Robinson and Witko, 1991). Many states require that residuals to be landfilled be a minimum of 20% solids and be able to pass the TCLP and also a paint filter test to determine whether there are any free liquids emanating from the residuals.

The RCRA and CERCLA guidelines for land disposal of residuals specify that disposal activities must be conducted in a manner that minimizes impact to the environment. Specifically, RCRA regulates the residuals quality and disposal activity, and CERCLA provides for the remediation of improperly operated or abandoned disposal sites. RCRA also encourages recycling of materials as an alternative to ultimate disposal.

Land application of residuals is loosely regulated by USEPA under 40 CFR 257, which deals primarily with wastewater treatment residuals as opposed to water treatment residuals.

Wastewater Disposal Regulations. The type of regulation applicable to wastewater disposal depends on the ultimate disposal plan. Discharge to an existing sewer would not trigger federal permitting, but would involve approval by the appropriate sewer authority. Federal approval would not be required for discharge to a septic system, but state or local approval would probably be required.

However, discharge to a nearby watercourse may require an NPDES individual permit from USEPA, and it may also be required that the discharge receive treatment before discharge, depending on the particular state and the discharge characteristics. (In Massachusetts, New Hampshire, and Maine, for example, some water treatment plant discharges are exempt from the need to obtain an NPDES individual permit, and only a general permit may be required.) Whether the discharge would be allowed is also determined by the ability of the discharge to meet the water quality criteria for the receiving watercourse.

Air Emissions Regulations. Air emissions may be regulated under several programs, depending on the amount and type of emission. A number of emerging USEPA clean-air rules may affect the design and operation of a water treatment plant. For example, on a case-by-case basis, state and local fire codes could require installation of scrubbers at water treatment plants to control accidental releases of chlorine from gas feed systems. Also, combustion of fossil fuels to operate boilers and emergency engines can result in emissions that may have to be permitted and controlled in the future.

One pertinent air quality regulatory program is 40 CFR Part 68, Chemical Accident Prevention Provisions, which addresses accidental airborne releases of substances that pose a significant threat to the public or the environment. Another is the Title V Operating Permit Program under the Clean Air Act (40 CFR Part 70), which is triggered if certain emissions thresholds are exceeded. A Title V permit is required if emissions exceed the following thresholds in tons per year (tpy):

- 100 tpy of a criteria pollutant (SO_2, NOx, ozone, CO, lead)
- 10 tpy of a toxic compound (volatile organic compound, VOC)
- 25 tpy of total VOCs

In determining whether any of these thresholds are being exceeded, all emission sources at a water treatment plant must be totaled.

State and Local Regulations

Construction and operation of a new water treatment plant is likely to require approvals from a number of state and local regulatory authorities. In fact, because there is increased deregulation at the federal level, much of the regulatory burden now rests with the states. Most states have received federally delegated authority to issue permits and enforce various federally mandated programs.

If a local authority has no direct permitting responsibility, input is still provided through the state permitting process. For example, in Massachusetts, the local board of health in a community where land application of residuals is proposed is given the first opportunity to comment on a permit application submitted to the state before the state takes any formal action on the application.

Many states have their own wetlands and surface water regulations that will require a permit if a new plant will affect these resources during construction or operation. Because federal approvals are also often required for activities affecting these resources, lack of consistency between state and federal requirements is commonly an issue that must be anticipated and addressed during the permitting process.

At the local level, zoning approvals are often required, as well as approvals from the local planning board, building inspector, and so on. Many communities also have specially mapped districts, in addition to zoning, to control activities in floodplains, for example.

Regulatory Trends

The most notable trend in environmental regulations pertaining to water treatment plants is deregulation at the federal level and increased reliance on individual states and even communities to monitor plant construction and operations. One example of this is a recently drafted state-USEPA accord (called the National Environmental Performance Partnership System), which would dramatically scale back USEPA oversight of state environmental programs by setting new goals for environmental protection and giving states broad flexibility to meet them. However, whether implemented at the federal, state, or local level, regulatory programs protecting water bodies, wetlands, and air from pollution are likely to exist in one form or another.

Regardless of the specific requirements, to understand the maze of permits and approvals required for construction and operation, a permitting plan should be developed during the preliminary design phase of a water treatment plant. Otherwise, it is possible that permits could delay the project, which can increase project cost. Developing and implementing such a plan is the subject of the next section of this chapter.

INTEGRATION OF ENVIRONMENTAL ISSUES AND PERMITTING INTO PLANT DESIGN, CONSTRUCTION, AND OPERATION

With the increase in environmental awareness and regulatory requirements in recent years, a project that might have breezed through the permitting process 10 years ago may find itself lost in the permitting maze today. As a result, it is extremely important to establish and follow a carefully planned permitting strategy even before plant design begins. The strategy should also include consideration of environmental issues, because these are the driving forces behind the permitting process. The overall steps in development and implementation of a permitting strategy are presented in Table 24.5 and summarized below.

Developing a Permitting Plan

The proposed permitting strategy should be outlined in a written plan that identifies specific permits and permit interrelationships and defines the overall approach for obtaining permits. The plan should also establish the links between specific project activities or environmental issues and permit requirements (i.e., permit triggering activities). Finally, the permitting plan should present a schedule for obtaining permits, including the approximate lead time needed to prepare applications, appropriate dates for filing applications, and dates when permits are needed. Establishing a realistic schedule is particularly important when there is a state or federal consent order involved, which specifies interim dates for project activities such as bid dates, construction start, and completion.

The permitting plan also provides a good opportunity to review preliminary design decisions and to make appropriate design modifications in order to eliminate or reduce permitting requirements. Identifying potential modifications before design is finalized may avoid costly later design changes and project delays caused by a lengthy permitting process.

TABLE 24.5 Permitting Strategies

Permit approach
1. Identify all project components.
2. Assess the existing site conditions.
3. Identify all required permits and approvals.
4. Identify the major permitting issues.
5. Develop a project-specific permitting approach.
6. Compile all necessary project information and data (coordinated with design).
7. Prepare and submit permit applications.
8. Track the review process.
9. Negotiate the permit conditions.
10. Obtain project approval by the agencies.
11. Follow through with permit conditions.
12. Know permit expiration dates.

Carrying Out the Permitting Plan

The permitting plan should be updated throughout design as necessary and should be used as a checkpoint for design decisions. Using the schedule provided in the plan, applications should be prepared and submitted to the review agencies. However, the process does not end here. Tracking the applications through the agencies and the review process is just as critical as timely preparation and submittal of those applications. Knowing who the reviewer is and establishing early contact with that individual is the first step toward a smooth review process.

Once the application is submitted, maintaining contact with the reviewer is also important to ensure that the application is not buried under others and that technical questions are not left unanswered. If technical questions are answered during the application review process, they are less likely to appear as permit conditions. Providing evidence that storm water discharges are not likely to exceed water quality standards, for example, may eliminate a storm water monitoring requirement. Another approach to ensuring reasonable permit conditions is to actively negotiate the conditions with the agency before the permit is issued. This strategy may streamline the process by producing a set of permit conditions that are reasonable to both the applicant and the agency, thereby avoiding time-consuming appeals.

Compliance during Construction and Operation

Most construction-related permits become part of the contract documents so that the prospective contractor can accurately bid the project. The selected contractor then becomes a party to permit compliance. However, relying solely on a contractor to meet the terms of a permit is not advisable because it is the applicant or project owner who is ultimately responsible for permit compliance.

As a result, the permit tracking system should not end with permit issuance, but should follow through construction and operation, as applicable. Even a construction-related permit may have conditions that extend beyond construction completion. Maintaining catch basins and other storm water controls is one example of a condition that may be effective beyond construction.

The permit tracking system should also take into account permit expiration dates and regulatory changes. If, for example, a permit program is revised, a new application may be required before the existing permit expires. When a permit expires, an extension is often easier to obtain than the original permit as long as regulatory changes have not occurred and conditions in the original permit have been met, although a smooth process is not necessarily guaranteed.

Conclusions

The key to successful permitting is to establish a strategy early in the project and implement it throughout design, construction, and operation. Obviously, some revisions to the strategy will be necessary as the project progresses, but a permitting plan can still provide a road map to help streamline and track the permitting process.

BIBLIOGRAPHY

Camp Dresser & McKee, Inc. *Wachusett Reservoir Water Treatment Plan. Final Environmental Impact Report.* Cambridge, Mass.: CDM, 1995.

Camp Dresser & McKee, Inc. *Taunton River Water Supply EIR.* Cambridge, Mass: CDM, 1996.

Hadden, Deborah, and William Keough. *Predicting Public Water Supply Well Operation, AWWA Annual Conference.* Denver, Colo.: American Water Works Association, 1994.

Robinson, M. P. and J. B. Witco. *Overview of Issues and Current State-of-the-Art Water Treatment Plant Waste Management Programs.* AWWA Conference Proceedings, 1991

Stanley, S. J. and D. W. Smith. "Modeling Chlorinated Discharges from Water Treatment Plants: a Case Study." *Canada Journal of Civil Engineering* 18:985, 1991.

Turner, A., and A. Chu. "Chlorine Toxicity as a Function of Environmental Variables and Species Tolerance." In *Water Chlorination: Environmental Impact and Health Effects.* Vol. 4, book 2. Ann Arbor, Mich.: Ann Arbor Science, 1983.

Vicory, A. H., and L. Weaver. "Controlling Discharges of Water Plant Wastes to the Ohio River." *Journal AWWA* 4(84):122, 1984.

CHAPTER 25
OPERATIONS AND MAINTENANCE

This chapter presents operations and maintenance (O&M) factors that are important in design and start-up of a water treatment plant.

OPERATIONS AND MAINTENANCE CONSIDERATIONS DURING PLANT DESIGN

While the engineering design team is considering the unit processes and structural details of a new plant, they must also constantly keep in mind the considerations of how best to make the plant easy, efficient, and economical to operate and maintain.

Design Considerations for Plant Operations

Some of the more important considerations of plant operations that must be considered are the power and control systems, the process systems, plant automation, plant security, and safety.

Power and Control Systems. Adequate electrical power and good control systems are extremely important to the successful operation of a modern water treatment plant. One of the main concerns with electrical power is reliability of the power source. Electrical equipment must be protected from damage from condensation, flooding, dust, and dirt. Control issues center around the location of controls and the level of automation. The following is a list of electrical power and control issues that should be considered by the designers:

- If possible, control systems should be provided to operate in both automatic and manual modes. Facilities that have automatic, semiautomatic, and manual control strategies provide full operator flexibility and are not labor intensive.

- Local manual control panels should preferably be placed at central sites that allow operators to perform manual operating duties while being able to visually observe equipment operation. This is important when dealing with filters, chemical feed systems, and sludge pumping systems.

- The use of manually operated valves on automatic systems should be minimized. Manually operated valves defeat the purpose of an automatic system, which is to minimize the labor required to operate the system.
- All valves in the plant should be provided with number tags. If valves are not properly identified, it becomes difficult to develop easy-to-follow standard operating procedures (SOPs). If tags are not identified by numbers, a lengthy written description must be provided to accurately locate the valve.
- The design of motor control centers and control stations should provide adequate room for O&M access.
- Motor control centers and control stations must not be located in areas that are subject to flooding.
- Motor control centers and control stations that are located within rooms must be provided with entrances and exits that are readily accessible.
- Mechanical equipment with electric motors should be elevated on pedestals or pads. Keeping equipment elevated above the floor reduces the possibility of damage and simplifies maintenance around the equipment.
- If possible, critical motor control centers should be decentralized. For example, if there are two sedimentation basins, it is preferable to furnish a separate control center for each one.
- Sample taps and flowmeters should be provided on all sidestreams to facilitate sampling and accounting of plant water uses.
- Flooding alarms should be installed near the floor in all equipment areas that are subject to flooding.

Process Systems. The following process areas in a water treatment plant have O&M issues that designers should consider.

Source Water Intakes. Intake facilities should be designed to be as accessible as possible to plant staff. If source water pumps are located more than a short distance from the main plant, they should be fully automated and monitored from a main control room.

Rapid Mix Basins. Considerations should be given in the design of rapid mix basins to providing variable-speed mixers, which will allow the plant operator to vary the intensity of mixing. This will generally allow better process control when there are changes in source water quality, water temperature, and flow rates.

The method of introducing coagulants and coagulant aids into the rapid mix basin is important to the overall performance of the basin. They should be introduced quickly and uniformly with the source water as it enters the basin. The location of the point of application and how it is applied can have a significant impact on treatment efficiency.

Flocculation Basins. Paddles or mixers in the flocculation basin should be provided with variable-speed drives so that operators have the ability to change the mixing energy to meet different water conditions and thus produce an effluent with the best settling characteristics.

Flocculation basins should be designed with sump drains and sluice gates between baffle walls to make it easier to drain and wash the basins. Sumps should be placed at the effluent end or lowest point in the basin floor.

Sedimentation Basins. The installation of automated sludge removal in sedimentation basins is usually desirable for reducing maintenance costs. The sludge valves should be opened and closed by a timer set by the operator to maintain the optimum sludge level in the basin. Providing a davit next to the sedimentation basins and installing permanent ladders in the basins help facilitate operations and maintenance activities.

Recarbonation Basins. Many recarbonation basins are located indoors, especially in northern climates. Enclosed recarbonation basins must be provided with exhaust and fresh air supply fans that are operated by one switch located at the enclosure entrance. There must also be two self-contained breathing devices located next to the enclosure entrance, as well as carbon dioxide analyzers and alarms to continuously monitor carbon dioxide levels in the enclosure.

Filters. Filters are process units that should preferably have local control. Remote control may be provided to a central point, such as the operations control room, but local control at each filter is necessary to facilitate O&M functions. Filter backwashing should be observed by operators at the filter, at least part of the time, or problems that occur in the filters may not be detected for extended periods of time. Typical problems that may be observed are formation of mudballs, cracking, jetting, incomplete backwashing, broken nozzles, and broken underdrain blocks. In addition, if the filters are provided with fully automatic functions, they should also have semiautomatic and manual functions for flexibility of O&M.

Provisions should be made for continuously monitoring filter head loss and effluent turbidity at each filter because they are key parameters that tell the operator how the filter is performing. Filter operating data should be recorded on chart recorders or by logging the data on a computer. Adequate records of filter operation will be important to the plant operator in determining how the filters are performing with respect to plant filtered water goals. The recorded data will also be important in the future, when the data can be reviewed for making long-term process decisions and future upgrades to the water system.

Controls must be provided for varying the backwash water flow rate and extending or shortening the backwashing cycle. Operators will need to control the backwash water flow rate and adjust it to the optimum rate as water temperatures change. This is especially critical for surface water plants located in northern climates where wide variations in water temperature can significantly change the optimum backwash rate. The operator should be able to make changes in the backwash cycle both at the central control panel, if one has been provided, and at the control panel in front of each filter.

Chemical Feed Systems. Chemical feed systems normally comprise storage tanks, day tanks, transfer pumps, suction and discharge piping, chemical injection points, and adjustable metering pumps. The following are features that should be included in chemical feed systems to assist operations and maintenance functions:

- Provide a calibration cylinder for each chemical feed pump.
- Provide a pulsation dampener on the suction and discharge sides of each chemical feed pump.
- Provide adequate pipe supports for metering pump suction and discharge piping. The supports should allow for movement of piping caused by the pulsation effect of the pumps.
- Whenever possible, install chemical distribution piping in troughs with removable grating for ease of access and repair.
- Provide frequent unions or flanged fittings in chemical piping for easy disassembly for repair or replacement.
- Construct adequate roadways adjacent to the plant to allow bulk chemical delivery trucks to turn completely around near the chemical unloading stations.
- Provide chemical storage tanks with monitoring devices, and allow ample work space around them for chemical testing.
- Provide for adequate mixing of chemicals with process water by feeding them through ejectors or by using motorized or static mixers.

- Liquid chemical systems should generally be selected over dry bulk chemical systems unless the transportation costs of the liquid chemical are prohibitive. Liquid chemicals are normally easier to store and feed, and there are no dust problems to contend with. In addition, liquid chemical feed systems usually have a lower capital cost and are simpler to operate and maintain.
- If any chemicals are to be purchased in bags or drums, the plant construction contract should require the contractor to furnish proper dollies, platform trucks, hydraulic lift trucks, hydraulic pallet trucks, or forklifts for handling the chemicals.
- If dry bulk chemical systems are to be used, a dust control system should be provided that is large enough to handle dust created by the vacuum unloading system or blower on the delivery truck. The blowers on bulk delivery trucks usually deliver approximately 700 ft^3/min (1,190 m^3/h) of air at up to 10 psig (69 kPa). The dust control system must have more capacity than this to contain the dust from the unloading process.
- Avoid installing chemical solution piping overhead wherever possible.
- Avoid using threaded stainless steel (typically 316 or 304) pipe. Welded or flanged stainless steel pipe is much easier to install and maintain than threaded. Installing threaded stainless steel so that it is leak free is extremely difficult, and the threads gall easily. If it is suitable for the application, use plastic pipe instead.

Cathodic Protection Systems. All cathodic protection systems should be clearly identified so that they can be monitored by operations staff. Location markers should be specified for locating cathodic protection anodes, cables, and monitoring stations for buried pipe. The contract for purchase of cathodic systems should specify vendor training for O&M personnel.

Computer Programs to Assist Operations and Maintenance. Numerous software packages are available to assist in the operations, maintenance, and administration of a water treatment facility. Various types of software can be purchased to run on personal computers (PCs), large mainframes, and mini-computers.

Operations Software and Hardware. Operations software usually implies process control software. Several off-the-shelf programs are available to optimize settling rates, chemical dosing, filtration rates, and backwashing. These programs will assist the operators in optimizing thier processes but are not the only controlling factor; the operator will still need to apply general process history, knowledge, and skill. Most of the off-the-shelf programs can be customized for a specific facility. Plant-specific process control software can also be designed by an outside consultant or someone on the owner's staff.

Hardware necessary to operate the process control programs should include a high-speed PC with as much random access memory (RAM) and hard disk memory as possible. The amount of RAM and hard disk memory available on new computers is growing rapidly every year. The reason the changes occur is because of technological advances that are needed to accommodate new, more advanced software.

If data will be accessed from other computers, the main computer for the treatment plant should have a modem with proper communication software. The computer monitor should be at least 17 inches, and preferably 21 inches, to minimize eye strain and allow for proper inspection of on-screen data. If a distributed control system (DCS) is used, operations software can be incorporated into the same system.

Maintenance Software and Hardware. Off-the-shelf computerized maintenance management systems (CMMS) can easily be tested and purchased for use with the facility maintenance program. To ensure that the program will be well suited for its uses, the program selected should be designed for the water treatment industry and not customized from another trade. A good CMMS will be capable of inventorying all equipment, scheduling preventive maintenance tasks, balancing workloads, controlling spare parts, inventorying and tracking consumables, tracking corrective maintenance, printing work orders, recording run times, and following labor hours. The program selected should be easy to use and customize.

The hardware necessary to operate the CMMS should include a computer similar to the one used for process control. If graphics abilities are included in the CMMS package, the maximum amount of RAM should be ordered with the PC. The hard disk storage should be the maximum amount that is currently available.

Laboratory Software and Hardware. Laboratory software and hardware are highly specialized. The software is usually for data collection. In some cases, the analytical equipment itself will have the software and storage capabilities to analyze, record, and maintain data. One laboratory computer should also have software loaded for word processing and spreadsheets.

The PC for the laboratory should be big enough and fast enough to handle off-the-shelf word processing and spreadsheet programs. Special equipment could include the ability to download data from the analytical equipment to the PC. Laboratory data can be collected, evaluated, and stored by the DCS, if a DCS is used.

Administration Software and Hardware. The uses for administration software include spreadsheets for budgets and data. Spreadsheets can be used to set up and monitor all of the facility's budgets. The use of spreadsheets is extremely beneficial for budgetary "what if" scenarios. Spreadsheets are also useful for manipulating and evaluating operations data. Scheduling programs are available for looking at working and training schedules. Treatment plant word processing requirements may include form letters, mailing labels, public relations, public presentations, and manual production and editing. Several companies produce training software for educating water system personnel.

Plant Security Considerations. Security at water treatment plants is an important issue. Controlling access to the plant is important from the standpoint of public safety, as well as ensuring the safety of personnel working at the plant. Recommendations to help make a plant more secure are:

- Provide a system of locks at the plant so that a master key or card can open all of them. Personnel at the plant can be issued keys or cards that give them different levels of access. For example, the plant manager would have a master key or card providing access to all secured areas, O&M supervisors would have access to most secured areas, and all other staff would have access only to the areas that are necessary to do their work.
- Install motion detectors inside the perimeter fence to activate a signal in the central control room to alert plant personnel of unauthorized entry into the plant area.
- Provide adequate lighting for all outside areas of the plant for security and safety.
- Supply O&M personnel with two-way radios for in-plant communications.

Safety Considerations. When designing a water treatment plant, designers must comply with the Uniform Building Code, Uniform Fire Code, National Fire Protection Association standards, state and federal OSHA standards, and other codes and regulations. By designing the plant according to these codes and regulations, the plant should be a safe place to work, but the designer should also consider the following items that may not be addressed in the codes and regulations:

- At least two self-contained breathing apparatuses (SCBA) should be provided outside each area that can contain toxic gases or be depleted of oxygen (e.g., chlorine feed room, chlorine storage room, indoor recarbonation basin).
- Noise arresters or earplugs should be provided at the entrance to areas that have high noise levels (e.g., pump rooms, diesel-generator room, or near air compressors).
- Plant telephones or intercoms should be installed in areas accessible to controls and equipment, but away from noise. If it is not possible to find a quiet area, install materials to control noise around communications equipment.
- All tunnels or galleries should have an exit at least every 150 ft (46 m).
- Hose bibs should be provided at intervals of no more than 50 ft (15 m) in tunnels or galleries.
- Electrical receptacles should be installed at no more than 100 ft (30 m) intervals in tunnels or galleries.
- The plans should specify that all drains must be tested before plant acceptance.
- It should be specified that all chemical and process piping be color coded and also labeled (for color-blind staff).
- Confined spaces must be given special considerations to meet new regulations. Special gear and gantries may have to be installed, as well as signs identifying the confined entry requirements. It is generally wise to use confined-space experts to review designs and ensure that all requirements are met.

Design Reviews. During the design of a water treatment plant, the plans should be reviewed by persons who have several years of experience in operations and maintenance. This type of review can be beneficial in making the plant user-friendly to O&M personnel. These people could be on the designer's staff, or personnel from the owner's staff.

Design Considerations for Plant Maintenance

If a water treatment plant is easy and efficient to maintain, it follows that it should also be economical to maintain. Equipment should be located so that it can be easily installed, maintained, and removed. Areas where equipment is installed must be well lighted, must have easy access to power, must be well ventilated, and must have moisture kept to a minimum.

Excess Equipment. Having "excess" equipment in a treatment plant is a lot like having an automobile with many extras that are not essential, but nice to have. From a maintenance perspective, the more extras there are, the more things there are to maintain.

An example of excess equipment in a water treatment plant would be a small plant that could just as well be operated manually, but has been designed and constructed with sophisticated automatic controls. Although the plant does not really need the automatic controls, the plant owner is obligated to keep them functioning properly, possibly at relatively great expense.

Another example of excess equipment would be a pumping facility in which no more than three pumps of the same size are required to operate to meet current demand, but the plant has been designed and constructed with six pumps. If one pump is counted as standby, there are now two extra pumps that have no real purpose, but must be maintained. Instead of installing all of the equipment to meet a future demand, the facility should have been designed with only the number of pumps required at present, but with equipment space allocated for the additional pumps.

Accessibility of Equipment. A common problem in the design of many water treatment plants is that insufficient room has been provided to work on equipment. Convenient access to all equipment is necessary for proper maintenance. The layout of basins and buildings should be designed with careful consideration of all necessary maintenance activities. This applies to elevated equipment as well as equipment installed on the floor. There should be space to place ladders or scaffolds to gain access to all elevated equipment.

A good rule of thumb is to provide a minimum of 3 ft (1 m) of clear space around each item of equipment that requires maintenance. When laying out areas such as pump stations, designers should remember to allow for associated appurtenances such as seal water piping, control panels, electrical conduit, and ventilation ducts. It is common for designers to show mechanical, electrical, and building utilities on separate drawings. So, unless all drawings are related to each other, it is possible for an inexperienced designer to provide what appears to be adequate clearance around mechanical and electrical equipment without taking into account the space taken up by ancillary items.

Maintenance personnel should be able to get to all equipment without climbing over other equipment or crawling over or under piping. Equipment should be placed so that it can be disassembled and reassembled in place or can be removed and replaced without removing other equipment or lifting it over other equipment. The pumps shown in Figure 25.1 are examples of equipment with good accessibility.

Equipment should be located so that grease fittings, oil fill and drain lines, oil filters, and air filters are easy to reach. Drive belts, drive chains, and guards should be easy to remove and replace. If equipment is mounted high above the floor, the area around it should be unencumbered so that a ladder can be safely placed to perform maintenance.

Piping with good accessibility is shown in Figure 25.2. The piping is layered on the supports with enough distance between the layers to repair, remove, and replace the piping.

Manways into tanks and vaults need to be sized so that maintenance personnel can comfortably enter the portholes wearing SCBA. Portholes should be located far enough away from walls, ceilings, and columns so that personnel can enter and exit without difficulty.

FIGURE 25.1 Pumping equipment with adequate access.

FIGURE 25.2 Process piping supports.

Installation of Cranes and Hoists.   Water treatment plants should be designed to include installation of cranes and hoists for maintenance of heavy equipment. A bridge crane, monorail, or hoist beam is often provided in larger plants. For equipment that can be lifted with a 2-ton (1,800 kg) chain hoist, lifting rings can be strategically installed in the building structure above each unit so that a chain hoist can be easily attached for use. The pump shown in Figure 25.3 has a monorail and chain hoist located directly above to facilitate removal and replacement of the pump and motor.

Power and Lighting Considerations.   Two important items in performing maintenance are having power receptacles located near where the work is to be performed and having adequate lighting in the work area. Power receptacles should be located so that maintenance personnel need to carry only a 50 ft (15 m) extension cord to obtain power tools and test equipment. Adequate lighting is absolutely necessary to perform maintenance efficiently and safely.

Tunnel and Gallery Considerations.   Galleries in water treatment plants are underground areas where equipment, process piping, and utility piping are located. Tunnels are underground areas or corridors where only process and utility piping is located.

FIGURE 25.3 Pump with monorail and chain hoist.

Galleries should be designed so that there is a clear passageway and adequate space to install, maintain, and remove the equipment. There should be hatches and elevators so that equipment can easily be taken in and out of the galleries. All piping in galleries and tunnels should be located so that there is access for maintenance.

Pipe chases and tunnels are areas where lighting is usually inadequate for maintenance. In order to conserve on energy during normal circumstances, half of the lighting should be put on a separate switch for normal conditions, but extra lighting will be available when maintenance is performed.

Galleries and tunnels are generally prone to collecting water. Wet spots on floors caused by condensation and leaks pose a significant slipping hazard, and biological growths that form due to wet and humid conditions may make floors even more slippery. It is essential to provide good drainage to minimize wet floors. Galleries and tunnels should have floor drains, gutters, pitched floors, and sumps with automatic pumps in order to keep floors as dry as possible.

There should also be a ventilation system for galleries and tunnels that will provide enough air circulation to dry the floors if moisture does occur, as well as ventilating the area. In plants where there are times when the water temperature will be considerably colder than ambient air temperature, it can be expected that there will be considerable condensation from piping in galleries and tunnels. This can sometimes be alleviated by insulating the piping. The other solution to keep the area dry is to install a dehumidification system.

Maintenance Shop Considerations. The design of the treatment plant maintenance shop should take into account the level of maintenance that will be done by plant personnel and the number of persons who might be working in the shop. Types of work that should be considered when designing the shop are:

- Overhaul of major equipment
- Vehicle maintenance
- Machining (lathe, press, punch, shears, milling cutters, drill press, etc.)
- Sheet metal work
- Welding (arc, submerged arc, cutting torch, etc.)
- Certified equipment calibration
- Tool storage
- Spare parts storage

If it is possible that complete overhaul of major equipment will be performed by plant personnel, the shop should be large and well equipped and should have large hoists and cranes to handle the equipment. The doors should be large enough to accommodate the largest equipment.

If vehicle maintenance is to be performed in the maintenance shop, some of the special requirements include:

- A vehicle lift and lubrication pit
- Tire changing equipment
- Ventilation equipment for removal of exhaust fumes
- Engine analyzing equipment
- Equipment for holding spent lubricants for disposal

Certified calibration of instrumentation and test equipment requires special equipment and a clean isolated environment to work in. It is important to have test equipment recertified to National Bureau of Standards (NBS) criteria.

In all but the smallest maintenance shops, secure rooms or cages should be provided for tool and spare part storage. The rooms must be sized to contain all of the tools and spare parts used by the plant and to provide enough extra space for expansion.

PREPARING A PLANT FOR START-UP

The designer must start making preparations for plant start-up far in advance to make sure that all permits, funding, staffing, and equipment checkout is completed when the plant is ready to begin operation (see Chapter 27).

Obtaining Operating Permits

There are numerous federal, state, and local government regulations and permitting requirements that must be complied with before a water treatment plant can be placed in operation. This is an area of concern not only at the beginning of the design of a water treatment plant, but also once the plant is constructed and placed in operation. Permits and regulations are discussed in detail in Chapter 24.

Preparing an Operations and Maintenance Budget

Several O&M budgets should be considered during the design phase of a water treatment facility. Initial considerations taken by the designer regarding these budgets can help the owner minimize ongoing O&M expenses. These budgets include staffing, chemicals, utilities, ongoing training, equipment, and other relevant O&M budgets. The fiscal effect of each of these budgets on the owner should be carefully considered by the designer.

Staffing Budget. Initial staffing costs for a newly constructed facility include recruitment costs such as costs for advertising the availability of the positions, interviewing costs, travel expenses, initial indoctrination and training, and possibly relocation expenses. The costs for recruitment of the higher management may include the services of a professional firm to locate potential candidates.

Once the facility is initially staffed, the staffing budget must consider employee wages, fringe benefits, and the continuing costs of restaffing due to employee turnover. Fringe benefit costs generally range from 30% to 50% of salaries. Employee turnover can be influenced by many factors, including the area unemployment rate and wages and benefits provided by other employers in the area. The turnover rate can be estimated using historical administrative records from the owner or from other water treatment plants in the area.

If the facility's employees will belong to one or more unions, labor rates should be established with the union before the new facility goes on line so that the rates can be accurately applied to the O&M budget.

Chemicals Budget. The chemicals budget for a water treatment facility can be a significant portion of the total budget. To project the chemical budget, it is necessary to list all chemicals to be used in the treatment processes and the projected quantities to be used annually. If possible, three price quotes should be obtained for each chemical, including shipping costs, along with the availability from each supplier.

Utilities Budget. The budget for the utilities used by a water treatment facility is usually significant, with electrical consumption usually being the single largest operating cost. If the electric utility offers special off-peak rates, power use should be closely monitored. This means, if possible, backwashing or using other electrically intensive equipment during off-peak periods rather than peak periods. Other means of electrical savings can include the use of motors with higher power factors and efficiency ratings. The electric utility should be contacted to discuss other means of electrical savings. Other utilities to be considered in the budget include gas, sewer, and phones.

Training Budget. Training is not only necessary for all employees who are initially hired to start up the plant, but should continue during the lifetime of the facility. Most states have mandated training requirements such as certification and safety. Training courses may or may not be provided locally. The training costs that should be included in the budget include salary for employees while they are attending training, as well as travel, registration, and training materials. It may also include the costs of hiring a professional trainer to provide on-the-job staff training.

Training includes new employee orientation and current employee proficiency training. Because the water supply industry is highly technical and ever changing, every opportunity should be taken to send facility employees to attend seminars, workshops, and conferences to upgrade their skills and to keep up professional contacts.

Equipment Budget. Equipment costs should be budgeted in several separate categories. Smaller new equipment will be purchased during the operation of the facility to replace broken equipment, add new types of equipment, and to upgrade the facility. These costs tend to increase with the age of the plant.

There should be a spare parts category in the budget to cover costs of maintaining an inventory of critical and hard-to-locate parts. Another budget category is consumables necessary to keep equipment in operation, such as oil, grease, nuts, and bolts. Finally, although some tools will be initially purchased, a budget allowance should be included for constantly replacing and upgrading the tools required for plant maintenance.

Large equipment is usually a budgeted capital cost. This could include such items as vehicles, construction equipment, new pumps and generators, and high capacity process equipment. Replacement of large equipment is usually a one-time expense, so it is budgeted by setting aside a certain amount annually in an equipment fund to purchase new equipment or replace old equipment that has failed or is a planned replacement because it is either out of date or no longer economically practical to maintain.

Other Budget Items. Other items that might have to be considered in the overall O&M budgets include payments to outside contractors for jobs such as sludge hauling, grounds keeping, painting, and maintenance of HVAC systems if these jobs are not performed by plant staff. The costs of public relations projects such as flyers, brochures, posters, and newsletters should be planned into the future. Other items that should be provided for in the budget include personnel awards, professional journals, professional memberships, uniforms, office supplies, and office equipment.

Preparing a Staffing Plan

The designer must begin considering plant staffing early in the plant design, both in terms of creating a design that can be efficiently operated by a minimum staff, and also to decide what degree of automation should be included in the design.

Staffing Criteria. Following are some general statements regarding how the design of a facility corresponds to the staffing requirements.

- The more automated a water treatment plant is, the fewer personnel are required to operate the plant. Although a large fully automated plant will require significantly fewer personnel for operation than the same size plant that is not automated, the automated plant may require additional instrumentation technicians to handle the increased repair and preventive maintenance of the special equipment.
- The more processes there are in a water treatment plant, the more personnel will be required to operate and maintain it. For instance, a direct filtration and chlorine disinfection plant would require fewer personnel to operate and maintain than the same

size conventional filtration plant with lime-soda softening, recarbonation, ozonation, chlorination, and solids dewatering.
- The more spread out a water treatment plant and distribution system are, the more personnel are required to operate and maintain them. For instance, water systems that have their source of supply, treatment plant, booster stations, and reservoirs all distributed many miles apart will have significantly greater personnel costs than a water system that is relatively compact.

The number of estimated personnel it would take to operate and maintain conventional water treatment plants with 5 and 50 mgd (19 and 190 ML per day) capacities is shown in Table 25.1. The estimated number of personnel are listed for semiautomated and fully automated facilities.

During the preliminary design stage, the designer should begin looking at the level of staffing required to operate and maintain the water treatment plant. This is valuable in helping the owner decide on which of several alternative designs is most suitable. One of the issues that a staffing estimate resolves is whether it is economically feasible to go to a fully automated plant, or whether it is better to use less automation and more operating personnel. A fully automated plant normally represents a higher capital cost and lower annual O&M cost; a less automated plant generally has a lower capital cost and high O&M cost.

The owner is usually responsible for staffing the new facility, but the designer will usually help estimate the staffing needs. In some instances, the owner may even have the designer locate, interview, and hire the staff. The designer needs to be aware of the job descriptions, job needs, and special consideration for each key position in order to ensure that the right people are selected for the right job and the right number of people are on hand to operate and maintain the plant once it is ready to begin operation.

Supervisory Positions. In the following discussions, supervisory positions are based on a medium to large water treatment plant.

Plant Manager. The plant manager's primary goals are to ensure that the quality of the final product meets all state and federal requirements and ensure that this is accomplished cost-effectively. It is the plant manager's responsibility to direct the management, operations, maintenance, and laboratory personnel toward this overall goal. The plant manager's office should be centrally located within the facility to allow easy access to plant supervisory personnel.

The plant manager should preferably be hired early in the design phase and included in meetings and discussions between the owner and designer regarding process design, specifications, construction, operations, and maintenance. During construction of the plant, the plant manager should be on site to become familiar with the plant as it is being constructed and also to help with on-site inspection during construction.

Operations Supervisor. The operations supervisor is the person directly responsible for controlling plant treatment processes. Duties include scheduling operator shifts, supervising operators, controlling the treatment processes, ensuring that sampling and monitoring are performed, providing operator training, and reporting on plant operations to the plant manager. The operations supervisor's office should be located centrally within the facility and should be easily accessible to operations personnel and close to plant processes.

Maintenance Supervisor. The maintenance supervisor oversees all mechanical, electrical, instrumentation, building, and grounds maintenance activities for the plant. Duties include scheduling and supervising preventive and corrective maintenance tasks and reporting on plant maintenance to the plant manager. Typically, the maintenance supervisor has an office located within the maintenance facility for better communication with maintenance personnel.

Laboratory Supervisor. The laboratory supervisor is responsible for supervising all process control and analyses required by regulations. The supervisor's office should preferably be located next to or in the laboratory. The plant laboratory is an integral part of operations; therefore the laboratory itself should be located near the central operations control area within the plant so that samples can easily be transported to it and the operations supervisor has easy access to the process-related analytical data.

Operations Personnel. Operations personnel are responsible for the day-to-day operation of the plant. They take samples, make process modifications, operate equipment, take and record operating data, make rounds visually inspecting processes and equipment, make entries into daily logs, keep chemical inventories, and perform various other duties related to operating the plant to meet finished water quality goals. There are usually one or two control room operators whose duties include monitoring indications of the various instruments in the control room and making changes in the processes as they are required. The plant may have other operators assigned to areas around the plant whose duties are related to those areas (e.g., chemical feed building, filters and sludge dewatering).

Maintenance Personnel. Maintenance personnel are responsible for taking care of equipment and facilities at a water treatment plant. Their goal is to ensure that the plant equipment and facilities are kept in good condition, thereby extending the useful life of the equipment and facilities. To accomplish that goal, maintenance personnel must be well trained and motivated. In addition, they should have sufficient tools, spare parts, and a well-planned and implemented maintenance management system. Duties include maintenance of mechanical, electrical, and instrumentation systems, as well as maintenance of buildings and grounds. Examples of each type of maintenance are given below:

- *Mechanical maintenance.* Plant maintenance technicians are responsible for preventive and corrective maintenance on mechanical equipment. This equipment includes air compressors, elevators, bridge cranes, pumps, valves, and piping. Plant maintenance technicians are also responsible for all routine equipment alignment but usually do not perform work involving electrical power and instrumentation and control (I&C).

TABLE 25.1 Estimated Staffing Requirements for 50 and 5 mgd Water Treatment Plants

Position	50 mgd (190 m³/d) water treatment plant		5 mgd (19 m³/d) water treatment plant	
	Semiautomatic	Fully automatic	Semiautomatic	Fully automatic
Plant manager	1	1	1	1
Operations supervisor	1	1	0	0
Maintenance supervisor	1	1	0	0
Operator	15	5	5	1
Mechanical technician	3	3	1	1
Electronics technician	2	2	1*	1*
Instrument technician	2	3	0	1
Laboratory technician	2	2	1	1
Buildings maintenance	1	1	1†	1†
Grounds maintenance	1	1	0	0

*This position is split between electrical and instrumentation duties.
†This position is split between janitorial and grounds-keeping duties.

- *Electrical maintenance.* Electricians are responsible for preventive and corrective maintenance on electrical equipment. This includes electric motors, motor control centers, emergency generators, uninterruptible power systems, unit substations, protective relays, lighting, distribution panels, and lighting panels. Electricians usually do not perform work on mechanical and I&C equipment. Electricians may be required to do equipment alignments if they have had to remove and replace or uncouple an electric motor in the course of their work.
- *Instrumentation maintenance.* Instrumentation technicians are responsible for preventive and corrective maintenance on I&C equipment. This equipment includes electrical, pneumatic, hydraulic, and thermal control systems. Examples of types of equipment electronics technicians will work on are equipment control panels, low-voltage control systems compressed air control systems, and programmable logic controllers (PLCs). Electronics technicians usually do not perform work on electrical power and mechanical equipment.
- *Building maintenance.* Plant personnel responsible for building maintenance perform work to ensure that the buildings are kept clean and in good repair and that the environment in the buildings is comfortable. Building maintenance work includes but is not limited to painting, masonry, concrete, glazing, carpentry, plumbing, HVAC systems, security systems, and cleaning.
- *Grounds maintenance.* Plant grounds personnel are responsible for maintaining all of the outside area of a treatment plant. Grounds maintenance work includes but is not limited to seeding and cutting lawns, planting and pruning shrubbery and trees, installing and repairing fences, installing and maintaining irrigation systems, fertilizing, roadway repair, and snow removal.

Laboratory Personnel. Laboratory personnel perform both chemical and biological testing of water. They perform process control and regulatory testing on samples collected by operations or by laboratory personnel. Results of data generated by laboratory staff are used by operations to control equipment and processes to meet plant water quality goals.

Plant Start-Up Preparations

Plans for staring up a new water treatment plant should be initiated once construction is well under way. This is accomplished through meetings with those who will be directly involved with start-up. O&M manuals must be produced to aid operations and maintenance personnel in running a new facility in an efficient and cost-effective manner. These manuals should cover everything from the designer's intent for the plant processes to a suggested list of spare parts that should be ordered and kept on hand.

Training of all employees is an essential part of the successful start-up of a new treatment plant. Even if new employees have previous experience in the water supply field, they will still need training in the operation of the specific equipment and processes in a new plant.

As a water treatment plant nears the end of construction, it is time to start equipment checkout. Equipment checkout should be thoughtfully and carefully planned.

BIBLIOGRAPHY

American Water Works Association. *Automation and Instrumentation, M2.* Denver, Colo.: American Water Works Association, 1977.

Baalim, W. A., G. Cleasby, G. Logsdon, and M. A. Allen. "Assessing Treatment Plant Performance." *Journal AWWA* 85(9):34, 1993.

California Department of Health Services, Sanitary Engineering Branch, and U.S. Environmental Protection Agency, Office of Drinking Water. *Water Treatment Plant Operation.* Vol. 1, 2nd ed. Sacramento, Calif.: California State University, 1990, p. 211.

Instrument Society of America. 1995. *Dictionary of Measurement and Control: Guidelines for Quality, Safety, and Productivity.* 3rd ed. Research Triangle Park, N.C.: Instrument Society of America, 1995.

Water Pollution Control Federation. *Plant Maintenance Program, Manual of Practice 0-3, Operations and Maintenance.* Washington, D.C.: WPCF, 1982.

CHAPTER 26
CONSTRUCTION COSTS

It has been said that construction cost estimating is more of an art than a science. This would be true only if the application of science did not require the use of sound judgment and experience. Because most process design engineers have little exposure to cost estimating, either in their formal education courses or internal training at work, most have a somewhat skeptical attitude to the process of developing a reliable and accurate cost estimate. Although a thorough understanding of the mechanical elements of a water treatment plant would appear to be the most important aspect in estimating its cost, it is only one discipline of a multidisciplined project.

All cost estimates contain a certain amount of experienced judgment or educated guesswork concerning the various cost elements that make up the estimate. An order-of-magnitude or preliminary cost estimate may be based completely on judgment; a definitive level estimate may require judgment only for those cost elements with historical cost fluctuations.

As an example, an order-of-magnitude estimate may require sound judgment in identifying a previous project with similar scope of work characteristics factored to estimate the cost of a new project. A definitive level estimate may require judgment only concerning specialized concrete forming labor with historically wide labor cost fluctuations, with all other elements being based on verified vendor quotes or historical cost data with minimal deviations.

The various types of cost estimates are based on the level of project design that defines the known scope of work. The scope of work can range from only a daily treatment flow rate and process to a complete set of plans and specifications for construction. The cost estimate to be used also reflects a certain contingency and accuracy range based on the scope of work reflected by the level of design.

LEVEL OF ESTIMATES

Three basic levels of cost estimates are identified by AACE International (formerly American Association of Cost Engineers): order-of-magnitude, preliminary, and definitive. A fourth type of estimate called a conceptual level has also been identified, and is most commonly developed between the order-of-magnitude and preliminary level estimates. As their names imply, these estimates correspond to the various phases of design through which a project progresses.

Order-of-Magnitude Estimates

An order-of-magnitude estimate is usually based on previous project data factored up or down to reflect the size and treatment capacity of the new project. Because of the numerous types and variations of water treatment plant (WTP) processes, it is recommended that a similar plant process type be used for factoring. Where previous project cost data may not reflect the same process facilities for the entire new plant, the estimator can itemize those facilities that are applicable and use them in factoring the cost in a cut-and-paste application. For those facilities for which no historical cost data are available, a separate cost estimate needs to be developed.

In addition, cost indexes for the location of the new plant can be compared with cost data used for the previous job, and the factored cost adjusted accordingly. In general, factoring of previous data is usually accurate for new construction only when there are favorable geotechnical conditions. Where the new project involves modifications or expansion of existing facilities, or where geotechnical issues need to be addressed, factoring of previous data alone does not provide accurate cost estimates of the new project.

Conceptual Costs

A conceptual level cost estimate is based on a preliminary scope of work that may reflect some equipment and structure sizing. This type of estimate is most commonly used for projects that involve modifications to existing facilities. It is similar to an order-of-magnitude estimate but requires additional scope identification to cost the modified facilities. It is also used where a more accurate cost estimate is required before the predesign stage, or where unusual site or geotechnical issues need to be addressed.

The scope of work required to modify existing facilities is one of the biggest unknowns in preparing a preliminary cost estimate for a treatment plant upgrade. However, if the required basic process mechanical elements can be identified, factors can be applied to these elements that will generate reasonable cost estimates for the related disciplines of work. The most common application of this method is to apply factors to the new process equipment to generate costs for the major work elements such as mechanical piping, electrical, and instrumentation and control.

To address structural or geotechnical issues, personnel experienced in these disciplines should be consulted. Although a thorough analysis of these discipline requirements is not practical, past experience with similar installations and an understanding of the work scope and existing conditions allow an experienced estimator to develop costs for these items.

Preliminary Design Estimates

The preliminary design cost estimate is considered by many as the most important estimate prepared for a project. This estimate is used to verify the basic concept of the project and usually determines the project's financial viability. It also forms the baseline cost of the project with which all future cost impacts of the evolving design will be compared.

A preliminary design cost estimate is based on the predesign concept of the project. The major facilities of the project are identified and the footprints for the structures are sized. In addition, major equipment elements are identified and sized, usually based on specifications for a specific manufacturer. Preliminary geotechnical information should be available in order to accurately evaluate structural considerations for the project. Major process mechanical piping and electrical equipment is also identified and sized.

The purpose of the preliminary design estimate is to obtain as much information as possible regarding the scope of the project. Most of this information is identified in a preliminary design report or other similar document. However, other pertinent information that could influence the cost estimate may not be identified in this report. Some of the information not usually identified in a preliminary design report are site or existing facility access restrictions, sequencing of new and existing process elements for existing plant operations, and identification of temporary power sources for early phase facility start-ups. Information of this type is usually known by the various design discipline engineers who prepared the preliminary design report.

Communication with the various design discipline engineers is essential in preparing an accurate preliminary design cost estimate. At this point in the project, they are the most knowledgeable with respect to their design elements and often are aware of special considerations that could influence the cost estimate. It usually requires an experienced cost estimator with past treatment plant experience to know what additional information is required to accurately reflect project cost.

As stated previously, the preliminary design cost estimate is usually identified as the baseline cost estimate for comparison with all future design considerations. This concept is also known as the design-to-cost principle. As the various facilities of the project proceed through final design, revised cost estimates based on the latest design information are prepared for comparison with the baseline cost estimate. Variations in these costs that exceed a predetermined amount (usually 15%) require an evaluation to determine the cause of the variance.

Definitive Level Estimate

A definitive level cost estimate is based on the project's final design documents. Also known as the engineer's estimate, this estimate represents the scope of work depicted by the contract documents from which construction contractors base their contract price to build the project.

At this cost estimate level, it is important to verify all major equipment costs by vendor quotations that include all start-up and training requirements. Major process piping material, including fittings and valves, should also be verified with vendor quotations. Major electrical components and process instrumentation and control systems (PICS) should have costs assigned and confirmed by vendors. Local labor factors and market conditions should be analyzed for application to the project. These are just a few of the numerous items to consider when preparing a definitive level cost estimate. It is recommended that a senior level cost estimator with previous definitive level cost estimating experience review the cost estimate to ensure the inclusion of all items, in addition to overall estimate quality control.

After bids have been received and the construction contract awarded, the definitive level estimate can be used as a basis of comparison to review the contractor's schedule of values for partial payment applications. Therefore it is essential that the engineer's estimate be prepared in the same format requested of the contractor in the bid documents.

ESTIMATING METHODOLOGIES

Most cost estimates, whether produced by a specialized cost estimating consultant or a water treatment plant design firm, are based to a certain degree on historical cost information. Where historical cost information is not available, other estimating factors must be used.

Historical Cost Data

Most firms that produce cost estimates have access to past water treatment plant bid data and other cost information generated during the construction of various projects. This information can be useful for estimating the costs of future projects if the data are compiled in a manner applicable to other projects.

One of the more significant and accurate sources of treatment plant cost data is the schedule of values required to be submitted by the construction contractor for partial payment applications. The contractor is required to submit a detailed cost breakdown of the bid proposal to enable assessment of the value of the work performed as the job progresses. This cost information is usually broken down by the Construction Specification Institute (CSI) format for the major facilities that make up the project.

Based on this past information, a database is usually set up to compile historical treatment plant costs. This database can be divided into process facilities and subdivided by CSI specification section. These sections can be compared with total process facility costs and total treatment plant costs, resulting in patterns of percentage allowances for each CSI specification section. This information can be useful for developing conceptual or preliminary cost estimates where only structural and major equipment elements are identified.

An example of this estimating application is a chlorine contact basin and pump station where only the civil, structural, and major process equipment elements of the facility are identified. These elements constitute the known costs and make up a certain percentage of the total facility cost based on historical cost data. Total facility cost can then be projected from the known cost percentage. Other unknown elements of the facility can be computed based on historical cost percentages applied to the projected facility total (Table 26.1). This application can also be used to verify detailed cost data that may be suspected of not being totally inclusive of the scope of work.

Cost Estimating Factors

Where historical cost data are unavailable or irrelevant to a new project, other cost estimating techniques can be employed. Because these other cost techniques are not as reliable as historical cost information, their use is usually confined to conceptual or preliminary design level cost estimates.

One such method is applying a designated percentage factor to the known cost of the major process equipment for a project. This percentage application method can be used to estimate all other cost factors of a project, as shown in Table 26.2.

Another cost estimating factor is applying the six-tenths rule. This method is primarily used to estimate costs of facilities based on a known cost and capacity of a similar facility. The capacity ratio of the known facility is multiplied by its cost and factored by six-tenths to account for the fixed costs associated with the construction of any facility. For example: A filter building facility with a capacity of 12 mgd has a construction cost of $18 million. Estimate the cost of a similar 15 mgd facility.

$$\text{Cost} = 18 \times \frac{(15)^{0.6}}{12} = \$20.58 \text{ million}$$

Although it is suggested that these cost factors be confined to use with order-of-magnitude or conceptual level cost estimates, they can also be used in conjunction with preliminary or definitive level estimates for minor items that do not significantly affect total facility cost. Using percentage factors for items such as site restoration, painting and finishes, and small piping less than 6 in. (150 mm) in diameter is appropriate because these costs historically do not exceed 10% of the total facility cost.

TABLE 26.1 Historical Cost Records

Item of work	Quantity	Unit	Unit cost	Total cost
Excavation and haul	24,000.0	CY	$10.00	$240,000
Steel sheet piling	23,000.0	SF	$10.00	$230,000
Dewatering—three wells with pumps	15.0	Mo	$7,500.00	$112,500
16″ sq. precast conc. piles	34,000.0	VF	$20.00	$680,000
Granular backfill	2,100.0	CY	$25.00	$52,500
30″ base slab	3,140.0	CY	$200.00	$628,000
20″ walls	1,325.0	CY	$300.00	$397,500
12″ walls	790.0	CY	$500.00	$395,000
8″ walls	42.0	CY	$600.00	$25,200
12″ elev. slabs	160.0	CY	$400.00	$64,000
72″ × 72″ sluice gates	4.0	Ea	$23,000.00	$92,000
Subtotal	75.0%			$2,916,700
General requirements	5.0%		$3,888,933	$194,000
Metals	5.0%		$3,888,933	$194,000
Finishes	2.5%		$3,888,933	$97,000
Mechanical/Misc. Equip.	10.0%		$3,888,933	$389,000
Electrical/I&C	2.5%		$3,888,933	$97,000
Total	100.0%			$3,887,700
Contingency	15.0%		$3,887,700	$583,000
Total cost, chlorine contact basin				$4,470,700
Excavation and haul	2,600.0	CY	$10.00	$26,000
Steel sheet piling	8,800.0	SF	$10.00	$88,000
Dewatering—one well with pump	9.0	Mo	$2,500.00	$22,500
16″ sq. precast conc. piles	4,000.0	VF	$25.00	$100,000
Granular backfill	700.0	CY	$25.00	$17,500
30″ wetwell slab	270.0	CY	$200.00	$54,000
20″ wetwell walls	300.0	CY	$300.00	$90,000
12″ weir walls	24.0	CY	$500.00	$12,000
24″ sq. columns	12.0	CY	$800.00	$9,600
12″ elev. slabs	86.0	CY	$400.00	$34,400
Electrical rm. slab on grade	60.0	CY	$200.00	$12,000
Building	3,500.0	SF	$100.00	$350,000
Overhead crane	1.0	LS	$25,000.00	$25,000
72″ × 72″ sluice gates	2.0	Ea	$23,000.00	$46,000
12,400 GPM pumps—CS	2.0	Ea	$22,500.00	$45,000
24,600 GPM pump—CS	1.0	Ea	$56,500.00	$55,500
24,600 GPM pump—AS w/AFD	1.0	Ea	$70,500.00	$70,500
24,600 GPM pumps—AS w/AFD	2.0	Ea	$86,000.00	$172,000
Subtotal	52.5%			$1,230,000
General requirements	5.0%		$2,342,857	$117,000
Metals	5.0%		$2,342,857	$117,000
Finishes	2.5%		$2,342,857	$59,000
Mechanical/misc. equip.	20.0%		$2,342,857	$469,000
Electrical/I&C	15.0%		$2,342,857	$351,000
Total	100.0%			$2,343,000
Contingency	15.0%		$2,343,000	$351,000
Total cost, pump station				$2,694,000

CY = cubic yards; SF = square feet; Mo = month; Ea = each; VF = vertical feet; LS = lump sum; CS = constant speed; AS = adjustable speed; AFD = adjustable frequency drive; I&C = instrumentation and control.

TABLE 26.2 Percentage of Total Facility Cost Based on Historical Cost Data

Equipment cost	100.0%
Equipment installation	50.0%
Process mechanical piping	65.0%
Instrumentation and control	20.0%
Electrical	10.0%
Buildings	20.0%
Yard improvements	10.0%
Service facilities	70.0%
Engineering and supervision	35.0%
Project management and overhead	40.0%
Total percentage of equipment cost	420.0%
Subtotal percent of project cost for above	80.0%
Additional project cost elements	
Misc. and unidentified equipment	10.0%
Misc. and unidentified process mechanical	5.0%
Misc. and unidentified electrical/I&C	5.0%
Percent of total project cost	100.0%

Source: Peters and Timmerhaus. *Plant Design and Economics for Chemical Engineers.* 3rd ed.

SPECIAL COST CONSIDERATIONS

Many construction projects have special circumstances that must be considered when preparing cost estimates, such as special soil or site conditions.

Geotechnical and Site Constraints

Locations for new water treatment plants, and to a lesser extent existing plants, are usually confined to the secondary real estate areas. The economic reality is that most municipalities cannot afford to purchase prime real estate that reflects good soil conditions and location. Therefore the construction costs for these facilities will usually reflect additional costs to build water treatment facilities in less than ideal soil conditions or at locations far removed from the water distribution systems.

One of the cost factors to be considered in the siting of any new water treatment plant is the proximity of the plant to its primary water source. Most land adjacent to any water body with soil conditions advantageous for construction has already been claimed for development, particularly in urban areas. Therefore the remaining land will probably require extensive geotechnical considerations for facility construction.

Costs for geotechnical site investigation, structural design of special foundation support, and permitting of wetlands mitigation are just a few of the possible additional engineering costs required for plant siting. The additional construction costs include groundwater dewatering and control, installation of piling systems for structure support if soil conditions are not satisfactory, and construction of new wetland areas if the new facility site requires filling in of existing wetlands. These additional costs are unique to specific sites and are not usually reflected in most historical cost data and therefore must be added to any factored cost data used in order-of-magnitude or preliminary level estimates.

Remote Site Locations

There are occasions when the geotechnical and site constraint criteria for new plant facilities are so extensive at the selected site, or the cost of the land so prohibitively expensive, that construction of new plant facilities at a location far removed from the distribution system may be more economical. This situation may result in additional construction costs also not typically reflected in historical cost data.

Transportation to the site of all construction material will be affected, with material from local suppliers being the most pronounced. Costs for imported granular fill material from local quarries, ready-mix concrete, and other materials that are small in quantity but require full truckloads for delivery are examples of construction materials that will increase in cost because of additional transportation costs.

Labor costs will also increase to a certain extent, based on the distance of the project site from normal work locations. If the location is far from normal work areas, travel time may have to be paid to the workers as an incentive to travel a further distance to work. If excess excavated material is required to be hauled off site, additional labor and trucking costs will be incurred.

These are just a few of the possible additional costs that would be incurred in the construction of a new plant at a remote site. A thorough review of the scope of work by an experienced cost estimator or construction manager should be performed to identify the extent of these special remote-site cost impacts.

Other Considerations

Consideration must also be given to special costs that are caused by various federal regulations and requirements. Some examples are indicated below.

Americans with Disabilities Act (ADA). Recent regulations are affecting the costs for new and renovated water treatment plants with respect to requirements for accessibility to public facilities. Some common additional costs can be categorized in the following areas:

- Additional parking areas for handicapped persons
- Reconfiguration of door openings to be fully automated
- Rehabilitation of existing lavatory layouts
- Addition of exterior ramps for handicapped persons
- For multistory facilities, possible addition of an elevator

State Sales Tax. Most public-financed water treatment plant projects are exempt from state and local sales tax. Where this is the case, a tax exemption certificate will be provided by the public agency. However, not all states follow this policy. In particular, any project for the federal government (such as the Corps of Engineers, the Park Service, and the Fish and Wildlife Service) will not be exempt from sales tax. The sales tax policy of the state and local municipal government where the project is to be constructed should be reviewed if there is a question of sales tax inclusion into the project cost. Where sales tax is applicable, it is usually applied only to the material cost portion of the project.

Hazardous Waste Audit and Removal. Recent trending in the market indicates that with future funding being levelized or reduced, the option to modify or rehabilitate existing facilities is another special cost consideration for the estimator. Because of the limited estimating manuals and available cost information, a cost estimator should request that an audit be conducted on the existing facility in the following areas:

- Lead paint—primarily on windows and walls
- Insulation material—particular attention to the presence of asbestos on piping systems, roofing systems, boiler packages, and ductwork systems
- Chimney stacks—condition, particularly the lack of liners
- Oil tanks and chemical tanks—particularly if there are any residuals or free products
- Flooring systems—types of materials and adhesives
- Underground storage tanks—the possibility of contaminated soil
- Transformers—any oil-filled transformers with polychlorinated biphenyls (PCBs)
- Ballasts in light fixtures—general condition
- Fireproofing of steel framing systems—presence of asbestos

A local hazardous waste remediation contractor should be consulted for a proper estimate and subsequent cost evaluation by the owner.

FINALIZING THE COST ESTIMATE

As the design of a facility progresses, the contingency allowance should be adjusted, and consideration must also be made of whether the scope of the job has changed since it was initiated.

Contingencies

Contingency is the application of a factor to the cost estimate that is intended to account for those items not specifically identified in the scope of work but which historically are found to be required. After completion of the estimate, a thorough review of the estimate must be made to access the completeness of the scope of work represented by the design information.

For order-of-magnitude or preliminary design level estimates, the scope of work presented by the design reports and communication with design personnel is usually insufficient to reflect the full scope of work for the project. For definitive level design estimates, the scope of work is highly detailed and sufficient to allow a contractor to actually build the project. In other words, the contingency applied to an estimate reflects the completeness of the scope of work as presented by the design documents for the particular level of design. A high contingency is generally applied to an estimate with a low degree of confidence in the scope of work, and a low contingency would be applicable to a detailed design document.

Recommended contingency factors that are applied to the various types of estimates are listed in Table 26.3. The theory is that, as the level of project detail increases, the estimator's confidence in the scope of work reflected by the design information also increases. As more of the unknown elements of the project become identified, the contingency for these unknown elements can be reduced. The direct or known cost of the project will increase,

TABLE 26.3 Level of Cost Estimates

Type of cost estimate	Level of accuracy	Recommended contingency
Order-of-magnitude	+50% to −30%	20% to 30%
Conceptual	+40% to −20%	20% to 15%
Preliminary design	+30% to −15%	15% to 10%
Definitive	+15% to −5%	10% to 5%

Note: The level of accuracy percentages listed above are interpreted as comparisons of the low bid with the established cost estimate. As an example, a low bid of 15% above to 5% below can be expected in comparison with a definitive level cost estimate.

but the contingency applied to the cost for unknown elements will decrease. The goal is for these increases and decreases to offset each other, thus keeping the total project cost within the parameters of the original cost estimate. However, when cost estimates between different levels of design are not within the same cost parameters, one of the more common explanations is "scope creep."

Trending or Scope Creep Identification

Design changes frequently occur between the preliminary design and final design phases. It is during this phase that predesign concepts are finalized and presented to the client for final approval, or conceptually identified items are more clearly defined. Clients who did not fully understand the concepts or details previously discussed are now presented with a detailed system or concept that may not be exactly what they had in mind. In addition, designers may also want to "tweak" their design based on some new information or product data for their systems. The result is usually a change from the preliminary design concept to incorporate new ideas into the project.

The scope of work for the project now changes from what was reflected by the preliminary level or baseline cost estimate. Although many designers will identify these changes to the client, and many clients will acknowledge these as changes, a formal process should be put in place to document these scope changes when and if they cause an increase in the cost estimate. A procedure similar to the change order process found in most construction contracts is now being used by many firms to document these design scope changes for future reference. At a minimum it should include the following:

- Identification of the design change and scope
- Reason for the design change
- Effect on project baseline cost and schedule

In addition, a formal document prepared by the designer reflecting the above information should be presented to the client for review and approval.

Cost Indexes

Several cost indexes are used in the municipal water market. The one that appears to be the most accepted and regionalized is the Construction Cost Index published by Engineering News Record (ENR-CCI), a McGraw-Hill weekly publication of the engineering and construction industry. The ENR-CCI reflects the cost of the basic commodities that affect the construction cost of any heavy construction or civil works project. The cost index is regionalized for 20 major U.S. cities. For those locations not in close proximity to any of the 20 listed cities, interpolation can be performed between nearby cities.

Cost Escalation

Although it is not as significant to construction costs as it was in the late 1970s and early 1980s, escalation of costs because of inflation over the life of the project is still a cost that should be taken into consideration. The most common method used is to base the cost estimate on current cost data and then escalate this cost to the midpoint of construction. This is usually computed by calculating the compounded percentage rate of material and labor cost increases to the construction midpoint, and applying this percentage to the current construction cost.

BIBLIOGRAPHY

Jelen, F. C., and J. N. Black. *Cost and Optimization Engineering.* 2nd ed. New York: McGraw-Hill, 1983.

Peters, M. S., and K. D. Timmerhaus. *Plant Design and Economics for Chemical Engineers.* 3rd ed. New York: McGraw-Hill, 1982.

CHAPTER 27
OPERATOR TRAINING AND PLANT START-UP

The design engineer's provisions for new, upgraded, or expanded water treatment facilities should go beyond the design and specification of equipment, steel, and concrete to include less tangible aspects of the total job: consideration of adequate staffing for the new facilities, personnel preparation, reference documentation, and start-up of new facilities. These aspects should be part of the total package of services provided by the engineer to comprehensively meet the client's needs. The engineer can then turn over facilities knowing that everything possible has been done to provide the client a complete, functional unit.

TRAINING AND START-UP CONSIDERATIONS

Categories covered in this chapter that both engineer and owner must take into consideration during design and construction phases are design-phase training considerations and construction-phase training, start-up, and post–start-up considerations.

Training Considerations

Operator training is a joint responsibility shared by the engineer, contractor, vendors, and owner. Equipment suppliers (vendors), under the general responsibility of the contractor, should train plant staff in the operations and maintenance (O&M) of equipment and systems they have provided. The content of that training should be defined by the engineer in the design specifications. During construction the engineer should oversee vendor training of operators as part of start-up planning, including providing systems training on the functioning of the unit processes and treatment system as a whole. In short, the contractor or vendors should provide the training on the parts, and the engineer on the sum of the parts. The owner's responsibility is to adequately fund training and start-up programs, allocate sufficient personnel to receive training, and provide suitable training facilities.

Reference and training materials complement personnel training. These may include AWWA publications, vendors' O&M manuals, engineer's O&M manual, design specifications, record drawings, and records of operator training sessions. These latter records may be a particularly important source of firsthand, field-verified knowledge that can provide future operators with information that would otherwise have to be learned through experience.

Videotaping training sessions provides an effective means of training future staff, but effective video training requires more time and effort than simply recording a trainer delivering a session. All too often, videotaping training sessions without off-line editing results in "talking-head" videos, which are boring, have poor sound and picture quality, and have gaps between significant subject matter.

Start-Up Considerations

Start-up is the process of placing new facilities into service, ideally after construction is completed. More realistically, however, start-up is usually a series of steps that are neither as smooth nor as coordinated as intended. Planning for start-up should be conducted as part of the design phase to identify construction-related aspects that might become troublesome issues during both construction and start-up. Early in the design phase, consideration must be given to the effect of construction activities on existing facilities, such as the need for tie-ins and the effects of shutdowns. These considerations should be written into the specifications and drawings to the extent possible. Similarly, start-up planning during construction should include developing start-up plans and close coordination between contractor, engineer, and owner.

Not all design and construction projects involve all activities described in this chapter. Upgraded or expanded facilities require careful planning to maintain existing operations during construction. Completely new plants, on the other hand, present different challenges if the client's staff is unprepared or must be hired with no previous experience of stepping in and running new facilities. In all cases, planning for effective training and start-up establishes the preconditions for successful treatment operations.

DESIGN-PHASE TRAINING

The design phase of a project is the time to incorporate vendor training and operations and maintenance manual requirements into contract documents.

Vendor Training Requirements

Vendor training means the equipment manufacturers or suppliers provide instruction to plant operators on operations and maintenance of equipment. Vendor training is usually specific to a piece of equipment or a subsystem provided by the vendor and generally does not cover an entire treatment process. Effective vendor training should follow a prescribed lesson plan and be conducted by a qualified instructor under controlled conditions.

To ensure that vendor training is as effective as possible, the engineer must define the requirements for vendor training and include these in the specifications. Table 27.1 lists provisions to be included in specifications.

General Training. Specifications should include general training requirements for distribution by the contractor to the vendors. These requirements typically include a lesson plan and submittal requirements. Inclusion of other general training requirements may also be appropriate. Job descriptions for all trade groups who are to receive training and general videotaping requirements may be included.

TABLE 27.1 Vendor Training Requirements

1. General conditions
   General conditions include preconditions to training (e.g., equipment has been installed and tested), need for audiovisual equipment, and training scheduling (e.g., no sooner than 30 days before training).

2. Instructor's qualifications
   The instructor's qualifications and experience should be acceptable to the engineer. Qualifications can be evaluated by requiring the contractor to submit information on the instructor for approval in advance of scheduling the training.

3. Submittal requirements
   Training lesson plans, handouts, and other documents should be submitted for approval before the training.

4. Lesson plan
   A lesson plan that outlines the training content should be submitted for approval. The lesson plan should identify the topics to be covered, trainee trade, and duration of training. Comprehensive lesson plans may include such things as:
   - Pretests and posttests
   - Target trade(s)
   - Purpose and specific learning objectives
   - Assumed trainee knowledge requirements (prerequisites)
   - Lesson plan contents, including duration of each phase
   - Identification of training aids, handouts, overheads, and slides and their use in the lesson plan
   - Location of training (classroom and hands-on training)
   - Audiovisual equipment requirements

5. Minimum required training days
   Each piece of equipment and associated specification section should be identified and the minimum number of training days specified. The owner should be consulted to develop this requirement. Also, the videotaping requirements, if any, and the duration of videotaping sessions should be defined.

Specific Training. Equipment sections of specifications typically include specific training requirements, such as:

- Vendor trainer qualifications
- Number of days allotted for training
- Specific content of training, if beyond the scope of the general requirements

Vendor Operations and Maintenance Manual Requirements

A vendor may supply a single piece of equipment, a subsystem, or an entire unit treatment process. In each case, the vendor should be required by the specifications to supply written O&M characteristics covering equipment supplied. Vendor-supplied information should be supplemented by the contractor with additional information and submitted to the engineer for approval. This section defines the type of information required from the vendor and the contractor so that useful and site-specific vendor O&M manuals are provided to the owner's staff.

Vendors' responsibility is limited to the equipment and systems each has provided. Thus, if a pump vendor has supplied only the pump and none of the controls, the vendor's responsibility is to define the operations and maintenance characteristics of the pump only. Vendors normally provide recommendations for installation and mounting equipment. The contractor ultimately controls this aspect of the job unless the design specifications require the vendor to check, inspect, test, or approve equipment after it is installed.

Specifications should require the contractor to provide O&M manuals for each system or piece of equipment having its own specification section. The contractor will supplement information provided by the vendor with any additional information required by the specifications.

Documentation should include equipment design expectations and controls, for example, pump curves, showing the design point. Ladder diagrams clearly identify control logic for equipment. Design loadings; quality characteristics; operating parameters; and detailed start-up, shutdown, and monitoring procedures must also be included.

Table 27.2 lists typical vendor O&M manual requirements. These requirements fall on the contractor to satisfy because of limitations in the vendors' responsibilities, as discussed above. These requirements should be considered a minimum, and additional information may be needed in specific situations.

Vendor O&M Manual Format. Widespread use of computers allows the engineer to consider alternative formats besides the standard three-ring binder O&M manual. Although this type of format may still be most useful to maintenance personnel in the field, engineers should also consider requiring the contractor to submit the vendor

TABLE 27.2 Vendor Operations and Maintenance Manual Contents

1. Project name, contractor, subcontractor (if any), client, consulting engineer, equipment name, and installation location (normally included on title page)
2. Name, address, and phone number of the manufacturer and the manufacturer's local representative; spare parts ordering information
3. Complete equipment identifying information, including model number, serial number, design parameters; functional information (such as pump curves) and operating limitations
4. Detailed operating instructions, including purpose and operational description; start-up, shutdown, and operational troubleshooting instructions; normal operating conditions or characteristics; abnormal or emergency operating conditions; safety shutdown and other shutdown conditions
5. Preventive maintenance instructions, including all information required to keep the equipment properly lubricated, adjusted, serviced, and maintained for efficient operation throughout its design life; recommended preventive maintenance schedule; lubricant type and lubrication requirements; comparative table of alternate lubricants from different manufacturers; list of required maintenance tools and specialized service equipment; maintenance troubleshooting instructions; illustrations as needed for clarity; corrective maintenance procedures and illustrations
6. Manufacturer's recommended spare parts list
7. Equipment warranty period
8. Approved shop drawings showing wiring, control logic, and installation details
9. Safe operating conditions; safe maintenance conditions; other safety information as may be required (e.g., material safety data sheets)
10. Information required by current regulations of appropriate government agencies

O&M information via computerized files. Computerized formats are more compact, they do not wear out as printed manuals do, they can be easily improved and modified, and copies of file pages can easily be printed.

If a hard copy format is desired, the engineer needs to decide whether to incorporate all O&M manuals into binder volumes or to require the contractor to submit a separate binder for each vendor's equipment. Separate binders are more portable for field use, but they require more storage space. In all cases, multiple copies of the vendor O&M manuals should be required. Table 27.3 lists format items that the engineer should consider when specifying O&M manual requirements.

Manual Submittal Schedule.   Vendor O&M manuals must be submitted as shop drawings for approval. The engineer should ensure by provisions in the specifications that the contractor will provide O&M submittals in a timely manner for approval, well in advance of start-up and vendor training. Because the system O&M manual prepared by the engineer will use some information from vendors' manuals, the latter should be submitted within a limited time after approval of the equipment shop drawing; 60 or 90 days is common.

CONSTRUCTION-PHASE TRAINING, START-UP, AND POST–START-UP

Start-up of a facility is the culmination of all preceding design and construction activities. Ideally, start-up should proceed only after all facilities have been tested and accepted, with staff properly trained, but this is often not the case. When facility expansion or upgrading is involved, start-up usually takes place in stages, with some areas of the site completed and in service while other areas are still under construction. This results in a difficult environment for plant staff and contractor alike, because both must coexist under less than ideal conditions. When start-up must be fragmented, start-up planning is an especially effective tool for coordinating activities of all parties. Planning start-up focuses attention on the responsibilities and activities of all parties

A number of elements are important when planning for start-up, including:

- Coordination of construction with existing operations (see Chapter 24)
- Vendor and systems training
- Facility operations and maintenance manual

TABLE 27.3 Vendor Operations and Maintenance Manual Format Requirements

1. Title page
2. Table of contents
3. Enclosed in a three-ring loose-leaf or triple-post binder with stiff covers
4. Page size, $8\frac{1}{2}$ by 11 inches; high rag content; larger drawings to be folded and bound so that they can be unfolded without removal from the manual, or drawings contained in bound pouch; binding margin of $1\frac{1}{2}$ inches; reinforced binding edge
5. Text to be typewritten (or computer-printed) originals, or permanent copies
6. Dividers and indexed tabs to identify major divisions of the manual, such as operating instructions and maintenance information
7. Number of manual copies to be provided

- Roles and responsibilities
- Start-up planning and initiation
- Record keeping
- Post–start-up activities

Vendor and Systems Training

Training required by the expansion or new construction of a water treatment facility can be extensive. A facility may be increased in size and complexity, with concurrent changes in operations and maintenance requirements. New systems and equipment may be state of the art, requiring higher skill levels on the part of facility staff for proper operation. To meet these challenges, staff members must acquire technical, organizational, and managerial skills to operate and maintain facilities in a cost-effective, efficient manner.

Before start-up, facility staff should be given both vendor training and systems training. Vendor training typically focuses on individual equipment, and systems training encompasses a system's operation, process monitoring, and control. Systems training is provided by the engineer, preferably by operations specialists, who not only are familiar with the design, intent, and operational requirements of the facility, but bring real-life experiences to compliment their instruction.

Vendor Training. Vendor training is typically delivered by vendors' field representatives during equipment testing, usually just before start-up. The use of knowledgeable proctors to coordinate and oversee vendor classes and verify that vendors convey appropriate information is an effective means to enhance vendor training class quality.

Vendor training tends to focus on equipment maintenance because vendors' representatives are normally field maintenance personnel with extensive mechanical experience on particular equipment. As a result, their training may emphasize mechanical maintenance issues and leave gaps in operations training.

Unless an entire system is supplied by a single vendor, the instructor generally only knows details of a particular piece of equipment and may not fully understand its role as part of a total process. Systems training by the engineer provides the "big picture" for facility staff, whereas vendor training normally provides only a piece of the picture.

Classroom quality control is extremely important because such details as class location, seating arrangements, lighting, background noise, use of training aids, instructor control over the class, instructor voice projection, and ability to follow the lesson plan can affect the quality of instruction. A proctor can assist the instructor with classroom arrangements.

Training Setup. Vendor training should be scheduled with plant staff far enough in advance to allow time to plan around staff needs and to ensure that the appropriate personnel are available on the proposed date. Specifications should require sufficient advance notice by the contractor, normally 15 to 30 days before the proposed training date. If the contractor's proposed training date is not acceptable because plant staff are not available, the contractor should be advised of alternate acceptable training dates.

Tentative training dates can be handled orally with plant staff, but they should also be informed by letter of the final training date. A copy of this notice should also be transmitted to the contractor. This ensures that the formal written notification has been provided to all parties. Plant staff should be reminded several days before and again one or two days before the scheduled training to ensure maximum participation.

The training location should meet minimum requirements of trainer and trainees. Issues such as adequate seating and desk space, lighting, ability to use overhead and slide projectors, availability of paper and pencils for note taking, and audio considerations should all be taken into account by the training coordinator. The trainer should be notified if projection equipment is available on site.

Lesson Plan and Trainer's Qualifications. The lesson plan and written trainer's qualifications provide the documentation to evaluate before training delivery whether training will be effective. The trainer's qualifications may appear impressive on paper, citing years of experience in field applications, however the quality of training also depends on the training abilities of the trainer. This is often not apparent until training begins, and the engineer can only hope that documented ability of the trainer will be matched by the trainer in person.

Proctoring Training. One of the best methods of ensuring that training will be effective is to provide a specialist familiar with training techniques and practices to witness training. Witnessing training, called proctoring, involves taking attendance, guiding the trainer if needed, making notes of key information, and preparing a brief letter report.

The proctor should use the lesson plan as a checklist and guide to verify that proposed training topics are covered. It is essential that all the topics be covered, and the order of coverage should generally follow the lesson plan. If the trainer deviates from the lesson, the proctor should ensure that training quality is not compromised.

Trainers who do not properly cover subjects or are not effectively presenting the subjects may need some guidance or coaching by the proctor. Guidance can be phrased in the form of questions by the proctor in order to provide greater information or clarity without embarrassing the trainer. If the proctor's questions and suggestions are ineffective in bringing out information adequately, the proctor, as a representative of the engineer, has the authority to terminate the training.

The proctor should take attendance, noting whether trainees were present for the entire training period. It may be important to track the duration of training in hours, because the training hours may be used as continuing education credits in states where periodic retraining is required to maintain an operator's certification. A training-received verification form should be developed, documenting the instruction with sign-off spaces to be completed by the instructor, engineer, owner, and contractor at completion of training.

The proctor should also make abbreviated notes of the training. In some instances the trainer will cover information about equipment operations and maintenance that is not included in the vendor's O&M literature. This type of information is especially valuable to plant staff and should be carefully recorded. It should later be transmitted to plant staff in printed form so that all staff members may take advantage of this experience.

Videotaped Vendor Training Sessions. Videotapes specially prepared for training purposes can be useful tools for continuing training of plant staff, and construction specifications may require that tapes on certain topics be delivered by the contractor. The common availability of video recording equipment, however, can mislead people into believing that virtually any training session can be taped and used as an effective training tool. This is normally not the case unless the video session is carefully planned and prepared and the tapes are edited to produce a clear, interesting video.

A videotape of a training session delivered in a classroom or field setting, for example, picks up all of the extraneous background distractions that go on in a training session and all of the actions of the trainer, including misspoken words, hesitations, interruptions, and pauses. Unless the tape is heavily edited and processed, noise intrusion may distract the audience from the subject matter. Furthermore, live training is more interesting than videotapes, and the audience is so used to the fast pace of broadcast television that even a well-planned, well-edited tape will be of little use unless the tape moves fast enough to retain the trainees' interest. Requirements for good videotaping, therefore, include the following:

1. Meet with the trainer in advance to plan the videotaping session and define the purpose of the training, topics to be covered, depth of coverage, training location limit, and material to be covered so that the final tape lasts preferably no more than 20 minutes.
2. Proctor the training session for the plant staff to determine appropriateness of coverage for the videotaping session. Develop a feel for the trainer's effectiveness when appearing on tape.
3. Tape a training session delivered for taping only without an audience present. The trainer should repeat when needed to obtain clearly and confidently delivered information.
4. Edit and assemble the tape; add a title, credits, and date; send to the trainer's company, if required, for approval.
5. Copy the approved tape as needed and deliver to the client; retain the original footage and edited master.

This sequence should also be followed when producing videotapes of engineer systems training.

Systems Training. Systems (process) training by the engineer provides facility staff with knowledge they need to operate equipment and understand new processes. Systems training helps staff relate the "small picture," such as individual equipment or peripheral systems, to the "big picture," such as an entire treatment process. Systems training is most effective when created and delivered using the following concepts:

- Prepare technically accurate information that places process information and procedures into context.
- Work closely with treatment plant staff to identify specific operational concerns and incorporate concerns into training.
- Vary delivery techniques and materials to enliven the presentation.
- Use graphics extensively.
- Communicate effectively and interact with students during training to maintain interest and improve student knowledge retention.
- Solicit and respect trainees' views and use their feedback constructively to improve the instructor's delivery.
- Relate trainees' responsibilities to overall plant function and regulatory compliance.
- Limit individual training sessions to less than three hours of contact time, with one or two breaks.
- Use hands-on demonstrations or walk-throughs to accompany or follow classroom sessions.

Training Materials Development. Materials used in systems training must meet intended audience needs and clearly describe specific equipment, instrumentation, processes, and controls. Training materials can be most effectively developed subsequent to or concurrently with the preparation of the facility operations and maintenance manual. The instructor has contract documents, shop drawings, manufacturers' O&M manuals, existing standard operating procedures (if any), and reference materials to draw on for information. The instructor must distill and simplify material and present information in an interesting and practical way. Training materials must be technically sound, user oriented, and effective. Formal lesson plans, handouts, and other instructional resources should be prepared for most systems training.

All training materials should be performance based, keyed clearly to learning and performance objectives. Evaluation methods should focus on direct and indirect indicators of performance by trainees, along with more conventional trainer evaluation.

Systems Training Delivery.   Systems overview training for all plant staff should be conducted several months before start-up to provide an overview of new treatment systems and the project as a whole. This training presents an opportunity for staff to voice concerns about future operations that inevitably develop as start-up gets closer. Some topics covered by this overview training include:

- New process systems, locations, function, and purpose
- The projected schedule of vendor training and identification of equipment for which training will be provided
- A description of the start-up schedule
- Answers to any questions or concerns that staff members have regarding the project

Specific systems training by the engineer, conducted just before start-up and after vendor training, completes the formal start-up training program. This instruction emphasizes hands-on aspects and procedural instruction to teach how and why the system operates. This training is considered performance based, and it should teach each staff member the skills necessary to achieve the best possible equipment and systems performance.

In addition, the initial month after start-up is an excellent opportunity for further instruction and confirmation of skills. Post–start-up training may be conducted either formally or informally by the engineer.

Facility Operations and Maintenance Manual

The facility O&M manual prepared by the engineer provides complete guidance to the client's plant and engineering staff on facility operation, with contents prepared to meet the needs of technical staff and plant operators. This dual audience, with its diversity of technical background and needs, imposes a challenge for the manual writer to prepare material both technically complete and comprehensible. Because plant operators rely on practical information as it applies to plant operations and maintenance, procedural information can be directed more to this group. Theory and design intent may be written more with technical personnel in mind.

Manual Preparation Schedule.   The O&M manual should be prepared before construction is completed in order to support preparation of operator training materials. Client staff should be involved at an early stage in developing the O&M manual to assist with format and content. Later, during start-up and initial facility operation, information will be gathered and procedures developed that will modify information in the O&M manual. This material should be carefully recorded so that the manual can be updated based on actual operating conditions. A final manual can then be issued that accurately reflects operating characteristics of the facility.

Computer-Based Manuals.   The traditional hard copy O&M manual is gradually being replaced by computer-based, electronic manuals in various formats. Computer manuals have a variety of characteristics that make them superior in many respects to paper manuals. Some of these advantages are ease and speed of access to information, ability to readily update the manual, and the capability of storing graphic information. Internet-based applications are linking electronic O&M documentation to a variety of information sources, including Internet addresses, document imaging, other reference databases, and maintenance management and laboratory information management systems.

Another consideration is the likelihood that manufacturers' literature will become available in digital formats such as CD-ROM and through the Internet. Computerized O&M manuals can be expected to become more common, and the engineer should give special consideration to developing the capability of the O&M manual to link to these other references.

The engineer must carefully consider factors that go into making a computer manual. It must generally be an entirely new document, not simply a reformatting of electronic files used to develop paper manuals. Considerations include the software used to develop the manual to make sure it is user-friendly. Input from the client's staff at an early stage of manual development is especially critical when preparing a computer manual.

O&M Manual Contents. Whether the O&M manual is computer-based or not, the engineer should determine the regulatory requirements over the contents of the manual. In addition to any such regulatory requirements, the manual should contain the following information:

- Design intent for each process and the overall plant
- Design parameters
- Water quality criteria and regulatory requirements
- Process operations and control
- Operational procedures
- Laboratory procedures
- Maintenance information, in addition to manufacturers' literature (preferably a complete maintenance management system)

A detailed listing of suggested manual contents is included in Table 27.4.

Start-Up Planning and Initiation

When an existing plant is upgraded, two types of start-ups commonly occur—interim and final. Both can substantially affect O&M practices and procedures, and each requires extensive planning and coordination with existing operations. Interim start-up occurs during critical construction phases as existing systems are being upgraded and new systems are brought into service on an interim basis. Final start-up occurs when the system is substantially complete.

Interim start-up planning is a part of construction sequencing. For example, to keep a plant operational, a minimum number of filters must be in service at any given time to meet water demands, and a plan to maintain plant operations is developed to do this.

All parties must collectively plan for and implement process changes to the facility during each start-up. Several process and operations issues must be considered:

- New equipment and control systems will be placed in service, such as tanks, chemical feed equipment, control valves, instrumentation, and pumping equipment. This equipment requires frequent monitoring and adjustment, which must be accounted for in the start-up plan.
- New flow patterns, channels, and tank configurations may be used, and new systems may also be placed into service.
- Contingencies should respond to potential emergencies, such as major equipment failure or instrumentation calibration and control problems.
- Additional operations personnel may be required to monitor and operate equipment during the critical initial phases of a start-up. It is commonly necessary to schedule

TABLE 27.4 Recommended Facility Operations and Maintenance Manual Requirements

1. Introduction and process overview
   Plant capacity, source water supply, and distribution system overview
   Treatment process overview
   Plant hydraulics
   Source water quality

2. Water quality criteria and regulations
   Water quality criteria
   Microbiological, organic, and inorganic contaminants
   Aesthetic qualities
   Regulations (such as federal Safe Drinking Water Act requirements)
   Nonregulatory quality issues
   Sampling and testing

3. Process chapters (each process)
   Design intent and criteria, process purpose
   Process and equipment description, operation, and control (standard operating procedures)
   Start-up, shutdown, monitoring, and troubleshooting
   Special safety considerations

4. Maintenance
   Organization of equipment manufacturers' literature
   Equipment identification system
   Maintenance and lubrication frequency tables
   Maintenance management system (work orders, maintenance history records, schedules, inventories, purchase orders)
   Standard maintenance procedures
   Warranties

5. Laboratory
   Monitoring program
   Quality assurance/quality control and chain of custody procedures
   List of analytical procedures and references
   Safety and materials safety data sheets (MSDS) for laboratory chemicals
   Inventories
   Ordering information and procedures

6. Safety
   Supervisory and staff responsibilities
   Safety equipment
   Plantwide safety procedures
   Confined space entry procedures
   MSDS for process chemicals

7. Utilities
   Electrical, gas, and water service
   Heating, ventilating, and air-conditioning systems

8. Records and reports
   Record-keeping needs
   Reporting to regulatory bodies

special personnel for 24-hour monitoring during start-up phases of new systems and equipment. Similarly, the contractor may need to schedule additional personnel and ensure that the vendor is on site and available during start-up.

Start-Up Plan. Start-up plans should include elements listed in Table 27.5. A formal start-up plan should be prepared, generally in the following sequence:

- Identify facilities to be placed into service and their functional objectives.
- Review, outline, and develop procedures with facility operations and maintenance personnel to maintain treatment process integrity.
- Identify specific process and equipment conversion sequences.
- Identify contractor preconditions to start-up, including satisfactory functional test documentation; vendor training completion; accepted vendor O&M documentation, turned over to the plant staff; and accessible and fully operational equipment.
- List additional preconditions, such as systems training requirements; necessary documentation availability; and related systems operability, including safety and auxiliary equipment.
- Establish process monitoring and control set points.
- Define owner, contractor, and engineer staff responsibilities, including emergencies and shift coverage.
- Identify other resources required for start-up, such as emergency requirements, special record-keeping forms, and specialized support or monitoring equipment.
- Establish the start-up schedule.
- Develop specific system start-up sequences for each unit process, including detailed procedures and operating responsibilities.

Start-Up Workshop. A highly useful tool for working out details of a start-up plan is the use of workshops to bring responsible parties together. The resident engineer, contractor, client staff supervisors and engineer, and start-up manager normally should attend these meetings. The start-up manager should prepare draft documents for consideration, and agreed-on revisions should be incorporated into revised drafts. Ultimately, a final plan will be developed.

Start-Up Sequence. From an operational point of view, start-up should ideally proceed in the following sequence:

1. Hire needed plant staff.
2. Complete vendor and system training.
3. Prepare and revise a start-up plan.
4. Complete construction.
5. Test, check out, and accept new facilities.
6. Stock needed chemicals and prepare for their use.
7. Notify regulatory agencies, as appropriate.
8. Conduct a walk-through of the facilities that are to be started before start-up.
9. Proceed with start-up.

TABLE 27.5 Start-Up Plan Elements

1. Overall objective of the plan
2. Facilities to be started
3. Sequence of events
4. Responsibilities of each party
5. Initial operating conditions and parameters
6. Intended final operating conditions and parameters
7. Laboratory requirements or arrangement for outside laboratory services
8. Operating procedures
9. Sampling and monitoring requirements

This may be the sequence for a small project that is completed and then placed into service, but for a large project that proceeds in phases, this sequence takes place for each phase in a less than ideal manner. New plant staff, for example, are often not hired until start-up begins, because of owner's budget limitations, and they have to be trained on site while some of the facilities are placed into service. It is also sometimes necessary to start facilities that are not complete, and the contractor is responsible for finish work to be completed at a later time. Such disruptions may cause conflict between the contractor and plant staff.

Start-Up/Operations Services Manager. An effective functional arrangement for coordinating start-up is the employment of a start-up/operations services manager. This person's responsibilities include not only start-up, but all operations services elements of a project. This function is particularly important when start-up is piecemeal and the resident engineer does not have time to coordinate training and start-up in addition to managing construction activities.

An effective operations services manager should establish good rapport with contractor and plant staff. Because conflicts or differences in opinion among these parties are almost inevitable, the manager should be able to work with the parties to resolve problems. Close communication and cooperation with the resident engineer are essential.

Record Keeping.

Record keeping during start-up is often a chore because it requires careful recording of process conditions that may change frequently as the contractor works toward start-up. Nevertheless, good record keeping during start-up is important because it:

- Defines what was done at a particular time and how the process responded to given conditions, which may be a useful guide for future process operation
- Establishes responsibilities for actions that may later be relevant to damage claims
- Defines the initiation of equipment warranties
- Identifies process operating problems and solutions
- Identifies equipment problems and unfinished work that need to be resolved by the contractor

Post–Start-Up Considerations

Post–start-up activities depend on scheduling start-up and the engineer's responsibility for follow-up work. If part of the new facilities have been started up, the engineer will remain on site and available to assist plant staff in process troubleshooting, even though the engineer's contract does not specifically define this as a contract function. If a start-up/operations services manager is employed, this person normally will assist the owner.

When a new facility has been completed and start-up is over, post–start-up services provided by the engineer depend on the engineer's contract with the owner. Budget conditions may limit services the engineer can provide. However, post–start-up operational and troubleshooting assistance is so important that the engineer should seek to write a provision for this service into the contract.

CHAPTER 28
PILOT PLANT DESIGN AND CONSTRUCTION

A pilot plant consists of equipment and materials used to simulate a full-scale process or processes. It is built to collect process design and operations data, but it is constructed on a smaller scale for ease of operation, installation, and manipulation. A pilot plant's size can vary from bench-scale equipment to systems that can treat millions of gallons per day.

Each unit treatment process has a limited scale-down factor. If this limit is exceeded, the pilot unit process will no longer simulate the full-scale process. Most scale-down limits have been established empirically from previous pilot studies. For example, gravity filter operations can be simulated using filter columns as small as 4 in. (10 cm) in diameter. However, backwashing studies using the same size filter do not provide quality data. This is primarily because the excessive ratio of sidewall area to surface area creates friction between the media and sidewall, short-circuiting washwater.

Pilot plant design and operations should be kept as simple as possible. Pilot plants should be operated manually to maintain control of the processes, to determine idiosyncrasies associated with unit processes and individual equipment pieces, and to rectify small problems as they occur. Problems not manually corrected could jeopardize the relevance of significant amounts of data. For example, flow can be controlled using rate controllers and variable-speed drives on pumps. With small flows, in-line flowmeter accuracy can be off by as much as 25% or more. Without a manual check of flow rate, all data collected would have a significant factor of error, which, if left unchecked, could result in a critical final design flaw. Automatic data collection is effective for establishing trends and monitoring operational and water quality data during the absence of operators.

Pilot plant design and operations must also be flexible because these plants are used to investigate the unknown and to stress the limits of a treatment process. During a study it is common to change process variables and even to change processes themselves. Flexibility in design allows the operator to quickly connect and disconnect unit processes or by-pass processes. Modular design allows for complete processes to be quickly removed or inserted into the treatment train, reducing downtime. For example, if the pilot plant is not operational for significant periods of time, specific water quality events important to the study might be missed.

Although the pilot plant comprises the tools used in performing a study, it is the design of the study that dictates much of the design characteristics of the pilot plant.

PILOT PLANT STUDIES

Before describing the construction of pilot plants, some of the common variables encountered in a pilot plant study should be discussed.

Purposes of a Pilot Plant Study

Pilot plant studies are common, and their use has become almost necessary to meet regulatory requirements at minimum construction costs. Pilot plants are generally used to:

- Compare alternative treatment processes
- Solve treatment process problems by investigating alternative process modifications
- Investigate new treatment processes
- Demonstrate confidence in recommended treatment processes
- Meet regulatory requirements
- Establish design criteria

Engineers generally turn to pilot plant studies to compare alternative treatment processes before design, construction, or plant modification. One significant advantage of using pilot plant studies in new facility design is that pilot plants allow the engineers to use site-specific design criteria rather than having to use generic or rule-of-thumb conservative design criteria. This advantage can result in significant cost savings without sacrificing treatment quality.

New treatment processes are usually investigated at the pilot scale to establish a success rate before their recommendation. For example, the concern for controlling the formation of disinfection by-products has led to using ozone as a preoxidant and disinfectant in place of chlorine and chlorine dioxide. Although ozone has been used for many years in Europe, it has only recently been commonly used in North America. Pilot plants and pilot studies have allowed ozone to be compared as an alternative oxidant and disinfectant. Ozone concentrations, contact times, and dissolution techniques are investigated at the pilot scale and proven practical before full-scale plant design is started. In many instances, pilot plant studies produce results that provide for regulatory agencies the confidence to approve the use of new technologies and for municipalities confidence to accept them.

Pilot studies are often used to demonstrate treatment processes over extended periods of time and under varying water quality conditions. Seasonal fluctuations due to lake turnover and algae blooms in reservoirs, daily water quality changes in rivers due to biological respiration, runoff events, and industrial discharges are some of the occurrences that can disrupt treatment processes. If water sources are likely to experience any of these events, it may be necessary to study their effects on various treatment processes for extended periods of time. Regulatory agencies may require operating a pilot plant for a year or more before granting permission to use new treatment technologies.

It is important to have a complete understanding of source water quality before establishing a pilot plant study. Historical data can be used to some extent to determine the stability of supply and the need for extended operations. Exceptionally stable supplies may require pilot testing for a short period of time during each season. All water quality episodes that happen in the supply should be determined and investigated as part of the study. For example, if a study is conducted for only two to three weeks per season on a large reservoir, there is a chance that an algae bloom occurrence may be missed. Also, depending on the depth of the intake, an isotherm may pass through the water at the intake level for short periods of time causing wide swings in pH, which may severely affect the coagulation process.

Another common use of pilot plant studies is to determine how best to correct full-scale unit process problems without shutting down the full-scale process. Once the comparability between the pilot plant process and the full-scale process has been established, individual process parameters can be isolated and tested for comparison with the full-scale process. For example, if floc carryover through a clarification tank is a problem, a pilot plant can be used to test how the problem can best be alleviated by testing various combinations of coagulant type and dosage, mixing times and energies, and flocculation time. Testing these variables on a full plant basis can be dangerous in that some changes could result in complete failure of the process or production of water of unsatisfactory quality.

Pilot Plant Study Checklist

In developing a pilot plant study, the following checklist of items must be considered before the design, construction, or acquisition of a pilot plant. Each item in the checklist will help focus on the type and size of pilot plant required.

- Define the purpose of the study.
- Identify the end product of the study.
- Collect all available background and historical data concerning water source and water quality.
- Compile a list of water treatment processes that should be considered.
- Establish the minimum and maximum design constraints of pilot plant processes.
- Define the length of operation of the study.
- Determine who will operate the pilot plant.
- Decide where the pilot plant will be located.
- Determine the flexibility required for process changes during the study.
- Decide the locations and frequency of sampling and analyses.

PILOT PLANT DESIGN

Pilot plant design generally consists of two steps: running bench-scale tests to obtain preliminary data, and then final design based on the unit processes to be studied.

Bench-Scale Testing

Bench-scale tests are often conducted before designing a pilot plant to establish the feasibility of various treatment processes.

Purpose of Bench-Scale Tests. In the absence of existing treatability data or treatment plant operational data, bench-scale tests can be run to help select the pilot-scale treatment processes. Bench-scale tests are typically batch tests; the results are not capable of showing dynamics found in the continuous flow process stream. Bench-scale tests can, however, provide immediate and short-term results of significant value. Results from bench-scale tests can also be incorporated into design criteria for the processes.

Bench-scale equipment can be arranged to simulate an existing treatment plant operation, proposed modifications, or entirely new processes. Once the correlation between the bench-scale test and the full-scale process has been established, bench-scale apparatus can be used in assessing various treatment options. For example, rapid mixing, flocculation, and clarification can be accomplished in one vessel by simulating various mixing times and energy inputs. Clarified or flocculated water can then be passed through filter paper, membranes, or a carbon column to simulate further treatment.

Bench-scale tests are generally conducted concurrently with the pilot plant study to help determine proper pilot plant protocols. Preliminary selection of treatment chemicals, mixing times and energies, settling rates, flotation rates, and media selection can be estimated using bench-scale tests.

Bench-Scale Equipment. The most common types of equipment used for bench-scale tests for potable water include:

- Six-paddle stirrer with variable-speed mixers (Figure 28.1)
- Set of 2L plastic jars with sample taps and pinch valves
- Supply of 15 cm Whatman filter paper with approximately 0.5 to 10 μm porosity
- A 4 in. (10 cm) filter funnel and a 500 ml graduated flask
- Vacuum filter apparatus with 0.5 μm glass fiber filters
- Analytical equipment, including a turbidimeter, pH meter, spectrophotometer, and colorimeter

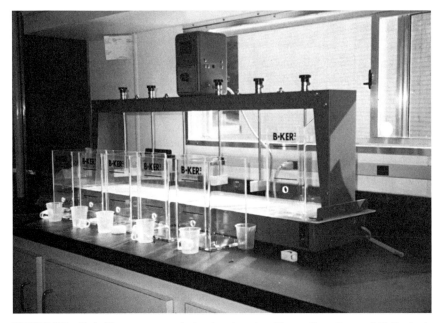

FIGURE 28.1 Typical jar test apparatus for bench-scale testing. *(Source: Camp Dresser & McKee Inc.)*

Other types of bench-scale equipment are available to test for settling, carbon adsorption, dissolved-air flotation, and solids dewatering characteristics.

Bench-scale tests are generally conducted by adding various treatment chemicals directly to the jars while mixing. Treatment chemicals can be prepared in advance and added by syringe or pipette to the jar at the appropriate time. Rapid mixing is usually simulated for 0.5 to 1.0 min at full rpm. Flocculation can be simulated by varying the rpm from 80 to 20 over an extended period of time, typically from 5 to 20 min. The floc is then allowed to settle for 30 to 60 min without mixing.

Samples are periodically collected from the sample tap on the jar to determine the settleability of the floc. Floc suitable for conventional settling will readily settle in less than 5 min, or sometimes even during the slow flocculation step. Floc suitable for flotation will remain suspended for over 30 min. Pin-size floc suitable for direct filtration will be barely visible and will remain suspended.

Samples collected from the jar after settling can be filtered through filter paper and analyzed for particulate removal and color reduction. True color can be used as a surrogate measurement for natural organic matter reduction. If samples are to be analyzed for total or dissolved organic carbon reduction, glass fiber filters should be used so as not to contaminate the samples with organic paper fibers.

General Design Considerations

Principal issues that must be addressed early in pilot plant design are the various hydraulic issues and provision of adequate electrical power to operate the equipment that will simulate processes provided in the full-scale plant.

Hydraulic Requirements. Several hydraulic concerns should be addressed before pilot plant construction or installation. The first is where the source water should come from and how it is to be supplied to the pilot plant. It is important that the source water supplied to the pilot plant be representative of the water that will be used in the treatment plant being designed. In many cases, providing a continuous supply of source water requires modifications to existing structures or running a long length of pipe from the source to the pilot plant.

The method of disposal of both treated and untreated water from the pilot plant must also be determined. It is often difficult to locate gravity drain lines from the pilot plant, and it may be necessary to arrange for a sump or collection tank with pumps to remove the wastewater to a sewer or drain.

If possible, the pilot plant should be designed so that pumping will not be required between the unit processes. In most cases, water being processed contains solids, and running the water through a pump could change the characteristics of the solids. This could have a significant adverse impact on pilot plant results. In addition, every pump introduces another potential source of failure. Staged platforms are usually used to support tanks and other hydraulic structures so that flow is by gravity and the number of pumps is kept to a minimum.

Electrical Power Requirements. The electrical loads required by the pilot plant must be determined in advance to avoid on-site problems and delays. The total electrical draw (amps) and voltage requirements must be compared with available on-site electrical facilities. If special electrical work is required to supply power to the pilot plant, it may be necessary to obtain a local permit and hire a licensed electrician. One way of minimizing outside labor charges for electrical work is to provide a power panel with the pilot plant prewired to all electrical equipment. In this way, it is necessary to provide only a single source of power to the panel at the site.

Pilot plants designed for overseas installation will probably require special consideration of the electrical design. The voltage and hertz of power available differ from country to country, and it may be necessary to provide a transformer as part of the pilot plant equipment package.

Design for Study of Unit Processes

Unit processes most often studied by using a pilot plant to obtain improved design information are:

- Preoxidation
- Coagulation
- Flocculation
- Clarification
- Filtration and granular activated carbon GAC contactors
- Ozonation contactors
- Solids collection and treatment

Preoxidation. Preoxidation is commonly used for biological growth control through treatment plant source water piping and structures, as a first-stage treatment for color removal, for taste and odor reduction, for reduction of natural or manufactured organic matter, for metal precipitation, for disinfection credit, and for coagulation enhancement. The most common chemicals used are:

- Chlorine gas or sodium hypochlorite
- Chlorine dioxide
- Potassium permanganate
- Ozone

Preoxidation to be used in the full-scale plant must be simulated in the pilot plant, but chlorine gas, chlorine dioxide, and ozone will generally be expensive, difficult to feed and control, and potentially dangerous to use. Sodium hypochlorite is recommended for most pilot studies because it is safe to use and easy to feed and control. Sodium hypochlorite is commonly available in 5.25% and 12.5% solutions and can be easily pumped into the process stream. Fumes associated with hypochlorite solutions are not generally a major concern.

Potassium permanganate may also be used as a preoxidant for a pilot plant. It is purchased as a dry powder, and a batch of solution can be prepared with a known concentration. This solution can then be injected into the process stream with a metering pump. Metering the powder directly to the process stream is not practical because of the low flows used in most pilot plants.

Depending on the application, some extended reaction time must be provided in the process train when using chlorine or permanganate. The optimum chlorine concentration and reaction time can be simulated in jar tests before the pilot study. The total retention time may be longer than 30 min for some reactions. Some agitation is required during the retention time to guarantee that the oxidant is dispersed throughout the process stream. This mixing can be accomplished by directly injecting chlorine into the pipeline ahead of several elbows or bends, or by using a static mixer, venturi injector, or mixing in a tank.

Polyethylene tanks or drums can be used to provide contact time in a pilot plant. To eliminate the short-circuiting of flow that would occur in one large tank, proper contact time can be ensured by installing several smaller tanks piped in series. By connecting piping to the threaded bulkhead fittings in the tanks and using pipe unions, tanks can be easily added and eliminated from the series. Changing the flow through a process can also control the contact time; however, other processes downstream may then be adversely impacted or restricted.

Ozone is a strong oxidant and an unstable gas that has to be generated on site. Its reaction time is typically much faster than other oxidants. If ozone is to be fed in the pilot plant, it can be directly injected into the process stream through a venturi injector or applied through a diffuser into a contact tank. Ozone injection can be accomplished by drawing off approximately 10% of the total process stream and pumping the flow through a venturi injector, creating a vacuum. The venturi is connected so that the vacuum draws the ozone gas and injects it into the partial process stream, and then the solution that has been formed is returned to the full process stream. The dosage is controlled by varying the flow of gas and the concentration of the gas.

Coagulation. Coagulation is the process in which an electrolyte is introduced to the process stream to reduce the net electrical repulsive forces of suspended particle surfaces. Coagulation aids in removing those substances associated with turbidity, including colloidal solids, clay particles, suspended and some dissolved organic matter, bacteria, algae, color, and taste- and odor-causing compounds. Coagulation depends on the selection of a coagulant, applied coagulant dosage, pH, temperature, mixing energy, and time.

Coagulant application requires facilities and equipment to store and deliver the chemical to the process stream. Many suppliers of coagulant chemicals provide quantities of their products for a pilot plant study at no charge other than those associated with shipping. Containers larger than 5 gal are heavy and may be difficult to receive and move around the pilot plant location.

Day tanks generally function as a control mechanism to measure the volume of chemical used. The day tank can be as small as a 500 ml beaker and as large as a 50 gal drum. The use of day tanks requires the pilot plant operator to frequently mix fresh chemicals for plant use.

If possible, coagulant chemicals should be fed at full strength. Once a coagulant has been diluted, hydrolysis immediately starts to take place. The metals begin to precipitate out of solution and render the coagulant less effective, which can have a significant impact on the performance of the coagulant. The design size of the process stream in the pilot plant and the selection of the coagulant feed pumps are critical for the coagulation process. The smaller the process stream, the more difficult it is to meter the coagulant accurately.

The coagulation process is almost instantaneous if a complete mixing system with sufficient energy is supplied. In-line static mixers can be used instead of mixing tanks and paddles. An in-line static mixer imparts a tremendous amount of energy in the short period of only about 1 or 2 s. If static mixers are used, it is recommended that chemicals be fed with a constant rate (peristaltic) chemical feed pump. If pulse-type feed pumps are used, chemicals are injected only intermittently into the system. Peristaltic pumps are excellent for this type of application. There are peristaltic pumps that can reliably feed as little as 0.1 ml/min.

If a mixing tank is provided, a minimum hydraulic detention time of 15 s should be provided. Rapid mix times longer than 1 to 2 min may lead to floc shear. A mixing system should be designed to impart a velocity gradient (G) value of approximately 300 to 500 s^{-1}.

The pH of the water being treated is critical in the control of the coagulation process. If source water pH varies, continuous pH monitoring may be required. If source water pH is stable, periodic grab samples usually suffice. A streaming current or zeta-potential meter may be considered to monitor the coagulation process if source water quality changes frequently. A sample tap should be located immediately before and after the coagulation process.

Flocculation. Flocculation imparts energy to the process stream to force agitation, which causes very small suspended particles to collide and agglomerate into larger, heavier particles, or flocs. The optimum size of the floc particles varies depending on the downstream processes. For direct filtration, a fine, dense floc is required; for dissolved air flotation, a small floc is required. For best settling, a large, heavy floc is generally optimum. Mixing time and energy control the behavior of the flocculation process. The flocculation process is usually designed to provide mixing with a decreasing level of energy over time.

Flocculation time of at least 10 min is usually required for most treatment systems, and some conventional settling plants find that optimum flocculation times as long as 30 min may be required. The determination of the optimum flocculation time is often one of the primary goals of a pilot plant study; therefore the flocculation tanks should be designed to accommodate wide variations in flow and retention time. To avoid short-circuiting through the flocculation process, the tank or tanks should be constructed to provide for at least three separate hydraulic zones. This can be accomplished by providing baffle walls between zones or by providing two or more tanks in series. The time through the process can be varied somewhat by controlling the flow through the tanks, as long as other processes are not affected by the flow change. Provisions for bypassing hydraulic zones or installing or removing tanks in series may be a better method for controlling the time of flocculation.

Mixing in the flocculation tanks can be accommodated by installing vertical shaft paddle mixers designed to impart various energies to the flocculation zones. Variable-speed drives should be provided for the simulation of flocculation. As an alternative, the speed of the mixer can be determined by visual inspection or with the use of a tachometer or strobe light. Flocculation velocity gradients between 10 and 80 s^{-1} are most common. Flocculation tanks should preferably be rectangular to promote complete mixing. If circular tanks are used, some type of stators should be incorporated to guarantee mixing.

Provisions for sampling should be provided throughout the flocculation process, specifically at points between mixing zones. Grab samples are probably best to evaluate the formation of floc. Velocities through piping should be kept low so as not to shear floc and present unrepresentative samples.

Clarification. Clarification removes much of the floc before filtration. Most clarification processes require sedimentation of the flocculated particles. Recently, dissolved air flotation has gained some popularity for its ability to remove significant amounts of floc before filtration, lower surface area requirements, and less upsets such as those associated with temperature gradients or hydraulic surges. Other clarification processes include the use of sludge blankets and roughing filters. These processes can be effective but are usually pilot tested by equipment suppliers with their own standard equipment. Conventional sedimentation remains the most common form of clarification process.

Sedimentation. Full-scale sedimentation basins have typically been divided into four specific zones:

- The inlet zone provides a smooth transition from the influent flow to the uniform steady flow desired in the sedimentation zone.
- The sedimentation zone provides volume and surface area for sedimentation to take place.
- The sludge zone receives and stores the settled floc particles.
- The outlet zone provides a smooth transition from the sedimentation zone to the effluent flow.

These four zones have equal importance at the pilot scale as well. The inlet zone must be designed to provide for a smooth transition from the flocculation process to the clarification process. High velocities and energy losses associated with small piping and fittings, valves, and orifices can shear and break flocculated particles, thus reducing the sedimentation tank efficiency. An open-channel zone providing laminar flow should be provided ahead of the chamber where upflow begins.

The sedimentation zone must be designed to provide the necessary volume and surface area to minimize floc carryover and maintain steady-state operations for extended periods of time. Simply scaling down the size of the tank to match the process flow requirements would result in significant short-circuiting through the tank with little sedimentation occurring. When pilot testing sedimentation, it is recommended that standard inclined tubes or plates be used. Even when using tubes or plates, it is recommended that a safety factor of 1.5 to 2.0 be applied in designing the surface area requirements. A design overflow rate of 0.5 gpm/ft^2 (0.1 m/h), based on the surface area of the sedimentation tank, has been used successfully.

To eliminate short-circuiting through the sedimentation tank, it is critical to provide for at least 4 to 6 in. (5 to 15 cm) of standing water over the top of the tubes or plates. In addition, collector pipes or perforated pipes should be laid across the top of the clarifier to evenly distribute the flow across the entire basin.

The sludge zone of the sedimentation tank must be large enough to collect solids over extended periods of time. Continuous sludge collection constantly stirs up and resuspends solids, which greatly reduces the effectiveness of the process. Three to four vertical feet of storage volume should be provided to allow for the storage of the solids. The bottom of the tank can be angled in such a manner as to direct the solids to a central place for collection, with a drain valve provided for the removal of the solids. If the drain line is to be run continuously, it should be provided with a flowmeter.

The outlet of the sedimentation basin should be designed similar to the inlet so that excessive velocities and energy drops are not encountered. Flocculated particles that pass through the sedimentation basin can be sheared to a size where they may later pass through a filter. Aggressive pumping of settled water can also shear floc, so low-speed positive displacement pumps should be used if pumping is necessary.

Process flows of 5 to 15 gpm (3 to 9 L/s) can incorporate conventional sedimentation relatively easily. Flows less than 2 to 3 gpm (1 to 2 L/s) make short-circuiting an intolerable problem. On the other hand, flows greater than 20 gpm (13 L/s) result in requirements for large tanks. Temporary pilot plant sites typically have constraints that restrict the size of tanks that can be used. These constraints are usually associated with open space, access, and structural limitations.

Dissolved Air Flotation. The dissolved air flotation (DAF) process requires the use of a clarification tank in which the flocculated particles are drawn to the surface by their attachment to tiny air bubbles; they are then removed from the surface and piped to waste or provided with further treatment (Figure 28.2). Overflow rates can range from 3 to 10 gpm/ft^2 (8 to 24 m/h) of surface area. DAF also requires a sidestream process whereby approximately 10% of the total process flow is injected with air, pumped through a pressure vessel at approximately 60 to 100 psig (414 to 689 kPa) to saturate it with air, and delivered to the process stream at the inlet to the DAF tank. At the tank inlet, the pressure in the sidestream is dropped to near-atmospheric conditions and the air in solution is released in the form of many tiny bubbles.

The DAF tank requires an inlet zone for smooth transition from the flocculation process, but the flows do not have to be as slow as those for sedimentation. The sidestream is injected into the process stream in the pipes directly ahead of the tank. A baffle wall inclined at approximately 60 degrees from horizontal, pointed in the direction of flow, is required immediately downstream from the inlet of the tank in order to direct the flow upward. The flow is then directed to the bottom of the tank for discharge.

FIGURE 28.2 Parallel DAF tanks with mechanical scrapers (tanks 50 gpm with overflow rate of 6 gpm/f^2). *(Source: Camp Dresser & McKee Inc.)*

Perforated collector pipes located a short distance off of the tank bottom can be used for collecting the DAF effluent and directing it to an adjustable weir that allows for maintaining a constant head in the tank at various flow rates. An inclined plate at the outlet end of the tank is provided as a "beach" for the ease of collecting the floating solids ("float"). Removal of the solids can be accomplished by backing up the flow and flooding the solids over the beach or by providing mechanical scraping of the solids toward the beach. Volume requirements for solids collection can be estimated by assuming the float material generated to be approximately 0.5% to 1.0% and 2% to 5% percent solids if collected hydraulically or mechanically, respectively.

The sidestream system requires piping and valves, centrifugal pump, rotameter, air compressor and receiver tank, solenoid valve, venturi injector, saturation tank with level switch and pressure relief, and discharge valves (Figure 28.3).

Filtration and GAC Contactors

Filtration is accomplished by passing water through a porous medium for the removal of suspended particles. GAC contactors are designed for adsorbing soluble organic matter onto the media. Most filters and contactors are designed to operate in a downflow manner, but up-flow filters are occasionally used ahead of another filter in series. Filters and GAC contactors can be pilot tested using similar material and equipment.

The process of filtration has been successfully simulated using filter columns as small as 4 in. (10 cm) in diameter. The use of smaller columns is not recommended because there is an excessive sidewall-to-surface ratio, which results in significant short-circuiting during the filter operation and excessive bed expansion or compaction caused by media-sidewall friction. In addition, the small flows required by small columns are difficult to control

PILOT PLANT DESIGN AND CONSTRUCTION 739

FIGURE 28.3 Author shown adjusting flow to a DAF recycle stream pump adjusted manually based on data from flowmeters and pressure gauges. *(Source: Camp Dresser & McKee Inc.)*

and maintain. The flow requirements for additional processes and for monitoring can help determine the size selection of the filter column, but in general, a 6 in. (15 cm) column is recommended.

The most popular material for construction of filter columns is clear polyvinyl chloride (PVC) because it allows the operator to visually inspect the bed while it is in operation. Visual inspection can help identify the level of media, formation of mud balls, excessive floc accumulating on the surface of the media, uneven distribution of media after excessive backwashing, and accumulation of media fines near the surface. Columns can be constructed in one section or in short sections with flanges that are bolted together. Inlets, outlets, overflows, underdrains, and manometer taps are installed relatively easily through the sidewall of the column (Figure 28.4).

The underdrain system can be attached directly to the bottom flange of the column. If a nozzle is used, the flange can be drilled and tapped to accommodate the thread size of the nozzle. If a porous plate is used, the flange can be cut to accommodate the plate, or a plate can be sandwiched between two flanges with the use of gasket material so that it will be watertight. Filter columns can either be supported off of the ground on a stand or hung from a structural wall using steel angle iron and plates. Mobile pilot plants are constructed with individual filters built into cabinets with casters that allow for movement from one site to another (Figure 28.5).

FIGURE 28.4 Top section of filter column preassembled with quick-connect unions, flanged sections bolted together. *(Source: Camp Dresser & McKee Inc.)*

FIGURE 28.5 Mobile field plant modules with filter assembly units mounted on casters, complete with filter, flow controls, turbidimeter, and head loss monitoring. *(Source: Camp Dresser & McKee Inc.)*

Flow control through a filter column can be controlled by several methods. One common method is by use of a constant-head tank or altitude valve. Electronic flowmeters can also be used in conjunction with a modulating valve or variable-speed pump to control the flow through the filter. When flow control is provided by a pump located ahead of a filter, there will be some pressure through the filter, which will impair filter performance if the pressure is excessive. This usually requires the installation of pressure sensors to monitor the pressure drop through the filter. In addition, the energy imparted to the water by the pump may change the character of any floc that has passed through the clarification process.

If possible, gravity feed to and through the filter is recommended for minimum adverse effects. The flow can be controlled at the filter outlet through a valve or pump. When using a pump, it should be a positive displacement pump that is not affected by the suction head. Because the filter will experience head losses approaching 8 to 10 ft (2.4 to 3 m), the available suction head will vary by the same amount.

Because of the low flow rate required when filtering through a small column, available flowmeters may not be accurate, so a rotameter is often installed on the influent line to provide visual inspection of the flow rate. In addition, a sample point should be provided at the filter effluent so that the flow rate can be checked by diverting the flow into a graduated cylinder while the time is measured with a stopwatch.

Filter backwashing facilities should be provided to clean the filters at the end of a filter run. These facilities should include a storage reservoir for clean water and a pump. The pump should be sized to provide water to the filter at a rate of up to 25 gpm/ft^2 (61 m/h) after head losses through the pipe, media, and column have been considered.

If air scour is also to be used, oil-free air should be provided if there are to be multiple backwashings. Air scour can be controlled through a needle valve–rotameter combination. The compressed air is usually delivered at a rate of 2 to 5 ft^3/min (0.9 to 2.4 L/s) for a period of 2 to 5 min before backwashing. If the pilot plant is to be set up at an existing filtration plant, there may be sources of backwash water and air available on site, but care should be taken to determine whether the available backwash water is chlorinated. The use of chlorinated water through GAC media can inactivate biological growth and reduce the adsorptive capacity of the carbon.

If several types or grades of media and media depths are to be investigated in the same filter column over time, several backwash outlets should be provided for different bed expansion depths. Sample taps should be provided according to the media configuration to be investigated. The taps should be installed to measure head loss at the media surface, at interfaces between media types, and in the bottom layer. Manometer tubes or pressure sensors can be used to measure the head loss through the filter column.

The backwash process cannot be simulated accurately enough for collecting design data when using 4 to 6 in. (10 to 15 cm) diameter filters. Filters with a 2 ft (60 cm) minimum diameter have been shown to be effective in simulating the backwashing process. However, at this size, large volumes of water are required to operate the filter and the collection of backwash solids requires several days of storage volume (Figure 28.6).

Ozonation Contactors. Ozone can be used for preoxidation, as well as for disinfection after clarification or after filtration. Ozone generation requires special equipment for providing dry compressed air, as well as the ozone generator itself (Figure 28.7).

Many sizes and types of ozone generators are available, and they are generally classified by the pounds of ozone produced. Most pilot plants can use generators that produce less than 2 lb/day (0.9 kg/day). The size of the generator can be determined by approximating the required dosage in advance and by knowing the process stream rate. These units are typically run with conventional 120 V, 60 Hz power requirements. Larger generators may require 220 to 480 V power.

Some generators come with desiccant air dryers built within their chassis. Others require independent air preparation equipment. Separate air compressors may be required ahead of the desiccant dryer if not provided as part of the unit. The air compressor should be able to provide oil-free air to the generator. Particulate filters and hydrocarbon filters should be installed between the compressor and ozone generator.

Small oxygen generators are also available for providing a clean oxygen stream, which allows an ozone generator to create more ozone than when using compressed air. They can produce oxygen streams that are up to 95% pure. The size of the air stream required can be approximated by knowing the percent weight ozone gas that can be produced by the generator and by computing the quantity of ozone required per day.

The required quantity of ozone feed-gas and off-gas must be determined in order to establish the transfer efficiency of the ozone contactor. Several types of ozone monitors are available for measuring the concentration of ozone in air, and they all work

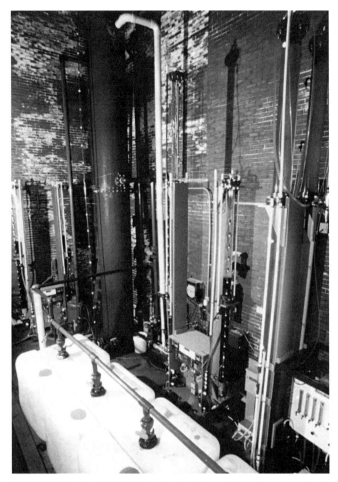

FIGURE 28.6 Manifolded backwash collection, settling and storage tanks in foreground. Large-scale filter (2 ft diameter) in background generates backwash solids. *(Source: Camp Dresser & McKee Inc.)*

primarily by measuring the adsorption of ultraviolet light by the ozonated air stream. Some monitors require a zero reference gas flow, as well as the ozonated gas flow. Rotameters are generally used to provide an accurate measurement of the volume of ozone gas directed to the monitor and to the feed point in order to establish the dosage and transfer efficiency. The rotameters must be made primarily of glass and stainless steel, and gasket material should be made of Teflon or Hypalon to resist attack by the ozone gas.

Measuring the quantity of off-gas concentration is relatively difficult for small pilot plant systems because most monitors require greater air flows than are typically provided to small ozone contactors for accurate measurement. Also, the off-gas contains high concentrations of water vapor, which can destroy or inhibit the optical materials inside a monitor. To minimize the adverse impacts to the monitoring system, the off-gas can be measured periodically instead of continuously.

FIGURE 28.7 Ozone disinfection module includes oxygen generator, ozone generator, contact columns, pumps, flow meters, residual and feed gas monitors, and power supply panel. *(Source: Camp Dresser & McKee Inc.)*

Ozone contactors for pilot plants with process flows less than 10 gpm (0.6 L/s) are typically made of clear PVC. Although PVC is attacked to some extent by ozone, most pilot studies are short term and intermittent, so the material will hold up for the duration of the test. The benefit of using clear PVC is the visibility it provides to the operator. The relative volume of gas and the size of the bubbles can be monitored, and a diffuser blockage can be noticed through visual inspection. Most contactors are designed to provide a countercurrent flow whereby the water is lifted to the top of the ozone contactor and allowed to fall, while the ozone gas is supplied to the base of the column and allowed to rise. This technique provides the necessary mixing action to induce the transfer of ozone into the liquid stream. Small porous-stone diffusers are generally used to disperse the ozone, and they can be attached to the bottom flange of the contactor.

Additional ozone contact time can be simulated by placing additional columns in series or by allowing the ozonated water to pass through one or more storage tanks. Efforts should be made to allow for smooth flow transitions between contact tanks because turbulence, hydraulic jumps, and free falls will drive ozone out of solution and reduce the contactor efficiency. In some cases a reduction of ozone residual may be required as part of the process. Excessive ozone residual, for example, can interfere with filtration and can consume chlorine. In such cases, the added turbulence can be used to reduce the ozone residual, and free falls of about 2 ft (0.6 m) have been shown to be effective in reducing ozone residual.

The contact time in an ozone contact tank can also be altered to change the contact time, but the variation in flow that can be achieved may be limited by minimum gas-to-liquid relationships and other process flow requirements. All ozone contactors should be covered and provided with an off-gas blower or fan to direct the vapors away from operators.

Sample ports should be provided throughout the contact tank system for the measurement of ozone residual. At a minimum, sample taps should be provided between each contact column or tank. In-line continuous ozone residual analyzers are available. Some work by continuously passing ozonated water across an ozone-specific membrane, and others work by agitating the ozonated water stream and stripping the ozone out of solution. The gas concentration is then measured by UV adsorption.

The depth of the water in the contact chamber will have a significant impact on the transfer of ozone into the water stream. Most full-scale plants use contactors that are 18 to 25 ft (5.5 to 7.6 m) deep. These depths may not be possible at the pilot plant scale, but the columns should be made as deep as possible.

In addition to the conventional bubble diffuser system, direct injection or sidestream ozone injection is available. These techniques may be more attractive where high ozone demands exist because the transfer is almost instantaneous.

Ozone can also be injected into the process stream through a venturi injector before a contact tank. A sidestream of approximately 10% of total flow is drawn off and pumped to a pressure of 20 to 35 psig (138 to 241 kPa) and then passed through an efficient, low-energy venturi injector, where a vacuum is formed. Ozone gas is drawn in by the vacuum of the injector, and the turbulence created by the flow creates a foam of ozonated gas and water. If the injector is mounted adjacent to the process flow, the foam can be returned to the process stream, where the mass transfer takes place.

Solids Collection and Treatment. Solids generated by the treatment of water have become a significant issue for most municipalities and water utilities. Pilot studies for the sole purpose of optimizing the solids treatment train are gaining popularity.

Proper pilot testing of mechanical thickening and dewatering of settled solids and backwash water solids require a significant amount of sludge, in the range of several hundred gallons, for a continuous run. In general, it is best to first see if there are vendors who will analyze solids, perform jar tests, and perform studies using mechanical dewatering equipment in their laboratories. They can also help in selecting the types of polymers and optimum dosages for thickening and dewatering. Some vendors also have small pressure vessels and membranes that can simulate dewatering methods.

It is important to not change the solids characteristic of sludge before it is tested. If solids generated from clarification, flotation, and backwashing must be pumped, positive displacement pumps should be used and operated at low speeds. The number of times the materials are transferred or moved should be minimized, and the material should be stored in a manner that will deter it from going anaerobic or becoming contaminated in any way.

On-site pilot testing of mechanical dewatering equipment is commonly used to establish design criteria. Solids from a full-scale plant are provided on a daily basis to operate a pilot-scale dewatering operation. Before beginning an on-site study, samples of the solids to be treated should be sent to the vendor whose equipment will be used for preliminary evaluation. This will minimize start-up time and operational errors on site. The transfer of the solids to the pilot plant and disposal of the solids and supernatant are usually the responsibility of the facility owner.

Lagoons, land drying, and freeze-thaw drying can be simulated at a fairly small scale in small, easily constructed drying beds. The beds should have drains installed to capture the filtrate and the sidewalls designed so that the volume reduction of the solids due to dewatering can be easily measured. Covers can be made of plywood for protection from rain and snow if necessary (Figure 28.8).

FIGURE 28.8 Water treatment residuals freeze-thaw beds with sand/gravel underdrain and leachate collection system included. Covers made of plywood are not shown. *(Source: Camp Dresser & McKee Inc.)*

PILOT PLANT CONSTRUCTION

Some of the options available for obtaining the facilities to run pilot studies are:

- The owner can design and build the pilot plant.
- The owner can design the plant and hire a contractor build it.
- A contractor can be engaged to both design and construct the plant (turnkey operation).
- An independent engineer can be hired to design the plant, with either in-house or contract construction.

- The owner can rent or lease portions of a plant, including specific unit processes that are difficult to build or are patented.
- A complete package pilot plant can be leased for the period of the pilot studies.

Much of the decision making is inevitably determined by the available money and time, as well as the end goals for the project.

Alternative Types of Pilot Plants

Pilot plants generally fall under two distinct categories of permanent and temporary units. Permanent pilot plants require the owner to purchase the pilot plant equipment and are more common with the construction of new water treatment plants. A permanent pilot plant offers the treatment plant operator the opportunity to fine-tune treatment techniques without adversely impacting the full-scale plant. Successful modifications identified through pilot tests provide the confidence to make the modification to the full-scale process. Without the pilot plant, operators commonly hesitate to recommend changes for fear of worsening the treatment.

The approach to permanent pilot plant construction differs significantly from temporary plant construction. Permanent pilot plants have to meet local building, plumbing, electrical, and fire codes. In addition, bidding laws may require extensive engineering services to provide design drawings and specifications before bids can be taken and the plant constructed.

In locating a permanent pilot plant, it is extremely important to identify the area and height requirements, source water and drainage availability, and electrical requirements. For instance, laying pipes along the floor surface is not usually acceptable for permanent applications. Any changes to the main treatment plant facility to accommodate the pilot plant should be determined to be part of the pilot plant cost.

Temporary pilot plants offer the owner the same opportunities to test treatment options, but usually at a fraction of the cost of building a permanent installation. Temporary pilot plants can be constructed in many ways. Modular pilot plants are gaining in popularity and make it easy for vendors to move their equipment from one site to another. Start-up time is reduced because the intricate connections move with the plant. The only on-site work is limited to source water piping, electrical supply modifications, drainage piping, and process-to-process piping.

The customer user can lease or rent the equipment for the time necessary to perform the evaluations, and when the tests are completed, the equipment can be removed and the area where it was located can be returned to its previous condition. Also, the use of modular plants more readily allows for process changes during the study compared with permanent installations.

Process Drawings and Specifications

Process drawings are required for every pilot plant installation. The number and type of drawings and detail required for the drawings vary with the type of plant selected and change according to how the plant is obtained.

At a minimum, a floor plan or site layout drawing should be created. This drawing identifies exactly where the process equipment is to be located, keeping in mind space and access requirements, the source water source, electrical supply, and available drainage (Figure 28.9). A process schematic or process and instrumentation diagram (PID) should also be created to provide the level of detail required for a complete operational pilot plant (Figure 28.10). The level

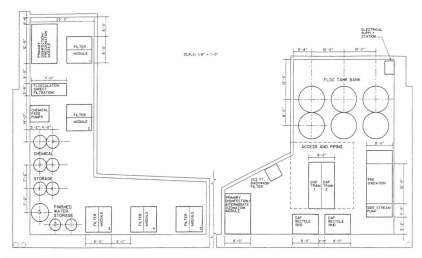

FIGURE 28.9 Pilot plant floor plan.

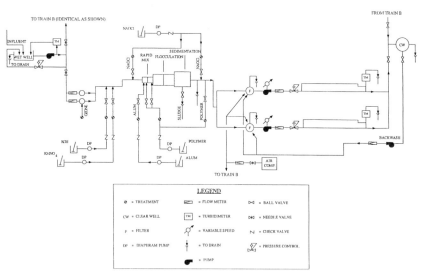

FIGURE 28.10 Pilot plant schematic of train A.

of detail provided by a process schematic cannot typically be provided on the floor plan. For simple plants, the schematic can include details of the mechanical process, electrical, instrumentation, and plumbing design criteria. More complicated plants may require individual function drawings.

An electrical single-line diagram is usually required to identify the numbers, types, and sizes of motor loads and other electrical demands, including those for instrumentation, controls, lighting, and lab equipment.

Specifications should also be produced for all pilot plants. These usually consist of a simple yet comprehensive design criteria table that lists the equipment and materials associated with each unit process.

A permanent pilot plant may require a full set of design drawings, including architectural, structural, mechanical, electrical, plumbing, and HVAC, along with a full set of specifications. If a permanent pilot plant is to be built within an existing facility, the existing conditions must be determined in detail so that the installation of new water lines, electrical conduit, and drain lines will not conflict with existing pipe and conduit. The structural limitations of floors may have to be considered when deciding on the location of large storage tanks or other heavy equipment.

Installation and Start-Up

When planning the installation of a pilot plant, the time required for installation and start-up is often either forgotten or underestimated. Careful and proper installation and start-up of the pilot plant will inevitably produce benefits during the pilot plant study. On the other hand, improper installation or start-up can cause many problems, including the collection of useless data.

The use of modular equipment reduces the installation and start-up efforts but does not eliminate it entirely. Site-specific constraints and minor process modifications can have major impacts. The use of equipment from different vendors, equipment built in different countries, and different construction materials, flow requirements, system pressures, and electrical demands are some of the problems that can cause installation nightmares.

Start-up time should include time for troubleshooting and repairs. Any weekend plumber knows how smart water can be when it comes to finding leaks. The shipping of equipment often produces some form of damage as well. Long periods of nonuse can also have an adverse impact on equipment.

Whenever possible, the people who are designated to operate the pilot plant should help with the installation and start-up. The contractors who install the plant often know the strengths and weaknesses of the equipment and materials that are provided, and this information can be valuable to the operator.

Operations and maintenance manuals for each piece of equipment and unit process should be available to the user for proper start-up, and these should be studied by the operators during the installation and start-up. Once the study has begun, time is usually limited and operators may not take the time to learn how to properly maintain the equipment.

Each unit process should be independently tested for proper hydraulic, electrical, and mechanical before testing the entire process train. Once all connections have been made, the entire system can be hydraulically tested by passing source water through it. Each chemical feed system should be calibrated under the flow and pressure conditions likely to be experienced before start-up, and preliminary jar tests can be conducted before start-up to identify likely chemical selections and dosages.

Consideration must also be given to the experience level of the installers and operator because this will affect the time of installation and start-up. When hiring engineers or contractors for the design, construction, and installation of a pilot plant, their experience with similar projects should be investigated before awarding the contract. Selection of the lowest bid with lowest experience can spell disaster in the long run.

BIBLIOGRAPHY

Thompson, J. C. "Overview of Pilot Plant Studies." Presented at the 1982 Annual Conference of the American Water Works Association, Denver, Colo.

American Society of Civil Engineers and American Water Works Association. *Water Treatment Plant Design.* 2nd ed. New York: McGraw-Hill, 1990.

American Water Works Association. *Water Quality and Treatment.* 4th ed. New York: McGraw-Hill, 1990.

Daniel, Philippe, Michael Zafer, and Paul Meyerhofer. *Effectively Planning Pilot Studies.* Walnut Creek, Calif.: Camp Dresser and McKee Inc., 1992.

APPENDIX A
PROPERTIES AND CHARACTERISTICS OF WATER TREATMENT CHEMICALS

ACTIVATED ALUMINA

See Aluminum oxide.

ACTIVATED CARBON

Activated carbon is a form of charcoal that has acquired the property of adsorbing various substances from water through treatment by a carefully controlled combustion process. It is available in two forms, powdered and granulated, and it has high affinity for adsorbing chlorine and taste- and odor-causing substances.

Activated carbon particles have a large surface area. The grade of activated carbon used for water treatment has a specific surface area ranging from 500 to 600 m^2/g. The particles appear to be solid, but are actually honeycombed with minute "tunnels" or pores on the molecular order of size. Pore dimensions are expressed in angstroms (Å). The materials responsible for taste and odor in water, such as the products of industrial waste, sewage, and plant and animal organisms, are thought to be adsorbed in pores not greater than 20 Å in diameter. In the highly activated carbon, about 75% of the surface exists as pores less than 20 Å in diameter.

Activated carbon is prepared from coal, wood, coconut char, petroleum coke, lignite, charred peach pits, and other carbonaceous materials. The carbons used in the water industry are prepared principally from paper char, hardwood charcoal, or lignite.

Granular Activated Carbon

Granular activated carbon (GAC) is used by passing the water through a relatively thick bed of carbon. Common applications are by replacing some or all of the granular media in a filter or using a separate tank filled with the carbon. It can be ground and screened to any desired size. In the size used in some water treatment plants, 100% will pass an 8-mesh screen, and at least 90% will be retained on a 12-mesh screen. Other sizes commonly available have effective sizes of 0.8 to 0.9 mm and 0.55 to 0.65 mm. GAC is covered in AWWA Standard B604.

Powdered Activated Carbon

Powdered activated carbon (PAC) is a very finely ground material; more than 90% passes a 300-mesh screen. PAC is used by adding it to the water to be treated at a point where it will be subsequently removed by sedimentation and filtration. PAC is covered in AWWA Standard B600.

ALUMINUM OXIDE

Aluminum oxide (Al_2O_3) is also known as activated alumina. It is a highly porous, granular material with a preferential adsorptive capacity for moisture from gases, vapors, and some liquids. One of the principal uses in the waterworks field is for removal of excess levels of fluorides from drinking water by percolating the water through beds of alumina. The beds can be arranged as either pressure or gravity filters, and piped for backwashing and regeneration.

When the alumina becomes saturated with fluorides, it must be regenerated. This is accomplished by first backwashing with water in order to remove the accumulated solids, and then backwashing with a weak caustic solution to remove fluorides. The residual caustic is then neutralized with a weak acid, followed by water rinses.

Activated alumina is available in granules ranging in size from a powder to approximately 1.5 in. (38 mm) in diameter. The 14 to 8 and the 8 to $\frac{1}{4}$ U.S. Standard Sieve mesh are the sizes most generally used.

ALUMINUM SULFATE

By far the most common coagulating agent used for water treatment is aluminum sulfate (alum). Most commercial grades of alum have the formula $Al_2(SO_4)_3 \cdot 14H_2O$. It is available in lump, ground, or liquid form. Ground aluminum sulfate for use in dry feed machines should be of such size that not less than 90% will pass a National Bureau of Standards no. 10 sieve, and 100% will pass a no. 4 sieve. It is a grayish-white crystalline solid completely soluble in water with a tendency to absorb moisture from the air. Under such conditions, it tends to deliquesce to a fine white powder. This does not affect its efficiency but requires a slight change in the chemical feed rate.

Liquid alum is a clear, amber-colored liquid sometimes called 50% alum. This is because a gallon of liquid alum weighs 11.2 lb (0.5 g) and contains 5.4 lb (0.2 g) of dry aluminum sulfate. Actually, it usually contains 8.5% or more available water-soluble alumina (Al_2O_3), as compared with the 17% Al_2O_3 available in dry alum. Alum is covered in AWWA Standard B403.

AMMONIA, ANHYDROUS

Ammonia is used in water treatment to add to chlorinated water so that chloramines will be formed. In the gaseous state, ammonia is colorless and about 0.6 times as heavy as air.

The liquid, also colorless, is about 0.68 times as heavy as water. Unconfined liquid ammonia rapidly vaporizes to gas. The temperature, pressure, and density characteristics of anhydrous ammonia are shown in Figure A.1.

AMMONIA, AQUA

See Ammonium hydroxide.

AMMONIUM ALUMINUM SULFATE

Also referred to as ammonium alum and crystal alum, ammonium aluminum sulfate has the form of dry, colorless, and odorless crystals or white powder. Although it has many industrial applications, it is not often used in water treatment because it has coagulation properties similar to alum but is more expensive. A characteristic unique to alum, though, is the residual ammonium ion after the coagulation process, which can be used to form chloramines.

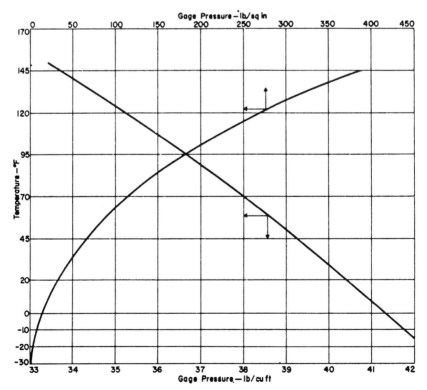

FIGURE A.1 Relationship of pressure–temperature to density–temperature in anhydrous ammonia.

AMMONIUM HYDROXIDE

Aqua ammonia is the ammonium hydroxide of commerce. At 60° F, it consists of a stable solution of 29.4% NH_3, with a specific gravity of 0.8974 and a density of 7.48 lb/gal (0.9 g/ml). Its boiling and freezing points are 84° F and −107° F, respectively. Its vapor pressure varies from about 4.5 lb/in.^{2} absolute (psia) (31 kPa) at 32° F (0° C) to 50 psia (344.5 kPa) at 104° F (40° C).

AMMONIUM SULFATE

Like ammonia, ammonium sulfate is sometimes used as a source of ammonia for the formation of chloramines for disinfection. Because it can vary widely in purity, purchase specifications regarding quality should be carefully written and should conform to the AWWA standard. Being a manufacturing by-product, it should be carefully examined for the presence of heavy metals. It is a white, crystalline solid that is readily soluble in water. Ammonium sulfate is covered in AWWA Standard B302.

BENTONITE

Bentonite is a form of clay that swells on absorbing moisture. It is sometimes used as a coagulant aid in waters of very low turbidity to create a nucleus around which floc will form and to add weight to the floc. Many variations of bentonite are available, depending on their source, but the primary constituents are silicone and aluminum oxides.

CALCIUM CARBONATE

Calcium carbonate is also known as unburned lime, limestone, calcite, whiting, chalk, and precipitated chalk. It is widely used as an agricultural lime and is also used in neutralization, stabilization, or corrosion control in water treatment. The chemical is commercially available in a number of forms, both prepared and naturally occurring. It is only slightly soluble in water. Calcium carbonate is also the main precipitant generated as a by-product in lime softening.

CALCIUM CHLORIDE

Calcium chloride ($CaCl_2$) is manufactured as white, odorless, deliquescent flakes or pellets with a minimum strength of 77%. It is used in water treatment as a coagulant and sludge conditioner. Calcium chloride is covered in AWWA Standard B550.

CALCIUM HYDROXIDE

See Lime.

CALCIUM HYPOCHLORITE

Calcium hypochlorite is an off-white chemical that is furnished in granular, free-flowing, or compressed-tablet form. Commercial high-test calcium hypochlorite products, such as HTH, contain at least 70% available chlorine and have from less than 3% up to about 5% lime residual. All commercial calcium hypochlorites also contain minor amounts of other impurities, such as calcium carbonate and other insoluble substances. In hard water, an additional amount of calcium carbonate is formed.

Although a highly active oxidizer, calcium hypochlorite is relatively stable. Under normal storage conditions, commercial preparations lose about 3% to 5% of their available chlorine content in a year. But contact with water or the atmosphere induces a pronounced increase in the decomposition rate and greatly increases the reaction rate with organic materials.

Decomposition is exothermic and proceeds rapidly if any part of the material is heated to 350° F (176° C), yielding oxygen and chlorine, as well as a powdery dust that has an irritating action. Heat is also evolved, which further supports and increases the decomposition rate. The granular form is essentially nonhydroscopic and resists moist caking tendencies when properly stored. Though calcium hypochlorite is a stable, nonflammable material that cannot be ignited, contact with heat, acids, combustible, organic, or oxidizable materials may cause fire.

Granular calcium hypochlorite is readily soluble in water, varying from about 21.5 g/100 ml at 32° F (0° C) to 23.4 g/100 ml at 104° F (40° C). Tablet forms dissolve more slowly than the granular materials and can provide a fairly steady source of available chlorine over an 18- to 24-hour period. Small tablets are often used for disinfection of new water mains. Large tablets are commonly used in swimming pool and sewage disinfection. Hypochlorites are covered in AWWA Standard B300.

CALCIUM OXIDE

See Lime.

CARBON

See Activated carbon.

CARBON DIOXIDE

Carbon dioxide (CO_2) gas is commonly used in water treatment to lower pH and stabilize lime-softened water. Once stabilized, further precipitation of soluble calcium carbonate is less likely to occur. In the past, CO_2 was commonly produced on site by the burning of carbonaceous material such as natural gas. It is now typically manufactured at an industrial plant and delivered to the treatment plant, where it is stored in liquid form in a refrigerated tank under pressure.

Carbon dioxide is a clear, colorless, and odorless gas with a high specific gravity. When confined, carbon dioxide tends to seek the lowest level possible. In water, its solution produces carbonic acid. Carbon dioxide functions as an aid to coagulation, primarily by adjusting the pH or solubilities so that proper coagulation is obtained. Carbon dioxide is covered in AWWA Standard B510.

CHLORINATED LIME

Chlorinated lime is sometimes referred to as bleaching powder. It has been used as a disinfectant in treating water supplies and more particularly for swimming pools. It is a white powder prepared by chlorinating slaked lime and decomposes in water, releasing 39% available chlorine for disinfecting action.

CHLORINE

In the gaseous state, chlorine is greenish yellow in color and about 2.48 times as heavy as air. The liquid is amber colored and about 1.44 times as heavy as water. Unconfined liquid chlorine rapidly vaporizes to gas with 1 volume of liquid yielding about 450 volumes of gas.

Chlorine is only slightly soluble in water, its maximum solubility being approximately 1% at 50° F (10° C). At lower temperatures, chlorine combines with water forming chlorine "ice," a crystalline hydrate ($Cl_2 \cdot 8H_2O$).

The temperature-pressure characteristics of chlorine are shown in Figure A.2. Chlorine confined in a container may exist as a gas, a liquid, or a mixture of both. Thus any consideration of liquid chlorine includes consideration of gaseous chlorine. The vapor pressure of chlorine in a container is a function of temperature and is independent of the contained volume of chlorine; therefore gauge pressure does not necessarily indicate its state.

The volume-temperature characteristics of chlorine in a container loaded to the limit authorized by Department of Transportation (DOT) regulations are shown in Figure A.3. If a container is filled to this limit, it becomes completely full of liquid at approximately 154° F (67° C) and at temperatures beyond that point, excessive pressure that could result in rupture of the container. Safety devices are provided on cylinders and containers to relieve the excessive pressures that accompany excessive temperature elevations.

Under specific conditions, chlorine reacts with most elements, sometimes with extreme rapidity. The reaction with hydrocarbons, alcohols, and ethers can be explosive. Although chlorine itself is neither explosive nor flammable, it is capable of supporting combustion of certain substances. It should, accordingly, be handled and stored away from other compressed gases (such as anhydrous ammonia) and kept apart from turpentine, ether, finely divided metals, and hydrocarbons or other flammable materials. Chlorine is covered in AWWA Standard B301.

CHLORINE DIOXIDE

Under atmospheric conditions, chlorine dioxide is a yellow to red, unpleasant-smelling, irritating, unstable gas. It is produced by mixing a chlorine solution (chlorine gas dissolved in water or sodium hypochlorite and hydrochloric acid) with a solution of sodium chlorite (typically 25%). Because chlorine dioxide decomposes very quickly, it is usually generated on site before its application.

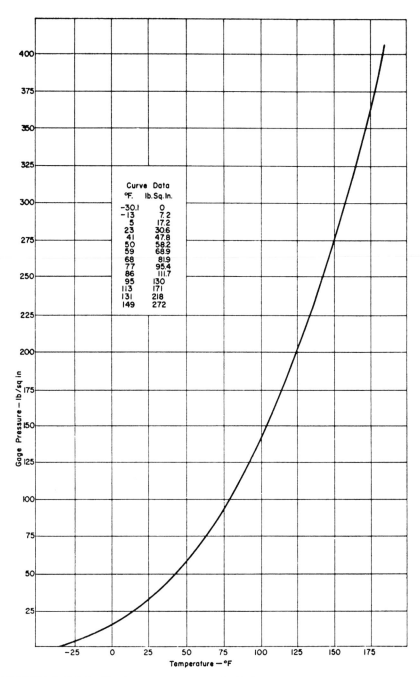

FIGURE A.2 Relationship of temperature to pressure in chlorine.

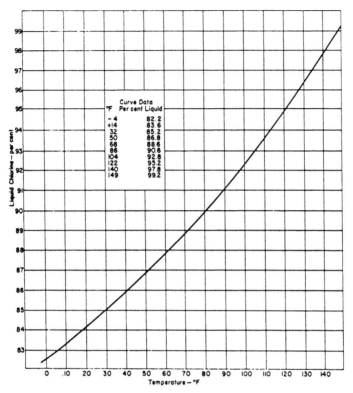

FIGURE A.3 Relationship of volume to temperature in chlorine.

To generate chlorine dioxide, the chlorine and sodium chlorite solutions are injected and mixed together and passed through a media-packed reaction column to promote mixing and a near-complete reaction (greater than 95%) to generate chlorine dioxide. Various types and designs of chlorine dioxide generators are manufactured. This material is a powerful oxidizing agent and disinfectant, used primarily in reducing or eliminating tastes and odors.

COPPER SULFATE

Copper sulfate is available in blue crystalline granules or powder. It is used most often in water treatment to control algae in impounding reservoirs. It is sometimes applied as a powder, distributed by dusting on the surface of the water. Another technique is to drag burlap bags of copper sulfate crystals through the water, effecting solution and distribution at the same time. Another method of application is to meter the copper sulfate solution into a narrow control channel at the entrance to the impounding reservoir. This chemical and its solution are poisonous and corrosive. Copper sulfate is covered in AWWA Standard B602.

DISODIUM PHOSPHATE

See Phosphates.

EPI-DMA POLYAMINES

Epichlorohydrin dimethylamine (EPI-DMA) polyamines are polyquaternary polymers produced from the reaction of dimethylamine and epichlorohydrin. Polymers may contain residual amounts of these chemicals and their by-products, so the chemical should be tested for impurities. The product is shipped in drums or bulk as an aqueous solution typically ranging from 10% to 65% by weight active polymer. EPI-DMA polyamines are covered in AWWA Standard B452.

FERRIC CHLORIDE, LIQUID

Ferric chloride is commonly used as a coagulant in potable water, sewage, and industrial waste treatment. It is usually prepared as a liquid of 37% to 40% solution, and is shipped in 55 gal (208 L) drums or 4,000 to 8,000 gal (15,000 to 30,000 L), rubber-lined tank trucks. Ferric chloride is fed about the same as alum, but because it is more corrosive than alum, special handling materials must be used. Ferric chloride is covered in AWWA Standard B407.

FERRIC SULFATE

Ferric sulfate is produced to some degree when ferrous sulfate is chlorinated. It is available on the chemical market as a reddish-brown deliquescent solid. Ferric sulfate is soluble in water, producing a somewhat acidic solution. Ferric sulfate is covered in AWWA Standard B406.

FERROUS SULFATE

The combination of ferrous sulfate and lime forms an effective coagulant for the clarification of turbid water. Ferrous sulfate itself is a greenish-white crystalline solid that is obtained as a by-product of other chemical processes.

The AWWA standard for ferrous sulfate gives size standards for both its granular and lump forms. The standard-size granular ferrous sulfate has a tendency to cake and arch in dry feeder hoppers or storage bins. Material of a finer crystal size that contains only five molecules of water crystallization ($FeSO_4 \cdot 5H_2O$) may be obtained. This material has less tendency to cake or arch. Ferrous sulfate is also available in a liquid form. Ferrous sulfate is covered in AWWA Standard B402.

FLUOROSILICIC ACID

Fluorosilicic acid (formerly called hydrofluosilicic acid) is commonly fed at a very low dose rate to drinking water supplies to increase the fluoride level in the water. Pure acid is a colorless, transparent, fuming, corrosive liquid, with a pungent odor and an irritating action on the skin. The commercial product used in water treatment is a 20% to 30% water solution of H_2SiF_6 that has a straw or slightly reddish color owing to dissolved iron. In the amounts used in water fluoridation, the iron presents no problem. Fluorosilicic acid is covered in AWWA Standard B703.

HYDRATED LIME

See Lime.

HYDROCHLORIC ACID

Hydrochloric acid is a clear or slightly yellow fuming, pungent liquid. It may contain iron or arsenic and is poisonous. If it is to be used in a potable water supply, care should be taken to obtain the purified form (USP). To reduce the fuming characteristics, it should be diluted (by adding the acid to water) to approximately 20% HCl. It is not often used in the treating of potable water supplies because it increases the chlorine content of the water.

HYDROFLUOSILICIC ACID

See Fluorosilicic acid.

HYDROGEN PEROXIDE

Hydrogen peroxide is an aqueous solution prepared in solution strengths ranging from 30% to 70%. In water treatment it is used as a catalyst with ozone for oxidation. It is shipped in 6 to 55 gal (23 to 208 L) polyethylene drums, tank trucks, and railroad tank cars.

HYPOCHLORITES

See Calcium hypochlorite and Sodium hypochlorite.

LIME

The most common forms of lime in water treatment are quicklime and hydrated lime. Other forms of lime may also be found in specific areas. Quicklime and hydrate lime are covered in AWWA Standard B202.

Quicklime

Quicklime (CaO) results from the calcination of limestone or an equivalent, such as dried water-softening sludge of suitable analysis, and consists essentially of calcium oxide in natural association with a lesser amount of magnesium oxide. The revised AWWA standard for quicklime and hydrated lime gives detailed recommendations on impurities and particle sizes. It recommends a minimum of 90% available calcium oxide content. Quicklime is available in any particle size required by a given installation, and size is usually specified in each contract. Quicklime has some tendency to air-slake, so exposure to outside air should be kept to a minimum.

Hydrated Lime

Hydrated lime, $Ca(OH)_2$, is a very finely divided powder resulting from the hydration of quicklime with enough water to satisfy its chemical affinity. Commercial hydrated lime consists essentially of calcium hydroxide or a mixture of calcium hydroxide and magnesium hydroxide, depending on the type of quicklime slaked. Hydrated lime is a powder that should be uniform in size and should have a minimum available calcium oxide content of 68.0%. Hydrated lime tends to absorb carbon dioxide from the atmosphere, even in the slurry form; therefore its exposure to outside air should be carefully controlled.

Dolomitic Hydrated Lime

This lime is produced by hydrating dolomitic quicklime. When normal hydrating conditions are present, only the calcium oxide is hydrated and the magnesium oxide remains unchanged. This is referred to as monohydration.

Dolomitic Lime

Dolomitic lime contains from 35% to 40% magnesium oxide. When the magnesium content exceeds the calcium content, the chemical is terminated *dolomitic magnesite*. If the magnesium oxide is greater than 5% but less than 35%, it is termed *magnesia quicklime*.

MONOSODIUM PHOSPHATE

See Phosphates.

OXYGEN, LIQUID

Liquid oxygen is pale blue, extremely cold, and nonflammable. It is a strong oxidizer, and as such, supports combustion. Oxygen will react with nearly all organic materials and metals. Materials that burn easily in air usually burn vigorously in oxygen. In water treatment pure oxygen is used to generate a higher concentration of ozone that can be generated with prepared air.

OZONE

Ozone is a faintly blue gas with a pungent odor. It is an unstable form of oxygen composed of three-atom molecules that break down readily to normal oxygen and nascent oxygen. The latter is a powerful oxidizing agent and has germicidal action. Ozone is usually produced with on-site generators by passing high-voltage electricity through dry atmospheric air or pure oxygen between stationary electrodes. This process converts a small percentage of the oxygen in the air into ozone. It is usually injected into the water to be treated in a highly baffled mixing chamber. The ozone residual is determined by use of the indigo blue test.

PHOSPHATES

Many different forms of phosphates are marketed for corrosion control, preventing scale formation, and sequestering unwanted precipitants. Generic forms of these compounds and compound groups include sodium hexametaphosphate, tricalcium phosphate, monosodium phosphate, disodium phosphate, trisodium phosphate, sodium tripolyphosphate, polyphosphates, zinc orthophosphates, bimetallic zinc orthophosphates, and silicate phosphates. These products are available in both liquid and powder forms. They often have a very low pH requiring special care in handling and storage. Phosphates are covered in AWWA Standards B502, B503, B504, and B505.

PHOSPHORIC ACID

Phosphoric acid (H_3PO_4) is a colorless and odorless water-white liquid used for corrosion and pH control. It is shipped in polyethylene drums, railroad tank cars, and tank trucks at concentrations ranging from 70% to 85%.

POLYALUMINUM CHLORIDE

Polyaluminum chloride (PAC) is a clear, pale amber liquid used in place of an inorganic coagulant. It is shipped in bulk and 55 gal drums. Polyaluminum chloride is covered in AWWA Standard B408.

POLYDADMAC

Polydiallyldimethylammonium chloride (polyDADMAC) is a synthetic cationic organic polyelectrolyte used as a coagulant. PolyDADMAC is usually sold as an aqueous solution between 10% and 50% by weight of active polymer in drums or bulk. PolyDADMAC is covered in AWWA Standard B451.

POLYMERS

Polymers are synthetic organic chemicals typically possessing high molecular weight. They are available in positively charged (cationic), negatively charged (anionic), or neutral (near zero net charge) forms and are manufactured as both powders and liquids. They are used as primary coagulants, coagulant aids, flocculants, filter aids, filter backwash water conditioning, and solids thickening and dewatering. Compared with other chemicals routinely used in water treatment, polymer dosages often are much lower.

When a polymer in powder form is to be used, a stock solution must first be prepared. Some liquid polymers must also be prepared in a stock solution before using, although other liquid products can be fed directly to the process with dilution water. In preparing the stock solutions, warm water up to 100° F (38° C) should be used to hasten the dissolving process, and a relatively long mixing time is usually recommended.

To protect the effectiveness of the polymer, there are typically upper and lower limits to the strength of the solution. Once the stock solution is ready to apply, it is further diluted to help ensure complete mixing with the process water. Because polymers are highly variable in their performance on various types of water, bench or pilot scale studies are necessary to make an appropriate selection.

POTASSIUM HYDROXIDE

Potassium hydroxide, also known as potash and caustic potash, is used in lime softening and for pH adjustment. It is available in both solid and liquid forms. As an unground solid it is shipped in 700 to 725 lb (318 to 329 kg) steel drums. Flaked, ground, crystalline, or beaded materials are shipped in drums, kegs, barrels, polyethylene-lined bags, or fiberboard drums. The liquid form typically contains 45% to 52% by weight and is shipped in carboys, steel drums, tank trucks, and railroad tank cars.

Considerable heat is generated when this product is mixed with water. As a result, when making solutions, protective clothing is required, and the material should only be added to water with constant stirring. Potassium hydroxide is covered in AWWA Standard B511.

POTASSIUM HYPOCHLORITE

Potassium hypochlorite is also referred to as potassium hypochlorite bleach. It has similar properties and is used as a substitute to sodium hypochlorite when the use of sodium must be avoided or minimized. As a stock solution, it is typically prepared at 10% strength. Heat, sunlight, and certain metallic contamination will affect the hypochlorite stability, so the solution should be stored away from heat and sunlight in containers of appropriate material.

POTASSIUM PERMANGANATE

This material is in the form of black or purple crystals or pellets with a blue metallic sheen. It is highly soluble in water, which allows easy application. It is usually prepared in dilute solution (1% to 4%) as needed for application. Potassium permanganate is covered in AWWA Standard B603.

QUICKLIME

See Lime.

SODA ASH

See Sodium carbonate.

SODIUM ALUMINATE

Sodium aluminate is sometimes used as an auxiliary coagulant for the removal of fine turbidity and color in soft, low-pH water. The solid form is a white or brown powder containing 70% to 80% $Na_2Al_2O_4$. The solution form is a concentrated solution containing approximately 32% $Na_2Al_2O_4$. This material is readily soluble in water, producing a noncorrosive solution. Sodium aluminate is covered in AWWA Standard B405.

SODIUM BISULFITE

Sodium bisulfite, also called sodium metabisulfite, is used as a dechlorination agent. The principal constituent of this product is sodium pyrosulfite, or sodium metabisulfite ($Na_2S_2O_5$). These white, crystalline powders are readily soluble in water. Sodium metabisulfite is covered in AWWA Standard B601.

SODIUM CARBONATE

Sodium carbonate, commonly called soda ash (Na_2CO_3), is commonly used in the water softening process. It is a white powder or granular material containing well above 99% sodium carbonate, or 58% as a sodium oxide equivalent. Although the product is available in two different grades, light and dense, they differ only in bulk density, size, and appearance. Quality and other properties are the same. Packaging is in 50 and 100 lb (23 to 46 kg) bags and as bulk in tank trucks and rail cars. Soda ash is covered in AWWA Standard B201.

SODIUM CHLORIDE

Sodium chloride is used in water treatment to recharge cation resins of ion exchange treatment units. The AWWA standard for sodium chloride requires that the material be homogeneous and in a dry granular form, white, grayish white, pink, brown, or brownish white in color. The size requirements state that rock salt must be of such fineness that it passes a no. 3 sieve and 95% of it is retained on a no. 7 sieve. Evaporated salt must be of such fineness that at least 85% is retained on a no. 7 sieve. The material must also dissolve rapidly without packing. The solution formed by dissolving the salt in distilled water is to have a phenolphthalein alkalinity of zero and a hydrogen-ion concentration (pH) no higher than 8.0. Sodium chloride is covered in AWWA Standard B200.

SODIUM CHLORITE

Sodium chlorite is a dry, flaked salt with a powerful oxidizing nature. It is stable when sealed or in solution, but it is highly combustible in the presence of organic material. Technical-grade sodium chlorite (approximately 80% dry weight $NaClO_2$) is an orange-colored flaked salt that is a powerful oxidizing agent.

Sodium chlorite in contact with acid will react rapidly to evolve chlorine dioxide gas. When heated above 347° F (175° C), sodium chlorite will decompose rapidly, liberating oxygen and evolving sufficient heat to make the decomposition self-sustaining. If this decomposition is confined, as in a closed container, the effect is explosive. Therefore it should be protected at all times from exposure to heat.

Sodium chlorite dissolves easily in water at ordinary temperatures to form an orange-brown solution that is chemically stable under ordinary conditions of temperature and pressure. Sodium chlorite is available from suppliers in dry form or as a solution. Because of the hazardous nature of dry sodium chlorite, liquid solutions are the preferred chemical form. For most applications, sodium chlorite solutions of 25% by weight should be used. This concentration has been selected by the industry because it has the lowest freezing point of sodium chlorite solutions, 5° F ($-15°$ C). If dry sodium chlorite is to be used, special precautions are required for storage, handling, dust control, and worker safety.

For most chlorine dioxide generation systems, a low-alkalinity (<10 g/L as $CaCO_3$) sodium chlorite should be used. Sodium chlorite is covered in AWWA Standard B303.

SODIUM FLUORIDE

Sodium fluoride is produced by neutralizing hydrofluoric acid with either sodium carbonate or sodium hydroxide. It is available as crystals or white powder. It is highly soluble in water, which enhances its use in the fluoridation of water supplies. Sodium fluoride is covered in AWWA Standard B701.

SODIUM HEXAMETAPHOSPHATE

This is one of several forms of what are known as glassy phosphates. It is used in water softening and boiler water treatment and is available from manufacturers in both powdered and granular form. It is also used as a sequestering agent in municipal water supplies and is highly soluble in water. See Phosphates and Sodium polyphosphate.

SODIUM HYDROXIDE

Sodium hydroxide (NaOH) is also known as caustic soda. It is available as a solid in flake form, but more commonly in liquid solution at a concentration of 50% as NaOH. The specific gravity of 50% liquid caustic is 1.525, or a density of 12.7 lb/gal (1,522 g/L). Because it is a powerfully alkaline corrosive chemical, extreme care must be used when handling caustic soda. Body contact will result in immediate burning of tissue. Sodium hydroxide is covered in AWWA Standard B501.

SODIUM HYPOCHLORITE

Commercial sodium hypochlorite, or liquid bleach, usually contains 12.5% to 17% available chlorine at the time of manufacture and is available only in liquid form. It is typically used for disinfection at smaller water treatment plants, where use of chlorine gas is a safety concern, and for swimming pool water disinfection.

All NaOCl solutions are unstable to some degree and deteriorate more rapidly than calcium hypochlorite. The effect can be minimized by care in the manufacturing processes, controlling the alkalinity of the solution, and storing away from heat and sunlight. Greatest stability is attained with a pH close to 11.0 and in the absence of heavy metal cations. Hypochlorites are covered in AWWA Standard B300.

SODIUM METABISULFITE

See Sodium bisulfite.

SODIUM POLYPHOSPHATE

See Sodium hexametaphosphate.

SODIUM SILICATE

Sodium silicate (liquid) in the form known as activated silica is used as a coagulant aid. The chief advantage of activated silica is its ability to toughen the floc. Activated silica may be prepared in a number of ways. Sodium silicate solution is available as a 40° Baumé liquid containing approximately 30% SiO_2. Sodium silicate is covered in AWWA Standard B404.

SODIUM FLUOROSILICATE

Sodium fluorosilicate was previously called sodium silicofluoride in the United States, but the name has been changed to conform with international usage. This material is a white, free-flowing, odorless, crystalline powder. It is produced by neutralizing hydrofluosilicic acid with soda ash (Na_2CO_3) or sodium hydroxide (NaOH) and evaporating the resulting solution. Sodium silicofluoride is the most extensively used compound in the fluoridation of water supplies because of its availability and low cost. Its disadvantages lie in its very low solubility, 60 gal of water being required to dissolve 1 lb of sodium fluorosilicate. Sodium fluorosilicate is covered in AWWA Standard B702.

SODIUM SULFITE

Sodium sulfite is similar in appearance and use to sodium bisulfite.

SODIUM TRIPOLYPHOSPHATE

See Phosphates.

SULFUR DIOXIDE

Sulfur dioxide is used in water treatment for dechlorination. It is produced in North America by the combustion of sulfur in special burners, by burning pyrites, or as a by-product of smelting operations. In the gaseous state, sulfur dioxide is colorless and about 2.26 times as heavy as air. The liquid, also colorless, is about 1.44 times as heavy as water. Unconfined liquid sulfur dioxide rapidly vaporizes to gas.

Sulfur dioxide is about 20 times as soluble as chlorine in water. At 32° F (0° C), up to 20% by weight will dissolve to form a weak solution of sulfurous acid. Because its vapor pressure increases with increasing temperature, its solubility decreases with increasing temperature; at 80° F (27° C) it is soluble to less than 10%. Sulfur dioxide is covered in AWWA Standard B512.

SULFURIC ACID

Sulfuric acid (H_2SO_4) is a strong, corrosive, dense, oily liquid, lightly colored or dark brown, depending on purity. For use in water treatment it should be of the USP grade and free of heavy metals. It is available in a number of grades, containing from 60% to 90% H_2SO_4.

QUALITY OF PURCHASED CHEMICALS

Chemicals can be purchased in several grades. Contaminants from manufacturing or by-products can be present in poorer-quality grades. The presence of some contaminants, such as heavy metal ions, could cause water quality problems in finished water or in the disposal of treatment plant residuals. The presence of iron, nickel, copper, cobalt, and others will cause sodium hypochlorite to decompose at a faster rate. For that reason, appropriate specifications for chemical quality should be used for all purchases. AWWA has prepared standards for purchasing certain water treatment chemicals to help ensure that quality is acceptable. Table A.1 provides an alphabetical listing of commonly used water treatment chemicals. Table A.2 lists the process use of common water treatment chemicals.

TABLE A.1 Alphabetical Listing of Commonly Used Water Treatment Chemicals

Chemical name and formula	Common or trade name	Shipping containers	Suitable storage materials	Available forms/descriptions	Density	Solubility, lb/gal	Commercial strength, percent	Additional characteristics and properties
Activated carbon, powdered carbon	Aqua Nuchor, Hydrodarco, Herite	Bags, bulk	Dry: iron, steel; wet: rubber and silicon linings, type 316 stainless steel	Black granules, powder	15 to 30 lb/ft^3	Insoluble (suspension used)		1 lb/gal suspension used for storage and handling
Aluminum oxide, Al_2O_3	Activated alumina	Bags, drums	Iron, steel	Powder granules (up to 1 in. in diameter)		Insoluble	100	
Aluminum sulfate, $Al_2(SO_4)_3 \cdot 14H_2O$ (dry)	Alum, filter alum, sulfate of alumina	100 to 200 lb bags, 300 to 400 lb barrels, bulk (carloads), tank truck, 228-36 tank car	Dry: iron, steel; wet: stainless steel, rubber, plastic	Ivory colored; powder, granule, lump	38 to 45 lb/ft^3 60 to 63 lb/ft^3 62 to 67 lb/ft^3	6.2(60° F)	17 as Al_2O_3 dry	pH of 1% solution: 3.4
Aluminum sulfate (liquid)	50% alum	Tank cars and tank trucks	FRP†, PE‡; type 316 stainless steel, rubber linings	Liquid	11.2 lb/gal	—	8.5 as Al_2O_3	Freezing point _4_F
Ammonium aluminum sulfate, $Al_2(SO_4)_3(NH_4)_2$–$SO_4 \cdot 24H_2O$	Ammonia alum, crystal alum	100 lb bags, barrels, bulk	FRP, PE, type 316 stainless steel, rubber linings	Colorless crystals or white powder	65 to 75 lb/ft^3	0.3 (32° F)	99	pH of 1% solution: 3.5
Ammonium hydroxide, NH_4OH	Ammonia water, ammonium hydrate, aqua ammonia	Carboys, 750 lb drums, bulk	Glass lining, steel, iron, FRP, PE	Colorless liquid	7.48 lb/gal	Complete	29.4 (NH3) max 26° Baumé	pH 14; Freezing point = −107° F
Ammonium silicofluoride, $(NH_4)_2SiF_6$	Ammonium fluorsilicate	100 and 400 lb drums	Steel, iron, FRP, PE	White crystals	65 to 70 lb/ft^3	1.7 (63° F)	100	White, free-flowing solid
Ammonium sulfate, $(NH_4)_2SO_4$	Sulfate of ammonia	50 and 100 lb bags, 725 lb drums	FRP, PE, ceramic and rubber linings; iron (dry)	White or brown crystal	70 lb/ft^3	6.3 (68° F)	>99	Cakes in dry feed; add $CaSO_4$ for free flow

TABLE A.1 Alphabetical Listing of Commonly Used Water Treatment Chemicals

Chemical	Other names	Shipping containers	Available forms	Weight	Solubility	Purity (%)	Remarks
Anhydrous ammonia, NH_3	Ammonia	50, 100, 150 lb cylinders, bulk tank cars, and trucks	Iron, steel, FRP, PE	Colorless gas	38.6 lb/ft³	3.9 (32° F) 3.1 (60° F)	99.9+ (NH_3)	
Bentonite	Colloidal clay, volclay, wilkinite	100 lb bags, bulk		Powder, pellet, mixed sizes	60 lb/ft³	Insoluble (colloidal solution used)		Free flowing, nonabrasive
Calcium fluoride, CaF_2	Fluorspar	Bags, drums, barrels, hopper cars, trucks	Steel, iron, FRP, PE	Powder		Very slight	85 (CaF_2), less than 5 (SiO_2)	
Calcium hydroxide, $Ca(OH)_2$	Hydrated lime, slaked lime	50 lb bags, bulk	FRP, PE, iron, steel, rubber lining	White powder, light, dense	28 to 36 lb/ft³	0.14 (68° F) 0.12 (90° F)	85 to 99 ($Ca(OH)_2$) 63 to 73 (CaO)	Hopper agitation required for dry feed of light form
Calcium hypochlorite, $Ca(OCl)_2 \cdot 4H_2O$	HTH, perchloron, pittchlor	5 lb cans; 100, 300, 800 lb drums	Glass, plastic, and rubber linings, FRP, PE	White granule, powder, tablet	52.5 lb/ft³	1.5 at 25° C	65 (available Cl_2)	1 to 3 (available Cl_2 solution used)
Calcium oxide, CaO	Burnt lime, chemical lime, quicklime, unslaked lime	80 and 100 lb bags, bulk	FRP, PE, iron, steel, rubber	Lump, pebble, granule	35 to 71 lb/ft³	Slaked to form hydrated lime	75 to 99 (CaO)	pH of saturated solution, on detention time temp. amount of water critical for efficient slaking
Carbon dioxide, liquid CO_2	Carbonic anhydride	Bulk	Carbon steel (dry); type 316 stainless steel (solution)	Liquid		0.012 at 25° C	99.5	Solution is acid
Chlorinated lime, CaO, $2CaOCl_2 \cdot 3H_2O$	Bleaching powder, chloride of lime	100, 300, 800 lb drums	Glass and rubber lining, FRP, PE	White powder	48 lb/ft³		25 to 37 (available Cl_2)	Deteriorates
Chlorine, Cl_2	Chlorine gas, liquid chlorine	100, 150 lb cylinders; 1-ton tanks; 16-, 30-, 55-ton tank cars	Shipping containers	Greenish-yellow, liquefied gas under pressure	91.7 lb/ft³	0.07 (60° F) 0.04 (100° F)	99.8 (Cl_2)	Forms HCl and HOCl when mixed with water

TABLE A.1 Alphabetical Listing of Commonly Used Water Treatment Chemicals (*Continued*)

Chemical name and formula	Common or trade name	Shipping containers	Suitable storage materials[a]	Available forms/ descriptions	Density	Solubility, lb/gal	Commercial strength, percent	Additional characteristics and properties
Chlorine dioxide, ClO_2	Chlorine dioxide	Generated as used	Glass, PVC, and rubber linings; FRP; PE	Greenish-yellow gas		0.02 (30 mu)	26.3 (available Cl_2)	Explosive under certain conditions
Copper sulfate, $CuSO_4 \cdot 5H_2O$	Blue vitriol, blue stone	100 lb bags, 450 lb barrels, drums	FRP, PE, silicon lining, iron, stainless steel	Crystal, lump, powder	75 to 90 lb/ft^3 73 to 80 lb/ft^3 60 to 64 lb/ft^3	1.6 (32° F) 2.2 (68° F) 2.6 (86° F)	99 ($CuSO_4$)	25% solution pH approx. 3.0
Disodium phosphate, anhydrous $Na_2HPO_4 \cdot 12H_2O$	Basic sodium phosphate, DSP, secondary sodium phosphate	100 and 300 lb drums, 50 and 100 lb bags	Cast iron, steel, FRP, PE	White crystal, granular or powder	60 to 64 lb/ft^3	0.4 (32° F) 6.4 (86° F)	64.3 (PO_4) 48 (P_2O_5)	Precipitates Ca, Mg; pH of 1% solution = 9.1; solubility is 11 g/100 g at 25° C (77° F)
Ferric chloride, $FeCl_3$ (33% to 45% solution)	Ferrichlor, iron chloride	55 gal drums, bulk	Glass, PVC, and rubber linings, FRP, PE	Dark brown syrupy liquid	11.9 lb/gal (40%)	Complete	37 to 45 ($FeCl_3$) 20 to 21 (Fe)	
Ferric chloride, $FeCl_3 \cdot 6H_2O$	Crystal ferric chloride	300 lb barrels	Keep in original containers	Yellow-brown lump			59 to 61 ($FeCl_3$), 20 to 21 (Fe)	Hygroscopic (store lumps and powder in tight container), no dry feed; optimum pH. 4.0 to 11.0
Ferric chloride $FeCl_3$	Anhydrous ferric chloride	500 lb casks; 100, 300, 400 lb kegs; 65, 135, 250 lb drums	Keep in original containers	Greenish-black powder or crystals	175 lb/ft^3		98 ($FeCl_3$) 34 (Fe)	
Ferric sulfate, $Fe_2(SO_4)_3 \cdot 9H_2O$	Ferrifloc, ferrisul	100 to 175 lb bags, 400 to 425 lb drums	Glass, plastic, and rubber linings, FRP, PE, type 316 stainless steel	Red-brown powder 70 or granule 72	60 to 70 lb/ft^3	Soluble in 2 to 4 parts cold water	90 to 94 [as Fe $(SO_4)_3$] 25 to 26 (Fe)	Mildly hygroscopic coagulant at pH 3.5 to 11.0

TABLE A.1 Alphabetical Listing of Commonly Used Water Treatment Chemicals

Ferrous sulfate, $FeSO_4 \cdot 7H_2O$	Copperas, green vitriol	Bags, barrels, bulk	Glass, plastic, and rubber linings, FRP, PE, type 316 stainless steel	Green crystal, granule, lump	63 to 66 lb/ft^3		55 ($FeSO_4$) 20 (Fe)	Hygroscopic; cakes in storage; optimum pH 8.5 to 11.0
Fluorosilicic acid, H_2SiF_6	Fluorosilicic acid	Rubber-lined drums, trucks, or railroad tank cars	Rubber-lined steel, PE	Liquid	−0.4		35 (approx.)	Corrosive, etches glass
Hydrogen fluoride, HF	Hydrofluoric acid	Steel drums, tank cars	Steel, FRP, PE	Liquid		Approx. 1.2 (68° F)	70 (HF)	Below 60% steel cannot be used
Oxygen, liquid	LOX	Dewars, cylinders, truck and rail tankers	Steel	Pale blue liquid	9.52 lb/gal at 68° F and 1 atm	3.16% by volume at 25° C	99.5	Prevent LOX from contacting grease, oil, asphalt, or other combustibles.
Ozone, O_3	Ozone	Generated at site of application		Colorless gas				
Phosphoric acid, H_3PO_4		Polyethylene drums, bulk	FRP, epoxy, rubber lining, polypropylene, type 316 stainless steel	Watery white liquid	13.1 lb/gal at 75% solution		75/80/85	Freezing point: 0.5° F at 75% 40.2° F at 80% 70.0° F at 85%
Polyaluminum chloride, $Al_{13}(OH)_{20}(SO_4)_2Cl_{15}$	SternPac	55 gal drums and bulk	FRP, PE, type 316 stainless steel, rubber linings	Pale amber liquid	10.0 lb/gal		10.3 (Al_2O_3)	Freezing point = −12° C
Potassium aluminum sulfate, $K_2SO_4 \cdot Al_2(SO_4)_3 \cdot 24H_2O$	Potash alum, potassium alum	Bags, lead-lined bulk (carloads)	FRP, PE, ceramic and rubber linings	Lump, granule, powder	60 to 67 lb/ft^3	0.5 (32° F) 1.0 (68° F) 1.4 (86° F)	10 to 11 (Al_2O_3)	Low, even solubility; pH of 1% solution, 3.5
Potassium permanganate, $KMnO_4$	Purple salt	Bulk, barrels, drums	Iron, steel, FRP, PE	Purple crystals	90 to 105 lb/ft^3	Infinite	100	Danger of explosion in contact organic matters

TABLE A.1 Alphabetical Listing of Commonly Used Water Treatment Chemicals (Continued)

Chemical name and formula	Common or trade name	Shipping containers	Suitable storage materials	Available forms/descriptions	Density	Solubility, lb/gal	Commercial strength, percent	Additional characteristics and properties
Pyrosodium sulfite	Sodium metabisulfite $Na_2S_2O_5$	Bags, drums, barrels	Iron, steel, FRP, PE	White crystalline powder		Complete in water	Dry 67 (SO_2), sol 33.3 (SO_2)	Sulfurous odor
Sodium aluminate, $Na_2OAl_2O_3$	Soda alum	100 to 150 lb bags, 250 to 440 lb drums, solution	Iron, FRP, PE, rubber, steel	Brown powder liquid (27° Baumé)	50 to 60 lb/ft³	3.0 (68° F) 3.3 (86° F)	70 to 80 (Na_2) Al_2O_4 min. 32 Na_2 Al_2O_4	Hopper agitation required for dry feed; very hygroscopic
Sodium carbonate, Na_2CO_3	Soda ash	Bags, barrels, bulk (carloads), trucks	Iron, rubber lining, steel, FRP, PE	White powder, extra light, light, dense	31.2 to 56.2 (light); 56.2 to 68.7 (dense) lb/ft³	1.5 (68° F) 2.3 (86° F)	99.4 (Na_2CO_3) 57.9 (Na_2O)	Hopper agitation required for dry feed of light and extra-light forms; pH of 1% solution, 11.3
Sodium chloride, NaCl	Common salt, salt	Bags, barrels, bulk (carloads)	Bronze, FRP, PE, rubber	Rock, fine	50 to 60 58 to 78 lb/ft³	2.9 (32° F) 3.0 (68° F)	98 (NaCl)	Absorbs moisture
Sodium chlorite, $NaClO_2$	ADOX dry	100 lb drums	Metals (avoid cellulose materials)	Light orange powder, flake or crystals	53 to 56 lb/ft³	3.5 (68° F)	80 ($NaClO_2$) 30 (available Cl_2)	Generates ClO_2 at pH 3.0; explosive
Sodium fluoride, NaF	Fluoride	Bags, barrels, fiber drums, kegs	Iron, steel, FRP, PE	Nile blue or white powder, light, dense	50 lb/ft³ 75 lb/ft³	0.35 (most temps.)	90 to 95 (NaF)	pH of 4% solution, 6.6
Sodium hexametaphosphate, $Na(PO_3)_6$	Calgon, glassy phosphate, vitreous phosphate	100 lb bags	Rubber linings, plastics, type 316 stainless steel	Crystal, flake, powder	47 lb/ft³	1 to 4.2	66 (P_2O_5 unadjusted)	pH of 0.25% solution, 6.0 to 8.3
Sodium hydroxide, NaOH	Caustic soda, soda lye	100 to 700 lb drums; bulk (trucks, tankcars)	Carbon steel, polypropylene, FRP, rubber lining	Flake, lump, liquid	95.5 lb/ft³; 12.8 lb/gal for 50% solution	2.4 (32° F) 4.4 (68° F) 4.8 (104° F)	98.9 (NaOH) 74 to 76 (NaO_2)	Solid, hygroscopic; pH of 1% solution = 12.9; freezing point of 50% solution = 53° F

TABLE A.1 Alphabetical Listing of Commonly Used Water Treatment Chemicals (*Continued*)

Sodium hypochlorite, NaOCl	Sodium hypochlorite	5, 13, 50 gal carboys; 1,300 to 2,000 gal tank trucks	Ceramic, glass, plastic, and rubber linings; FRP, PE	Light yellow liquid		12 to 15 (available Cl_2)	Unstable
Sodium silicate, Na_2OSiO_2	Water glass	Drums, bulk (tank trucks, tank cars)	Cast iron, rubber lining, steel, FRP, PE	Opaque, viscous liquid		38 to 42° Baumé	Complete	Variable ratio of Na_2O to SiO_2; pH of 1% solution, 12.3
Sodium fluorosilicate, Na_2SiF_6	Sodium silicofluoride	Bags, barrels, fiber drums	Cast iron, rubber lining, steel, FRP, PE	Nile blue or yellowish white powder	72 lb/ft^3	0.03 (32° F) 0.06 (68° F) 0.12 (140° F)	99 (Na_2)	pH of 1% solution, 5.3
Sodium sulfite, Na_2SO_3	Sulfite	Bags, drums, barrels	Cast iron, rubber lining, steel, FRP, PE	White crystalline powder	80 to 90 lb/ft^3	Complete in water	23 (SO_2)	Sulfurous taste and odor
Sulfur dioxide, SO_2	Sulfurous acid anhydride	100 to 150 lb steel cylinders, ton containers, tank cars, tank trucks	Shipping container	Colorless gas		20% at 32° F, complete in water	99 (SO_2)	Irritating gas
Sulfuric acid, H_2SO_4	Oil of vitriol, vitriol	Bottles, carboys, drums, trucks, tank cars	FRP, PE porcelain, glass, and rubber linings	Solution	81.4 lb/ft^3 (59.3 Baumé)	Complete	77(59.3 Baumé)	Approx. pH of 0.5% solution, 1.2
Tetrasodium pyrophosphate, $Na_4P_2O_7 \cdot 10H_2O$	Alkaline sodium, pyrophosphate, TSPP	125 lb kegs, 200 lb bags, 300 lb barrels	Cast iron, steel, plastics	White powder	68 lb/ft^3	0.6 (80° F) 3.3 (212° F)	53 (P_2O_5)	pH of 1% solution, 10.8
Tricalcium phosphate	Fluorex	Bags, drums, bulk, barrels	Cast iron, steel, plastics	Granular	Variable	Insoluble		Also available as white powder
Trisodium phosphate, $Na_3PO_4 \cdot 12H_2O$	Normal sodium phosphate, tertiary sodium phosphate, TSP	125 lb kegs, 200 lb bags, 325 lb barrels	Cast iron, steel, plastics	Crystal—course, medium, standard	56 lb/ft^3 58 lb/ft^3 61 lb/ft^3	0.1 (32° F) 13.0 (158° F)	19 (P_2O_5)	pH of 1% solution, 11.9

*Always contact chemical suppliers to select best materials for handling.
†FRP = fiber-reinforced plastic.
‡PE = polyethylene.

TABLE A.2 Process Use of Common Water Treatment Chemicals

Process	Chemicals	AWWA Standard
Coagulants and coagulant aids	Aluminum sulfate	B403-93
	Bentonite	
	Calcium carbonate	
	Calcium hydroxide	
	Calcium oxide	
	EPI-DMA polyamines	B452-90
	Ferric chloride	B407-93
	Ferric sulfate	B406-92
	Ferrous sulfate	B402-90
	Polyaluminum chloride	B408-93
	PolyDADMAC	B451-92
	Polymers	
	Sodium aluminate	B405-94
	Sodium silicate	B404-92
Dechlorination	Activated carbon, granular	B604-90
	Ion-exchange resins	
	Sodium bisulfite	
	Sodium metabisulfite	B601-93
	Sulfur dioxide	B512-91
Disinfection and chlorination	Anhydrous ammonia	
	Ammonium hydroxide	
	Ammonium sulfate	B302-90
	Calcium hypochlorite	B300-92
	Chlorinated lime	
	Chlorine	B301-92
	Chlorine dioxide (sodium chlorite + Cl_2)	
	Ozone	
	Sodium chlorite	B303-88
	Sodium hypochlorite	B300-92
Fluoridation and fluoride adjustment	Activated alumina (aluminum oxide)	
	Calcium fluoride	
	Fluorosilicic acid	B703-94
	Hydrogen fluoride	
	Sodium fluoride	B704-94
	Sodium fluorosilicate	B702-94
Mineral oxidation	Chlorine	B301-92
	Chlorine dioxide (sodium chlorite + Cl_2)	
	Ozone	
	Potassium permanganate	B603-93

TABLE A.2 Process Use of Common Water Treatment Chemicals

Process	Chemicals	AWWA Standard
Organics adsorption oxidation	Activated carbon, granular	B604-90
	Chlorine dioxide	
	Ozone	
	Potassium permanganate	B603-93
pH adjustment, stabilization, and corrosion control	Calcium carbonate	
	Calcium chloride	B550-90
	Calcium hydroxide	B202-93
	Calcium oxide	B202-93
	Carbon dioxide	B510-89
	Disodium phosphate	B505-88
	Hydrochloric acid	
	Monosodium phosphate	B504-88
	Phosphoric acid	
	Potassium hydroxide	B511-90
	Sodium carbonate	
	Sodium hexametaphosphate	B502-88
	Sodium hydroxide	B501-93
	Sodium polyphosphate	B502-88
	Sodium silicate	
	Sodium tripolyphosphate	B503-89
	Sulfuric acid	
Softening	Calcium hydroxide	B202-93
	Calcium oxide	B202-93
	Sodium carbonate	B201-92
	Sodium chloride	B200-93
	Sodium hydroxide	
Taste and odor control	Activated carbon, powdered	B600-90
	Activated carbon, granular	B604-90
	Chlorine	B301-92
	Chlorine dioxide (Sodium chlorite + Cl_2)	
	Copper sulfate	B602-91
	Ozone	
	Potassium permanganate	B603-93

APPENDIX B
ABBREVIATIONS COMMONLY USED IN THE WATER INDUSTRY

A

acre-foot: **acre-ft**
acutely hazardous material: **AHM**
Advanced Notice of Proposed Rule Making: **ANPRM**
aggressiveness index: **AI**
alkylbenzene sulfonate: **ABS**
alternating current: **AC**
American Association of Cost Engineers: **AACE**
American Association of State Highway and Transportation Officials: **AASHTO**
American Chemical Society: **ACS**
American Concrete Institute: **ACI**
American Gas Association: **AGA**
American Institute of Chemical Engineers: **AIChE**
American Institute of Electrical Engineers: **AIEE**
American National Standards Institute: **ANSI**
American Public Health Association: **APHA**
American Public Works Association: **APWA**
American Society of Civil Engineers: **ASCE**
American Society of Heating, Refrigeration and Air Conditioning: **ASHRAE**
American Society of Mechanical Engineers: **ASME**
American Society for Testing Materials: **ASTM**
American Water Works Association: **AWWA**
American Water Works Association Research Foundation: **AWWARF**
Americans with Disabilities Act: **ADA**
ampere: **A**
ampere-hour: **A·h**
answer: **ans.**
area: **A**
asbestos cement: **A-C**

asbestos-containing material: **ACM**
assimilable organic carbon: **AOC**
Association of Metropolitan Water Agencies: **AMWA**
Association of State Drinking Water Administrators: **ASDWA**
Association of State Sanitary Engineers: **ASSE**
atomic absorption spectrophotometry: **AAS**
atomic emission spectroscopy: **AES**
atomic weight: **at wt**
available filter head: **AFH**
avoirdupois: **avdp** or **avoir.**

B

background organic matter: **BOM**
barrel: **bbl** or **brl**
base (as the "length" of a triangle): **B** or **b**
becquerel (metric equivalent of curie): **Bq**
benzene-toluene-xylene (the volatiles measured in gasoline): **BTX**
best available technology: **BAT**
billion: means 10^9 to Americans, 10^{12} to Europeans
billion electron volts: **BeV**
billion gallons: **bil gal**
billion gallons per day: **bgd**
biochemical oxygen demand or biological oxygen demand: **BOD**
biodegradable organic matter: **BDOM**
biologically enhanced activated carbon: **BEAC**
board feet (feet board measure): **fbm**
brake horsepower: **bhp**
British thermal unit: **Btu** or **BTU**
bushel: **bu**

C

calculation error: **CE**
carbon usage rate: **CUR**
cast iron: **CI**
cathode ray tube: **CRT**
Centers for Disease Control: **CDC**
centimeter: **cm**
central processing unit: **CPU**
chemical oxygen demand: **COD**
chloramines: includes **monochloramine (NH_2Cl), dichloramine ($NHCl_2$),** and **trichloramine (NCl_3)**
chlorinated polyvinyl chloride: **CPVC**
chloroplatinate units (color indicator): **cpu**
Clean Air Act: **CAA**
Clean Water Act: **CWA**
Code of Federal Regulations: **CFR**

colony-forming units: **cfu**
color units: **cu**
community water system: **CWS**
Comprehensive Environmental Response, Compensation, and Liability Act: **CERCLA**
computerized maintenance management system: **CMMS**
Consumer Products Safety Commission: **CPSC**
contact time (disinfectant with water): *T*
counts per minute: **cpm**
cubic feet: **ft^3** or **cu ft**
cubic feet per hour: **ft^3/h** or **cu ft/h**
cubic feet per minute: **ft^3/min** or **cu ft/min**
cubic feet per second: **ft^3/s** or **cu ft/s**
cubic inches: **in.^{3}**
cubic meters: **m^3**
cubic yards: **yd^3** or **cu yd**
curie: **Ci**
cycles per second: **cps** (1 cps = 1 Hz)

D

deciliter: **dl**
Department of Agriculture: **DOA**
Department of Commerce: **DOC**
Department of Interior: **DOI**
Department of Transportation: **DOT**
diameter: **D**
diatomaceous earth (filtration): **DE**
dibromochloropropane: **DBCP**
dichlorodiphenyltrichloroethane: **DDT**
differential pressure cell: **DP cell**
differential pressure meter: **DP meter**
direct current: **DC** or **dc**
disinfectant by-products: **DBPs**
disinfectant concentration × time: *CT* or **C × T**
disinfectant–disinfection by-products: **D-DBP** or **D/DBP**
dissolved air flotation: **DAF**
dissolved organic carbon: **DOC**
dissolved organic halogen: **DOX**
dissolved oxygen: **DO**
distributed control system: **DCS**
Ductile Iron Pipe Research Association: **DIPRA**

E

effective size (of granular media): **E.S.**
efficiency: **eff.**
electrodialysis: **ED**
electrodialysis reversal: **EDR**

electrolytic conductivity detector: **ElCD**
electromotive force: **emf**
electronic data processing: **EDP**
electron volts: **eV**
Emergency Planning and Community Right-to-Know Act: **EPCRA**
empty bed contact time: **EBCT**
Endangered Species Act: **ESA**
Engineers Joint Council: **EJC**
Enhanced Surface Water Treatment Rule: **ESWTR**
environmental impact statement: **EIS**
Environmental Protection Agency (US): **USEPA**
epichlorohydrin dimethylamine: **EPI-DMA**
ethylenediaminetetraacetic acid: **EDTA**
ethylene dibromide: **EDB**

F

Federal Emergency Management Agency: **FEMA**
Federal Insecticide, Fungicide, and Rodenticide Act: **FIFRA**
feet: **ft**
feet per hour: **ft/h**
feet per minute: **ft/min**
feet per second: **ft/s**
fiberglass reinforced plastic: **FRP**
fiscal year: **FY**
flow rate: **Q**
foot-pound: **ft-lb**

G

gallon: **gal**
gallons per capita per day: **gpcd**
gallons per day: **gpd**
gallons per hour: **gph**
gallons per minute: **gpm**
gallons per minute per square foot: **gpm/ft^2**
gallons per second: **gps**
gallons per square foot: **gal/ft^2**
gas chromatography: **GC**
gas chromatography-mass spectrometry: **GC-MS**
General Accounting Office: **GAO**
gigaliter: **GL**
Government Printing Office: **GPO**
grains per gallon: **gpg**
granular activated carbon: **GAC**
greater than: **>**
greater than or equal to: **≥**
ground water under the direct influence of surface water: **GWUI**

H

haloacetic acid: **HAA**
haloacetic nitriles: **HANs**
heating, ventilating, and air conditioning: **HVAC**
hectare: **ha**
height: **h** or **H**
hertz: **Hz**
heterotrophic plate count (replaces *standard plate count*): **HPC**
high-density, cross-linked polyethylene: **HDXLPE**
high-density polyethylene: **HDPE**
high-intensity discharge (lighting): **HID**
high-pressure liquid chromatography: **HPLC**
highway-addressable remote transducer: **HART**
horse power: **HP**
hour: **h** or **hr**
hundred (centum): **C**
hundredweight: **cwt**
hydraulic grade line: **HGL**
hydraulic grade line elevation: **HGLE**
hydraulic retention time: **HRT**

I

imperial: **imp**
inch: **in.**
inches per minute: **in./min**
inches per second: **in./s**
Information Collection Rule: **ICR**
inorganic chemical: **IOC**
inside diameter: **ID**
Institute of Electrical and Electronic Engineers: **IEEE**
Instrument Society of America: **ISA**
instrumentation and control: **I&C**
integrally molded flanged outlets: **IMFO**
International Research Center: **IRC**
Interstate Commerce Commision: **ICC**
iron pipe size: **IP**

J

Jackson turbidity units: **Jtu**
joule: **J**

K

kilogram: **kg**
kilohertz (kilocycles): **kHz**
kilometer: **km**
kilovolt-ampere: **KVA**
kilowatt: **kW**
kilowatt-hours: **kW·h**

L

Langelier saturation index: **LSI**
length: **l** or **L**
less than: **<**
less than or equal to: **≤**
linear feet: **lin ft**
liquid oxygen: **LOX**
liquefied petroleum gas: **LPG**
liter: **L**
local emergency planning committee: **LEPC**
locally unacceptable land use: **LULU**
local operator interface (control system): **LOI**
lowest observed adverse effect level: **LOAEL**
lumen: **lm**

M

mass spectrometry: **MS**
mass transfer zone: **MTZ**
material safety data sheet: **MSDS**
2-methylisoborneol: **MIB**
maximum contaminant level: **MCL**
maximum contaminant level goal: **MCLG**
maximum permissible concentration: **MPC**
maximum tolerated dose: **MTD**
mean sea level: **MSL**
megahertz (megacycles): **MHz**
megaliter: **ML**
membrane filter: **MF**
meter: **m**
microfiltration: **MF**
mile: **mi**
miles per hour: **mph**
mille (thousand): **M**
milliampere direct current: **mADC**
milliamperes: **mA**
milligram: **mg**
milliliter: **ml**

millimeter: **mm**
million electron volts: **MeV**
million gallons: **mil gal**
million gallons per day: **mgd**
minute: **min**
mixing intensity: **G**
mole: **mol**
molecular weight: **mol wt**
molecular weight cutoff: **MWCO**
moles per liter: **mol/L**
month: **mo**
most probable number: **MPN**
motor starting current: **IM**

N

nanofiltration: **NF**
nanometer: **nm**
National Academy of Science: **NAS**
National Electrical Code: **NEC**
National Electrical Manufacturers Association: **NEMA**
National Environmental Policy Act: **NEPA**
National Fire Protection Association: **NFPA**
National Institute of Occupational Safety and Health: **NIOSH**
National Pollution Discharge Elimination System: **NPDES**
National Sanitation Foundation: **NSF International**
natural organic matter: **NOM**
nephelometric turbidity units: **ntu**
net positive suction head: **NPSH**
New England Water Works Association: **NEWWA**
nitrogen (gaseous): **GAN**
nitrogen (liquid): **LIN**
N,N-diethyl-p-phenylenediamine (used to determine presence of chlorine): **DPD**
nominal pipe size: **NPS**
not equal to: $\neq$

O

Occupational Safety and Health Administration: **OSHA**
ohm: Ω
one thousand: **M**
operations and maintenance: **O & M**
Organization for Economic Cooperation and Development: **OECD**
ounce: **oz**
outside diameter: **OD**
oxidation-reduction potential: **ORP**
oxygen (gas): **GOX**
oxygen (liquid): **LOX**
ozone: O_3

P

package: **pkg.**
parts per billion: **ppb** or **μg/L** (preferred)
parts per million: **ppm** or **mg/L** (preferred)
parts per trillion: **ppt** or **ng/L** (preferred)
perchloroethylene: **PCE**
perimeter: **P**
personal computer: **PC**
picocuries: **pCi**
picocuries per liter: **pCi/L**
point of entry (water treatment device): **POE**
point of use (water treatment device): **POU**
polyaromatic hydrocarbon: **PAH**
polychlorinated biphenyl: **PCB**
polydiallyldimethylammonium chloride: **PolyDADMAC**
polytetrafluoroethylene: **PFTE** (generic name for Teflon)
polyvinyl chloride: **PVC**
Portland Cement Association: **PCA**
pound: **lb**
pounds per day: **lb/d**
pounds per square foot: **lb/ft^2**
pounds per square inch: **psi**
pounds per square inch absolute: **psia**
pounds per square inch gauge: **psig**
powdered activated carbon: **PAC**
practical quantitation level: **PQL**
presence-absence (coliform test): **P-A**
pressure reducing valve: **PRV**
pressure-swing adsorption (manufacture of O_2): **PSA**
process and instrumentation drawing: **P&ID**
process hazard analysis: **PHA**
process safety management program: **PSM**
programmable logic controller: **PLC**
proportional, plus integral, plus derivative (control system): **PID**
publicly owned treatment works: **POTW**

Q

quantity (flow rate): **Q**

R

radius (of a circle): **r**
random access memory: **RAM**
redox potential: **pE**
regulatory negotiations: **reg neg**
reinforced thermoset plastic: **RTP**

relay ladder logic (computer): **RLL**
reliable detection limit: **RDL**
remote terminal unit: **RTU**
resistance temperature detector: **RTD**
Resource Conservation and Recovery Act: **RCRA**
reverse osmosis: **RO**
revolutions per minute: **rpm**
revolutions per second: **rps**
risk management program: **RMP**
roentgen equivalent, mammal: **rem**

S

Safe Drinking Water Act: **SDWA**
second: **s**
second feet (cubic feet per second): **ft^3/s**
self-contained breathing apparatus: **SCBA**
specific gravity: **sp gr**
specific heat: **sp ht**
square foot: **ft^2** or **sq ft**
square inch: **in.^{2}** or **sq in.**
square meters: **m^2**
square miles: **mi^2** or **sq mi**
square yards: **yd^2** or **sq yd**
standard plate count: see *heterotrophic plate count*
standard query language: **SQL**
state emergency response commission: **SERC**
streaming current detector: **SCD**
Superfund Amendments and Reauthorization Act: **SARA**
supervisory control and data acquisition: **SCADA**
Surface Water Treatment Rule: **SWTR**
synthetic organic chemical: **SOC**

T

temperature: **T** or **temp.**
tetrachloroethylene (perchloroethylene): **PCE**
tetrafluoroethylene: **TFE**
thermal-swing adsorption (manufacture of O_2): **TSA**
thermocouple: **TC**
thousand: **thous**
thousand (mille): **M**
threshold odor number: **TON**
ton: **T**
total dissolved solids: **TDS**
total organic carbon: **TOC**
total organic halide: **TOX**
total suspended solids: **tss**

total trihalomethanes: **TTHMs**
toxic characteristic leaching procedure: **TCLP**
Toxic Substances Control Act: **TSCA**
transient voltage surge suppression: **TVSS**
trichloroethylene (or trichloroethene): **TCE**
trihalomethane: **THM**
trihalomethane formation potential: **THMFP**
true color units: **b**
turbidity units: see *nephelometric turbidity units*

U

ultrafiltration: **UF**
ultra high-molecular-weight (plastic): **UHMW**
ultraviolet: **UV**
underground injection control: **UIC**
uniformity coefficient: **uc**
uniformity coefficient (of granular media): **U.C.**
uninterruptible power system: **UPS**
unreasonable risk to health: **URTH**
U.S. Environmental Protection Agency: **USEPA**
U.S. Public Health Service: **USPHS**

V

vacuum-swing adsorption (manufacture of O_2): **VSA**
variable frequency drive: **VFD**
variable speed drive: **VSD**
volatile organic chemical: **VOC**
volt: **V**
volt drop: **VD**
volt-amperes:**VA**
volts alternating current: **VAC**
volts direct current: **VDC**
volume: **V** or **vol.**

W

Water Environment Federation (formerly Water Pollution Control Federation): **WEF**
water treatment plant: **WTP**
watt: **W**
week: **wk.**
weight: **wt**
width: **w** or **W**
World Health Organization: **WHO**

X

xenon: **Xe**

Y

yard: **yd**
year: **yr**

Z

zeta potential: **zp**

INDEX

INDEX NOTE: An *f.* following a page number refers to a figure; a *t.* refers to a table.

AACE International, 705
Abbreviations, 777–787
Activated carbon, 751. *See also* Granular activated carbon, Powdered activated carbon
ADA. *See* Americans with Disabilities Act
Adsorption clarifiers, 149, 150*f.*
Aeration, 61
  cascade aerators, 62
  cone aerators, 62, 63*f.*
  design of diffused air equipment, 72–73
  diffuser aerators, 63, 83–84, 84*f.*
  diffusion-type equipment, 63–64
  draft-tube aerators, 64
  emerging technologies, 65–69
  equilibrium conditions, 69–70
  equipment types, 61–69
  Henry's law, 71, 72*t.*
  in-well, 64
  low-profile, diffused air system, 65–66, 66*f.*
  maxi-strip system, 67, 67*f.*
  mechanical, 65
  multiple-tray aerators, 62, 74–75
  packed columns, 62–63, 64*f.*, 75–83
  pilot testing, 83–85
  pressure aerators, 65
  principles, 69–71
  process design, 71–83
  rate of achievement, 70
  rate of transfer, 70–71, 72*t.*
  rotor-strip unit, 67–68, 69*f.*
  saturation value, 70
  significance of films, 70
  spray aerators, 62, 73–74
  submerged aerators, 65
  surface aerators, 65
  uses, 61
  waterfall equipment, 62–63
Air emissions
  environmental impact, 680–681
  regulations, 684–685
Air stripping, 61, 84*f.*
  equilibrium conditions, 69–70
  Henry's law, 71, 72*t.*
  principles, 69–71
  rate of achievement, 70
  rate of transfer, 70–71, 72*t.*
  saturation value, 70
  significance of films, 70
  and volatile organic chemicals, 61, 62–63, 64, 64*f.*, 67, 67*f.*
Algae
  and slow sand filtration, 195, 196*t.*
Alum
  recovery, 497
  sludge, 488–489
  storage, 428, 429*f.*
Aluminum oxide, 752
Aluminum sulfate, 752
American National Standards Institute, 532, 546
  rapid sand filtration standards, 187
American Water Works Association
  background data for master plan, 8
  Construction Contract Administration manual, 330
  hardness water quality goals, 282
  rapid sand filtration standards, 186
Americans with Disabilities Act, 25, 532, 533, 534–535, 557, 663, 711
Ammonia, 752–753, 753*f.*
  in chloramination, 241–242
  handling and storage, 422
  removal by ion exchange process, 305–306
Ammonium aluminum sulfate, 753
Ammonium hydroxide, 754
Ammonium sulfate, 754
Architectural design, 26–27, 531
  administrative facilities, 535, 536*f.*
  aesthetics, 532
  Americans with Disabilities Act, 532, 533, 534–535
  client relations, 534
  codes and standards, 532–533
  colors of finish materials, 546
  contexturalism, 532
  covered or uncovered process units, 544
  employee facilities, 535–538, 537*f.*
  and environmental concerns, 533, 545
  flexibility in gender-specific facilities, 536–537
  interior finishes, 546
  and laboratory equipment needs, 539
  and laboratory operations, 540
  and laboratory space needs, 539–540, 541*f.*
  and laboratory standards, 538–539
  lobby and reception area, 535
  locker rooms, 537, 537*f.*
  lunchrooms, 537–538
  maintenance facilities, 541–544, 543*f.*
  maintenance storage, 542–544
  and maintenance tools, 542
  office area, 535
  organizations offering design standards and recommendations, 547–551
  panelized building modules, 545
  and plant arrangement, 668
  pre-engineered structures, 544–545
  repair shops, 541–542, 543*f.*
  site planning, 533–534
  surface coatings, 546

INDEX

Arsenic, 310–311
  removal by activated alumina, 312
  removal by strongly basic ion exchange resins, 311–312
Available filter head, 470, 479

Backwashing
  control systems, 589–590
  granular activated carbon, 389, 391f., 403–404
  and hydraulic design, 481–482
  microfiltration, 365
  and pilot plants, 742, 743f.
  process and instrumentation diagram, 601, 601f.
  rapid sand filtration, 163–168, 163f., 166f., 168f.
  ultrafiltration, 365
Bacteria
  treatment options, 14t.
Barium, 306
  removal by combined softening and decationization, 307
  removal by hydrogen-form weak-acid dealkalizers, 307
  removal by softeners, 306–307
BAT. See Best available technology
Belt-type gravimetric feeders, 435
Bench-scale testing
  equipment, 732–733, 732f.
  membrane processes, 374
  process residuals, 527–528
  purpose, 731–732
Bentonite, 754
Best available technology, 12
Bid administration, 29
Bourdon tubes, 579
Bromate, 12
Building codes, 464, 532, 693

Calcium carbonate, 754
Calcium chloride, 754
Calcium chlorite
  feed system, 439
Calcium hypochlorite, 232, 755
Carbon dioxide, 369, 755–756
  handling and storage, 422
Carbon slurry, 442–445, 444f.
Cathodic protection systems, 692
Caustic soda
  in lime softening, 285
  storage, 428, 430f., 431f.
Change orders, 29–30
Chemical mixing
  air mixing, 94–97
  hydraulic mixing, 97–98, 98t.
  in-line blenders, 93–94
  in-line jet mixers, 94, 95f., 96f.
  intensity, 90
  mechanical in-line blenders, 94, 96f.
  mechanical mixers, 92f., 93f.
  rapid mixing design criteria, 92, 93, 94t.
  static in-line blenders, 94, 97f.
  temperature effects, 91, 91t.
  types of rapid mixing systems, 91–98

Chemicals. See also Chemical mixing, Coagulant chemicals, Inorganic chemicals, Organic chemicals, Volatile organic chemicals
  ancillary feed equipment, 448–452
  and building codes, 464
  bulk storage and handling, 467
  carrying water, 454
  centrifugal pumps, 442
  conditioning water for, 454
  and confined spaces, 461–462
  containment and treatment systems, 464–465
  and cross connections, 454–455
  day tanks, 449
  delivery, handling, and storage, 418–433
  diffusion and diffusers, 455–456, 456f.
  dilution, 453
  displacement or decanting feeders, 453
  dissolving tanks, 453
  dry chemical handling and storage, 423–425
  dry feed equipment, 435–439
  dust containment, 425
  eductors, 442
  and environmental impact planning, 678–679
  evaporators, 422–423, 423f.
  eye washes, 463
  fire and explosion prevention, 465–466
  and fire codes, 464
  flushing feeders, 453–454
  gas feed equipment, 433–435
  gravity orifice feeders, 453
  inventory size, 418
  liquefied gas receiving and storage, 419–423
  liquid chemical effective densities, 426t.
  liquid chemical handling and storage, 426–433
  liquid feed equipment, 439–456, 441f.
  list of, by process type, 774t.–775t.
  list of, with properties and characteristics, 753–767
  list of, with water industry features, 768t.–773t.
  Material Safety Data Sheets, 456, 461, 461t.
  new ones for water treatment, 467
  NIOSH Pocket Guide to Chemical Hazards, 456
  NSF listings, 461
  O&M design issues for feed systems, 691–692
  permits, 683–684
  piping and conduits, 449–452, 450t.–451t.
  piston and diaphragm pumps, 440
  polymer feed systems, 448
  pot feeders, 453
  progressive cavity pumps, 440
  protective clothing, 462
  pumps, 440, 442
  receiving shipments, 419
  recent trends, 466–467
  regulations, 459–460, 460t.
  respirators and masks, 462–463
  rotameters, 452
  safety design considerations, 456–466, 457t.–458t.
  safety equipment, 462–463
  safety screens, 463
  safety showers, 463

Chemicals (*Cont.*)
  single application points, 666
  size of storage and feed systems, 417–418, 418*t*.
  slurry feed systems, 442–447
  spill containment, 427–428
  toxic gas protection, 463
  transfer pumps, 448–449
  utility water, 453, 454–455
  valves and appurtenances, 440–442, 441*f*.
  ventilation requirements, 464
Chick, Harriette, 221, 230
Chick-Watson theory, 230
Chloramine and chloramination, 232–234, 234*t*.
  ammonium sulfate systems, 242
  anhydrous ammonia systems, 241–242
  aqueous ammonia systems, 242
  chlorine/ammonia feed ratio, 241
  and SWTR, 241
  systems, 241–242
Chlorinated lime, 756
Chlorine and chlorination, 230–232, 756, 757*f*., 758*f*.
  analyzers and detectors, 581–582
  calcium hypochlorite, 232
  chemistry, 232, 233*f*.
  chloramination, 232–234, 234*t*.
  chloramine systems, 241–242
  control methods, 245, 246*f*.
  control systems, 591
  cylinders, 419, 420*f*., 421
  density, 237, 239*f*.
  free residuals, 232, 234*t*., 235*f*.
  gas, 232
  gas feeders, 236, 237*f*., 238*f*., 433–434, 434*f*.
  gas piping, 236–239, 435
  gas systems, 234–240
  Handbook on Chlorination, 232
  handling and storage, 419–422, 420*f*.
  hypochlorous acid, 232
  initial mixing, 239–240
  leak detectors, 582
  materials for handling, 240, 240*t*.
  new gas leak safety requirements, 466
  on-site generation, 244–245
  residual analyzers, 247, 248*f*., 582
  sample point location, 245–247
  scrubbers, 465
  sodium hypochlorite, 232
  sodium hypochlorite systems, 242–244
  system design for SWTR compliance, 247–249
  and Uniform Fire Code, 232
  viscosity, 239
Chlorine dioxide, 249, 756–758
  aqueous chlorine-sodium chlorite system, 251–252, 253*f*.
  disinfection, 249–250
  excess chlorine requirement, 251, 252*f*.
  gas chlorine-sodium chlorite system, 252, 253*f*.
  generation, 251–253
  sodium chlorite, 254
  system schematic, 250, 250*f*.
Chlorine Institute, 421
Chlorites, 12

Circular sedimentation basins, 128, 129*f*.
  bottom slopes, 131–132
  dimensions, 128–129
  flocculation zone, 130, 130*f*.
  inlet design, 129–130
  outlet design, 130–131
  sludge hoppers, 131
  sludge removal, 131–132
  turntable collector drives, 131
Civil/mechanical process design, 25–26
Clarification, 111
  basin depth and velocities, 114
  circular basins. *See* Circular sedimentation basins
  contact, 149–151, 150*f*.
  conventional design, 111–132
  detention time, 113
  dissolved air flotation, 145–149, 146*f*.
  high-rate, 132–137
  number of tanks, 114
  overflow rates, 113, 113*t*.
  pilot plants, 736–738
  plate settlers, 134–137, 135*f*., 136*f*.
  rectangular basins. *See* Rectangular sedimentation basins
  sedimentation basins, 111, 112*f*.
  sedimentation theory, 112
  sludge blanket, 140–145, 142*f*., 143*f*.
  solids contact/slurry recirculation units, 137–140, 138*f*., 140*f*.
  square basins, 132
  tube settlers, 132–134, 133*f*.
Clean Air Act, 643
Clean Water Act, 494–495, 683, 684
Coagulant chemicals, 88–89
  defined, 88
Coagulation, 88
  adjustment of pH, 90
  chemical selection, 89
  chemicals, 88–89
  coagulant recovery, 497
  control systems, 591–592
  defined, 87
  flocculation aids, 89
  multipurpose effects, 16*t*.
  pilot plants, 735–736
  sludge, 488–489
Color
  and diatomaceous earth filtration, 209
  removal by ion exchange processes, 317
  and slow sand filtration, 195
  treatment options, 14*t*.
Conceptual estimates, 706
Conductivity sensors, 584
Construction, 28
  administration, 29
  bid administration, 29
  coordination of design disciplines, 29
  owner approval of change orders, 29
  regulatory approval of change orders, 29–30
Construction costs. *See* Estimates and estimating
Construction Specifications Institute, 602–603

INDEX

Contact clarification, 149, 150*f.*
 adsorption clarifiers, 149, 150*f.*
 application, 149
 design criteria, 149
 inlet conditions, 151
 operation, 150
 outlet conditions, 151
 sludge removal, 151
Contaminant Rules, 10
Control systems. *See also* Instrumentation
 adaptive gain, 588
 auxiliary equipment, 599–600
 centralized, 594
 chlorination, 591
 coagulation, 591–592
 and communications equipment, 600
 computer-aided operations and maintenance, 598
 control center locations, 595–596
 distributed, 594–595
 drawings and diagrams, 600–602
 expert systems, 596–597
 filter backwash, 589–590
 flocculation, 591–592
 flow, 589, 590*f.*
 future trends, 603
 fuzzy logic, 588–589
 integral-only, 587
 linked computer applications, 598
 local control panels, 596, 597*f.*
 model-based, 588
 neural networks, 597–598
 nonlinear control, 588
 O&M design issues, 689–690
 open loop, 586
 operator interface locations, 595–596
 pH, 591
 pressure, 589
 and programmable logic controllers, 592–593, 593*f.*
 proportional, 586–587
 proportional plus integral, 587
 proportional plus integral plus derivative, 588
 proportional-speed floating, 587–588
 remote terminal units, 593–594, 594*f.*
 SCADA systems, 595–596
 signal error sources, 585, 586*t.*
 software, 596–598
 and surge suppression, 600
 two-position/on-off, 586
 unattended plant operation, 596
 and uninterruptible power systems, 599–600
Coordination
 of design disciplines, 25–27, 29
 of design disciplines during construction, 29
 with regulatory agencies, 21
 with support disciplines, 23–24
Coordination of design disciplines, 25–27
Copper sulfate, 758
Costs. *See* Estimates and estimating
Cryptosporidium, 12
 and diatomaceous earth filtration, 193, 219
 and slow sand filtration, 193, 196
 and Surface Water Treatment Rule, 223

DDBPR. *See* Disinfectant-Disinfection By-Product Rule
Dealkalization, 309
 by dual strong acid columns, 310, 311*f.*
 pH effects, 309, 310
 with weak acid resins regenerated with acid, 309–310
Definitive estimates, 707
Desalination. *See* Electrodialysis/electrodialysis reversal
Design. *See also* Architectural design, Civil/mechanical process design, Electrical design, Hydraulic design, Instrumentation design, Mechanical design, Plant arrangement, Structural design, Water treatment engineering
 changes, 713
 development, 23
 documents, 24–25
 final, 24–28
 organizations offering design standards and recommendations, 547–551
 preliminary, 21–24
 reviews, 27–28
 specifications, 602–603
 standards, 25
Design-build approach, 3
Desorption, 61
Diatomaceous earth filtration, 16, 207–208
 ancillary facilities, 217–219
 circular leaf, 211, 211*f.*
 and color, 209
 and *Cryptosporidium,* 193, 219
 design considerations, 210–212
 development, 207
 and *Entamoeba histolytica,* 207
 and entrapment, 193
 filter construction, 210–211
 filter elements, 211–212, 211*f.,* 212*f.*
 filter leaf backing screen, 213
 filter leaf binding frame closures, 214
 filter leaf design, 212–214, 213*f.*
 filter leaf drainage chambers, 212–213
 filter leaf outlet connections, 214, 215*f.*
 filter septum, 213
 filter types, 210
 fixed filter elements, 211
 flat leaf filter, 210*f.,* 213*f.*
 and *Giardia,* 193
 handling and delivery, 217–218
 and iron, 209–210
 and manganese, 209–210
 operation, 208–210, 208*f.*
 porosity control, 208–209, 209*f.*
 pressure filter systems, 214–216, 216*f.*
 recovery system, 218–219, 219*f.*
 rectangular leaf, 211, 211*f.*
 rotating filter elements, 211–212
 slurry conveyance, 218
 slurry storage, 218
 source of diatomaceous earth, 208
 steps, 208, 208*f.*
 supplementary treatment, 209–210

Diatomaceous earth filtration (*Cont.*)
  system controls, 219
  tubular filters, 211*f.*
  and turbidity, 193
  vacuum filter leaves, 214, 214*f.*, 215*f.*
  vacuum filter systems, 214–215, 216–217, 217*f.*
  waste disposal, 218–219
Direct filtration
  defined, 88
Disinfectant-Disinfection By-Product Rule, 10, 11–12, 13*t.*, 228
Disinfection. *See also* Chloramine and chloramination, Chlorine and chlorination, Chlorine dioxide, Oxidation, Ozone and ozonation
  and baffling conditions, 225*f.*, 226*f.*, 227*t.*
  Chick-Watson theory, 230
  contact requirements, 223–227
  determining contact time, 224–226
  groundwater rule, 228
  historical overview, 221–222
  monitoring requirements, 227, 228*t.*
  physical constants, 221*t.*
  and Surface Water Treatment Rule, 222–228, 223*t.*
Disinfection by-products, 227–228
Dissolved air flotation, 145, 146*f.*
  basin dimensions, 147
  chain and flight skimmer, 148
  design criteria, 147
  effluent design, 147
  floated solids removal, 148–149
  hydraulic removal, 149
  influent design, 147
  mechanical skimming, 148
  oscillating skimmer, 148
  pilot plants, 737, 738*f.*, 739
  recycle system, 146–147
  rotating skimmer, 148–149
  theory and operation, 145–146
Dissolved oxygen
  and slow sand filtration, 196
Dolomitic hydrated lime, 761
Dolomitic lime, 761
Drawings and specifications, 24

ED/EDR. *See* Electrodialysis/electrodialysis reversal
Electrical design, 27
Electrical systems
  adequate power outlets, 696
  adjustable-speed drives, 617
  autotransformers, 616
  capacitors, 615
  circuit breakers, 612, 613
  direct current starters, 617
  double-ended, 609, 611*f.*
  equipment selection, 612–614
  fault protective device maintenance, 618
  fault protective devices, 612–613
  fuses, 612
  generator maintenance, 618
  generators, 606, 607
  high-mast lighting, 613
  induction motors, 615
  lighting fixtures, 613–614, 696
  lightning rods, 614
  loop feed, 607, 610*f.*
  maintenance, 617–619
  main-tie-main, 609, 611*f.*
  motor starting and voltage drop, 615–617
  National Electrical Code, 605
  National Fire Protection Association standard, 605
  O&M design issues, 689–690
  one-line diagrams, 607, 608*f.*
  pilot plants power requirements, 733–734
  power distribution reliability, 607–611
  power factor correction, 614–615
  power supply reliability, 606–607, 629–630
  power system arresters, 614
  primary selective, 607–608, 610*f.*
  protective device settings, 613
  radial, 607, 609*f.*
  reduced voltage motor starters, 616–617
  relay testing, 618
  reliability, 605–611
  secondary selective, 608–611, 611*f.*
  signal system arresters, 614
  siting and proximity to utility services, 646
  slip drives, 617
  solid-state starters, 617
  starters, 615–617
  surge capacitors, 614
  surge protection, 600, 614
  switch gear maintenance, 618
  synchronous motors, 615
  synchronous starters, 617
  testing, 617–618
  transformer maintenance, 619
  transformers, 613
  variable-frequency drives, 617, 626
  voltage drop, 615–616
  voltage selection, 611–612
  wire and cable testing, 618
  wound rotor regenerative starters, 617
  wye-delta starters, 616–617
Electrodialysis/electrodialysis reversal, 337, 338
  back diffusion, 367
  current leakage, 367
  desalination ratio, 367
  differential pressure, 368
  electric current, 366
  electrode compartments, 368
  hydrogen sulfide control, 354
  instrumentation and control, 371
  iron and manganese control, 354
  limiting current density, 367
  membrane configuration, 341
  membrane types, 367
  operating mode, 368
  organics control, 354
  and pH, 355, 368
  posttreatment disinfection, 369
  posttreatment for corrosion control, 369
  power consumption, 368

product water quality, 367
recovery, 367
residuals disposal, 371
scaling control, 347
and silt density index, 347, 347*t.*
solute removal, 367
staging, 368
temperature, 367
temperature control, 355
and turbidity, 347, 347*t.*
typical applications, 339*t.*
Emergency Planning and Community Right-to-Know Act, 459, 460*t.*
Endangered Species Act, 644
Engineering. *See* Water treatment engineering
Engineering News Record Construction Cost Index, 713–714
Enhanced coagulation
defined, 87–88
Enhanced Surface Water Treatment Rule, 11, 12
ENR-CCI. *See* Engineering News Record Construction Cost Index
Entamoeba histolytica
and diatomaceous earth filtration, 207
Environmental impact
air emissions, 680–681
and architectural design, 533, 545
chemical handling and storage, 678–679
determining major issues, 675
environmental impact statements, 683
environmental review steps, 673, 674*t.*
and facility construction, 676
federal regulations, 682–685
groundwater withdrawals, 677–678
and landscaping, 677
and plant construction, 673–677, 674*t.*, 676*t.*
and plant operations, 677–681
of process residuals, 526–527
residuals treatment and disposal, 679–680
and site preparation, 675
and site selection, 642–644
stormwater management, 680
surface water withdrawals, 677
wastewater treatment and disposal, 680
Environmental Protection Agency. *See* U.S. Environmental Protection Agency
EPA. *See* U.S. Environmental Protection Agency
EPI-DMA polyamines, 759
Equipment
pre-engineered vs. custom-built, 22–23
start-up, 30
Estimates and estimating, 24, 705
and Americans with Disabilities Act, 711
bid documents, 28
conceptual, 706
contingency factor, 712–713, 713*t.*
and cost escalation, 714
cost estimating factors, 708, 710*t.*
cost indexes, 713–714
definitive, 707
and design changes ("scope creep"), 713
final cost estimates, 28

finalizing, 712–714
financial feasibility evaluation, 23
geotechnical and site issues, 710–711
historical cost data, 708, 709*t.*
levels, 705–707
methodologies, 707–710
order-of-magnitude, 706
percentage application method, 708, 710*t.*
preliminary, 706–707
and remote sites, 711
and sales tax, 711
six-tenths rule, 708
special cost considerations, 710–712
ESWTR. *See* Enhanced Surface Water Treatment Rule
Evaporators, 422–423, 423*f.*
Existing facilities
upgrading or expanding, 22
Expert systems, 596–597

Faraday's law, 366
Ferric chloride, 759
Ferric sulfate, 759
Ferrous sulfate, 759
Filtration, 1, 153. *See also* Backwashing, Diatomaceous earth filtration, Granular activated carbon, Membrane processes, Powdered activated carbon, Rapid sand filtration, Slow sand filtration
and lime softening, 294–295
methods, 15, 16
multipurpose effects, 16*t.*
O&M design issues, 691
pilot plants, 738–742, 740*f.*, 741*f.*
waste discharge, 482
Final design
architectural design, 26–27
bid documents, 28
civil/mechanical process design, 25–26
coordination of design disciplines, 25–27
design documents, 24–25
design reviews, 27–28
design standards, 25
drawings and specifications, 24
electrical design, 27
final cost estimates, 28
final documents, 28
geotechnical investigation, 26
hydraulic design, 26
instrumentation design, 27
mechanical design, 27
owner reviews, 28
project control, 24–25
regulatory agency reviews, 28
schedules, 25
site surveying, 26
structural design, 27
value engineering, 28
Financial feasibility evaluation, 23
Flocculation, 98
aids, 88
buoyant coarse media flocculator, 109*f.*

794 INDEX

Flocculation (*Cont.*)
  compartments, 98, 99f.
  contact, 106–107, 108f., 109f.
  defined, 88
  design criteria, 100t., 102t., 690
  downflow gravel-bed flocculator, 108f.
  end baffles, 106
  energy requirements, 101, 102t.
  facilities, 101f.
  flocculator types, 102–107
  horizontal shaft paddle flocculator, 99f.
  hydraulic, 103–106, 105f.
  incidental, 98–100
  in lime softening, 288
  maze and baffle flocculators, 103, 105f.
  mechanical flocculators, 99f., 101f., 102–103, 103f., 104f.
  oscillating flocculator, 104f.
  pilot plants, 736
  proprietary systems, 106, 107f.
  sludge recirculation, 106
  solids contact, 106
  time, 100–101
  upflow gravel-bed flocculator, 108f.
  vertical paddle flocculator, 103f.
Floods and flooding
  and intakes, 39
  and plant arrangement, 667–668
  and site selection, 644
Flowmeters
  differential pressure transmitters, 571–572
  electromagnetic, 572–573, 572f., 573f.
  orifice plates, 571
  piping requirements, 575, 576t.
  pitot tubes, 571, 571f.
  propeller, 575
  turbine, 575
  ultrasonic, 574
  venturi, 570
  vortex-shedding, 574
  weirs and flumes, 574
Fluoride
  removal by ion exchange processes, 314
  residual measurement instrumentation, 583–584
Fluorosilicic acid, 759–760
Foundations, 558
  deep, 560
  and geotechnical report, 558–559
  lateral soil pressures, 560
  mat, 559
  spread footings, 559
Free chlorine residuals, 232, 234t., 235f.
Freeboard, 471

Gas adsorption, 61
Geotechnical investigation, 26
Giardia
  and diatomaceous earth filtration, 193
  and rapid sand filtration, 184
  and slow sand filtration, 193, 196
  and Surface Water Treatment Rule, 222–223
Granular activated carbon, 16, 157, 377–378, 751–752. *See also* Powdered activated carbon
  abrasion resistance, 379
  addition to slow sand filtration, 205–206, 206t.
  adsorber configuration, 392–393
  adsorber maintenance, 405
  adsorber vessel design suggestions, 406–407
  adsorber volume and bed depth, 388–389
  adsorption beds, 392–393
  ash content, 379
  backwashing, 389, 391f., 403–404
  biological activity, 404–405, 406
  breakthrough, 386–388, 388f.
  capping slow sand filters, 396–397
  carbon loss, 405–406
  carbon usage rate, 389–392, 393, 410
  carbon weight, 378–379
  circular adsorbers, 400–401, 401t.
  construction materials, 408
  controlling carbon depth, 403–405
  design considerations, 386–392
  dry feed, 438
  durability, 379
  empty bed contact time, 386, 388
  expanded beds, 399
  facility sizing, 393
  as filter-adsorber, 385, 385f.
  fixed beds in parallel, 399
  fixed beds in series, 398–399
  hydraulic loading rate, 389, 390f.
  iodine number, 378
  location options, 385–386, 385f.
  maintenance requirements, 405–406
  manufacturing, 380
  mesh sizes, 386, 387t.
  moisture content, 379
  molasses number, 378
  multipurpose effects, 16t.
  open or enclosed adsorbers, 400, 400t.
  operating problems, 403
  particle size, 379
  physical properties, 378–379
  pilot plants, 738–742
  postfiltration adsorber configurations, 397–400, 398t.
  postfiltration adsorber design, 400–403
  postfiltration adsorbers, 385–386, 385f.
  as prefilter adsorber, 385, 385f.
  pressure or gravity flow adsorbers, 400, 400t.
  properties, 386, 387t.
  pulsed beds, 399–400
  rapid small-scale column test, 389
  reactivation rate, 410
  rectangular adsorbers, 401–403, 402f.
  regeneration, 408–414, 409f.
  regeneration by-products, 412–413
  regeneration furnaces, 410–412, 411t., 413f., 414f.
  for removal of VOCs after packed column stripping, 82, 82f.
  removing carbon from adsorbers, 404
  retrofitting GAC filters, 393–396, 397f.
  sandwich layer in slow sand filtration, 397
  storage and transport facility design suggestions, 407–408
  support systems, 394–396

INDEX **795**

thermal regeneration, 410
transfer equipment maintenance, 405
transporting hot regenerated carbon, 414
underdrains, 394–395, 395f., 396f., 406
washwater troughs, 393–394, 394f.
Gravimetric feeders, 435
Groundwater, 1
  disinfection rule, 228
  under the direct influence of surface water, 222
Guiding Principles for Chemical Accident Prevention, Preparedness, and Response, 678

Haloacetic acid, 11, 12
  treatment options, 14t.
Handbook on Chlorination, 232
Hardness. *See also* Ion exchange processes, Lime softening, Membrane processes
  AWWA goals, 282
  carbonate, 281
  defined, 281
  degrees of, 282
  hot-process softening, 281
  noncarbonate, 281
  treatment options, 14t.
Hazardous substances, 456, 643–644, 667, 712
Head, 470
Head loss
  in channels and troughs, 477
  in filter media, 475–476
  in filter piping, 476–477
  in gates and ports, 477
  in pipelines, 473–474
  in valves and fittings, 474–475
  in weirs, 477–478
Height of transfer unit (HTU), 76
Henry's law, 71, 72t.
High-rate clarification, 132
  plate settlers, 134–137, 135f., 136f.
  tube settlers, 132–134, 133f.
Historical cost data, 708, 709t.
Hydrated lime, 761
Hydraulic design, 26
Hydraulic grade line, 470, 471–472, 473f.
Hydraulic grade line elevation, 470
Hydraulics
  available filter head, 470, 479
  computer programs, 479
  and filter backwashing, 481–482
  and filter effluent piping, 480–481
  and filter waste discharge, 482
  flow distribution, 479–480
  freeboard, 471
  head, 470
  head loss, 473–478
  hydraulic grade line, 470, 471–472, 473f.
  hydraulic grade line elevation, 470
  maximum plant flow, 470
  and plant arrangement, 665–666
  preliminary design, 469–472
  preliminary profile 471–472
  and pumps, 482
  and rapid mixing basins, 481
Hydrochloric acid, 760

Hydrogen peroxide, 467, 760
Hydrogen sulfide, 369
Hypochlorous acid, 232

ICR. *See* Information Collection Rule
Information Collection Rule, 10
Inorganic chemicals
  treatment options, 14t.
Instrumentation, 567. *See also* Control systems
  analyzers and detectors, 581–584
  categorical schematic for treatment platn, 568f.
  conductivity sensors, 584
  controllers for efficiency in energy, materials, and labor, 570
  design, 27
  design specifications, 602–603
  drawings and diagrams, 600–602
  elementary control diagrams, 602
  emergency response, 570
  equipment monitoring, 569
  flowmeters, 570–575
  fluoride residual measurement, 583–584
  future trends, 603
  hydraulic controls, 569
  hydraulic monitoring, 568–569
  instrument loop diagrams, 602
  interconnection diagrams, 602
  level switches and transmitters, 575–579
  membrane processes, 371
  pH measurement, 584
  pressure switches and transmitters, 579
  process and instrumentation diagrams (P&IDs), 600–601, 601f.
  process controls, 569
  process monitoring, 569
  sample requirements for analytical measurement, 584
  signal error sources, 585, 586t.
  streaming current detectors, 584
  temperature switches and transmitters, 581
  water production controls, 569
Intakes, 31
  and anchor ice, 52
  capacity, 36
  cellular design, 44
  chemical treatment considerations, 50, 51t.
  components, 32, 32f.
  conduits, 46–48, 50t.
  crib, 35, 40f., 46, 48f.
  design, 41–56, 690
  and dynamic ice, 51, 52t.
  exposed, 33
  fish protection, 54, 55f.
  floating, 33, 36f.
  flood considerations, 39
  flotation considerations, 48
  and frazil ice, 51–53, 52t., 53f., 54, 54t.
  geotechnical considerations, 49–50
  hydraulic criteria, 41, 44t.
  hydraulically balanced inlet cones, 46, 49f.
  ice design considerations, 51–54
  icing control methods, 54
  inlet works and ports, 44–46, 45t., 46t.

796 INDEX

Intakes (*Cont.*)
  lake and reservoir, 39–41, 42*f.*, 46, 48*f.*
  location considerations, 37, 37*t.*, 39, 41
  mechanical considerations, 50
  movable carriage, 35, 36*f.*
  pump stations, 48
  purpose of, 31
  and quagga mussels, 55
  racks, 56–57, 57*t.*
  reliability and redundancy, 37
  river, 37–39, 38*f.*, 40*f.*
  screens, 56–57, 57*f.*, 57*t.*, 58*f.*, 59
  shore wells, 48
  silt and bed load considerations, 39
  siphon-well, 33, 35*f.*
  and static ice, 51, 52*t.*
  submerged, 33, 35, 43*f.*, 46, 49*f.*
  submergence design, 44
  and surface ice, 51
  towers, 33, 34*f.*, 42*f.*, 45, 47*f.*
  types of, 33–35, 33*t.*, 37–41
  ventilation systems, 50
  vertical velocity cap, 54, 55*f.*
  and water quality, 41
  and zebra mussels, 54–55, 56*t.*
Integrated treatment systems, 2
International Research Center, 198
Ion exchange processes, 299–300
  acid and caustic cycles, 325
  adsorbents, 303
  ammonia removal, 305–306
  anion processes, 308–317
  arsenic removal, 310–312
  automation, 329
  barium removal, 306–307
  brine wastes, 491–492, 492*t.*
  cation processes, 304–308
  centralized systems, 301
  chemical fouling, 304
  chemical storage and dilution, 327–328
  co-flow exchangers, 319–321, 320*f.*
  color removal, 317
  consultants, 330
  countercurrent regeneration exchangers, 322–323, 322*f.*
  dealkalization, 309–310
  degasification, 319
  demineralization, 318,324, 324*f.*
  distributors, 327
  equipment design, 318–329
  external regeneration, 325
  fiberglass tanks, 326
  and flow rates, 318
  fluoride removal, 314
  future trends, 330–331
  gel and macroporous resins, 302–303, 303*t.*
  inorganic zeolites, 299, 303
  media depth, 318–319
  monitoring, 328–329
  nitrate removal, 312–314
  operating capacity, 305
  oxidation and resin degradation, 304

  packed-bed exchangers, 324, 324*f.*
  piping design, 327
  point-of-use systems, 301
  radium removal, 307–308
  regeneration frequency, 318–319
  regeneration reactions, 300–301
  resin capacity, 300
  resin degradation, 303–304
  resin qualities, 302–303, 302*t.*
  resin regeneration, 324–325
  resin replacement monitoring, 329
  resins, 302–304, 302*t.*
  salt regeneration, 324–325
  salt selection, 305
  selenium removal, 314–315
  and source water quality, 301
  split-flow exchangers, 321–322, 321*f.*
  steel tanks, 326
  sulfate removal, 312
  suppliers, 330
  tanks, 326
  THM precursor removal, 317
  TOC removal, 317
  underdrains, 326–327
  uranium removal, 315–316
  vessel design, 325–327
  waste collection and disposal, 328
Iron
  control in membrane processes, 354
  and diatomaceous earth filtration, 209–210
  iron coagulant recovery, 497
  oxidation, 229, 230*t.*, 230*f.*
  precipitate sludge, 489–490
  pretreatment for membrane processes, 354
  and slow sand filtration, 196
  treatment options, 14*t.*

Koch, Robert, 221

Laboratory facilities
  and architectural design, 538–540, 541*f.*
  personnel, 703
  and plant arrangement, 669–670
Laboratory supervisor, 702
Lagoons
  in plant arrangement, 667
  sludge, 493, 495, 505–506, 506*f.*
Landscaping, 668, 677
Langelier Saturation Index, 348–349, 348*f.*
Lead and Copper Rule, 10
Level switches
  conductance probe, 577
  float-type, 576–577
Level transmitters
  continuous purge bubbler, 578
  differential pressure (for use in closed vessels), 579
  periodic purge bubbler, 578
  radio frequency capacitance, 579
  submersible hydrostatic, 577–578, 577*f.*
  ultrasonic, 578–579

Lime, 760
　dolomitic, 761
　dolomitic hydrated, 761
　hydrated, 761
　quicklime, 761
　slakers, 436
　slurry, 445–447
Lime softening
　aeration, 286
　caustic soda use, 285
　chemical feed considerations, 295
　chemical reactions, 284
　chemical requirements, 284–285
　chemistry, 283–285
　conventional softening basins, 287–289
　design considerations, 292–295
　encrustation problems, 293
　excess lime process, 283, 291–292, 291f.
　and filtration, 294–295
　and flocculation, 288
　future trends, 295–296
　lime recovery, 497–498
　lime-soda ash process, 283
　pellet reactors, 295
　presedimentation, 286
　pretreatment, 286
　recarbonation, 287, 291–292
　recarbonation basins, 289, 691
　residuals production, 293–294
　residuals recycling and particle-size distribution, 288, 288f.
　residuals removal and handling, 293
　single-stage lime process, 283, 287–290, 287f.
　solids contact softening basins, 289, 290f.
　split-treatment, 285, 292, 292f.
　types of processes, 283, 283t.
　U.S. plants, 283
Liquid oxygen, 467, 761
Loss-in-weight gravimetric feeders, 435
LSI. *See* Langelier Saturation Index

Maintenance. *See also* Operations and maintenance
　building, 703
　electrical systems, 617–619, 703
　facilities design, 541–544, 543f., 698
　granular activated carbon, 405–406
　grounds, 703
　instrumentation, 703
　mechanical, 702
　personnel, 701, 702–703
　and plant arrangement, 669
　slow sand filtration, 199–201, 200f.
　storage, 542–544
　supervisor, 701
　tools, 542
Manganese
　control in membrane processes, 354
　and diatomaceous earth filtration, 209–210
　oxidation, 229, 229t., 230f.
　precipitate sludge, 489–490
　pretreatment for membrane processes, 354

　and slow sand filtration, 196
　treatment options, 14t.
Master plan, 7
　and alternative source waters, 9
　alternative treatment options, 8, 9
　background data, 8
　issues, 7–8
　main considerations, 8
　periods covered, 9
　and proposed rules, 12–13
　and regulatory considerations, 8–9
Material Safety Data Sheets, 456, 461, 461t.
Maximum contaminant levels, 1
Maximum plant flow, 470
McCabe-Thiele diagram, 85
MCLs. *See* Maximum contaminant levels
Mechanical design, 27
Membrane processes, 16, 335. *See also* Electrodialysis/electrodialysis reversal, Microfiltration, Nanofiltration, Reverse osmosis, Ultrafiltration
　applied pressure, 341
　bench testing, 374
　brine wastes, 490–491
　building design, 372
　calcium carbonate control, 348–349
　concentrate pressure, 341
　concentration of sparingly soluble salts and silica, 352, 352t.
　construction materials, 373
　differential pressure (delta-P), 341
　electrical voltage-driven, 337
　electrodialysis, 337, 338, 339t., 341
　electrodialysis reversal, 337, 338, 341
　feed pressure, 341
　feed water, 344, 345t.
　flat-sheet membranes, 338
　flux, 341
　hollow-fiber membranes, 338–341
　hydrogen sulfide control, 354
　instrumentation and control, 371
　interstage pressure, 341
　iron and manganese control, 354
　membrane cleaning systems, 370
　membrane composition, 338
　membrane configurations, 338–341
　microbial control, 353
　microfiltration, 336, 339–341, 339t.
　nanofiltration, 336, 339t., 356–364
　net driving pressure, 343
　operations and maintenance design, 372
　organics control, 354
　osmotic pressure, 341–342
　pH control, 354
　pilot testing, 374
　posttreatment disinfection, 369
　posttreatment for corrosion control, 369
　pressure-driven, 336–337, 336f., 337t., 338
　pretreatment, 344–353, 346f.
　process reliability and redundancy, 374
　product pressure, 343
　recovery, 343

798 INDEX

Membrane processes (*Cont.*)
  removal of gases and VOCs, 369
  residuals disposal, 371
  reverse osmosis, 336, 338–339, 339*t.*, 340*f.*, 356–364
  scaling control, 347
  schematic, 344*f.*
  selecting, 337–338, 339*t.*
  silica control, 349–351, 351*f.*
  and silt density index, 347, 347*t.*
  solute passage, 343
  solute rejection, 343
  spiral-wound modules, 338
  sulfate scale control, 349, 350*f.*
  suspended solids control, 346–347
  system components, 344–355, 344*f.*
  temperature control, 355
  terminology, 341–343
  transmembrane osmotic pressure, 342
  transmembrane pressure, 343
  and turbidity, 347, 347*t.*
  types of, 335–341
  ultrafiltration, 336, 339, 339*t.*, 341*f.*
  vendor contact, 374
Metropolitan Water District of Southern California, 33–35, 34*f.*, 35*f.*
MF. *See* Microfiltration
Microfiltration, 336
  backwashing, 366
  cleaning system, 366
  cross-flow velocity, 366
  flux, 365
  instrumentation and control, 371
  membrane configuration, 339–341
  membrane integrity tests, 366
  membrane module arrays, 366
  pumping requirement, 365
  recovery, 365
  solute rejection, 365
  temperature, 365
  temperature control, 355
  typical applications, 339*t.*
Milwaukee, Wisconsin, 46, 48*f.*
Mixing
  defined, 87
Monroe Reservoir, 45, 47*f.*

Nanofiltration, 336
  automatic flushing systems, 364
  blending membrane permeate and bypass water, 362–363
  concentrate staging, 361–362, 361*f.*
  concentrate stream solubility criteria, 352*t.*
  design equations, 356
  energy recovery, 364
  feed pressure and recovery, 356, 357*t.*
  feed pressure requirement, 360
  flux, 359
  flux decline curve, 359, 361*f.*
  hydraulics, 358
  hydrogen sulfide control, 354
  instrumentation and control, 371
  iron and manganese control, 354
  membrane module arrays and staging, 361–362
  membrane type, 358
  microbial control, 353
  operating pH, 359
  organics control, 354
  performance modeling and system design, 364
  permeate drawback tank, 364
  permeate staging, 362, 362*f.*
  permeate water quality, 358
  and pH control, 355
  posttreatment for corrosion control, 369
  pretreatment, 346*f.*
  recovery considerations, 358
  residuals disposal, 371
  scaling, 358
  scaling control, 347
  and silt density index, 347, 347*t.*
  solute rejection and passage, 358
  source water use, 358
  sulfate scale control, 349, 350*f.*
  temperature, 359
  temperature control, 355
  and turbidity, 347, 347*t.*
  typical applications, 339*t.*
National Environmental Policy Act, 683
National Fire Protection Association, 464, 532, 538, 693
  electrical standard, 605
National Historic Preservation Act, 644
National Pollution Discharge Elimination System, 683, 684
  and rapid sand filtration, 186
National Register of Historical Places, 644
National Science Foundation
  rapid sand filtration standards, 187
National Secondary Drinking Water standards, 222
Net positive suction head, 449
Neural networks, 597–598
NF. *See* Nanofiltration
NIOSH Pocket Guide to Chemical Hazards, 456
Nitrates, 312
  removal by selective resins in ion exchange processes, 313–314
  removal by type 1 and type 2 resins in ion exchange processes, 312–313, 313*f.*
North Pennsylvania Water Authority, 64
NPDES. *See* National Pollution Discharge Elimination System
NPSH. *See* Net positive suction head
Nutrients
  and slow sand filtration, 196

Occupational Safety and Health Administration, 25, 187, 459–460, 538, 539, 663–664, 683, 693
Operations and maintenance. *See also* Maintenance
  accessibility of equipment, 695, 695*f.*, 696*f.*
  budget, 699–700
  chemicals budget, 699
  computer-aided, 598, 631
  computer assistance software, 692–693
  control system design issues, 689–690

cranes and hoists, 696, 697f.
design considerations for maintenance, 694–698
design considerations for operations, 689–694
design review, 694
designing adequate power and lighting outlets, 696
designing to facilitate, 631
and excess equipment, 694
equipment budget, 700
facility O&M manual, 723–724, 725t.
laboratory supervisor, 702
maintenance supervisor, 701
membrane processes, 372
operating permits, 699
operations supervisor, 701
plant manager, 701
plant security design issues, 693
power system design issues, 689–690
preparing for plant start-up, 698–703
process system design issues, 690–692
and safety, 632
safety design issues, 693–694
staffing budget, 699
staffing plan, 700–703
staffing requirements, 701, 702t.
start-up/operations services manager, 727
supervisory positions, 701–702
training, 631
training budget, 699–700
tunnels and galleries, 696–698
and up-to-date drawings, 631
utilities budget, 699
vendor manuals, 717–719, 718t., 719t.
Operations supervisor, 701
Operator training, 30, 715–716
  in construction phase, 719
  in design phase, 716
  systems training, 722–723
  training materials development, 721, 722–723
  vendor manuals, 717–719, 718t., 719t.
  vendor training, 716–717, 717t., 720–722
Organic chemicals
  treatment options, 14t.
Organic compounds, 1
Organizations. See also AACE International, American National Standards Institute, American Water Works Association, Construction Specifications Institute, International Research Center, National Fire Protection Association, National Science Foundation, Occupational Safety and Health Administration, Portland Cement Association, U.S. Army Corps of Engineers, U.S. Bureau of Reclamation, U.S. Environmental Protection Agency
  offering design standards and recommendations, 547–551
OSHA. See Occupational Safety and Health Administration
Owner
  approval of construction change orders, 29
  final design reviews, 28

Oxidation, 228
  chlorination, 230–241
  iron and manganese removal, 229, 229t., 230t., 230f.
  preoxidation pilot plants, 734–735
  taste and odor control, 229, 231f.
Oxygen
  handling and storage, 422
  liquid, 467, 761
Ozone and ozonation, 16, 254, 762
  adsorptive air separation, 271–273
  aftercoolers, 264
  air feed systems, 260–261, 262f., 263f.
  cell evaluation, 275–276, 276t.
  chemistry, 254–256, 255f., 256f.
  chilled water driers, 264–265
  coalescing filters, 267, 267t.
  compressor selection, 263–264
  contactor selection, 274, 274t.
  contactors, 273–277
  contaminants, 259–260
  conventional horizontal tube generator, 257, 258f.
  cryogenic air separation, 268–271, 270f.
  desiccant drying, 265–266
  in diatomaceous earth recovery, 219
  direct expansion driers, 264
  direct injection contactors, 276–277
  external heat regeneration, 266
  feed gas, 259–260
  final feed gas, 267
  fine bubble contactors, 274–276
  gas compressors, 261–264
  gas filtration, 266–267
  generation, 256–259, 257f., 258f.
  high purity oxygen, 267–268
  internal heat regeneration, 266
  liquid ring compressors, 261
  liquid storage tanks, 267–268
  Lowther plate generator, 257, 258f.
  mass flow rate, 260
  moisture content, 259
  off-gas disposal, 277–278
  on-site generation of oxygen, 268–273
  Otto plate generator, 257, 258f.
  oxygen concentration, 260
  packed column contactors, 276
  particulates, 259–260
  pilot plants, 742–745, 744f.
  preozonation with slow sand filtration, 205
  pressure, 260
  pressure swing regeneration, 265–266
  raw gas filtration, 266–267
  refrigerative drying, 264–265
  rotary lobe compressors, 263
  rotary screw compressors, 261–263
  system construction materials, 278
  systems, 256–273
  temperature, 260
  transfer efficiency factors, 273
  turbine contactors, 276
  vaporizers, 268, 269t.

P&IDs. *See* Process and instrumentation diagrams
Packed columns, 62–63, 64f., 79f.
  air and water flow requirements, 75–76
  air blowers, 79
  column design, 76, 77f., 78f.
  design considerations, 78
  distributors, 78, 80f.
  exhaust emission considerations, 81–83
  fouling, 83
  height of transfer unit (HTU), 76
  packing, 78–79, 81f.
  pilot testing, 84–85
  site considerations, 79–81
Particle counters, 583
Particulates
  treatment options, 14t.
Partnership Program for Safe Water, 12
Pasteur, Louis, 221
Percentage application estimating method, 708, 710t.
Permits. *See* Project permitting
Pesticides, 1
pH
  control systems, 591
  measurement instrumentation, 584
Phosphates, 467, 762
Phosphoric acid, 467, 762
Pilot plants, 729
  backwash set-up, 742, 743f.
  and bench-scale testing, 731–733
  construction, 746–749
  for clarification, 736–738
  for coagulation, 735–736
  design, 731–745
  for dissolved air flotation, 737, 738f., 739
  electrical power requirements, 733–734
  for filtration, 738–742, 740f., 741f.
  for flocculation, 736
  floor plan, 748f.
  for GAC contactors, 738–742
  hydraulic requirements, 733
  installation, 749
  modular, 747
  options for designing, building, leasing, 746–747
  for ozonation, 742–745, 744f.
  for preoxidation, 734–735
  permanent, 747
  process drawings and specifications, 747–749, 748f.
  schematic, 748f.
  sedimentation, 736–737
  start-up, 749
  temporary, 747
  for treatment residuals, 745, 746f.
Pilot testing
  aeration, 83–85
  checklist, 731
  membrane processes, 374
  packed columns, 84–85
  process residuals, 528
  purposes, 730–731
  slow sand filtration, 204

Pipes and piping
  and air entrainment, 480–481
  ANSI color coding, 546
  butterfly valve alignment, 480
  in chemical feed systems, 449–452, 450t.–451t.
  compound 90-degree bends, 480
  conveyance of water between process units, 666
  filter effluent, 480–481
  and flowmeters, 575, 576t.
  for gas in chlorination, 236–239
  head loss, 473–474, 476–477
  for ion exchange processes, 327
  net positive suction head, 449
  pipe lateral underdrains in rapid sand filtration, 172–173, 172f.
  in slow sand filtration, 202, 202f.
Plant arrangement, 655. *See also* Site selection
  access and facilities for visitors, 664–665
  access for disabled persons, 663
  access for equipment maintenance, 663
  architectural considerations, 668
  basic layout types, 655–661, 657f.
  building orientation, 668
  campus layout, 657f., 658
  climate considerations, 668
  compact layout, 657f., 658–661, 659f., 660t., 662f.
  convenience of routine operations, 663
  conveyance of water between process units, 666
  factors impacting, 655, 656t.
  finished water storage, 667
  and flexibility of operations, 661
  and flooding, 667–668
  and hazardous chemical location, 667
  and hydraulic profile, 665–666
  laboratory facilties, 669–670
  lagoons, 667
  landscaping, 668
  linear layout, 656–658, 657f.
  maintenance facilities, 669
  noise control, 663–664
  operational issues, 661–665
  possible capacity expansion, 669
  possible process additions, 669
  provision for future changes, 668–669
  and redundancy, 661
  separation of filtered and nonfiltered water, 666
  single application points for chemicals, 666
  site security, 665, 693
  source water storage, 667
  support facilities, 669–670
  technical issues, 665–668
Plant manager, 701
Plant siting analysis, 21–22. *See also* Site selection
Plant start-up, 715, 716. *See also* Operator training
  equipment start-up, 30
  facility O&M manual, 723–724, 725t.
  final, 724
  initiation, 724–726
  interim, 724
  manager, 727
  operations and maintenance preparations, 698–703
  operator training, 30

plan elements, 726, 727t.
planning, 724–726
post–start-up activities, 728
record keeping, 727
sequence, 726–727
workshops, 726
Plate settlers, 134, 135f., 136f.
  basin dimensions, 135
  design criteria, 135
  inlet design, 136
  outlet design, 136–137
  solids removal, 137
  theory, 134–135
Polyaluminum chloride, 467, 762
PolyDADMAC, 762
Polymers, 467, 763
  in sludge conditioning, 518–519, 519f.
Portland Cement Association, 562
Potassium hydroxide, 763
Potassium hypochlorite, 763
Potassium permanganate, 229, 231f., 763
  volumetric feeder, 436, 437f.
Powdered activated carbon, 377–378, 752
  application equipment, 382–384, 383f.
  application methods, 382
  black water, 384
  characteristics, 380
  chemical handling, 384
  delivery, 382
  dry feed, 378, 382, 438
  explosive possibilities of carbon dust, 384
  interaction with other treatment chemicals, 384
  manufacturers' specifications, 380t.
  manufacturing, 380
  operating problems, 384–385
  selecting application point, 380–381, 381t.
  slurry, 378, 382, 383–384, 383f., 384–385
Preliminary design
  coordinating with regulatory agencies, 21
  coordination with support disciplines, 23–24
  cost estimates, 24
  design development, 23
  financial feasibility evaluation, 23
  plant siting analysis, 21–22
  pre-engineered or custom-built equipment, 22–23
  upgrading or expanding existing facilities, 22
Preliminary estimates, 706–707
Pressure switches and transmitters
  accessories, 581
  Bourdon tubes, 579
  capacitive, 579–580, 580f.
Primary Drinking Water Standards, 10, 11t.
Privatization, 3
Process and instrumentation diagrams, 600–601, 601f.
Process residuals
  alum recovery, 497
  alum sludge, 488–489
  bench testing to select handling processes, 527–528
  brine wastes and disposal, 490–492, 494
  coagulant recovery, 497
  coagulant sludges, 488–489
  contingency planning, 525–526, 525t.

deep well injection, 495–496
development of handling system, 485, 487f.
disposal limitations, 522
disposal permits and regulations, 494–496, 684
disposal process selection, 526–528
environmental impacts, 526–527, 679–680
equipment sizing, 525
estimating solids production, 524–525
filter backwash water sludge, 487–488
handling and transportation, 498, 516–519, 517f., 518f., 519f.
information sources, 528
ion exchange process waste, 491–492, 492t.
iron and manganese precipitates, 489–490
iron coagulant recovery, 497
iron salts coagulation sludge, 489
lime recovery, 497–498
mass balance, 523–524
membrane process waste, 490–491
minimizing sludge quantities, 496
pilot plants, 745, 746f.
pilot testing to select handling processes, 528
quantity and quality data, 522–523
recycling or disposing of liquid wastes, 519–520, 520f.
resource recovery potential, 523
sludge dewatering, 498–499, 502–515
sludge disposal, 492–494
sludge drying, 498–499, 515–516
sludge thickening, 498–499, 499–502, 521, 521t.
sludge types, 486–490
softening-plant sludge, 490
toxic characteristic leaching procedure, 679
types of, 485, 486f.
Programmable logic controllers, 592–593, 593f., 626, 630
  future trends, 603, 626
Project control, 24–25
Project permitting, 681, 682t.
  air emissions regulations, 684–685
  for chemical delivery and handling, 683–684
  compliance, 687
  for construction, 682–683
  environmental impact statements, 683
  federal environmental regulations, 682–685
  National Environmental Policy Act, 683
  National Pollution Discharge Elimination System, 683
  for operations, 683–685, 699
  and regulatory trends, 685
  for residuals disposal, 684
  state and local regulations, 685
  strategy, 686–687, 686t.
  from U.S. Army Corps of Engineers, 683
  for wastewater disposal, 684
Pumps and pumping
  centrifugal pumps, 442
  for chemicals, 440, 442
  and hydraulic design, 482
  intakes pump stations, 48
  microfiltration requirement, 365
  piston and diaphragm pumps, 440

Pumps and pumping (*Cont.*)
  progressive cavity pumps, 440
  slurry pumps, 447
  transfer pumps, 448–449
  ultrafiltration requirement, 365

Quagga mussels, 55
Quicklime, 761

Radium
  removal by ion exchange processes, 307–308
Rapid mixing
  design criteria, 92, 93, 94*t.*, 690
  and head loss, 481
  types of systems, 91–98
Rapid sand filtration. *See also* Slow sand filtration
  air binding, 189
  ANSI-NSF standards, 187
  auxiliary equipment, 182–184
  AWWA standards, 186
  block underdrains, 173, 173*f.*, 174*f.*, 175*f.*, 176*f.*
  coarse-to-fine media, 156–158
  conduits, 182–183, 182*f.*
  constant-rate control with constant water level and influent splitting, 160, 161*f.*
  constant-rate control with rate-of-flow controller, 159–160, 160*f.*
  constant-rate control with varying water levels and influent flow splitting, 160–161, 162*f.*
  continuous backwash, 165–166
  declining-rate control, 161, 162*f.*
  dual-media beds, 156–157
  false-bottom underdrains, 175–176, 177*f.*, 178*f.*
  filter aid polymers, 155
  filter box depth, 170–171
  filter configuration, 168–171, 170*f.*
  filter design criteria, 184–187
  filter efficiency, 187–188
  filter media, 155–158
  filter media replacement, 190
  filter media washing, 161–168
  filter operational control, 159–161
  filter performance monitoring, 178
  filter ripening, 187
  filter wash rate, 181
  filtration rate, 158, 179–180
  fine-to-coarse media, 156
  flocculation and coagulation, 154
  granular activated carbon, 157
  head loss monitoring, 180
  inadequate filter washing, 188–189
  inadequate pretreatment, 188
  length of filter run, 181
  mechanism, 153–154
  media depth and size relationship, 158, 159*f.*
  multipurpose effects, 16*t.*
  and National Pollution Discharge Elimination System, 186
  net filter production, 188
  number of filters, 168
  operation and control, 181
  particle counting, 179
  pilot filters, 181
  pipe lateral underdrains, 172–173, 172*f.*
  porous-bottom underdrains, 176–178
  pretreatment, 154–155
  regulatory agency standards, 184–196
  restart after shutdown, 189–190
  self-backwashing systems, 163–164, 163*f.*
  size of filters, 168
  support gravel upset, 189
  and Surface Water Treatment Rule, 184–185
  turbidity monitoring, 178–179
  underdrain systems, 171–178
  ungraded media, 156
  uniform media, 158
  upflow water backwash with air scour, 164–165, 167–168
  upflow water backwash with surface wash, 165, 166*f.*, 168*f.*
  upflow water backwash without auxiliary scour, 164, 167
  valves, 183
  wash rates, 166–168, 167*f.*, 168*f.*
  washwater sources, 162–164
  washwater troughs, 183–184, 185*f.*, 186*f.*
Recommended Standards for Water Works. *See* Ten States Standards
Rectangular sedimentation basins
  carriage-type collectors, 120
  chain-and-flight collectors, 118*f.*, 119*f.*
  circular collector equipment, 119–120
  cross collectors, 121, 121*f.*
  cross hoppers, 121–122, 122*f.*
  dimensions, 114–115
  floating bridge-type hydraulic removal systems, 124–128, 127*f.*
  indexing grid systems, 123–124, 123*f.*
  inlet zone, 116–117, 116*f.*
  manual solids removal, 117–118
  mechanical solids removal, 118–128
  outlet design, 117
  track-mounted hydraulic removal systems, 124, 125*f.*–126*f.*
  traditional desludging equipment, 118–119, 118*f.*, 119*f.*
  two-tray, 115*f.*
  underflow control, 128
Redundancy, 621, 629*t.*, 661
Regulations, 2
  current trends, 685
  finalized, 10
  and project permitting, 685
  proposed rules, 10–13
  stages of development, 9–10
Regulatory agencies
  approval of construction change orders, 29–30
  design reviews, 28
  and preliminary design, 21
Reliability, 621
  A-B syndrome, 624, 624*f.*, 625*f.*
  of chemical storage, 630
  components affecting, 623–624
  disaster considerations, 627

failure analysis, 623
flexibility of process units, 622–623
gravity flow or pumping, 624–625
and human-caused disasters, 627
of individual components, 626
of individual unit processes, 624, 624f.
multiple processes performing same function, 622
of multiple treatment units, 622
and natural disasters, 627
and operations and maintenance, 631
overall, 623
of power supply, 629–630
of process performance, 630
of process train, 624, 625f.
of processes, 628–629, 629t.
and redundancy, 621, 629t.
reserve process capacity, 622
of source water supply, 627–628
Remote terminal units, 593–594, 594f.
future trends, 603
Residuals. See Free chlorine residuals, Process residuals
Resource Conservation and Recovery Act, 495, 684
Reverse osmosis, 336, 338–339
automatic flushing systems, 364
blending membrane permeate and bypass water, 363
concentrate staging, 361–362, 362f.
concentrate stream solubility criteria, 353t.
design equations, 357
energy recovery, 364
feed pressure and recovery, 357, 357t.
feed pressure requirement, 360
flux, 359
flux decline curve, 359, 361f.
hydraulics, 358
hydrogen sulfide control, 354
instrumentation and control, 371
iron and manganese control, 354
membrane configurations, 338–339, 340f.
membrane module arrays and staging, 361–362
membrane type, 358
microbial control, 353
operating pH, 359
organics control, 354
performance modeling and system design, 364
permeate drawback tank, 364
permeate staging, 362, 362f.
permeate water quality, 358
and pH control, 355
posttreatment for corrosion control, 369
pretreatment, 346f.
recovery considerations, 358
residuals disposal, 371
scaling, 358
scaling control, 347
silica control, 351
and silt density index, 347, 347t.
solute rejection and passage, 358
source water use, 358
sulfate scale control, 349, 350f.
temperature, 359

temperature control, 355
and turbidity, 347, 347t.
typical applications, 339t.
Reynold's number, 237
RO. See Reverse osmosis

Safe Drinking Water Act, 9, 10, 12, 495, 538
Safety. See also Occupational Safety and Health Administration
chemical design considerations, 456–466, 457t.–458t.
chemical safety equipment, 462–463
chlorine gas leak requirements, 466
Material Safety Data Sheets, 456, 461, 461t.
and operations and maintenance, 632
operations and maintenance design issues, 693–694
safety showers, 463
screens, 463
SCADA systems, 595–596, 630
Schedules, 25
Schmutzdecke, 194
Scope creep, 713
SDSI. See Stiff and Davis Saturation Index
Secondary Drinking Water Standards, 10, 12t.
Sedimentation
pilot plants, 736–737
Sedimentation basins
circular, 128–132
O&M design issues, 690
rectangular, 114–128
square, 132
Sedimentation theory, 112
Seismic design, 560
and equipment, 561
and soil liquefaction possibility, 560
and tanks, 561
Selenium, 314
removal by ion exchange processes, 314–315
Shasta Dam, 45
Silt density index
and membrane processes, 347, 347t.
Site selection. See also Plant arrangement
aesthetic issues, 642–643
air quality concerns, 643
appropriate criteria, 654
archaeological and historical considerations, 644
buffer area, 642
and capital costs, 648
coarse screening, 651, 652–653
compensation measures, 651
criteria, 638–639, 640t.
criteria weighting, 654
drainage, 646, 647
ecological issues, 644
elevation and pumping considerations, 644–645
environmental issues, 642–644
evaluation methodology, 651–655
factors involved, 636, 637t.
financial considerations, 648–651
floodplain considerations, 644
and future expansion, 642

804 INDEX

Site selection. (*Cont.*)
  hazardous chemical concerns, 643–644
  institutional issues, 647–648
  land acquisition and ownership, 647–648
  and local politics, 648
  minimum area, 639–642, 641*f.*, 641*t.*
  mitigation measures, 651
  multiattribute analysis techniques, 651, 653
  multiple sites, 636
  net present value, 649
  net present value and inflation, 650
  noise level concerns, 643
  and operating costs, 648
  opportunity costs, 648–649
  overlay mapping, 651–652
  permits and approvals, 648
  and plant arrangement planning, 635
  proximity to distribution system, 645
  proximity to utility services, 646
  proximity to water source, 645
  public participation, 636–638
  and public response, 635–636
  risk and uncertainty analysis, 654–655
  seismic potential, 647
  sieving techniques, 651–655
  size requirements, 639–642, 641*f.*, 641*t.*
  steps, 636–639, 638*t.*
  sunk costs, 649
  and taxes, 648
  technical issues, 644–646
  topography and soil, 646–647
  traffic and transportation issues, 644
  transportation and site access, 645–646
  wetland considerations, 644
Site surveying, 26
Six-tenths rule, 708
Slow sand filtration, 16, 193–194. *See also* Rapid sand filtration
  access ramp, 199, 200*f.*
  adding a granular activated carbon cap, 396–397
  adding a granular activated carbon sandwich layer, 397, 398*f.*
  and algae, 195, 196*t.*
  automation, 203
  and color, 195
  and *Cryptosporidium*, 193, 196
  and dissolved oxygen, 196
  effluent water quality, 196, 197*t.*
  and entrapment, 193
  filter bed, 197*f.*
  filter box, 197–198
  filter design, 196–203
  filter draining, 201–202, 202f
  filter fabric, 206
  filter harrowing, 206–207, 207*f.*
  filter rates, 203
  filter-to-waste system, 201
  flow control, 202
  and *Giardia*, 193, 196
  granular activated carbon addition, 205–206, 206*t.*
  historical overview, 193–194
  and iron, 196
  lighting, 201
  maintenance and cleaning access, 199–201, 200*f.*
  and manganese, 196
  new developments, 205–207
  and nutrients, 196
  pilot testing, 204
  piping gallery, 202, 202*f.*
  preozonation, 205
  recording and monitoring, 203
  roofing system, 201
  roughing filters, 205
  sand characteristics and depth, 198
  schmutzdecke, 194
  ship's doors, 199, 201
  source water quality, 194–196, 195*t.*
  source water storage, 198–199
  support system, 198
  and Surface Water Treatment Rule, 194
  treatment mechanism, 194
  and turbidity, 193, 194–195
  underdrain, 198, 199*f.*
  ventilation, 201
Sludge
  alum sludge, 488–489
  belt presses, 508–511, 512*f.*
  burial in a landfill, 493–494, 495
  centrifuges, 511–513, 513*f.*, 514*f.*, 515*f.*
  coagulant sludges, 488–489
  dewatering, 498–499, 502–515
  discharge to a lagoon, 493, 495
  discharge to a natural waterway, 492–493, 494–495
  discharge to a sanitary sewer, 493, 495
  disposal alternatives, 492–494
  drying, 498–499, 515–516
  drying beds, 503–504
  filter backwash water sludge, 487–488
  filter presses, 507–508, 508*f.*, 509*f.*, 510*f.*
  flotation thickening, 499–501
  freeze-assisted sand beds, 504
  gravity belt thickeners, 501, 502*f.*
  gravity thickening, 499, 500*f.*
  handling and transportation, 498, 516–519, 517*f.*, 518*f.*, 519*f.*
  high-concentration wastes, 517–518
  iron and manganese precipitates, 489–490
  iron salts coagulation sludge, 489
  lagoons, 505–506, 506*f.*
  land application, 494, 495
  low-concentration wastes, 517
  mechanical dewatering, 506–515
  mechanical thickening, 501–502, 502*f.*
  minimizing quantities, 496
  natural dewatering, 502–506, 503*f.*
  polymers added for conditioning, 518–519, 519*f.*
  softening-plant sludge, 490
  solar drying beds, 504–505, 505*f.*
  thickening, 498–499, 499–502, 521, 521*t.*
  types, 486–490
  vacuum filtration, 513–515, 516*f.*
  vacuum-assisted drying beds, 505
  wedgewire beds, 505

Sludge blanket clarification, 140–141, 142f., 143f.
  design criteria, 141
  inlet design, 144–145
  outlet design, 145
  pulsed blanket clarifier, 141–144, 143f.
  sludge removal, 145
Slurry. *See also* Solids contact/slurry recirculation units
  carbon, 442–445, 444f.
  diatomaceous earth filtration, 218
  dipper-wheel feeders, 447
  feed systems, 442–447
  flow control valves, 447
  lime, 445–447
  powdered activated carbon, 378, 382, 383–384, 383f., 384–385
  pumps, 447
  splitter boxes, 447
Snow, John, 221
Soda ash
  feed system, 439
  in lime softening process, 283
Sodium aluminate, 764
Sodium bisulfite, 764
Sodium carbonate, 764. *See also* Soda ash
Sodium chloride, 764–765
  feed system, 439
Sodium chlorite, 254, 765
  feed system, 438–439
Sodium fluoride, 765
Sodium fluorosilicate, 766
Sodium hexametaphosphate, 765
Sodium hydroxide, 765–766
Sodium hypochlorite, 232, 242, 766
  feed, 244
  increased use of, 466
  purchasing, 242–243
  storage, 243–244, 430–433, 432f., 433t.
Sodium silicate, 766
Sodium sulfite, 767
Softening. *See* Water softening
Solids contact clarifiers
  defined, 88
Solids contact/slurry recirculation units, 137, 138f.
  accelerator design, 139–140, 140f.
  effluent design, 139
  equipment, 139
  flocculation/contact zone, 138
  influent design, 139
  operation and design criteria, 137–139
  solids removal, 139
  surface loading rate, 138
Specifications, 602–603
Square sedimentation basins, 132
Stiff and Davis Saturation Index, 348–349, 349f.
Stormwater management, 680
Streaming current detectors, 584
Stripping, 61
Structural design, 27
  and aluminum, 555
  basins, 562–563
  of buildings, 563–565
  channels, 562–563
  codes and standards, 554
  conduits, 562–563
  and concrete, 554–555
  and construction below water table, 560
  corrosion protection, 558
  design team, 554
  efficiency of structures, 556
  and fiberglass, 555
  foundations, 558–560
  and freezing, 557
  and geotechnical report, 558–559
  interaction between structures, 556
  jointing of structures, 556–557
  lateral soil pressures, 560
  layout of structures, 555–556
  loading criteria, 562–563
  and masonry, 555
  mat foundations, 559
  material selection, 554–555
  and public access considerations, 557
  responsibilities of structural engineer, 553
  seismic design, 560–562
  and soil liquefaction possibility, 560
  spread footings, 559
  and stainless steel, 555
  and steel, 555
  vaults, 562–563
  watertightness, 557–558
Sulfates
  removal by ion exchange processes, 312
Sulfur dioxide, 767
Sulfuric acid, 767
Supervisory control and data acquisition systems. *See* SCADA systems
Surface Water Treatment Rule, 2, 10, 222–223
  and baffling conditions, 225f., 226f., 227t.
  and chloramination, 241
  determining contact time, 224–226
  disinfectant contact requirements, 223–227
  disinfectant monitoring requirements, 227, 228t.
  disinfection requirements, 223, 223t.
  and groundwater under the direct influence of surface water, 222
  proposed changes, 227
  and rapid sand filtration, 184–185
  and slow sand filtration, 194
Surge protection, 600, 614

Tanks
  chemical dissolving, 453
  clarification, 114
  day, 449
  ion exchange processes, 326
  ozone storage, 267–268
  permeate drawback, 364
  and seismic design, 561
Taste and odor
  oxidation, 229, 231f.
  treatment options, 14t.
Taxes, 648, 711
TCLP. *See* Toxic characteristic leaching procedure

Technology Transfer Handbook: Management of Water Treatment Plant Residuals, 528
Temperature switches and transmitters, 581
Ten States Standards, 184–185, 198, 289
Total Coliform Rule, 2, 10
Total organic carbon
 removal by ion exchange processes, 317
Total trihalomethanes, 1, 11, 12
Toxic characteristic leaching procedure, 679
Treatment processes, 13–15, 14*t*.–15*t*.
 baseline filtration trains, 15–16, 17*f*., 18*f*., 19*f*.
 and chemicals used, 774*t*.–775*t*.
 comparison and evaluation, 16–18
 filtration methods, 15
 monitoring and control, 107
 multiple capabilities, 15, 16*t*.
 O&M design issues, 690–692
 treatment train alternatives, 15–16
Trihalomethanes, 50
 removal of precursors by ion exchange processes, 317
 treatment options, 14*t*.
Tube settlers, 132–133, 133*f*.
 design criteria, 134
 effluent design, 134
 inlet conditions, 134
 solids removal, 134
Turbidimeters
 surface scatter, 582–583
 transmittance, 583
Turbidity
 and diatomaceous earth filtration, 193
 and membrane processes, 347, 347*t*.
 and slow sand filtration, 193, 194–195
 treatment options, 14*t*.
 turbidimeters and particle counters, 582–583

U.S. Army Corps of Engineers, 683
U.S. Bureau of Reclamation, 562
U.S. Environmental Protection Agency, 2, 459
 background data for master plan, 8
 development of rules, 9–10
 groundwater disinfection rule, 228
 National Secondary Drinking Water standards, 222
 Risk Management Program, 683
 and slow sand filtration, 194
 Surface Water Treatment Rule, 222–227
 Underground Injection Control, 495
 wastewater plant power requirements, 606
UF. *See* Ultrafiltration
Ultrafiltration, 336
 backwashing, 366
 cleaning system, 366
 cross-flow velocity, 366
 flux, 365

instrumentation and control, 371
membrane configuration, 339, 341*f*.
membrane integrity tests, 366
membrane module arrays, 366
pumping requirement, 365
recovery, 365
solute rejection, 365
temperature, 365
temperature control, 355
typical applications, 339*t*.
Underground Injection Control, 495
Uniform Building Code, 693
Uniform Fire Code, 232, 693
Uninterruptible power systems, 599–600, 630
Uranium, 315–316
 removal by ion exchange processes, 316, 316*t*.
USEPA. *See* U.S. Environmental Protection Agency

Value engineering, 24, 28
Viruses
 treatment options, 14*t*.
VOCs. *See* Volatile organic chemicals
Volatile organic chemicals, 1, 643
 and air stripping, 61, 62–63, 64, 64*f*., 67, 67*f*.
 removal after membrane processes, 369
 treatment options, 14*t*.
Volumetric feeders, 435–436

Wastewater
 disposal permits, 684
 environmental impact of treatment and disposal, 680
 plant power requirements, 606
Water hardness. *See* Hardness, Lime softening
Water Quality and Treatment, 283, 292, 299, 417
Water softening. *See also* Ion exchange processes, Lime softening, Membrane processes
 benefits, 282–283
 hot-process, 281
 sludge, 490
Water treatment chemicals. *See* Chemicals
Water treatment engineering
 design-build approach, 3
 historical overview, 1–2
 list of abbreviations used, 777–787
 multidiscipline teams, 3
 new technologies, 2–3
 and organic compounds, 1
 privatization, 3
 project delivery approaches, 3
 project development, 4
Water Treatment Plant Waste Management, 528
Watson, H. E., 230

Zebra mussels, 54–55, 56*t*.